Sustainable Development Goals Series

The **Sustainable Development Goals Series** is Springer Nature's inaugural cross-imprint book series that addresses and supports the United Nations' seventeen Sustainable Development Goals. The series fosters comprehensive research focused on these global targets and endeavours to address some of society's greatest grand challenges. The SDGs are inherently multidisciplinary, and they bring people working across different fields together and working towards a common goal. In this spirit, the Sustainable Development Goals series is the first at Springer Nature to publish books under both the Springer and Palgrave Macmillan imprints, bringing the strengths of our imprints together.

The Sustainable Development Goals Series is organized into eighteen subseries: one subseries based around each of the seventeen respective Sustainable Development Goals, and an eighteenth subseries, "Connecting the Goals," which serves as a home for volumes addressing multiple goals or studying the SDGs as a whole. Each subseries is guided by an expert Subseries Advisor with years or decades of experience studying and addressing core components of their respective Goal.

The SDG Series has a remit as broad as the SDGs themselves, and contributions are welcome from scientists, academics, policymakers, and researchers working in fields related to any of the seventeen goals. If you are interested in contributing a monograph or curated volume to the series, please contact the Publishers: Zachary Romano [Springer; zachary.romano@springer.com] and Rachael Ballard [Palgrave Macmillan; rachael.ballard@palgrave.com].

More information about this series at http://www.springer.com/series/15486

Baleshwar Thakur
Rajiv R. Thakur
Srikumar Chattopadhyay
Rajesh K. Abhay
Editors

Resource Management, Sustainable Development and Governance

Indian and International Perspectives

Springer

Editors
Baleshwar Thakur
Department of Geography
Delhi School of Economics
University of Delhi
New Delhi, Delhi, India

Srikumar Chattopadhyay
Resources Analysis Division
National Centre for Earth
Science Studies
Thiruvananthapuram, Kerala, India

Rajiv R. Thakur
Department of Geosciences
Missouri State University
West Plains, MO, USA

Rajesh K. Abhay
Department of Geography
Dyal Singh College, University of Delhi
New Delhi, Delhi, India

ISSN 2523-3084 ISSN 2523-3092 (electronic)
Sustainable Development Goals Series
ISBN 978-3-030-85838-4 ISBN 978-3-030-85839-1 (eBook)
https://doi.org/10.1007/978-3-030-85839-1

This Springer imprint is published by the registered company Springer Nature Switzerland AG
The registered company address is: Gewerbestrasse 11, 6330 Cham, Switzerland

Preface

The studies in this volume focus on the lessons that emerge from the diversity of experiences in the realm of resource management, sustainable development, and governance both in India and in other countries. More specifically, it sheds light on the connections and interdependence between resource management, sustainable development, and governance in the context of India and many other countries. They also complement what can be learned from other experiences of challenges and opportunities in natural resource management. This volume is the outcome of the need for a more intense engagement from a critical perspective between the intrinsic relationship of resource management, sustainable development, and governance as circumstances evolve between human interactions and their ecosystems, causing dynamic changes that warrants attention. The volume is organized as a festschrift in recognition of the contributions of Bruce Mitchell FRSC, Distinguished Professor Emeritus, University of Waterloo, Canada, to various sub-fields of geography like environment management, environmental analysis, natural resource management, sustainability, and water governance. The volume also reflects the deep commitment of the editors to the dynamics of integrated resource management and sustainability. To that end, this research handbook is also motivated by the need to examine the same natural resource management and governance issues through the lens of UN-SDGs, which has emerged as the dominant framework of understanding and action to preserve the planet.

The experience of resource management in both India and other developing countries, as well as some developed countries, has recently undergone dynamic changes under the impact of climate change and associated vulnerabilities. Therefore, governance at different scales and places need to transition by developing innovative and adaptive practices, and techniques to protect their regions and better cope with new and worsened risks even as the population expands and so does growth in carbon emissions which threaten the natural environment. The authors come together in their recommendations that there is a need to act now, together, and differently adopting an integrated sustainable natural resource management approach as the dominant inertia. Hence, stakeholders need to understand the ever-evolving interactions between natural and social processes, which will determine how natural resources will be governed, produced, managed, and allocated to society.

In this context, the volume is designed for senior undergraduates and graduate students, research scholars, and practitioners. The research has been presented in an accessible way, but some of the processes and ideas that they discuss are inevitably complicated. Case studies examine and or investigate issues in resource management at multiple scales of ecosystem processes, and is not, therefore, focused exclusively at processes on the national scale. For example, the editors believe that integrated natural resource management has as much to contribute to thinking about how gender roles in the household play out within the spaces of rural livelihoods and human interaction with the local ecosystems, as it does in understanding the natural resource dynamics at other scales. The editors of the volume do not prescribe a particular quantitative or qualitative technique as a method of analysis, rather emphasize on the stimulation of the reader with critical perspectives and arguments. For example, in thinking about how tribal women's role and responsibilities in rural Uttarakhand is structured, we are more interested in inviting the reader to consider and think about the processes that lie behind such dynamics than in explaining how to demonstrate statistically that such patterns exist. That said, several authors use geospatial techniques and tools to supplement qualitative analysis and understanding. Finally, the editors and authors do not seek to establish impervious boundaries between resource management and other cognate fields concerned with social, cultural, and political processes. Our vision of the volume's theme is porous, and we sincerely take the need to view the ecosystem as embedded in other spheres of life. For example, we see resource consumption not 'just' as a livelihood act, but also as a political engagement interacting and responding to policy geography.

The book takes the form of a series of linked chapters within well-defined themes and contemporary debates that draw upon, and showcase some of the best research in resource management and governance. The chapter contributors have analyzed the spatial, temporal, ecological, environmental, economic, political, and social relationships between resource management, sustainable development, and governance. We see each of these as issues, rather than just phenomena, i.e. they are processes to be debated rather than factual realities to be described. This then, is not a conventional research handbook. Rather our aim is to develop well-grounded arguments from the perspective of integrated resource management perspectives, not necessarily to present simplifications of multiple viewpoints or collection of facts and data.

The contributors include early faculty and research scholars (including those outside formal academia) as well as established geographers. In different ways, all of us are concerned with conceptual foundations of resource management and its governance, the need to deconstruct the issues, evaluate the human dimensions of the same, and examine local, regional, and national responses, and emerging challenges and opportunities. Our disciplinary interest, training, and engagements include a broad diversity, i.e. geography, environment, regional science, and forestry to include a few. But in general, the approach and work are spatial in nature, and our interventions, analyses, and insights are specific to contexts, but these also have wider relevance. We would like to offer clarification that this volume has been prepared specifically, under Springer's *Sustainable Development Goals Series*. While individual research has acknowledged the support from respective funding agencies, the volume as a whole was not funded by any individual or institution.

New Delhi, Delhi, India — Baleshwar Thakur
West Plains, MO, USA — Rajiv R. Thakur
Thiruvananthapuram, Kerala, India — Srikumar Chattopadhyay
New Delhi, Delhi, India — Rajesh K. Abhay

Acknowledgements

This book could not have been prepared without the generous contribution of many individuals and organizations. The editors wish to acknowledge their special debt to Bruce Mitchell, University of Waterloo, and to Robert Doe, Zach Romano, and Zoe Kennedy at Springer who have contributed to shaping the volume's vision and outcome. Errors of commission and omission are the sole responsibility of the editors. The editors thank the chapter author(s) for preparing original contributions for the volume. The volume benefited greatly from intellectual advice and consultations of several experts who gave their time freely but wanted to remain anonymous. The editors also recognize that the volume benefitted enormously from the editorial support of Dr. Patricia Thakur, Adjunct Faculty, Missouri State University, and the generous support and encouragement from our respective spouses (Chanda, Patricia, Mahamaya, and Mamta) that allowed us to spend endless hours away from our loved ones. The second editor would like to acknowledge the support and encouragement of colleagues (including Dr. Dennis Lancaster, Interim Chancellor, and Dr. Michael Orf, Interim Academic Dean, who facilitated book project-related travel funding to India in the summer of 2021) at Missouri State University and in particular the untiring support of colleagues at the library—Neva Parrot, Shanon Conlon, and Sophia Skinner.

New Delhi, India — Baleshwar Thakur
West Plains, Missouri, USA — Rajiv R. Thakur
Thiruvananthapuram, Kerala, India — Srikumar Chattopadhyay
New Delhi, India — Rajesh K. Abhay

Contents

Part I Introduction

1 **Resource Management, Sustainable Development, and Governance: Introduction and Overview** 3
Baleshwar Thakur, Srikumar Chattopadhyay, Rajiv R. Thakur, and Rajesh K. Abhay

2 **Bruce Mitchell: Professional Career and Contributions** 13
Baleshwar Thakur, Rajesh K. Abhay, Rajiv R. Thakur, and Srikumar Chattopadhyay

Part II Understanding Conceptual Foundations

3 **Early and Modern Paradigms in Natural Resource Management: Global and Indian Experiences** 27
Baleshwar Thakur and Rajiv R. Thakur

4 **Approaching the Collaborative "Turn" in Water Governance: A Critical Reappraisal** 59
Nigel Watson and Rob C. de Loë

5 **Disaster Risk Governance and Management: An Asian Perspective** 75
Indrajit Pal and Jayant K. Routray

6 **Reshaping Natural Resource Management in China** 89
Bing Xue and Wanxia Ren

7 **Regional Sustainable Development and Natural Resource Decision-Making in India: Methods and Implications** 105
Sudhir K. Thakur

Part III Unpacking Problems

8 **Rethinking Resettlement as a Development Opportunity: Need for Good Practices** 135
Vinita Mathur and Gaurav Sikka

9 Will the Water Revolution Be Centralized? Investigating the "Downscale" and "Upscale" Challenges of Urban Rainwater Harvesting 143
Georgina Drew

10 Rethinking Capacity Building in Water Governance: Factors Influencing Risk Interpretation and Decision-Making in Delhi 159
Shabana Khan

Part IV How Resource Management, Sustainable Development and Governance Work: Case Studies

11 Regional Environmental Governance: Analytical Framework for Resource Conservation in the Eastern Ghats, India 179
Kapil Kumar Gavsker

12 Groundwater Sustainability in Haryana: Challenges to Governance 193
Inder Jeet

13 Wetland Resources in the Brahmaputra Valley, Assam: Characteristics, Use, and Sustainable Development 205
A. K. Bhagabati and Nityananda Deka

14 Impact of Physical Factors on Transboundary Water Management and Governance in the Kosi Basin 225
Ramashray Prasad

15 Governance and Management of Teesta River Water Resources: A Geopolitical Appraisal 249
Sudeepta Adhikari and Subinita Kamle

16 Governance Issues for Sustainable Water Management in Rapti River Basin, Uttar Pradesh 267
Narendra Kumar Rana and Neha Singh

Part V Exploring Human Dimensions

17 Social Transformation, Ecosystem Services, and Resource Sustainability in Nepal Hills 283
Krishna Prasad Poudel

18 Determinants of Land Use Dynamics and Its Ecological Implications in India: A State-Level Analysis 299
Shahab Fazal, Nasrin Banu, and S. K. Azharuddin

19 Land Use Change and Its Impact on Ecosystem Services: Food, Livelihood, and Health Security in Kumaon Himalayas 319
Prakash C. Tiwari and Bhagwati Joshi

20 Impact of Land Use Change on Livelihood Options: A Case Study of Upper Pasolgad Watershed, Uttarakhand 331
Bindhy Wasini Pandey, Abhay Shankar Prasad, and Jitendra Kumar Mahto

21 Rural Livelihood and Women: Glimpses from an Indian Tribal Village 347
Purva Yadav, Shreya Akarshna, and Anuradha Shankar

22 Land Degradation and Agricultural Sustainability in Kendujhar District, Odisha 365
Rajesh K. Abhay and Punyatoya Patra

23 Diversification of Agriculture Through Medicinal Plants in Kangra District, Himachal Pradesh 377
Nitu and R. B. Singh

Part VI Response to National, Regional, and Global Change

24 Sanitation Challenges and Policy Options in Developing Countries: A Critical Review 399
Nuthan Maharaj and Brij Maharaj

25 Solid Waste Management for Environmental Sustainability in India 415
Surya Tewari

26 Social Impact Assessment of Indian Water and Allied Policies and Programs 429
Dipankar Roy, Shobha Kumari, Akhilesh Kumar Mishra, and Suresh Chand Rai

27 Challenges and Opportunities Towards Management of Solid Wastes in Indian Cities: Beyond the Rhetoric of Convenience 445
Ruchira Ghosh and M. Satish Kumar

28 Natural Resource Evaluation for Ecotourism and Geotourism Destination in Hong Kong 461
Lawal M. Marafa

29 Climate Change Knowledge: Comparison of People's and Scientists' Perceptions in Western Himalayas 475
Pallavi V. Das

30 Climate Change Modeling for Ecosystem Services in Western Himalayan District: A Methodological Approach Towards Sustainability Science 487
Swarnima Singh

31 Urban Sustainability and Good Governance in Delhi Metropolitan Region 507
Shweta Rani

32 Investigating the Variability in Sanitation and Sustainability Issues: Evidence from Resettlement Colonies of Delhi 519
Pooja Yadav and Subhash Anand

33 Water Governance in Thiruvananthapuram City, Kerala: Existing Practices and Alternative Paradigms 537
Srikumar Chattopadhyay and K. N. Harilal

34 Bhitarkanika Wildlife Sanctuary, Odisha: People's Attitude Towards Its *Conservation* 557
S. C. Rai and Arpita Panda

35 Socio-Economic Development in Highway Corridor Zones of Bihar: Case Study of Kuchai Kot-Muzaffarpur Section. 563
Krishna Kumar and Anjan Sen

Part VII Future Directions

36 Sustainable Resource Governance: Lessons for the Future 583
Rajiv R. Thakur, Baleshwar Thakur, Srikumar Chattopadhyay, and Rajesh K. Abhay

Index. 589

Contributors

Rajesh K. Abhay Department of Geography, Dyal Singh College, University of Delhi, New Delhi, India

Sudeepta Adhikari Department of Geography, Patna University, Patna, India

Shreya Akarshna Centre for the Study of Regional Development, School of Social Sciences, Jawaharlal Nehru University, New Delhi, India

S. K. Azharuddin Department of Geography, Aligarh Muslim University, Aligarh, India

Subhash Anand Department of Geography, Delhi School of Economics, University of Delhi, Delhi, India

Nasrin Banu Department of Geography, Aligarh Muslim University, Aligarh, India

A. K. Bhagabati Department of Geography, Gauhati University, Guwahati, India

Srikumar Chattopadhyay National Centre for Earth Science Studies, Thiruvananthapuram, India

Pallavi V. Das Department of History, Lakehead University, Orillia, ON, Canada

Nityananda Deka Department of Geography, Gauhati University, Guwahati, India

Georgina Drew Anthropology and Development Studies, The University of Adelaide, Adelaide, Australia

Shahab Fazal Department of Geography, Aligarh Muslim University, Aligarh, India

Kapil Kumar Gavsker Department of Geography, Institute of Science, Banaras Hindu University, Varanasi, India

Ruchira Ghosh Centre for Sustainable Technologies, Belfast School of Architecture and the Built Environment, Ulster University, Jordanstown, UK

K. N. Harilal Centre for Development Studies and Member, State Planning Board, Government of Kerala, Thiruvananthapuram, India

Inder Jeet Department of Geography, Maharshi Dayanand University, Rohtak, India

Bhagwati Joshi Department of Geography, Government Post-Graduate College, Rudrapur, India

Subinita Kamle Department of Geography, Patna University, Patna, India

Shabana Khan Indian Research Academy, New Delhi, India

Krishna Kumar Discipline of Geography, School of Sciences, Indira Gandhi National Open University (IGNOU), New Delhi, India

M. Satish Kumar School of Natural and Built Environment, Queen's University Belfast, Belfast, UK

Shobha Kumari Department of Geography, Institute of Science, Bananas Hindu University, Varanasi, India

Rob C. de Loë School of Environment, Resources and Sustainability, University of Waterloo, Waterloo, ON, Canada

Brij Maharaj University of KwaZulu-Natal, Durban, South Africa

Nuthan Maharaj Economic Development Unit, Durban, South Africa

Jitendra Kumar Mahto Department of Geography, Delhi School of Economics, University of Delhi, Delhi, India

Lawal M. Marafa Centre for Environmental Policy and Resource Management, Department of Geography and Resource Management, The Chinese University of Hong Kong, Sha Tin, N.T., Hong Kong

Vinita Mathur Department of Geography, Dyal Singh College, University of Delhi, New Delhi, India

Akhilesh Kumar Mishra Department of Geography, Delhi School of Economics, University of Delhi, Delhi, India

Nitu Indian Institute of Public Administration, Delhi, India

Indrajit Pal Department of Development and Sustainability, Asian Institute of Technology, Pathumthani, Thailand

Arpita Panda Department of Geography, Delhi School of Economics, University of Delhi, Delhi, India

Bindhy Wasini Pandey Department of Geography, Delhi School of Economics, University of Delhi, Delhi, India

Punyatoya Patra Department of Geography, Aditi Mahavidyalaya, University of Delhi, Delhi, India

Krishna Prasad Poudel Department of Geography, Central Department of Education, Tribhuvan University, Kathmandu, Nepal

Ramashray Prasad Department of Geography, B.R. Ambedkar College, University of Delhi, Delhi, India

Abhay Shankar Prasad Department of Geography, Dyal Singh College, University of Delhi, New Delhi, India

Suresh Chand Rai Department of Geography, Delhi School of Economics, University of Delhi, Delhi, India

Narendra Kumar Rana Department of Geography, Institute of Science, Banaras Hindu University, Varanasi, India

Shweta Rani Department of Geography, Dyal Singh College, University of Delhi, New Delhi, India

Wanxia Ren Institute of Applied Ecology, Chinese Academy of Sciences, Shenyang, China

Jayant K. Routray Department of Development and Sustainability, Asian Institute of Technology, Pathumthani, Thailand

Dipankar Roy Department of Geography, Institute of Science, Bananas Hindu University, Varanasi, India

Anjan Sen Department of Geography, Delhi School of Economics, University of Delhi, Delhi, India

Gaurav Sikka Department of Geography, L.N. Mithila University, Darbhanga, India

Neha Singh Department of Geography, H. R. P. G. College, Khalilabad, Sant Kabir Nagar, India

R. B. Singh Department of Geography, Delhi School of Economics, University of Delhi, Delhi, India

Swarnima Singh Department of Geography, Deen Dayal Upadhyay Gorakhpur University, Gorakhpur, India

Surya Tewari Institute for Studies in Industrial Development, New Delhi, India

Baleshwar Thakur Department of Geography, Delhi School of Economics, University of Delhi, Delhi, India

Rajiv R. Thakur Department of Geosciences, Missouri State University, West Plains, MO, USA

Sudhir K. Thakur Department of Finance, Insurance and Real Estate, College of Business Administration, California State University, Sacramento, CA, USA

Prakash C. Tiwari Department of Geography, Kumaun University, Nainital, India

Pooja Yadav Department of Geography, Delhi School of Economics, University of Delhi, Delhi, India

Purva Yadav Centre for the Study of Regional Development, School of Social Sciences, Jawaharlal Nehru University, New Delhi, India

Nigel Watson Lancaster Environment Centre, Lancaster University, Lancaster, UK

Bing Xue Institute of Applied Ecology, Chinese Academy of Sciences, Shenyang, China

About the Editors

Baleshwar Thakur is a former Professor and Chair of the Department of Geography, Delhi School of Economics, University of Delhi, and former Vice-Chancellor of L.N. Mithila University, Darbhanga (Bihar, India). He has approximately 57 years teaching and research experience in resource management, urban development and planning, and geographical thought and research methods. He has authored and co-authored 37 journal articles and 82 book chapters, edited and co-edited 33 books, and authored a book. He was a Canadian Commonwealth Scholar, University of Waterloo, Canada (1970–1972), Fulbright Visiting Professor, University of Akron, USA (1992), and Shastri Indo-Canadian Fellow, York University, Canada (2004). He was President of the Earth System Sciences of Indian Science Congress Association (2000), National Association of Geographers, India (2002), and the Institute of Indian Geographers (2012). He is a recipient of the Distinguished Scholar Award (2005) from the Regional Development and Planning Specialty Group of the American Association of Geographers (AAG).

Rajiv R. Thakur is currently an Associate Professor of Geography at Missouri State University, West Plains, Missouri, USA. He has a doctoral degree from Indiana State University and master's in Geography from the University of Akron. He has previously taught at the University of Tennessee, Knoxville (2011–2013) and the University of South Alabama, Mobile (2008–2011). He is an Economic Geographer with expertise in regional development and planning and has emerging interests that reside at the intersection of economics and the environment. He has co-authored *Research Design and Proposal Writing in Spatial Science* (with Jay D. Gatrell, Gregory D. Bierly, and Ryan R. Jensen), lead edited *Urban and Regional Planning and Development: Twentieth Century Forms and Twenty First Century Transformations* (with Ashok K. Dutt, Sudhir K. Thakur, and George Pomeroy), and co-edited *Spatial Diversity and Dynamics in Resources and Urban Development* (with Ashok K. Dutt, Allen G. Noble, Frank J. Costa, Sudhir K. Thakur, and H. S. Sharma). He was Chair of the West Lakes Division of the AAG (2016–2017) and Chair of the Regional Development and Planning Specialty Group of AAG (2011–2012) and is recipient of National Science Foundation Robert Noyce Scholarship (2006–2008).

Srikumar Chattopadhyay is ICSSR National Fellow, former ICSSR Senior Fellow, and former Scientist at the National Centre of Earth Science Studies, Thiruvananthapuram, Kerala, India. He was Post-Doctoral Fulbright Fellow at East-West Centre, Hawaii, USA. He specializes in resource and environmental management, sustainable development, and local level planning. He has authored and edited 7 books and published over 100 research papers. He was the President of Institute of Indian Geographers (2015) and is the recipient of R.N. Dubey Memorial Award (2001).

Rajesh K. Abhay is currently an Assistant Professor of Geography at Dyal Singh College, University of Delhi, and was previously Assistant Professor and faculty at several colleges of the University of Delhi. His doctoral and graduate degrees are from the Delhi School of Economics, University of Delhi. His research expertise includes natural resource management and sustainable resource development with the application of geospatial technologies. He has authored and edited 3 books and published 23 research papers. He is a recipient of the Professor NP Aiyyar Young Geographer Award (2014) of the National Association of Geographers, India.

Part I

Introduction

1 Resource Management, Sustainable Development, and Governance: Introduction and Overview

Baleshwar Thakur, Srikumar Chattopadhyay, Rajiv R. Thakur, and Rajesh K. Abhay

Abstract

This volume is a festschrift in honor of Bruce Mitchell, Distinguished Professor Emeritus, University of Waterloo and celebrates his contribution to the theme "resource management, sustainable development and governance." The volume is published under the *Sustainable Development Goals* (SDGs) series of Springer as chapter contributions are engaged through the lens of SDGs. This collection of studies from scholars around the world considers critical emerging issues in natural resource management not only in India but elsewhere in the context of increased population growth and economic growth which has led to the intense use of resources and in many cases has challenged the resilience limits of resource systems. The volume identifies three systems: climate change, rate of biodiversity loss, and human interference with the nitrogen cycle, as critical to sustainable preservation of ecosystems in places. Sustainable resource management is dependent on resource governance given the demands of formal and nonformal actors and other stakeholders in decision-making, planning, and resource management for sustainable development. The volume ends with several proposals for future research directions.

Keywords

Ecosystem · Governance · India · Paradigm Shifts · Resilience · Resource Management · Sustainable Development · Sustainable Development Goals

Resource is a techno-economic concept. Elements of a geo-system, when transformed into utilizable material form resources. The knowledge base, labor, technological capacity, and economy are the prime governing factors in this process of transformation. While these basic premises are true, there is finitude (finite nature of resources), entropy, and complex ecological interdependence which combine to provide biophysical limits for resource utilization to growth (Daly 1987). To

B. Thakur
Department of Geography, Delhi School of Economics, University of Delhi, Delhi, India

S. Chattopadhyay
National Centre for Earth Science Studies, Thiruvananthapuram, Kerala, India

R. R. Thakur (✉)
Department of Geosciences, Missouri State University, West Plains, MO, USA
e-mail: rajivthakur@missouristate.edu

R. K. Abhay
Department of Geography, Dyal Singh College, University of Delhi, New Delhi, India
e-mail: rkabhay@dsc.du.ac.in

B. Thakur et al. (eds.), *Resource Management, Sustainable Development and Governance*, Sustainable Development Goals Series, https://doi.org/10.1007/978-3-030-85839-1_1

meet the demand of increasing population pressure and socioeconomic growth there had been intense use of resources and in many cases the resilience limits of resource systems have been surpassed. Rockstrom et al. (2009) of the Stockholm Resilience Centre conducted a study to assess safe operating space for humanity. This study identified nine indicators such as climate change, rate of biodiversity loss, nitrogen cycle, phosphorus cycle, stratospheric ozone depletion, ocean acidification, global freshwater use, change in land use, atmospheric aerosol loading, and chemical pollution for analysis. Out of these nine, three systems (climate change, rate of biodiversity loss, and human interference with the nitrogen cycle) have already crossed their respective limits. Planetary boundaries are tightly coupled and not mutually exclusive therefore transgression in one boundary may lead to serious risk in case of other boundaries. Millennium Ecosystem Assessment (MEA) report (2005) has indicated that 15 out of 24 ecosystem services examined are being degraded or used unsustainably. The apparent gains in economic development and growth have been achieved at the growing costs in the form of degradation of many ecosystem services and an increased risk of nonlinear changes. Situations in the developing countries are further complicated due to the high rate of population growth, increasing poverty, overdependence on land and water resources for economic betterment, and limited opportunity to shift surplus labor from primary to nonprimary sectors. With progress of human society, concentration of large number of people/activities in certain places, and technological advancement, the nature-human interaction frame has just not widened, but it has become nonlinear, complex, and multidimensional and is making profound changes in the ecosystem. Excessive drawdown of resources and unequal distribution of development are civilizational challenge and there is global concern about prudent resource management for smooth sailing of the spaceship-Our Earth.

The worldwide concern about these issues is perhaps well epitomized in the report of the World Commission on Environment and Development (WCED 1987)—*Our Common Future*, in which the fundamental interrelationships between environment and development and the unsustainability of current practices have been clearly spelt out. As an alternative, the concept of sustainable development has been introduced. Since then, there were Earth Summit in 1992 at Rio, World Summit on Sustainable Development at Johannesburg in 2002, and Rio+20 summit in 2012 and several other global meets to address the crisis of development and devise the way for sustainable development, a term gained wide currency even among the politicians. The WCED defined sustainable development as "Development that meets todays need without compromising the ability of the future generations to meet their needs" (WCED 1987). It is proposed to factor in intergenerational equity with the development process and at the same time the importance of meeting the need for present generation is also stressed. The sustainable development is construed as a process operating within the framework of economic, social, and ecological boundaries; however, the challenge is how to operationalize this concept and deliver.

It is now globally argued that resource management science is passing through a crisis (Holling et al. 1998). In some parlance, it has been opined that sustainability is neither a realistic goal nor a useful concept, while some other authors wandered about utility of scientific research in designing policies for sustainable management of resources (Ludwig et al. 1993). Such arguments primarily emanate from an attempt to discuss sustainability following a reductionist linear approach and trying to work out sustainable use of a particular sector of natural resources like forestry, fisheries, agriculture, etc., and formulate policy accordingly. System's perspective is often missing. The Millennium Ecosystem Assessment (MEA) report (2005) approach the sustainability issue with a set of four scenarios (Chattopadhyay and Franke 2006):

- *Global Orchestration*—Trade and liberalization dominate, ecological problems are treated reactively, poverty and inequality are reduced, and heavy investment in education and infrastructure.

- *Order from Strength*—Nations and regions focus on their own problems. Security, protection, little attention to public goods, and a reactive approach to ecological problems.
- *Adapting Mosaic*—Regional watershed ecosystems are the focus. Local institutions are strengthened and local management strategies are developed. A proactive approach to management of ecological systems.
- *Techno Garden*—Globally connected, highly managed ecosystems. A proactive approach.

A society can position itself in the context of these four scenarios and design its own course of action. In 2000, UN General Assembly identified eight goals, known as Millennium Development Goals (MDG) for the countries to steer their development process and achieve certain milestones by 2015. The goals were as follows:

- Goal 1: Eradicate Extreme Poverty and Hunger.
- Goal 2: Achieve Universal Primary Education.
- Goal 3: Promote Gender Equality and Empower Women.
- Goal 4: Reduce Child Mortality.
- Goal 5: Improve Maternal Health.
- Goal 6: Combat HIV/AIDS, Malaria, and TB.
- Goal 7: Ensure Environmental Sustainability.
- Goal 8: Develop Global Partnership for Development.

Many of the countries could make significant progress. India performed well in some sectors and in some other sectors it must progress further (Government of India 2017a, b). Nevertheless, to continue with the momentum of MDG, the World adopted a set of new agenda and targets under "Sustainable Development Goals: Transforming our World by 2030" in 2015 to complete unfinished agenda under MDG and proceed further.

The post 2015 development agenda strongly advocated for the sustainable management of natural resources to achieve sustainable development goals. Sustainable resource management in a globalized economy warrants actions at different scales from local to global, and relevant policy formulation and management practices require a detailed scientific information base and new strategies to use resources. This calls for new institutional capacity and governance arrangements (Bringezu et al. 2016). The issue of governance is of paramount importance in striving for sustainable development.

Resource governance entails a range of political, social, economic, and administrative systems that are in place to develop and manage resources in a sustainable manner. The emphasis is on providing space for formal and nonformal actors and all other stakeholders in decision-making, planning, and resource management for sustainable development. Governance is commonly defined as "the interactions among structures, processes, and traditions that determine how power and responsibilities are exercised, how decisions are taken, and how citizens or other stakeholders have their say" (Graham et al. 2003). The institutions pertain to formal laws, rules, and regulations as well as informal norms and customary practices that guide the behavior of individuals and groups with respect to environment/natural resources. Effective institutional interventions would be those that account for this complexity of interests and interactions and aim for a sustainable outcome.

As there are competing demands and multiplicity in management authorities, the challenge of resource governance is to resolve conflicts among techno-scientific, market, policy administration, ecological, and sociopolitical actors. Besides, there are the issues of property rights, decision-makers, geographical scale, and beneficiary likely to figure in devising appropriate resource governance system. There is also a need to look beyond the immediate vicinity and consider the broader territory, establish a relationship with the surrounding areas, evolve reciprocity with the hinterlands, and operate in the frame of comanagement with other administrative units. One of the key factors in governance is interdependence and interaction among diverse actors from different territories at multiple governmental scale (Davidson et al. 2006). This will call for integration and follow system approach that looks at the resource system from provenance to consumption and market. The drivers of resource use and abuse are location-specific, so the insights on resource management should arise from local-level experiences. A careful analysis of the resource governance system, its actors, interests, values, and processes in each locality is necessary to bring out required change in the present governance practices.

Going beyond the "instrumental and idealistic" notion of governance and an attempt to depoliticization as nowadays being advocated in some parlance, it is necessary to strike a balance among different aspects of resource management activities, ensure convergence between research and practice in resource management science, and help the society to evolve a proper governance system through democratic means of debate and stakeholder participation in policymaking. This requires multiple level interactions between government and all other stakeholders/actors, all of whom must be drawn into continuous dialog and negotiations, which may turn into conflicts and uncertainties; however, all these as part of democratic resource governance may progressively lead to concentrative process in order to reach some kind of agreement to move on to implementation and evolve iterative multilevel governance processes that continually progress through social learning and create a broader conceptual space for wide-ranging debate. Prudent resource management practices therefore cut across ecological, economic, social, and governance dimensions (Fig. 1.1). It is multi-scale and nonlinear.

Our attempt, in this volume, is to flag some of these issues through scholarly chapters with empirical data. We requested established and well-accomplished researchers to contribute. Altogether there are 36 articles covering 23 topics contributed by 60 authors. Water emerged as an important topic of discussion. It has been examined from various perspectives. Land and land use dynamics have been dealt by several authors. Among the emergent fields, there are papers on climate change, sanitation and solid waste management, disaster risk, and social impact assessment. Most of the papers are empirical studies based on microlevel field data highlighting local conditions. Theme of the volume runs as common thread among all the papers. However, considering thrust of the papers and principal argument advanced in these chapters we have organized all 36 papers into seven sections (Table 1.1).

1.1 Part I Introduction

In what follows, we highlight some of the salient points in all the contributed chapters. Part-1 which is introductory in nature has two chapters. The introductory chapter by the editors provides an overview of the conceptual aspects of resource

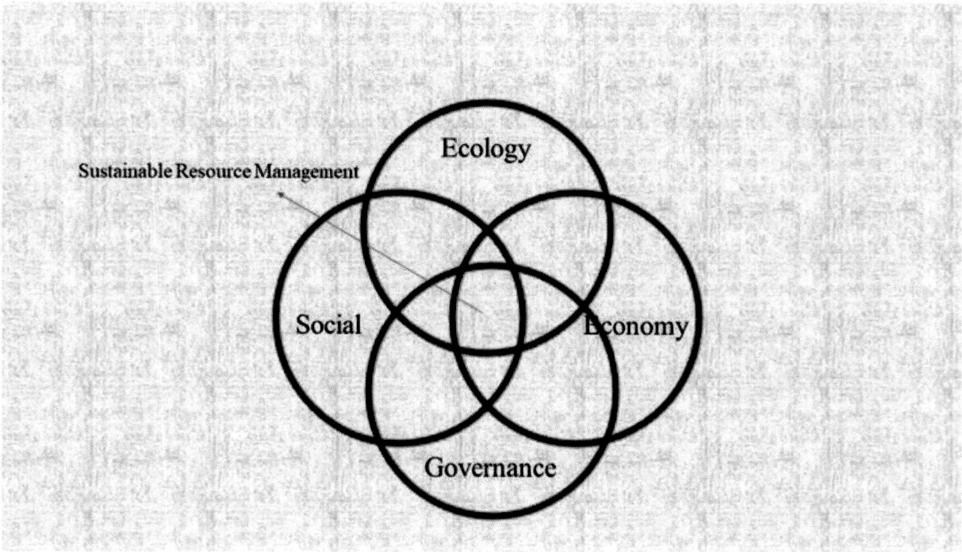

Fig. 1.1 Dimensions of Sustainable Resource Management

Table 1.1 Content organization

Section-1	Introduction (2 papers)
Section-2	Understanding Conceptual Foundations (5)
Section-3	Unpacking Problems (3)
Section-4	How Resource Management, Sustainable Development, and Governance Works: Case Studies (6)
Section-5	Exploring Human Dimensions (7)
Section-6	Response to National, Regional, and Global Change (12)
Section-7	Future Directions (1)

management, sustainable development, and governance. The second chapter in this section is a review of the illustrious career of Bruce Mitchell, Distinguished Professor Emeritus, University of Waterloo.

1.2 Part II Understanding Conceptual Foundations

The focus of part two is to develop an understanding of the conceptual foundations and consists of five chapters drawing on changing paradigms, collaborative turn in water governance, disaster risk governance and management, and regional sustainable development. More specifically, chapter 3 by Baleshwar Thakur and Rajiv R. Thakur engages with paradigm shifts in sustainable resource management both in India and globally. Resource management practices are undergoing change in India and abroad. The demand-supply and commodity mode of resource management are no more viable as has been evident from large-scale environmental degradation associated with all resource management. On the one hand, nonrenewable resources are getting exhausted and on the other hand renewable or environmental resources are under stress resulting in loss of production potential. Technological advancement has equipped the society in precision resource management at the same time it has opened the avenue for alternative resource management and material substitution. Sustainable development, which is the professed goal warrants a paradigm shift in resource management. There are several challenges. Nevertheless, a knowledge-based society, towards which we are moving has the potential to address the challenges and transit to sustainability. The next chapter in this section on "Approaching the Collaborative 'Turn' In Water Governance: A Critical Re-Appraisal," by Nigel Watson and Rob de Loë critically examines the signs for a "turn" in water management and governance towards collaborative approaches and institutional arrangements, drawing on recent literature, and their own original research and practical experiences of working with collaborative institutions and groups. They focused on collaboration in the water sector, where this approach to governance has become particularly significant. Lessons are important as collaboration is considered as a new tool to overcome many of the problems concerning water resource management. The third chapter in this section by Indrajit Pal and Jayant K. Routray on "Disaster Risk Governance and Management: An Asian Perspective" examines the "disaster risk governance" framework of various policy interventions and governance mechanisms that have been developed to improve the resiliency and sustainability of communities and reduce their vulnerability to natural disasters in the Asian region. The next two chapters deal with China and India, respectively. In Chap. 4, Bing Xue and Wanxia Ren discuss reshaping natural resource management as a key component for meeting the challenges of transitioning to sustainable development in China. In doing so, the authors review current institutional changes for natural resource management and identify key stakeholders in the governance system. In the next chapter, Sudhir K. Thakur critically engages with methodologies adopted in regional sustainable development and natural resources decision-making and raises questions about (a) the relationship between natural resources, economic progress, and sustainable development, (b) considers alternative methods to natural resources decision-making, and (c) visualization of natural resource distribution in India.

1.3 Part III Unpacking Problems

Part three contains three chapters which unpacks critical issues such as resettlement and displacement, challenges associated with urban rainwater harvesting, and capacity building in water governance. Chapter 8 on "Rethinking Resettlement as A Development Opportunity: Need for Good Practices," by Vinita Mathur and Gaurav Sikka presents some of the "good practices" in resettlement and compensation planning as observed in the case of Sardar Sarovar Project in India. It is argued that as one size does not fit all, so what may be a "good practice" in one circumstance might not be so good in another situation. There is a need for creating a shelf of "best practices" which can be treated as tools for guiding proper resettlement and achieving the aim of inclusive development. Chapter 9 by Georgina Drew on "Will the water revolution be decentralized?" traces progress in debates over urban rainwater harvesting, as well as the uptake in rainwater harvesting practices, that have taken place since the publication of the "A Water Harvesting Manual for Urban Areas," in 2003. Drawing from a selection of documents and case studies, the author argues that several disincentives persist that either deter people from taking up the clarion call of household-level rainwater harvesting, or that prevent them from doing it altogether. It has been suggested that successful moves towards decentralized urban rainwater harvesting and water management require enhanced centralized cooperation and capacity building. The next chapter in this section by Shabana Khan is on "Rethinking Capacity Building in Water Governance? This study is based on stake holders" interview which examines the interplay of risk interpretation and decision-making in the current water-governance system in Delhi by using the Risk Interpretation and Action (RIA) framework. The results highlight the need to rethink capacity building in terms of preparing varied stakeholders for their greater engagements and participation in the development of effective water governance.

1.4 Part IV How Resource Management, Sustainable Development, and Governance Work: Case Studies

Composed of six chapters, part four of the book addresses how resource management, sustainable development, and governance work through several case studies. In Chap. 11, Kapil Gavsker discusses how regional environmental governance can be a reality and an effective strategy by addressing some crucial issues towards natural resource conservation and their sustainable uses. In the context of the Eastern Ghats of India, he analyses the role of space, stakeholders, institutional structure, and socio-ethnic elements in dealing with regional environmental challenges across this ecosystem. This chapter also critically examines the contemporary development process and practices and attempts to offer a general framework to deal with regional environmental challenges. Chapter 12 on "Groundwater sustainability in Haryana," by Inder Jeet enquires the evolution, trends, present state of groundwater development, management, and governance in the state of Haryana. Some indicators like groundwater level and groundwater quality have been adopted to measure groundwater sustainability. This chapter concludes that small landholdings and intensive agricultural and government policies are the main causative factors of groundwater exploitation in Haryana. The next chapter is on "Wetland resources in the Brahmaputra Valley, Assam: Present status and development prospects" where A. K. Bhagabati and N. Deka study the distribution of wetlands, their status as natural water bodies in the Brahmaputra Valley and presents an inventory of the water and biological resources available in the wetland environments. It assesses threats and pressures on the wetlands and finally suggests some workable strategies and action plans for their sustainable development. The present relevance of the traditional knowledge systems associated with the wetland ecosystems among different tribal and non-tribal communities is also examined in the changing environmental contexts. The last three chapters in this section are concerned with the spatial dynamics of transboundary river basins. In Chap. 14 Ramashray Prasad focuses on "Transboundary water management and governance problems in Kosi Basin," where he identifies various concerning physical issues and factors related to water in the monsoonal regime of Kosi River basin in North Bihar. This basin is located completely in the plains where the water is supplied by the Himalayan catchment lying in Nepal. Many of the problems of Kosi River basin have their genesis lying in the provenance region, over which the Government of India has little control. It is a geopolitical issue and therefore requires intergovernmental cooperation. Moving from Kosi river basin to Teesta River basin, the issues and challenges are nearly the same. Here, in Chap. 15, Sudeepta Adhikari and Subinita Kamle, present their study titled "Governance and Management of Teesta River Water Resources: A Geopolitical Appraisal." River water management is a great challenge as it involves multiple stakeholders. This chapter discusses geopolitical dimension of Teesta River water management. Geopolitical issue assumes great significance in the present context as Teesta water management is a bilateral issue between India and Bangladesh. This chapter also deals with resource use conflicts and strife over inclusive control. The last chapter in this section is concerned with governance issues for sustainable water management in Rapti River basin where Narendra K. Rana and Neha Singh present an empirical study highlighting how integration is difficult in case of a river shared by two riparian nations and identified the complexity caused by multiple stakeholders at the basin level. The study also identifies a number of governance issues like, management of floodplains and its resources, compliance to flood forecasting and warning, public utility management within the active channel zones, annual maintenance of river banks, illegal sand mining, integration of development schemes within the context of floodplain environment, livelihood issues, and the incorporation of community expectations that need to be prioritized for sustainable water management at basin scale at microlevel.

1.5 Part V Exploring Human Dimensions

Part five consists of seven chapters that explore human dimensions and their role in ecosystem services, land use change dynamics, livelihood, and their impact on sustainable land management. Chapter 17 by Krishna Prasad Poudel on "Social Transformation, Ecosystem Services and Resource Sustainability in Nepal Hills" has tried to investigate impacts of social transformation on ecosystem services and sustainability of the resource supply in the hilly region. This chapter is based on an intensive field study in three settlements, i.e., Taksar of Syangja district, Machhapuchhre of Kaski district and Bandipur of Tanahun district from the mid-hills of Nepal. These three villages represent three different ecosystems. With the modern intervention on infrastructural development, education, employment opportunities, diversification on economic activities, and social transformations have been observed. Social transformations and ecosystem services are closely linked. The second chapter in this section by Shahab Fazal, Nasrin Banu, and S. K. Azharuddin titled "Determinants of Land Use Dynamics and its Ecological Implications in India: A State Level Analysis" examines land use dynamics during the period from 1990–1991 to 2010–2011, and its ecological implications, by budgeting different category of land use in India and among states. The study brings out that India is passing through a critical phase of land transformation. The net sown area is decreasing, along with land under pastures, and miscellaneous trees, etc. Urban growth impacts land allocation under the agricultural sector. This change may affect the agrarian economy and lead to ecological challenges.

In this section, the next three chapters focus on the varying dynamics of ecosystem change in the state of Uttarakhand. Chapter 19 is on "Land Use Change and Its Impact on Ecosystem Services: Food, Livelihood, and Health Security in Kumaon Himalayas" by Prakash C. Tiwari and Bhagwati Joshi and presents an illustration of Upper Kosi catchment in Kumaon Himalaya of India. Using remote sensing data, and a combination of qualitative and quantitative methods, this chapter has detected land use change and identified the impact of these changes on population growth. Socioeconomic fallout due to waning of ecosystem services as linked to land use change have also been discussed. In Chap. 20 Bindhy Wasini Pandey, Abhay Shankar Prasad, and Jitendra Kumar Mahto discuss "Impact of Land Use Changes on Livelihood Options: A Case Study of Upper Pasolgad Watershed, Uttarakhand." This case study in a hilly watershed highlights how land use change affects livelihood options. It considers several biophysical parameters including climate change and anthropogenic factors to assess the impact of change. Management of common property resources and sustainable livelihood are intertwined and need due care for watershed development program. Chapter 21 also has its focus on Uttarakhand however with a difference. This chapter on "Rural livelihoods and women: Glimpses from an Indian tribal village" by Purva Yadav, Shreya Akarshna, and Anuradha Shankar draws on insights from a small tribal village called Audali in Uttarakhand, India. The dominant Tharu tribe who migrated to this area from the state of Rajasthan centuries back, finds their livelihood transformed because of changing development process. Role of women became paramount at the household and community level with the changing socioeconomic profile of the household and the village. This chapter also deals with the experiences of the group of motivated Tharu women in this changed milieu.

Different from the experience of Uttarakhand, in Chap. 22, Rajesh K. Abhay and Punyatoya Patra make a compelling case for resilience approach in the long-term as they deconstruct the process of land degradation associated with traditional agricultural practices, deforestation, shifting cultivation, and mining activities in Kendujhar District of Odisha. In the last chapter of this section Nitu and R. B. Singh's study explores the livelihood situation in changing socioeconomic environments of Kangra district of Himachal Pradesh and suggests introduction of medicinal plants as part of crop diversification. The challenges of inequitable use of water and soil resources can no more be addressed through traditional crops. Human capital development through training is considered as an essential input to overcome present agricultural problems.

1.6 Part VI Response to National, Regional, and Global Change

Part six containing 12 chapters looks at a variety of sustainable resource governance challenges at different scales. Chapters examine the response to both policy and process as well as emerging opportunities in the context of climate change, population, and economic growth as well as their resultant impact on ecosystems.

In the first chapter of the book Nuthan Maharaj and Brij Maharaj critically review "Sanitation Challenges and Policy Options in Developing Countries" and present the endemic sanitation challenges experienced by developing countries. They also address the issues with specific reference to women, children, and the disabled. The influence of sanitation problems on the realization of the Millennium Developmental Goals is also analyzed and finally it assesses different policy options to attain sustainable development goals. In a related vein, Surya Tewari's Chap. 25 on "Solid waste management for environmental sustainability in India," dealt with problems of solid wastes. An emerging problem across the world, solid waste management is a major challenge to achieve environmental sustainability. The author's focus is on levels of waste generated and handled at the level of states/Union Territories and cities/towns in the country. The best practices at the country and cross-country level have been documented along with critical evaluation of the new Municipality Solid Waste Rules (2015). Chapter 26 of this volume authored by Dipankar Roy, Shobha Kumari, Akhilesh Kumar Mishra, S C Rai is unique as the author's study the dynamics of "Social Impacts Assessment of Indian Water and

Allied Policies and Programs." Social impact assessment (SIA) is having a noteworthy degree of independent applicability apart from being a subset of an environmental impact assessment (EIA) scheme. The Indian water resource sector is a specific example of the defected practice of SIA and consequent visible problems of inefficiency and underperformance. The focus of this chapter is on the link between poor SIA planning and underperformance of the sector. It has discussed the existing scope of SIA in Indian water resource management during both pre-feasibility and post-development stages. Chapter 27 by Ruchira Ghosh and M Satish Kumar is both compelling in its arguments and provides an elaborate review of challenges in solid waste management in the context of India given how human inflow to cities has made landfill sites supersaturated. The authors of this chapter argue that given the basket of opportunities within waste management in India, while the bandwagon of recycling and reuse is successful, unfortunately, "reduction" has lost its vitality. They articulate the role of smart cities framework as critical in making decentralized management approach, vis-à-vis, the significance of the informal sector and opportunities for women and children addressing livelihood, health, and hygiene. The fifth chapter in this section by Lawal M. Marafa brings out the significance of natural resource evaluation for ecotourism and geotourism in the context of Hong Kong. The author attempts to evaluate and assess the natural resource base and explore a potential site in Hong Kong for ecotourism and geotourism. The study reports a simple and effective method in identifying and assessing resources for ecotourism, geotourism, and nature-based tourism on Tung Ping Chau that can be replicated elsewhere. Such a methodology can identify attractions and inventory relevant resources for sustainable use. The study is expected to help formulate recommendations on planning and management for the sustainable development of ecotourism and geotourism in Tung Ping Chau (TPC). In Chap. 29 Pallavi V. Das draws our attention to the socio-economic impact of climate change in a section of the Western Himalayas. Das focuses on how apple farmers have been adapting in this region to climate change. Unlike the popular perception, both climate scientist and farmer perception of climate change are nearly the same. Das's chapter is evident that stakeholder's perception if honed can be used as adaptive strategy in the wake of climate change in the Himalayan region. In Chap. 30 Swarnima Singh draws on PRECIS data model to study ecosystem services due to changing climate in Kangra district of Himachal Pradesh in India. The meteorological data across western Himalayas are examined and analyzed. There is a visible increase in temperature. North-eastern and south-western parts of the district depict much significant variation in the mean minimum and mean maximum temperature. Both quantity and quality of ecosystem services were found to be affected. In Chap. 31 Shweta Rani documents the intricate relationships between urban infrastructure and the development of the urban environment influencing its sustainability. It is suggested that a proper understanding of this relationship coupled with good governance practices is necessary to address many of the problems of the Delhi Metropolitan Region (DMR). Besides, some of the best practices in good governance pertaining to this mega city have also been highlighted. The development and deployment of sanitation has become a global phenomenon. In Chap. 32, Pooja Yadav and Subhash Anand focus on the present status of sanitation in four resettlement colonies of Delhi. It is found that the services are not adequate and there is scope to improve the situation further. In Chap. 33 Srikumar Chattopadhyay and K N Harilal review water governance in Thiruvananthapuram city of Kerala. They highlight challenges such as overcoming spatial differentiation in service delivery, providing quality service, and devising measures for source sustainability as emerging concerns of urban water management in Thiruvananthapuram city. Despite a strong commitment to decentralization, the water management is fragmented and centralized with little role left for the city authority. Applicability of integrated urban water management concept has been examined. It is suggested that participatory polycentric governance may be adopted to address emerging challenges. In Chap. 34 S.C. Rai and Arpita Panda in studying Bhitarkanika Wildlife Sanctuary, Odisha, layout that Mangrove forests are one of the most productive and biodiverse wetlands. They argue that the Bhitarkanika ecosystem, a Ramsar site, in Odisha, is under stress. The continued exploitation of mangroves has led to habitat loss, changes in species composition, loss of biodiversity, and shift in dominance and survival ability. The study suggests that ecosystem services provided by the mangrove are of immense economic and livelihood value to the coastal communities. Chapter 35 which is also the last in this section by Krishna Kumar and Anjan Sen is unique in design, composition as well as content. This chapter examines Kuchai Kot-Muzaffarpur Section of East-West Highway Corridor to understand how highways have provided better connectivity and accessibility to the urban area to avail better health and educational facilities. This chapter demonstrates that highway corridor is playing an important role in the development of socioeconomic status of the region. Highway development also contributes to increased land price and triggers change in settlement pattern. Although highway generates growth impulses, it also fuels spatial inequality.

1.7 Part VI Future Directions

In this last chapter of the volume, the editors not only summarize the findings of individual chapters but also address the integration of sustainable resource management and governance as well as make recommendations for future research.

Thus, this volume provides global and local level information. We expect that this volume will serve as reference material for academicians, professionals, and students. Location-specific data brought out in some of the chapters will be helpful to the researchers of the locality. Many papers have raised new research questions, which will generate interests among researchers and students.

References

Bringezu S, Potǒcnik J, Schandl H, Lu Y, Ramaswami A, Swilling M, Suh S (2016) Multi-scale governance of sustainable natural resource use—challenges and opportunities for monitoring and institutional development at the national and global level. Sustainability 8:778. https://doi.org/10.3390/su8080778. www.mdpi.com/journal/sustainability

Chattopadhyay S, Franke R (2006) Striving for sustainability- environmental stress and democratic initiatives in Kerala. New Delhi: Concept

Daly H (1987) The steady-state economy: alternative to growth-Mania. Population environment balance, monograph, Lousiana State University, USA, April 1987

Davidson J, Lockwood M, Curtis A, Stratford E, Griffith R (2006) Governance principles for regional natural resource management. Project Report No 1, Pathways to good practice in regional NRM governance Project overview, Land and Water, Australia

Government of India (2017a) Millennium Development Goals-Final Country Report India. Central Statistical Office, Ministry of Statistics and Programme Implementation, New Delhi: Government of India. www.mospi.gov.in

Government of India (2017b) Millennium Development Goals-Final Country Report India. Central Statistical Office, Ministry of Statistics and Programme Implementation, New Delhi: Government of India. www.mospi.gov.in

Graham J, Amos B, Plumptre T (2003) Governance Principles for Protected Areas in the 21st Century. The Fifth World Parks Congress, Durban, South Africa. Ottawa: Institute on Governance, Parks Canada, and the Canadian International Development Agency

Holling CS, Berkes F, Folke C (1998) Science, Sustainability and Resource management. In: Berkes F, Folke C (eds) Linking social and ecological systems- management practices and social mechanisms for building resilience. Cambridge University Press, Cambridge, pp 342–362

Ludwig D, Hilborn R, Walters C (1993) Uncertainty, resource exploitation and conservation: lessons from history. Science 260(17):36

Rockstrom J, Steffen W, Noone K et al (2009) A safe operating space for humanity. Nature 461(7263):472–475

WCED (1987) Report of the world commission on environment and development: our common future. OUP, Oxford

2 Bruce Mitchell: Professional Career and Contributions

Baleshwar Thakur, Rajesh K. Abhay,
Rajiv R. Thakur, and Srikumar Chattopadhyay

Abstract

This chapter is a celebration of Bruce Mitchell's distinguished career as a geographer who has been at the University of Waterloo, Canada for more than five decades. Bruce is an internationally recognized geographer whose specialization is in the field of resource and environmental management. As a widely traveled scholar, his work in geography contributed to various subfields like environmental management, environmental analysis, natural resource management, sustainability, among others. His scholarly and professional engagements took him to many countries including the USA, Europe, India, Australia, Indonesia, China, and Hong Kong in a variety of distinguished capacity. Mitchell's academic footprint earned him recognition in the form of several awards, leadership positions, and honors. His distinguished career is an inspiration to his students, colleagues, and collaborators worldwide. Bruce Mitchell, currently Distinguished Professor Emeritus has left an indelible impact on several generations of geographers in Canada and abroad and will continue to be recognized as a prolific researcher with his stellar contributions to resource and environmental management, especially water resources and in providing recommendations for overcoming challenges in implementing various policies and plans in distant places.

Keywords

Bruce Mitchell · Natural Resource Management · University of Waterloo · Water Resources

2.1 Background

Bruce Mitchell (Fig. 2.1) is an internationally recognized geographer in the field of resource and environmental management. Born in 1944 at Prince Rupert, British Columbia, Bruce Mitchell attended Prince Rupert Senior Secondary School before heading to the University of British Columbia where his love for geography took shape. He was 22 years when he completed his

B. Thakur
Department of Geography, Delhi School of Economics, University of Delhi, Delhi, India

R. K. Abhay
Department of Geography, Dyal Singh College, University of Delhi, New Delhi, India

R. R. Thakur (✉)
Department of Geosciences, Missouri State University, West Plains, MO, USA
e-mail: rajivthakur@missouristate.edu

S. Chattopadhyay
National Centre of Earth Science Studies, Thiruvananthapuram, Kerala, India

B. Thakur et al. (eds.), *Resource Management, Sustainable Development and Governance*, Sustainable Development Goals Series, https://doi.org/10.1007/978-3-030-85839-1_2

Fig. 2.1 Professor Bruce Mitchell (02/12/1944)

Bachelors in Geography and by 23, he had a master's degree in Geography, with the help of the Government of British Columbia Scholarship and University of British Columbia Graduate Fellowship, respectively. His Master's dissertation titled "The Water Component of the Industrial Location Problem: British Columbia's Pulp and Paper Industry," at the Department of Geography, University of British Columbia, perhaps the first attempt in getting the specialization in the field of water resource management with Professor John D. Chapman as his advisor. As a recipient of the prestigious Leverhulme Research Fellowship, Mitchell moved from northeast Pacific location to northeast Atlantic to complete his Ph.D. from the University of Liverpool. He took only 2 years to complete his doctoral thesis, which he worked on "Decision-Making for Water Supply in England and Wales." He did this under the guidance of Professor Stanley Gregory and Professor Edward S. Simpson.

2.2 At Home in Waterloo

Upon finishing his doctoral studies at the University of Liverpool in 1969, Bruce joined the University of Waterloo in southeastern Ontario, Canada as Assistant Professor in the Department of Geography. A man of academic excellence, Mitchell remained here and achieved all his academic accomplishments. He was promoted to the position of Associate Professorship after 4 years (1973) and to Professor after another 4 years in the Department of Geography and Environmental Management (1979). It shows a steep academic development achieved at the age of just 35 years. He held the responsibilities of Chair at the Department of Geography during 1986–1990 and served the Department for more than four and a half decades until his retirement in November 2015.

It is said, a teacher never retires. The University of Waterloo recognized Bruce for his lifelong achievements and bestowed on him the title of *Professor Emeritus* (2015–2016) in the Department of Geography and Environmental Management (GEM). Bruce became Distinguished Professor Emeritus at the University of Waterloo and continues to be active with his scholarship.

Mitchell's career is not limited to academic activities only. He has shown his resourceful capabilities in the form of various administrative responsibilities given to him at the University of Waterloo. He was appointed Associate Dean of Graduate Studies in the Faculty of Environmental Studies (1983–1984) and later was appointed Dean of Graduate Studies and Research in the Faculty of Environmental Studies (1984–1986). He also held the responsibilities of Associate Vice President, Associate Provost, Interim Vice-Provost, at the University of Waterloo (Table 2.1).

2.3 Engagements Outside Waterloo

Mitchell's academic footprint took him outside Waterloo. His influence was felt in Europe, Asia, and Australia. He was Visiting Professor at many universities of Europe. He was the invited Visiting Professor at the School of Geography, University of Leeds, England (1981), and joined the Department of Geography, at the University of Edinburgh in Scotland from October to December in 1982.

Mitchell traveled widely in Asia. He visited India, Indonesia, China, and Hong Kong for his academic and scholarly exchange. In particular,

Table 2.1 Administrative Positions Held by Professor Mitchell at University of Waterloo

Term	Positions held
1983–1984	Associate Dean of Graduate Studies, Faculty of Environmental Studies, University of Waterloo
1984–1986	Associate Dean of Graduate Studies and Research, Faculty of Environmental Studies, University of Waterloo
1986–1990	Chair, Department of Geography, University of Waterloo
1998–2003	Associate Vice President Academic, University of Waterloo
2008–2009	Interim Associate Vice President International, University of Waterloo, July–June
2009	Interim Vice President Academic and Provost, University of Waterloo, July and August
2010	Associate Provost, Academic and Student Affairs, University of Waterloo
2010–2013	Associate Provost, Resources, University of Waterloo
2011–2012	Chair, Confucius Institute Board, Renison University College and University of Waterloo
2012–2013	Interim Associate Provost, Human Resources, September–June

Source: Compiled by the editors

he was invited Visiting Professor in the Department of Geography, University of Madras during winter of 1986. China attracted the attention of Bruce where his engagement brought him back in various capacities. He returned as Visiting Professor in the Department of Geo and Ocean Studies at Nanjing University, Nanjing (1989) and the Wei Lun Visiting Professor at the Chinese University of Hong Kong (2001). He was Honorary Professor at the Dalian University of Technology (2005); Xi'an Jiao Tong University, Xi'an (2009); and at Shenyang University, Shenyang (2011). In addition to these, he was Concurrent Professor at the Nanjing University, Nanjing (2004) and Advisory Professor at the Shanghai Jiao Tong University, Shanghai from 2015 to date. In Australia, he was the invited Visiting Research Fellow in the Department of Geography and Planning, University of New England, Armidale, N.S.W. (1987) and as University Visiting Fellow at the Centre for Water Research, University of Western Australia, Perth (1991). Besides, he was also Visiting Professor in the Department of Geographical Sciences and Planning, University of Queensland, Brisbane (1998).

Besides, the academic positions at several university abroad, he also held advisory and consultative positions. To name a few, he has been the invited Water Advisor at the Centre for Environmental Studies, Universitas Gadjah Mada, Yogyakarta, Indonesia (1996) and Invited Specialist at the Hong Kong Council of Accreditation of Academic and Vocational Qualifications (2010). Mitchell was recognized for his expertise both at home and abroad.

2.4 Awards, Honors, and Recognition

Mitchell's work at various levels, from micro to macro and from local to regional, recognized him with honors and awards. The Canadian Association of Geographers recognized his works in the field of geography and presented him "Award for Scholarly Distinction in Geography" in 1994. The University of Waterloo also gave him Distinguished Teacher Award (1996). He was the recipient of the "Service to Ontario Geography Award," by Canadian Association of Geographers, Ontario Division (1999) and a "Distinguished Service Award" by the Canadian Water Resources Association (2001). In 2008, he was felicitated by the "Massey Medal," by the Royal Canadian Geographical Society.

His contributions to research in water resources led him to become the President of the Canadian Water Resources Association. He served as consultant or advisor to Provincial and Federal Government in Canada as well as the Organization for Economic Cooperation and Development (OECD), the National Academy of Sciences and the World Bank, the International Joint Commission, the Government of Western Australia, Asian Institute of Technology, Bangkok, and numerous private consulting firms in Canada.

2.5 Professional Contribution

Mitchell has been teaching at the University of Waterloo for the past five decades. He is known internationally for his outstanding contributions to resource management and environmental planning. He is one of the most prolific researcher and publisher among Canadian geographers. His work covered every aspect of geography related to resource and environmental management, with particular emphasis on integrated management and policy and program evaluation and has published in most major journals of geography in Canada, the UK, the United States, Europe, Australia, and India.

As Distinguished Professor Emeritus in the Department of Geography and Environmental Management at the University of Waterloo, Mitchell's research focused on resource and environmental management, especially water, and overcoming challenges in implementing various policies and plans.

2.6 Major Contributions

Mitchell's five decades of research has contributed to the various subfields of geography like environment management, environmental analysis, natural resource management, sustainability and sustainable development, human geography, and soil erosion. His contribution in the field of geography is discussed in the following sections.

2.6.1 Natural Resource Management

Mitchell received his higher education in geography at the University of British Columbia and the University of Liverpool. At these institutions, his focus of study was resource management particularly, with respect to water. Consequently, research publications focused on the same. His academic journey started with the publication of his first paper on water pollution control with reference to British Columbia (Mitchell 1968). He made a review of the functions of Pollution Control Board and highlighted how water functions as a carrier to waste. After this paper, he published several papers on resource management. Besides, water pollution, he also published on water quality management (Li et al. 2011) and heavy metal pollution (Ren et al. 2014). Mitchell wrote a small paper on Middle East Water which argued that water is a key issue for the peace, stability, and progress in the development of Middle East countries (Downey and Mitchell 1993).

In 1990, the paper on "Resource Management in Geography: Progress and Opportunities" outlined significant research areas. These were related to forecasting, demand management, technological hazards, noxious facilities, remote sensing and GIS, environmental monitoring and reporting, global change, sustainable development and conservation strategies, and integrated resource management. It was recommended that these are the areas for geographers and others where they should invest time and energy in the years ahead. Mitchell also wrote on several innovations in resource management with special attention to their institutional dimensions (Mitchell 1993).

Within the field of natural resource management, Mitchell kept his focus on water resources since the beginning. Besides writing on water pollution, he along with coauthors also wrote on water use information and sustainable water management with reference to Ontario (Vandierendonck and Mitchell 1997). The paper answers the question that whether there is an adequate database in Ontario to allow to understand "water use" with confidence so that sustainable water management decisions can be made. The study found that there is no uniform, readily accessible database on water use in Ontario. Data about agricultural and rural domestic water use are not being collected or maintained on a systematic basis. There appear to be some serious inaccuracies and inconsistencies in some water use databases, leading to recommendations regarding a protocol for water use data collection, a system for water use information management, and more attention to variability of water use and non-withdrawal uses of water. Professor Mitchell also contributed to forecasting water use (Mitchell and Leighton 1977).

Water management has been his favorite area of research which is quite visible in his works done elsewhere in England and Wales, Israel, Nigeria, and Ontario. In England and Wales, he worked on institutional framework for water management (Mitchell 1970), performance of water system under master plan has been completed in Israel (Kay and Mitchell 1998), and participatory issues in water management have been addressed in the Ontario study (Hofmann and Mitchell 1995). Integrated land and water management has also been suggested in the study of Sokoto-Rima River in northwestern Nigeria.

Water efficiency and water security are other areas related to water which Mitchell contributed. In a paper, Heath and Mitchell (2002) addressed the importance of education in water efficiency programs and reviewed the results of a study undertaken in the Regional Municipality of Waterloo in Ontario to assess the effects of education on participation in voluntary efficiency initiatives.

Integrated resource management (IRM) is the area in which Mitchell worked intensively. It involves the coordinated use and management of land, water, vegetation, and other natural resources in the context of a river basin. He analyzed this approach in Hunter Valley, Australia. The Hunter Valley Conservation Trust had been established in 1950 to address problems of flooding and land degradation. Its activities and performance were evaluated relative to integrated resource management, focusing upon context, legitimation, functions, structures, processes, and mechanisms, as well as organizational culture and participant attitudes (Mitchell and Pigram 1989). The study on IRM related to institutional arrangements regarding nitrate pollution in England (Watson et al. 1996; Mitchell 2005a; b) has highlighted various issues related to the problem. The study found that the emergence of the problem of nitrate pollution has drawn attention to the need for integrated management of land and water resources. It is widely believed that improved resource and environmental management outcomes would occur if those responsible took a long-term view, considered whole systems rather than their components in isolation, and coordinated and integrated their activities with each other. There has been substantial action in many countries to achieve these ideals but with mixed success (Hooper et al. 1999). Shrubsole et al. (2017) analyzed IWRM in Canada where various issues and challenges have been identified to implement integrated water resource management (IWRM). IWRM has also been linked with water risk management. The paper argued that IWRM is one means to achieve effectiveness and efficiency through systematic coordination and collaboration. Effective integration (IWRM) and innovative governance can improve implementation, and thereby enhance capacity to address water risk management (Mitchell 2015).

The role of governance in IWRM has also been emphasized by Mitchell (2013). It was argued that without an appropriate governance system, the scientific and technological aspects of water-related work are unlikely to be effectively implemented at the policy and management levels. Mitchell's contribution can also be seen in the field of river basin management and planning. The contribution can be seen in the form of conceptualizing the river basin management and associated problems and opportunities, particularly, in Canada (Mitchell 1980a, b, 1983a, b, 1986).

2.6.2 Environmental Management

Mitchell's major interest besides, natural resource management, is environmental management. Within environmental management, he has also published on environmental justice and environmental impact assessment. One of the important papers on environment has been published with respect to Bali Sustainable Development Project (Bater et al. 2000). This paper provides a self-evaluation regarding the interactive learning approach applied in the Bali Sustainable Development Project (BSDP). The BSDP was part of a larger institutional capacity building and human resource development project supported by the Canadian International Development Agency. Experiences related to contextual and substantive aspects are reviewed. Particular attention is given to the need for incorporating culture into sustainable development strategies, developing an iterative and adaptive approach, and capitalizing on synergistic opportunities. Another paper on participatory partnerships has also been published with respect to engaging and empowering for enhancing environmental management and quality of life (Mitchell 2005a; b). The paper argues that when partnerships with civil society are created, members of the public become engaged in defining and solving problems, and as a result become empowered through enhanced understanding of substantive problems, and of the processes used by society to deal with them.

There is a wide range of papers published by Mitchell in the field of environmental impact assessment (EIA). They are very relevant with respect to environmental conditions of that time as well as of the current evolving scenario. The approaches to EIA of the Canadian federal and Ontario provincial governments were examined by Mitchell (1976a; b). According to Diduck and Mitchell (2003a) policymakers and scholars have shown increased interest in the learning outcomes of resource and environmental management initiatives. Using a transformative framework and a qualitative methodology, Diduck and Mitchell (2003a) investigated the learning outcomes from involvement in an Environmental Assessment (EA) of a major hog processing facility in Brandon, Canada. The implications for EA process design and the pursuit of key social objectives of sustainability have also been examined in the paper (Diduck and Mitchell 2003b).

2.6.3 Environmental Justice

Environmental Justice is another field of interest for Bruce. Environmental justice refers to the right to a safe, healthy, productive, and sustainable environment for all, in which environment is seen in its totality, and includes ecological, physical, social, political, esthetic, and economic components (Draper and Mitchell 2001). Various issues, responses, strategies, and actions have also been analyzed in environmental justice particularly with reference to Canada (Mitchell 2001). Interesting concepts like LULUs (Locally Unwanted Land Uses) and NIMBYs (Not in My BackYard) have been conceptualized and their relationship with environmental justice (Mitchell 2011). The paper explored the ways in which governments in North America have interpreted and used environmental justice as one means to address issues related to LULUs and NIMBYs and practical experiences or examples from Canada for LULUs management.

2.6.4 Sustainable Development

Mitchell's stay in Indonesia in the 1990s involved him in sustainable development activities. He wrote many papers particularly on the rural development and institutional framework. Developing and implementing sustainable development strategies involve overcoming many institutional obstacles. Some key institutional challenges which most countries must handle to achieve sustainable development, such as realizing cross-sectoral integration, combining top-down and bottom-up planning, and creating credibility for new policy initiatives have been highlighted with reference to Bali, Indonesia. In the paper, strengths and weaknesses related to institutional challenges were identified and assessed, and their general implications were considered. It was also stated that there is a need to develop sustainable development strategies which reflect the conditions and needs specific to the targeted region, and to address different national and regional perspectives (Mitchell 1994a; b; c). Institutional issues have also been assessed in Sokoto State, Nigeria. In Nigeria, as in most countries, institutional barriers to successful rural development exist and need to be addressed. In Sokoto State, a variable and unpredictable Sahelian climate in combination with conditions of high population growth, low levels of income, and environmental degradation, create and intensify institutional challenges. The study examined the sustainability of government initiatives regarding the provision of agricultural inputs, assistance, and extension services to farmers, and the extent to which agency efforts are coordinated and integrated, particularly regarding water management. It was found that Sokoto State has many institutional barriers to sustainable rural development which need to be addressed (Cridland et al. 1995).

Mitchell has also used a "stress-capability framework" to examine the problems and opportunities for sustainable development at the village level in Bali, Indonesia. He argued that Balinese culture incorporates a traditional form of local government which emphasizes cooperation, consensus building, and balance. These aspects provide a strong foundation for sustainable development initiatives. At the same time, many decisions are being taken external to the villages, and even to Bali, which may lead to problems for development initiatives (Mitchell 1994a, b, c). Knight et al. (1997) also linked sustainable development with tourism and coastal management. They argued that distinctive landscape and culture in Bali offers a microcosm to test the concept of sustainable development. Bali encountered significant challenges in the promotion of policies to encourage vigorous economic development, while simultaneously enhancing traditional culture and protecting the integrity of the natural environment. An integrated approach is recommended to protect and rehabilitate linked coastal ecosystems with the context of cultural tourism policies.

2.6.5 Disaster Management

Mitchell was also concerned about disaster management particularly with reference to flood disasters. He extensively wrote on flood, its management, institutional issues particularly with reference to Ontario Province of Canada. In 1978, Mitchell published a paper on physical adjustments and institutional arrangements to flood in Ontario (Mitchell et al. 1978). This paper describes and evaluates physical adjustments to and institutional arrangements for the flood hazard in Bridgeport (Kitchener), Galt (Cambridge), Paris, Brantford, and New Hamburg within the Grand River Watershed, Ontario. The paper finds out four major trends concerning community adjustments to flood hazard. It has been seen that municipalities have been tardy in providing the information needed to make a flood-warning system effective. Further, it has been found out that a recently created Citizen's Advisory Committee has been partially successful due to differing perceptions as to what should be its role in the decision-making process. In another paper, the impact of flood has also been analyzed (Babcock and Mitchell 1980). The study is an examination of the relationship between flooding and property values for an urban community in south-western Ontario which has a lengthy history of flooding. Further, peoples' perception of the effect of flooding on property values is also analyzed. The study concluded that differences in flood risk and flood experience did not adversely affect actual or perceived long-term property values (Babcock and Mitchell 1980). The history of flood damages in Ontario has also been analyzed (Shrubsole et al. 1993).

Floodplain management is also an issue which has been addressed by Mitchell and his colleagues. Bennett and Mitchell (1983) analyzed floodplain management in Cambridge city in Ontario with reference to land acquisition and preservation of historic buildings. Floodplain policy and practice in the Credit River Watershed in Ontario has also been analyzed which revealed that although there has been an increase in the use of nonstructural adjustments, municipalities and residents favor structural adjustments to reduce flood hazard at the community level (Veale and Mitchell 1983). An analysis of the existing and proposed policy in the Grand and Credit River Watersheds related to floodplain management has also been done by Mitchell and Gardner (1980). This section concludes that Mitchell has extensively worked on flood and its management particularly in Ontario Province.

2.7 Other Significant Contributions

2.7.1 Soil Erosion

For Bruce soil erosion was a major theme of research while in Indonesia. His work recognizes that soil erosion in Indonesia has occurred at an alarming rate. Here, a management challenge arises not only because of the lack of understanding of soil erosion processes and lack of data but also due to difficulties in the application of soil erosion models. Kusumandari and Mitchell (1997) tried to measure erosion rates and sediment yields in a watershed in West Java using the Agricultural Non-point Source Pollution (AGNPS) model, to document rates of erosion from forestry and agro-forestry land uses in the basin, and to compare soil erosion rates calculated by the AGNPS and Universal Soil Loss Equation (USLE) models. With McNairn and Mitchell (1991), Mitchell analyzed farmer's perception of soil erosion and economic incentives for conservation tillage in south-western Ontario. Farmers gave a wide range of opinions on the best type of economic incentive. The study suggested that the best combination of educational and economic incentives should be applied to encourage conservation adoption.

2.7.2 Fishing

Fish habitats are being damaged or permanently destroyed through industrial development. Gross overcapacity exists in both the primary and secondary sectors of the fishing industry, which is primarily attributable to the common property status of the resource. Mitchell touched these issues through his research and publications. He wrote on the linkages between politics, fish, and international resource management (Mitchell 1976a, b). He also tried to address the issue of overfishing through the discussion of resource conflict and policy changes (King and Mitchell 1984). The paper argues that fish habitats are being damaged or permanently destroyed through industrial development. Gross overcapacity exists in both the primary and secondary sectors of the fishing industry, which is primarily attributable to the common property status of the resource. The paper suggested that in the short run, there is a need to develop approaches that are implementable in the context of diverse conflicts and vested interests.

2.7.3 Geography, Geographers, and Ethics

Mitchell has not only contributed to various subdisciplines of geography, but he was also concerned about Geography and Geographers and ethics in geographical research. According to Mitchell and Draper (1983) ethical dilemmas involving issues of harm-benefit, privacy, deception, and sponsor relations are encountered frequently in geographical research. Geographers have not always been sensitive to ethical issues, nor have they always been able to reconcile their obligations to understanding and knowledge with those of respecting the dignity and integrity of research subjects. They further state that if geographers became more aware of the issues, principles, and strategies associated with ethics, they should be able to anticipate and respond to these problems in a more systematic manner than they usually do now (Mitchell 1983a; b). Besides, ethics in research and geographer's role, Mitchell was also concerned and addressed the critical issues of geographic education and career opportunities for students of geography.

2.8 Contribution in Development of Models in Geography

Decision-making process in resource management and their final outcomes have caused tensions and conflicts throughout the world. Mitchell provided a framework to improve decision-making. This framework is referred to as the "RESPECT model" and includes the principles of research, equity, sustainability, participatory decision-making, education, communication, and trust. It was stated that the lack of consideration of these principles is the root of tensions in a case study of groundwater extraction in Ontario, Canada. The paper concluded by suggesting that if decisions followed the principles of the RESPECT model, the tensions of the study area could have been reduced or not have occurred and that the RESPECT model could be a useful tool in many water management decision-making processes. Mitchell also wrote on "Models of Resource Management" which evaluated the role and applications of models in both resource analysis and resource management. These models were grouped into biophysical, economic, cultural, and integrative. Mitchell noted that such models have yet to lead the generation of a geographical theory of natural resources or of a more limited theory applicable to a specific open problem in resource management (Mitchell 2003).

2.9 Resource Analysis to Environmental Change to Management

The worldwide recognized his book "Geography and Resource Analysis" outlined the contributions of geographers, and those allied disciplines. The first edition came in 1979 and the second in 1989. This book interrelates research traditions in geography, basic research issues, and natural resource policy concerns. Mitchell discusses how resource analysis has been influenced by recent developments in geography. Among the issues dealt with are resource inventory and allocation, perception and attitudes to resource use, landscape evaluation, concepts of carrying capacity, natural hazards, environmental impact assessment, institutional constraints and opportunities in resource management, resource evaluation, policy formulation, and decision-making. These issues are illustrated with examples of resource problems from Europe, North America, Africa, the Soviet Union, Australia, New Zealand, and Asia.

Environmental policies remain an area of controversy in Canada and around the world. The book on "Environmental Change and Challenge: A Canadian Perspective" (1998–2016 in five editions) is written as an introductory text for students in environmental science or environmental studies. The case studies range from well-known national controversies to local cases using alternative dispute resolution methods to mediate a solution among affected parties. The book evaluates the role of various actors as the question of liability and compensation. Another recognized textbook by Mitchell on "Resource and Environmental Management" served as the standard environmental science course after the first edition published in 1997. Now, 20 years later, an updated third edition (Mitchell 2019) allows for the inclusion of recent developments in resources and management of environment. The book highlights the theories and concepts of environmental resource management in detail.

Today, at the age of 75, Bruce Mitchell is active, alert, and agile as before and is busy writing and editing books and delivering lectures. He is currently working on various research projects as Distinguished Professor Emeritus in the Department of Geography and Environmental Management, University of Waterloo. The editors of the current volume put on record and pay tribute to a man who has helped in treading the path of success in academic as well as related professional fields.

References

Babcock M, Mitchell B (1980) Impact of flood hazard on residential property values in Galt (Cambridge), Ontario. Water Resour Bull 16(3):532–537

Bater JH, Gertler L, Haryadi KD, Martopo S, Mitchell B, Wall G (2000) Capacity building for environmental management in Indonesia. J Transl Manag Dev 4(3–4):107–134

Bennett WJ, Mitchell B (1983) Floodplain management: land acquisition versus preservation of historic buildings in Cambridge, Ontario, Canada. Environ Manag 7(4):327–337

Cridland W, Adeniyi P, Bullock R, Mitchell B (1995) Sustainable rural development in Sokoto State, Nigeria: an assessment of institutional arrangements and management practices. Nat Res Forum 19(3):203–214

Diduck A, Mitchell B (2003a) Learning, public involvement, and environmental assessment: Canadian Case Study. JEAPM 5(3):1–26

Diduck A, Mitchell B (2003b) Curing' affected communities and publics. In: Lawrence D (ed) Environmental impact assessment: practical solutions to recurrent problems. John Wiley, Toronto, pp 268–269

Downey TJ, Mitchell B (1993) Middle east water: acute or chronic problem? Water Int 18(1):1–4

Draper D, Mitchell B (2001) Environmental justice considerations in Canada. Can Geogr 45(1):93–98

Hofmann N, Mitchell B (1995) Evolving towards participatory water management: the permit to take water program and commercial water bottling in Ontario. Can Water Resour J 10(2):91–100

Hooper B, McDonald GT, Mitchell B (1999) Facilitating integrated resource and environmental management. J Environ Plan Manag 42(5):747–766

Kay PA, Mitchell B (1998) Performance of Israel's water system under a new master plan: post-audit and implications for the future. Int J Water Resour Dev 14(1):109–119

King P, Mitchell B (1984) Resource conflict, policy change and practice in canadian fisheries management. Geoforum 15(3):419–432

Knight D, Mitchell B, Wall G (1997) Bali: sustainable development, tourism and coastal management. Ambio 26(2):90–96

Kusumandari A, Mitchell B (1997) Soil erosion and sediment yield in forest and agro-forestry areas in West Java, Indonesia. J Soil Water Conserv 52(5):376–380

Li S, Guo W, Mitchell B (2011) Evaluation of water quality and management of Hongze Lake and Gaoyou Lake Along the Grand Canal in Eastern China. Environ Monit Assess 176(1–4):373–384

McNairn HE, Mitchell B (1991) Farmers' perceptions of soil erosion and economic incentives for conservation tillage. Can Water Resour J 16(4):307–316

Mitchell B (1968) Water pollution control in British Columbia. Effluent Water Treat J 8:295–296

Mitchell B (1970) The institutional framework for water management in England and Wales. Nat Resour J 10(3):566–589

Mitchell B (1976a) Environmental impact assessment: Canadian approaches. Area 8(2):127–131

Mitchell B (1976b) Politics, fish and international resource management: the British-Icelandic Cod war. Geogr Rev 66(2):127–138

Mitchell B (1980a) Models of resource management. Prog Hum Geogr 4(1):32–56

Mitchell B (1980b) Unified river basin management? Can Water Resour J 5(2):87–95

Mitchell B (1983a) Comprehensive river basin planning in Canada: problems and opportunities. Water Int 8(4):146–153

Mitchell B (1983b) Ethical issues in geography - a Canadian perspective. Zeitschrift der Gesellschaft fur Kanada - Studien 3(2):49–62

Mitchell B (1986) Integrated river basin management: Canadian experiences. In: Hydrology and water resources symposium 1986: River Basin Management, Barton, A.C.T., Institution of Engineers, Australia, pp. 140–147

Mitchell B (1993) Institutional innovations in resource management. In: Foster HD (ed) Advances in resource management: Tributes to W. R. Derrick Sewell. Belhaven Press, London, pp 1–17

Mitchell B (1994a) Addressing "Edge" problems in land and water management in The Sokoto-Rima River Basin, Nigeria. Geoforum 25(2):133–143

Mitchell B (1994b) Institutional obstacles to sustainable development in Bali, Indonesia. Singap J Trop Geogr 15(2):145–156

Mitchell B (1994c) Sustainable development at The Village Level in Bali, Indonesia. Hum Ecol 22(2):189–211

Mitchell B (2001) Environmental justice in Canada: issues, responses, strategies and actions. Zeitschrift für Kanada-Studien 21(1):24–43

Mitchell B (2003) Models of resource management. In: Thakur B (ed) Perspectives in resource management in developing countries, vol 1. Concept Publishing Company, New Delhi, pp 102–129

Mitchell B (2005a) Integrated water resource management, institutional arrangements, and land-use planning. Environ Plan A 37(8):1335–1352

Mitchell B (2005b) Participatory partnerships: engaging and empowering to enhance environmental management and quality of life? Soc Indic Res 71(1–3):123–144

Mitchell B (2011) Lulus, nimbys, and environmental justice. In: Fung T, Lesbivel SH, Lam K-C (eds) Facility siting in the Asia-pacific: perspectives on knowledge production and application. Chinese University Press, Shatin, pp 57–83

Mitchell B (2013) Resource and environmental management. Routledge

Mitchell B (2015) Water risk management, Governance, IWRM and Implementation. In: Fra Paleo U (ed) Risk governance: the articulation of hazard, politics and ecology. Springer, Dordrecht, pp 317–335

Mitchell B (2019) Resource and environmental management, 3rd edn. Oxford University Press, New York

Mitchell B, Draper D (1983) Ethics in geographical research. Prof Geogr 35(1):9–17

Mitchell B, Gardner J (1980) Floodplain management in Ontario: an analysis of existing and proposed policy in the grand and credit river watersheds. J Environ Manag 11(2):119–132

Mitchell B, Pigram JJ (1989) Integrated resource management and the hunter valley conservation trust, N.S.W., Australia. Appl Geogr 9(3):196–211

Mitchell B, Gardner J, Cook R, Veale B (1978) Physical adjustments and institutional arrangements for the urban flood hazard in the Grand River Watershed, Ontario. Can Water Resour J 3(2):31–60

Mitchell B, Leighton PH (1977) A comparison of multivariate and trend forecasting Estimates with actual water use 1. JAWRA Journal of the American Water Resources Association 13.4: 817–824

Ren W, Xue B, Geng Y, Sun L, Ma Z, Zhang Y, Mitchell B, Zhang L (2014) Inventorying heavy metal pollution in redeveloped brownfield and its policy contribution: case study from Tiexi District, Shenyang, China. Land Use Policy 38:138–146

Shrubsole D, Kreutzwiser R, Mitchell B, Dickinson T, Joy D (1993) The History of flood damages in Ontario. Can Water Resour J 18(2):133–143

Shrubsole, D., Walters, D., Veale, B., & Mitchell, B. (2017). Integrated Water Resources Management in Canada: the experience of watershed agencies. Int J Water Resour Develop 33(3), 349–359.

Vandierendonck M, Mitchell B (1997) Water use information for sustainable water management: building blocks and The Ontario Situation. Can Water Resour J 22(4):395–415

Veale BJ, Mitchell B (1983) Floodplain policy and practice in the credit river watershed, Ontario. Can Water Resour J 8(3):39–57

Watson N, Mitchell B, Mulamoottil G (1996) Integrated resource management: institutional arrangements regarding nitrate pollution in England. J Environ Plan Manag 39(1):45–64

Part II

Understanding Conceptual Foundations

3 Early and Modern Paradigms in Natural Resource Management: Global and Indian Experiences

Baleshwar Thakur and Rajiv R. Thakur

Abstract

In this chapter, the authors present: (1) India's resource management problems, (2) review paradigm concept, (3) analyze evolution and application of identified paradigms and sub-paradigms, and (4) compare global and Indian experiences. This chapter reviews 12 stratified purposive sample of books and major journal articles in natural resources management between 1935 and 2005. They are focused on the United States, Canada, the UK, and India. The study is divided into Early (1935–1975) and Modern period (1976–2005). Identification of four paradigms and eight sub-paradigms was accomplished by stratification. The methodology of this chapter is qualitative because the analysis is based on purposive selection of one book/journal article from each paradigm and sub-paradigm. The Early Period was divided into ecological, integrated, economic and technological-I, and Modern Period into decision-making and technological-II. Both technological-I and II were further divided into four sub-paradigms each. In addition, the study adopted any four of seven criteria for the selection of paradigm developer and exemplar book/journal article. Our study demonstrates that: (1) technological-I and II have been the prominent uninterrupted sub-paradigms during the past 45 years, (2) dominance of non-technological paradigms in the Early Period and technological sub-paradigms in the Modern Period. (3) Sir A. Tansley, Gilbert White, Piers Blaikie in the Early Period and Bruce Mitchell in the Modern Period as the nontechnological paradigm developer. On the other hand, the technological paradigm developers in the Early Period are R.P. Misra, David Ebdon, D.W. Rhind, T.M. Lillesand, and R.W. Kiefer; and P. J. Taylor, John Campbell, J.R. Jensen, C.P. Lo, and A.K. W. Yeung in the Modern Period.

Keywords

Decision-making · Exemplar · India · Natural Resource Management · Paradigms · Technological

3.1 Introduction

Nature-society interaction producing natural resources for the satisfaction of human wants is fraught with many issues and challenges. For instance, recently, interest has focused on unequal

B. Thakur
Department of Geography, Delhi School of Economics, University of Delhi, Delhi, India

R. R. Thakur (✉)
Department of Geosciences, Missouri State University, West Plains, MO, USA
e-mail: rajivthakur@missouristate.edu

B. Thakur et al. (eds.), *Resource Management, Sustainable Development and Governance*, Sustainable Development Goals Series, https://doi.org/10.1007/978-3-030-85839-1_3

land access and redistribution; land degradation and desertification; soil erosion, waterlogging, and salinization; deforestation, shifting cultivation, and loss of biodiversity; water scarcity; declining water quality and groundwater depletion; transboundary water resources and interlinking of rivers; increasing hazards and degrading atmosphere and inadequacy of minerals and energy crisis. Multidisciplinary research is essential for addressing many of the arising issues in natural resource management as its domain cannot be adequately addressed through a single research discipline (Janssen and Goldsworthy 1996). However, what multidisciplinary research signifies and how it can be put into practice is not always clear to the resource managers, academicians, and practitioners. The current study presents and discusses the review paper under four headings: (1) resource management problems in India, (2) concepts of paradigm, (3) evolution and application of identified paradigms and sub-paradigms, and (4) comparison of global and Indian experiences.

3.2 Resource Management Problems in India

India is the seventh largest country in the world and the second most populous. However, the land-man ratio in India is not as favorable as in many other countries like Australia, Canada, Argentina, the USA, Chile, Denmark, and Mexico. Conversely, the land-man ratio is more favorable in India than Japan, the Netherlands, Egypt, the United Kingdom, Israel, and China (Thakur 2007). India has many physiographic divisions like mountains, plateaus, hills, lowlands, and valleys. Mountains over 7000 ft high constitute 11% of the country's area, plateaus below 4000 ft comprise 28% of the total surface area, hill land between 1000 ft and 7000 ft embrace 18%, 43% under 1000 ft comprise lowlands, and nearly 95% of it is suitable for cultivation (Noble 1972). A notable example of research in this area is by Dayal (1972), who concludes that about two-thirds is topographically usable. In this context, the diverse behavioral response, as well as human-environment relationships, to the use of land resources in each ecological setting are distinct. The literature survey of resource management problems in India identifies the following eight issues:

3.2.1 Unequal Land Access and Redistribution

Agricultural land is considered as the base for economic growth and development in Indian villages. It provides day-to-day commodities like food, fodder, fuel, and fiber to the people. Unequal accessibility is a major obstacle in the power structure in rural India. Land access is defined as the process by which people gain rights to occupy and utilize land on a temporary or permanent basis (Thakur and Sinha 2017). Dominance refers to control or exercise authority over land. The village having two types of caste dominance is known as dominant caste and nondominant caste. Dominant caste is the base of economic and social power structure. They are the rural elites and occupy the best land in the village, i.e., most fertile land (Thakur and Das 2013). Srinivas (1959) explained the concept of dominant caste which wields economic and social power and owes the major portion of agricultural land. On the other hand, nondominant caste possesses little amount of land or are landless or agricultural laborers. They possess only marginal and low fertility land. The *usar* and *banjar* lands are generally in the possession of nondominating castes who are poor (Thakur and Sinha 2017).

Consequently, the Government of India adopted land reform policies since independence. The policies were at the top of the agenda in each Five-Year Plan. Land reform measures involved the following: (1) remove hindrance to increase agricultural production, (2) eliminate exploitation and social injustices, and (3) provide security for the tiller of soil (Thakur 2007). More importantly, the crux of the policies was: (1) land to the tiller, (2) redistribution of land in favor of landless and small cultivators, and (3) tenancy reforms. The 1950s–1970s have been a period of expansion in approaches taken in this regard. The *Bhoodan movement* launched by Acharya Vinoba Bhave was born in the mid-1950s. By and large, the decades from 1955 to 1970 saw the achievement of Vinoba Bhave, who walked through the country covering 80,000 km (49,710 miles) for 14 years and obtained 42 lakh acres (4.2 million acres) of land as *gramdan* (Natarajan 1995). Such land was made available free for distribution to the tillers. Sociologists, anthropologists, geographers, and economists have worked on land problems, land access, and poverty in India.

3.2.2 Land Degradation and Desertification

Land degradation and desertification appear frequently in Indian literature. Land degradation is defined as the reduction in the status of land quality. The causes of land degradation are complex and diverse. Thus, land degradation signifies a loss in land productivity which restricts the land's productive capacity and reduces its ability to sustain yields or reduce economic returns (Chisholm and Dumsday 1987). The degraded land is the damaging threat which affects the physical, chemical, and biological status of land and decreases its usefulness. Johnson and Lewis (1995) argue that land degradation is a human-induced process and describes human interference as the fundamental cause of degradation. All land is not equal, some land is better than others. The total degraded land of India is 146.82 m ha or approximately 45% of the country's total geographical area. Today, at least 2 out of every 3 acres of land is in poor health, one-third of the total is almost completely unproductive, another one-third is partially productive, and it is only the remaining one-third which is in good health (Thakur 2020). During the past three decades, several publications have appeared in India. Kumar and Bhandari (1993) investigated socio-economic aspects of land degradation in sand dune ecosystem in Churu and Jhunjhunu district. Patra (2006) studied land's susceptibility, classified it on its characteristics, measured land variations, causes, and trends of land degradation, and proposed guidelines for their regeneration in the Northern Highlands of Odisha. Abhay (2015) developed Susceptibility-Resilience Matrix (SRM) for identifying critical areas with respect to land degradation in Kendujhar Plateau to improve land management practices.

The progressive downgrading of land is known as desertification. Therein, a particular concern is the impact of climate change, environmental degradation, unsuitable land use practices, improper soil, and water management, reduced productivity of land, land abuse, and overgrazing (Sharma 2020). Concerns have also been expressed regarding the expanding desert from core to periphery. It happens most rapidly in times of drought and varies a great deal from place to place. Desertification, which is a decrease in the productivity of the land under arid, semi-arid, and dry subhumid climates and which may eventually lead to desert-like conditions, is prevailing in Rajasthan. The Thar Desert of Rajasthan has advanced 1500 km eastwards in the last 8000 years. In the last 50 years, it has encroached upon 13,000 hectares of land in Rajasthan, Gujarat, and Haryana (Valdiya 1987). Dhabaria, Kar, Sharma, and Singh pioneered the geographic research in Rajasthan on expanding desert to northeast Rajasthan with the help of Remote Sensing and GIS methods. Sharma (2020) argued that the decision-makers in the center have now realized the serious magnitude of desertification and the ecological crisis in the Aravalli range. Thus, land degradation and desertification in India are as much a socioeconomic problem as they are a biophysical problem. Geographers and geoscientists have contributed more to the study of processes, patterns, and managerial policies to land degradation and desertification in India.

3.2.3 Soil Erosion, Waterlogging, and Salinization

Soil erosion is described as the carrying away of soil. It is the theft of soil by natural elements like water, wind, waves, and glacier. Erosivity and erodibility are important factors of soil erosion. It occurs in many parts of India and is regarded as the most serious and widespread form of land degradation. In India, soil erosion has emerged as one of the important resource management problems which has received considerable attention. Out of the total geographical area of 329 million ha of India, about 173 million ha are subjected to varying degrees and forms of soil erosion (Soil and Water Conservation Annual Report 2000). Soil erosion occurs in diverse geographical settings, ranging from Jammu and Kashmir to Kerala, from Narmada valley to Barak valley and is also heavily concentrated from Chambal valley in Central India to Chotanagpur in the northeast Peninsular plateau. Thus, gullies and ravines are localized within a triangle whose nodes are Surat, Jammu, and Murshidabad. Physical geographers in India have made significant contributions in locating four major areas of ravines and gullies within this triangle (Ahmad 1970; Sharma 1980). They are: (1) The Yamuna-Chambal ravine Zone, (2) The Gujarat ravine Zone, (3) The Punjab Shivalik Foothill Zones, and (4) The Chotanagpur Zone (Thakur 2020). As alluded to earlier, many techniques to combat soil erosion have been introduced by the Government in India to hold the soil in place and may not be drifted from original place. For example, they are improved agricultural practices, contour terracing on the hill slopes, planting of trees in desert regions, afforestation in the upper watersheds in the Himalayas, check dams, gully plugging, land shaping, water harvesting, and control of shifting cultivation (Thakur 2007). Taken together, in India, soil scientists, agricultural scientists, geographers, and geologists have worked on soil erosion, waterlogging, and salinization.

Waterlogging is another problem the country is facing today. In India, 6–10 million ha of infertile land suffer from waterlogging. Many factors contribute to waterlogging. They include overuse of water in irrigated areas, seepage of water from canals and distributaries, lack of land development, adoption of unsuitable cropping patterns, poor water use practices, and poor conjunctive use of groundwater facilities (Bowonder *et al.* 1987). A large part of Punjab, Haryana, Rajasthan, and Gujarat has developed serious waterlogging problems. Resource geographers, in India, have demonstrated that due to excess irrigation, waterlogging chokes plants and kill them.

Soil salinization is another problem arising from canal irrigation. Livernash (1995) offered an understanding of soil salinization that when water table is a meter or so below the soil, water flows to the surface, evaporates, and leaves salt deposits. Research has shown that, currently, about 3 million ha of land are influenced by salinity (Livernash 1995).

3.2.4 Deforestation and Shifting Cultivation

Forests are renewable resources. They are prominent geographic features of India (Thakur 2007). In 2017, India covered 21.54% under forest. They contain most of the terrestrial biodiversity and provide livelihoods to over a million people. Forests have a major role in climate change mitigation, in economic development of the country, moderating local climate, regulate stream flow, reduce soil erosion, control floods, enhance the quality of environment, act as carbon sinks and improve human values and spiritual strength (Thakur 2007). Thus, forests provide numerous benefits to society and tribal and forest dwelling communities. Haigh (1998) stated that deforestation is an environmental problem that threatens the survival of the entire ecosystem. It is defined as the complete removal of tree formations and their replacement with other land uses. Most importantly, it reduces biodiversity, increases global warming, and expands deserts.

The history of deforestation in India is associated with political kingdoms. During the rule of Ashoka and Akbar, India was rich in forests. The colonial government accelerated the process of deforestation for approximately 200 years by placing demand on forest-based commodities (Gadgil 1990). For example, teak forests along the Malabar coast were overexploited for British Navy, sandalwood trees of Karnataka went to Europe, railway expansion (1853–1891) used Himalayan deodars, and Garhwal and Kumaon

Himalayan trees were cleared for agriculture (Thakur 2007). Since, India's independence, deforestation has been witnessing alarming spread across the country. At the end of the twentieth century, the situation changed completely. Between 1951 and 1980, about 4.3 million ha of forest land had been distracted for non-forest uses. As a result, factors like decreasing forest cover have put stress on the forest resources in the tribal areas. Studies also show that during 1987–1995, forest cover declined by 1893 km^2 in the north-eastern region due to jhum cultivation, fire, and flood, illicit falling, and growth of wood-based industries (Thakur 2007). The years 2015–2017 were monumental for forest cover increase in Andhra Pradesh (2141 km^2), Karnataka (1101 km^2), and Kerala (1043 km^2) followed by Odisha, Assam, Telangana, Rajasthan, Himachal Pradesh, Uttar Pradesh, Jammu and Kashmir, and Manipur. Secondly, during the same year, major part of loss in forest cover occurred in northeastern region due to shortening of shifting cultivation cycle along with biotic pressure, particularly, in Mizoram, Nagaland, and Arunachal Pradesh (Rani and Thakur 2020). The important causes of deforestation are the following: (1) rapidly increasing population, (2) conversion of forest to agricultural land, (3) source of fuel, (4) irrigation and power projects, (5) construction of road, and (6) deforestation due to medical resources.

Shifting cultivation, also known as slash and burn, is a popular method of farming in northeastern region. It literally means changing the position and direction of farming sites over sloping land. It is a temporary use of rainforest land for agriculture by cutting and burning trees and bushes. Their ash serves as manure, the soil becomes exhausted and fertility of the site declines after 3–4 years (Shafi 2006). Then the farming sites are abandoned, and fresh sites are selected. Thus, decline in fertility followed by degraded land and selection of fresh site and repeated practice on new land are characterized by soil erosion and deforestation. Different terms are used to connote this cultivation in different parts of India like *jhum* in Assam, *anam* in Kerala, and *podu* in Odisha (Singh and Saroha 2014). Ecologists, geographers, meteorologists, and environmentalists have contributed to the study of deforestation problems in India.

3.2.5 Water Scarcity, Water Quality, and Groundwater Depletion

Lack of adequate water is one of the serious resource problems for human survival in India. Such problems vary from urban to rural and region to region. Water scarcity refers to situations in which water resources available for producing output are insufficient to satisfy human needs (Mehta 2003). Water scarcity is the result of imbalance between water availability and increasing water demand, rapid urban population growth, rapidly growing human population, inadequate access to water, inefficiency of public water supply, agricultural water use efficiency, water pricing policies, water loss, augmenting water supplies, challenges in water management, and changing lifestyle both in urban and rural areas. There is a thirst for water in mega cities, million cities, and even cities in India. The megacities, Delhi, Mumbai, Kolkata, Chennai, Bengaluru (formerly Bangalore), and Hyderabad form the apex of Indian urban system and are characterized by a strong diversified economy.

The biggest problem associated with megacities and million cities dynamism is the provision of an adequate and safe quality supply in the face of greatly accelerated demand. Due to the phenomenal growth of the population, water supply did not keep pace with demand. There is a constant demand-supply gap of about a thousand million liters of water every day. Some 120 hill towns located at elevations of 1500–3000 m above sea level were established by the British on the lower mountain ranges of the Himalayas. Until now, the hill towns have depended on its springs and gravity sources of water. However, water supply is seriously threatened because of deforestation and dying springs. The inadequacy in water supply is accentuated by demands in peak tourist season, lack of additional source of water, lack of proper infrastructure, irregular and low voltage of power supply, insufficient storage capacity of water, and inadequate distribution system of water, as well as inadequate capacity of pumps. In Lucknow, a city with 4 million people, water is available only 10 h each day (Roy 1994). In Rajkot, a city with a population of 0.7 million, the piped water runs for only 20 min (Bhatt 1994). Mathur (2004) provides a conceptualization that distinguishes water scarcity in Lucknow district covering urban and rural parts. The author has also investigated the impact of deteriorating water quality on water scarcity and has shown an increase in tendencies toward socioeconomic and political factors. In another study, Abhay (2012) has provided insights by demonstrating that all the income groups in sampled village in the Sikar district, Rajasthan are dissatisfied with water supply in rural areas. He has also demonstrated an empirical basis for water use for irrigation in rural areas and for domestic use in urban areas.

Concerns about the deteriorating quality of water have spurred an increasing level of research in India. Water quality indicates the suitability of water to sustain various uses or processes (Trivedi 2008). Many studies have addressed the issues of increase in population, unchecked urbanization, unplanned industrialization, and fertilizers in agricultural development of the quality of water both in urban and rural areas. The Central Board for the Prevention and Control of Water Pollution was constituted in 1984 to study water pollution in both the Ganga and the Yamuna basins (Thakur 2007). Pollution in the Ganga River has been caused by pilgrimage activities, scattered ashes of cremated bodies over the river, and generations of sewage from the cities located on the banks of the Ganga in Uttar Pradesh, Bihar, and West Bengal (Livernash 1995). These practices contaminate river water. In 1985, and afterward, the Ganga Action Plan and Yamuna Action Plan were created to clean up the polluted section of the river for which millions of rupees have been pumped by the center, but this is not helping the river (Narain *et al.* 2018). Another management problem is related to megacities. For example, in Delhi, the physiochemical analysis of drinking water shows the quality of water deteriorates with distance from the Yamuna River. The area falling to the east of the Yamuna exhibits a contrast in water quality (Balaseetha 1986). As monsoon arrives, Mumbai's water supply gets contaminated and sewage seeps through the pipelines and contaminates drinking water. Much of the drinking water in Kolkata is contaminated with human excrement and the arsenic contamination problem which has been reported from time to time. Poor drinking water quality and irregular garbage disposal have become important issues for the people of Chennai.

Groundwater depletion is emerging rapidly in the wake of freshwater requirements for irrigation, domestic need, and industrial demand in the country. Groundwater is contributing significantly to the development of Indian agriculture for attaining food security. Agriculture is the primary user of groundwater. Wells and tube wells are the main ways by which irrigation is practiced. Thus, groundwater depletion is defined as the sinking of water table at a fast rate to utilize freshwater for different purposes. There has been a sudden increase in groundwater withdrawals in northwestern, western, and southern India for meeting the requirements. The number of agricultural pumps used in India over the past 50 years has proliferated from 5000 in the 1950s to more than 6 million along with 60,000 deep tube wells by the early twenty-first century. A scientific research by Jeet (2001) addresses the magnitude and rate of groundwater depletion and the impact of factors, especially mechanical lifts, amount of rainfall, drainage characteristics, groundwater quality, rice cultivation, and spacing of tube wells on the depletion. Finally, the study investigates the dynamics of groundwater depletion and its management in eastern Haryana including the districts of Ambala, Kurukshetra, Yamunanagar, Karnal, Panipat, Sonipat, Faridabad, Gurgaon, Rewari, and Mahendragarh.

3.2.6 Transboundary Water Resources and Interlinking of Rivers

The distribution of potable water in India is uneven. Its unevenness is threatened with perceptible climate change. The rise in the population of India has been unprecedented. The modern human lifestyle has created a situation in which, the water quality has gone down leading to further lowering of per capita potable water availability. Life without sufficient usable and consumable water cannot be imagined. The importance of water is not hidden to anyone. Sustainable development goal for clean water and sanitation explains the need of its urgency to achieve the target by 2030.

The flow of water knows no political boundaries. It runs along the slope under the force of the earth's gravity. Therefore, basin/watershed-wise management of water is the best way, but regions and administrative units are governed by different administrators who perceive the challenges differently. From the perspective of administrators, water management may be understood in two different ways—transboundary and Inter-State Water Disputes (ISWD). The first one is concerned with the management of water running through two or more than two states (nations) where mutual agreement is to be worked out under the international water laws. The Nile River in Africa and the Jordan River in West Asia are some examples of very well-managed water resources and agreed upon to share the benefits among stakeholder countries. It has happened due to maturity displayed by the concerned nations. More than 263 watersheds and 300 aquifers across political boundaries of two or more than two countries have been identified over the globe. This accounts for 45% of the earth's area, 40% of the world population, and about 80% of the entire global flow of the river. Despite all these challenges, water scarcity is a big issue all over the world. It is a great challenge to reach an agreement for proper implementation of water management plans for the betterment of mankind. It must be managed in such a way so that there are no negative impacts on our environment and a sustainable approach is applied. Before benefits sharing agreement is reached, inventory of water resources of the basin is considered at the first stage like collection of hydrological data, analysis, estimation, evaluation, basin coverage area, cultural aspects of water, and its use by the community of the countries, etc. Involvement of local-level community leader, NGO's administrator, basin-level authorities, etc., are involved to get feedback and then water and other related benefits are agreed to share with the concerned nations.

The second one is concerned with those rivers which fall within the same nation but fall within different jurisdictions, i.e., state/ province. Their approach is simply to serve the need of the constituency they are representing. This sort of attitude to get more and more water, even at the cost of others, becomes the focal point. It has led to interstate water disputes. The burning example of this type of problem in India is: (1) Cauvery water dispute between Tamil Nadu, Karnataka, and Kerala, (2) The Krishna water dispute between Maharashtra, Karnataka, and Andhra Pradesh, (3) The Godavari water dispute between Andhra Pradesh, Madhya Pradesh, Chhattisgarh, Odisha, and Karnataka, (4) The Narmada water dispute between Gujarat, Maharashtra, Madhya Pradesh, and Rajasthan, (5) The Ravi and Beas river water dispute between Punjab, Haryana, Himachal Pradesh, Rajasthan, Jammu and Kashmir, and Delhi, and (6) The Satluj-Yamuna link canal dispute between Punjab, Haryana, and Rajasthan in India. Generally, this kind of dispute is referred to river tribunal and the aggrieved parties present their case and, finally, it is resolved. Once the resolve is not acceptable to the party, legal procedure is followed within the jurisdiction of a sovereign country. The primary emphasis is to reach an understanding about the share of benefits of water and its related outcome for the better life of the people of their constituency.

The interlinking of rivers (ILR) in India is a mega-project. It is a miracle solution to water scarcity and monsoon floods. The National Water Development Agency (NWDA) was established in 1982 on the pattern of "Ganga-Cauvery Link Canal" (1972) and "Garland Canal" (1977). The main objective of NWDA is to connect the surplus water region of northern, eastern, and north-eastern parts with the dryer region of southern and western parts of the country. The interlinking of rivers contains Himalayan and Peninsular components. ILR is the Government of India's proposal to link 37 rivers through 31 links (14 in the Himalayan and 17 in the Peninsular), dozens of large dams and thousands of kilometers of canals making it the largest water project in the world (Thakur and Abhay 2020; Bandyopadhyay and Parveen, 2004). The Project's estimated cost is Rs. 560,000 crores (5,600,000 million) which seems a huge investment. However, there are many benefits of the project, for example, (1) transportation of surplus water from the north and east to deficit south and west, (2) flood problem in North Bihar and Assam will be solved, (3) 34,000 MW hydroelectricity would be produced, (4) will provide inland navigation, (5) drought-affected areas will get water for drinking and irrigation purposes, (6) interstate water disputes will be resolved, and (7) metros and megacities water needs will be satisfied. The Project has been criticized on account of mammoth investment of public money, huge power, and technology in lifting the water from surplus to deficit basins, destruction of ecology due to construction of big dams and canals, displacement due to reservoir construction, and consultation with neighboring countries.

3.2.7 Increasing Hazards and Degrading Atmosphere

This section describes the state of natural hazards in India. Hazard brings loss of life, property, and livestock. It is generally considered as a process causing an extreme event. On the other hand, disaster is an abrupt and unexpected calamity with serious consequences. Kapur's (2010) book provides an overview of India's disasterscape and identifies those disasters are now considered to be socially constructed. The Himalayas, monsoon rainfall, and complex river systems surrounded by oceans on three sides have produced half a dozen natural hazards in the country. On an average 57% geographical area of India is vulnerable to earthquakes, 28% to droughts, 12% to floods, and 8% to cyclones (Samra et al. 2006). Most importantly, the distribution, occurrence, frequency, magnitude, and duration of flood have been influenced by heavy or prolonged rainfall, uncontrolled deforestation, silting of rivers, advent of embankments, faulty land practices, and increasing anthropogenic pressure. The impacts of flood are devastating in North Bihar between rivers Gandak and Teesta and further in the Brahmaputra valley. Drought is the outcome of abnormal water deficiency (Thakur 2003a, b). The principal cause of drought is erratic behavior of the monsoon, excessive withdrawal of groundwater, fall in water table, and deforestation. Thus, the drought-prone areas cover parts of Rajasthan, Gujarat, Madhya Pradesh, Chhattisgarh, Odisha, Karnataka, and Maharashtra. The country experienced 1966 and 1987 two major droughts which affected agriculture, livestock, water resources, and livelihood security of the people. Cyclones are characterized by strong winds, heavy rains, devastating storms, resultant floods, and high storm tides (Thakur 2003a, b). The Indian Ocean is the major cyclone origin region which affects the east coast of India. In fact, more cyclones are formed in the Bay of Bengal than the Arabian Sea. Coastal regions covering West Bengal, Odisha, and Andhra Pradesh experience destructive winds. Such winds in 1999 were known as "super cyclone" with casualties more than 10,000 in Odisha. India is located on the boundary of Tibetan and Indian plates. Earthquakes are caused by the collision of the two plates and the subduction of the Indian plate beneath the Tibetan plate (Thakur 2003a, b). The country has been divided into five seismic zones by the Geological Survey of India: Zone-I, Zone-II, Zone-III, Zone-IV, and Zone-V based on the severity of the earthquakes. For instance, Zone-I is seismically least active, while Zone-V is seismically more active. Based on the distribution of earthquakes, the country has been divided into the Himalayan Zone, the Indo-Gangetic Zone, and the Peninsular Zone. Landslide is a form of mass movement on steep mountain slopes containing rock debris, mud, and soil. Landslides triggering factors are heavy rainfall, cloud burst, deforestation, road construction, and earthquake-induced materials. The geology and topography of the slope also play an important role when the materials move down. The important examples are the Himalayas, Western Ghats, Nilgiri Hills, Meghalaya Plateau, and Arakan-Yoma belt of northeast India (Bagchi, Kumar and Sharma 2011). The sub-Himalayas,

Rishikesh-Badrinath Highway, Alaknanda tragedy of 1970, and Malpa tragedy of 1998 are famous.

Tsunami has been identified in India recently as the deadliest natural disaster. A massive tsunami was struck on December 26, 2004, along the eastern coast of India including Indonesia, Thailand, the Andaman and Nicobar Islands, Sri Lanka, the Maldives, and eastern Africa along the Indian Ocean. It is an earthquake-related hazard under the sea which is caused by the tectonic displacement of the seabed and volcanic eruptions. "Tsunami waves occur all of a sudden due to seismic activities in the bed of the sea, which comes without any warning" (Sinha 2014). "Tsunami waves travel at a speed of 725–800 km/h, swells up to 15 m high and breaks like a wall" (Sinha 2014). Research has shown that many scientific studies have increased our understanding about coastal morphological change of affected areas, coastal ecosystems, sea water intrusion, damage caused by tsunami, and destruction of coconut and other trees (Karan and Subbiah 2011).

The following section presents changing characteristics of urban atmospheric resources and their causative factors. The atmosphere is made up of a mixture of gases. Nitrogen (78%), Oxygen (21%), and carbon dioxide (0.03%) by volume for providing fertilizers for plants, for life to living beings, and chief source of carbon for plants. Air is ubiquitous but is a troubled and polluted resource. It is becoming increasingly perceived in terms of resources. It is a vital flow resource that lasts indefinitely. In four megacities, industries and automobiles are the important pollutants. Badarpur and Indraprastha thermal power stations are important in the National Capital Region of Delhi. Together they emit 180 tonnes of fly ash and 70 tonnes of sulfur dioxide everyday (Agarwal *et al.* 1982). Most importantly, Delhi has the highest number of cycle trips and Mumbai has the highest number of walking trips. Although cars meet less than 10% of the travel demand of Kolkata, but the city has banned bicycles on 174 roads to make way for cars (Narain *et al.* 2018). In Mumbai, several cotton mills are located near the docks on the eastern part of the island between Thane Creek and Mahim Bay (Dutt and Geib 1998). The most common public transport in megacities is the bus service because they allow greater flexibility, greater geographical coverage and accessibility, cost-effective travel, and space efficiency (Narain *et al.* 2018).

3.2.8 Mineral Problems and Energy Crises

Minerals are essential and the base for industrial development of the country. India is rich in metallic and nonmetallic minerals. The country produces as many as 64 minerals, 4 fuel minerals, 11 metallic minerals, and 49 nonmetallic industrial minerals (Lahiry 1997). Iron, manganese, copper, and bauxite are metallic minerals, while mica, limestone, marble, granite, sand, and gravel are nonmetallic minerals. The country is self-sufficient in coal and limestone and has a surplus of iron-ore, manganese, mica, kyanite, and chromite (Dutt and Geib 1998). It is also emphasized that the country is deficient in petroleum, copper, zinc, and tin and must import them (Dutt and Geib 1998; Mukerjee 1992).

India has probably the world's largest reserves of iron-ore in the Dharwar and Cuddapa rocks of the peninsular plateau. They are principally located in Singhbhum district in Jharkhand and Kendujhar, Sundargarh, and Mayurbhanj districts in Odisha. It is known as the iron-belt which contains high-quality ore. The country contains abundant manganese reserves and is the second largest exporter of manganese in the world. The noteworthy deposits of manganese exist in Nagpur-Bhandara in Maharashtra and Balaghat-Chindwara in Madhya Pradesh. Bauxite, a raw material utilized in the aluminum industry is adequate. It is mined principally in the Lohardaga region of Jharkhand. Mica and limestone are important nonmetallic minerals in India. India produces 75% of the world's mica. Mica deposits are located principally, in Giridih-Kodarma region of Jharkhand. It is mainly used in the electrical industry. It is the larger exporter in the world and exports to the USA, Japan, Russia, Germany, France, Belgium, and Holland.

Limestone is found in the Vindhyan quarries of the Sone valley, Andhra Pradesh, Madhya Pradesh, Ajmer, and Jabalpur. It is used in cement, steel making, and chemical industries.

India is facing an energy crisis. The unequal distribution of energy resources and the imbalance between demand for energy and power supply is acute. However, more recently, India has made progress in reducing the gap between supply and demand of energy resources. Commercial energy sources consist of coal, petroleum, hydroelectricity, and nuclear power. India is among the 10 top countries with vast reserves of coal. However, the quality of Indian coal is not good enough. Coal is, principally, found in two geological formations in India: Lower Gondwana and the Tertiary (Thakur 1998). The coal deposits in both the formations are distributed unevenly. More than 80% of the reserves are in Jharkhand, Odisha, Chhattisgarh, West Bengal, Madhya Pradesh, and Andhra Pradesh. Movement of coal within India emanates from these areas to northwestern, western, and southern regions. Hence, transport of coal over long distances involves a heavy strain on transport charges to the consumers situated in remote areas (Dayal 1958).

After coal, petroleum is the second most important source of energy. Petroleum occurs in sedimentary rocks of marine origin and is called "Liquid Gold." India contributes about 1.5% of the total petroleum production of the World, but its known resources are limited. The production of petroleum was 105 lakh tonnes (10.5 million) in 1980 which increased to 324 lakh tonnes (32.4 million) in 2000 and which further increased to 380 lakh tones (38 million) in 2010 (Singh and Saroha 2014). Petroleum is obtained from four regions: (1) Bombay High, (2) Brahmaputra Valley, (3) Gujarat Coast, and (4) The Eastern Coast. The Bombay High region is the biggest and most important oilfield located 176 km northwest of Mumbai. It is the leading producing area since 1976. The Brahmaputra Valley is the oldest oilfield in India which contains Surma Valley, Dibrugarh, Sibsagar, Digboi, and Naharkatia. The Gujarat Coast is the second largest oil producing area in the country with Ankleshwar, Cambay region, and Ahmadabad region. The Eastern coast has many oilfields discovered in deltaic regions of Kaveri, Krishna, Godavari, and Mahanadi rivers. India is not an oil exporter as about 60% of oil requirements are met through imports. Saudi Arabia, Iraq, Iran, Kuwait, and Bahrain are chief suppliers of petroleum to India. A network of pipelines has been laid down connecting oilwells to 21 refineries and transport petroleum products to consumer markets.

After coal and petroleum, hydropower is the third important source of energy in the country. It is a renewable energy resource which is mobile, clean to use, can be used precisely in the quantity needed, and is easy to transmit over short distances, but long transmissions entail higher costs and greater loss. Hydropower supplied 21% of electricity in 2000 less than coal or petroleum but more than nuclear power. It is a cheap source of energy which requires perennial character of rivers with sufficient flow of water throughout the year. Surface water resources in the country fluctuate greatly between seasons. Hydropower energy also depends on undulating topography, suitable site for dam construction, fixing turbines, and demand for energy. It is also generated in a location with vertical drop in the valley, steeply sloping, and with waterfall. The most important hydropower projects are the Damodar Valley Project in Jharkhand, the Bhakra-Nangal Project in Punjab, the Hirakud Project in Odisha, the Koyna Project in Maharashtra, the Mettur Dam in Tamil Nadu, the Periyar Project in Kerala, the Rihand Project in Uttar Pradesh, the Tehri Dam in Uttarakhand, and the Sardar Sarovar Dam in Gujarat. Thus, the hydropower sources predominate in the coal deficient north and south in the country (Dutt and Geib 1998).

Due to poverty in petroleum resources, India has investigated nuclear energy as the dependable source. Although nuclear energy contributes only 2% of our total power generation, India has the vast potential for future energy development. Nuclear energy is created from uranium and thorium minerals. India has limited uranium resources and is found in the Singhbhum district in Jharkhand, Aravalli range in Rajasthan (Jaipur and Udaipur), and Kanyakumari district in Tamil

Nadu. India is advantageous to have the world's vast reserves of thorium (4.5 lakh tons or 0.45 million) in the monazite sands along the coast of Kerala (Chavara), Manaviarurichi in Tamil Nadu, and in the Ranchi plateau. Thus, the abundant availability of nuclear minerals provided the country an advantage in the exploitation of nuclear energy for power production. It coincided with the arrival and development of nuclear technology in India. In 1960, the Atomic Energy Commission drew up program to harness nuclear technology for power requirement of India. India's first nuclear power station at Tarapur near Mumbai with 420 MW was started in 1969 followed by Kota in Rajasthan in 1973 with a capacity of 440 MW, Kalpakam near Chennai in 1984 with a capacity of 235 MW, Narora in Uttar Pradesh in 1991 with a capacity of 440 MW, Kakrapara in Gujarat in 1993 with 220 MW, Kaiga in Karnataka in 2000 with 220 MW, and Rawatbhata in Rajasthan in 2000 with 220 MW (Thakur 1998), Kaiga in Karnataka in 2007 and 2011 with 660 MW, Kundankulam in Tamil Nadu in 2014 and 2017 with 2000 MW, Tarapur unit-III in 2005 and 2006 with 1080 MW, Rawatbhata in Rajasthan in 2000 (III and IV Unit) with 220 MW, and (V and VI Unit) with 2010 MW.

3.3 Concept of Paradigm

Paradigm has emerged as the dominant ways of scientific thinking within a discipline. The term paradigm is used to denote a universally recognized set of assumptions and procedures which serve to define both subjects and methods of scientific enquiry (Stoddart 1981). It has been widely accepted in philosophical and historiographical discussion in the social sciences, especially in sociology, anthropology, economics, psychology, and geography (Stoddart 1981). The concept gained sudden popularity in geography during 1960s and 1970s. Kuhn has become as familiar in geography as Humboldt and Hartshorne. Haggett (1979) defines paradigm as a supermodel. Strictly speaking, paradigm suggests scientists and researchers where to look because we need more epistemologically rigorous foundations for research and a more open discussion of values before we can identify a credible core for the discipline of geography (Buttimer 1981). More attention has also been paid to what scientists may expect to discover in their field.

Kuhn (1962 and 1970) in his classic work "The Structure of Scientific Revolutions" defined paradigm "as the entire constellation of beliefs, values, techniques shared by the members of a given community," Kuhn's concept is a key to understanding both formal structure of investigation and the interpretation of change in the history of science (Stoddart 1981). Kuhn developed a model of paradigm for the philosophy of science. According to him, the growth of knowledge depends on a paradigm shift which passes through four phases: anomalies, crises, revolution, and normal science. The Kuhnian model of paradigm shift has been used to examine the evolution of human geographic thought (Harvey and Holly 1989). This he calls an evolutionary model of paradigm development. In course of development of science, from time to time, goals, subjects, and problems appear less satisfactory, old problems lose significance, and the prevailing paradigm is replaced by a new paradigm. Thus, research moves into new areas, ultimately into a period of normal science which accumulates theoretical knowledge and empirical data that fall within the dominant paradigm (Kuhike 2006). The discipline then makes progress with same stages repeated. At a certain point in the development of science, when it fails to make progress or does not provide solutions for contemporary problems, clusters of new ideas emerge and challenge assumptions taken previously (Kuhike 2006). Holt-Jensen (2009) argued that the most basic function of Kuhn's paradigm is an exemplar which is a concrete problem solution within a discipline that serves as a model for successive scientists. Harvey and Holly (1989) present exemplars in geography, such as Ratzel's Anthropogeographie (1882, 1891), Vidal de la Blache's Tableau de la Geographie de la France (1903), Sauer's Morphology of Landscape

(1919–1922), Hartshorne's Nature of Geography (1939), and Schaefer's Exceptionalism in Geography (1953).

Kuhn stimulated a large literature on paradigm beliefs, which became a powerful toolkit for researchers. For example, Bird (1975) and Holt-Jensen (2009) advocated that an active science will be in a state of permanent revolution. Barnes (1982) in his book Kuhn and Social Science similarly argued that Kuhn's model is descriptive, not prescriptive. Johnston (2000) stressed that Kuhn's is a normative view of science, not a positive one.

3.4 Materials and Methods

The paradigm analysis is relatively new in natural resource management. Harvey and Holly (1989) substantiated Kuhn's model with reservations. In our exploratory study, we decided to use secondary data sources to investigate the status of published data to investigate the status of published literature on paradigms in natural resource management. The units of analysis are books and peer reviewed journal articles. This chapter was written during the early phase of the Covid-19 pandemic, as a result, the challenges were obvious. Thus, our project is a descriptive longitudinal study of a small sample based on the availability of books and journals. However, limitations of this chapter are the exclusion of book chapter materials. Most books selected under review were published in English language and are geographically focused on the United States, Canada, the UK, and India. The selected study period was approximately 1935–2005 because many conceptual and methodological changes have taken place in natural resource management during this time. This period saw conceptual evolution from ecosystem concept, political economy and political ecology, sustainable development and resilience, climate change, governance and institution, behavioral revolution, perception and attitudes, descriptive and inferential statistics, multivariate statistics, spatial statistics, geographic information systems, computer-aided cartography, aerial photography, remote sensing imagery, geoinformatics, Internet, disaster mapping, to qualitative data analysis, and map use and analysis.

We have adopted a three-phase approach for the selection of books and journals. A stratified purposive sampling was conducted. In the first phase, the 1935–2005 period was divided into Early and Modern periods. The second phase consisted of stratification into ecological, integrated, economic and technological-I (early period), and decision-making and technological-II (modern period). Their temporal extent is provided in Table 3.1. The third phase comprised purposive selection of one book/journal from each paradigm. The following four criteria out of seven mentioned below were selected for the recommendation of exemplar in each paradigm: (1) classic work or renowned book, (2) development of model/theory, (3) development and use of appropriate method, (4) name of publisher, (5) number of editions, (6) translation into different languages (minimum four), and (7) adjustment of time.

3.5 Early Period

The contribution of the early period (1935–1975) and its seven paradigms have been important with respect to the accumulation of knowledge. The early period begins from 1935 and is divided into ecological, integrated, economic, and technological-I (Table 3.2). The technological-I is further subdivided into cartography, descriptive and inferential statistics, computer-aided cartography, and remote sensing and are termed as sub-paradigms.

3.5.1 Ecological Paradigm

This is the oldest paradigm in the history of natural resource management was developed by British botanist Sir Arthur Tansley in 1935. He authored a seminal paper which suggested the concept of ecosystem. Tansley advocated for the rational and efficient management of natural resources through knowledge of dynamics of several interacting and interdependent elements.

Table 3.1 Stratification and Sampling Frame of Books/Journals for Samples

	Period	Paradigms	Sampling Frame (Name of the Books/Journals)
1.	Early (1935–1975)	Ecological	Sir A. Tansley (1935) "The Use and abuse of vegetational concepts and terms," Ecology E.A. Odum (1971) Fundamentals of Ecology, W.B. Saunders Co. P. Robbins (2004) Political Ecology: A Critical Introduction, Blackwell
		Integrated	G.F. White (1945) Human Adjustment to Floods, Research Paper No. 29, University of Chicago. B. Mitchell (1994) "Addressing edge problems in land and water management in Nigeria," Geoforum. S.K. Saha (1979) River basin planning in the Damodar Valley of India," Geographical Review A.K. Biswas (2008) "Management of Ganges-Brahmaputra-Meghna System: Way forward.
		Economic	P. Blaikie (1985) Political Economy of Soil Erosion in Developing Countries, Longman T. O'Riordan (1971) Perspectives on Resource Management, Pion. A.C. Mohapatra (2013) "Political economy of natural resources and geography," Transactions, IIG.
		Technological-I: Cartography	Elements of Practical Geography, Kalyani Publishers Fundamentals of Cartography (with R.B. Singh, B. Misra, and A. Pandey) Second Revised and Enlarged Edition, Concept
		Descriptive and Inferential Statistics	D. Ebdon (1977) Statistics in Geography: A Practical Approach, Basil Blackwell. John Silk (1979) Statistical Concepts in Geography, George Allen and Unwin. S. Siegel (1956) Non-parametric Statistics for the Behavioural Sciences, McGraw-Hill.
		Computer-Aided Cartography	"Computers and spatial analysis: Extensions of geographic research," Geoforum. D.W. Rhind (1977) "Computer-aided cartography," Transactions, Institute of British Geographers, New Series. Computer Applications in Geography, Wiley.
		Remote Sensing	T.M. Lillesand and R.W. Kiefer (1979) Remote Sensing and Image Interpretation, Wiley. E.C. Barrett and L.F. Curtis (1976) Introduction to Environmental Remote Sensing, Wiley.

(continued)

Table 3.1 (continued)

	Period	Paradigms	Sampling Frame (Name of the Books/Journals)
2.	Modern (1976–2005)	Decision-Making	B. Mitchell (1979) Geography and Resource Analysis, Longman. W.R.D. Sewell and Ian Burton (eds) (1971) Perceptions and Attitudes in Resources Management, Information Canada. I. Burton (1971) The social role of attitude and perception studies in W.R.D. Sewell and I. Burton (eds) Perceptions and Attitudes in Resources Management, Information Canada. T. O'Riordan (1971) Perspectives on Resource Management, Pion.
		Technological-II (Spatial Analysis)	Quantitative Methods in Geography: An Introduction to Spatial Analysis, Houghton Mifflin Company. An Introduction to Quantitative Analysis in Human Geography, McGraw-Hill.
		Map Use and Analysis	J. Campbell (1991) Map Use and Analysis, McGraw-Hill. Statistical Mapping and the Presentation of Statistics, Edward Arnold. P.C. Muehreke (1992) Map Use: Reading, Analysis and Interpretation, J.P. Publications.
		Remote Sensing of the Environment	Introduction to Remote Sensing, Guilford Press. J.R. Jensen (2000) Remote Sensing of the Environment: An Earth Resource Perspective, Pearson Education. Remote Sensing and Its Applications, Universities Press.
		Geographic Information Systems	C.P. Lo and A.K.W. Yeung (2002) Concepts and Techniques of Geographic Information Systems, Prentice-Hall of India. Geographic Information Systems and Science, Wiley.

Source: Compiled by the Authors

He stressed the use of ecosystem to express ideas that populations of plants and animals interact with each other and their nonliving environment in a systematic way. We recommend his publication as an exemplar in the ecological paradigm because it is a classic work, famous for model/theory published in Ecology journal with adjustment of time. Odum was well-known among ecologists and being an influential ecologist, he had a significant control on geography. He attracted substantial attention among geographers both globally and in India. Odum (1971) envisaged that any unit including all organisms in each area interact with the physical environment so that a flow of energy leads to clearly defined trophic structure, biotic diversity, and material cycles within the system. This paradigm is divided into three subcategories: (1) ecological interlinkages, (2) biosphere reserve, and (3) political ecology.

3.5.1.1 Ecological Interlinkages

Ecological interlinkages demonstrate flows between different physical parameters of the ecosystem. One of the earliest and most important works between different physical parameters of the ecosystem was by Burman (2003). She has offered some reflections on ecology and natural

Table 3.2 Period, Paradigms, and Developer of Paradigm

Period	Paradigms and Sub-Paradigms	Time Period	Developer of Paradigm
Early (1935–1975)	*Ecological*	1935–2005	A. Tansley, E.A. Odum
	Integrated	1945–2005	Gilbert White, Bruce Mitchell
	Economic	1985–2005	Piers Blaikie, T. O'Riordan
	Technological-I		
	a. Cartography	1969–2005	R. P. Misra
	b. Descriptive and Inferential statistics	1977–2005	David Ebdon
	c. Computer-aided cartography	1977–2005	D.W. Rhind
	d. Remote Sensing	1979–2005	T.M. Lillesand and R.W. Kiefer
Modern (1976–2005)	*Decision-Making*	1979–2005	Bruce Mitchell, W.R.D. Sewell, I. Burton, P.J. Taylor, M. Yeates
	Technological-II		
	a. Spatial Analysis	1977–2005	J. Campbell, P.C. Muehreke
	b. Map Use and Analysis	1991–contd.	J.R. Jensen
	c. Remote Sensing of the Environment	2000–contd.	
	d. Geographic Information Systems	2002–contd.	C.P. Lo and A.K.W Yeung P.A. Longley M.F. Goodchild D.J. Maguire D.W. Rhind

Source: Compiled by the Authors

resource management from a system perspective. The systems approach provides a basis for finding out a connection between the physical system, natural resources, and the human usages of the resources (Burman 2003). The author investigates methodologies selected by traditional systems of agriculture like grazing systems and utilization along slopes by adapting the village structure to their optimum level (Thakur 2003a, b). The significant attribute of these traditional systems is the adaptation of the humans to get the most out of each situation in terms of locality, product, time, and available technology (Thakur 2003a, b). It is indicated that the concept of development began to change in the second half of the twentieth century in underdeveloped countries.

3.5.1.2 Biosphere Reserve

Biosphere reserve concept has evolved from ecology for natural resource management in the mid-1970s under the UNESCO's Man and Biosphere Programme (MAB). It aims at the conservation of ecosystems by preserving representative samples of significant ecosystems, original habitats of domesticated plants and animals, and remnant populations of rare and endangered species (Thakur 2003a, b). It protects national parks, protected areas, sanctuaries, and biodiversity. Most importantly, and probably with the passage of time, attention has also focused on promoting sustainable resource development and ensuring people's participation. The article by Batisse (2003a) in Nature and Resources (1986) on "Developing and Focusing the Biosphere Reserve Concept" is reprinted in Thakur (Thakur 2003a, b) and deals with the evolution, present content, and formulation of the concept. In another paper on "Development and Implementation of the Biosphere Reserve Concept and its applicability," Batisse (2003b) reprinted in deals with three roles of the concept, that is, development, logistic, and conservation in the light of present experience with the application Thakur (Thakur 2003a, b).

3.5.1.3 Political Ecology

Political ecology is defined as empirical, research-based explorations to explain linkages in the condition and change of social/environmental systems, with explicit consideration of relations of power (Robbins 2004). The concern of geographers to political ecology has been caused by deep investigation between society, environmental problems, and political relationships in parts of Asia, Africa, and Latin America (Blaikie and Brookfield 1987; Bryant 1992). This approach to ecological paradigm is interdisciplinary in nature which is shared by geography, anthropology, sociology, development studies, and environmental studies. Political ecology rests upon three assumptions, first, neo-Malthusian arguments that overpopulation was the primary cause of resource depletion and environmental degradation; second access to and control over natural resources, whether by class, gender, ethnicity, or race and its social and environmental outcomes; and third, emphasis on history—both environmental history and the institutional history of resource management (Yeh 2006). Yeh (2006) also noted that: (1) the field focused on rural land uses, (2) failed to address gender and power, and (3) too much emphasis on poverty as a cause of environmental problems. Finally, Yeh (2006) comments that political ecology lacks empirical and theoretical coherence; also, it neglects ecological aspects of environmental phenomena. Thus, it can be stated that political ecology as an approach to resource management has reached a crossroads in its development.

3.5.2 Integrated Paradigm

This paradigm of natural resource management primarily deals with the integration of natural resources. Integration is concerned with the combination of parts or elements of resource system to produce a harmonious whole and to coordinate diverse elements (Thakur 2003a, b). Land, soil, water, crops, natural vegetation, and landscape are often integrated through a spatial approach. The discipline of geography integrates because it is a spatial science. Water acts as a lubricant to unite and integrate various elements of resource systems to sustain its ecological unity and integrity within the watershed of the river basin. Teclaff (1967) recommends the river basin as the appropriate unit for the development and management of resources because of physical unity in the basin. Thus, most notably, the river basin forms a biogeographical system which is generally called process-response system (Thakur 2003a, b). Toward this end, it is argued that the use of river basin water has progressed from single purpose to multipurpose uses, such as irrigation, flood control, navigation, and generation of hydel power due to the development of engineering techniques.

Since the middle of the twentieth century, the scholarly leadership in geography has come from those who have focused on natural resource management and hazards. Gilbert White has drawn from other fields of enquiry, for example, hydrology, water resources engineering, and economics and has provided a spatial perspective on the management of natural resources. He became increasingly involved with a large volume of research, scholarly publications, and contributions to public policy. He addressed national and international problems and the great global issues of our time. As a graduate student at the University of Chicago in the late 1930s, White studied the Mississippi River basin's water resources for the Federal Government and proposed a radical idea (Kayastha 2003). A reliance on dams, levees, and channel improvements to reduce losses was the focus of his highly influential dissertation at the University of Chicago. White (1945) wrote that "floods are acts of God, but flood losses are largely acts of man." White helped shape the United States policies on floodplain, water use, and natural disasters for more than six decades (Kates and Burton 2008). It has since been called the most influential work ever written by an American geographer. White (1945) is known worldwide as the "father of floodplain management." White's work on natural hazards changed the way people deal with and made the world safer for people to inhabit. His most notable work involved the identification and classification of adjustment mechanism for flooding in the USA. White (1945) has listed eight types of human adjustments: (1) elevating land, (2) abating floods by land (watershed) treatment, (3) protecting against floods by levees and dams, (4) providing emergency warning and evacuation, (5) making structural change in buildings and transportation, (6) changing land use to reduce vulnerability, (7) distributing relief, and (8) taking out insurance. White concluded that public policy should consider all possible adjustments to use one floodplain over others. Thus, his book deserves to be an exemplar not only in integrated paradigm but also a founder of natural resource management. His publication on flood adjustment published by the University of Chicago Press, developed a model on flood plain management with adjustment of time.

Bruce Mitchell (1986b) backgrounds the concept of integrated resource management by incorporating elements like multiplicity of purposes and means, combining and intermingling of various resource sectors, and as a device for social and economic change. Mitchell (1986a) is also of the view that for more than 55 years now, watershed has been used in different regions of Canada as the basis for water management. Mitchell and Pigram (1989) addressed the problems of land degradation and flooding in the Hunter Valley in Australia. Similarly, in a path-breaking work in a developing country, Mitchell (1994) has made significant contributions which have examined Dublin principles and made suggestions that there should be an integrated approach in water resource management in the Sokoto-Rima River basin, Nigeria.

Saha (1979) analyzed the present situation of river basin planning undertaken 31 years ago in the Damodar Valley region. The Damodar watershed spreads across Jharkhand and West Bengal. The Damodar Valley Corporation (DVC) is essentially a multipurpose project on the Damodar River and its tributaries designed by TVA engineers. The recurrence of annual devastating flood caused by the Damodar River in West Bengal was notorious from 1823 to 1943. The multipurpose project was therefore created in 1948 by the Act of Indian Parliament in Central Jharkhand. It was the first Indian attempt to put an integrated resource management concept into practice (Saha 1979). The objectives of the project were to construct seven multipurpose dams with a purpose to regulate flood water, generate hydropower, provide irrigation water, afforestation on slopes, and create navigation channel. The study shows that in 1979 (Saha 1979) only four dams were built with achievements of flood control, power generation, and irrigation water to 4 lakh ha (0.4 million) of agricultural land. Finally, there is a need to undertake studies on intra-regional flows generated by individual program, the degree of regional leakage or linkages between the rural-agricultural sector and the high-technology-based urban of the economy (Saha 1979).

Gupta and Mitra (2004) have made significant contributions to the Subarnarekha River basin located in northeastern peninsular plateau in India. The river originates near Ranchi and drains Jharkhand, West Bengal, and Odisha states before reaching the Bay of Bengal. The basin is marked by inadequate water supply, unequal distribution of freshwater, conflict among competing water uses, freshwater used by industry (small-engineering and manufacturing, steel industry, chemical, forest, etc.), widespread poverty, increasing water scarcity, and high pollution load. Water scarcity, water quality, and water use conflicts are three distinguishing characteristics of this basin. Thus, a comprehensive integrated management approach was started in 1982–1983 on the Subarnarekha Multipurpose Project to meet expectations and benefits with financial assistance from the World Bank. The project was expected to be completed in 2012–2013.

Mismanagement of water resources has contributed to increasing poverty and deteriorating quality of life in the Ganges-Brahmaputra-Meghna (GBM) basin. Biswas (2008) has formulated a framework for the sustainable development of the GBM basin to enhance the quality of life through human development, environmental conservation, and efficient governance. The basin covers about 1.75 million km^2 stretching across five countries (Bangladesh, Bhutan, China, India, and Nepal). It is characterized by endemic poverty. It contains rich natural endowments like arable land, water, and energy. The development and utilization of the above resources in an efficient manner have never been sought by the five countries due to difference, mistrust, and lack of goodwill (Biswas 2008). They can be developed only through collaborative efforts. The author offers the following components for managing sustainable development for the future based on an integrated paradigm. Such win-win paths to sustainable development are flood management, drought mitigation, hydropower generation, agricultural development, human development, environmental conservation, efficient governance, improving the quality of life of poor people, dry season flow augmentation and water sharing, water quality improvement, inland navigation, watershed management, and domestic water supply. Finally, Biswas (2008) presents long-term regional vision for sustainable development: (1) transferring water-related technology for caring irrigation efficiency, water storage, disaster management, and pollution control; (2) vigilance against potential instability and inequality in income distribution; (3) establishing good governance; (4) community participation in water development; (5) water-based regional cooperation for achieving peace, stability, and enhancing quality of life; (6) holistic, multidisciplinary and integrative approach following macro, meso, and micro policies within each country and their coordination, and that (7) being a water-rich region, if properly harnessed, they can be an important source of development to improve the quality of life of millions of poor people in this basin.

3.5.3 Economic Paradigm

This paradigm is traditional and has a long history with numerous interpretations. It is concerned with how people, either individually or in groups, attempt to allocate scarce resources to meet their unlimited wants through the process of production, substitution, and exchange (Goodall 1987). Also, it is a system of activities and administration through which a society uses its resource, to produce wealth is the aim of this paradigm (Goodall 1987). The paradigm thus emphasizes the consistency of rules which allow values of meaning to be determined with certainty. It also stresses the relation among social-structural formations, economic forces, and political power relations that determine resource development. This paradigm identifies two significant aspects of resource development which are of concern: (a) resource allocation and (b) political economy.

3.5.3.1 Resource Allocation

The objective of resource allocation is to achieve economic efficiency by minimizing production costs (labor and capital) and maximizing monetary profits. Under resource allocation, the issue of scarcity, choice, and optimization has received considerable attention in this paradigm. Omara-Ojungu (1992) discusses that rational allocation of resources is possible in a free and competitive market economy. The recent economic literature is voted to analyzing the assumption of resource allocation which rests upon the assumption of efficiency and the concept of maximization of net returns (O'Riordan 1971). Net returns are maximized when the marginal value of goods and services consumed just equals the marginal costs involved in producing them (O'Riordan 1971).

3.5.3.2 Political Economy

In the English-speaking world of the eighteenth and nineteenth centuries, numerous celebrated economists emerged who contributed to the field of political economy. The classical period of economics is usually defined by the work of Adam Smith's The Wealth of Nations (1776), David Ricardo's The Principles of Political Economy and Taxation (1912), T.R. Malthus's an Essay on the Principles of Population (1798), J.S. Mills's Principles of Political Economy (1848), and Karl Marx's Capital, Volume 1, in 1776. It was only by the 1980s that the political economy as we know it today emerged as a popular concept. It first emerged in human geography in the late 1960s (Barnes 1982) and the term became more pervasive and diffused in the mid-1980s (Barnes 1982). This approach also includes historical-structural stance incorporating within its framework, structural-institutional forces, and individual or household responses in the wider political economy (Thakur 2003a, b). This approach has been rarely applied to developing nations, especially on the economic condition of the farmers. Blaikie's (1985) research presents a different view in the case of Nepal.

Piers Blaikie (1942–), a British geographer, is internationally renowned for the study of environment, agrarian change, and political ecology. Blaikie's book (1985) *The Political Economy of Soil Erosion in Developing Countries* is a classic acclaimed worldwide. He argued that soil erosion in developing countries is because of political economy on the economic conditions of farmers and not due to mismanagement, overpopulation, or physical processes. This has got wider acceptance among social scientists and many of them are working in this direction set by Blaikie. Blaikie emphasized on the social elements, that is, the social relations of productions under which land is used are key elements in the understanding and explanation of soil erosion (Thakur 2003a, b). Blaikie (1985) provided a clear rationale that erosion occurs in several social contexts: peasants and pastoral groups employing family labor, peasants and pastoral groups working under exploitative class relations, centrally planned economies, and advanced capitalist economies. However, the occurrence of soil erosion is dependent on a few physical and social variables, it is probably not agreeable to endeavor a single theory of soil erosion (Thakur 2003a, b). Nonetheless, the quality of work has been very high and Blaikie attempts to theorize substructures in a theory of soil erosion. It is therefore recommended Blaikie's (1985) book on *The Political Economy of Soil Erosion in Developing Countries* as the exemplar in economic paradigm because of high-quality work on soil erosion, developed theory of soil erosion, published by Longman and with adjustment of time. Apart from this work, some of his other significant works are: (i) Blaikie and Brookfield (1987) Land Degradation and Society and (ii) Blaikie *et al.* (1994) At Risk: Natural Hazards, Peoples Vulnerability and Disasters, London: Routledge.

In his Presidential Address delivered at IIG, Mohapatra (2013) analyzed political economy of natural resources and geography. He has provided a critical overview of current research on political economy and geography in India. It is argued that geographers working within this paradigm are confronted with challenges of combining geographical occurrence of natural resources with how accessed and consumed are political. Of concern is Mohapatra's (2013) observation is the concept of equity and justice. They are both fundamental because without equity there shall be no justice and without justice there is no polity. Most importantly, geographers are actively providing bases for policies on natural resources, such as price equalization, mineral royalty, urban land, affordable houses, naxalism in tribal belt, problems of biological diversity, effective conservation for biosphere reserves, and free from inequities and injustice.

3.5.4 Technological Paradigm-I

The close linkage between technology and resources has received attention by Thakur (2003a, b). This has been one of the most prominent uninterrupted paradigms for the past five decades in natural resource management. Very simply defined, technology is the capacity by which man extends his energy over his surroundings (Thakur 2003a, b). Some of the prominent emerging achievements of technology, by and large, are modifying weather, sustaining men in space, reaching the ocean depths, harnessing tides, Internet, submarine cables, global telecommunication, and blasting cities to ruins. The appraisal of natural resources is completed by aerial photographs, remote sensing, GIS, and most importantly, space technology which has brought revolution in this field. Finally, we have developed and classified technological paradigm into four sub-paradigms which will be elaborated under the following section: (1) cartography sub-paradigm, (2) descriptive and inferential statistics sub-paradigm, (3) computer-aided cartography sub-paradigm, and (4) remote sensing sub-paradigm. We shall consider each of them in turn.

3.5.4.1 Cartography Sub-Paradigm

The cartographic analysis became the techniques of representations, such as dots, choropleths, isopleths, bar, and pie charts to highlight the spatial pattern of distribution of phenomenon. Perhaps more importantly data is stored in maps. Cartography sub-paradigm provides a technical service to geography and geographers. It is the first attempt in technological paradigm-I which deals with drafting and drawing maps. Cartography, 60 years ago, was not only a science of representing spherical earth on a flat surface but also an art of creating maps (Misra 1969). Misra (1969) observed that maps were not only informative and educative but also artistic. Cartography gives a formal system for the presentation and communication of spatial data (Goodall 1987). Cartography is defined as the art, science, and technology of making maps. Johnston *et al.* (1981) succinctly summarize views on maps as the focus of cartography, map as models, source of map-making is the real world, map using signs, map characterized as mapmaker and scale, direction, projection, and coordinates.

R.P. Misra (1930–2021) is a well-known geographer, urban and regional planner, and Gandhian scholar. He had a good training in cartography while he was an M.A. student in Geography at Allahabad University (1952–1954). Misra joined the National Atlas and Thematic Mapping Organization (NATMO), Ministry of Scientific and Cultural Affairs (1957–1961). He obtained his doctorate degree at the University of Maryland, USA under the Fulbright Program (1961–1964). His interest in cartography continued at Maryland, too. The early graduate centers of cartographic instruction and research emerged at the University of Kansas, Washington, Wisconsin-Madison, Ohio State, Syracuse, South Carolina, and Maryland, among others. Upon return from the USA, Misra headed the Department of Geography, University of Mysore as Reader (1964–1968) and Professor (1968–1975). He used his excellent training and insights gained at Allahabad, NATMO, Maryland, and Mysore to produce a detailed, comprehensive, and remarkably valuable book on Fundamentals of Cartography (1969), First Revised and Enlarged Edition (1989) and Second Revised and Enlarged Edition (2014), New Delhi: Concept Publishing Company. According to the author "drawing maps by hand became obsolete, surveying whether ground or aerial was no longer the only major source of information." The book demonstrates part two on Cartographic Information and part four on the Impact of Information Technology on Cartography. In essence, because of the merit, the book is recommended as an exemplar in Cartography. It is a renowned textbook published by Popular Publisher, with three revised editions, incorporating remote sensing, GIS, GPS and computer mapping, and adjustment of time.

3.5.4.2 Descriptive and Inferential Statistics Sub-Paradigm

Descriptive and inferential statistical measures have been introduced to analyze distribution, production, and their changing spatial patterns which started in the early 1960s. The techniques are central tendency, standard deviation, correlation, regression, nearest neighbor, and quadrat methods on interval and ratio data. Inferential statistics have been developed and introduced to solve the problems of nominal and ordinal data by using Chi square (χ^2), Spearman rank correlation and other tests. Cartography has undergone massive changes over the last four decades.

In resource management, statistics deals with collection, classification, analysis, and presentation of numerical data for dissemination of statistical information. Resource managers are concerned with analyzing natural resources distribution, production, consumption, prediction, and their changing spatial patterns and processes. Consequently, it is possible that we may use many statistical measures and procedures to present quantitative information on maps which started in the early 1960s.

The important characteristics of descriptive or parametric statistics are: (1) the observations must be drawn from distributed population, (2) the observations must be independent and random, and (3) variables must be measurable on interval and ratio scale (Siegel 1956). Techniques like central tendency, standard deviation, quartile deviation, correlation, regression, variance, ratio, percentages, proportions, nearest neighbor, and quadrat methods are the basic tools of univariate or parametric statistics for the description and analysis of spatial data.

By contrast, in the last 60 years, an alternative branch known as nonparametric statistics or inferential statistics, or distribution-free methods have been developed to unravel the problem of nominal and ordinal data (Silk 1979). Nonparametric methods do not make the assumptions of normality like parametric, focus is to relax traditional parametric assumptions, focus on the difference between the medians can be used under general conditions, easier to understand than *t*-test or variance they replace, simple in computation and provide easy tests. However, they are less robust than their counterparts. Many recent studies in human geography including natural resource management have employed techniques like Chi square (χ^2), Spearman rank correlation, the K-S test, Mann–Whitney U test, and Kruskal–Wallis H test. It is also emphasized that nominal and ordinal levels of data which are not truly numerical are analyzed by nonparametric techniques, and interval and ratio levels of data which are numerical are investigated by parametric techniques.

David Ebdon, University of Nottingham, is a well-known name in the field of statistical geography. His book "Statistics in Geography: A Practical Approach" published by Basil Blackwell in 1977 is well recognized among the community (Ebdon 1977). He has discussed statistical concepts, sampling, comparisons, trends, relationships description, and spatial statistics in a very simple way to understand, describe, examine, analyze, and interpret in a meaningful manner. The edge of the book lays in easy understanding, articulating presentation, illustrative description, effortless flow of writing, intersecting with nice figures and tables make the learner engrossed into it. Numerous examples and exercises related to different aspects of the content in a satisfactory way enhance the quality of book and its acceptability among students, teachers, and researchers alike. The author has succeeded in conducting different types of tests based on sample and universe with clear diagrams and illustrations as well as has done justice in explaining the study of comparisons, trends, and relationships like correlation and regression very eloquently. The beauty of this book is due to the simplification of difficult and complex topics in a very simplified fashion. The second edition of this book appeared in 1985. Overall, its popularity can be gauged through several reprints of this book which is still available today—more than 40 years of its first edition. We therefore recommend this book as an exemplar in the field of descriptive and inferential statistics.

3.5.4.3 Computer-Aided Cartography Sub-Paradigm

One of the most stimulating developments in the 1970s has been the growth of computer-aided cartography. Its performance in map production, rapid data analysis, and accuracy has been impressive.

Rhind (1943–) is a renowned geographer, computer-aided cartographer, and GIS expert. He earned his Ph.D. in geomorphology from the University of Edinburgh in 1969. He taught at Durham and London and served as Vice-Chancellor of Tity University. He published more than 15 books, numerous articles, and many research projects to his credit. He has worked on computer-aided mapping. His book on "Computer Mapping of Drift Lithology from Borehole Records," was published in 1973 by Her Majesty's Stationery Office, London. His work on computer cartography and GIS continued in several volumes of books and research papers after the 1970s. Among his academic works, "Computer-aided cartography," published (in 1977) in Transactions, Institute of British Geographers, New Series 2(1), 71–97, is well accepted and acclaimed recognition in the field of modern cartography. His paper is a turning point in computer cartography which can very well be traced from the initial year of 1950 through the development of line printer, cathode ray tube (CRT), tabulating equipment, etc. By the beginning of 1970s, meteorologists, geologists, geophysicists, geochemists, plant ecologists, and many other earth scientists started using computer for their illustration. In computer-aided, cartography, several terms are used, for example, digital, automated, computer-aided, and cartography. The development in the field of computer has witnessed enormous change since 1970s when the concept of computer cartography was initiated by Rhind. We therefore recommend Rhind's paper on computer-aided cartography as an exemplar in the field because it is classic, innovative, and was published in leading international journals, original contribution, and adjustment of time.

3.5.4.4 Remote Sensing Sub-Paradigm

Remote sensing is a multidisciplinary tool, technique, and technology used by geographers and spatial scientists as well as practitioners from across disciplines. It represents a major technological step forward in the gathering of data about the spatial distribution of the earth's resources. It acquires data by satellite about the Earth and human action (Goodall 1987). It is uncommon and incomparable as it obtains basic biophysical information unlike cartography, GIS, and statistics which depends on information that are already in existence and accessible. The distance between sensors attached to Earth-orbiting satellite and the remotely located objects varies for about hundreds of kilometers. Thus, "remote sensing is defined as the art, science and technology of obtaining reliable information about the Earth and its environment through the process of acquiring, recording, analysing and interpreting imagery" (Sheng 2011). The Earth-orbiting satellite collects data from sensors attached to spacecraft. The satellite operates and revolves around the Earth, making 14 revolutions a day, cover once in every 18 days at a height of 920 km with spatial coverage of 180x180 km swath (Haggett 1979). Haggett (1979) provides two commendable examples of relevance and significant advances in digital image processing. For example, radar imagery is independent of solar illumination and is unaffected by darkness, cloud cover, or rain; second, radar imagery provides greater detail of the terrain also. In other words, cloud covered humid tropics and regions of polar nights can always be scanned. Haack (1982) describes 11 characteristics of satellite spatial information related to resource inventory and appraisal. They are: (1) worldwide coverage, (2) repetitive coverage, (3) synoptic view, (4) uniformity over time, (5) uniformity over large areas, (6) multispectral, (7) digital, (8) planimetric, (9) readily available, (10) easily usable, and (11) inexpensive. Imagery has wide range of application areas, such as cartography, agriculture, environmental studies, forestry, geography, geology, hydrology, meteorology, military, oceanography, and urban and regional planning.

The development of science and technology has increased knowledge horizon and reduced time and distance, allowing humans interdisciplinary interactions. Important among them is Remote Sensing which is outbound to any discipline. A classical book—Remote Sensing and Image Interpretation is written by Thomas M. Lillesand and Ralph W. Kiefer published for the first time in 1979 from New York: Wiley. Both authors are from engineering background by education and profession. Lillesand obtained a degree in Civil Engineering (1969), master's in science (1970), and Ph.D. (1973) all from Wisconsin-Madison. He has been named as honorary member of the American Society for Photogrammetry in Remote Sensing (ASPRS), given for distinguished services to ASPRS, which is the highest award for its member. He is a Professor Emeritus in the Department of Forest Ecology and Management at Nelson Institute for Environmental Studies and the Department of Civil and Environmental Engineering. He has authored and coauthored more than 200 professional publications. Kiefer is also a Professor Emeritus in the Department of Civil Engineering, University of Wisconsin-Madison with more than 35 Publications.

The book Remote Sensing and Image Interpretation is a first-hand systematic account equally accepted by students, planners, professionals, researchers, and interested persons of the subject from different disciplines. In fact, this book has broken the barrier of the subject and became interdisciplinary in acceptance. The book is very comprehensive and up-to-date about remote sensing and image interpretation. The book emphasizes aerial photogrammetry and photographic interpretation, as well as equal weightage, is assigned to satellite digital data acquisition, restoration, enhancement, classification, analysis, and interpretation. With advancing knowledge and research in the concerned area, this subject is bound to change which has been taken up in the successive additions (1979, 1987,

1994, 1999, 2007, and 2015). With major changes, this book is translated into several languages which tells its success story. We therefore recommend this book as an exemplar in the field of remote sensing because of its six editions published in several languages, a classic textbook and published by Wiley.

3.6 Modern Period

The modern period extends from 1976 to 2005 and is classified into decision-making paradigm and technological paradigm-II. Both paradigms entered the period through revolution and have a deep impact on the field. Decision-making was a move toward process geography while technological-II paradigm was about pattern recognition, use of satellite, geographic information systems, information technology, and electronic computers to speed up the processing, recording, and analysis of data.

3.6.1 Decision-Making Paradigm

The decision-making paradigm emerged as an important component of human geography in the last quarter century. It is a significant part of behavioral geography which is broader in scope. Much of the research in resource management decision-making has been concerned with understanding the geography of space and place and the way people experience and perceive the environment. Goodall (1987) has observed that "decision-making is the process whereby alternative courses of action are evaluated." It thus deals with spatial choice and patterns of spatial behavior. Choice is defined as a search among alternative and decisions made by individuals is governed by the utility, values, and attractiveness assigned to alternatives (Thakur 2003a, b). Mitchell (1979) suggested that mind mediates between the environment and behavior. Geographers are now recognized outside their own discipline for providing insights into management strategies, input into the planning process, allocation of resources, shaping spatial decision policy, improving citizen participation, and improving resource use (Thakur 2003a, b).

Burton (1971) has noted that there has been a growing body of research literature on perception and attitudes in decision-making. Perception is the ability of mind to apprehend objects through sight, hearing, smell, and touch. On the other hand, attitude is organized feelings and beliefs which influence behavior. In response to the above developments, the following section briefly presents a few studies on resource management decision-making process.

Mitchell (1944–) is a renowned Professor of Geography with specialization in various subfields of geography but his contribution in behavioral geography especially decision-making is well appreciated. He started his academic career in 1969 as Assistant Professor and became Professor in 1979 in the Department of Geography, University of Waterloo. He served the University in different administrative capacities like Chairman of the Department of Geography, Associate Dean of Undergraduate Studies, and Associate Provost of Academic and Student Affairs. Several awards are to his credit, for example, scholarly distinction in geography from Canadian Association of Geographers, Distinguished Service Award from the Canadian Water Resources Association, and Distinguished Teacher Award from the University of Waterloo, etc. He has published more than 25 books and over 140 research papers. Many of his books are translated into different languages like Chinese, French, Indonesian, and Spanish.

Mitchell's important publications on resource management decision-making are as follows. Mitchell (1969) identified and analyzed problems associated with creation and implementation of national water supply policy in England and Wales. He focussed upon determining spatial and temporal guidelines as well as an operational framework for decision-making. Mitchell (1971) identified catchment areas of upland water supply lakes and reservoirs in the Lake District. The author also recognized conflicting viewpoints and lay citizens wondering who is right and who is wrong. The study shows that experts are not the best judges in resource decisions. Mitchell (1976) illustrated that the incremental model provided a close approximation of a fisheries management program in British Columbia. It is also pointed out that fishery managers implemented a four-stage program to increase Salmon fisherman's incomes, reduce capitalization, and improve management of resources. Mitchell's book on Geography and Resource Analysis (1979) was published by Longman. The book provides a thorough coverage regarding policy needs and problems in resource analysis and management. It is an outstanding contribution to the field of geographical approach to resource analysis and management. It deals with resource management model and theory and how they have been influenced by recent development in geography. The chief features of the book are lively preface and introduction; worthwhile text covering role of attitudes, perceptions, and behavior in resource management; landscape evaluation, carrying capacity, natural hazards, policy models, and resource management, and institution setting. Interestingly, in recent years, the book has become a flag-bearer of resource management decision-making. It is powerful and articulate. We recommend the book as an exemplar in the field of decision-making paradigm because it is a renowned textbook, characterized by model building, translated into four languages, and published by Longman with and two-editions.

Hoffmann and Mitchell (1998) suggested a framework to improve decision-making in the context of groundwater extraction in Ontario arising from tensions in decision-making processes. They have developed RESPECT model including the principles of research, equity, sustainability, participatory decision-making, education, communication, and trust. It is recommended that it is likely to make major contributions to reduce tensions, when the situations arise in other problem areas. "Geographic modelling for resource management is in its infancy" and such models "have yet to lead to the generation of geographical theory of natural resources" (Mitchell 1980). The second book by Mitchell (1995) deals with physical basis of resource analysis, political culture of resource management, and resource policy. Most chapters include heartland-hinterland resource issues, sustainable development, wildlife, and parks institutional content. Finally, the book presents insight into the debate between traditional economics and ecosystem analysis. An interesting book has been published by Mitchell on Resource and Environmental Management in (1997, Longman), second edition (2002, Prentice-Hall) and third edition (2019, Oxford University Press). The book is organized into chapters covering ecosystem approach, sustainable development and resilience, governance, social learning organization, and stakeholders, and partnership.

Sewell and Burton (1971) edited a scholarly and timely book on Perceptions and Attitude in Resource Management. The editors approached the subjects from the paradigm of decision-making which is characterized by conceptual and methodological approaches, public involvement in the planning process, and selected case studies from Canada and the USA. Probably, the most interesting idea in the book suggested by the authors is perception and attitude matrices questionnaire formulation, methods to examine perception of air pollution, and public perception toward water supply. Thus, it is concluded that the community should participate in resource management decisions.

Burton (1971) has provided information on the serious role of perception and attitude in resource management decision-making. It is notable that they provide input into the planning process and policy implementation. Increasingly, researchers have developed ways to enhance the role of public participation, for example, Mitchell (1989) addressed significant ways at people's participation program: (1) degree of public involvement, (2) which public should be consulted, (3) at what point public opinion should be sought, and (4) components of good participation program. Sewell (1973) constructed a descriptive model of policy-making progress. He found weaknesses in the past procedure for evaluation to procure data for decision-makers. The paper contributes by developing methodologies by acceleration in resource management.

3.6.2 Technological Paradigm-II

The technological paradigm-II exhibited a high level of mathematical and statistical analytical techniques. They fostered a methodological trend toward model building. In the realm of model and theory development, qualitative data analysis has been particularly influential in increasing our understanding of resource management decisions. Finally, we have classified technological paradigm-II into four categories which will be elaborated under the following section: (1) spatial analysis sub-paradigm, (2) map use analysis sub-paradigm, (3) Remote Sensing of the environment sub-paradigm, and (4) Geographic Information Systems sub-paradigm. We shall consider all, in turn.

3.6.2.1 Spatial Analysis Sub-Paradigm

The fundamental spatial concepts include location, distance, direction, shape, pattern, and distribution. Spatial analysis raises the question on why are spatial distributions structured the way they are? It adopts a rigorous way for describing spatial patterns and attempts to identify factors controlling patterns of distribution (Goodall 1987). Thus, this paradigm attempts to identify, describe, and interpret regularities of geographical variables, relating such regularities to the process which produces them (Abler, Adams and Gould 1971). It seeks to explain patterns of human behavior and their spatial expression (Mayhew 2015). Techniques like correlation, regression, factor analysis, point-pattern analysis, quadrat analysis, gravity model, network analysis, trend-surface analysis, and location-allocation model would answer the above questions.

Peter Taylor (1944–), a British geographer, received his B.A. in geography in 1966, and Ph.D. in 1970 from the University of Liverpool. Before moving to Loughborough University as Professor of Geography in 1995, he taught at Newcastle University for 27 years. He was a Senior Lecturer in Geography at the University of Newcastle upon Tyne in 1986. He was elected a Fellow of the British Academy in 2004 and awarded an Honorary Doctorate by Oulu University in 2006. His main research interests are theoretical and methodological concerns in electoral geography, world system analysis, globalization, and World cities, relationship between World cities, spatial analysis, and space and place.

Taylor's first book on Quantitative Methods in Geography: An Introduction to Spatial Analysis published in 1977 by Houghton Mifflin Company, Boston is a comprehensive textbook. The book is divided into seven chapters and covers descriptive, inferential statistics, multivariate, pattern analysis, flow models, and allocation models. Most chapters have a lively writing style. The text is well written, logically organized, and is supported by an extensive list of references. The book is thought-provoking, intelligently written, path-breaking, and authoritative. Materials are well articulated. Tables, graphs, and maps give readers valuable insights. Thus, this is a classic book with a good image not only in the West but in the East also. Chapter four (point-pattern analysis), five (areal Association), six (factorial ecology and analysis of variance), and seven (applied spatial analysis covering gravity model and location-allocation models) are beautifully expressed with real-world examples. Thus, the authors recommend this book as an exemplar, in the spatial analysis sub-paradigm, because it is a renowned textbook, published by a popular publisher, contributes methodologically, and evolved with time. His other works are: (1986) A World in Crisis and Geographical Perspectives with RJ Johnston, Basil, Blackwell; (2004) World City Network: A Global Urban Analysis, Routledge; (2006) Political Geography: World Economy, Nation-State, Locality, Prentice-Hall, 5th Edition with C. Flint; (2006) Cities in Globalisation: Practices Policies and Theories, Routledge, edited with B. Derudder, P. Sacy and F. Witlox.

3.6.2.2 Map Use Analysis Sub-Paradigm

Maps are the basic tools and database of geography. The last four decades have witnessed a growing interest in the study of map use and analysis. Its study has emerged fast with the nuances of satellite imageries, geographic information systems, computer cartography, global positioning system, and land information system. Maps are viewed as physical, economic, resource, and cultural artifacts on the earth's surface. Johnston *et al.* (2000) have observed that maps have helped to reconstruct field systems, land use change, providing evidence in boundary disputes, coastal change, ecological and vegetation change, patterns of roads, and spacing between towns. Campbell (1991) suggested that alignments of streams, lakes, crests of hills and the floors of valleys, and patterns of soil erosion can also be analyzed using maps. This paradigm is about the world of maps. This views map as a broad framework and suggests mapping to be a model of displaying spatial information. It is a vibrant field because in the recent past there has been a cartographic revolution bringing tremendous change in its approaches.

John Campbell (1932–) is an imminent geographer in the United States. He received his B.A., M.A., and Ph.D. in Geography from the University of Washington. He taught for 22 years at the University of Wisconsin, Parkside from where he retired in 1996 and was elevated to Professor Emeritus in Geography. Subsequently, he became Adjunct Professor of Geography at the University of New Mexico, Albuquerque and resides in Santa Fe. His teaching, research as well as publications focus on the broad themes of cartography, GIS, economic geography, industrial location, and conservation of natural resources.

Campbell's book on Map Use and Analysis (1991) was published by McGraw-Hill Higher Education. The book covers a broad range of map-related topics. These topics include map projections, scale, terrain and interpretation, topographic features, shape and point patterns, aerial photography, remote sensing, computer-assisted cartography, digital map, and GIS are attractive and readable. The chapters are eclectic in nature which adds depth to highlight the different aspects of the relationships among scale, projections, computer mapping, and image interpretation. The salient features of the book are: (1) the focus is on map use rather than map reproduction, (2) it is a comprehensible overview of the world of map, and (3) the appearance is clean and well balanced. Chapters on aerial photography, remote sensing, computer-assisted cartography, and GIS are modern techniques devoted to map users. Finally, a glossary and six appendices provide quick points of reference for the researcher. In addition, the book is not only well written but also well illustrated by maps, diagrams, charts, tables, graphs, and photographs which are presented in an attractive manner. It is a valuable contribution to the growing literature on map use and analysis. Thus, Campbell's (1991) book on Map Use and Analysis is recommended as an exemplar in the Map Use and Map Analysis sub-paradigm as it is a comprehensive textbook, published by McGraw-Hill, five editions and perfect adjustment of time.

3.6.2.3 Remote Sensing of the Environment Sub-Paradigm

Remote Sensing technology through image processing helps in planning and management of natural resource. Remote Sensing is used for a variety of national and international Earth resource monitoring programs (Jensen 2000). It has potential, usefulness in the management of soils, forest, water, mineral, geology, land use, clouds, snow covering river basins, deserts, and marine areas. Moreover, remote sensor systems provide a wealth of data that is relatively inexpensive and easily usable.

Jensen (1949) is a very well-known name in the field of environmental remote sensing. He completed his M.Sc. in 1972 and Ph.D. in 1976 from the University of California, Los Angeles in geography. He became Professor in the Department of Geography, University of South Carolina, Columbia. Jensen is a distinguished expert in remote sensing and digital image processing. He is a member of the American Society for Photogrammetry and Remote Sensing (ASPRS). He also served at ASPRS as its President during 1995–1996. Several books and numerous research papers are to his credit.

Jensen's book on *Remote Sensing of the Environment: An Earth Resource Perspective*, first published in 2000 by Pearson Education, introduces the fundamentals of remote sensing from an earth resource perspective. Its second edition came in 2007 with the addition of two chapters on "ground remote sensing" and "light detection and ranging" (LiDAR). Both editions have several reprints conveying the demand among students, teachers, and related professionals. The author emphasizes that biophysical or socioeconomic information about any area can very well be extracted by using remote sensing. This book provides a comprehensive wide-ranging up-to-date analysis of the fundamentals of remote sensing and its applications. It deals with monitoring and managing resources of the earth in-depth. The book is balanced in presentation of both theory and application of remote sensing, particularly, with respect to the study of resources found in different regions of the world. The conceptual parts are very well illustrated with the diagrammatic presentation and written in a simple way to understand thoroughly. The beauty of this book lies in providing contemporary developments in the field of remote sensing and its applications in the study of natural resources. Thus, we recommend this book as an exemplar in remote sensing of the environment sub-paradigm because it is a popular textbook, published by Pearson Education (several editions), and methodological contribution.

3.6.2.4 Geographic Information Systems Sub-Paradigm

Burrough (1986) defined the geographic information system as "a powerful set of tools for collecting, storing, retrieving, as well as transforming, and displaying spatial data from the real world." The first mention of GIS occurred in the literature in the mid-1960s, but massive growth began only in 1980 with the introduction of Super Mini-Computers by manufacturers, such as Digital Equipment Corporation and Prime Computers (Johnston *et al.* 2000). It is a rapidly growing field. Here, information is its heart. It has revolutionized the mapping process. Geographic Information Systems is multidisciplinary. It is used by a variety of academic disciplines, such as resource management, environmental studies, remote sensing, computer science, forestry, social science, surveying, and landscape architecture, as well as practitioners. Perhaps more significantly, Abler (1988) also viewed GIS as a set of processes and procedures that approach the heart of modern geography.

Lo (1949–2007) was an internationally acknowledged geographer, who made an inspiring contribution by using remote sensing and GIS and studied urban and human-environment interactions. He did his M.A. in Geography in 1966 from the University of Hong Kong and Ph.D. from the University of Glasgow in the UK and completed it in 1971. He was Associate Professor till 1984 at the University of Hong Kong before joining the University of Georgia and at this university, he became Professor in 1988. Many awards and recognition are to his credit. For his outstanding contribution to geographic remote sensing, Lo was recognized by Association of American Geographers in 2001 with Research Honour award, by Southeast Division of the Association of American Geographers he was conferred Lifetime Achievement Award in 2002 and 2005, respectively. He has authored and coauthored 11 books and more than a 100 papers. Dr. Albert Yeung is associated with Ontario Police College, Aylmer, Ontario, Canada. He is a GIS practitioner with over 35 years of work experience in the academic and public sectors as well as the private consulting industry in Canada.

The book *Concepts and Techniques of Geographic Information Systems* by Lo and Yeung published by Prentice-Hall of India, for the first time in 2002, is a milestone in GIS studies. It focuses on concepts and techniques required for GIS professionals. This book provides up-to-date coverage of the concepts and techniques concerning GIS and its applications in various walks of life and research. The authors have tried to strike a balance between theory and practice of GIS. The book emphasizes on the integration of GIS and information technology, and it discovers new spatial analysis techniques or landscape matrix and discusses about geovisualization. It examines new terrain data acquisition by light detection and ranging (LiDAR) and covers emerging technology in solving day-to-day problems. The authors have given numerous real-world experiences of using and implementing it. The book is very well utilized by students, researchers, and professionals. The second edition of this book was published in 2007 before the senior author passed away. Several reprints of this book are in the hands of students and professionals as well as in libraries world globally. We therefore recommend this book as an exemplar in geographic information systems sub-paradigm because it is a renowned textbook not only in the West but also in the East, published by Prentice-Hall of India, comprises concepts and techniques with several editions.

3.7 Global and Indian Experiences

This chapter makes a significant contribution to the history of paradigm development at global and Indian level between 1935 and 2005. The study demonstrates contrasting evolving patterns raised by this work on paradigm development in natural resource management at the global and Indian scale. Global paradigm development experience and challenges are of practical as well as academic relevance to resource managers in India. What factors promote the development of paradigm in a discipline, whether global (the USA, Canada, the UK) or Indian: (1) stage of development of discipline, (2) quality of publications, (3) number of qualified scholars for paradigm formation, (4) quantitatively talented practitioners, (5) quantitative training of a new generation, (6) availability of standard computers, and (7) role of infrastructures, such as quality of teaching, research laboratory, training in systems, behavioral, quantitative, qualitative, and mixed methods, and departments offering advanced instructions. Since they differ in western countries and Indian institutions, the results are bound to be different. The research reveals that the important paradigm developers at the global level are F. Berkes, C. Holling, R.E. Grumbine, N.H. Gundersen, G.H. Norton, G.H. Walker, Bryan Walker, David Salt, T. O'Riorden, T.F. Saarinen, W.R.D. Sewell, R.W. Kates, Ian Burton, Gilbert White, Paul Robbins, T. Birkenholtz, E.P. Odum, P.A. Longley, M.F. Goodchild, D.J. Maguire, N. Hoffman, A. Burrough, B. Barnes, E. Yeh, C.J. Barrow, N. Sud, M.C. Morales, and P. Woodhouse. Model construction has been a significant feature of paradigm development by R.W. Kates, Bruce Mitchell, W.R.D. Sewell, Kasperson, T. O'Riorden, Burton and Kates, and Burton, Kates, and White. Many of the research papers and books are conceptually and methodologically strong. Theoretical contributions are many. Technologically, more advanced in the western countries in terms of computer mapping, satellite imageries, GIS, and GPS. Policy implications are the important features by many experts. More emphasis is placed on systems and behavioral approaches because they provide explanations.

The research also reveals that in India, the paradigm developers are P. Dayal, M. Shafi, S.L. Kayastha, Moonis Raza, Jayant Bandyopadhyay, T. Shah, M. Dinesh Kumar, Savitri Burman, M.H. Qureshi, H.S. Sharma, H.N. Misra, Abha Lakshmi Singh, P.S. Tiwari, S.C. Rai, P.C. Tiwari, A.C. Mohapatra, S.K. Chattopadhyay, Kuntala Lahiri-Dutt, R.B. Singh, Inder Jeet, Sahab Fazal, A. Rahman, B.W. Pandey, Punyatoya Patra, Rajesh Kumar Abhay, Narendra Kumar Rana, and S.K. Sharma. However, we find serious theoretical shortcomings, lack of model construction, conceptually and methodologically weak, and theoretical contributions are few. Little emphasis is placed on systems and behavioral approaches. Surprisingly, technologically more advanced in many universities of India on computer mapping, imageries, geographical information systems, and global positioning system. Many of the Indian resource managers have contributed to integrated paradigm, for example, S.K. Saha, A.K. Biswas, D.B. Gupta, H.N. Misra, S.K. Chattopadhyay, among others.

3.8 Conclusion

This study on paradigm concept and its applicability in the field of resource management in India is the first of its kind. The study was conducted over a 70-year period and was time-consuming. It is a longitudinal, descriptive, analytical study from 1935 to 2005. This period is characterized by revolutions and conceptual as well as methodological developments in natural resource management, such as ecology, economics, geology, environmental studies, geography, anthropology, and psychology. This chapter has analyzed the achievements and experiences of 12 scholars from different fields of resource management.

The study has adopted three-phase approach to select prime publications in the field during the abovementioned period which Kuhn calls exemplar. The approach consisted of small samples of books and major journal articles with international readership for this period. The selected books under review were published in English language focused on the USA, Canada, the UK, and India. In the first phase, the period was divided into early (1960–1975) and modern (1976–2005). A stratified purposive sampling was conducted. The second phase consisted of stratification into 4 paradigms and 8 sub-paradigms. The third phase comprised of a purposive selection of one book/journal article from each paradigm and sub-paradigm.

The Early Period consisted of ecological, integrated, economic, technological-I paradigms and Modern Period into decision-making and Technological-II. Both Technological-I and Technological-II have been divided into four sub-paradigms each. The research adopted any four of the following seven criteria for the selection of paradigm developers as the exemplar of the book. They were: (1) classic book or renowned textbook, (2) development of model and theory, (3) development and use of appropriate method, (4) name of the publisher, (5) number of editions, (6) translation into different languages (minimum four), and (7) adjustment to time.

Our analysis empirically demonstrated: first, the main findings are that Technological-I and II have been the most prominent uninterrupted sub-paradigms for the past 45 years. It is important to note that cartography, descriptive and inferential statistics, computer-aided cartography, remote sensing, spatial analysis, map use and analysis, remote sensing of the environment, and GIS sub-paradigms are the emerging fields of analysis, and they are worthy of further exploration, including space technology. Second, the research has revealed that a three-phase approach with seven criteria has improved identifying accuracy considerably related to paradigms, sub-paradigms, and developer of paradigms. Third, interestingly, this research has contributed knowledge by suggesting the dominance of non-technological paradigm in the early period and ruling technological sub-paradigms in the modern period. Fourth, the important non-technological paradigm developers are Sir A. Tansley, Gilbert White, and Piers Blaikie in the Early Period and Bruce Mitchell in the Modern Period. On the other hand, the important technological paradigm developers in the early period are R. P. Misra, David Ebdon, D. W. Rhind and T. M. Lillesand and R.W. Kiefer; and P. J. Taylor, J. Campbell, J.R. Jensen and C. P. Lo and A. K. W. Yeung in the modern period. Fifth, not surprisingly, the study has contributed to the theory and practice of paradigm literature by identifying, classifying, and using qualitative inquiry. The results have significance for the emerging field of paradigm analysis. Sixth, there were some limitations of this work. We faced several problems in completing this chapter during the COVID-19 pandemic. It was difficult to procure appropriate books and journals articles for evaluation. Admittedly, it was time-consuming and a laborious task.

Further, research needs to be carried out in the following two directions: first, there is a need to extend the analysis to contemporary period (2006–continued), and to compare the content of early and modern with the contemporary period. A more in-depth analysis is worthwhile to investigate and develop paradigm analysis to create meaningful data and analyze data by suitable techniques. Second, attention should be given to the role of resource management problems on ruling and developing paradigms. Their relationships are incompletely known especially between Technological Paradigm-III and non-technological paradigms.

Acknowledgments The authors are sincerely thankful to Bruce Mitchell for his thoughtful feedback in strengthening this chapter for which we are deeply grateful to him. We also extend gratitude to Patricia Thakur, Ramashray Prasad, Rajesh Kumar Abhay, Bhawan Singh Dahiya, Sudhir Thakur, Jitender Saroha, Richa Goel, Rajrani Kalra, Inder Jeet, Punyatoya Patra, and Chanda Thakur for their diverse support.

References

Abhay RK (2012) Water scarcity management in Rajasthan: a case study of Sikar District. Lambert Academy, Saarbrücken

Abhay RK (2015) Land Degradation and its Impact on Agricultural Production in The Kendujhar Plateau. Ann Natl Assoc Geogr India 35(1):71–88

Abler RF (1988) Awards, rewards, and excellence: keeping geography alive and well. Prof Geogr 40:135–140

Abler RF, Adams JS, Gould PR (1971) Spatial organization: the geographers view of the world. Prentice-Hall, Englewood Cliffs

Agarwal A, Chopra R, Sharma K (1982) The state of India's environment 1982: A Citizen's Report. Centre For Science and Environment, New Delhi

Ahmad E (1970) Soil erosion in India. Asia Publishing House, New Delhi

Bagchi J, Kumar R, Sharma S (2011) Variability of landslides incidences in India. Indian J Geosci 65(3):223–228

Balaseetha R (1986) Variation in drinking water quality in Delhi: a spatial analysis. M. Phil. Dissertation, Department of Geography, Delhi School of Economics, Delhi University, Delhi

Bandyopadhyay J, Parveen S (2004) Interlinking of rivers in India: assessing the justifications. Econ Polit Wkly 39:5307–5316

Barnes B (1982) T. S. Kuhn and social science. Macmillan, London

Batisse M (2003a) Developing and focusing the biosphere reserve concept. In: Thakur B (ed) Perspectives in resource management in developing countries, Resource management: theory and techniques, vol 1. Concept, New Delhi, pp 160–177

Batisse M (2003b) Development and implementation of the biosphere reserve concept and its applicability to coastal regions. In: Thakur B (ed) Perspectives in resource management in developing countries, Resource management: theory and techniques, vol 1. Concept, New Delhi, pp 178–194

Bhatt A (1994) Rajkot: Chronic Scarcity. The Hindu Survey of The Environment May 31. pp. 113–117

Bird JH (1975) Methodological implications for geography from the philosophy of K. R. Popper. Scott Geogr Magaz 91:153–163

Biswas AK (2008) Management of ganges-brahmputra-meghna system: way forward. In: Varis O, Tortajada G, Biswas AK (eds) Management of trans-boundary rivers and lakes. Springer, Berlin, pp 143–164

Blaikie P et al (1994) At risk: natural hazards, people's vulnerability, and disasters. Routledge, London

Blaikie P, Brookfield H (1987) Land degradation and society. Methuen, London

Blaikie P (1985) The political economy of soil erosion in developing countries. Longman, London

Bowonder B, Ramana KV, Ravi C, Srinivas C (1987) Land use, waterlogging and irrigation management. Land Use Policy 4:331–341

Bryant RL (1992) Political ecology: an emerging research agenda in third world studies. Polit Geogr 11:12–36

Burman SG (2003) Ecological interlinkages in resource management and development. In: Thakur B (ed) Perspectives in resource management in developing countries, Resource management: theory and techniques, vol 1. Concept, New Delhi, pp 147–159

Burrough P (1986) Principles of geographical information systems for land resource assessment. Clarendon, Oxford

Burton I (1971) The social role of attitude and perception studies. In: Sewell WRD, I B (eds) Perceptions and attitudes in resources management. Information Canada, Ottawa, pp 1–6

Buttimer A (1981) On people, paradigms, and progress in geography. In: Stoddart DR (ed) Geography, ideology, and social concern. Blackwell, Oxford, pp 70–80

Campbell J (1991) Map use and analysis. McGraw Hill, New York

Chisholm A, Dumsday R (eds) (1987) Land degradation: problems and policies. Cambridge University, Cambridge

Dayal P (1958) The acute coal zone of India. Natl Geogr 2:1–6

Dayal P (1972) Agricultural and industrial potentialities. In: Dutt AK (ed) India: resources, potentialities, and planning. Kendall/Hunt, Dubuque, IA., pp 38–71

Dutt AK, Geib MM (1998) Atlas of South Asia: a geographic analysis by countries. Science Publishers: Enfield (USA) and Oxford and IBH Publishing, New Delhi

Ebdon D (1977) Statistics in geography: a practical approach. Basil Blackwell, Oxford

Gadgil M (1990) India's deforestation: patterns and processes. Soc Nat Resour 3:131–143

Goodall B (1987) The penguin dictionary of human geography. Penguin Books, Harmondsworth

Gupta DB, Mitra S (2004) Sustaining Subarnarekha River Basin. Int J Water Resour Dev 20(3):431–444

Haack BN (1982) LANDSAT: a tool for development. World Dev 10:899–909

Haggett P (1979) Geography: a modern synthesis. Harper and Row, New York

Haigh MJ (1998) Deforestation in the Himalaya. In: Robert N (ed) The changing global environment. Blackwell, Oxford

Harvey ME, Holly BP (1989) Paradigm, philosophy, and geographic thought. In: Harvey ME, Holly BP (eds)

Themes in geographic thought. Rawat Publications, Jaipur
Hoffmann N, Mitchell B (1998) The RESPECT Model: evaluating decision-making approaches in water management. Water Policy 1:341–355
Holt-Jensen A (2009) Geography: history and concepts: a student's guide, 4th edn. Sage, Thousand Oaks
Janssen W, Goldsworthy P (1996) Multidisciplinary research for natural resource management: conceptual and practical implications. Agric Syst 51:259–279
Jeet I (2001) Water resource management: a case study of groundwater depletion in Haryana. Manisha Publications, New Delhi
Jensen JR (2000) Remote sensing of the environment: an earth resource perspective. Pearson Education, Delhi
Johnson DL, Lewis LA (1995) Land degradation: creation and destruction. Blackwell, Cambridge
Johnston RJ (2000) Geography and geographers: anglo-american human geography since. Arnold, London, p 1945
Johnston RJ et al (eds) (1981) The dictionary of human geography. Basil Blackwell, Oxford
Johnston RJ, Gregory D, Pratt G, Watts M (eds) (2000) The dictionary of human geography, 4th edn. Blackwell, Oxford
Kapur A (2010) Vulnerable India: a geographical study of disasters. Sage, New Delhi
Karan PP, Subbiah SP (eds) (2011) The indian ocean tsunami: the global response to a natural disaster. Cambridge University Press of India, Delhi
Kates RW, Burton I (2008) Gilbert F. White, 1911–2006: local legacies, national achievements, and global visions. Ann Assoc Am Geogr 98(2):479–486
Kayastha SL (2003) Prof. Gilbert F. White: An appreciation. Natl Geogr J India 49(1):63–64
Kuhike O (2006) Paradigm. In: Warf B (ed) Encyclopaedia of human geography. Sage, Thousand Oaks, pp 349–350
Kuhn TS (1962) The structure of scientific revolutions. University Of Chicago Press, Chicago
Kuhn TS (1970) The structure of scientific revolutions, 2nd edn. University of Chicago Press, Chicago
Kumar M, Bhandari MM (1993) Human use of sand dune ecosystem in the semi-arid zones of Rajasthan desert, India. Land Degrad Rehabil 4(1):21–36
Lillesand TM, Kiefer RW (2004) Remote sensing and image interpretation. Wiley, Singapore
Livernash R (1995) "India" World Resources 1994-95. Oxford University Press, Delhi, pp 83–106
Lo CP, Yeung AKW (2002) Concepts and techniques of geographic information systems. Prentice-Hall of India, New Delhi
Lahiry SC (1997) Mineral development and environment. Yojana (August):88–92
Mathur V (2004) The Political Ecology of Water Scarcity in Lucknow District. Ph.D. Dissertation, Department of Geography, University of Delhi, Delhi
Mayhew S (2015) Oxford dictionary of geography, 5th edn. Oxford University, Oxford
Mehta L (2003) Contexts and constructions of water scarcity. Econ Polit Wkly 29:5066–5072
Misra RP (1969) Fundamentals of cartography. (With Contributions from Singh RB, Misra B, Pandey A). In: 2nd Revised and Enlarged Edition. Concept, New Delhi
Mitchell B (1969) Decision-Making for Water Supply in England and Wales. Unpublished Ph.D. Dissertation. Department of Geography, Liverpool University, Liverpool
Mitchell B (1971) Water in England and Wales: Supply, Transfer, And Management. Research Paper No. 9. Liverpool University, Liverpool
Mitchell B (1976) Decision-making and consultation in resource management: The B.C. Salmon Fishery. J Environ Manage 4:211–223
Mitchell B (1979) Geography and resource analysis. Longman, London
Mitchell B (1980) Models of resource management. Prog Hum Geogr 4(1):32–52
Mitchell B (1986a) Integrated river basin management: canadian experiences. The Institution of Engineers, Australia, pp 3–7
Mitchell B (1986b) The evaluation of integrated resource management. In: Lang R (ed) Integrated approaches in resource planning and management. University Of Calgary Press, Calgary, pp 13–26
Mitchell B (1994) Addressing "Edge" problems in land and water management in the Sokota-Rima River Basin, Nigeria. Geoforum 25(2):133–143
Mitchell B (2019) Resource and environmental management, 3rd edn. Oxford University Press, New York
Mitchell B (ed) (1995) resource and environmental management in Canada: addressing conflict and uncertainty. Oxford University Press, New York
Mitchell B, Pigram JJ (1989) Integrated water resource management and the hunter valley conservation trust, NSW, Australia. Appl Geogr 9:196–211
Mohapatra AC (2013) Political economy of natural resources and geography. Trans Inst Indian Geogr 35(2):149–156
Mukerjee NK (1992) Tin Resources of India and National Development. Presidential Address, Earth System Sciences, Indian Science Congress, MS University, Baroda
Narain S et al (2018) State of India's environment. In: Down to earth annual. Centre for Science and Environment, New Delhi
Natarajan KM (1995) No time to lose. Kurukshetra 44(October):109–111
Noble AG (1972) Physical basis of resources. In: Dutt AK (ed) India: Resources, potentialities, and planning. Kendall/Hunt, Dubuque, IA, pp 13–37
O'Riordan T (1971) perspectives on resource management. Pion, London
Odum EP (1971) Fundamentals of ecology. W.B Saunders Co., Philadelphia
Omara-Ojungu PH (1992) Resource management in developing countries. Longman Scientific & Technical. Harlow: Longman

Patra P (2006) Land degradation in tropical highlands. Rajesh Publications, New Delhi

Rani S, Thakur B (2020) Regional patterns of deforestation in India. In: Thakur B (ed) Perspectives in resource management in developing countries, Bioresource depletion and conservation, vol 9. Concept, New Delhi

Rhind DW (1977) Computer-aided cartography. *transactions*. Inst Br Geogr New Series 2:71–97

Robbins P (2004) Political ecology: a critical introduction. Blackwell Publication, Malden

Roy PK (1994) Lucknow: slow death of a water source. The Hindu Survey of The Environment May 31. pp. 119–122

Saha SK (1979) River basin planning in the damodar valley of India. Geogr Rev 69:273–287

Samra JS, Singh G, Dagar JC (eds) (2006) Drought management strategies in India. Indian Council of Agricultural Research, New Delhi

Sewell WRD (1973) Broadening the approach to evaluation in resource management decision-making. J Environ Manage 1:33–60

Sewell WRD, Burton I (eds) (1971) Perceptions and attitudes in resources management. Information Canada, Ottawa

Shafi M (2006) Agricultural geography. Pearson Education, Delhi

Sharma HS (1980) Ravine erosion in India. Concept, New Delhi

Sharma HS (2020) Monitoring desertification in rajasthan: status, processes, and causes. In: Sharma HS, Padmaja S (Eds). Land resources: processes, degradation, and evaluation, NAGI Resource Series 3, (Series Editor, R.P. Misra). Concept, New Delhi. pp. 215–230

Sharma HS, Padmaja S (Eds) (2020) Land resources: processes, degradation, and evaluation, NAGI Resource Series 3, (Series Editor: R.P. Misra). Concept, New Delhi

Sheng Y (2011) Remote sensing. In: Agnew JA, Livingstone DN (eds) Geographical knowledge. Sage, Los Angeles, pp 171–184

Siegel S (1956) Non-parametric statistics for the behavioural sciences. McGraw Hill, New York

Silk J (1979) Statistical concepts in geography. George Allen and Unwin, London

Singh S, Saroha J (2014) Geography of India. Access Publishing, New Delhi

Sinha BN (2014) Tsunami: the killer waves. In: Mishra BK, Panda GK (eds) Natural hazards and disaster management. Artech, Bhubaneswar, pp 235–249

Soil and Water Conservation Annual Report 1999-2000 (2000) Central Soil and Water Conservation Research and Training Institute, Dehradun

Srinivas MN (1959) The dominant caste in Rampura. Am Anthropol 61(1):1–16

Stoddart DR (ed) (1981) Geography, ideology, and social concern. Basil Blackwell, Oxford

Tansley A (1935) The use and abuse of vegetational concepts and terms. Ecology 16:284–307

Teclaff LA (1967) The river basin in history and law. Martinus Nijjhaff, The Hague

Thakur B (1998) The energy resources of India: problems and planning. *Reg Symbiosis 6:75–91*

Thakur B (2003a) Coping with natural hazards: a geographer's agenda. Ann Natl Assoc Geogr India 23:1–20

Thakur B (2007) Emerging trends in India's resource development: implications for sustainable development. In: Thakur B (ed) Perspectives in resource management in developing countries, Part-1: population, resources, and development, vol 2. Concept, New Delhi, pp 3–87

Thakur B, Das A (2013) People, resource, and space: changing land and caste relations in rural Bihar. Natl Geogr J India 51:221–236

Thakur B, Sinha P (2017) Rural poverty and unequal access to land in Masauri block in Patna district. In: Thakur B, Sharma HS, Misra S, Chattopadhyay SK, Singh S (eds) Regional development: theory and practice, Development challenges, vol 5. Concept, New Delhi, pp 133–162

Thakur B (2020) Depletion and degradation of land resources in India. In: Sharma HS, Padmaja S (Eds). Land resources: processes, degradation, and evaluation. NAGI Resource Series, Vol. 3, (Series Editor, R.P. Misra). Concept, New Delhi. pp. 169–214

Thakur B (ed) (2003b) Perspectives in resource management in developing countries, Resource management: theory and techniques, vol 1. Concept, New Delhi, pp 1–45

Thakur B, Abhay RK (2020) Feasibility of linking indian rivers. In: Thakur B (ed) Perspectives in resource management in developing countries, Water utilization and management, vol 7. Concept, New Delhi

Trivedi RC (2008) Water quality management in India. In: Kanjlia VK, Gupta AC (Eds) Proceeding's 3rd International Conference, Water Quality Management, 6–8 February, Nagpur, Central Board of Irrigation and Power, Chanakyapuri. New Delhi, pp. 1–22

Valdiya KS (1987) Environmental geology: the Indian experience. Tata McGraw Hill Book Company, New Delhi

White GF (1945) Human Adjustment to Floods: A Geographical Approach to The Flood Problems in The United States. Research Paper No. 29. Department Of Geography, University of Chicago, Chicago

Yeh E (2006) Political ecology. In: Warf B (ed) Encyclopaedia of human geography. Sage, Thousand Oaks, pp 360–362

Check for updates

4 Approaching the Collaborative "Turn" in Water Governance: A Critical Reappraisal

Nigel Watson and Rob C. de Loë

Abstract

Since its inception as a professional and academic field, natural resources management and governance has been influenced by many different factors including, but not limited to, prevailing economic conditions, social values, political beliefs, and technology. In addition, the field has been shaped by many different ideas and concepts which have been taken up and advocated by professional groups and have become part of the dominant paradigm for resources policy and practice during different eras. Beginning in the early 1990s, "collaboration" emerged as one of the most popular and influential ideas of the present time — to the extent that collaboration among governments, private corporations, not-for-profit organizations, civic groups, and other actors is becoming an increasingly advocated solution for water problems at a variety of spatial scales. Unfortunately, the popularity of collaboration has not been matched with scrutiny of its strengths or weaknesses. In this chapter, we critically examine the signs for a "turn" in water management and governance towards collaborative approaches and institutional arrangements, drawing on recent literature, and our own original research and practical experiences of working with collaborative institutions and groups. Our focus is on collaboration in the water sector, where this approach to governance has become particularly important.

Keywords

Collaborative Turn · Water Governance · Water Management

4.1 Introduction

Complex, inter-connected, and seemingly intractable problems involving natural resources and the environment require innovative responses and, where possible, similarly creative and novel solutions. For at least the last several decades, geographers and scholars in allied disciplines have known this to be the case. Throughout his entire career, Bruce Mitchell has been a key figure and prolific writer on a wide range of matters related to innovative environmental problem-solving and approaches to decision-making. In particular, he has led efforts to understand and improve institutional and organizational arrange-

N. Watson (✉)
Lancaster Environment Centre, Lancaster University, Lancaster, UK
e-mail: n.watson1@lancaster.ac.uk

R. C. de Loë
School of Environment, Resources and Sustainability, University of Waterloo, Waterloo, ON, Canada
e-mail: rdeloe@uwaterloo.ca

B. Thakur et al. (eds.), *Resource Management, Sustainable Development and Governance*, Sustainable Development Goals Series, https://doi.org/10.1007/978-3-030-85839-1_4

ments for integrated resource management and governance, sustainability, and resilience. In this regard, Bruce Mitchell exemplifies the ethos that an academic should be driven by both intellectual curiosity and a desire to make the world a better place for present and future generations.

Regarding institutional and organizational arrangements, in the last 20 years increased attention has been drawn by policymakers and academics to innovative approaches involving collaboration. Whether the challenges relate to climate change, natural or technological hazards, resource use or consumption, or any other type of problem involving the environment, it seems that almost everyone believes that "collaboration" is necessary for developing viable policy responses and implementing effective solutions. The emergence of collaboration as a favored mode of operation is not unique in this regard, and other ideas such as public participation and grassroots or bottom-up decision-making have all had their moments in the spotlight. Equally notable however is how the "turn" towards collaboration has spread so widely and rapidly, to the extent that collaborative working may now actually constitute a new natural resources management paradigm (Holley et al. 2012). However, the paradigmatic shift and turn towards collaboration appear to be happening without very much explicit inquiry or critical reflection apparently having taken place. There is ample literature advocating greater collaboration as an answer to the major resource and environmental challenges faced by society, but considerably less published research critically examining the concept or challenging its underlying norms and assumptions. In much the same way that interagency cooperation became a myth for resource management in the 1980s (Dorcey 1987), it appears that collaborative resource and environmental management has become an idea that many people choose to firmly believe in and to advocate, but not necessarily based on a full and careful appreciation of what is likely involved, the important factors or conditions which can influence success and potential risks and limitations, in addition to potential positive impacts and benefits. Furthermore, collaboration is increasingly applied to resource planning and governance, in addition to management, and many other policy areas such as public health, social care, national and international security, education, transportation, and urban, rural, and regional development. As such, the early twenty-first century appears to be an era defined by cross-scale interagency and multiparty collaboration, in much the same way that the twentieth century was defined as an era of big government and industrial conglomeration.

Given this clear increase in the popularity of collaboration, and a rise in the number of applications, the time has come for a detailed and systematic examination and critical appraisal of collaboration, considering literature from both inside and beyond the field of natural resources, and insights from practical experiences. A fully comprehensive and detailed assessment of collaboration is beyond the scope of this chapter. Nevertheless, our goal here is to add to the ongoing scholarly debate and appraisal of collaboration by shedding some further light on collaborative processes and arrangements. Through this short contribution, we hope to encourage deeper and more critical understanding of collaboration in the spirit of supporting future developments and successful applications. This chapter is organized as follows. First, attention is given to examining the meaning of collaboration and defining characteristics which can help to differentiate collaboration from other types of interaction, along with potential benefits and costs. In the second section, the focus shifts towards the nature of collaboration as a process and identifying key insights from the literature regarding important elements and phases, and factors which are important influences on the impacts and overall success of collaborative initiatives and arrangements. In the third section, some of the insights from the literature are used to examine two case studies of collaborative resource management, one from the UK and one from Canada. This chapter concludes by highlighting key lessons from the combination of literature and case studies, and by considering potential future implications for policymakers, resource managers, and scholars interested in improving understanding of collaboration and developing and applying effective practical approaches.

Box 4.1 Sustainable Development Goals

The sustainable development goals (SDGs) have emerged as constructive allies in navigating water governance. Achievement of SDG 6 (water and sanitation) or water management and governance more generally involves not only the collaboration between institutions (government, private corporations, nonprofits, and civic groups) but also articulating the sensitive interrelations among land, water, and food across spatial scales. Using two case studies of collaborative resource management from the UK and Canada, this chapter takes a critical look at how good governance can promote accountability, transparency, efficiency, and rule of law at all scales, ensuring the availability and access to safe drinking water and sanitation for all, focusing on the sustainable management of water resources, wastewater, and ecosystems, as well as acknowledging the significance of an enabling environment. As the authors focus on collaboration in water governance, they examine the meaning of collaboration, its characteristics, nature, and the process that determines collaboration itself. Their findings are bold and humbling as their recommendation hinges on the fact that while collaboration is significant, it is not a panacea for water governance. Collaboration is an iterative process and requires the coming together of different stakeholders over time to trust the leadership at different scales. In so much, sustainable management of water resources and access to safe water and sanitation are essential for unlocking economic growth and productivity and provide significant leverage for existing investments in health and education.

4.2 Definitions, Benefits, and Costs of Collaboration

The term "collaboration" is derived from the Latin words *com* and *laborare,* and at a basic level means "to work together". While a useful starting point, such a broad definition includes an infinite number of possibilities and does not provide sufficient conceptual clarity to differentiate collaboration from other common forms of work-related interaction, such as, for example, consultation, cooperation, or coordination. A related challenge is that collaboration has tended to be defined and examined within specific contexts, such as planning (for example, Booher and Innes 2002; Frame et al. 2004), governance (for example, Ansell and Gash 2007; Vangen et al. 2015), or management (for example, Conley and Moote 2003; Bidwell and Ryan 2006; Bonnell and Koontz 2007). As a result, explicit operational definitions of collaboration per se are often absent, and therefore it is not always clear whether research findings are directly attributable to the nature of the collaborative process and associated arrangements under investigation, or perhaps to other factors or variables related to the planning, governance, or management context.

Of the literature that does provide an explicit definition of collaboration, the work of Barbara Gray, Professor of Organisational Behaviour at Pennsylvania State University, is among the most widely cited (Gray 1985, 1989). Gray (1985) defined collaboration as: "(1) The pooling of appreciations and/or tangible resources, e.g., information, money, labor, etc., (2) by two or more stakeholders, (3) to solve a set of problems which neither can solve individually".

Building on her original definition, Barbara Gray (1989) later described collaboration as a process "through which parties who see different aspects of a problem can constructively explore their differences and search for solutions that go beyond their own limited view of what is possible". Thompson and Perry (2006) provided a further definition highlighting the need for participants to create means to enable their own interactions: "Collaboration is a process in which autonomous actors interact through formal and informal negotiation, jointly creating rules and structures governing their relationships and ways to act or decide on the issues that brought them together; it is a process involving shared norms and mutually beneficial interactions".

Potential benefits arising from collaboration have been described in a variety of ways. For example, Hudson et al. (1999) define collaborative capacity as the level of activity or degree of change a collaborative relationship can sustain without any partner losing a sense of security in the relationship. Sense of security includes the tangible resources which are central to the collaborative endeavor, and the autonomy and relative strength of the individual actors. Huxham (1996) described "collaborative advantage" as a synergy among organizations when something unusually creative is produced that no organization could have produced on its own and when each organization, through the collaboration, can better achieve its own objectives than it otherwise could by acting alone. In contrast, Booher and Innes (2002) refer to "network power" as an important outcome and benefit of collaboration: "Network power emerges from communication and collaboration among individuals, public and private agencies, and businesses in society. Network power emerges as diverse participants in a network focus on a common task and develop shared meanings and common heuristics that guide their action. The power grows as these players identify and build their interdependencies to create new potential. In the process, innovations and novel responses to environmental stresses can emerge. These innovations, in turn, make possible adaptive change and constructive joint action".

Evaluations of collaboration in the environmental domain have pointed to various kinds of benefits. For instance, after evaluating resource planning in British Columbia, Frame et al. (2004) concluded that collaboration was an effective means of resolving environmental conflict, and that collaboration produced significant additional benefits, such as improved stakeholder relations, skills, and knowledge. van Tol et al. (2015) highlighted practical ways in which collaboration allows for diverse types of knowledge to be brought together. Importantly, however, people almost never collaborate to improve stakeholder knowledge, relations, and skills. Instead, they collaborate to address shared environmental problems. Evidence that collaboration produces better environmental outcomes than other governance modes is rare, in part because of three challenges. Koontz and Thomas (2006) revealed: (1) gathering data that measures environmental outcomes rather than only outputs, (2) allowing for long time horizons between the implementation of collaborative outputs and environmental change outcomes, and (3) designing research protocols that untangle the effects of multiple interacting variables that shape environmental change. Thus, studies, such as the one reported by Scott (2015) are important. In a study of collaborative governance and environmental quality in 357 watersheds, Scott (2015) found evidence of positive impacts on water chemistry and in-stream habitat conditions, particularly where groups were engaged in management activities rather than collaborative planning or policymaking.

Practical evaluations of the benefits of collaboration are necessary, but in our view not sufficient. Collaborative processes frequently are created in situations where more established governance mechanisms have failed (Watson et al. 2009; Watson 2015). As a result, the people designing collaborative processes—or simply "failing" into collaboration—may give insufficient attention to basic concerns, such as legitimacy, accountability, and fairness. We are far from the first people to raise these kinds of concerns. To illustrate, McClosky (2000) challenged many of the basic assumptions common in the collaboration literature relating to fairness, quality of decisions, and the legitimacy of the processes. Similarly, Bryan (2004) noted that critics of collaborative approaches to natural resources argue that decision-making processes that devolve government authority, hand responsibility for management to unauthorized groups, ignore or circumvent existing laws and regulations, and/or preclude the rights of citizens to participate through administrative channels should be avoided. Proponents of collaboration argue that such problems and limitations can be avoided by giving careful attention to process design and understanding the various factors and conditions which are likely to promote effective, equitable, and efficient collaborative decision-making. However, the extent to which good process design can mitigate power imbalances among the actors engaged in collaboration is poorly understood in the field. In a systematic review of collaboration literature linked to water governance, Brisbois and de Loë (2015) found that power was generally ignored and usually poorly theorized when it was acknowledged. We explore these kinds of concerns in more detail in the next section.

4.3 Understanding Collaboration as a Process

Most authors concur that collaboration is a deeper and more richly joined process compared to cooperation or coordination, and that the collaboration most often involves several iterative cycles of decision-making, each of which is likely to include a series of recognizable stages or phases. Building on the earlier work of McCann (1983) and Gray (1985, 1989), Watson (2004) proposed a process-based conceptual model for collaboration to describe how participants engage in an iterative cycle of problem-setting, direction-setting, structuring, and producing outputs and outcomes. Throughout the cycle or multiple cycles, participants can be expected to continually evaluate the benefits and cost arising from the collaboration to determine whether to increase, maintain, or reduce their future level of commitment to the joint venture.

In a different approach, Ansell and Gash (2007) developed a model of collaborative governance which drew attention to starting conditions, institutional design, facilitative leadership, and a collaborative process that included five key variables: commitment, shared understanding, intermediate outcomes, face-to-face dialog, and trust-building, all of which contribute to the generation of outcomes. Ring and Van de Ven (1994) similarly characterized collaboration as cyclical and iterative rather than a linear process, and drew attention to the relationships among negotiation, commitment, and implementation and the idea that participants make ongoing assessments to determine whether collective actions are executed in a reciprocal fashion and whether to expand their commitments, seek corrective measures or reduce their involvement. Thompson and Perry (2006) noted that collaboration may be driven by self-interest (termed "classic liberalism") or commitment to collective needs and shared preferences (termed "civic republicanism") and highlight five key elements for a collaborative process: governance, administration, organizational autonomy, mutuality, and norms of trust and reciprocity.

There is also discussion in the literature regarding key conditions and factors which are likely to encourage successful and resilient collaboration. For example, Levesque et al. (2017) found that collaboration related to the conservation of vernal (i.e., temporary) water pools had worked well where power had become distributed among members, trust had formed across core interests, and social learning had resulted in shared understanding and joint solutions. In addition, agreement regarding rules for group membership, organizational and social norms of inclusion and openness, and the use of small working groups had contributed to the observed patterns of power, trust, and learning. Referring to experiences of collaboration regarding social work, Perrault et al. (2011) identified four success factors: attention to informal connections and member relationships; developing trust, respect and understanding; having learning as a purpose; and sharing leadership. Vangen et al. (2015) argued that all collaborations require effective governance and make a distinction between "collaborative governance" and "governing collaboration". Three key design elements were identified for effective governance of collaboration: (1) structure — providing connections among the participating individuals and organizations and links with other collaborative groups; (2) processes — enabling communication, sharing, and decision-making; and (3) actors — ensuring the involvement of actors with power and sufficient knowledge to influence and enact the collaborative agenda and agreement.

Various dimensions related to power and power relations have been noted in the literature as particularly important influences on collaboration. In their discussion of network power, Booher and Innes (2002) introduced the acronym "DIAD" to highlight the need for diversity, interdependence, and authentic dialog. Booher and Innes suggested that authentic dialog enables all agents to speak in an open and informed way about their interests and understandings and ensures that all are listened to and taken seriously by other participants. Without such dialog, it is suggested that understandings and meanings will not become truly shared, members will not identify with a common system or community, opportunities for reciprocity will be missed, important information about the problem will not surface and creative solutions are much less likely to emerge. Referring to water governance, Brisbois and de Loë (2015) drew attention to three key forms of power which are likely to be important in many collaborative situations. First, instrumental power is overt and involves the use of resources to influence policy outcomes in competition with others. Second, structural power involves the ability to shape policy agendas and to privilege dominant views or preferences over those of more marginalized groups. Third, discursive power represents the ability to manipulate the wants and desires of others. A key point to emerge from this discussion is that the phenomenon of power is more complex than is often appreciated and that providing broad and balanced representation at the decision-making table alone is unlikely to be sufficient to ensure effective collaboration and equitable decisions.

In summary, organizations and groups choose to collaborate to deal with overlapping interests and responsibilities and to enhance their problem-solving capabilities. Regarding natural resources and the environment, collaboration is an increasingly preferred approach, particularly in complex and cross-scale institutional settings characterized by multiple public and private sector actors with interdependent needs and limited individual authority and power to act. A defining feature and key quality associated with collaboration is that typically participants give up some of their independence and make joint decisions in pursuit of mutually identified and agreed superordinate goals or objectives. As such, collaborative relationships are different from cooperative relationships and interorganizational arrangements based on coordination, where levels of interaction are likely to be lower, participants act more autonomously, and attention tends to be focused on the achievement of separate rather than joint objectives.

In the literature, collaboration is described as an iterative and dynamic process in which participants engage in learning and negotiation to develop shared understandings of their situation and the problems or challenges, they face. Assuming that agreement on such matters is reached, the process may then lead to joint commitments, formal and informal agreements, policies, plans, and actions. To achieve any of this, most collaborative groups are likely to need their own governance arrangements to address matters such as representation and membership, decision rules, finances, monitoring and reporting, and accountability. Furthermore, collaboration may occur in relation to governance, planning or management, and combined initiatives involving different regimes and functions in more complex cross-scale collaborative arrangements. Because collaboration depends on mutual agreement, there are no fixed rules other than those developed by the participants and, because emphasis is often placed on learning, collaborative arrangements can be expected to evolve over time. In this regard, collaborators are likely to continually reevaluate the benefits and costs of their participation and, consequently, increase, maintain, or decrease their involvement as the collaborative process develops and unfolds.

Given the often complex, dynamic, or even "messy" nature of collaboration, it is not at all surprising that authors and commentators have generally avoided detailed prescriptions, protocols, and exact models. Nevertheless, previous research does suggest that, for collaborative initiatives and arrangements to be effective and enduring, particular attention should be given to

(1) antecedents and motives; (2) power and power relations; (3) dialog and trust; (4) interdependency and reciprocity; and (5) outputs and outcomes. These matters are further explored in the following section, where attention is turned to learning from experiences regarding two examples where a collaborative approach has developed to address concerns and problems related to land and water resources.

4.4 Learning from Experience

4.4.1 Case Study 1: Collaborative Catchment Management in the UK

Loweswater is a small lake of around 0.6 km^2 with a surrounding upland catchment area of 7.6 km^2 in the county of Cumbria, Northwest England. Situated in an isolated area designated as a "quiet valley" by the Lake District National Park Authority (LDNPA), Loweswater contrasts sharply with the other larger iconic lakes in the region, such as Windermere and Coniston which attract hundreds of thousands of visitors each year. The lake at Loweswater is owned by the National Trust (NT), which is a charitable and membership-based conservation organization established in 1895. The surrounding lowland pasture and upland fells are incorporated within eight separate landholdings, each farmed by tenants for sheep and cattle production. Land within the catchment has low agricultural potential due to the landscape and climate, and additional inputs of organic and inorganic nutrients are needed to maintain the pastures and to produce sufficient animal feedstocks for the winter. There are around 50 permanent residents in the catchment area, including retirees, small business owners who provide camping facilities, rental cottages and hotel accommodation for visitors, and around ten farming families. During the summer months, many thousands of people pass through and stay in the area while visiting the Lake District. In addition to the LDNPA and NT, other organizations with interests in Loweswater include the Environment Agency (EA) which has statutory responsibilities for pollution control and river basin management, and Natural England (NE) which administers many of the voluntary and payment-based environmental improvement schemes available to farmers in selected areas.

Antecedents and Motives: deteriorating water quality at Loweswater first began to attract local concern in the 1990s because of increased prevalence of blue-green algal blooms, particularly during the colder winter months when such problems would not usually be expected. The NT posted signs around the perimeter of the lake warning of the dangers to people and animals encountering potentially toxic cyanobacteria. Despite the controversy and increased local concern, little if any action was taken until 2000 when the EU Water Framework Directive was adopted in England and Wales and responsibility for implementation was handed to the EA. Initial assessments by the EA indicated that Loweswater was unlikely to meet the requirements for "Good Status" and notices were issued to several local farmers warning of the possibility of prosecution if they were found to be causing or contributing to the problem. Relations among the farmers, residents, and staff from the EA, NT, and NE became increasingly strained. Members of the farming community believed they were being blamed for a problem that they had not necessarily caused and were anxious to avoid costly legal proceedings and stricter regulations regarding land use and management. Similarly, as the owner of the lake, the NT wanted to improve relationships with the local community. Meanwhile, the EA was facing reductions in finances and was increasingly interested in locally led, voluntary approaches rather than expensive regulatory action.

Keen to show that the problem was being taken seriously, in 2002 the local farming community established a ten-member Loweswater Improvement Group (LIP) to enhance practical understanding and strengthen relationships with the various agencies. Next, in 2003/04, the NT and NE jointly funded research by the Centre for Ecology and Hydrology (CEH) to test soil samples and investigate the impacts of farming practices on the lake. Around the same time, several local property owners took the step of installing new septic tanks to deal with human waste and some of the farmers invested in new slurry stores and drainage systems to separate clean rainwater from dirty yard water.

Despite improved relationships and some positive action, there were still many uncertainties and different opinions regarding the causes of the algal blooms on the lake, possible links to past and present farming methods and other nutrient sources, and ways to effectively mitigate the problem. Recognizing the opportunity, and with the agreement of local farmers and agencies, researchers from CEH and Lancaster University proposed a project to enable a diverse range of actors to collaborate in conducting new research to improve understanding and management of the catchment. Funding for the 3-year project was awarded through the joint UK Research Council (RCUK) Rural Economy and Land Use Programme (RELU) in 2007.

Power and Power Relations: given the diverse range of actors involved with Loweswater, and the history of weak organizational relationships and acrimony regarding the lake, a key challenge was to ensure that groups or interests would not dominate and therefore undermine the collaborative process. To address this concern, Latour and Weibel's (2005) concept of a "knowledge collective" was applied to create an open forum which served as a social, community-based, mechanism to bring people together to exchange information, explore and evaluate different claims and beliefs regarding the catchment and, where possible, develop shared understandings of the situation. The knowledge collective was named by the group as the Loweswater Care Project (LCP) and to ensure as much as possible that individual groups or people did not try to dominate or control the process, five guiding principles were adopted:

1. There is not a single correct or self-evident way of understanding or explaining the environment.
2. All forms of knowledge and expertise need to be debated.
3. Uncertainties regarding knowledge and understanding should be recognized.
4. Making new connections among organizations, system variables, processes, and bodies of knowledge are valuable.
5. Doubt and questioning should be applied to claims and arguments from all fields, including science and politics.

The five principles were applied through a series of 15 open-forum meetings held every 2–3 months in the village hall at Loweswater over 3 years. Each meeting was held in the evening to enable as many people as possible to attend, lasting for between 3 and 4 h and attracting 25–35 people. The format worked well, and the meetings were attended by a diverse mix of people from inside and beyond the catchment area, including representatives for the EA, LDNPA, NE, and NT, farmers and local business owners, residents, and social and natural scientists from CEH and Lancaster University. Three factors appeared to help ensure the LCP was reasonably effective in dealing with power and power relations: holding meetings in the community, rather than in agency premises; carefully following principles designed to encourage open debate; and focusing explicitly on knowledge production rather than management decision-making, which helped to lower the stakes and reduce the risks for the participants. Nevertheless, the LCP was not an entirely effective mechanism. For example, some residents, farmers, organizations, and groups did not attend at all, and some of the people who did participate remained uncomfortable with being part of a collective and would have preferred a more conventional committee-style structure and process.

Dialog and Trust: several design elements were included in the collaborative project to promote dialog and build trust. Prior to the start in 2007, researchers from CEH and Lancaster University had gradually developed strong relationships with some of the local farmers and had provided technical input and support to the predecessor LIP group since 2003/04. In addition, researchers undertook an initial scoping study to gauge potential interest and gather support, before preparing a proposal for a full-scale project. Subsequently, one of the researchers spent several days visiting and interviewing local people, in part to gather background information but also to build interpersonal relationships. The first few meetings of the LCP were used to explain the approach to collaboration that the researchers were recommending, to establish aims for the group and to agree the procedures to be followed.

To encourage open and inclusive dialog during the meetings, a range of relevant topics were identified for discussion, and over the 3-year period the LCP examined and debated diverse matters including local cultural and economic history, past and current farming practices, recreational fishing on the lake, tourism, domestic waste, and septic tanks, catchment hydrology and geomorphology, and institutional arrangements related to the governance and management of land and water in the catchment. In addition to measures intended to encourage open dialog, several other elements were introduced to develop trust among the participants. For example, a widely respected and trusted local farmer was hired as a part-time researcher to help with communication, facilitate access to land for research, and act as an intermediary. In another effort to promote trust, an LCP meeting was organized at which representatives for all the key agencies gave presentations and answered questions from the participants regarding statutory responsibilities and powers, policies, resources, and capabilities with regard to the catchment and lake. Nevertheless, there were inevitable disagreements and heated exchanges at some of the gatherings, some participants stopped attending after the first year, and the farmers were often reluctant to disclose details in public and held several of their own meetings outside of the LCP forum.

Interdependency and Reciprocity: according to advice in the literature, one of the keys to collaboration is to create a strong sense of interdependence and similarly strong reciprocal relationships among the participants. In the Loweswater project, one of the guiding principles concerned the importance and value of making new connections. However, in practice, finding agreement on relevant systems, boundaries and knowledge proved to be one of the most difficult challenges and the issue was still was not resolved at the end of the 3-year period. To illustrate, some participants viewed the blue-green algae as a complex problem and believed that the LCP needed to focus on developing a detailed understanding of lake nutrient cycles and ecological processes to eradicate or at least reduce algal blooms on the lake. In contrast, other participants regarded the algae to be a symptom of a much larger problem regarding the functioning of the catchment area, including the management and drainage of land for farming, other business purposes, and conservation. To a degree, the interplay among the different views and preferences was an important feature of the collaborative and iterative processes and, in general, participants were willing to compromise and accept there are multiple ways of framing resource-related problems. One innovation which proved to be useful for handling such differences was related to the funding awarded by RCUK. The research grant included an additional budget of £35,000 for initially unspecified expenditure to enable the LCP to commission additional research on important matters that might arise during the project. The flexibility around some of the funding enabled members of the LCP to suggest additional research and therefore accommodate a range of different priorities and interests. Two small-scale studies, one on attitudes to tourism and the other on the use of septic tanks, were subsequently carried out by LCP members from the local community. Three additional pieces of research, on land use change and lake sediments, phosphorous applications on farmland and flows to the lake, and macroscale water movements in the catchment were conducted by local people working with some of the scientists from the project. As such, attempts were made to establish two-way reciprocal relationships among scientists and laypeople and to avoid situations where scientists were perceived to be the only experts. While these kinds of approaches and strategies were positively received by the LCP at the time, it is also true that some members felt disillusioned and were disappointed at the end of the 3-year project because answers for the specific problem of blue-green algae in the lake had not been found.

Outputs and Outcomes: given the often considerable additional time, effort, and resources required for collaborative initiatives, it is entirely reasonable that questions arise regarding the impacts relative to other approaches potentially available to organizations and groups. In attempting to evaluate the impacts and benefits of collaboration at Loweswater, it is important to keep in mind that the funding awarded by RCUK, and the subsequent work was aimed at improving knowledge and understanding and did not include financial resources for changes or improvements to the management of land and water in the catchment. Nevertheless, it was hoped that practical action might occur because of new information and evidence generated from the LCP and the associated collaborative research.

Prior to the project, one of the main challenges related to Loweswater concerned the small size of the lake and surrounding catchment, which meant that the area was not a high priority for government and public agencies with the ability to support improvements in management. However, the LCP helped to draw attention from policymakers at a time when national interest in catchment management and community-based action was increasing. The LCP provided legitimacy for management of the catchment which had previously been lacking, and the various technical reports produced during the project provided credible fresh evidence indicating that changes to land use and farm management could have positive impacts on lake water quality and help to reduce the incidence of algal blooms.

By the end of 2010, many of the local and agency-based LCP participants were keen for the initiative to evolve to a new phase where practical action could be taken. However, the researchers from CEH and Lancaster University felt it was not appropriate for them to lead a new phase, given the RCUK funding had ended, and that attention was starting to move more towards catchment management and away from research and knowledge production. As a result, two members of the Loweswater community acted on behalf of the LCP and applied for new funding under the England Catchment Restoration Fund (CRF) and were awarded £275,000 to address the causes and effects of nutrient enrichment of the lake. Following this success, proposals were created to take the LCP forward without the involvement of CEH or Lancaster University but, unfortunately, none of the remaining organizations and groups were willing to accept responsibility or take the lead because of perceived legal difficulties and financial risks. With the collapse of the LCP, an eight-person Management Committee was formed by a small group of local farmers, residents, and representatives for the EA, NT, and NE in association with an umbrella organization called the West Cumbria Rivers Trust to administer the CRF grant. Subsequently, the Committee awarded funding to four of the eight farms in the catchment for improvements to animal slurry stores, roofing for farmyards, infrastructure to separate rainwater from dirty water, fencing-off land close to streams, stream restoration work, shared farm machinery to reduce soil compaction and improve aeration, new farm buildings outside of the catchment, plus a program operated by two residents and staff from the NT to monitor lake water quality. Initial monitoring results indicate improvements in water chemistry but less clear changes in algal populations, with some species increasing in numbers while others have declined.

Overall, collaboration has been an important and valuable aspect of the development of catchment management at Loweswater. Collaboration among farmers, residents, conservation organizations, public agencies, and researchers has enhanced credibility and legitimacy, produced new evidence and knowledge, and enabled practical action to be taken. However, the LCP did not result in a lasting governance or management transformation, and collaboration appears to have been subsequently replaced with more conventional and less inclusive approaches to decision-making.

4.4.2 Case Study 2: Source Water Protection in Ontario

The Province of Ontario is the most populous of Canada's ten provinces and three territories. Most of its residents (approximately 8 in 10) are served by drinking water treatment and distribution systems that draw their source water from surface and ground sources. The provincial government establishes the regulatory and policy framework under which drinking water provision occurs, while responsibility for operating drinking water treatment and distribution systems lies primarily with municipalities and public utilities (Water Strategy Expert Panel 2005).

The recent history of drinking water provision in Ontario has two main eras: before a waterborne disease outbreak in the Town of Walkerton, Ontario (population 4800), in May 2000, and after. The former era is characterized by widespread complacency about drinking water safety. The regulatory environment had been weakened through a series of changes, budgets had been reduced and technical staff in provincial agencies responsible for monitoring and regulating systems cut back dramatically (Kreutzwiser 1998). Contamination of the water supply of the Town of Walkerton in May 2000 by *E. coli* O157:H7 and *Campylobacter jejuniin* resulted in seven deaths and over 2300 cases of illness. In response, on the advice of a wide-ranging public inquiry, the provincial government strengthened the regulatory system and instituted a multi-barrier approach to drinking water safety. Importantly, a key feature of the new, post-Walkerton drinking water system in Ontario is source water protection organized at the watershed (catchment) scale and based on a collaborative planning model. Under the new *Clean Water Act* (Statutes of Ontario 2006), 19 collaborative bodies organized around watershed boundaries have played a crucial role in developing drinking water source protection plans and the policies that are being implemented at the local level to protect these sources.

Antecedents and Motives: The challenge of protecting drinking water sources is universal, and countries around the world have taken different approaches. For instance, in the United States, the federal government created the overarching rules under its *Safe Drinking Water Act*; states and local governments play key roles in developing and implementing plans (de Loë and Lukovich 2004). In England, source water protection takes place through a host of legal and policy instruments that are coordinated by government and nongovernment organizations operating at the national, regional, and local levels (Watson et al. 2009). Ontario has a long history of watershed management through its watershed-based conservation authorities but has been a relative latecomer to drinking water source protection. In the aftermath of the Walkerton tragedy, the commission of inquiry charged with analyzing the causes of the outbreak and then, proposing a way forward was presented with numerous ideas and models for how to build a robust system for protecting drinking water sources. Considering the deep cuts to the capacity of the provincial government that had taken place throughout the 1990s, a top-down planning approach would not have been feasible. Fortunately, conservation authorities were ready and willing to play key supporting roles. These locally created watershed-based organizations that have existed in Ontario since 1946 covered most of settled southern Ontario and were well positioned in terms of technical capacity to support a more bottom-up, collaborative approach to source water protection.

The system that exists, today, illustrates how collaboration can be inserted into a regulatory environment that already includes a host of actors with implementation responsibilities. The *Clean Water Act* established the framework for source water protection. Collaborative source protection committees were formed in each of 19 source protection jurisdictions based on the boundaries of one or more existing conservation authorities. Source protection committees have a chair and members representing municipalities (one-third); agricultural, commercial, or industrial interests (one-third); and environmental, health, and other sectors, including the public (one-third). In addition to these members, committees are required to seek members from First Nations (Indigenous) communities located within their area of jurisdiction. Responsibilities and operating procedures for these committees are specified in a provincial regulation made under the *Clean Water Act*. Nonetheless, the committees are meant to function as true collaborative bodies.

Power and Power Relations: Power takes many forms, including instrumental (the power to force others to do one's will); structural (the power to set agendas); and discursive (the ability to shape social norms and values) (Lukes 2005). Collaboration poses challenges for those interested in the way power is used and distributed in society. Power imbalances within collaborative processes are an obvious area of concern. For instance, industry is crucial to Ontario's economy, and a source of many threats to source waters. Representatives of large industries are members of source protection committees in jurisdictions where these industries operate. It would be naïve to believe that the financial and technical power firms hold, let alone their elite access to decision-makers in government, disappears once they participate in a collaborative process. Brisbois and de Loë (2016a) found that large industries were enthusiastic participants in collaborative source protection committees in Ontario but had a limited commitment to engage in shared learning and to re-examine values and interests (two key benchmarks for successful collaboration). Indeed, firms benefited from participating in these processes through having opportunities to influence other participants and to learn about their perspectives, while still being able to exert influence directly on policymakers through elite-level relationships and lobbying that took place entirely outside of the collaborative processes.

Concerns for power did not exist solely within the collaborative committees. In considering the motives of the provincial government, Brisbois and de Loë (2016b) found that the provincial government exerted considerable power over the collaborative processes from agenda setting through to implementation. As explained previously, the provincial government made the rules. In part this simply reflected the fact that the provincial government is responsible for drinking water safety in Ontario, and, thus, relied heavily on its traditional command-and-control mechanisms despite creating these collaborative processes. Command-and-control and rigid bureaucratic structures are common characteristics of the state. However, Brisbois and de Loë (2016b) also found that the state used its power over the collaborative processes to protect economic actors; this reflected a concern for jobs and growth that is also widely shared by governments around the world.

Dialog and Trust: Concerns relating to open dialog and trust among participants flow in multiple directions in the source protection planning process in Ontario. Not only is source water protection a highly technical activity, but also it tends to be extremely political because protecting water usually means changing land use practices. Activities that have been legal and accepted in the past — from spreading manure on fields to fertilize crops, to using salt on roads in winter to protect drivers from accidents — may become unacceptable in some areas due to the need for source water protection.

Concerns, such as these naturally surfaced during the process of plan and policy development. They were especially prominent in the context of farming, and the involvement of the farm sector. As a proportion of the population in Ontario, farmers represent only approximately 2%—yet they own and manage approximately one-third of the land in southern Ontario (OMAFRA 2012). Farmers in Ontario have long recognized that normal farm practices affect air and water quality. Thus, the sector, with the support of the provincial government, had a history of using voluntary stewardship programs and measures to protect water resources (Armitage 2001). The Walkerton crisis triggered a prescriptive, regulatory approach to source water protection that disregarded "historical practices and shared understandings, especially in rural areas with long agricultural traditions" (Ferreyra et al. 2008). Thus, the potential existed for conflict and unproductive relationships on the committees.

Through intensive behind-the-scenes dialog, consultation, and capacity building, a constructive approach to including farmers in source protection planning efforts was developed. A provincial farm sector organization undertook a deliberate program of capacity building to ensure that farmers could participate effectively in the multi-actor source protection committees. The farm sector, in general, and individual farmers, then viewed engaging with other source protection committee members and sharing their general and local knowledge about agriculture as key priorities (Simpson et al. 2015). Trust developed through processes of knowledge coproduction, and through farm participants in the collaborative committees demonstrating the value of agricultural knowledge for the work of source protection planning. In many cases, farm representatives were able to challenge technical knowledge provided by specialists supporting the work of the collaborative committees (Simpson et al. 2015).

Interdependency and Reciprocity: Collaboration can be an effective strategy for drinking water source protection because threats to water quality have multiple historical and contemporary sources linked to a vast range of practices and behaviors on the landscape, and because responsibility and authority for dealing with those threats tend to be widely shared. This is certainly the case in Ontario, where threats to drinking water sources identified in Ontario Regulation 385/08 of the *Clean Water Act* include agricultural practices, such as manure storage and fertilizer application; routine winter road maintenance activities, such as storing snow and handling, storying and applying road salt; industrial practices related to the handling and storage of fuel and organic solvents; and activities that involve taking water without returning it to the source. These activities implicate a host of actors, including farmers, business owners, municipalities, and provincial and federal government agencies. As a result, it is well recognized and accepted in Ontario that planning for source water protection benefits from collaborative approaches where those who are causing the threat and/or responsible for dealing with the threat share knowledge and resources, take account of the impacts of their activities on each other, and mobilize the tools available for dealing with threats to source waters in a coordinated fashion.

In Ontario, recognition of these interdependencies exists not only within each source protection jurisdiction, but also among them. Within each of the 19 source protection jurisdictions, policies to respond to identified threats had to be developed on an extremely tight timeline. Because they are based on watersheds, the boundaries of the regions divide municipalities. Some municipalities are in multiple regions, and each region contains multiple municipalities. Hence, a shared policymaking challenge existed in each region—with the collaborative source protection committees at the center. In response to this challenge, Conservation Ontario, the umbrella organization that represents the conservation authorities that provide core technical support for source protection planning, partnered with the Ontario Ministry of the Environment, which is responsible for source protection planning. They developed several tools designed to speed up the policymaking process: a catalog of existing work; a policy web forum; and a policy database (de Loë et al. 2016). Policy planners in the 19 regions used these tools to facilitate rapid, concurrent policy development that supported the work of the collaborative source protection committees. The ability to "reuse" or share policies developed for one of the source protection jurisdictions in another was variable, and most successful in cases where the challenge was not generic and not defined by extremely local circumstances or contexts (e.g., policies relating to the development of signage); in contrast, policies for taking water without returning it to the same aquifer were more difficult to share—yet sharing did occur among neighboring source protection jurisdictions (de Loë et al. 2016).

Outputs and Outcomes: Ensuring that clear and relevant outputs and outcomes emerge from a collaborative process is a central challenge. In collaborations that form entirely through bottom-up processes, benchmarks for success in terms of outputs and outcomes can be unclear. In the case of Ontario's source protection committees, outputs and outcomes were, for the most part, clear and measurable. Under the *Clean Water Act*, the committees were required to lead the characterization of water resources, including the creation of water budgets; the delineation of vulnerable areas; and the identification of threats to drinking water. With this foundational work complete, they were required to create source protection plans and related policies that use a range of tools to protect source waters in vulnerable areas. Outputs were thus tangible products, such as water budgets, source protection plans, and policies. Much of the actual work of developing these complex products was undertaken by staff from conservation authorities, municipalities, and consulting firms. Nonetheless, the collaborative source protection committees played a crucial steering role. As of the time this chapter was written, source protection plans had been completed and approved for all 19 regions, and implementation was underway.

Important outcomes from the collaborative approach to source water protection that has been established in Ontario are already evident. First and foremost, while no comprehensive assessment has yet been completed, few would disagree that Ontario's drinking water sources are better protected now than they were in 2000. An enormous effort has been made to improve understanding of water resources, especially groundwater systems, and to identify threats and vulnerable areas. Many people and organizations in conservation authorities, municipalities, universities, and other organizations were involved, but the collaborative source protection committees played an important leadership role in ensuring this work was completed. A less tangible, but by no means insignificant, outcome of the collaborative process used is the strengthened networks and relationships that have developed not only within the source protection regions, but among them, and among the diverse actors who play a role in ensuring that sources of drinking water in Ontario are protected from contamination.

4.5 Lessons and Implications

A review of literature and exploration of two experiences in the UK and Canada provides some useful insights regarding circumstances in which collaborations tend to emerge, how they develop, their benefits and costs, and factors which seem to affect performance and success. Critically examining experiences related to the management of a small catchment area and the protection of source water has revealed some valuable lessons, which have implications for organizations and individuals attempting to collaborate, or perhaps considering collaboration as a possible future approach.

Below, we draw attention to six key lessons which have emerged from our analysis, and comment on some of the potential implications for policy and practice. First, collaboration is clearly not a panacea for all resource-related problems or management situations. Public, private, and third-sector organizations, and groups often have clear legal rights and responsibilities which cannot be ignored or over-ridden. When existing rights and responsibilities are not fully understood and acknowledged, and therefore potentially under threat, it is unlikely that a collaborative initiative will even get off the ground. This implies that one of the first steps in considering the potential for collaboration should be a careful analysis of existing institutional arrangements to identify areas where there is genuine interdependency and scope for reciprocal, mutually beneficial relationships.

Second, as shown by both cases studies, collaboration does not happen overnight and often emerges from a lengthy period of interaction and discussion among actors and stakeholders. This may also involve failed attempts by individual organizations or smaller groups of actors to solve problems, eventually leading them to realize that collaboration may be necessary. Collaboration cannot be forced. Judgments need to be made regarding whether conditions are favorable, and the time is right to propose collaboration as a way forward. Even in cases where the collaborative process is formed in a top-down manner, as occurred in the case of source water protection in Ontario, participation ultimately is voluntary.

Third, collaboration requires credible, trusted leadership to work well. In the case of Loweswater, this came from university academics with established relationships in the area. In Ontario, leadership originated from multiple directions, including the provincial government, which established the framework and provided substantial resources to support foundational technical work; the conservation authorities, that provided technical support; and the various organizations that provided support in other ways. The active capacity-building support provided by farm sector organizations to the farmers who served on the collaborative committees in Ontario is a striking example of the importance of informal or unofficial leadership.

Fourth, skepticism is often associated with collaboration and collaborative initiatives. Not only is such skepticism or wariness warranted but can also be constructive. It has long been argued by some that collaboration can be used inappropriately by governments to share the burden of work, without adjusting the balance of power and authority, or simply to create the appearance of action. Especially in situations where collaborative processes are situated in an already complex institutional landscape — which described virtually every water-related example — great care is needed to integrate collaboration and the outputs of collaborative processes into existing governance systems. Collaboration needs to be based on a clear and explicit power-sharing arrangement that is seen to be fair and is accepted by the participants. At the same time, it is crucial that participants understand that collaboration cannot erase power differentials, and indeed that some parties engaged in collaboration have considerably more power than others (and can draw on that power to effect the changes they desire outside of the collaborative process).

Fifth, collaboration should not be entered lightly. A commitment to long-term effort is required. Collaborations are easily derailed and require heavy organizational investment to ensure collective understanding of the problem, to allow for dialogue, and to permit learning. A clear vision and clear goals are essential. In the context of water-related collaborations, governments that support collaborations (whether bottom-up examples or ones they themselves have formed) must be consistent in their support and clear about their purpose.

Finally, collaboration can generate a range of benefits related to knowledge sharing, joint learning, trust, and conflict resolution. However, these in themselves are largely incidental benefits, and experience suggests the primary motive is often the desire or need to improve resource and environmental conditions. This reinforces the fact that collaborations work best when there is a clear, shared understanding of the mutual problems faced by the participants. In the absence of a shared problem that cannot be effectively addressed by participants acting alone, collaborations are unlikely to be able to maintain their momentum and will falter.

References

Ansell C, Gash A (2007) Collaborative governance in theory and practice. J Public Adm Res Theory 18:543–571

Armitage DG (2001) Ontario farmer's commitment to the natural environment (prepared on behalf of the Ontario farm environmental Coalition for the Walkerton Commission). Ontario Farm Environmental Coalition, Guelph, ON

Bidwell RD, Ryan CM (2006) Collaborative partnership design: the implications of organizational affiliation for watershed partnerships. Soc Nat Resour 19(9):827–843

Bryan TA (2004) Tragedy averted: the promise of collaboration. Soc Nat Resour 17(10):881–896

Bonnell JE, Koontz TM (2007) Stumbling forward: the organizational challenges of building and sustaining collaborative watershed management. Soc Nat Resour 20(2):153–167

Booher DE, Innes JE (2002) Network power in collaborative planning. J Plan Educ Res 21(3):221–236

Brisbois MC, de Loë RC (2015) Power in collaborative approaches to governance for water: a systematic review. Soc Nat Resour 29(7):775–790

Brisbois MC, de Loë RC (2016a) Natural resource industry involvement in collaboration for water governance: influence on processes and outcomes in Canada. J Environ Plan Manag 60(5):883

Brisbois MC, de Loë RC (2016b) State roles and motivations in collaborative approaches to water governance: a power theory-based analysis. Geoforum 74:202–212

Conley A, Moote MA (2003) Evaluating collaborative natural resources management. Soc Nat Resour 16(5):371–386

de Loë RC, Lukovich D (2004) Groundwater protection on Long Island, New York: a study in management capacity. J Environ Plan Manag 47(4):517–539

de Loë RC, Murray D, Michaels S, Plummer R (2016) Policy transfer among regional-level organizations: insights from source water protection in Ontario. Environ Manag 58(1):31–47

Dorcey AHJ (1987) The myth of interagency cooperation in water resources management. Can Water Resour J 12(2):17–26

Ferreyra C, de Loë RC, Kreutzwiser RD (2008) Imagined communities, contested watersheds: challenges to integrated water resources management in agricultural areas. J Rural Stud 24(3):304–321

Frame TM, Gunton T, Day JC (2004) The role of collaboration in environmental management: an evaluation of land and resource planning in British Columbia. J Environ Plan Manag 47(1):59–82

Gray B (1985) Conditions facilitating inter-organizational collaboration. Hum Relat 38(10):911–936

Gray B (1989) Collaborating: finding common ground for multi-party problems. Jossey-Bass, San Francisco, CA

Holley C, Gunningham N, Shearing C (2012) The new environmental governance. Routledge, London

Hudson, B, Hardy B, Henwood M, Wistow G (1999) In pursuit of inter-agency collaboration in the public sector: what is the contribution of theory and research? Public Manag: Int J Res Theory 1(2):235–260.

Koontz TM, Thomas CW (2006) What do we know and need to know about the environmental outcomes of collaborative management? Public Adm Rev 66:111–121.

Kreutzwiser RD (1998) Water resources management: the changing landscape in Ontario. In: Needham RD (ed) Coping with the world around us: changing approaches to land use, resources, and environment. Department of Geography University of Waterloo, Waterloo, ON, pp 135–148

Latour B, Weibel P (eds) (2005) Making things public: atmospheres of democracy. MIT Press, Cambridge MA

Levesque VR, Calhoun AJK, Bell KP, Johnson TR (2017) Turning contention into collaboration: engaging power, trust, and learning in collaborative networks. Soc Nat Resour 30(2):245–260

Lukes S (ed) (2005) Power: a radical view, 2nd edn. Palgrave Macmillan, Basingstoke

McCann JE (1983) Design guidelines for social problem-solving interventions. J Appl Behav Sci 19(2):177–192

McClosky M (2000) Problems with using collaboration to shape environmental public policy. Valparaiso UniV Law Rev 34:423–434

OMAFRA (Ontario Ministry of Agriculture, Food and Rural Affairs) (2012) 2011 census of agriculture. Ontario Ministry of Agriculture, Food and Rural Affairs, Strategic Policy Branch, Guelph, Ontario, Canada: Queen's Printer of Ontario. www.omafra.gov.on.ca/english/stats/county/southern_ontario.htm (downloaded 19 November 2012)

Perrault E, McClelland R, Austin C, Sieppert J (2011) Working together in collaborations: successful process factors for community collaboration. Adm Soc Work 35(3):282–298

Ring PS, Van de Ven AH (1994) Development processes of cooperative inter-organizational relationships. Acad Manag Rev 19(1):90–118

Scott T (2015) Does collaboration make any difference? Linking collaborative governance to environmental outcomes. J Policy Anal Manage 34(3):537–566

Simpson H, de Loë R, Andrey J (2015) Vernacular knowledge and water management: towards the integration of expert science and local knowledge in Ontario, Canada. Water Altern 8(3):77–97

Thompson AM, Perry JL (2006) Collaboration processes: inside the black box. Public Adm Rev 66(S1):20–32

Vangen S, Hayes JP, Cornforth C (2015) Governing cross-sector, inter-organizational collaborations. Public Manag Rev 17(9):1237–1260

van Tol SE, de Loe R, Plummer R (2015) How knowledge is used in collaborative environmental governance: water classification in New Brunswick, Canada. J Environ Plan Manag 58(3):423–444

Water Strategy Expert Panel (2005) Watertight: the case for change in Ontario's water and wastewater sector. Queen's Printer for Ontario, Toronto, ON

Watson NM (2004) Integrated River basin management: a case for collaboration. Int J River Basin Manag 2(4):243–257

Watson N, Deeming H, Treffny R (2009) Beyond bureaucracy? Assessing institutional change in the governance of water in England. Water Altern 2(3):448–460

Watson N (2015) Adaptation through collaboration: emerging institutional arrangements for catchment management and governance in England. Int J Water Governance 3:55–80

5 Disaster Risk Governance and Management: An Asian Perspective

Indrajit Pal and Jayant K. Routray

Abstract

Disasters are becoming routines in the present world and the loss of life, property, infrastructure, economy, causing hindrance to growth and development. While improving the technologies and strengthening of facilities are important for disaster prevention, the enhancement and implementation of proactive disaster risk governance for long-term mitigation and preparedness are increasingly emphasized for effective disaster prevention and response. The natural disasters in the Asian region have evidenced a large number of administrative and structural ineffectiveness and inefficiency in many crucial fields, such as governance, education, health, social security, environment, and emergency management. To complement traditional structural measures, greater emphasis has been given to developing strong interorganizational collaboration and governance systems that aim to not only reduce vulnerability to disasters through planning but also improve the response mechanism following a disaster. The Sendai Framework for Disaster Risk Reduction (SFDRR) 2015–2030 considers economic, structural, legal, social, cultural, educational, environmental, and technological dimensions of disaster risk reduction (DRR). However, the political and institutional dimensions play a significant role in the context of the framework of implementation as it crosses across several stakeholders, and, therefore, can impact substantially on the expected outcomes. In this context, the present study examines the "disaster risk governance" framework of various policy interventions and governance mechanisms that has been developed to improve the resiliency and sustainability of communities and reduce their vulnerability to natural disasters in the Asian region.

Keywords

Disaster management · Disaster risk reduction (DRR) · Risk governance · Sustainable development · The Sendai Framework for Disaster Risk Reduction (SFDRR)

5.1 Introduction

There are several attributes of governance that interplay with reference to before, during, and after situation of any disaster. These attributes also unfold several governance lessons evidenced through response mechanism of the recent catas-

I. Pal · J. K. Routray (✉)
Department of Development and Sustainability, Asian Institute of Technology, Khlong Luang, Pathumthani, Thailand
e-mail: indrajit-pal@ait.asia

B. Thakur et al. (eds.), *Resource Management, Sustainable Development and Governance*, Sustainable Development Goals Series, https://doi.org/10.1007/978-3-030-85839-1_5

trophes in the context of Asian countries and more particularly in Nepal, Myanmar, Thailand, India, and Japan. There have been profound implications for the governance of disaster risks with the continuous increase in number, scale, and intensity of natural disasters over the last decade. As a result, most of the countries have established national disaster management agencies to coordinate, manage, and respond to future disasters along with the resource management for DRR efforts. The absence of strong national and local institutions for dealing with disaster risks is a major contributing factor to increased vulnerability to natural disasters in developing countries (Anderson 1995; Smith 2001). United Nations International Strategy for Disaster Reduction (UNISDR 2004) recognises the need to decentralize and institutionalize risk mitigation at the community and local levels. A well-organized, coordinated, and decentralized partnership between national and local disaster management institutions, with decentralized access to communication, information, decision-making, and the control of resources facilitate the local governance of disaster risks (Bollin 2003; UNISDR 2004).

The Asia-Pacific region by virtue of its geo-climatic location is the most disaster-prone region in the world. Nearly 2 million people across the region were killed between 1970 and 2011, which represents 75% of the global disaster fatalities. Frequency of occurrence of natural disasters are also increased many folds due to the climate anomalies and extreme weather events. More than a billion people have been exposed to hydro-meteorological events since 2000, and this number is expected to rise due to the increased frequency and intensity of extreme weather events.

Annual Disaster Statistical Review 2012 describes that in the last 10 years five countries most hit by natural disasters were China, the United States, the Philippines, India, and Indonesia. In 2012 alone, Asia accounted for nearly 65% of global disaster victims (Guha-Sapir et al. 2012).

In 2014, the number of hydrometeorological or climate-induced disasters was equivalent to the annual 2004–2013 average. Though the number of geophysical (17) and hydrometeorological disasters (65) were respectively, the third lowest and the lowest since 2004 and showed a decrease of 19.0% and 21.7%, respectively, as compared to their annual 2004–2013 average. On the other hand, the number of meteorological disasters (57) was the highest since 2004 and showed a 21% increase when compared to its annual 2004–2013 average. The number of victims in Asia in 2014 (97.8 million persons) was far below the 2004–2013 annual average (160.7 million persons) and a decrease was observed for all disaster types, except climatological disasters of which the number of victims (31.7 million persons) increased in 2014 by almost 20%, but 87% of these victims suffered from one drought in China (Guha-Sapir et al. 2015).

Disaster Risk Management and Reduction include the systematic development and application of policies, strategies, and practices to prevent or prepare for hazards, or to mitigate their adverse effects (UNISDR 2010). The Department for International Development (DFID) (2005) has divided DRR measures into five categories: policy and planning measures; physical coping and/or adaptive measures; physical preventative measures; and community capacity building measures (DFID 2005). Four ways of reducing vulnerability, which can be used for disaster risk governance, have also been identified by McEntire et al. 2010. These are engineering methods—focusing on ways to increase resistance through construction practices in the built-environment; physical science methods—stressing exposure to hazards and risk reduction in unsafe environments; structural methods—concentrating on socioeconomic factors and demographic characteristics with a focus on cultural and traditional perceptions of vulnerability; and organizational dimensions, which focus on the effectiveness of preparedness, response, recovery, and management operations (McEntire et al. 2010). Governance is the exercise of political, economic, and administrative authority in the management of a country's affairs at all levels. National governments are expected to play a pivotal role in disaster risk management (DRM).

The governments in Asia and the Pacific region have developed a wide range of innovative solutions at the national level. The present research reviews the trends and patterns in developing governance and institutions in disaster risk management and recommends necessary actions to establish disaster risk governance system for developing countries, including mainstreaming disaster risk management (DRM) into development plans and policies.

It has been summarized at the 2010 International Recovery Forum by the Minister of a disaster prone South Asian nations that "Governance is everything." Even in case of disaster recovery context, governance is the overall process by which affected governments, organizations, institutions, and populations: (1) determine what is to be done, how it is to be done, and who it is to benefit and (2) apply themselves to implementing these decisions. Keeping in mind the dynamic and unpredictable environment, the impacts of these decisions and their implementation could be either for the better or the worse through drastically changing lives, social systems, economies, and the recovery process itself.

5.2 Disaster Governance: Concept and Principles

Not incidentally or accidentally but thoughtfully, disasters have been referred to as outcomes or impacts of hazardous environmental processes or events, technical or operational failures, faulty developmental practices, and poor civil administration or governance. Governance represents a set of tools and administrative environment wherein decision-makers and community leaders help them appreciate the importance and benefits of good policies, appropriate institutional and legislative systems at the national level as frameworks for the effective recovery plans and programs (UNISDR 2004). "Governance" is a function of actors, structures, and processes by which societies share power and make collectively binding decisions (Lebel et al. 2006; Marjolein and Renn 2011). Adopting a governance perspective entails giving attention to the distribution of political power both internal and external to the state (Goodwin 1998). The distribution of power between state and non-state actors has changed significantly over the last few decades, because of neoliberal economic and political restructuring. This is often referred to as the shift from "government to governance." This shift of thoughts and practice at policy, planning, and field actions is crucial for the decentralization and diversification of disaster management in the country, as the basis for mainstreaming DRR and climate change adaptation together into developmental planning and practices of government and communities. Nongovernmental organizations (NGOs) have found a new place within the neoliberal global order in terms of the "outwards" redistribution of state functions (Frewer 2013). NGOs are an extremely diverse group of independent nonprofit organizations out of the government operational functions (Lewis and Kanji 2009). The disaster risk reduction approach requires redefining the role of the governments as decentralization, capacity building, and involvement of other stakeholders towards self-reliance at the local level. Nevertheless, the role of the state/civil administration is always critical as governments and local institutions to be the most important set of actors in disaster management (Christoplos et al. 2001; Bulkeley and Jordan 2012).

The modern paradigm of disaster management—disaster risk reduction (DRR)—and convergence with environmental management, climate change adaptation, and socioeconomic security systems towards achieving the Sustainable Development Goals (SDGs) calls for a holistic vision of "good governance." In the neoliberal environment of reduced responsibility for the state, along with the established global platforms to implement the Hyogo Framework for Action, a new arena has been opened for a multitude of stakeholders to engage in disaster risk reduction. It is noted widely that some governments have successfully adopted and implemented disaster risk reduction (DRR) policies, while others are lagging especially at the level of local governments and field implementations

(Williams 2011). It is also evident that shortcomings in DRR are increasing because of the poor governance and due to lack of political will (Williams 2011). However, very little attention has been given to researching the processes of governance of DRR, such as the formulation of policy and the roles of various stakeholders in field implementation. In addition, there is a lack of evidence on the effectiveness of different governance systems (UNISDR 2011a; b). The governance mechanism is particularly hindered by the complex interplay of power and knowledge among diverse groups of stakeholders with irrational command over resources, roles, and responsibilities (Ojha et al. 2009). In DRR, this complex interplay of power and knowledge among diverse players gives rise to different governance approaches. For long-term recovery operations, it is particularly important to advocate for "good governance" at the local government level and in the policy programming. As the immediate public service provider and the interface with citizens, local governments are naturally placed in the best position to raise awareness of the citizens on resilient recovery (i.e., building back better approaches) and listen to their concerns. Recovery operations may fail if the communities are not properly informed and engaged.

5.3 Risk Governance

An overview of the concept of risk governance promises to offer a comprehensive means of integrating risk identification, assessment, management, and communication, is provided by Renn (2008). Risk governance denotes both the institutional structure and the policy process that guide and restrain collective activities of a group, society, or international community to regulate, reduce, or control risk problems (Klinke and Renn 2012). Foresight activities strengthen public awareness and risk perception among relevant stakeholders. Communication and participation are the key for successfully implementing risk governance.

To analyze various issues relating to disaster risk governance, it is necessary to try and understand the concepts involved and the definitions used. The terms Hazard, Disaster, Disaster Risk, Disaster Risk Management, and Disaster Risk Reduction are used in this chapter are in line with the UN International Strategy for Disaster Reduction (UNISDR) and in the Hyogo Framework of Action (HFA). In view of the significance of DRR, the definition in the UNISDR Terminology is reproduced below:

> Disaster Risk Reduction is the concept and practice of reducing disaster risks through systematic efforts to analyse and manage the causal factors of disasters, including through reduced exposure to hazards, lessened vulnerability of people and property, wise management of land and the environment, and improve preparedness for adverse events.

Further, "DRR includes early warning, improved governance, building up community and household resilience, and reducing the underlying risk factors while strengthening disaster preparedness" (ISDR 2004).

Understanding the dynamics, structures, and functionality of risk governance processes require a general and comprehensive conceptualization of procedural mechanisms and structural configurations. The literature on urban governance and pro-poor development indicates that there are both demand and supply constraints to achieving pro-poor governance. These can be applied effectively to governance issues related to adaptation and urban environmental management. On the supply side, municipal governments are constrained by several factors (Satterthwaite 2001; Huq et al. 2007), starting from weak local governance, legal mandates and restrictions, financial resources, and risk perceptions of the communities.

Demand-side constraints commonly include low-frequency nature of high-impact events, low levels of awareness about the changing nature of climate-related risks due to climate change, low levels of empowerment and mobilization by the groups of citizens most adversely impacted by climate shocks and stresses. In most cities at risk from floods, wealthier groups and formal enterprises do not face serious risk. For example, Mumbai has the resources capable of reducing risk from flooding, yet the costs are often borne by the low-income groups, who live in more risk-prone settlements (Huq et al. 2007). Despite being at serious risk of increased flooding Dhaka, Mumbai, and Shanghai have all attracted much private investment despite their vulnerability to storms and sea-level rise (De Sherbinin et al. 2007). The speed at which the economic "risk map" for cities will change is likely to be slower than the actual climate changes and risks (for example, well-documented risk of New Orleans did not cause entrepreneurs and residents to move) (Huq et al. 2007). A so-called moral hazard may also play a role, as government decision-makers may not invest in a perceived low-probability event because they assume that the international relief community would come to their rescue in the event of a significant disaster (De Sherbinin et al. 2007). Devas (2001) argues that the quality of pro-poor governance in urban settings is determined by the ability of the poor to influence political decision-makers. Broadly, these demand-side variables include issues of institutional design whereby low-income citizens can participate meaningfully in elections and an active civil society, which is able to advocate for the rights of the poor.

5.3.1 The International Scenario

Worldwide, during the last two decades from 1994 to 2013, floods were the most frequent natural disasters: 43% of the hazards occurring during the period were floods and 2.5 billion people were affected. During the same period, storms/cyclones were the second most frequent event, which killed 244,000 people and the damage caused was estimated at USD 936 billion. Earthquakes and tsunamis killed more people than all other types of disasters put together, with 750,000 deaths. Tsunamis are the deadliest disasters in nature, with 79 deaths per thousand persons affected, against 4 deaths per thousand affected persons in case of earth movements only. Droughts were only 5% of the disasters but affected over a billion people. The number of people dying directly due to droughts is not on the same scale as some other events but the resultant malnutrition and the aftereffects, which linger for a longer time, on food production, health, and availability of potable water. Disease follows droughts and takes a heavy toll (Guha-Sapir et al. 2016).

Disaster risk reduction received a boost in the last decade of the twentieth century with the designation of the 1990s as the International Decade for Natural Disaster Risk Reduction by the United Nations General Assembly, which was a landmark in the history of disaster risk reduction. Following this, three World Conferences focusing on disaster and climate risk management in the context of sustainable development were convened, indicating the priority attached to this matter. All three conferences were held in Japan, the first in Yokohama in 1994, the second in Kobe in 2005, and the third and recent one in Sendai in 2015. All participating countries made commitments to act as per the agreement during the Third World Conference at Sendai. To have a comprehensive picture, the framework that evolved during the Conference is stated in full below.

The Sendai Framework (UNISDR 2015) is a 15-year nonbinding agreement, which recognizes that the State has the primary role to reduce disaster risk, but that responsibility should be shared with other stakeholders including local government and the private sector. It aims for the following outcome, to quote:

> "The substantial reduction of disaster risk and losses in lives, livelihoods and health and in the economic, physical, social, cultural and environmental assets of persons, businesses, communities and countries." (UNISDR 2015)

The Sendai Framework emerged from 3 years of consultations and negotiations, supported, and coordinated by UNISDR, during which UN member states, NGOs, and other stakeholders made calls for an improved version of the existing Hyogo Framework, with a set of common standards, a comprehensive framework with achievable targets, and a legally based instrument for disaster risk reduction. Member states also emphasized the need to tackle disaster risk reduction and climate change adaptation when setting the Sustainable Development Goals, particularly considering an insufficient focus on risk reduction and resilience in the original Millennium Development Goals.

The Sendai Framework sets four specific priorities for action:

1. Understanding disaster risk.
2. Strengthening disaster risk governance to manage disaster risk.
3. Investing in disaster risk reduction for resilience, and
4. Enhancing disaster preparedness for effective response, and to "Build Back Better" in recovery, rehabilitation, and reconstruction.

To support the assessment of global progress in achieving the outcome and goal of the Sendai Framework, seven global targets have been agreed. The Framework for DRR has set seven significant targets during 2015–2030.

1. Substantial reduction of global mortality during disaster by 2030, aiming to lower average per 100,000 global mortality rates during the period 2020–2030 compared to 2005–2015.
2. Substantial reduction in the number of affected people globally by 2030 aiming to lower the average global figure per 100,000 between 2020 and 2030 compared to 2005–2015.
3. Reduction in direct disaster economic loss in relation to global Gross Domestic Product (GDP) by 2030.
4. Substantial reduction in disaster damage to critical infrastructure and disruption of basic services through developing their resilience by 2030.
5. Substantial enhancement of the number of countries with national and local disaster risk reduction strategies by 2020.
6. Substantial enhancement of international cooperation to developing countries through adequate and sustainable support to complement their national actions for implementation of this framework by 2030.
7. Substantial increase in the availability of and access to multi-hazard early warning systems and disaster risk information and assessments to the people by 2030.

5.3.2 Risk Governance in Asia

Natural hazards continue to pose a major threat to the entire world with prospects of even greater impacts on life and property in the future (Aini and Fakhrul-Razi 2010; Hayles 2010). "Natural" disasters cause the greatest impact on poor communities in developing countries (97% fatalities) that have the least resources to cope (Amin et al. 2008; World Bank 2001). Developing countries need to place DRM as a core element within the structure of the government to mainstream DRM into development policies and operations. The structure and quality of governance need to be improved at all levels from national to local governments by legally defining mandates and status. Also, public involvement is critical in all aspects of DRM (Governance; UNISDR 2011a; Wisner et al. 2004). Many developing countries need responsive, accountable, transparent, and efficient governance structures in DRM (UNDP 2010). Governance is the combined functional exercise of political, economic, and administrative authority in managing a country's affairs. Governance influences how income and assets are distributed to the people, and determines how the people protect themselves from hazards, and how they access support in recovery. Since many developing countries lack the administrative, organizational, financial, and political capacity to effectively cope with disasters, the poor become particularly vulnerable. Low-income countries have suffered only 9% of worldwide disasters since 1980, but suffered 48% of the fatalities (World Bank 2012).

The developing countries need to place DRM as a core element within the structure of the government. Countries with well-established institutions can decrease the number of affected people and economic losses from natural disasters, while mortality is increasing in countries with weak governance capacities (Cannon 2008a; b; Raschky 2008). The Hyogo Framework for Action (HFA), which was adopted at the World

Conference on Disaster Reduction at Kobe in 2005, defines "development, and strengthening of institutions, mechanisms and capacities to build resilience to hazards" as one of the strategic goals, and emphasizes the action of ensuring DRM as a national and a local priority with a strong institutional basis for implementation (UNISDR 2005). In Japan, the Disaster Countermeasures Basic Act, which was legislated in 1961, stipulates the DRM framework. The framework covers: (i) the roles and responsibilities of national and local governments and communities; (ii) the details of disaster management plans, institutions, and countermeasures; and (iii) platforms at national and local levels.

Each country should mainstream DRM into policy, planning, and management in all relevant sectors. Mainstreaming DRM has important implications for a country's growth and development agenda since disasters can pose serious obstacles to socioeconomic development. The principal strategic goal of the HFA is to effectively integrate disaster risk considerations into sustainable development policies, planning, programming, and financing at all levels of government. As recommended in Sect. 5.3.2, governments should develop a range of innovative programs to prevent increasing vulnerability during the economic development process. For example, Bangladesh's Outline Perspective Plan, produced by the Planning Commission, integrates DRM and climate change adaptation into national development strategies (UNESCAP and UNISDR 2012). In Japan, the government is reviewing DRM approaches by learning lessons from the Great East Japan Earthquake (GEJE) in 2011 and will mainstream DRM further in all relevant sectors by assessing risks and vulnerabilities and allocating necessary resources to prepare for disasters (Committee on Promoting Disaster Management of Central Disaster Management Council, 2012).

Box 5.1 Sustainable Development Goals

If anything, the authors of this chapter have echoed through their research that disaster risk reduction (DRR) cannot be achieved without sustainable development. SDG–13 (which is about climate action) urges the need to combat climate change and its impacts by regulating emissions and promoting renewable energy development. Disaster governance is an emerging concept in the disaster research literature that is closely related to risk governance and environmental governance. Disaster governance arrangements and challenges are shaped by forces such as globalization, world-system dynamics, social inequality, and sociodemographic trends. Further, the Sendai Framework on Disaster Risk Reduction (2015–2030) is an ambitious agreement that sets out the overall objective to substantially reduce disaster risk and losses in lives, livelihoods, and health and in the economic, physical, social, cultural, and environmental assets of persons, businesses, and communities. This chapter also displays how SDG-13 could be incorporated at different spatial scales in Asian countries from a disaster risk governance perspective to transform communities and societies. Disaster Risk Governance and Management is a multidimensional process including administrative, financial, and social challenges which can together mitigate disasters and help advance the goals of Agenda 2030.

A wide range of stakeholders must be coordinated since DRM concerns everyone. DRM requires a multi-sectoral approach, which covers urban development, infrastructure, water, education, health, and many other sectors. Single-sector development planning cannot address the complexity of problems caused by disasters, nor can such planning build resilient societies (World Bank 2012). For example, DRM plans should be linked with urban planning and DRM education at school, which are effective measures to decrease disaster casualties and damage. Since no single organization can have the ultimate responsibility for managing disaster risks, various stakeholders should share the responsibility. In addition to governmental organizations, the private sector and civil society play crucial roles in DRM. The private sector can contribute to mitigating disaster damage in a wide range of areas, such as the logistics of providing relief goods, payment of insurance claims, restoration of damaged infrastructure, and the continuation of banking services. Civil society organizations (CSOs) can respond to the various needs of affected people at the grassroots level. As discussed in Sect. 5.3.1, the government and CSOs can play strategic roles in creating safety net systems to protect vulnerable groups from disasters.

Each country should create a platform to coordinate various organizations at different levels. Inter-sector coordinating mechanisms are needed to properly design and implement DRM strategies. UNISDR defines the national platform as a nationally owned and led forum or a committee of multiple stakeholders. The national platform requires several elements (1) political, (2) technical, (3) participatory, and (4) resource mobilizing components to promote DRM (Gopalakrishnan and Okada 2003). As the complexity of society increases, different institutions and formal or informal groups can be effectively involved in DRM. As Sect. 5.3.3 emphasizes, national governments should strengthen linkages with local governments to guide and support local governments to promote DRM on the ground. Also, local governments should create coordination mechanisms at the local level in line with national policies.

5.4 National Level Actions on Risk Management Legislations in Asia

Countries in Asia and the Pacific have substantially improved institutions and policies in DRM. Mega-disasters, such as the Indian Ocean Tsunami, Cyclone Nargis, and the GEJE, became opportunities to strengthen risk governance in affected countries. Various countries have developed risk governance by creating focal point agencies, establishing national platforms, and promoting legislation. These steps have been taken in line with the HFA. The number of national platforms is increasing globally, and rose from 38 in 2007 to 73 in 2011. An increasing number of countries have been adopting or updating existing DRM legislation (UNISDR 2011b). Legislation can establish new agencies or empower existing agencies with new responsibilities as well as create budget lines and policy responsibilities (Pelling and Holloway 2006). Mega-scale incidents can profoundly change DRM institutions. Many countries have been continuously strengthening national DRM systems based on lessons learned from disasters in and outside the countries (Amini-Hosseini and Hosseinioon 2012; Ikeda 2012; Nishikawa 2010). The terrorist attacks of September 11, 2001, in the US generated a major governmental reorganization there and led to the creation of the Department of Homeland Security. Further changes in emergency management in the US followed the catastrophic hurricanes of 2004 and 2005. In Taipei, China, Typhoon Morakot initiated a change in national institutions. The National Fire Agency was transformed to the National Disaster Prevention and Protection Agency through the Disaster Prevention and Protection Act to expand the agencies of functions to include comprehensive countermeasures. Japan enacted the Disaster Countermeasure Basic Act in 1961 after a high tide disaster in Nagoya in 1959, which caused over 5000 deaths. The main driver of the latest version of Japan's DRM plan after the GEJE is the need to account for low-probability and high-impact compound hazards. Countries in Asia and the Pacific have taken legislative actions to establish focal point agencies within their central government structures and national platforms. Of the 61 countries and areas in the region, 30 have enacted national or central legislation that specifically deals with DRM (UNESCAP and UNISDR 2012). Following the Indian Ocean Tsunami in 2004, affected countries have strengthened their focal point agencies. Sri Lanka established the Disaster Management Ministry by newly creating a disaster management center and merging it with the meteorological agency. Thailand has strengthened the coordination roles of the Disaster Management Department in its government. Indonesia has created a national disaster management agency and local agencies throughout the country. Myanmar is planning to establish a new agency and enact a national disaster management law reflecting lessons from the Cyclone Nargis disaster in 2008. Vietnam and the Philippines, which are major disaster-prone countries in Asia, have strengthened existing legislation and institutions. Most countries in the Pacific have created national disaster management offices as standalone agencies (Hay 2009).

One of the most crucial issues is how the focal point agency should be positioned within the government to coordinate and lead various organizations. There is no sole model for institutions since each country has developed its institutions according to the disaster scale and type, socioeconomic conditions, and geography.

The agencies vary considerably across Asian countries in terms of their positions within the government, their mandates, and roles (Table 5.1). For example, floods are historically a major disaster and have often threatened national security in China, Japan, and Vietnam. These countries have accumulated experience in flood management and established strong institutions and developed countermeasures to mitigate flood damages (Ishiwatari, 2010). Communities have prepared mainly for floods as major disasters. The experts in flood management have led policies and countermeasures in DRM including other disasters. In other Asian countries that have conducted relief activities following disasters as main activities, relief organizations have become leading agencies within the governments. In Singapore, the focal point agency, the Civil Défense Force, has the main mandate of managing fire disasters and other urban disasters, and would probably not need drastic institutional reform to change it into a body for coordinating all organizations. This is because this island city-state has rarely suffered from natural mega-disasters, such as floods from major rivers and earthquakes, and has prepared mainly for man-made disasters in urban areas.

Three models of focal point agencies are in place in Asia and the Pacific (Fig. 5.1): (1) designated as a coordination agency without implementation role; (2) located in parallel with other line ministries in the government; and (3) developed from existing implementation organizations. Characteristics of these models are summarized in Table 5.2.

5.5 Humanitarian Response and Coordination

The emphasis on humanitarian actions, which is rightly given since nothing is more important than human lives, results in disaster response and relief getting primary importance by governments, NGOs, and media. Nobody can argue with the proposition that it is an essential measure of governance to do the utmost to first save lives while taking up the work of emergency response and relief in disasters situations. NGOs devoted to humanitarian activities also likewise apply their resources first to saving lives, and to associated functions of health, sanitation, drinking water, etc. Major resources, financial, human, and technological are thus applied towards strengthening the response mechanism. The Hyogo Conference in 2005 provided the much-needed focus on disaster risk reduction and helped to increase awareness of the increasing importance of risks associated with disasters. Since then, during the last two decades many countries took up measures aimed at reducing the risks attendant to disasters. How far these measures are adequate and whether there is a need for further clarity of the issues involved as well as action required merits discussion across a broad spectrum.

5.6 Discussion and Conclusion

Countries in Asia and the Pacific have substantially developed disaster risk governance by creating focal point agencies, establishing national platforms, pro-

Table 5.1 National Platform and Focal Point Agencies in Selected Asian Countries

	National Platform (Chair)	Focal Point Agency	Related Act	Model, Original Mandate
Southeast Asia				
Brunei Darussalam	–	Nat. DM Centre, Min. of Home Affairs	–	(i)
Cambodia	Nat. Com. for DM (PM)	Nat. Com. DM General Secretariat	Sub-decree No.35 ANK	(i)
Indonesia	–	Nat. DM Agency	DM Law	(i)
Lao PDR	Nat. DM Com.	Nat. DM Office, Min. Labor & Social Welfare		(iii) relief
Malaysia	DM & Relief Com. (Deputy PM)	National Security Division, PM Dep.	Nat. Security Council Directive No. 20, 1997	(i)
Myanmar	Central Com, on Nat. Dis. Prevention (PM)	Relief and Resettlement Dep. Min. of Social Welfare,	Rehabilitation Board Act. 1950, DM Law (draft)	(iii) relief
Philippines	Nat. Dis, Coordination Council (Defense Minister)	Office of Civil Defense, Dep. of Nat. Defense	Dis. Risk Reduction, Man. & Recovery Act	(i)
Singapore	–	Civil Defense Force, Min. of Home Affairs	Civil Defense Act 1986	(iii) S&R, fire
Thailand	Nat. Dis. Prevention & Mitigation Com. (PM or Deputy PM)	Dep. of Disaster Prevention & Mitigation, Min. of Interior	Dis. Prevention and Mitigation Act, 2007	(iii) S&R, fire
Viet Nam	Central Com. for Flood & Storm Control (PM)	DM Center, Dep. Dike Man. &Flood & Storm Control, Min. Agriculture Rural Development	Decree No 168, 1990	(iii) flood man.
South Asia				
Bangladesh	Nat. DM Council (PM)	DM Bureau, & Directorate of Relief and Rehabilitation, Min. of Food & DM	DM Act, 2008	(iii) relief
Bhutan	Nat. Com. for DM (Cabinet Minister)	DM Division, Min. of Home & Cultural Affairs	Nat. DM Act	(i)
India	Nat.DM Authority(PM)	Nat. Institute of DM, Min. Home Affairs	DM Act 2005	(i)
Maldives	–	DM Center, Min. of Defense	(Draft)	(i)
Nepal	Central Nat. Dis. Relief Com. (Home Minister)	DM Section & Nat. Emergency Operation Centre, Min. of Home Affairs	Dis. Relief Act 1982	(iii) relief
Pakistan	Nat. DM Commission (PM)	Nat. DM Authority	Nat. DM Ordinance 2006	(i)
Sri Lanka	Nat. Council for DM (President and PM)	DM Centre, Min. of DM	DM Act 2005	(ii)
East Asia				
Japan	Central DM Council (PM)	DM Office, Cabinet Office	Dis. Countermeasures Basic Act	(i)
Mongolia	State Emergency Commission (Deputy PM)	Nat. Emergency Man. Agency	Law on Dis. Protection 2003	(iii) S&R, fire
People's Republic of China	Nat. Commission for Dis. Reduction (Vice Premier of State Council)	Nat. Dis. Reduction Center, Min. of Civil Affairs	More than 30 laws and regulations	(iii) relief

(continued)

Table 5.1 (continued)

	National Platform (Chair)	Focal Point Agency	Related Act	Model, Original Mandate
Republic of Korea	Central Safety Man. Council (PM)	Nat. Emergency Man. Agency, Min. of Public Administration & Safety	Act on Dis. Risk Man. and Reduction	(iii) S&R, fire

Notes: *Com.*: Committee, *Dis.*: Disaster, *Dep.*: Department, *DM*: Disaster Management, *Man*: Management, *Min*: Ministry, *Nat.*: National, *PM*: Prime Minister, *S & R*: search and rescue.
(i): Designation as a coordination agency without implementation role; (ii): located in parallel with other line ministries in the government; (iii): developed from implementation organizations.

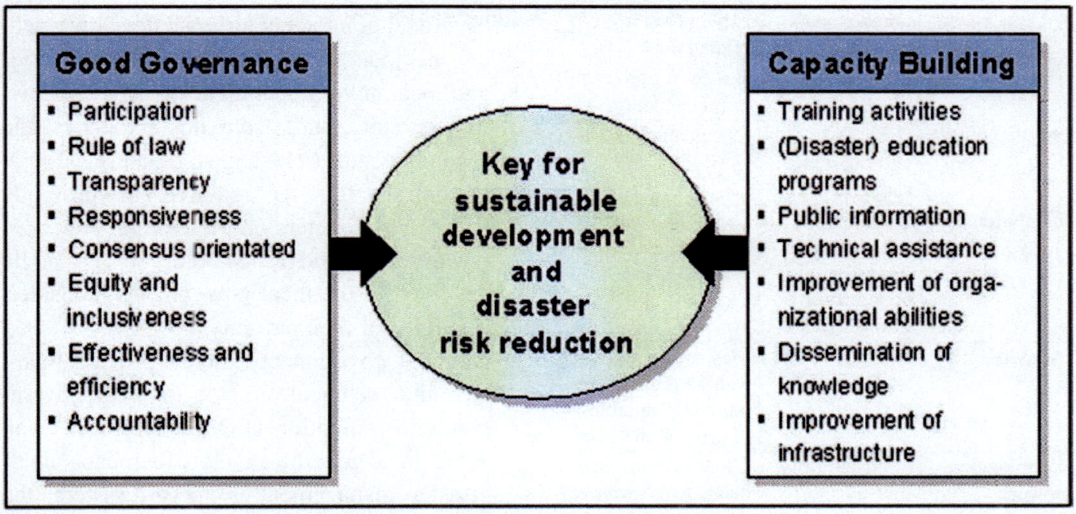

Fig. 5.1 Summarizes good governance and capacity building as a central component regarding the process and implementation of disaster risk management and sustainable development (IFS, 2006)

moting legislation in line with the HFA, and developing mechanisms for cooperation between national and local governments. Mega-disasters, such as the Indian Ocean Tsunami, Cyclone Nargis, and the GEJE, became opportunities for making these improvements. A designated coordinating agency with high authority in the government is theoretically required to neutrally coordinate and to lead organizations concerned with DRM, but this has not been easy to realize. In some countries implementing agencies have expanded their mandates for coordination. The focal point agencies have made various practical efforts in coordination, such as forming technical committees.

While devolution to local governments is needed to effectively respond to the needs of people affected by disasters on the ground, local governments typically face difficulties in taking on powers and budgets devolved from the national government because of their limited capacity. In some cases, national governments have established local offices or seconded staff to the local governments to promote DRM at the local level. Some national governments have also provided local governments with financial subsidies and technical support to guide DRM measures at the local level and place DRM as a core part of national development strategies and programs to mainstream DRM.

Firstly, strengthening the coordination role of the national government through an enhanced legal framework is currently pushing it in the following ways.

Table 5.2 Characteristics of Focal Point Agencies

	Designated coordination agency (i) and (ii)	Developed from implementing organizations (iii)
Background	Newly established or strengthened coordinating body	Long history of implementing agency in disaster management, such as fire management, relief, or engineering
Staff	Limited number and experience	Large number and experiences in related field
Budget	Limited	Budget for project implementation
Sustainability	Difficult without donor support	Sustainable
Institutional development	Often established as a new organization from scratch	Can utilize existing institutions and staff
Activity	Neutral	May overlook, or be biased on, one or more key variables of original mandate
Coordinating power	Need authority, such as chairmanship of state heads	Can utilize original mandate as leverage

- The designated coordinating agency with high status in the government should be set as the goal of establishing the focal point agency at the national level. This agency has authorities to formulate a vision and national policies, to allocate DRM-related budgets for government organizations, and to demand compliance and actions from the organizations.
- Since it is quite difficult to form such agency in practice, a gradual approach should be taken. A coordinating body could be newly established or strengthened as the focal point agency. Alternatively, an existing organization of disaster management, such as fire management or relief, could expand its functions to coordinate other organizations at the pre-disaster stage.
- The focal point agency should be situated under a leading body in the government, such as the cabinet office, President's office, home affairs ministry, or defense ministry. The coordination agency needs a high-profile status inside the government.

Secondly, by building up a flexible cooperation system among concerned organizations and with local governments:

- The focal point agency should develop practical methods of coordinating other ministries and local governments. These include receiving seconded staff from line ministries and seconding staff to local governments.
- Decentralization is required to promptly respond to disasters on the ground. However, power and budget should be gradually devolved to the local governments, considering limits of their capacity.
- National governments should provide financial and technical support to local governments in promoting DRM at the local level. When local governments lose their capacity of disaster management in mega-disasters, the national government needs to mobilize resources throughout the country to respond to the disaster.
- Overall, the disaster governance is in the evolutionary process in many Asian countries following some major incidents occurred over the past two decades. Countries prone to multi-hazards and disasters are more aware and conscious of improving the existing process and management system and gearing towards disaster governance following the basic principles of HFA and Sendai Framework. Depending on the type, nature, scale, and intensity of the disaster phenomena, it is imperative to work collectively from local to national and regional levels for strengthening the disaster governance beyond the national level for prevention as well as response to transboundary phenomena and extreme impacts. Sharing experience and best practices, data, early

warnings, and technology is no doubt very important in addition to interagency and inter-sectoral collaboration to follow Build Back Better both in structural and nonstructural measure issues.

References

Aini M, Fakhrul-Razi A (2010) Development of socio-technical disaster model. Saf Sci 48(10):1286–1295

Amin S, Cox M, Goldstein M (2008) Using data against disasters: overview and synthesis of lessons learned. In: Amin S, Goldstein M (eds) Data against natural disasters. The World Bank, Washington, DC, p 316

Amini-Hosseini K, Hosseinioon S (2012) Evaluation of recent developments in laws regulations for earthquake risk mitigation and management in Iran. Risk Hazards Crisis Public Policy 3(1):1–20

Anderson M (1995) Vulnerability to disaster and sustainable development: a general framework for assessing vulnerability. In: Mohan M, Clarke C (eds) Disaster prevention for sustainable development: economic and policy issues. World Bank, Washington, DC, pp 41–59

Bollin C (2003) Community-based disaster risk management. Experience gained in Central America, GTZ (Gesellschaft für Technische Zusammenarbeit), Eschborn

Bulkeley H, Jordan A (2012) Trans-national environmental governance new findings and emerging research agendas. Environ Plan 30(4):556–570

Cannon T (2008a) Reducing people's vulnerability to natural hazards: communities and resilience. Research paper 2008/34. United Nations University- World Institute for Development Economic Research, Helsinki

Cannon T (2008b) Vulnerability, "innocent" disasters and the imperative of cultural understanding. Disaster Prev Manag Int J 17(3):350–357

Christoplos I, Farrington J, Kidd AD (2001) Extension, poverty, and vulnerability: inception report of a study for the Neuchâtel initiative. ODI working paper no. 144. Overseas

DFID (Department for International Development) (2005) Natural disasters and disaster risk reduction measures: a desk review of costs and benefits. Draft Final Report, DFID, London

Frewer T (2013) Doing NGO work: the politics of being 'civil society' and promoting 'good governance' in Cambodia. Aust Geogr 44(1):97–114

Goodwin H (1998) Sustainable tourism and poverty elimination. A discussion paper for the Department for the Environment, transport and the regions and the Department for International Development

Gopalakrishnan C, Okada N (2003) Enhancing the effectiveness of integrated management for disaster risk: the case for institutional innovations. Paper presented at third international symposium on integrated disaster risk management, Kyoto, 3–5 July

Guha-Sapir D, Hoyois P, Below R (2016) Annual disaster statistical review 2015: the numbers and trends. CRED, Brussels

Guha-Sapir D, Vos F, Below R, Ponserre S (2012) Annual disaster statistical review 2012: the numbers and trends. Weblink: http://www.undp.org.cu/crmi/docs/cred-annualdisstats2010-rt-2011-en.pdf (Accessed on 3 May 2012)

Guha-Sapir D, Vos F, Below R, Ponserre S (2015) Annual disaster statistical review 2014: the numbers and trends. Weblink: http://www.cred.be/sites/default/files/ADSR_2014.pdf (accessed in 2015)

Hay JE (2009) Institutional and policy analysis of disaster risk reduction and climate change adaptation in Pacific Island countries. United Nations international system for disaster reduction (UNISDR) and the United Nations, Suva

Hayles C (2010) An examination of decision-making in post-disaster housing reconstruction. Int J Disaster Resil Built Environ 1(1):103–122

Huq SS, Kovats HR, Satterthwaite D (2007) Editorial: reducing risks to cities from disasters and climate change. Environ Urban 19(1):3–15

Ikeda M (2012) Knowledge note 2–2 disaster management plans. In: Learning from megadisasters. Weblink http://wbi.worldbank.org/wbi/document/disaster-management-plans

International Federation of Surveyors (2006) The Contribution of the Surveying Profession to Disaster Risk Management. FIG Publication No 38. FIG Working Group 8.4, International Federation of Surveyors (FIG), Frederiksberg, Denmark

International Strategy for Disaster Reduction (2004) Living with risk: a global review of disaster reduction initiatives, vol 1. United Nations Publications, Herndon, VA

Ishiwatari M (2010) "Chapter 14 Integrated management of urban flooding for climate change adaptation in developing countries", Shaw R, Pulhin JM, and Jacqueline Pereira, J. (Ed.) Climate Change Adaptation and Disaster Risk Reduction: Issues and Challenges (Community, Environment and Disaster Risk Management, Vol. 4), Emerald Group Publishing Limited, Bingley, pp. 305–323. https://doi.org/10.1108/S2040-7262(2010)0000004020

Klinke A, Renn O (2012) Adaptive and integrative governance on risk and uncertainty. J Risk Res 1(1):3–20

Lebel L, Anderies JM, Cambell B, Folke C, Hatfield-Dodds S, Hughes TP, Wilson J (2006) Governance and the capacity to manage resilience in regional social-ecological systems. Ecol Soc 11(1):19

Lewis D, Kanji N (2009) Non-governmental organizations and development. Routledge, London

McEntire DA, Crocker CG, Peters E (2010) Addressing vulnerability through an integrated approach. Int J Disaster Resil Built Environ 1(1):50–64

Devas N (2001) Does city governance matter for the urban poor? Int Plan Stud 6(4):393–408

Marjolein B.A. van Asselt & Ortwin Renn (2011) Risk governance, Journal of Risk Research, 14(4): 431–449, https://doi.org/10.1080/13669877.2011.553730

Nishikawa S (2010) From Yokohama strategy to Hyogo framework: sharing the Japanese experience of disaster risk management. Asian J Environ Disaster Manag 2(3):249–262

Ojha HR, Cameron J, Kumar C (2009) Deliberation or symbolic violence? The governance of community forestry in Nepal. Forest Policy Econ 11(5): 365–374

Pelling M, Holloway A (2006) Legislation for mainstreaming disaster risk reduction. Tearfund, Teddington

Raschky PA (2008) Institutions and the losses from natural disasters. Nat Hazards Earth Syst Sci 8:627–634

Renn O (2008) White Paper on Risk Governance: Toward an Integrative Framework. In: Renn O, Walker KD (eds) Global Risk Governance. International Risk Governance Council Bookseries, vol 1. Springer, Dordrecht. https://doi.org/10.1007/978-1-4020-6799-0_1

Satterthwaite D (2001) Reducing urban poverty: constraints on the effectiveness of aid agencies and development banks and some suggestions for change. Environ Urban 13(1):137–157

De Sherbinin A, Schiller A, Pulsipher A (2007) The vulnerability of global cities to climate hazards. Environ Urban 19(1):39–64

Smith K (2001) Environmental hazards: assessing risk and reducing disasters, 3rd edn. Routledge, London

UNDP (2010) Disaster risk reduction, governance, and mainstreaming. Weblink: http://www.undp.org/content/dam/undp/library/crisis%20prevention/disaster/4Disaster%20Risk%20Reduction%20-%20Governance.pdf (Accessed on 26 June 2012)

UNESCAP and UNISDR (2012) Reducing vulnerability and exposure to disasters. The Asia-Pacific Disaster Report 2012. Weblink: http://www.unescap.org/sites/default/files/publications/Asia-Pacific-Disaster-Report-2012_0.pdf

UNISDR (2004) Africa regional strategy for disaster risk reduction. International strategy for disaster reduction. Weblink: http://www.unisdr.org/files/13093_AFRICAREGIONALDRRSTRATEGYfullPDF.pdf (Accessed on 01 April 2010)

UNISDR (2005) Hyogo framework for Action 2005–2015. Building the resilience of nations and communities to disasters. Extract from the final report of the World Conference on Disaster Reduction (A/CONF.206/6). United Nations Office for Disaster Reduction (UNISDR). Weblink: http://www.unisdr.org/files/1037_hyogoframeworkforactionenglish.pdf. (Accessed on 12th July 2013)

UNISDR (2010) Local governments and disaster risk reduction. International strategy for disaster reduction. Weblink: http://www.unisdr.org/preventionweb/files/13627_LocalGovernmentsandDisasterRiskRedu.pdf (Accessed on 30 April 2010)

UNISDR (2011a) Global assessment report on disaster risk reduction', UNDRR, United Nations

UNISDR, 2011b, Global assessment report on disaster risk reduction. Revealing risk, redefining development. Geneva, Switzerland: United Nations international strategy for disaster reduction

UNISDR (2015) The human cost of weather-related disasters 1995–2015. UNISDR

Williams G (2011) Study on disaster risk reduction, decentralization, and political economy: the political economy of disaster risk reduction. Background report for the global assessment report on disaster risk reduction. UN/ISDR, Geneva

Wisner B, Blaikie P, Cannon T, Davis I (2004) At risk: natural hazards, people's vulnerability, and disasters, 2nd edn. Routledge, London

World Bank (2001) World development report 2000–2001. World Bank, Washington, DC

World Bank (2012) World development indicators 2012. World Development Indicators, World Bank, Washington, DC

6 Reshaping Natural Resource Management in China

Bing Xue and Wanxia Ren

Abstract

During the past several decades, China has achieved dramatic economic growth and has been the second largest economy in the world. China has also been the biggest carbon emitter and one of the most air, soil, and water polluted and biodiversity degraded regions in the world. Even though Chinese people earn better quality of life and life expectancy, more and more people are suffering from much more diseases and losses because of environmental pollution and global warming, largely due to its massive resources-based growth model affecting China's long-term sustainability, which has been questioned according to its performance in the fields of economic growth, environmental improvement, and social development. Today, rapid urbanization, pursuing affluent lifestyles and increased demand for resources and services are exerting increasing pressure on land, forest, sea, and natural resources in China and challenging sustainability at both national and international levels. Since 2012, a new green deal under the name of ecological civilization has been offered by the Chinese leadership. Therefore, reshaping natural resource management has been taken as a key component for meeting the challenges of transitioning to sustainable development mode in China. In this chapter, a clear and comprehensive understanding of China's natural resource consumption during the past decades are presented as well as the associated social, economic, and environmental impacts and temporal and spatial differences of China's resource consumption are investigated. In addition, a systemic review presents the changes to the institutional system for natural resource management at the national level and provincial level, followed by the identification of key stakeholders in the governance system, and, then, some typical local practices are identified and presented aiming to illustrate the integration of resource management, sustainable development, and livelihood strategy. Finally, policy implications are proposed.

Keywords

China · Natural resource management · Resources and environmental policy · Sustainable development

B. Xue (✉) · W. Ren
Institute of Applied Ecology, Chinese Academy of Sciences, Shenyang, China
e-mail: xuebing@iae.ac.cn

B. Thakur et al. (eds.), *Resource Management, Sustainable Development and Governance*, Sustainable Development Goals Series, https://doi.org/10.1007/978-3-030-85839-1_6

6.1 Resource Consumption in China

Natural resources in the form of materials, water, and energy, as well as the land available to us on Earth, are the basis of all living beings on our planet. Humans are also part of nature. Without the constant use of natural resources, neither economy nor society could function. However, the natural resource-based society is in danger of overexploitation and collapse. Due to the growth of the world population and continued high levels of resource consumption in the developed countries and rapid industrialization of countries, such as China and India, worldwide demand on natural resources and related pressures on the environment are steadily increasing.

Since the reform and opening policy of the late 1970s, China has achieved rapid economic growth and social development. For example, China's GDP was 364.52 (billions of Yuan) in 1978 and increased to 90,030.9 (billions of Yuan) in 2018. However, China's scale-driven economic development has led to inefficient natural resource utilization in the production process, as well as massive pollution emission during the past 40 years. From 1981 to 2018, China's GDP has increased more than 180 times, while the energy consumption in 2018 was 7.81 times of that in 1981.

Upon entering the new period, China's economy has been changing from speed-based development model to quality-based development model. The demand for natural resources has accordingly shifted from pursuing the quantity to the quality, which improving usage efficiency and promoting environmentally friendly processing ways.

6.1.1 Energy Consumption

China is the world's largest energy producer and consumer (BP Statistical Review 2019). In 2018, China's total primary energy production was 3.77

Box 6.1 Sustainable Development Goals

China is one of the major countries that has shown great enthusiasm in following Agenda 2030. Inherent in the new development and economic growth in China is the embedded attempt of poverty reduction with a new narrative of "green economy" and "green growth." The story of China's economic growth and rapid urbanization is one which embraces across the board all 17 UN Sustainable Development Goals. Discussing natural resource management in China in recent years, the authors discuss the Chinese version of sustainable development which includes being "open and inclusive" and "transformative and innovative." The Chinese version emphasizes respect for national sovereignty and diversity in development models. China as it reshapes its own natural resource management strategies has also become a firm proponent of "common but differentiated responsibilities." In other words, the Chinese leadership wants the world to respect its unique vantage point with respect to economic growth, climate change, and natural resource management. As China reshapes its own natural resource management, it is doing this through partnership with neighboring countries and regions in the developing world as it enters into collaborative effort for massive infrastructure development (gas pipelines, telecommunications, ports, airports, highways) initiatives through projects such as "Belt and Road Initiative" among others. In all this, one observes that China's economy has been changing from speed-based development model to quality-based development model. The demand for natural resources has accordingly shifted from pursuing quantity to quality, i.e., in environmentally friendly ways.

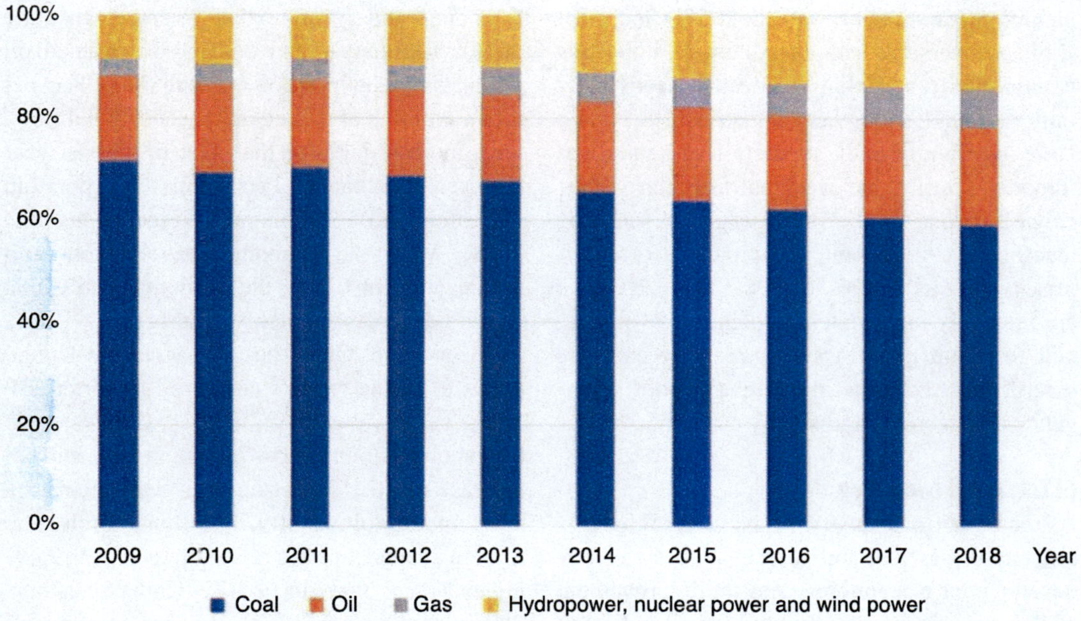

Fig. 6.1 Primary Energy Consumption Structure

billion tons of standard coal equivalents, average annual growth of 5.0% (Fig. 6.1); the total consumption was up to 4.64 billion tons of standard coal equivalents, increased by 3.3% and the energy self-sufficiency rate was 81.3%. In 2018, the energy consumption structure was composed of coal, accounting for 59.0%; oil, accounting for 18.9%; natural gas 7.8%; and other energy including hydropower, nuclear power, wind power, etc., which accounted for 14.3%. China's energy consumption structure has been improved continuously and the proportion of coal has continued to decline. In 2018, the proportion of coal consumption was 1.4% lower than that in 2017 and 12.6% points lower than that in 2009.

6.1.1.1 Natural Gas

Natural gas is cleaner and emits less greenhouse gases than fossil fuels, such as coal and oil. China's natural gas market has developed rapidly, and natural gas has become one of the most important energy sources. With the introduction of the policy of "campaign to prevent and control air pollution to make our skies blue again," increasing the proportion of natural gas consumption is an important way for China to achieve air quality standards and control greenhouse gas emissions.

Domestic natural gas has grown steadily in the past decade and imported natural gas has increased rapidly. In 2018, domestic natural gas production year-on-year growth was about 5.3%, lower than that in 2017 (9.7%). At the sectoral level, the year-on-year growth of gas consumption in urban, industries, power generation, and chemicals in 2018 increased by approximately 27%, 33%, 17%, and 26%, respectively. Simultaneously, the construction of China's pipeline interconnection is accelerating, including the West Second Line Guangnan Branch Main Line Zhangzhou Compressor Station Project, Sino-Russian East Line Natural Gas Pipeline Project Chang ling to Changchun Branch Line, Ordos-Anping-Cangzhou Gas Pipeline Phase I Project, and Shenzhen LNG to Large.

However, China's natural gas industry faces some problems. Firstly, China's exploration and production need to be improved, and the growth rate of natural gas production is far less than the growth rate of consumption. The growth rate of natural gas consumption in 2018 in China was about 16%, while the growth rate of domestic natural gas production was about 5%. Secondly, China's diversified supply system of importing natural gas from global perspective needs to be improved and strengthened, because there is a large uncertainty in the protection of natural gas imports. Thirdly, the weak infrastructure interconnection makes the initial capacity seriously insufficient. At present, the pipelines of enterprises, such as CNPC,[1] CPCC,[2] and CNOOC,[3] are relatively independent, which makes it difficult for them to form synergies between those enterprises, resulting in relatively poor emergency management and supply capacity.

6.1.1.2 Oil Industry

As the blood for industry, oil is essential for petrochemicals as fuel and source. With the steady development of economic and the improvement of living condition, the demand for oil will keep increasing in the following years. However, with a high level of external dependence on oil and the urgent need to tackle the environmental issue, China will continue to complete flexibly oil substitution in some key industrial fields.

The domestic production of oil in China has been decreasing since 2016, dropped to 189 million tons in 2018, decreased 1.3% year-on-year. The decline was 1.9% lower than the last year. China's resource endowment conditions are characterized as "rich coal, poor oil and gas." Proven oil reserve was only 1.73% of the world in 2017 (National Bureau of Statistics of China (NBS) 2012-2018). Meanwhile, the proven oil and gas resources are complex in accumulation, difficult to explore, and relatively expensive according to current mining technology. In addition, the international oil prices are relatively low compared with the domestic production cost, the oil production in China is lack of economic efficiency.

Despite the decline of domestic crude oil production, China's crude oil demand maintains a steady growth trend. Apparent consumption reached 625 million tons in 2018, at a growth rate of 7% (China National Petroleum Corporation Economic and Technological Research Institute 2018). The supply and demand for refined oil products maintain a low growth rate. The net import volume of refined oil reached 40.9 million tons, increased 12.4% than that in the last year (China National Petroleum Corporation Economic and Technological Research Institute 2018). With the economic development and urbanization in China, the oil demand in China would be increased quickly soon.

Since 2016, China has surpassed the United States to be the world's largest oil importer (BP Statistical Review 2019). In 2018, China's crude oil imports amounted to 462 million tons, with an increase of 10.1% year-on-year, equivalent to a daily import volume of 9.24 million barrels. The foreign dependence has raised up to 69.8%, 2.4% higher than that in 2017 (China Customs Administration n.d.). With the fast-growing oil demand resulting from economic development, a huge amount of imported oil in China would be increased continually in the future decades.

The crude oil processing volume reached 606 million tons in 2018, increasing 6.7% year-on-year, and the annual refined oil output was up to 365 million tons, increasing 1.8% year-on-year. At the same time, China imported net refined oil of 40.9 million tons, increasing 12.4% year-on-year. However, China is still facing the pressure of structural over capacity in the oil refining industry, and new refining capacity in China has taken more than half of the global net increased capacity.

6.1.1.3 Coal

Coal is a principal primary energy source in China and plays a key role in the economic growth of China. In 2018, China's coal consumption was 3.8225 billion tons, accounting for 50.54% of the global coal consumption. Why has China used so much coal every year? First, a large part of coal consumed in China has been used in thermal power generation. China's total thermal power generation reached 4979.47 billion kWh in 2018, which is more than the total power generation of the United States in a whole year. Another part of coal consumption is used to heat in winter, mostly used in Northeast China.

[1] China National Petroleum Corporation.

[2] China Petroleum and Chemical Corporation.

[3] China National Offshore Oil Corporation.

The third part of coal consumption is used in the steel industry. China is the world's largest steel producer, producing more than 50% of the world's steel. In 2018, China's crude steel production exceeded 900 million tons, accounting for 51.3% globally (Ministry of Natural Resources 2018), increasing 6.6% than the production of 870 million tons in 2017.

The price of power coal (23.0 MJ/kg) around Bohai Sea region was 569 Chinese yuan/ton at the end of 2018, decreased 1.6% year-on-year, 0.9% lower than that in the last year. Main business revenue was 2266.03 billion Chinese yuan, with an increase of 5.5% year-on-year, and the growth rate was 20.4% lower than that in the last year. At the same time, the total profit of China's coal industry reached 288.82 billion Chinese yuan, increasing 5.2% year-on-year, which was significantly lower than the growth rate of 290.5% in last year (UNEP 2013).

The substantial consumption of coal energy has caused severe environmental problems that have plagued several regions in China. Moreover, it has been the main cause of China ranking as the first in producing essential air pollutants and greenhouse gas emissions. In 2013, the State Council of China had promulgated the Air Pollution Prevention and Control Action Plan, which is its main purpose to control coal consumption in the long-term development of China. Based on the abundance of the proven reserves and its stability in supply, coal in China will continue to be the key component of the primary energy mix in the country at least over the next decades. Meanwhile, coal is responsible for a huge share of greenhouse gas (GHG) emissions and plays a key role in regional environmental pollution and global warming; hence, it is vital for China and the world how to improve and manage coal production and consumption in China to jointly promote economic development, improve environmental pollution, and mitigate global warming.

6.1.2 Steel Industry

China accounts for one-third of the world's total iron ore consumption and ranks the first in the world in iron ore import and crude steel output (Liu et al. 2003). The iron and steel industry are some of the basic industries in the process of economic development and its growth is an important indicator of industrialization. China has focused on the development of the iron and steel industry since the foundation of the People's Republic of China in 1949 because of the need for reconstruction and redevelopment of the destroyed country by war. In 1996, the amount of China's crude steel production exceeded 100 million tons for the first time, and then it is experiencing a rapid growth and has ranked first place consistently in the world for 15 years (IPCC 2007). The average annual growth rate of crude steel production has reached up to 18.5% in the first decade of the twenty-first century. In 2018, the output of the sector increased to 928 million tons, which accounted for 51.3% of the world's total steel production.

China's crude steel production is about 8 and 6 times larger than that of the United States and Japan, respectively. Consistent with the overall growth trend, export of China's iron and steel is also growing, and its structure is optimizing. China had changed into a net importer of steel products from a net exporter in 2006. From 2002 to 2011, exports of semi-finished and finished steel products had increased from 6.64 million tons to 47.9 million tons, increasing nearly 7 times (Ministry of Natural Resources 2018). With massive infrastructure development and rapid urbanization in China, iron and steel production will be increasing soon. However, the rapid development of the iron and steel industry has consumed a huge amount of energy in the sector as shown in Fig. 6.2.

6.1.3 Direct Resource Consumption Sectors

As Fig. 6.3 illustrates, in 2015, the construction industry accounted for almost 50% of natural resource consumption, followed by nonmetallic mineral production, food and tobacco production, transportation field, chemical products, and agriculture field. Natural resources are used mainly to construct buildings and infrastructure, such as roads, railways, and airports. Furthermore, we

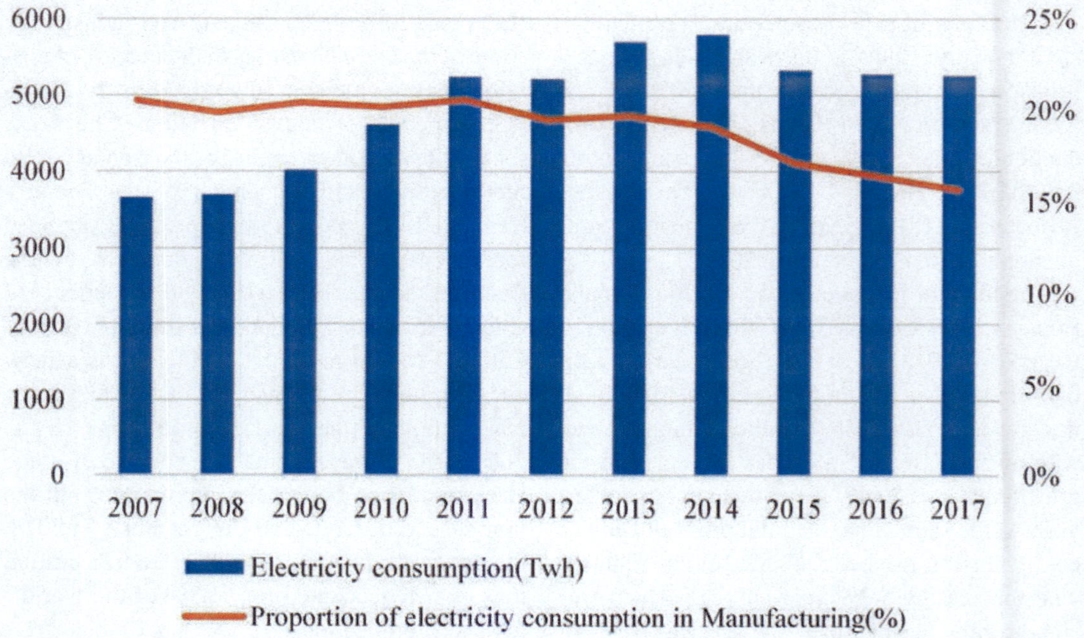

Fig. 6.2 Electricity consumption of China's iron and steel industry (Source: China Energy Statistical Yearbook (2018))

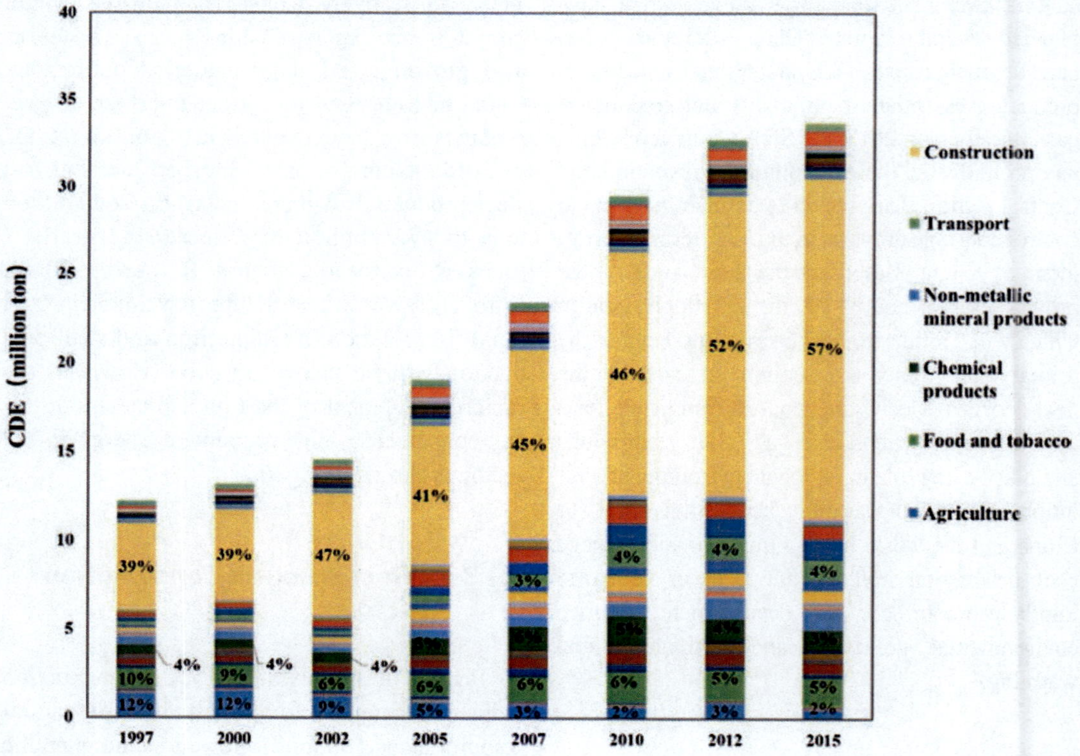

Fig. 6.3 China's top 5 sectors of direct resource consumption (1995–2015)

need electricity for lighting, cooling, and heating generated from various natural resources, such as oil, gas, and wood. Food and tobacco account for 5% of resource use in China. It includes food and beverages purchased in shops and consumed in hotels and restaurants. The food and tobacco industry require many resources, such as agricultural products, machinery, and energy, to produce these goods.

In addition, the retail network uses a lot of transport. All these resource uses are part of this consumption category. Transport contributes about 4% of resource consumption, including the fuel purchased for driving cars, the kerosene for airplanes and ferries, and the electricity for railways and public transport. All means of transport also require huge amounts of resources in manufacture: cars, ships, and airplanes need many metals, such as steel, aluminum, and copper, as well as plastics, glass, and textiles. These resource requirements are also included in this category. The remaining part of around 30% consists of a large variety of other consumption areas.

6.1.4 Pollutant Emissions

The way we use natural resources often provokes an irreversible change in ecological system. Extraction and processing of non-regenerative raw materials are often energy-intensive activities involving extensive interventions in ecosystems, which and resulting in air, soil, water pollution, and degradation of biodiversity. Even the extraction and production of renewable resources, involving extensive usage of energy, materials, chemicals, and water, are translated into different kinds of pollutants. Greenfield land is often transformed to create arable land, and, in some cases, whole ecosystems are destroyed in the use of natural resources.

Due to the rapid industrialization and urbanization, China is now the world's second largest resource consumption country, and at the same time it is the largest total pollutant emitter in the world (Geng and Doberstein 2008; Ou et al. 2017), which have affected severely China's long-term sustainable development (Lin and Wang 2014; Meng et al. 2019; Peters et al. 2010).

With the huge efforts implemented by multi-level Chinese governments, the pollutant emissions have been reduced rapidly, such as industrial sulfur oxide emissions and industrial wastewater discharges. Coal cleansing and reduction have made a great contribution to improving regional air pollution and mitigating global warming in China. These improvements are related closely to the installations of flue gas desulfurization systems in Chinese electricity plants, which has increased from 12% in 2005 to 82.6% in 2010. Emission controls implemented in the industrial processes have been proved to be effective to reduce primary $PM_{2.5}$ emissions (Ma et al. 2013). Energy mix has contributed to the emission growth (except primary PM2.5) during 2004–2007 and has offset emissions reduction during 2007–2011 (Ma et al. 2013). At current, the Chinese government has forecasted that the coal ratio in China's energy mix would decrease to 55% in 2020 based on the ratio of 68% in 2015 and the energy policy would contribute to reducing environmental pollution and adapting global warming.

6.2 Resource Management in China

China is facing complex challenges to manage natural resources to promote the sustainable development of China. No simple solution exists—but the policy reforms to reduce excess capacity and reflect suitable resource prices can help China to walk on the right path of synergy sustainability of economic, environmental, and social development.

6.2.1 State Level

Natural resource management in China has been implemented by adjusting the economic structure and changing the development path since 1950 (Fu 2008) (Table 6.1). During the past decades,

Table 6.1 Important policies and regulations of natural resources management in China

Date	Department	Regulation
12.22.1950	Administration Council of Central People's Government	Provisional Regulations of the People's of China on Mining
3.19.1986	Standing Committee of the 6th National People's Congress	Mineral Resource Law of the People's Republic of China
11.7.1992	28th meeting of the Standing Committee of the 7th National People's Congress	Mine Safety Low of the People's Republic of China
8.29.1996	21th meeting of the Committee of the 8th National People's Congress	Coal Law of the People's Republic of China
2.26.2016	19th meeting of the Standing Committee of the 12th National People's Congress	Law of the People's Republic of China on the Exploration and Development of Resources in Deep Seabed Areas

the reforms to streamline administration, delegate powers, and improve regulation and services have been deepened, the standards of approving mineral rights have been stricter than before, and the overcapacity has been improved. Cleanup of mineral licenses in national nature reserves also has been conducted very well.

The regional distribution of natural resources exploration has been optimized furtherly. The unified registration of rights to natural resources with reserves has been promoted, and the construction of natural reserve standard system has been advanced. The new natural resource planning has been issued and implemented, and the natural resources support capability to the national economy has been continuously improved.

The Chinese government had announced in December 2015 that Supply-Side Structural Reforms (SSSR) would help "guide the economy to a new normal mode." The 19th Party Congress, held in October 2017, had further elevated the SSSR as a main task in the national strategy of developing a modernized economy.

There are five focus areas under the SSSR: capacity reduction, housing inventory destocking, corporate deleveraging, reducing corporate costs, and improving "weak links" which contains a suit of measures aiming at industrial upgrading, infrastructure investment, etc., well-known as "three cuts, one reduction, and one improvement." In essence, the SSSR represents a departure of China's traditional demand-side stimulus policies to improve the quality of growth by managing the natural resources and addressing the structural imbalances in the economy.

Firstly, capacity cuts targeted two heavy industries of steel and coal that have experienced a rapid expansion in recent years. With the decline of GDP growth rate in China at the beginning of 2010, capacity utilization in the steel industry had also reduced from 79% in 2010 to less than 70% in 2015 (Fig. 6.4) and decreased from more than 90% in 2010 to 65% in 2015 for coal industry. Falling capacity utilization has been accompanied by a sizable deterioration in corporate profitability and debt-servicing capacity (Chen et al. 2018).

Capacity reduction has also been introduced in the efforts of the Chinese government's broad natural resource management. Severe national air smog since 2013 has aroused widespread concerns by government, scholars, and public, and has forced the central government to adopt more effective and tougher environmental management measures. In March 2016, Premier Li Keqiang announced measures to cut coal consumption and vehicle emissions as well as to punish government officials who ignore environmental crimes.

Emissions from Hebei, a province in North China with much more steel production than the United States, are a major source of Beijing's pollution (Chen et al. 2018). It is therefore not surprising that when the steel capacity reduction target was announced to be achieved in 2016, Hebei Province alone accounted for more than a third of the planned annual reduction.

In order to facilitate the implementation of these policies and measures, the State Council of China established a new department in 2018 (Fig. 6.5), named the Ministry of Natural

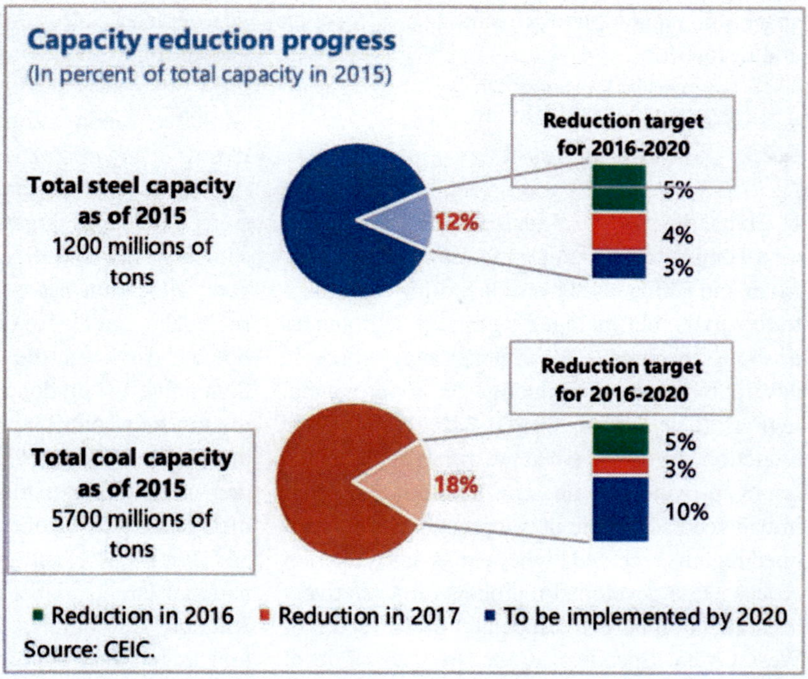

Fig. 6.4 Capacity reduction progress

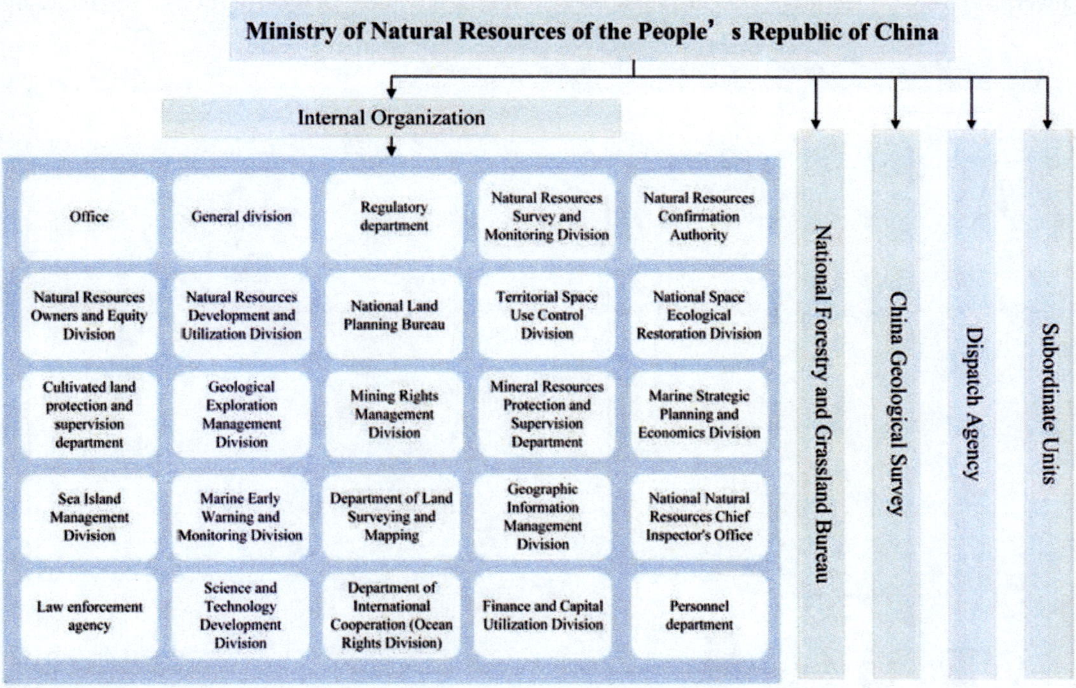

Fig. 6.5 Current structure of the Ministry of Natural Resources

Resources to integrate the responsibilities of the former Ministry of Land and Resources and other departments to uniformly exercise the responsibilities of the owners of natural resources assets owned by the whole people, and to uniformly exercise all the responsibilities of national land space use regulations and ecological protection and restoration.

6.2.2 Provincial Level

In China, the provincial region is the most important administrative unit in policy formulation. Mainland China has 31 provincial-level administrative units, but there are significant differences in both resource endowments and economic development levels among these provincial regions (Chen et al. 2014). Clear quantitative reduction targets have been distributed to the different provinces with consideration of their imbalanced economic development. More developed regions received higher emissions reduction request, less developed regions with a relatively less reduction burden, especially in the region of West China. Specifically, the volumes of local government-controlled target versus central government-controlled targets on chemical oxygen demand, sulfur dioxide, ammonia–nitrogen, and nitrogen oxides are 2.165 mt versus 0.124 mt, 2.004 mt versus 0.19 mt, 278,000 tons versus 14,000 tons, and 2.52 mt versus 0.246 mt, respectively (Zhang et al. 2011).

The sectoral and spatial allocation of the emissions reduction targets during 2011–2015 are shown in Fig. 6.6. Spatial difference of the targets' allocation has been considered due to the regional inequality of economic and social development such as the developed provinces of Shandong, Guangdong, Jiangsu, and Henan are the top four contributors of emission reductions, the subtotal amounts of their chemical oxygen demand, sulfur dioxide, ammonia–nitrogen, and nitrogen oxides reductions account for 35.7%, 35.73%, 34.24%, and 39.52%, respectively of the national targets; Tibet Autonomous Region and Xinjiang Autonomous Region have zero reduction targets and Hainan Province is allowed to increase 0.71% and 0.55% of nitrogen oxides and sulfur dioxide. Qinghai Province is the only

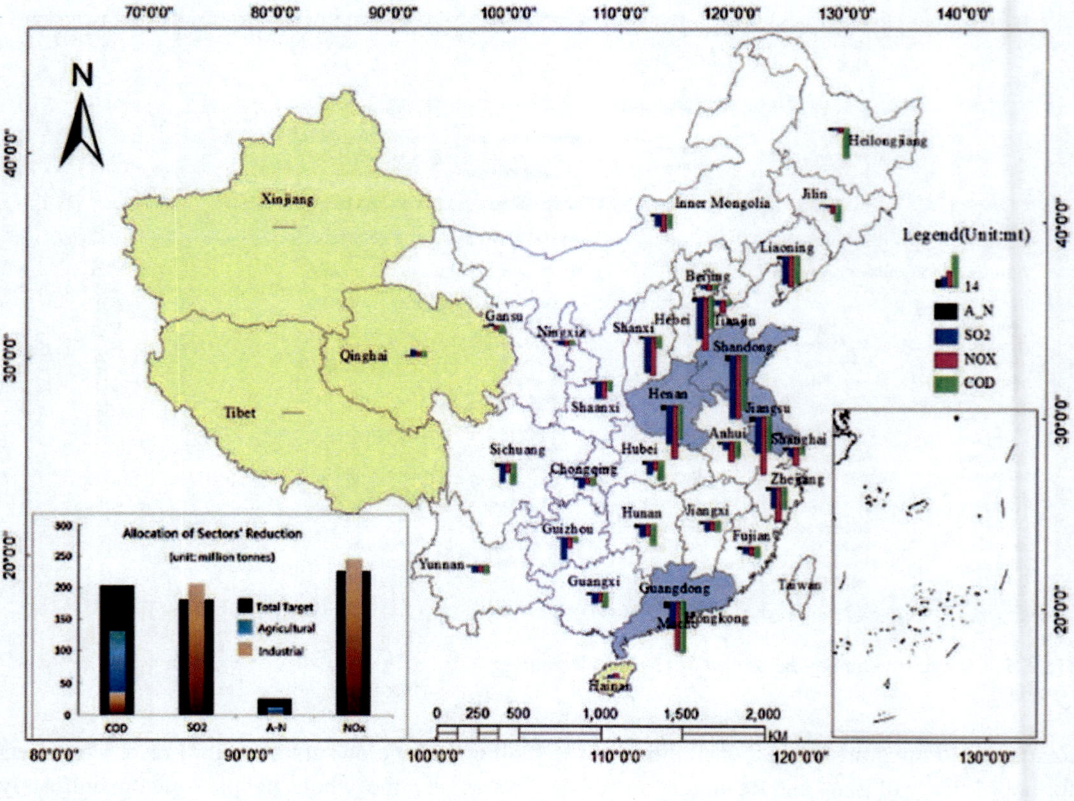

Fig. 6.6 Sectoral and spatial allocation of the emissions reduction targets during 2011–2015 (Source: Xue et al. 2014)

province to allow the increase of the above four pollutants with 0.88% for chemical oxygen demand, 1.30% for sulfur dioxide, 0.50% for ammonia–nitrogen, and 0.71% for nitrogen oxides (Zhang et al. 2011).

To manage natural resources and respond to climate change, low-carbon development in China had been raised by the National Development and Reform Commission[4] in 2010. Five provinces and eight cities had been chosen as the first batch of national low-carbon development pilot regions, which focuses on optimizing energy structure and improving energy efficiency.

For example, as one of the important and largest energy and industrial bases in China, Shanxi Province covers only 1/60 of the territory of China, but its coal production accounts for a quarter of China's total coal production, coke production of 2/5, and power generation of 1/17. To achieve its energy consumption reduction targets, Shanxi Province has been making and adopting key policies and measurements to integrated management of natural resources protect environment,[5] such as natural reserves, forest parks, spring areas, and Fenhe River, are prioritized for protection and any mining activity must be prohibited and every mine must be restricted for development or even closed in these areas.[6]

The integration of coal resources has focused on the two targets including economic sustainability and ecological improvement. The integration in Shanxi Province aims at pursuing the mechanization, the informatization, and the modernization of coal-mining scale extraction. To maximize coal resource utilization and technological improvement, Shanxi government made three plans, including closing small-scale mines, improving mining productivity, and merging mining enterprises (UNEP 2013).

In terms of Guangdong, to address the challenges of natural resource management and related air pollution, Guangdong Province and Hong Kong Special Administrative Region are playing leading roles in the natural resource's management in China. Back in 2003, the governments of Guangdong and Hong Kong had signed the agreement of the Pearl River Delta Regional Air Quality Management Plan to pursue regional collaborative pollution reductions. In 2012, the two governments had endorsed a successful management plan that would further reduce the air pollutant emissions up to 2020. Determining how to meet the emission reduction targets and improve resource efficiency while maintaining economic growth and the local living standard is huge challenge and key task for the two governments and policymakers.

6.3 Challenges and Suggestions

Although some achievements have been achieved, challenges still exist and need to be responded appropriately. We hereby discuss various challenges related to natural resource management in China and then, give our suggestions to better sustainable development in China.

6.3.1 Improving Reliability in Data Statistics System

The application of science and technology could significantly improve the outcomes of environmental problems by influencing several aspects of institutional changes (Xue and Ren 2012), while the data are the foundation underlying scientific-based policymaking. However, data reliability in China is always an issue for scientific communities, for example, the giga ton gap in carbon dioxide inventories between the aggregate of China's 30 provinces (excluding Tibet, Hong Kong, Macao, and Taiwan) and the nation in 2010 (Xue and Ren 2012), which can be explained as due to the differences in the statistical systems between national and provincial levels (Xue et al. 2014). The data discrepancy also exists in the national emissions statistical system; for instance, the total discharge

---

[4]NDRC, a ministry leveled agency in charge of planning issues for the whole country.

[5]Shanxi Provincial People's Government, 2017

[6]Leadership Office of Coal Mine Corporation Merger and Coal-Resource Integration of Shanxi, 2009.

of chemical oxygen demand in 2010 reported in the National Statistical Yearbook 2011 was 12.38 million tons (Ministry of Natural Resources 2018), while the data reported in ECER was 25.51 million tons (Schreifels et al. 2012), with a difference of nearly double times. Scientists and other stakeholders interested in this issue would thus feel uncertain because no official explanations for such a huge discrepancy have been released.

Although an official regulation on Environmental Management Statistics aiming to improve data's reliability and timeliness was released in November 2006 by the Ministry of Environmental Protection, the discrepancy between different data sources still exists, which indicates that this regulation should be modified, and more detailed statistical methods should be launched. Moreover, the "Measures for the Environmental Information Disclosure" has existed since 2008, but its effectiveness is limited (Yang et al. 2013). Different governmental agencies have different channels and methods for collecting and treating data, but neither of these agencies is subordinate to one another, nor can any of them play a leading role in clarifying data conflicts. In addition, no institutional channels exist for the public to express their concerns on the issue. For instance, both National Development and Reform Commission (NDRC) and Ministry of Environmental Protection are currently dealing with pollutant emissions reduction, but a lack of coordination and cooperation between the two ministries has led to significant errors and distortion in data. Apart from this, such a problem also exists between the central government and regional government, further worsening the issue at the national level.

To solve it, it will be rational for NDRC to play a leading role since it is the most important sector in China to coordinating various planning issues and can coordinate different ministries when conflicts arise among them. NDRC should invite experts to raise a unified data collection and treatment method on pollutant emissions reduction and prepare a national regulation based upon it, while other agencies should share their data with NDRC and join the discussion with related officials at NDRC so that data conflicts can be avoided. To reach this target, a round table system among relevant agencies should be established to ensure that all related aspects from different stakeholders are equally considered. It also creates an opportunity for them to exchange information and feedbacks, to obtain technological and institutional support, and to negotiate.

Companies must play key roles in the transformation of our societies towards a sustainable resource usage. Many enterprises do not know in detail how much energy and resources they purchase and what they cost. The potential for saving resources and money remains undetected and unexploited.

Production with high resource efficiency will also be an increasingly important factor to maintain the competitiveness of companies in the international markets. As prices for commodities and energy go up, using fewer natural resources is one key strategy to reduce the costs of production. Producers should also take a life cycle perspective in their production activities. This includes expanding their sphere of responsibility to the materials and intermediate products they purchase from their suppliers. Companies should set high environmental and social standards for purchasing raw materials, energy, and resource-efficient intermediate products.

6.3.2 Strategic Investments Towards Resource-Efficient Cites

Rapid urbanization, particularly in developing countries has resulted in economic, social, and environmental challenges. Cities in China face the triple challenges of providing basic services to urban poor people and, contemporarily, reducing resource consumption in wealthy areas and planning infrastructure and management systems for the growing population, all within the same urban context.

Cities can offer a gateway for a sustainable future. They differ in their initial carbon endowments due to climate, population and urban form, transport and built infrastructure, and economic structure. The main role of cities is to provide economic opportunities and better quality of life to their citizens (State Council 2011). Cities that opt for low-carbon transformation are also likely to experience economic and social gain as they will maximize low-carbon energy sources, enhance efficiency in delivering urban services, and, thus, become more efficient, competitive, livable, and sustainable. There can be a strong alignment between low-carbon growth and sustainable urban development objectives, as outlined in Table 6.2.

Cities have an opportunity to respond proactively to the future in terms of resilience to declining fossil fuel supplies and global temperatures. The future lies in an alternative mode of development that is not only carbon-frugal but also economically and socially inclusive by keeping their focus on sustainability.

Solutions to reduce the ecological impacts of cities have already existed (UNEP 2015). Cities must discard their linear metabolism and make the "cradle to grave" model redundant. Careful analysis of all incoming and outgoing flows can help to better understand how the natural resources, such as raw materials, and energy are consumed and transformed into wastes. Instead of considering the transport and disposal of waste towards peripheral areas subservient to the function of the city, production chains must be organized those wastes are perceived as resources that can be recycled, upgraded, and transformed within an interconnected network into new products.

A key component of the sustainable city will be the "circular metabolism" which discards the "take-make-dispose" thinking and follows the biomimetic approach, reflecting the notion that in nature all the interdependencies' feed into and benefit from one another (Geng and Doberstein 2008). Nature's best ideas can be studied, and then imitated in designs and processed to solve urban problems. Drastic changes must be made in the urban metabolism to reduce the pressure on natural resources, minimize the impacts on the environment, and provide better value to the society in the form of higher economic benefits, employment opportunities, delivery of goods and services, and improved quality of life.

6.3.3 Promoting Energy Efficiency

The rapid growth of Energy Conservation/Efficiency Frameworks adopted, or Acts/Laws promulgated by Chinese provincial governments reflects their recognition of its critical role in sustainable development, by bridging the demand-supply gap, improving international economic competitiveness, lowering import dependency, protecting against fluctuating energy prices, enhancing national security, and reducing the threats of global warming and climate change. Such laws are generally comprehensive in the sense that they target the different economic sectors and a range of energy policy measures, some

Table 6.2 Low-carbon growth and sustainable urban development objectives (Source: Baeumler et al. 2012)

Smart urban form and spatial development	Energy-efficient industry and buildings	Low-carbon vehicles and a public transport-oriented systems	Low-carbon waste management and other services
• Preserved agricultural land	• Reduced air pollution	• Reduced congestion	• Improved solid waste management
• Reduced contingent financial liabilities	• Improved energy security	• Reduced air pollution	• Reduced air pollution
• Improved rural land compensation and equity concerns	• Enhanced energy efficiency and industrial competitiveness	• Improved traffic safety	• Increased efficiency of water resource utilization and protection
• Limited encroachment into sensitive sites	• Increased resource efficiency in buildings and heating	• Increased urban livability	

mandatory and others voluntary in nature (State Council 2011).

Experiences in China show that energy performance standards and labeling schemes for appliances and equipment are a cost-effective policy tool for transforming markets and encouraging energy-efficient products. Governments usually follow a process of consensus and negotiation of standards that the industry can meet with reasonable increases in prices. Initially, a voluntary target is applied. As the market transformation proceeds, the targets are introduced as standards. However, standards may fail to induce sufficient energy efficiency improvements if they are largely based on negotiations with industry members without any explicit standard-setting method and if they are not revised periodically.

6.4 Conclusion and Policy Implications

To conclude, the downward trend of energy intensity in China is a good sign of the positive impacts of energy efficiency. However, we are still quite far from realizing the significant potential that energy efficiency offers, at costs below the rising fossil fuel prices. China still has a long way to go in the development process may consider taking a more aggressive attitude to tap energy efficiency to the fullest. This will enable China to achieve the developmental goals with the least adverse impacts on the environmental quality and global ecological system. According to Natural Resources Defense Council or NRDC, similar energy-saving initiatives have been adopted by China, as part of the Copenhagen Accord process, with formal commitments for a 40–45% reduction in CO_2 emissions per unit of GDP by 2020, compared to 2005 levels.

6.4.1 Approaching Co-Benefits through Policy Integration

Co-benefits approach refers to the development and implementation of policies and strategies that simultaneously contribute to tackling climate change and solving local environmental problems or vice versa (Guan et al. 2012). As the world's second largest economy and the heavy reliance on natural resources, China's high carbon emissions have become an important issue both in China and in the world (Liu et al. 2003). However, even great efforts have been made in China to reduce pollutant emissions, the intensity indicators such as greenhouse gas emissions per GDP are still increasing due to its rapid economic development. Consequently, to reduce the total carbon emissions by employing absolute emission reduction indicators (for example, total CO2 emission reduction) instead of relative emission reduction indicator should be taken and integrated into China's policymaking. Unfortunately, even CO2 emissions reduction was mentioned in the 13th planning of ECER, but without any clear quantitative reduction targets, which caused global arguments on China's responsibility in responding to global climate change. Therefore, a more appropriate approach that can address both pollutants emission reduction and climate change mitigation should be raised.

Co-benefits approach is suggested especially for most developing countries, especially for China, facing more challenges with limited natural resources. For instance, Alhajeri et al. (2011) demonstrate that there are co-benefits achieved by replacing the energy generated by coal with natural gas, including reductions of 24–71% in sulfur oxides emission, 16–82% in mercury emission, and 8.8–22% in carbon dioxide emission and water consumption also decreased by 4.4–8.7% (Alhajeri et al. 2011). Although the term of co-benefits has been recognized in political rhetoric at times, policy synergies of climate change and environmental quality often compete against each other in practice. Chinese policymakers need to develop a deeper appreciation for the multiple and far-reaching benefits of green growth by taking emission reduction policy as a fundamental principle for integrating absolute emissions reduction indicators-based climate policy consideration.

References

Alhajeri N, Donohoo P, Stillwell A, King C, Webster M, Webber M, Allen D (2011) Using market-based dispatching with environmental price signals to reduce emissions and water use at power plants in the Texas grid. Environ Res Lett 6:1–9

Baeumler A, Ijjasz-Vaquez E and Mehndiratta S (2012) Sustainable low-Carbon City development in China. The World Bank, Washington DC available at: https://siteresources.worldbank.org/EXTNEWSCHINESE/Resources/3196537-1202098669693/4635541-1335945747603/low_carbon_city_full_en.pdf

BP Statistical Review (2019) China's energy market in 2018. Available at: https://www.bp.com/content/dam/bp/business-sites/en/global/corporate/pdfs/energy-economics/statistical-review/bp-stats-review-2019-china-insights.pdf

Chen W, Yin X, Ma D (2014) A bottom-up analysis of China's iron and steel industrial energy consumption and CO2 emissions. Appl Energy 136:1174–1183. https://doi.org/10.1016/j.apenergy.2014.06.002

Chen L, Ding D, Mano R (2018) China's capacity reduction reform and its impact on producer prices. International monetary fund. Available at: https://www.imf.org/en/Publications/WP/Issues/2018/09/28/Chinas-Capacity-Reduction-Reform-and-Its-Impact-on-Producer-Prices-46223

China Customs Administration (n.d.). Available at: http://english.customs.gov.cn

China National Petroleum Corporation Economic and Technological Research Institute (2018) Domestic and Foreign Oil and Gas Industry Development Report. Available at: https://www.cnpc.com.cn/cnpc/ndbg/gywm_list.shtml

Fu B (2008) Blue skies for China. Science 321(5889):611

Geng Y, Doberstein B (2008) Greening government procurement in developing countries: a case of China. J Environ Manag 88:932–938. https://doi.org/10.1016/j.jenvman.2007.04.016

Guan D, Liu Z, Geng Y, Lindner S, Hubacek K (2012) The gigatonne gap in China's carbon dioxide inventories. Nat Clim Chang 2(6):672–675. https://doi.org/10.1038/nclimate1560

IPCC (2007) The physical science basis: contribution of working group I to the fourth assessment report of the intergovernmental panel on climate change. Cambridge University Press, Cambridge, pp 100–101

Lin B, Wang X (2014) Exploring energy efficiency in China's iron and steel industry: a stochastic frontier approach. Energy Policy 72:87–96. https://doi.org/10.1016/j.enpol.2014.04.043

Liu J, Ouyang ZH, Pimm SL, Raven PH, Wang X, Miao H, Han N (2003) Protecting China's biodiversity. Science 300(5623):1240–1241. https://doi.org/10.1126/science.1078868

Ma Z, Xue B, Geng Y, Ren W, Fujita T, Zhang Z, de Oliveira JPA, Jacques DA, Xi F (2013) Co-benefits analysis on climate change and environmental effects of wind-power: a case study from Xinjiang. Renew Energy 57:35–42. https://doi.org/10.1016/j.renene.2013.01.018

Meng J, Yang H, Yi K, Liu J, Guan D, Liu Z, Huang T (2019) The slowdown in global air-pollutant emission growth and driving factors. One Earth 1(1):138–148. https://doi.org/10.1016/j.oneear.2019.08.013

Ministry of Natural Resources (2018) China Mineral Resources. http://www.gov.cn/xinwen/2018-10/22/5333589/files/1bf146ddd1fd4ab998ba375aa0e98714.pdf

National Bureau of Statistics of China (NBS) (2012-2018) China Statistical Database, Available at: http://219.235.129.58/welcome.do

Ou J, Meng J, Zheng J, Mi Z, Bian Y, Yu X, Guan D (2017) Demand-driven air pollutant emissions for a fast-developing region in China. Appl Energy 204:131–142. https://doi.org/10.1016/j.apenergy.2017.06.112

Peters GP, Guan D, Hubacek KJ, Minx C, Weber CL (2010) Effects of China's economic growth. Science 328(5980):824–825. https://science.sciencemag.org/content/328/5980/824.2

Schreifels J, Fu Y, Wilson EJ (2012) Sulfur dioxide control in China: policy evolution during the 10th and 11th five-year plans and lessons for the future. Energy Policy 48:779–789. https://doi.org/10.1016/j.enpol.2012.06.015

State Council (2011) Comprehensive Work Program for Energy Conservation and Emission Reduction in 12th Five-year Period, Available at: http://www.gov.cn/zwgk/2011-09/07/content 1941731.htm (in Chinese)

UNEP (2013) Natural resource Management in Transition Setting. Available at: shttps://www.un.org/en/land-natural-resources-conflict/pdfs/UNDG-ECHA_NRM_guidance_Jan2013.pdf

UNEP (2015) Sustainable Consumption and Production-A Handbook for Policymakers. Available at: https://www.unenvironment.org/explore-topics/resource-efficiency/what-we-do/sustainable-consumption-and-production-policies

Xue B, Ren W (2012) China's uncertain CO2 emissions. Nat Clim Chang 2(11):762. https://doi.org/10.1038/nclimate1715

Xue B, Mitchell B, Geng Y, Ren W, Müller K, Ma Z, Tobias M (2014) A review on China's pollutant emissions reduction assessment. Ecol Indic 38:272–278. https://doi.org/10.1016/j.ecolind.2013.11.020

Yang L, Wu J, Shen P (2013) Roles of science in institutional changes: the case of desertification control in China. Environ Sci Pol 27:32–54. https://doi.org/10.1016/j.envsci.2012.10.017

Zhang J, Fu M, Geng Y, Tao J (2011) Energy saving and emission reduction: a project of coal-resource integration in Shanxi Province, China. Energy Policy 39(6):3029–3032. https://doi.org/10.1016/j.enpol.2011.03.026

7 Regional Sustainable Development and Natural Resource Decision-Making in India: Methods and Implications

Sudhir K. Thakur

Abstract

Natural resources are sources of agriculture output, input in the production process, and recreation amenity. Traditionally, economic growth has been measured by gross domestic product (GDP). This measure of economic development is misleading since natural resources such as agricultural land, forests, watersheds, aquifers, atmosphere, and ecosystems services are consumed in the process of development, but its consumption and depletion are not reflected in the GDP. Several of these resources can regenerate, but others consumed excessively will deplete to nonrenewable levels. Further, there are considerable regional variations in the regeneration, consumption, and depletion of natural capital in India. Does the depreciation of natural capital reflect in the national system of accounts with appropriate adjustments? Gundimeda (1998, 2001) has analyzed the accounting of forest resources for Maharashtra to compute environment-adjusted domestic product during 1993–1994. Another facet is the theme of sustainable development that states current consumption should not compromise the ability of production and consumption of future generations. Given this overview, this chapter addresses the following three overarching questions: (1) what is the relationship between natural resources, economic progress, and sustainable development? (2) what are the alternative methods to natural resources decision-making? (3) what are the patterns of natural resource regional variations in India during selected years 2006–2017? GeoDa is utilized for mapping data. Data for analysis are collected from Environmental Statistics of India and Central Statistical Organization, New Delhi.

Keywords

GeoDa · India · Natural Resource Decision-Making · Regional Sustainable Development

7.1 Introduction

Natural resources are the free resource endowments from nature, such as soil, water, air, minerals, forests, watersheds, aquifers, and ecosystem services, that contribute to economic development. Such resources are of two kinds: renewable and nonrenewable. The former regenerates on its own and is in perpetual supply and the latter is fixed in supply. An analogous concept views natural resources as stocks and *flows*. Stock resources

S. K. Thakur (✉)
Department of Finance, Insurance and Real Estate, College of Business Administration, California State University, Sacramento, CA, USA
e-mail: thakurs@csus.edu

B. Thakur et al. (eds.), *Resource Management, Sustainable Development and Governance*, Sustainable Development Goals Series, https://doi.org/10.1007/978-3-030-85839-1_7

are those, which have taken millions of years in its formation, and, hence, are treated as fixed in supply. Likewise, flow resources are those that are renewed in a short period of time (Rees 1985). Further, economic growth and natural resources are intrinsically linked. They are sources of agriculture output, input in the production process, foreign exchange, attraction of foreign capital, and sources of recreation (Burton and Kates 1965). Analysts have questioned gross domestic product (GDP) as a measure of economic growth. Natural resources are consumed, and its services are utilized in the process of development, and yet its consumption is not accounted for in the GDP. Several of these resources can regenerate, but those that are excessively consumed can deplete to nonrenewable levels, and, thus, would distort the measure of economic progress. Miller (1990) in this context suggested the computation of a Green GDP to take account of adjustments in GDP due to natural resource depletion.

Daly and Farley (2011) provide an exemplary understanding of biological, physical, and social sciences perspectives of ecology overlaid upon economic analysis. Ruth (2018) succinctly provides an understanding of the interdependence among economic and environmental change, and the environment providing services, such as energy, materials supply, and waste assimilative capacity. The environment and economy are intrinsically linked since they are interdependent and are considered part of a complex economic-ecological dynamic system. The economic and environmental interactions manifest as material and energy flow across system boundaries. Such interactions exemplify the transformations within the economy and environment that are assessed by material and energy flow accounts and ecological statistics (Fig. 7.1) (UN 2000). These processes occur under the governance of the laws of physical sciences, such as the laws of thermodynamics (Ruth 1993). Further, the notion of ecological *stability* and *resilience* is important to understand economic and environmental interdependence. Stability implies the resistance of a system from any departure from an equilibrium condition of steady-state and resilience implies the tendency of a system to retain its organization structure following perturbation (Mahendrarajah et al. 1999).

For long economists believed that the production of goods and services were dependent upon the accumulation of physical and human capital only. However, during the past few decades, this view has changed, and a third factor is included in the production process called the "natural capital." Natural capital is defined as the natural and environmental resource endowment available to the economic system, such as trees, fish, oil, and minerals (Barbier 2005). Thus, the total capital stock of an economy comprises physical, human, and natural capital. Table 7.1 shows the wealth share by global regions and income groups during 2014 (Lange et al. 2018). The largest share of natural capital to total wealth is noted in the Middle East, North Africa, Sub-Saharan Africa, and South Asia, and the least in Europe and Central Asia. The largest share of manufactured or produced capital is observed in Europe and Central Asia and East Asia and Pacific regions, while the least share is noticed in the Middle East, North Africa, and Sub-Sahara Africa. Analogously, the highest share of human capital (skilled and unskilled) is observed in East Asia and Pacific, Europe and Central Asia, and Sub-Sahara Africa, and the least in the Middle East and North Africa. Likewise, carbon assets, i.e., fossil fuel reserves are bestowed in the largest amount in the Middle East and North Africa and the least in East Asia and Pacific and Europe and Central Asia.

This geographical pattern raises an important question—do countries bequeathed with natural resources perform better in economic development? Auty (2001) proposed the *resource curse thesis* to examine the economic performance of resource abundant countries. He opined economic growth is inversely proportional to the share of natural resource rents and is evidenced with the poor performance of mineral-rich countries and debilitating state of affairs of oil-rich economies (Auty 2007). The resource curse thesis integrates neoliberal, political, institutional, and environmental factors to provide an explanation of the fate of such economies. A threat of the excessive dependence upon mineral extractions

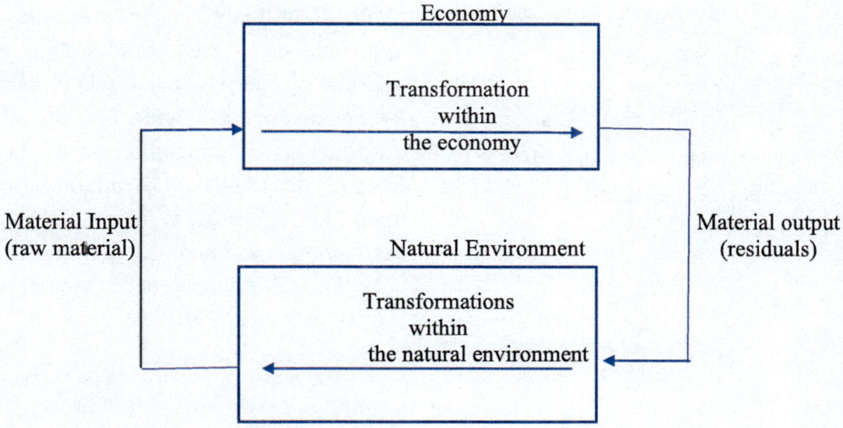

Fig. 7.1 Interrelationship between the economy and the natural environment. Source: Integrating Environment and Economic Accounting: An Operational Manual Figure 14, page 31 (UN 2000)

is due to depletion of resources, mineral sector marginalization with synthetic substitutes, low elasticity of primary sector products, and volatile nature of revenues from this sector (Auty 1993). Humphreys et al. (2007) investigate economies with oil and gas as natural resources and noticed that such resource wealth has led to lower growth rates, heightened volatility, excessive corruption, and devastating civil wars. In addition, Auty and Furlonge (2018) studied Trinidad and Tobago and Mauritius economies comparatively and provided a reconceptualization of the resource curse thesis that is consequential from the broader and more complex theory of rent curse. They proposed two development trajectories as an explanation of the rent curse theory, these are *low rent competitive diversification model* and *high rent staple trap model*. The former model suggests low rent incentivizes the elite to promote the national economy efficiently, as well as stimulates rapid and egalitarian economic growth with incremental democratization. The latter model motivates rent-seeking behavior for immediate consumption at the expense of long-term growth. Further, high rent constrains, competitive diversification, and regularly causes protracted growth collapses. In sum, both stylized models are nondeterministic explanations for understanding variants of rent curse applications.

Perloff and Wingo (1970) examined the role of natural resources in the national economic growth in the US. They analyzed three types of resources: agriculture, mineral, and amenity resources and exemplified the geographical variations in the resource endowment of places and its temporal significance. Further, they observed that less tangible amenity resources assume higher significance. Analogously, Carmignani and Chowdhury (2012) examined the relationship between natural resources and development for a sample of 92 countries across continents. Statistical analysis for the period 1970–2005 supports the presence of regional variations in the effect of natural resources on income, institutional quality, and schooling. The impact of natural resources on development in Sub-Sahara Africa is negative due to the combined effect of resources and the disease environment.

In addition, there are considerable regional variations in the regeneration, consumption, and depletion of natural capital in India. A pertinent question is whether the depreciation of natural capital reflects in the national and regional system of accounts with appropriate adjustments. Gundimeda (1998, 2001) and (Gundimeda et al. 2005, 2006) have analyzed the forest resource accounting in India at state levels and for Maharashtra to estimate environment-adjusted domestic product (EDP) during 1993–1994. Kadekoki (2004) has underscored the need for a continued emphasis on the methodological development for accounting natural resources. Further,

Table 7.1 Wealth Shares by Region and Income Group, 2014

Region	Carbon Assets (% of total wealth)	Natural Capital (% of total wealth)	Produced capital (% of total wealth)	Human capital (% of total wealth)
East Asia and Pacific	2	10	28	60
Europe and Central Asia	2	5	33	62
Latin America and the Caribbean	3	18	23	61
Middle East and North Africa	40	44	15	35
South Asia	3	25	26	51
Sub-Saharan Africa	9	36	16	50
Income group				
Low-income countries	1	47	14	41
Lower-middle-income countries	6	27	25	51
Upper middle-income countries	4	17	25	59
High income non-OECD countries	26	30	22	42
High income OECD countries	1	3	26	70

Source: The Changing Wealth of Nations (2018), World Bank (2018)

Chopra and Dayal (2009) have explored the theoretical, historical, evolutionary, and theme-based perspectives in the analysis of problems dealing with natural resource economics in India. Another aspect is the sustainable development of natural resources such that current consumption does not compromise the ability of production and consumption of future generations (Brundtland Commission 1987). Such a theme has been further explored utilizing an environmental economics approach and by incorporating the "first and second law of thermodynamics," "principles of ecological resilience," and "sustainability of the economy" (Thampapillai and Ruth 2019). Recently, the Government of India has pronounced upon the interrelation among natural resource accounting, sustainable development, climate change, and public-private sector participation (Pattanayak et al. 2020).

Box 7.1 Sustainable Development Goals

As the author examines regional sustainable development and natural resource decision-making in India, readers will notice the 17 SDGs all, directly or indirectly being influenced for successful implementation. In the Indian context, natural resource decision-making has several stakeholders in terms of institutions such as the Ministry of Environment, Forest, and Climate Change, the Ministry of Home Affairs, Defense, and the states and union territories. Further, with India's federal planning body *Niti Aayog,* various verticals have been created to provide vision and mission to different aspects of SDGs. One such vertical includes Natural Resources and Environment, which overlooks the relationship between natural resources, economic progress, and sustainable development. In its attempt to adhere to the UN SDGs (17 SDGs, 169 targets, and 231 indicators) the broader vision of this vertical lies in the formulation of strategies and policies for the sustainable management of natural resources which includes the maintenance of the clean, green, and healthy environment and mitigation of climate change. As the author reviews alternative methods to natural resource decision-making in the Indian context, he acknowledges the need on the part of decision-makers and policy analysts while delivering SDG strategies and plans to be mindful of robust research and data collection, knowledge sharing, capacity building, innovation and adaptation across spatial scales.

Given this overview, this chapter addresses the following three overarching research questions: (1) what is the relationship between natural resources, economic development, and regional sustainable development? (2) what are the alternative methods to natural resource decision-making? (3) what are the patterns of selected natural resources and its regional variations in India during the years 2006–2017? GeoDa software is utilized for map preparation. The remainder of this chapter is divided into five additional sections. The second section elaborates the research methodology adopted in this chapter; the third section reviews the relationship among natural resources, economic progress, and sustainable development at various geographical scales, followed by the fourth section, which reviews decision-making techniques to account for natural resources. The fifth section describes map patterns for selected natural resources and years in India. In addition, the last section concludes.

7.2 Research Methodology

The purpose of this chapter is to argue for the accounting of natural resources, due to its significance for regional sustainable development in India. Natural resources are consumed and depleted in the process of economic development but are unaccounted for in the computation of GDP. This suggests that the quantitative measurement of economic progress is inflated due to the noninclusion of natural resource depletion in the process of development. It also provides incorrect price and market signals for resource allocation and investment planning. This chapter collated data from the Compendium of Environmental Statistics and Environmental Statistics of India at the state-level data on three environmental indicators: total forest area by states for 2006 and 2016, total renewable installed power by states for 2007 and 2016, and net annual groundwater availability by states during 2013 and 2017. The size of resource endowment is computed as a percent of the total for nation and the resulting data are mapped for visualization.

7.3 Natural Resources and Sustainable Development: Literature Review

Natural resources and economic development are intrinsically linked in at least three different ways. The first is the role of natural capital in economic growth and development. Traditionally, economists have defined two major factors of production namely labor (skilled and unskilled) as well as capital or produced capital (machinery, building, infrastructure). The GDP measure subsumes the role of "nature" under the factor "land" and this process is an understatement, as it underestimates the contribution of natural capital in economic growth. Natural capital is regenerative but is also characterized by depletion and deterioration when over-used. It is conceptualized as such natural capital assets like aquifers, ocean fisheries, tropical forests, estuaries, and the associated ecological services it provides to earth, such as soil regeneration, recycle nutrients, assimilates waste, operates the hydrological cycle, and maintains the gaseous composition of the atmosphere (Dasgupta 2010a; b). Thus, the problem is when nations utilize natural capital in the production process; and its depletion is not accounted for loss of the national wealth. A second perspective is the relationship between environmental degradation and per capita income. It is postulated that first, environmental degradation will increase with economic growth, reach a maximum and then decline as per capita income continues to increase. The empirical verification of the environmental Kuznets curve (EKC) showed the rise, then decline in degradation with sustained increase in per capita income (Shafik 1994). The third perspective is to take cognizance of several stylized facts in understanding the role of natural capital in economic development (Barbier 2005): (1) many developing nations are highly dependent upon the consumption of natural resources for exports, (2) heavy dependence upon mineral resources has been associated with an inverse correlation with economic performance, (3) development pressures in several developing economies have led to land use change of forests, wetlands, and natural habitats for urban, commercial, and industrial development, and (4) a significant number of people in less developed nations reside in ecologically fragile areas that have been converted from poor quality upland and forest frontier areas for sustenance of livelihoods.

Each nation is endowed with a stock of natural capital in addition to human and produced capitals. The sum of the three constitutes the total capital stock of the economy. The consumption of these stocks brings economic progress and determines the level of social welfare to both the current and future generations. Thus, society must decide how much to consume today and bestow to the next generation so that they have at least the same level of natural capital. The Brundtland Commission in this context elaborated sustainable development as the "development that meets the needs of the present without compromising the ability of future generations to meet their own needs" (Brundtland Commission 1987). This notion raises the question, if current levels of natural capital are irreversibly depleted, then can the current generation compensate the future generation of the losses and by how much? Economists regarding sustainability have alluded to two views and they are referred to as *weak sustainability* and *strong sustainability* (Edward-Jones et al. 2000; Barbier 2005). The former contends that there is no inherent difference among the three types of stocks, and if the depletion of natural capital is substituted by human and produced capital, keeping the aggregate stock intact or increasing there is weak sustainability. Alternatively, proponents of strong sustainability reckon that human and manmade capital cannot substitute natural capital. They purport some forms of natural capital are essential for maximizing human welfare. This implies the non-substitutable natural capital needs to be preserved and not depleted and is called strong sustainability.

Table 7.2 shows the temporal variations in per capita wealth in India during 1995–2014. There was an increase in the total capital stock by 108% during the 19-year period 1995–2014. The largest increase was observed in produced capital (150.6%) followed by human capital (108.9%) during the same period in per capita dollar constant terms. Other natural assets that showed an increase during the same period were natural gas, metals, sub-soil assets, coal, and oil while a depletion was observed in forests (non-timber resources). It is important to note that natural capital accumulation was observed at the slowest pace during the period. The annual percent rate of growth in per capita total wealth was 2.38% during 1995–2000, 4.53% during 2000–2005, 5.58% during 2005–2010, and 4.68% during 2010–2014. This demonstrates a modest increase in sustainability during the period (Table 7.1). Figure 7.2 shows per capita wealth variations in India for selected natural resources and assets such as total wealth, produced capital, and natural capital. Human capital, and a decline in net foreign assets, during selected periods.

7.4 Sustainable Development

The concept of sustainable development (SD) is central in any debate on global sustainability. The term SD is conceptualized as an organizing principle that aids in meeting human development goals. This is accomplished in a way that the ability of the natural resource system is sustained for the provision of natural resources and ecosystem services. This section describes the *genesis of SD*, *SD as a complex system*, *measurement of economic progress*, and *economic analysis of sustainability*.

The notion of SD became important through an influential publication titled "Limits to Growth" by the Club of Rome (Meadows et al. 1972). The authors argued that continued patterns of economic growth would overshoot the carrying capacity of finite resources of the biosphere. Later, the Brundtland Commission (1987) conceptualized sustainable development as *development that meets the needs of the present without compromising the ability of future generations to meet their own needs*. The term "intergenerational" was acknowledged as significant at the Rio Earth Summit (1992). A critical implication of the notion was that "development today

Table 7.2 Per Capita Wealth in India: 1995–2014

Per Capita, constant 2014 USD	1995	2000	2005	2010	2014	Percent change (1995–2014) (%)
Total wealth	**8733**	**9773**	**11,989**	**15335**	**18211**	**108.5**
Produced capital	**2059**	**2340**	**2973**	**4189**	**5161**	**150.6**
Natural capital	**2619**	**2617**	**2776**	**4071**	**4739**	**80.9**
Forests, timber resources	25	22	18	30	37	47.6
Forests, non-timber resources	47	43	41	40	38	−19.1
Protected areas	83	88	82	95	94	13.9
Cropland	1450	1392	1279	1792	2036	40.4
Pastureland	803	855	925	1089	1429	77.9
Sub-soil assets	212	217	431	1026	1105	422.2
Oil	72	87	136	186	220	207.1
Natural gas	3	6	18	27	41	1101.3
Coal (all grades)	102	89	190	452	487	378.5
Metals and minerals	35	35	87	361	357	924.5
Human capital	**4192**	**4941**	**6386**	**7390**	**8755**	**108.9**
Net foreign assets	**−138**	**−125**	**−147**	**−315**	**−444**	**221.9**
Population	960,874,982	1,053,481,072	1,144,326,293	1,230,984,504	1,295,291,543	34.8

Source: The Changing Wealth of Nations (2018), World Bank (2018)

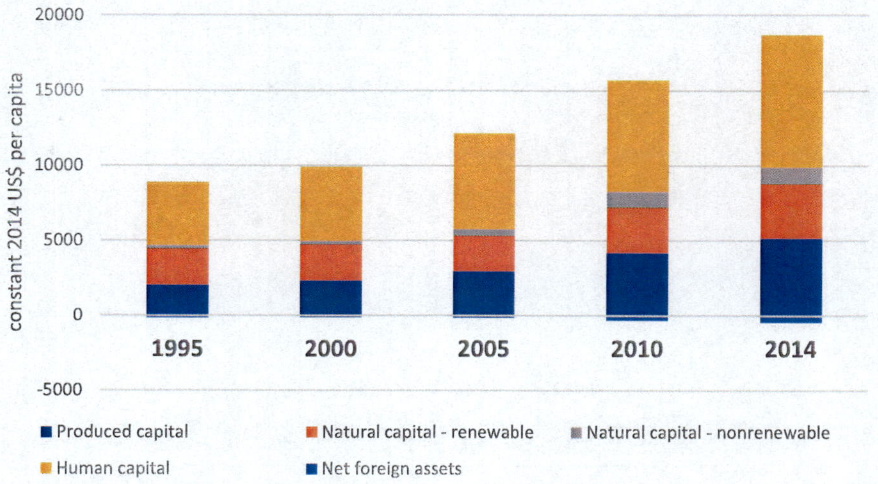

Fig. 7.2 Per Capita Wealth—India: 1995–2014. Source: The Changing Wealth of Nations (2018), World Bank (2018)

must not threaten the needs of the present and future generations. Subsequently, the idea of SD embraced a more comprehensive thinking linking 'economic development', 'social inclusion' and 'environmental sustainability'." Sachs (2015) posits the three pillars of sustainable development, as a component of an analytical understanding that endeavors to comprehend and predict the nonlinear interaction of the human and natural systems. SD is reckoned as a science of complex systems. A complex system is interpreted as a group of interacting components that together with its rules of interaction generates an interconnected whole. This system is characterized by interdependencies among its components, to the extent that the removal of any one element of the system destroys the system behavior to such an extent, that goes far beyond what is encapsulated in the element itself (Miller and Page 2007; Sachs 2015). Further, a sustainable complex system is characterized by emergent properties which imply the interaction of system components generates outcomes that is "more than the sum of its parts." In addition, a complex system is characterized by unexpected characteristics, indicating a small perturbation to the components of the ecosystem will generate a nonlinear outcome. Thus, SD includes four interacting system components, i.e., "global economy," "social interactions," "earth systems," and "governance" (Sachs 2015). The global economy encompasses the entire world at all geographical scales, i.e., household, village, local, regional, and national scales, which focusses on social interactions, such as trust, social network, inequality, and beliefs in communities. These component parts affect the earth system via ecological and environmental systems; and together these interacting parts influence the governance of the economy, business, and government. The four components of complex systems (economic, social, environment, and government) demonstrate persistently the emergent behavior and nonlinear dynamic properties of interaction between the natural and socioeconomic systems (Sachs 2015). SD exhibits properties of resilience implying first, it can absorb disturbance and remain with the same state; second, the extent to which it displays self-organizing behavior or the lack of, and, third, the magnitude until which the system shows learning and adaptation capabilities (Norberg and Cumming 2008). Further, SD is also viewed as a panarchy where natural and social systems and coupled natural and social systems are interlinked via "adaptive cycles of growth," "accumulation," "restructuring," and "renewal." Such systems exhibit two features, namely, hierarchy and its nested adaptive cycle capability displaying discontinuity marked by gaps and clumps (Gunderson and Holling 2002; Allen and Holling 2008). Analogously, Dasgupta and Maler (2004) characterize most ecosystems as nonconvex relative to the simplified assumption of convexity of economic systems, and, hence, the relationship of nature with economic development is marred with complexity.

Another dimension of the debate on SD is the measurement of economic progress. Clearly, the measurement of development utilizing GDP has been deceptive. A depletion of aquifers, coastal erosion, loss of wetland, air, soil, deforestation, scenic landscape, and loss of regional biodiversity are not accounted for in the GDP. Thus, natural resources are components of national wealth and are assets that deplete over time, and conventional accounting systems have inadequately accounted for the services rendered by ecosystems. Some of these renewable resources can regenerate but others will deplete and be lost forever. Thus, the GDP should be adjusted by deducting the depreciated assets from the GDP to accurately reflect the level of development. Incorrect estimates of GDP may convey faulty signals to policymakers for resource allocation (Miller 1990; Dasgupta 2010a). Therefore, this limitation led to the development of a new measure called the Index of Sustainable Economic Welfare (ISEW). The ISEW is a measure of economic progress that utilizes GDP and accounts for current stocks of natural resource assets. An increase in intergenerational social well-being over time elaborates the notion of SD. Further, a new measure of SD called the inclusive wealth index has been developed (Managi and Kumar 2018; UNU-IHDP 2018; 2012; Dasgupta 2010a). National wealth includes not only manufactured capital but also comprises human and natural capital, which is considered as the productive base of an economy. It is necessary for societies to consume natural resources for production in such a manner that the depleted natural capital can be regenerated for sustainable use. The Inclusive Wealth Report 2018 (UNU-IHDP 2018) concludes 44 of the 140 countries surveyed have suffered a decline in inclusive wealth per capita, although GDP per capita increased in almost all the countries surveyed. This observation surmises 31.4% of the countries in the sample were not on a sustainable path of development since 1998 although their GDP were growing.

Therefore, what should economic policymakers suggest to the government? Three economic principles are recommended: "polluter-pays-principle," "pollutee-pay-principle," and "payment for ecosystem services" (PES). The pollutee-pays-principle is an approach according to which the polluter shall bear the cost of measures to abate pollution to the extent of damage done to society or exceeding an optimal level of pollution. The tax is exemplified as a Pigouvian tax, which is equivalent to the marginal damage cost at the optimal level of pollution (Goodstein 2005; Dasgupta 2010a). Likewise, pollutee-pay-principle is the reverse of the former where the recipient of a clean environment is required to compensate the polluter not to pollute the environment. This idea is also represented as a payment for ecosystem services (PES). An example illustrating this principle is where land and forest owners should be provided incentives to engage in exchange for managing their land for providing ecological services. Although, it is difficult to conceive of a market for ecosystem, but, if it were a possibility then owners of ecological capital and beneficiaries of ecological services will be tempted to negotiate a transaction (Dasgupta 2010a; Tacconi 2012). The subsequent section elaborates upon the notion of regional sustainable development in the Indian context.

7.5 Regional Sustainable Development in India

Three interrelated terms need elaboration: *sustainable development*, *sustainable economic development*, and *regional sustainable development (RSD)*. First, SD implies the current generation's ability to fulfill its needs without conceding the ability of future generations to satisfy its own needs. This suggests recognition of the element of time in natural resource utilization and intergenerational equity. Second, economic development is opined, as sustainable if relative to the population a society's productive base does not diminish. The productive base consists of produced capital, human capital, and natural capital. Then, there is a form of equity missing from the discussion—spatial equity, i.e., equality in the use and access to natural resources across subnational regions to the lowest unit, i.e., across the globe, nations, regions, states, cities, and households. Thus, although the concept of SD is appealing it lacks an operational counterpart. Third, RSD implies societies grow in such a manner that it leaves the same level of resources for future generations while maximizing inter and intraregional equity in resource use. Therefore, the practical and applicable equivalent to the conception of SD is manifested in the view of RSD. Nijkamp et al. (1991a) and Nijkamp et al. (1991b) conceptualized RSD as:

> [a development which ensures that the regional population can attain an acceptable level of welfare-both at present and in the future-and that this regional development is compatible with ecological circumstances in the long-run while at the same time it tries to accomplish a globally sustainable development (Nijkamp et al. 1991b, p.3)]

Subsequently, RSD goals need to satisfy two objectives: first, the regional population should have a level of welfare, which is sustainable, and, second, the objectives of RSD should not counteract national or supra-regional SD goals. Analogously, if all regions have an attainable RSD, then the global SD is achievable. However, this framework of RSD cannot be generalized since regions would have specific social, economic, and environmental contextual factors and would imply that each region or system of regions would follow its own trajectory of development to attain the goals of RSD.

Khoshoo (1986) identified several priority areas and remedial actions that would steer the nation towards attaining national SD. Such actions epitomized in the context of Indian social, cultural, and economic milieu would pay dividends (Khoshoo 1986). He expressed the view that the benefits of economic planning had failed to reach the poor in India. A large proportion of the rural population fulfilled their everyday needs through biomass or biomass-related products. He advocated an ecosystem-based approach where the linkages among cropland, woodland, grassland, and rural energy were fostered for a rural sector-based SD. Table 7.3 shows the regional distribution of forest coverage in India during the period 1972–1975 and 2015–2016 for five periods. The forest coverage was 545,656 km^2 during 1972–1975, which increased to 767,419 km^2 during 2015–2016. There was a 40.6% increase in forest coverage during the 42 years period.

Forests provide both harvestable products that have direct use value such as timber, fuel lumber, industrial raw materials, construction materials, paper, fodder, fiber grasses, and a wide array of non-timber forest products derived from plants and animals. It also provides a range of ecosystem services, such as absorbing the excess carbon dioxide emitted from industrial societies, soil conservation and its protection from erosion, watershed protection and regulation of local water regime, protection of local ecosystem from floods, carbon sequestration, and provision of habitat for biodiversity. Further, forests have a pivotal role in conserving soil by holding soil constituents at the roots of the trees. It helps soils from erosion from wind or water flow. Similarly, forests protect soil from the impact of windblast and reduce wind speed. In addition, the wastes of forest biomass of plants and animals supply the organic nutrients to the soil for water retention. Thus, forests are important as a source of food, fiber, agricultural, and industrial raw materials as well as the nonmarket ecosystems services. Therefore, the perennial use and abuse of forests lead to the destruction and degradation of forests in many ways like unsustainable harvesting, cattle grazing, industrial wood demand, loss of regenerative capacity of forest growth, changes in species diversity, loss of biodiversity, and impact on soil, water, and climate (Sengupta 2001; Cutter and Renwick 2004). Forests have been degraded in India. The Faustmann's rotation model in this context would suggest the maximization of the present value of forestland through an infinite series of rotations, by a perpetual series of planting and harvesting. Such an approach would suggest when to optimally harvest forests (Amacher et al. 2009).

Forest coverage in India is regionally concentrated in eastern, central, and western

Table 7.3 Regional Distribution of Forest Coverage (percentage) in India

Region	1972–1975	1980–1982	2000	2010–2011	2015–2016
Northwestern Region	9.3	6.6	16.8	16.8	17.3
Western Region	14.8	13.6	15.8	15.9	16.0
Southern Region	4.6	4.5	4.4	4.4	4.5
Southeastern Region	9.0	8.8	9.2	9.2	9.3
Eastern Region	37.6	42.1	31.5	31.5	30.6
Central Region	24.6	24.3	22.3	22.2	22.3
Grand Total	100.0	100.0	100.0	100.0	100.0

Source: Compendium of Environmental Statistics India (2010–2016), Government of India; TN Khoshoo (1986)

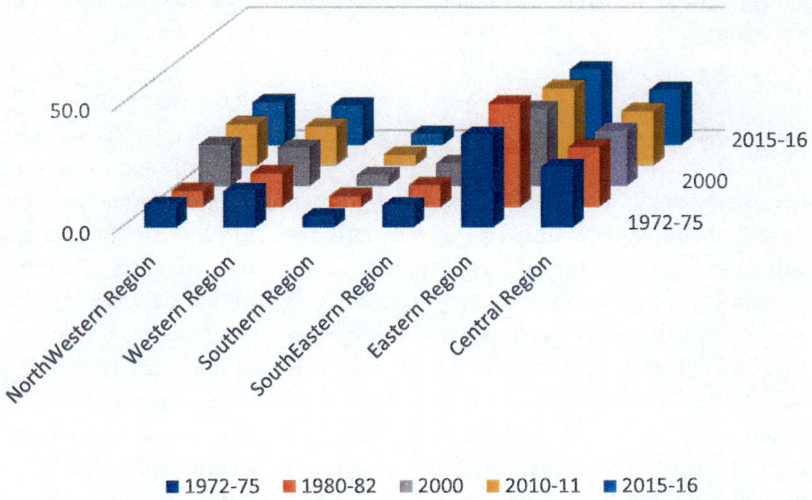

Fig. 7.3 Regional Distribution of Forest Coverage in India. Source: Compendium of Environmental Statistics India (2010–2016), Government of India (2016) and TN Khoshoo (1986)

regions (Table 7.3 and Fig. 7.3). It has increased in coverage in the northwestern region from 9.3 to 17.3% as well as the western region from 14.8 to 16.0% during 1972–1975 and 2015–2016. The southern and southeastern regions showed no increase during the entire period. The eastern region showed a 7.0% decrease during the period 1972–1975 and 2015–2016 as well as a 2.3% decrease in the central region. Thus, although forest coverage shows a 40% increase during 1972–1975 and 2015–2016, the regional spread shows an increase (northwest and western), decrease (eastern and central), and minuscule change (southeastern and eastern) in forest coverage. In sum, though forest coverage in India has increased during 1972–1975 to 2015–2016, it does not imply regional sustainability as the regional pattern demonstrates contrast and intermittent fluctuations in coverage.

7.6 Natural Resource Analysis and Decision-Making: Quantitative Measurement

Ecological economic analysis is at the cusp of three systems: economic, ecological, and social. This subdiscipline integrates these three systems at a few hierarchies and scales from the local, region, nation, and international (Edward-Jones et al. 2000). The real world is complex, and, thus, models are constructed to understand its nuances and intricacies. Models are conceptualized as an idealized and structured representation of the real world. To understand the analytical attributes of the models and its validity several methods are employed to corroborate the models. There are four types of modeling in ecological economics: conceptual, empirical studies, policy analysis, and modeling. Proops and Safonov (2004) have identified nine areas of modeling in ecological economics, and they are evolutionary models, input-output models, Neo-Austrian models, entropy in ecological models, thermodynamic models, multi-criteria analysis, agent-based models, and environmental Kuznets curve. Several methods have been utilized to understand the predictive performance or realism of assumptions of the models. This chapter will discuss four selected methods of decision-making in ecology and natural resource analysis, and these are cost-benefit analysis, environmental impact analysis, multi-criteria analysis, and natural resource accounting.

7.6.1 Cost-Benefit Analysis

The cost-benefit analysis (CBA) is a tool of development planning utilized for evaluating the full costs and benefits of major public sector investment projects in monetary terms. It is unlike the private sector where the major criterion for project appraisal is profit and loss accounting. The tool was originally applied to public sector and harbor projects in the US. Since then, it has been employed for project appraisal in such areas as urban air quality, irrigation dams, airport construction, highway maintenance, disposal of hazardous waste, choices among power generation schemes, and employment creation by alternate urban and regional policies. The underlying rationale for CBA analysis is contingent upon the theory of welfare economics whose objective is maximization of social welfare. This approach assumes the utilitarian principle, which enables the government in selecting those projects that maximize social welfare. The approach further depends on the Pareto efficiency criteria that states that those plans are better, if they make some (or all) people better off without making others worse off. However, such a criterion has a limitation since it is difficult to operationalize. Further, a widely applied variant of the Pareto principle is referred to as the Kaldor-Hicks's compensation principle. Further, if the beneficiaries of the selected project could in theory compensate the losers and remain better off, then the project benefits everyone in society (Edward-Jones et al. 2000).

The CBA is implemented in six stages: (1) project definition, (2) physical impacts of the project, (3) valuing impacts, (4) discounting of costs and benefits, (5) applying the net present value (NPV), and (6) performing sensitivity test (Little and Mirrlees 1974: Hanley 2001; Edward-Jones et al. 2000). In the first stage, the analyst defines the objectives, nature, and scope of the project, whose welfare are being considered, size of population, and duration of project. Assume a highway expansion project is under consideration that will clear large tracts of agricultural land and forests, but, in return reduce congestion and road accidents. Then, the second stage will entail quantifying the physical impacts of the project, such as consumption of materials, effects upon regional employment and its scale, and impacts on the physical environment, such as land conversion, wildlife habitat destruction, loss of agricultural land, and deforestation. In the third stage, all relevant effects will be expressed in monetary terms for aggregation purpose. This necessitates valuing the impacts in terms of marginal social cost and marginal social benefit. The shadow prices are estimates of marginal social cost/benefit when prices distort due to the presence of externalities and or government intervention. Under certain conditions, shadow prices will be essential to compute. Further, in the fourth stage when all the costs and benefits have been expressed in monetary terms, then, it is essential to convert them into its present value terms. This is necessary due to the time preference attribute of money. Therefore, due to discounting and impatience, it is necessary to discount costs and benefits. This can be performed by applying the discounting principle where the present value of a cost or benefit is computed by Eq. (7.1). The present value (PV) of a cost or benefit in time k, r is discount rate, and X_k is value of project after k years, and, then, present value can be computed using Eq. (7.1):

$$PV = \left[\frac{X_k}{(1+r)^k} \right] \tag{7.1}$$

In the penultimate stage, the CBA method assumes that the costs and benefits over the lifespan of the project (P) allow for the computation of an estimate called the net present value (NPV). The NPV can be expressed as Eq. (7.2) where k denotes time when the project begins and capital K as time horizon when the impact of the project comes to fruition. The term B_k is defined as benefits and C_k is defined as costs during period k. The social discount rate is defined as r and it should be noted that this rate has no relationship with the market rate of interest. If the sum of the monetary value of benefits outweighs the costs, then the proposed project, plan, or policy is meritorious and acceptable.

$$P=\sum_{k=0}^{k=K}\frac{\left(B_{k-}C_{k}\right)}{\left(1+r\right)^{k}}>0 \tag{7.2}$$

In the final stage, a sensitivity analysis is performed. This implies recalculating net present value when the value of certain critical parameters is perturbed. These parameters are discount rate, physical quantities and qualities of inputs, shadow prices of inputs, physical quantities and qualities of outputs, and timeline of project. Such a computation generates multiple estimates to create alternate scenarios.

The CBA methodology was utilized to assess the Ganga Action Plan (GAP) in India for evaluating ecological vulnerability and economic vitality. The river Ganges is one of the most polluted in India, and, thus, poses significant threats to human health and the environment. Markandaya and Murty (2000, 2004) conducted an analysis of the costs and benefits of cleaning the river Ganges. The contingent valuation method was employed for estimating user and nonuser benefits from the household surveys. For measuring the nonuser benefits a sample of 2000 household surveys were drawn from 10 cities (Delhi, Chennai, Baroda, Bengaluru, Hyderabad, Vijayawada, Kanpur, Lucknow, and Prayagraj) and for user benefits the sample was drawn from three cities (Prayagraj, Varanasi, and Kolkata) from the Ganga basin. In addition, they alluded to some of the emerging environmental and development issues in the river cleaning program. Further, they applied the market and nonmarket valuation of environmental goods to estimate the benefits such as user and nonuser benefits, health benefits to the poor residing along the river, and benefits to the agricultural farmers. The analysis estimated a net present social benefit of 10% social rate of discount and a high internal rate of return of 15%. The assessment of the Ganga Action Plan suggested several policy instruments to raise financing for cleaning efforts. These mechanisms were polluter pay principle, user pay principle (with government involvement), a user pay principle (without government involvement), and financing from tax collection (Markandaya and Murty 2004).

Similarly, interlinking of rivers (ILR) in India has been an ambitious civil engineering and inter-regional project aimed to develop and manage water resources at an estimated cost of $168 billion. The project's purpose was to link perennial rivers in Northern India, such as Ganga and Brahmaputra with Southern peninsular river system. The project's goal entailed to link Indian rivers through a network of reservoirs and canals for water management. The plan was expected to benefit the common person from the onslaught of floods and droughts by enhancing irrigation potential and recharging of acquirers. Such a gargantuan endeavor requires public debate and in-depth engineering, economic feasibility, and ecological analysis. The Ken-Betwa link a segment of ILR project would benefit Madhya Pradesh and Uttar Pradesh by providing irrigation needs, drinking water, and electricity consumption. The project has been assessed with respect to the pre-feasibility report available to the public (Alagh et al. 2006). The project utilized CBA for the decision analysis although it ignored the distributional aspect of costs and benefits. The CBA has estimated a ratio of 1.87 and an internal rate of return of 13%. In the construction of this project, large tracts of forestland will be lost, and the question is at what rates will lost forests be compensated? For instance, 45 km^2 of Panna National Park will be submerged which cannot be converted to other land uses by law.

Thus, a robust socio-ecological assessment needs to be made of the forestland it will lose, the tourist income it will forego and the sources of livelihood it will evade. Further, concerns related to costs of alignment of canal distribution system due to land elevation as well as resettlement and rehabilitation costs need to be resolved (Chopra 2006).

7.6.2 Environmental Impact Analysis

Environment impact analysis (EIA) is conceptualized as the minimization of the probability of potential unforeseen consequences (economic, environmental, and social impacts) associated with development projects, such as highway construction, airport, housing projects, and dam construction. EIA was first introduced in the USA in 1970 under the National Environmental Policy Act. Since then, EIA is a legal requirement for the evaluation of development projects. However, the identification of impacts is not easy and poses problems with decision-makers like identification of the impacts, and, second, the next best option once the impact has been identified. Edward-Jones et al. (2000) posit EIA as an aid to development policymaking. The methodology is articulated as a process of predicting the impact of a planned activity on the environment before the project is articulated. The execution of the EIA involves four stages: (1) project description, (2) evaluating the study and baseline assessment of environmental conditions, (3) estimation and assessment of impacts, and (4) identification of mitigating activities (Edward-Jones et al. 2000). The first stage involves collating general information on the site of the proposed project for vetting purposes. The second stage involves identifying a set of significant issues from a range of prospective problems. This process is important since not all probable issues can be studied and not all of them will be important. The third stage involves implementing baseline studies. This stage involves collecting data on issues that were identified in the second stage for a thorough and in-depth scientific analysis for projecting the impacts. Several techniques, such as "Leopold matrix," "environmental evaluation system," and "adaptive environmental assessment and management," can be utilized (Edward-Jones et al. 2000; Mitchel 1979). The Leopold matrix method is the analysis of the "specific activities" associated with the proposed project versus "attributes of the environment". As an example, if a new dam is constructed, it will have an environmental impact on the region. The "activities" will be dam construction, placing transmission line, reservoir filling, heavy metal discharge, relocation of inhabitants; and the affected attributes of the environment will be health, archeological artifacts, tourism, social and economic aspects, forestry, fishery, navigation, and aquatic plants among many others. An environmental evaluation system (EES) was developed as a methodology for improving the reliability and completeness of the EIA. The adaptive environmental assessment management (AEAM) technique is claimed to be responsive to changes in the decision-making and environmental domains.

EIA in India has undergone three phases of evolution: (1) the pre-1994 period when EIA analysis was performed only for selected government projects, (2) during 1994–2006 period a legal framework was utilized under the auspicious of the 1994 regulation, and (3) during the post-2006 period which enunciated the implementation of the modified EIA regulations under the direction of the Ministry of Environment and Forests (Nandimath 2009). Thus, the EIA has gained prominence in the evaluation of development projects in India. Apparently, public participation hearings in the Kullu district of Himachal were limited in the three-hydro development projects as ascertained (Sinclair and Diduck 2000). This was happening due to inaccessible information, ignorance of EIA procedures, and lack of institutional capacity, which suggested the embryonic stage of EIA in the Indian Himalaya region. In addition, rapid urbanization has led to an increasing demand of construction grade sand for regional development. An EIA of sand mining in the floodplain areas of three rivers in the Southwest coast of India was performed. The study was implemented for Kochi city and revealed sand mining activities have affected the health of the river ecosystem and degraded its overbank areas to a considerable extent. This concluded barring of mining activities in the deeper zones of river channels (Sreebha and Padmalal 2011). In addition, an environmental impact of the water quality parameters from a point source of the river Damodar in West Bengal was conducted for the period 2004–2007. The analysis performed an evaluation of the impact of mining and industrial effluents from the collieries and industries. The study suggested the presence of coliform, and, thus, needed proper mitigation strategies in the domestic water use from the river Damodar (Chatterjee et al. 2010).

7.6.3 Multi-Criteria Evaluation

Planners often confront goal maximization subject to alternate plans, multiple objectives, and sorting through several criteria. Let us suppose firefighters are planning to minimize human causalities due to forest fires. Heat spells, wind speed, and lack of rain can aggravate house fires and human causalities. In addition, fire departments work with limited budget, workforce, and thus need to prioritize resource allocation based on community vulnerability to save human lives. Thus, fire departments in such situations would develop alternate plans; assess the impact of each plan or the set of activities that will constitute a plan under conflicting and incommensurate criteria. The planner will, then, compare the impact of individual criterion to select the most desirable plan that fulfills the central objectives. A GIS analyst would use map overlay for integrating regional data for use in spatial decision-making. However, map overlay analysis might not perform well when confronted with more than four or five factors. Such a procedure will not allow for treating factors with unequal importance. A procedure that addresses this limitation is called multiple criteria evaluation (MCE). This technique allows weighing factors to reflect their relative importance. It is a robust methodology for exploring solutions to spatial decision-making problems that are ill-defined. Thus, MCE is conceptualized as a tool often utilized in management science, government, medicine as well as ecological problems, and other disciplines. The method utilizes data that are quantitative or qualitatively weighted, scores to reflect their importance according to a single or multiple set of objectives (Heywood et al. 2006) rank criteria. Further, this technique is implemented in four steps: (1) selection of criteria, where data layers are important to the problem identified, (2) standardization of criteria scores, where data layers are standardized for the sake of comparison, (3) allocation of weight, where numerical scores are allocated to place relative importance, and (4) MCE algorithm will multiply the standardized scores with weights for each data layer to determine the optimum location or outcome. MCE has become relatively popular due to methodological improvements in the structuring of the problem as well as advances in computing technology. It is configured similar to environmental impact analysis (EIA) but is relatively superior to cost-benefit analysis due to its restrictive nature. MCE has also gained popularity in addressing natural resource and ecological problems (Edward-Jones et al. 2000; Malczewski 1999).

This method has been utilized in the study of natural resource and ecological problems in India. Vadrevu et al. (2010) utilized MCE approach to assess the occurrence of forest fires at the district level in Andhra Pradesh (India). The analysis abetted in identifying potential hotspots such that forest protection measures can be implemented in advance. Droughts and floods are twin natural hazard problems that occur recurrently due to Monsoon rainfall and dry season spells. Shivaprasad Sharma et al. (2018) utilized MCE at the village level in Kopili river basin (Assam). They employed flood hazard layer from satellite data sets along with socioeconomic data sets, infrastructure, and land use vulnerabilities to generate flood risk zones. Such an analysis aids in alleviating hazard-based vulnerability due to floods. The local government can take measures to minimize the adverse impacts on human beings. Similarly, spatial MCE analysis was utilized to study earthquakes, vulnerability, and risk assessment in three locations in the

Delhi region. Data layers were collated for earthquake hazard component layer comprising seismic zone, peak ground acceleration, soil characteristics, liquefaction potential, geological characteristics, land use, proximity to fault, and epicenter. Likewise, data layers were also collected for physical vulnerability components like building information, number of stories, year-built range, area, occupancy, and construction type. Utilizing weighted overlay techniques, earthquake hazard, and vulnerability layers risk maps were generated. Such an analysis can be very useful in developing risk containment policies (Sinha et al. 2016). The next section elaborates the natural resource accounting technique.

7.6.4 Natural Resource Accounting

The environment interacts with the economy, contributes to the economic performance and thereby increases human welfare. It performs several functions, such as supply of natural resources for production and consumption activities, assimilates waste, and provides ecosystem services that are vital to human beings and earth's sustainability. In this context, conventional national balance sheet partially accounts for the functions provided by the environment. Only such functions are measured that are within the purview of market transactions and reflect important factors in welfare generation. The United Nations developed a System of National Accounts (SNA 1968) which provided a framework of systematic compilation and presentation of national economic data pertaining to the circular flow of national income and related macro-aggregates. Further, the United Nations Statistical Division articulated a System of Integrated Environmental and Economic Accounting (SEEA) that provides a structure to systematically collect data pertaining to the stocks and flows of environmental resources. This methodology was consistent with the SNA (1968) and exemplified the importance of assessing the environmental sustainability of economic performance. Further, ecological economic analysis provides an array of methods, i.e., "net present value" (Edward-Jones et al. 2000), "net price method" (Repetto et al. 1989), "user cost allowance" (El Serafy 1989), and "physical input-output accounts" (Proops and Safonov 2004).

First, most environmental assets do not have a market and so the market valuation principle rests on using the prices of the goods extracted from, or services provided by, or utilizing the future sales value and adjusting it by reducing extraction costs. Further, if the exploitation is spread over inter-temporally, then the flow of future sale proceeds is discounted (UN 2000). This can be expressed as Eq. (7.3) where the present value U_O of a natural resource is the sum of the expected net revenue flows N_tQ_t, discounted at nominal or real interest rate s. The interest rate is assumed constant for the life T of the asset. Also, N_t is defined as the total unit sales value of the resource less the production cost and Q_t is defined as the amount of resources exploited during the period t.

$$U_0 = \sum_{t=0}^{T} N_t Q_t / (1+s)^t \tag{7.3}$$

Second, the economist Robert Repetto defined the net price method as the difference between the average market value of the resource per unit (K_t) less its marginal exploitation cost per unit (L_t) including (operating costs, industry's previous investment, and capital equipment expenditure and normal rate of return). Then, the value of the natural resource is estimated as the product of the quantity of natural resource (R_t) and net price (N_t) (UN 2000). This can be written as Eq. (7.4) where U_t is the volume of a resource at the beginning of period t, U_t.

$$U_t = (K_t - L_t) R_t = N_t * R_t \tag{7.4}$$

Third, El Serafy (1989) developed the user cost method. This method addresses the question what proportion of the total income stream (S) of a natural resource is true income (Y)? Assume the life expectancy of the resource is (m) years and the social discount rate is (p). Equation (7.5) estimates the income from resource that is true income (Edward-Jones et al. 2000):

$$\frac{Y}{S} = 1 - \left[1/\left(1+p\right)^{m+1} \right] \quad (7.5)$$

Thus, it can be generalized the greater the life expectancy of a natural resources and higher the social discount rate, then higher will be the size of true income as a proportion of total income receipts from natural resources.

Finally, input-output (I-O) analysis is an accounting method for tracking all economic transactions for an economy at a given point in time and space. Tables describing the structure of an economy by recording transaction flows between industries in terms of purchase and sale represent it. In economic-ecological analysis, flows of materials and energy through networks of businesses in geographic proximity can be systematically recorded. These businesses share resources like water, energy, byproducts, and wastes. The analytical framework can be represented by the Leontief equation in an economic-environmental context. The traditional I-O table can be extended to identify the environmental costs and benefits associated with economic activities and distributed by sectors (Cumberland 1966; Isard et al. 1968) (Eq. 7.6):

$$X = \left(I - A\right)^{-1} Z \quad (7.6)$$

In Eq. (7.6) X is the n x 1 vector of outputs by industry, I is the n x n identity matrix, A is the n x n matrix of interindustry flow relationships, and Z is the n x 1 vector of final demand. The I-O analysis has made ecological studies methodologically rich due to three reasons. First, I-O analysis can be extended to the multi-sector and multiregional framework, incorporating numerous economic and ecological activities simultaneously along with their temporal and spatial interaction. Second, the I-O methodology can estimate direct and indirect effects, and, third, the method recognizes linkages, interdependence, and interaction among sectors and the evaluation of various alternatives.

There have been attempts to compile natural resource accounts for many countries. Repetto et al. (1989) applied the accounting methodology for integrating natural resource depletion into a revised national accounting system for Indonesia. They utilized timber, petroleum, and soil for the estimation purpose. They asserted Indonesia grew at an average rate of 7% during 1970–1974 and with an appropriate adjustment for the depletion of forests, soil, and oil the rate of growth was only 4%. Similarly, Saebo (1994) compiled Norwegian natural resource accounting system for forests, energy, and emissions during 1992–1993. In addition, the Dutch developed the National Accounting Matrix including Environmental Accounts (NAMEA), which developed accounts for environmental burdens in relation to economic development (Keuning et al. 1999).

India has been compiling environmental statistics since 1997 (CSO 1997). A systematic analysis was conducted under the supervision of Sir Partha Dasgupta utilizing a natural resource accounting framework to develop green adjusted GDP (Dasgupta Committee Report 2013). Subsequently, the Central Statistical Office (CSO) compiled the first official environmental economic accounts that portrayed asset accounts in physical terms for four categories of natural resources—forests, land, mineral, and water (CSO 2018). Prior to these efforts, several attempts were made to develop natural resource accounts in India at the state, multistate, and national levels. Dayal (2009) provides an in-depth review of natural resource accounting in India with respect to pollution, renewable, nonrenewable resources, and ecosystems. Gundimeda (1998, 2003) compiled forest-based resource accounts for the state of Maharashtra and integrated it into the system of national accounts (SNA) and System of Integrated Environmental and Economic Accounting (SEEA). The limitation of the study was that it considered only the monetary benefits provided by the forests and excluded the degradation of forests due to noninclusion of environmental services. Further, Gundimeda et al. (2006) and Atkinson and

Gundimeda (2005) estimated the value of forest resources in India's national and state accounts. The analysis revealed traditional measures undervalued the contribution of forests to regional income. The analysis showed the ratio of environmentally adjusted state domestic product (ESDP) as a ratio of adjusted net state domestic product (NSDP) was higher than 1, implying they are more sustainable than states with values less than 1. Several states in Northeastern India (Arunachal Pradesh, Manipur, Tripura, Meghalaya, Nagaland, and Sikkim) and southern states like Kerala and Tamil Nadu demonstrated values higher than 1. In addition, states such as Goa and Himachal Pradesh had values less than 1 and were dependent on tourism. Such states experienced stress on the forest sector and for other states with similar values specific contextual factors mattered, such as biodiversity values influence on local and regional climate, and impact on water resources, flood prevention, and drought control. Gundimeda's (2001) national scale analysis observes India's forest wealth, as extensive, yet net changes in wealth were not substantial in relation to GNP. However, when examined in the context of wealth tempering effect it is necessary to pay attention to additional savings to mitigate the effect of loss in forest values. Kumar (2012) accounted for inland wetland ecosystem loss for selected states in India (Gujarat, Jammu and Kashmir, Kerala, Rajasthan, and West Bengal) and estimated a loss of 0.46 million hectares during 1991–2001. This physical loss was equivalent to $1022 million-dollar loss. The air pollution levels were estimated in the twin cities of Hyderabad and Secunderabad in Andhra Pradesh in 2003. The damages from current levels of pollution were estimated to be equivalent to Rupees 6400 million [$ 87.48 million] (Murty and Gulati 2006). In addition, Tata Energy Research Institute (TERI) (2006) conducted an analysis of resource accounting in Madhya Pradesh and West Bengal for mineral resources utilizing the net present value and user cost allowance methods for computing value added by the coal sector to assess the cost of depletion. Their study estimated that in Madhya Pradesh depletion of coal varied from 2.5 and 6% assuming a 6 and 3% rate of discount, respectively during 2001–2002. The study determined resource rents were negative in West Bengal due to uneconomic mining practices. Analogously, Mali et al. (2016) estimated forest resource accounts in three forest subdivisions in Kerala state utilizing maintenance cost and reevaluation methods during 1999–2000 and 2006–2007. They observed positive revenue deficits in the forest sector in India including Kerala. Their study suggested forest resource accounting would decrease the deficit by integrating the value of unrecorded forest benefits and costs. Their analysis determined an increase in contributions of "forestry and logging sector" from 1.4% of GSDP to 1.5% of GSDP if the value of unrecorded forest benefits and carbon sequestration were accounted. The analysis was indicative of sustainable forest management in Kerala.

7.7 Description of Maps

Forests are important resources since they provide harvestable products that have direct use value as well as a wide array of ecosystem services. It is an important rural industry and a major environmental resource in India. Figure 7.4 shows total forest coverage in India in percentage (state total/national total*100) with five categories: very high, high, medium, low, and very low. Lighter color shows low values and darker color shows high values. Very high (darkest color) category consists of states, such as Madhya Pradesh, Andhra Pradesh, Maharashtra, Chhattisgarh, Odisha, Arunachal Pradesh, and Karnataka. The high category consists of states, such as Himachal Pradesh, Uttarakhand, Rajasthan, Assam, and Tamil Nadu.

The medium category consists of states, like Gujarat, Manipur, Mizoram, Uttar Pradesh, Meghalaya, and Nagaland. The low and very low categories include states, such as Bihar and Delhi. Forest coverage during the intervening period has not changed significantly. The forest coverage is highest in poorest parts of Northeastern India and richest states of western,

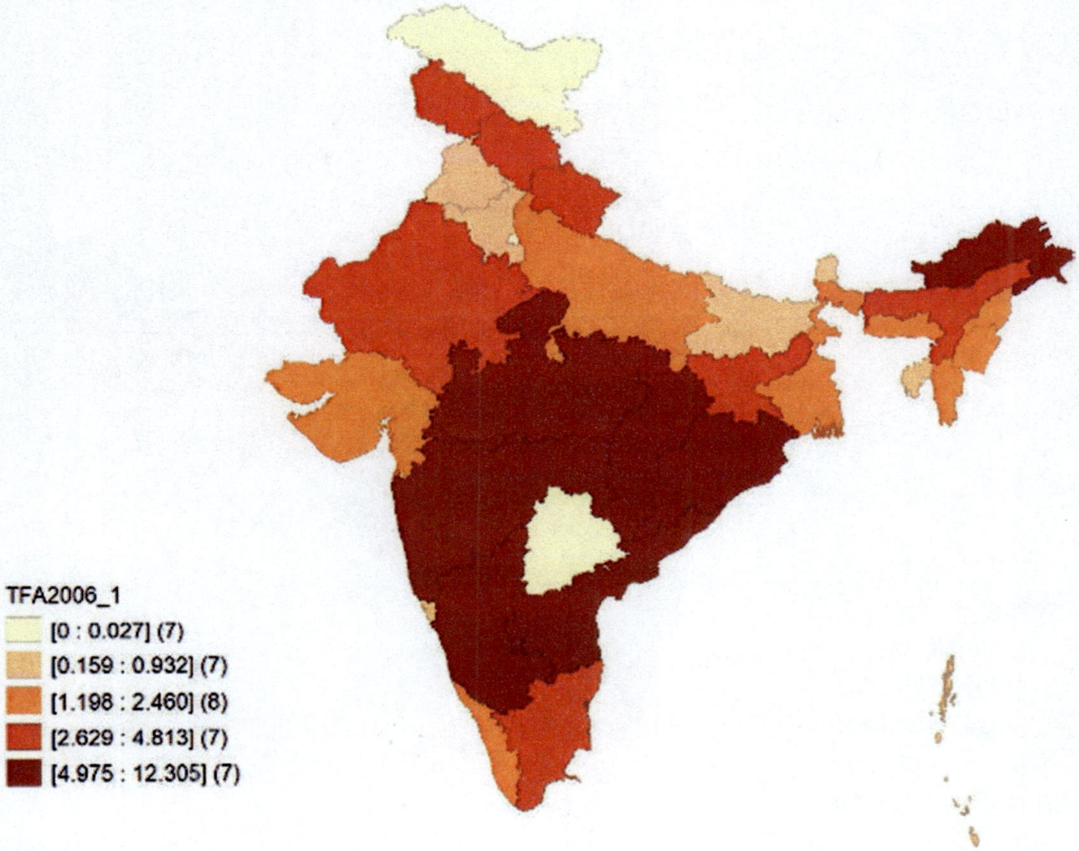

Fig. 7.4 Total Forest Area in India by States (km², %): 2006. Source: Compendium of Environmental Statistics, 2017, New Delhi

southern, and eastern India. The rich states utilize forest products as inputs for growth, while the poor and especially the tribal population's livelihood are dependent on the forest economy (Figs. 7.4 and 7.5).

India is one of the few countries with a commitment to a large production of energy from renewable resources. The Indian government has committed to a target of achieving 40% of its total electricity requirements from nonfossil fuel sources by 2030 in the Paris Agreement.

The government has also enunciated a target of installation of rooftop solar projects by 2022. Figures 7.6 and 7.7 show total renewable installed power in India at the state level during 2007 and 2016 and the map shows five categories: very high, high, medium, low, and very low. The maps show an east-west contrast with high percentage of renewable installed power in Western India. The top states in very high category are Tamil Nadu, Maharashtra, Karnataka, Gujarat, Andhra Pradesh, Rajasthan, and Panjab. Uttar Pradesh, Madhya Pradesh, Kerala, and West Bengal are in high category. Selected Northeastern states are in medium category, Tripura, Odisha, Manipur, Assam are in low category, and Telangana and Delhi are in very low category.

Groundwater resources are important for the agricultural sector, industrial, and residential use. It is estimated that 22% of groundwater resources are dried up or in the critical condition of being over-exploited. The government planners advise efficient use of water and suggest 10% of water should be saved for agricultural sector. They have advocated demand-side management in conjunction with supply-side management of water resources. The Central Groundwater Board (Government of India 2019)

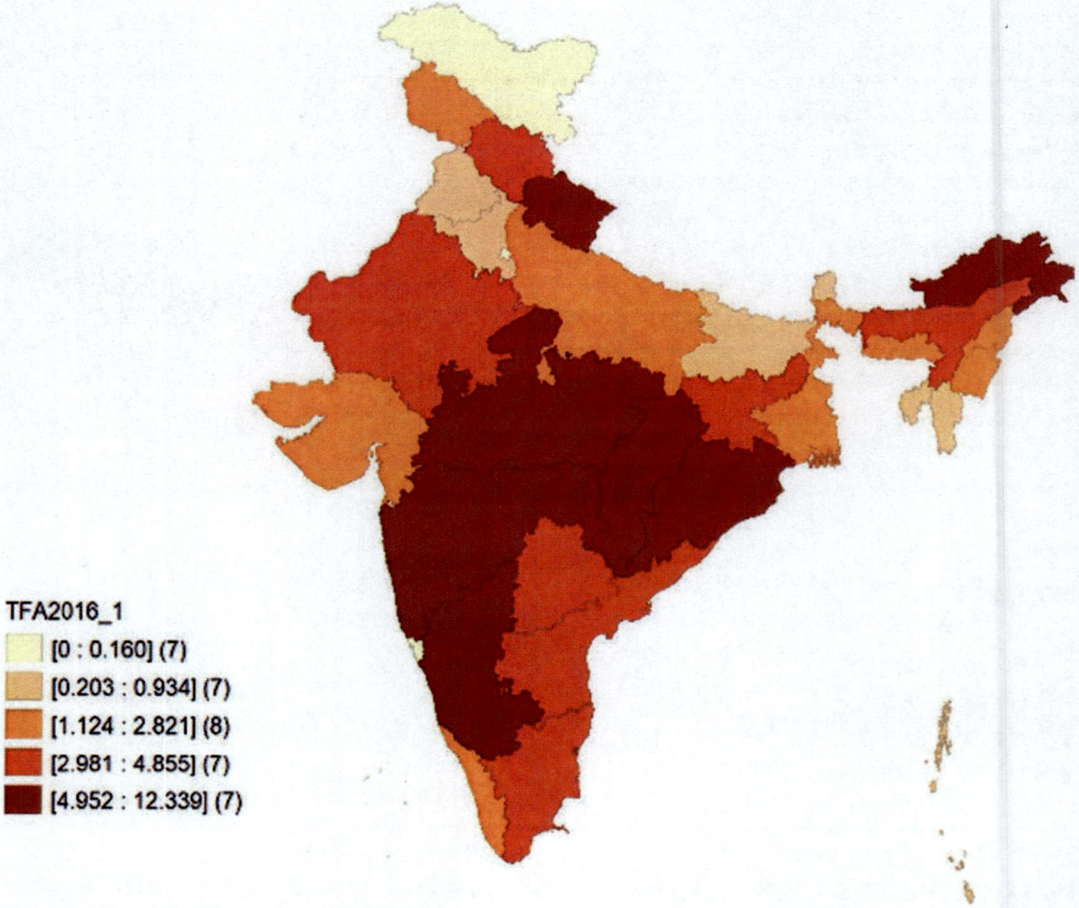

Fig. 7.5 Total Forest Area in India by States (km^2, %): 2016. Source: Compendium of Environmental Statistics, 2017, New Delhi

opined the over-exploited areas were mostly concentrated in parts of Punjab, Haryana, Delhi, and Western Uttar Pradesh where withdrawal exceeded recharge of aquifers. In addition, Rajasthan and Gujarat faced a grim situation due to arid climate. Further, parts of Karnataka, Andhra Pradesh, Telangana, and Tamil Nadu showed depleted groundwater due to inherent properties of crystalline aquifers.

Figures 7.8 and 7.9 show the net annual availability of groundwater resources in Indian states during 2013 and 2017 and the map shows five categories: very high, high, medium, low, and very low. Both maps show a similar pattern with a wedge that cuts through the middle of the western state of Maharashtra, Madhya Pradesh, Punjab, Uttar Pradesh, Bihar, West Bengal, and Assam. These are the states with a very high availability of groundwater. In the high category are the states of Gujarat, Tamil Nadu, Andhra Pradesh, and Odisha. The medium category consists of the states of Rajasthan, Haryana, Kerala, Arunachal Pradesh, Meghalaya, and Tripura. The states with less availability of groundwater are Delhi and Himachal Pradesh in the low category and Goa, Sikkim, and Mizoram in the very low category states.

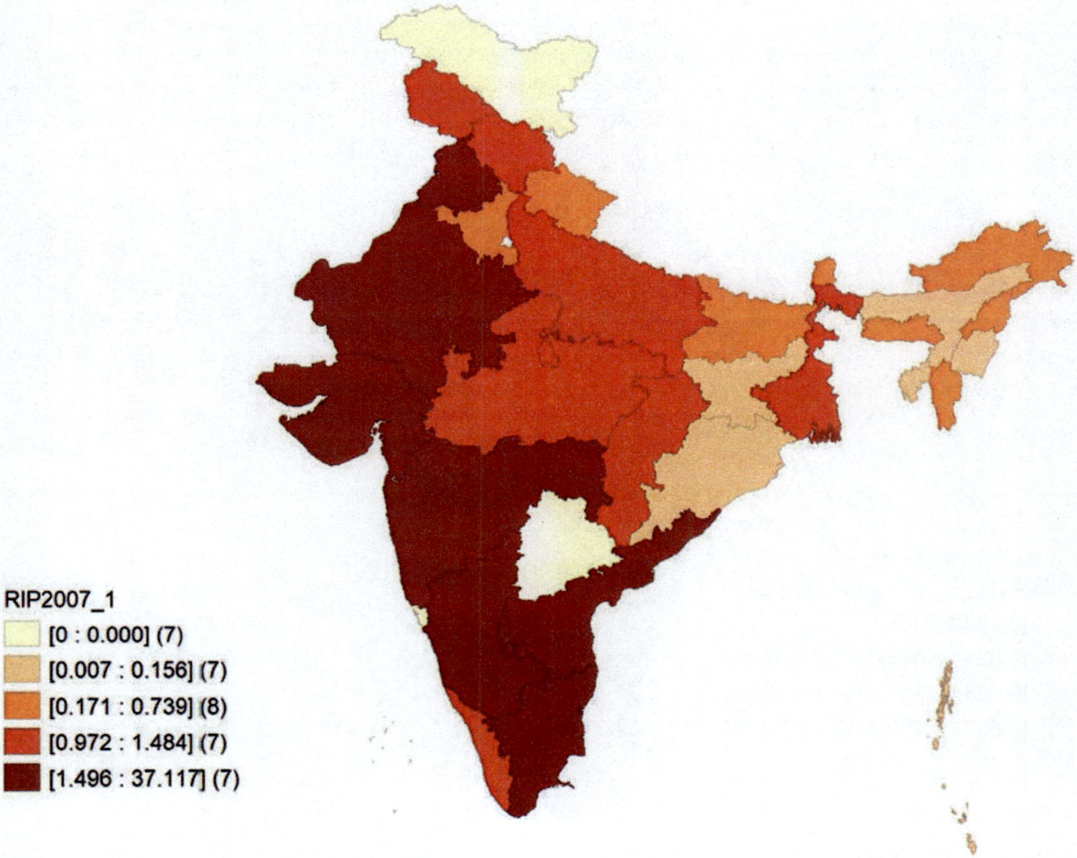

Fig. 7.6 Total Renewable Installed Power in India by States (MW, %): 2007. Source: Compendium of Environmental Statistics, 2008, New Delhi

7.8 Conclusion

This chapter examines three themes: first, the relationship between natural resources, development, and sustainable development; second, alternative methods of natural resource decision-making; and, third, visualization of natural resource distribution in India. Natural resources are important for economic development as it provides agricultural output, input for production process, sources of foreign exchange, and recreational amenity. It provides several kinds of ecological services, such as soil regeneration, recycling of nutrients, and maintains hydrological cycle. Natural resources are of two kinds: stocks and flows. Stocks are fixed in supply, are nonrenewable, whereas flows are renewable resources, and are available perennially. The economic and environmental systems are interdependent and are characterized by stability and resilience properties. Stability implies the resistance of a system from any departure from an equilibrium condition of steady-state and resilience implies the tendency of a system to retain its organization structure following perturbation. For a long time, economists believed physical and human capital were the two important factors of production. In addition, a third factor has become prominent and is called natural capital. It consists of the world's stock of natural wealth like tree, air, water, soil, and aquifers, among others. Thus, the total capital stock of an economy consists of human, physical, and natural capital. Auty (2007) examined the economic performance of resource, abundant economies, and opined economic growth is inversely proportional to the share of natural resource rents in the

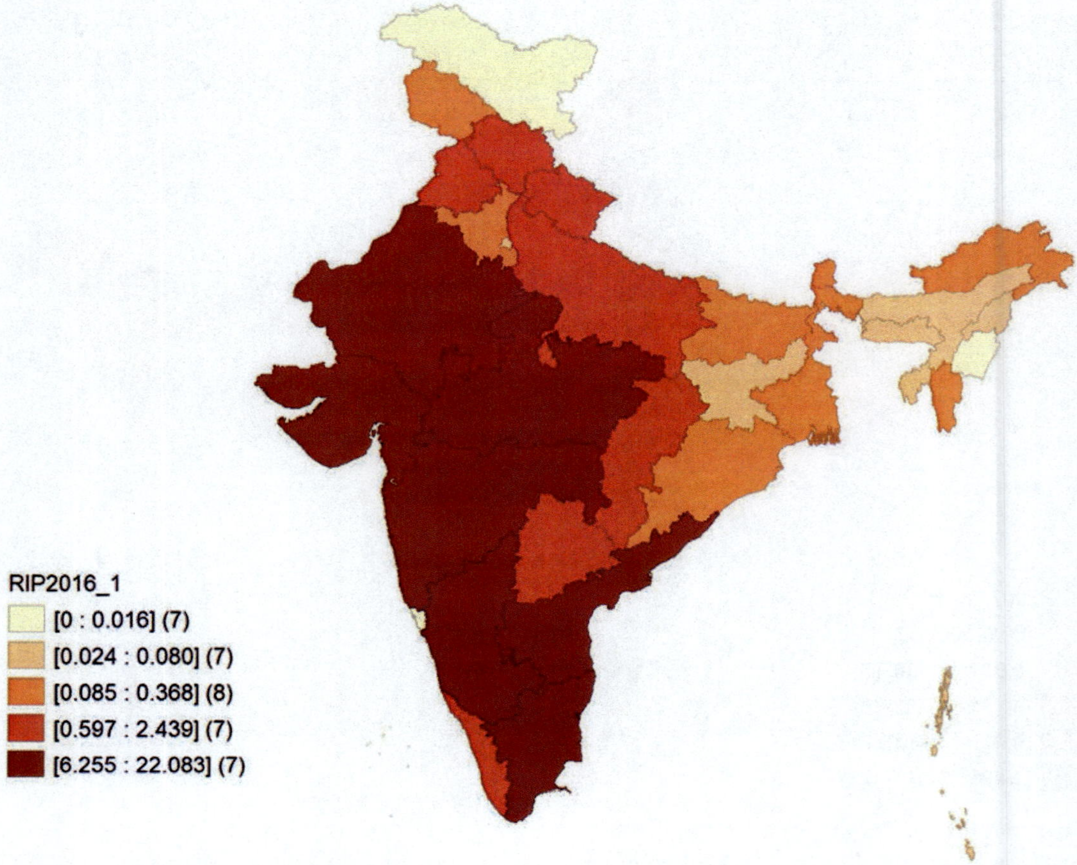

Fig. 7.7 Total Renewable Installed Power in India by States (MW, %): 2016. Source: Compendium of Environmental Statistics, 2017, New Delhi

development process. Debilitating performance of mineral- and oil-rich economies provide the evidence. In addition, natural capital is regenerative but also depletes and deteriorates when overused. Thus, utilization of natural capital in the production process, and its depletion and degradation lead to depletion, and then the GDP must be adjusted. Analogously, resource analysts provided evidence for the environment Kuznets's curve (EKC) that states environmental degradation will increase with development, reach a maximum, and then decline as per capita income continues to rise. Such an increasing trend of environmental degradation leads to the issue of sustainability.

Sustainable development is described as "development that meets the needs of the present without compromising the ability of future generations to meet their own needs." Two views provide alternative paths for sustainable development: weak and strong sustainability. The former contends that there is no inherent difference among the three types of stocks (human, produced, and natural capital), and if the depletion of natural capital is substituted by human and produced capital, keeping the aggregate stock intact or increasing there is weak sustainability. Alternatively, proponents of strong sustainability reckon that human and manmade capital cannot substitute natural capital. They purport some form of natural capital, such as biodiversity, hydrological cycle, forest reserve, unique environment, and natural habitat, are essential for human welfare. This implies the non-substitutable natural capital needs to be preserved and not depleted and is called strong sustainability. In

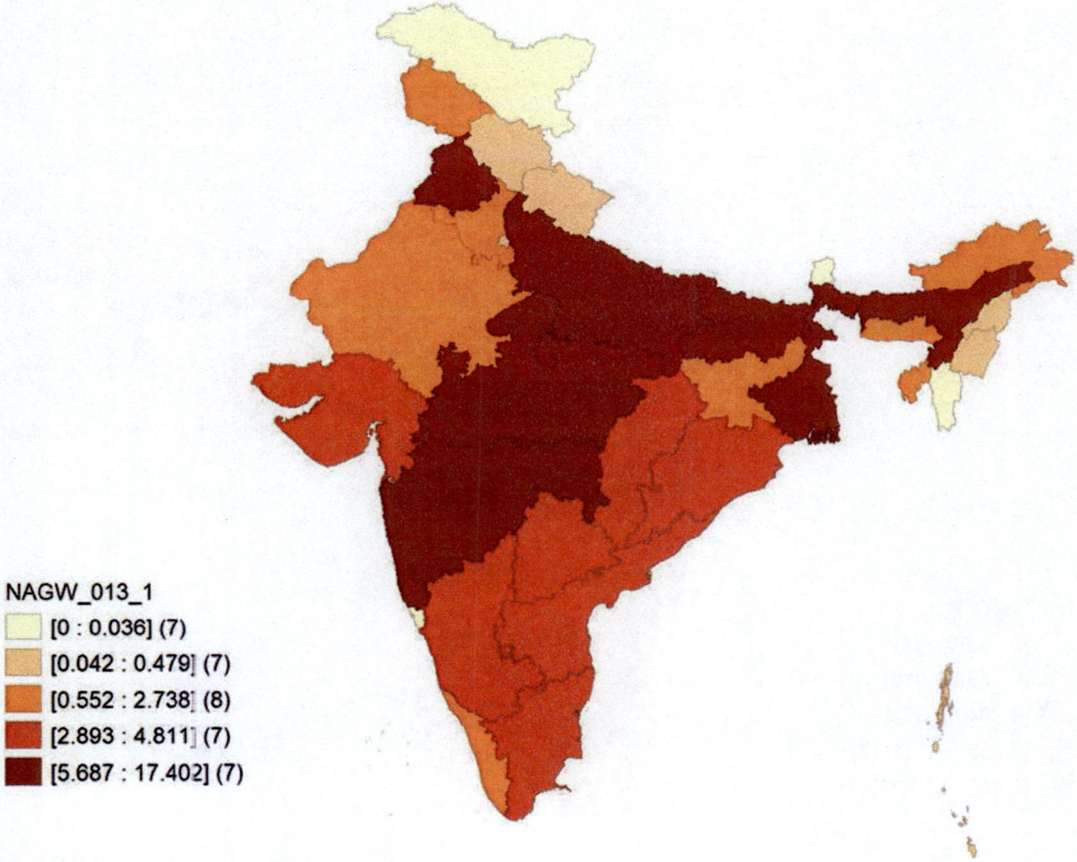

Fig. 7.8 Net Annual Ground Water Availability in India by States (Billion Cubic Meter/Year, %): 2013. Source: Environmental Statistics India, 2019, New Delhi

addition, it is essential to operationalize sustainable development in terms of regional sustainable development. RSD implies attaining social welfare for regional population both for the current and future generations subject to the attainment of sustainability at the global scale.

The second theme is the quantitative measurement of natural resource decision-making. There are four methods for natural resource evaluation: cost-benefit analysis, environmental impact assessment, multi-criteria analysis, and natural resource accounting. Cost-benefit analysis is a development analysis tool for evaluating public sector projects in monetary terms. Environment impact analysis is the minimization of the probability of potential unforeseen consequences (economic, environmental, and social impacts) associated with development projects. Multi-criteria analysis is a management science tool of goal maximization subject to multiple constraints. The method utilizes data that are quantitative or qualitatively weighted, and scores to reflect their importance according to a single or multiple set of objectives. It is also called map overlay analysis in spatial analysis. In addition, natural resource accounting is a balance sheet approach to account for the natural resource assets, its consumption, depletion, and regeneration.

Third, regional variations of natural resource indicators, such as forest cover, renewable installed power, and groundwater availability, show interesting trends. The forest coverage is highest in poorest parts of Northeastern India and richest states of western, southern, and eastern India. The rich states utilize forest products as

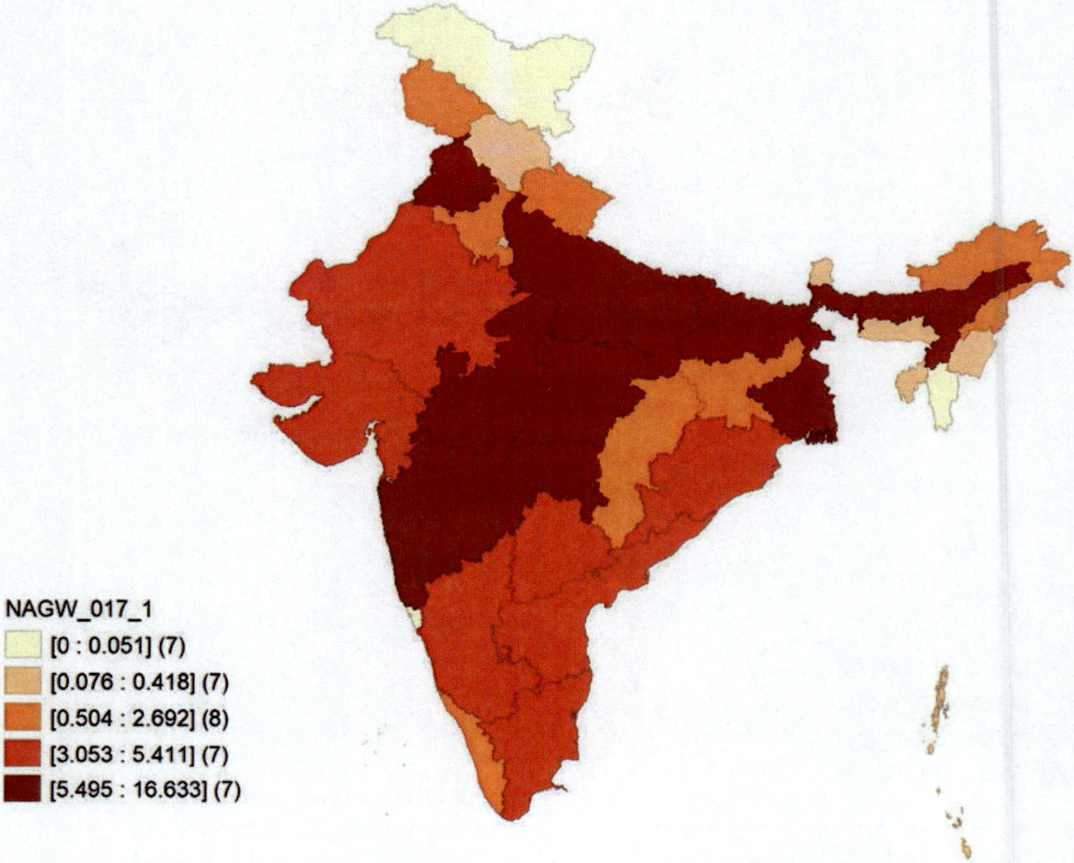

Fig. 7.9 Net Annual Ground Water Availability in India by States (Billion Cubic Meter/Year, %): 2017. Source: Environmental Statistics India, 2020, New Delhi

inputs for growth, while the poor and especially the tribal population's livelihood are dependent on the forest economy in Northeast and Central states. Likewise, renewable installed power shows an east-west contrast with high percentage of renewable installed power in Western India. In addition, the distribution of groundwater resources shows a wedge that cuts through the middle of the western, central, and eastern states. These are the states with very high availability of groundwater.

References

Alagh YK, Pangare G, Gujja B (eds) (2006) Interlinking of rivers in India; overview and Ken-Beta Link. Academic Foundation, New Delhi

Allen CR, Holling CS (2008) Discontinuities in ecosystems and other complex systems. Columbia University Press, New York

Amacher GS, Ollikainen M, Koskela E (2009) Economics of forest resources. The MIT Press, Cambridge, MA

Atkinson G, Gundimeda H (2005) Accounting for India's Forest Wealth. Ecol Econ 59:462–476

Auty RM (2007) Natural resources, capital accumulation, and the resource curse. Ecol Econ 61:627–634

Auty RM (ed) (2001) Resource abundance and economic development, UNU/WIDER studies in development economics. Oxford University Press, Oxford

Auty RM (1993) Sustaining development in mineral economies: the resource curse thesis. Routledge, New York

Auty RM, Furlonge H (2018) The rent curse: policy choice, and economic development. Oxford University Press, Oxford

Barbier EB (2005) Natural resources and economic development Cambridge University Press: Cambridge.

Brundtland Commission (1987) Our common future: the world commission on environment and development. Oxford University Press, New York

Burton I, Kates RW (1965) Readings in resource management and conservation. University of Chicago Press, Chicago

Carmignani F, Chowdhury A (2012) The geographical dimensions of the development effects of natural resources. Environ Resource Econ 52:479–498

Central Statistical Office (2018) Compendium of Environment Statistics, Ministry of Planning and Program Implementation, Government of India, New Delhi

Central Statistical Office (1997) Compendium of Environment Statistics, Ministry of Planning and Program Implementation, Government of India, New Delhi

Chatterjee SK, Bhattacharjee I, Chandra G (2010) Water quality assessment near an industrial site of Damodar River. India Environ Monit Assess 161:177–189

Chopra K, Dayal V (eds) (2009) Handbook of environmental economics in India. Oxford University Press, New Delhi

Chopra K (2006) The feasibility report of the ken-betwa link project: an analysis of assumptions and methodology. In: Alagh YK, Pangare G, Gujja B (eds) Interlinking of rivers in India: Overview and Ken-Betwa link. Academic Foundation, New Delhi

Cumberland J (1966) A Regional Interindustry model for analysis of development objectives. Papers Reg Sci Assoc 17:65–94

Cutter SL, Renwick WH (2004) Exploitation, conservation and preservation: a geographic perspective on natural resource use. John Wiley and Sons, Hoboken, NJ

Daly HE, Farley J (2011) Ecological economics: principles and applications, 2nd edn. Island Press, Washington

Dasgupta Partha Committee Report (2013) Green national accounts in India: a framework, Ministry of Statistics and Program Implementation, Government of India, New Delhi.

Dasgupta P (2010a) Nature's role in sustaining economic development. Philos Trans R Soc B 365:5–11

Dasgupta P (2010b) The concept of natural capital, Inter Academy Panel Biodiversity Conference on Integrating Ecosystem Services into Biodiversity Management, Royal Society, January 13–14

Dasgupta P, Maler KG (eds) (2004) The economics of non-convex ecosystems. Kluwer Academic Publishers, New York

Dayal V (2009) Recent development in natural resource accounting in India. In: Chopra K, Dayal V (eds) Oxford handbook of environmental economics. Oxford University Press, New Delhi, pp 121–136

Edward-Jones G, Davies B, Hussain S (2000) Ecological economics: an introduction. Blackwell Publishing, Malden, New Delhi

El Serafy S (1989) The proper calculation of income from depletable natural resources. In: Ahmad YJ, El Serafy S, Lutz E (eds) Environmental accounting for sustainable development. World Bank, Washington, D.C

Goodstein ES (2005) Economics and the Environment, 4th edn. John Wiley and Sons, Hoboken, NJ

Gunderson LH, Holling CS (2002) Panarchy: Understanding Transformations in Human and Natural Systems. Island Press, Washington, DC

Government of India (2019) Dynamic Groundwater Resources of India, 2017, Department of Water Resources, Central Groundwater Board, Faridabad

Government of India (2016) Compendium of Environment Statistics India, Ministry of Statistics and Program Implementation, Central Statistical Organization, New Delhi

Gundimeda H (2003) Integrated forest resources into the system of national accounts in Maharashtra, India. In: Perrings C, Vincent JR (eds) Natural resource accounting and economic development: theory and practice. Edward Elgar, Northampton, MA, pp 180–193

Gundimeda H (2001) Integrated environmental and economic accounting: an application to the forest resources in India. Environ Resource Econ 19:73–95

Gundimeda H (1998) Forest resource accounting: preliminary estimates for the state of Maharashtra. Dev Policy Rev 16:131–151

Gundimeda H, Sukhdev P, Sinha RK, Sanyal S (2006) Natural Resource Accounting for Indian States: Illustrating the case of forest resources. Ecol Econ 61:635–649

Gundimeda H, Sanyal S, Sinha R, Sukhdev P, (2005) The Value of Timber, Carbon, Fuelwood, and Non-Timber Forest Products in India's Forests. Monograph 1 for the Green Accounting for Indian States Project. Teri Press: Tata Energy and Resources Institute, New Delhi

Hanley N (2001) Cost-benefit analysis and environmental policy making. Environ Plan C 19(1):103–118

Heywood I, Cornelius S, Carver S (2006) An introduction to geographical information systems, 3rd edn. Person Prentice Hall, New York

Humphreys M, Sachs JD, Stiglitz JE (2007) Introduction: What is the problem with natural resource wealth? In: Humphreys M, Sachs JD, Stiglitz JE (eds) Escaping the resource case. Columbia University Press, New York, pp 1–20

Isard W, Bassett K, Choguill C, Furtado J, Izumita R, Kissin J, Romanoff E, Seyfarth R, Tatlock R (1968) On the Linkage of Socioeconomic and Ecological systems. Papers Reg Sci Assoc 21(1):79–99

Kadekoki GK (2004) Approaches to natural resource accounting in the Indian context. In: Kadekodi GK (ed) Environmental economics in practice: case studies from India. Oxford University Press, New Delhi, pp 332–372

Khoshoo TN (1986) Environmental priorities in India and sustainable development, Presidential Address, 73rd Session, January 3-7, Indian Science Congress, New Delhi.

Keuning JS, Dalen JV, Haan MD (1999) The Netherlands' NAME: presentation, usage and future extensions. Struct Chang Econ Dyn 10:15–37

Kumar P (2012) Measuring natural capital: accounting of inland wetland ecosystems from selected states of India. Econ Pol Wkly 47(22):77–84

Lange GM, Wodon Q, Carey K (2018) The changing wealth of nations: building a sustainable future. International Bank for Reconstruction and Development World Bank, Washington, DC

Little IMD, Mirrlees JA (1974) Project appraisal and planning for developing countries. Oxford and IBH Publishing Company, New Delhi

Mali KP, Singh K, Kotwal PC, Omprakash MD (2016) Strengthening forest resources valuation and accounting system: a case of forest resources in Kerala, India. In: Ghosh N, Mukhopadhyay P, Shah A (eds) Nature, economy and society: understanding the linkages. Springer, New Delhi, pp 87–112

Mahendrarajah S, Jakeman AJ, McAleer M (1999) Modeling change in integrated economic and environmental systems. John Wiley and Sons Limited, New York

Malczewski J (1999) GIS and multicriteria decision analysis. John Wiley and Sons Inc., New York

Managi S, Kumar P (2018) Inclusive wealth report 2018: measuring progress towards sustainability. Taylor, and Francis Group, London

Markandaya A, Murty MN (2004) Cost-benefit analysis of cleaning the ganges: some emerging environment and development issues. Environ Dev Econ 9(1):61–81

Markandaya A, Murty MN (2000) A cost-benefit analysis of the ganga action plan. Oxford University Press, New Delhi

Meadows DH, Randers J, Meadows DL, Behrens WW (1972) The limits to growth. Universe Books, New York

Miller J (1990) A Green GNP: Taking the Environment into Account, Dollars and Sense, November, pp. 6–8 and 22

Miller JH, Page SE (2007) Complex adaptive systems: an introduction to computational models of social life. Princeton University Press, Princeton

Mitchel B (1979) Geography and resource analysis. Longman, New York

Nandimath OV (2009) Oxford handbook of environmental decision-making in India. Oxford University Press, New Delhi

Murty MN, Gulati SC (2006) natural resource accounts for air and water pollution: case studies of Andhra Pradesh and Himachal Pradesh States of India. Institute of Economic Growth, New Delhi

Nijkamp P, Bergh CJM, Soeteman FJ (1991a) Regional Sustainable Development and Natural Resource Use, Proceedings of the World Bank Annual Conference on Development Economics 1990, International Bank for Reconstruction and Development, World Bank

Nijkamp P, Laschuit P, Soeteman F (1991b) Sustainable development in a regional system, Faculteit der Economische Wetenschappen en Econometrie Research Memorandum 1991–93, Vrije Universiteit: Amsterdam

Norberg J, Cumming GS (2008) Complexity theory for a sustainable future. Columbia University Press, New York

Pattanayak A, Srivastava KK, Sinha A (2020) Concept Paper on Natural Resource Accounting in India - An Initiative of GASAB, Office of the Comptroller and Auditor General of India, New Delhi

Perloff HS, Wingo L (1970) Natural resource endowment and regional economic growth. In: Burton I, Kates RW (eds) Readings in resource management and conservation. The University of Chicago Press, Chicago

Proops J, Safonov P (eds) (2004) Modelling in ecological economics. Edward Elgar, Northampton, MA

Rees J (1985) Natural resources: allocation, economics, and policy, 2nd edn. Routledge, New York

Repetto R, Magrath W, Wells M, Beer C, Rossini F (1989) Wasting assets: natural resources in the national income accounts. World Resources Institute, Washington, DC

Ruth M (2018) Advanced introduction to ecological economics. Edward Elgar Publishing, Northampton: MA

Ruth M (1993) Integrating economics, ecology and thermodynamics. Kluwer Academic Publishers, Boston

Saebo HV (1994) Natural Resource Accounting: The Norwegian Approach, Department of Economic Statistics, Division for Resource Accounts and Environment, Paper Presented at the UNEP/ECE/UNSTAT Workshop on Environmental and Natural Resource Accounting in Modra-Harmonia, Slovakia, 21–23 March, pp. 1–17

Sachs JD (2015) The age of sustainable development. Columbia University Press, New York

SenGupta R (2001) Ecology and economics: an approach to sustainable development. Oxford India Paperbacks, Oxford University Press, New Delhi

Shafik N (1994) Economic development and environmental quality: an econometric analysis. Oxford Econ Papers 64:757–773

Shivaprasad Sharma SV, Roy PS, Chakravarthi V, Srinivasa RG (2018) Flood risk assessment using multi-criteria analysis: a case study from Kopili River Basin, Assam, India, Geomatics. Nat Hazards Risk 9(1):79–93

Sinclair AJ, Diduck AP (2000) Public involvement in environment impact assessment: a case study of hydro development in Kullu District, Himachal Pradesh. India Impact Assess Project Appraisal 18(1):63–75

Sinha N, Priyanka N, Joshi PK (2016) Using spatial multi-criteria analysis and ranking tool (SMART) in earthquake risk assessment: a case study of Delhi Region, India Geomatics. Nat Hazards Risk 7(2):680–701

Sreebha S, Padmalal D (2011) Environmental impact assessment of sand mining from the small catchment rivers in the Southwestern coast of India: a case study. Environ Manag 47:130–140

System of National Accounts (SNA) (1968) A system of national accounts. Department of Economic and Social Affairs, United Nations, New York

Tacconi L (2012) Redefining payment for environmental services. Ecol Econ 73:29–36

Thampapillai DJ, Ruth M (2019) Environmental economics: concepts, methods and policies. Routledge, New York

The Energy and Resource Institute (TERI) (2006) Accounting for unsustainable mineral extraction in Madhya Pradesh and West Bengal. TERI, New Delhi

UNU-IHDP and UNEP (2018) Inclusive Wealth Report (2018) Measuring sustainability and well-being. Cambridge University Press, Cambridge

UNU-IHDP and UNEP (2012) Inclusive Wealth Report (2012) Measuring progress toward sustainability. Cambridge University Press, Cambridge

United Nations (2000) Integrated Environmental and Economic Accounting: An Operational Manual, Studies in Methods Series F, No. 78. United Nations Publications: New York

Vadrevu K, Eaturu A, Badarinath KVS (2010) Fire risk evaluation using multicriteria analysis-a case study. Environ Monit Assess 166:223–239

World Bank (2018) The Changing Wealth of Nations 2018: building a sustainable future. The International Bank for Reconstruction and Development, Washington, DC

Part III

Unpacking Problems

Check for updates

8

Rethinking Resettlement as a Development Opportunity: Need for Good Practices

Vinita Mathur and Gaurav Sikka

Abstract

The spatial displacement of people due to development projects has become a common phenomenon. This process of displacement has adverse socio-economic impacts on the affected people. It will not be an overstatement that these development projects in many cases have rendered a climate of deprivation and impoverishment. Therefore, a proper resettlement approach is the need of the hour, which also provides a development opportunity for the affected people. In this context, good practices in resettlement planning hold key to positive outcomes and proper management of consequences due to displacement. This chapter presents some of the "good practices" in relation to the impact assessment for large development projects, and in resettlement and compensation planning. However, as one size does not fit all so what may be a "good practice" in one circumstance might not be in another. Therefore, there may be more than one "best practices" for any resettlement project. In fact, best practices can be treated as a tool for guiding proper resettlement and achieving the aim of inclusive development. Moreover, the resettlement process needs to be considered as a development opportunity.

Keywords

Development · Displacement · Good Practices · Resettlement

8.1 Introduction

In the simplest sense, development should be raising the ability of all to realize their truest potential without any fear or obstacle. However, the development strategy adopted in post-independent India has involved the displacement of millions and created a gamut of deprivation and impoverished poor. In its quest for development, India built mega-projects like multipurpose dams, industrial projects, etc. The cost of this development was justified as being in the interest of the common good. It was evident that there could be no *vikas* (development) without *vinash* (destruction).

The spatial displacement of people due to development projects became a common phenomenon. However, no reliable official statistics on displacement and resettlement are available. This process of displacement has adverse socio-economic impacts on the affected people. It will not be an overstatement that these development

V. Mathur
Department of Geography, Dyal Singh College, University of Delhi, New Delhi, India

G. Sikka (✉)
Department of Geography, L.N. Mithila University, Darbhanga, India

B. Thakur et al. (eds.), *Resource Management, Sustainable Development and Governance*, Sustainable Development Goals Series, https://doi.org/10.1007/978-3-030-85839-1_8

projects in many cases have rendered a climate of deprivation and impoverishment. Cernea and Mathur (2008) add that impoverishment is the fundamental threat looming in every case of development-induced displacement and resettlement. Although the risks of impoverishment are known in advance, they are not effectively prevented, arrested, or mitigated (Cernea and Mathur 2008). The affected communities have undergone risks pertaining to landlessness, loss of livelihood, food insecurity, loss of common property resources, and cultural transformations (Cernea 1996, 1997). India's own record of failed resettlement and rehabilitation has been documented, among others, by Fernandes (2008) and Mathur and Marsden (1998). All this has manifested in aggravated pauperization of the already poor, by these development projects. Therefore, a proper resettlement approach is the need of the hour. A resettlement is required which is a development opportunity for the affected people. This is needed so that the displacement must not end in tragedy. It has been found that not enough is being done to channel resources for reconstructing the livelihood of the affected population. This is often due to lack of imagination in using the resources available, absence of guidelines, half-baked resettlement plans, and their improper implementation.

In this context, good practices in resettlement planning hold key to positive outcomes and proper management of consequences, arising out of displacement. This chapter argues in its initial section on sharing benefits with the affected community and highlights some of the best practices of benefit sharing adopted globally. Although large development projects are being justified on their developmental outcomes, the displaced are seldom their primary beneficiaries. As a matter of fact, much of the opposition to the projects is because, despite the sacrifices the affected people make for the project, they do not see any benefits coming their way from them. Therefore, sharing project-generated benefits could be an effective approach (Mathur 2009).

This chapter presents some of the global and local good practices in relation to the impact assessment for large development projects, and in resettlement and compensation planning. However, as one size does not fit all, what may be a "good practice" in one circumstance might not be so in another. Therefore, there is more than one "best practice" for any resettlement project. In fact, best practices can be treated as a tool for guiding proper resettlement and achieving the aim of inclusive development. Moreover, the resettlement process needs to be considered as a development opportunity. Before proceeding further, it is necessary to understand key concepts on spatial displacement and resettlement process through the following discussion.

Box 8.1 Sustainable Development Goals

Development projects most often create displacement and have unintentional but serious socioeconomic consequences for displaced people. Because of the complexities involved, these situations require knowledge, policies, and practices to manage it successfully, particularly, in the case of development projects. Also, for long, not a single indicator of the 2030 agenda for sustainable development mentioned a solution to the problems of refugees or displaced people. This gap has since been addressed and the United Nations (various commissions) have begun to consider refugees and displaced individuals as part of the SDG agenda. In a world, increasingly shaped by human actions resulting often in climate change, poverty, and conflicts SDGs cannot be achieved without taking into consideration the rights and needs of displaced people whether in India, China, the US, Brazil, or just about anywhere. In this context, the authors discuss "good practices" in some places across the developing and developed world. The authors also synthesize the conceptual idea that real development or sustainable development may only happen when resettlement policies are inclusive of displaced people and their needs.

8.2 Key Concepts

Displacement: When a certain population is forced out of their original homes and made to inhabit another area. Displacement is a process that uproots or forces to uproot the communities and individuals out of their homelands. The term "displaced persons" was coined by Russian-American Sociologist Eugene M. Kulischer in his book "The Displacement of Population" in Europe (1943). According to Terminski (2013), "displacement can be understood in two different ways: (1) eviction of people from their homeland without adequate compensation, guarantees or mechanism of social support; (2) the initial phase of a resettlement process associated with the physical relocation of people from their homes." Dams are a poignant case of displacement because they destroy the possibility of returning to the original home since the home itself no longer exists.

Compensation: Compensation can be understood as packages in cash or other forms, for persons directly or indirectly affected by the development interventions, as reparation for their losses, that includes their movable and immobile assets, livelihood, common property resource, and habitat (Hemadri et al. 2000). Compensation is what the owner(s) of a property forcefully acquired by the government receives in lieu of the property so acquired (Escudero 1988).

Resettlement: Resettlement refers to physical, preplanned relocation, combined with appropriate support mechanisms, including social support, in the new location. According to Robert Chambers, "resettlement is characterized by two main features: A movement of population; and an element of planning and control." According to the Encyclopaedia of World Environmental History, resettlement may be defined as "the process through which populations displaced from their habitat and/or economic activities relocated to another site and re-establish their productive activities, services, and community life." This definition strongly emphasizes that resettlement is a combination of physical relocation (displacement) with subsequent attempts to restore the displaced people's livelihood in the new place, it considers resettlement a holistic exercise, encompassing several aspects of life (Terminski 2013). Resettlement is a process of developing resettlement sites and relocation of the affected people at those sites. It also includes all the facilities as per the terms of compensation and resettlement guidelines.

Rehabilitation: Long-term support provided to resettlers in the form of packages, process, and policy provided in addition to the compensation package. Rehabilitation is needed to be done to ensure that the resettlers and their next generations can have a sustainable and better standard of living. Mostly, the second generation faces the effects of impoverishments or the benefits of rehabilitation over time. Figure 8.1 shows the overview of various dimensions with compensation being the immediate support provided in terms of cash, jobs, or any other suitable intervention. Resettlement is a physical relocation of people with a provision of immediate social support at the relocation sites. The dimension of rehabilitation is vast as shown here which has a provision of long-term support not only for the present generation but successive generations of the affected people.

8.3 Good Practices of Benefit Sharing

Projects and the project-affected persons see one another as separate groups and engaging in conflict. Both the project as well as the people are the losers in such circumstances. Projects lose due to people's protests and legal battles, and people lose because their concerns and requirements are ignored by the project authorities. A matter of fact is that people are important for the projects as they provide their lands. Land is an important factor of production; therefore, people who contribute their lands to big projects need to be recognized as the "stakeholders" in the project. People who give up their land, house, or livelihood base should be the first to benefit from the project (Council for Social Development 2008; Cernea and Mathur 2008; World Commission on Dams [WCD] 2000). Therefore,

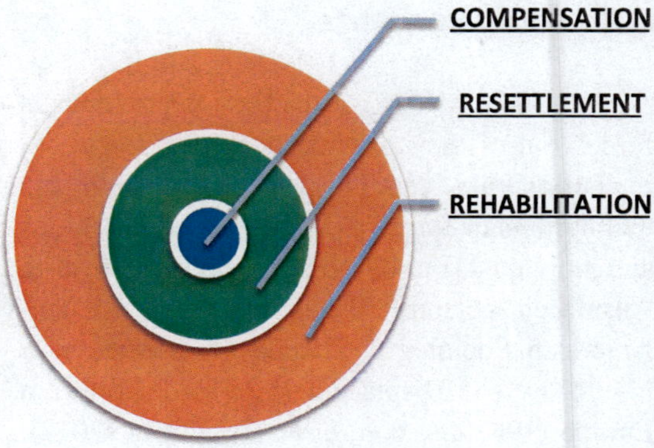

Fig. 8.1 Different dimensions of compensation, resettlement, and rehabilitation

involving affected people in development projects will reduce conflict situations and ensure a more equitable development process.

In the light of the above discussion, we have listed below some of the good practices followed by many countries in sharing the benefits accrued, with the affected communities.

8.3.1 Japan

When the three Jintsu-Gawa dams were built, the Japanese Government did not apply the expropriation law and instead leased the land required for the reservoirs from the owners. The payment of land lease was structured into twin financial transfers: a) first, onetime payment to the landowners leasing land to the project companies so that these farmers could develop alternative livelihood for themselves, b) second, a provision of regular rent payments for the leased land, paid to the farmers continuously for the entire life of the project.

In this practice the leased land, although now deep under waters of the reservoir, remains nevertheless, a continuous source of income for the affected persons and their subsequent generations.

8.3.2 Canada

Hydro Quebec is Canada's major power utility and when engaging in the James Bay project, faced resistance by the tribal Cree Indian population and indigenous Inuit people. The Canadian government adopted a strategy of partnering with the local indigenous communities. The project authority decided to enter into agreements with the affected community for equity sharing in the hydropower projects. In this agreement, the local indigenous communities will be considered as the investors in the project as they are contributing their lands. This benefit sharing is for the long-term and proportionate with their land share in the project. The equity was offered in addition to the compensation, paid to the local population for the loss of land and for assisting and sacrificing their livelihood activities in the new circumstances.

8.3.3 Brazil

Brazil has initiated many large hydro projects, comparable to India and China. The multiplication of large dam projects has led to large displacements. Due to the absence of any guidelines on resettlement and rehabilitation, the conditions were dismal. Affected people anarchically settled into slums around big towns. The Brazilian government took a politically important decision to redress this deplorable situation by amending the constitution. In 1988, Brazil introduced in its constitution a new principle of reinvesting a percentage of royalties from the hydropower in the resettlement areas. Subsequently, laws were enacted that defined the entitlements, specific amounts for royalties, and procedures for assuring regular allocations. The legislations explain the difference between royalties and financial compensations: the latter to be paid to the municipalities with area inundated by the reservoir. It also stipulated those royalties are to be paid throughout the lifeline of the power plants, to help provide for the long-term economic sustainability of the affected communities (Egre et al. 2008). Gomide (2004) in an assessment of this program mentions that annually, US$400 million is being paid as compensation and royalties through all the hydro projects in Brazil.

8.3.4 China

China's government in the 1990s announced a change from a resettlement policy primarily based on the physical transfer of people to a developmental resettlement policy. This was evident in the special policy guidelines for the Three Gorges Project in 1991 which focussed on "developmental resettlement." Cernea (2008) classifies China's innovative approaches on displacement and resettlement into four clusters as follows: (1) Measures to increase the restitutionary ex-ante financing for lost lands and assets were introduced, to upgrade the nation's compensation system for the affected communities, (2) Measures to supplement the ex-ante compensation with ex-post financing via targeted investments in area development. Under these measures, certain amount of the economic rent or other project benefits are transferred to the state agencies and their schemes, that are addressing the needs of the project affected persons, (3) Apart from the two measures mentioned above, the resettlers are given finance that can be used by them for consumption and/or for generating productive assets, and (4) Measures for correcting past under-compensation and compensation errors for displaced people. These measures include providing additional financing as an ex-post remedial investment.

The aim of mentioning the sharing benefits approaches was to demonstrate that sharing financial benefits with the affected communities is not utopian and idealistic. Moreover, it is a realistic and feasible practice, which is implemented by many countries. In the following section, we strive to highlight some of the cases where gender-inclusive attempts have been made in resettlement and rehabilitation.

8.4 Identifying the "Engendered" Good Practices

As gender has been virtually invisible in India's Resettlement and Rehabilitation policy formulation and its implementation (Thukral 1996; Mehta 2009), it is difficult to identify any best practice on gender-just distribution of benefits of its large dam projects. Nevertheless, some cases of gender inclusive attempts are produced here.

The first case is of some rural people who succeeded in securing land, it being one of the most important mechanisms for fighting off poverty and destitution. In Bodh Gaya, Bihar, landless laborer women protested the fact that land titles were only being distributed to men, even though they had been co-protestors for ownership rights to the land they had sown for many years. The women argued, "we are part of the struggle, so we should also get land" (Agarwal 1996). This incidence of the Bodh Gaya struggle in India stands out as an example of how women can wrest land rights for themselves in a hostile, gender-biased environment, given a movement's basic commitment to gender justice and a persistent effort to reduce gender inequities (Mehta and Srinivasan 2000).

Another instance of gender participatory consultation mechanism is that of the Shuikou project in China where women's participation was explicitly sought by implementing authorities at all levels: in the formulation of policy as well as in implementation (Operations Evaluation Department (OED) 1998). Authorities did not overlook women's substantial roles in the family's decision-making processes and ensured that they were involved in resettlement planning as well. Civil society was active in the towns of the project area with representation at the village level. This helped in identifying the needs of all the members of the community. Most laudable efforts were seen in this project. Unlike most other projects in the World where displaced persons are not adequately absorbed in the creation of new employment facilities in the command area, the workforce in several of the new factories established in this project area comprised resettled women. This was possible, primarily because the state encouraged "developmental" resettlement, which emphasized a productive base rather than "passive" compensation, and the fact that the land tenure system is based on collective ownership in China (Operations Evaluation Department (OED) 1998).

These inclusive mechanisms by the project authorities led to a marked increase in women's participation. As is commonly evident, it is the women who bear the brunt of the negative impacts of displacement and resettlement. This can be offset by an early inclusion of women into the planning and implementation process.

However, this requires both gender awareness and a commitment on the part of the planners to ensure that all members of the community have a say. Usually, implementers tend to interact with and operate via, male elites, who do not represent the community in its entirety. Often articulation of women's needs may not be in the interests of the male elites, given that addressing issues of gender justice call for a redistribution of already scarce resources, benefits, and privileges. Thus, a system where women leaders and women's networks are included in the negotiations with the state is desirable. It is important that women's associations are invested with authority, both within their communities and within wider regional and state processes. This will preempt situations where women are mere tokens in decision-making processes.

Given the deeply entrenched nature of gender inequality, one of the first things to disappear from a resettlement policy is the "women" (Shankaran, 2009). While the Gujarat Resettlement Policy only accepted "unmarried son" to be treated as a separate family for the purpose of compensation, Odisha Resettlement and Rehabilitation Policy 2006, in contrast, has taken great steps towards gender equality (Jena, 2006). It defines family in a progressive manner and mentions different categories which will be treated as a separate family for the purpose of extending rehabilitation benefits (Government of Orissa 2009).

These include a major son irrespective of marital status; an unmarried daughter/sister more than 30 years of age; a physically and mentally challenged person irrespective of age and sex; a minor orphan who has lost his/her parents; and a widow or a woman divorcee. Apart from the compensation benefits, this broad definition of families can help the affected families in getting employment in development projects as the policy mandates the project authorities to hire at least one person from the displaced families.

The Orissa R&R policy also mandates the constitution of a Rehabilitation and Periphery Development Advisory Committee (RPDAC) to ensure participation of the affected people in the design, implementation, and monitoring of R&R. The Orissa Policy states that "adequate representation will be given to women and indigenous communities in the committee" (Government of Orissa 2009). This good step ensures women's representation on the committee; however, "what is adequate" has not been defined clearly.

Another experience of Song Bong 4 Hydropower Project in Vietnam is noteworthy. The resettlement plan of the project has ensured equal participation of women at all stages. Women's voices were heard, and agreements were reached in resettlement consultations, women's role in site selection, women's contribution in design and management of infrastructure at sites were some of the pioneering features of gender inclusive approach adopted by these authorities. The intrinsic value of women's participation for women themselves and communities is reflected in the fact that women are continuing to meet on a regular basis even after their villages have been fully relocated (Asian Development Bank 2014). These meetings are avenues to discuss issues ranging from private to community matters at the resettlement sites.

Gender-related benefits in resettlement have also emerged in this project. Women have equal entitlement to compensation like joint titles in the name of both husband and wife, same individual rights have been guaranteed for households headed by a single man or a single woman, practice of paying cash compensation to both husbands and wives equally, transparently, and publicly. Besides, affected women have direct channels for grievance redressal, enjoy improved mobility, access to services like healthcare, and maximum opportunities to develop their skills and capacity.

8.5 Conclusion

Vision, wisdom, and efforts are the need of the hour for making resettlement a development opportunity. The best practices identified in this chapter are not an end, but a means to achieve real development of people by raising their abilities to their truest potential without any obstacles. The development, thus, achieved, will be inclusive, participatory, and empowering. When the resettlement is neglected and the resettlement policy is half-baked and implemented half-heartedly, the affected people are pushed into a deep crisis and end up impoverished in the process. The development projects should ensure that the resettlement policies are sensitive with the provisions of sustainable livelihood opportunities for the affected, consultation for both men and women in all the phases of the project, sharing project benefits with the affected, and an "inclusive" compensation entitlement. Lastly and most importantly, political will is required, which is necessary for investing in resettlement and rehabilitation—a process which must be considered as an opportunity for development.

References

Agarwal B (1996) A field of One's own: gender and land rights in South Asia. Cambridge University Press, Cambridge

Asian Development Bank (2014) Navigating gender inclusive resettlement: the experience of the song bong 4 hydropower project in Vietnam. Asian Development Bank, Manila

Cernea M, Mathur HM (2008) Can compensation prevent impoverishment: reforming resettlement through investments and benefit- sharing. Oxford University Press, New Delhi

Cernea MM (1996) Eight Main risks: impoverishment and social justice in resettlement. The World Bank (Environment Department) Paper, Washington, DC

Cernea MM (1997) The risks and reconstruction model for resettling displaced populations. World Dev 25:1569–1587

Cernea MM (2008) "Reforming Resettlement: Supplementing Compensation with Investments and Benefit Sharing" in India (2008) Social Development Report: Development and Displacement edited by Council for Social Development, Oxford University Press, New Delhi

Council for Social Development, India (2008) Social development report: development and displacement. Oxford University Press, New Delhi

Egre D, Roquet V, Durocher C (2008) Benefit sharing to supplement compensation in resource extractive activities: the case of multipurpose dams. In: Cernea MM, Mathur HM (eds) Can compensation prevent impoverishment? Oxford University Press, New Delhi, pp 171–190

Escudero C (1988) Involuntary resettlement in Bank-assisted projects: an introduction to legal issues. Legal Department, The World Bank, Washington, DC

Fernandes W (2008) Sixty years of development-induced displacement in India: scale, impacts and the search of alternatives, in India social development report 2008 - development and displacement. Oxford University Press, New Delhi

Gomide FLS (2004) Benefit sharing: experiences from Brazil. Presentation at the United Nations symposium on hydropower and sustainable development, Beijing, China

Government of Orissa (2009) Orissa resettlement and rehabilitation policy, 2006. Orissa Reference Manual, Bhubaneswar

Hemadri R, Mander H, Nagaraj V (2000) Dams, displacement, policy and law in India. In: Thematic review 1.3: displacement, resettlement, rehabilitation, reparation and development, contributing paper to the World Commission on Dams, World Commission on Dams, Cape Town

Jena M (2006) "Orissa: Draft Resettlement and Rehabilitation Policy", Economic and Political Weekly. 41(4):384–387

Mathur HM, Marsden D (1998) Development projects and impoverishment risk-resettling projects affected people in India. Oxford University Press, New Delhi

Mathur HM (2009) "Gender Biases in Resettlement Planning" in Displaced by Development: Confronting Marginalization and Gender Justice, edited by Lyla Mehta, New Delhi: SAGE Publications, 166–194

Mehta L (2009) The double bind: a gender analysis of forced displacement and resettlement. In: Mehta L (ed) Displaced by development: confronting margin-

alization and gender justice. SAGE Publications, New Delhi, pp 3–33
Mehta L, Srinivasan B (2000) Balancing pains and gains: a perspective paper on gender and dams. In: Adams W (ed). The social impact of large dams: equity and distributional issues. Paper prepared for the world commission on dams (WCD), WCD thematic review, social issues 1.1. World Commission on Dams, Cape Town
Operations Evaluation Department (OED) (1998) Recent Experiences with Involuntary Resettlement: China-Shuikou (And Yantan). No. 17539, World Bank, Washington, DC
Shankaran D (2009) Like a white flag? The gender dimensions of the Orissa state policy on resettlement and rehabilitation. In: Mehta L (ed) Displaced by development: confronting marginalization and gender justice. SAGE Publications, New Delhi, pp 225–245
Terminski B (2013) Development-induced displacement and resettlement: social problem and human rights issue. Research paper no. 9/2013, Geneva
Thukral EG (1996) Development, displacement and rehabilitation: locating gender. Econ Polit Wkly 31(24):1500–1503
World Commission on Dams [WCD] (2000) Dams and development: a new framework for decision making. Earthscan Publications, London

9 Will the Water Revolution Be Centralized? Investigating the "Downscale" and "Upscale" Challenges of Urban Rainwater Harvesting

Georgina Drew

Abstract

In 2003, the Centre for Science and Environment—one of India's most prominent environmental organizations—published a text urging city dwellers to take up the socially responsible act of catching rain where it falls, which is otherwise known as rainwater harvesting. The text argued that unless people are involved in urban rainwater harvesting at the household level, it would be "very difficult to meet the looming water crisis" that India confronts. Just how viable however are individual and household efforts for addressing the water crises on the horizon? This chapter takes up that question by looking at the progression in debates over urban rainwater harvesting, as well as the uptake in rainwater harvesting practices, that have taken place since the publication of the aforementioned manual. Drawing from a selection of documents and interviews, this chapter argues that several disincentives persist that either deter people from taking up the clarion call of household-level rainwater harvesting, or that prevent them from doing it altogether. This content shifts the onus of responsibility onto the centralized water system, and onto the municipal agencies charged with water management. Using a political ecology analysis that focuses on the scalar disparities of the water-power nexus, this chapter ultimately argues that urban rainwater harvesting requires enhanced centralized cooperation and capacitation to foster a viable integrated water resource management approach. At stake in this discussion is the fate of water self-sufficient Indian cities, and the viability of sustainable urban water management.

Keywords

Delhi · Political ecology · Sustainable urban water management · Urban rainwater harvesting · Water crisis

"We get a lot of rain, yet we do not have water. Why? Because we have not reflected enough on the value of the raindrop." (Centre for Science and Environment 2003, p. 3)

G. Drew (✉)
Anthropology and Development Studies, The University of Adelaide, Adelaide, SA, Australia
e-mail: georgina.drew@adelaide.edu.au

B. Thakur et al. (eds.), *Resource Management, Sustainable Development and Governance*, Sustainable Development Goals Series, https://doi.org/10.1007/978-3-030-85839-1_9

9.1 Reflecting on the Value of the Raindrop: An Introduction

In 2003, the Centre for Science and Environment—one of India's most prominent environmental organizations—published a text urging city dwellers to take up the socially responsible act of catching rain where it falls, which is otherwise known as rainwater harvesting. With the banal title, "A Water Harvesting Manual for Urban Areas," the text argued that unless people are involved in urban rainwater harvesting at the household level, it would be "very difficult to meet the looming water crisis" that India confronts (CSE 2003, p. 1). Just how viable however are individual and household efforts for addressing the urban water crises on the horizon? And what can municipalities and government organizations do to promote the upscaling of urban rainwater harvesting, be it at the household or the institutional level? This chapter takes up these questions while looking at the progression of debates over urban rainwater harvesting, as well as the uptake in rainwater harvesting practices, that have occurred in the city of New Delhi since the publication of the manual. To quickly foreshadow what is to follow, it appears that the turn-of-the-century calls for the widespread implementation of rainwater harvesting at the household level have not yet been fully realized. For reasons that will be explored, this observation shifts the onus of responsibility onto the centralized water system, and onto the municipal agencies charged with water management.

Within this discussion, the challenges of "upscaling" and "downscaling" rainwater harvesting are a central point of focus. The use of the term "upscaling" refers to the effort to expand the use of urban rainwater harvesting. The term "downscaling" reflects the need, and the difficulty, of promoting the technology's widespread implementation at the individual or household levels. Since it is the municipality and the government that is involved in promoting (and even potentially enforcing) urban rainwater harvesting at this smaller-scale level, the word "downscale" seems to fit better than the term "decentralized."[1] According to scholars, such as Elinor Ostrom and Arun Agrawal (2001), decentralization entails the democratic transfer of power over the management of resources closer to those who are most affected by management decisions (487). It is because of this definition that they state that decentralization is "synonymous with (the) redistribution of power, resources, and administrative capacities through different territorial units of a government and across local groups" (Ostrom and Agrawal 2001).

Box 9.1 Sustainable Development Goals

Rainwater harvesting is a traditional and sustainable method. This chapter presents a strong case for urban rainwater harvesting (URH) as a sustainable water resource management approach for the national capital region (NCR) of Delhi. To that end, SDG 6, 11, and 13, as well as the "land and water resources" vertical of ***Niti Aayog*** (the planning wing of the government of India) provides conceptual and policy support to the decision-making process. Indeed, in the context of burgeoning population growth and its consequent demand for urban water in NCR of Delhi, ensuring equitable access to water while focusing on the sustainable development of the region brings out issues of water governance in overcoming challenges and realizing the SDGs. The author's research which draws upon a wide selection of documents and interviews, clearly, brings out challenges of water governance, Vis-à-Vis, legal, policy, and institutional frameworks. Drew's work displays how an integrated water management strategy can contribute to household water security, but needs to be facilitated through programatic and strategic water governance efforts coordinated

[1] The Godyer Institute, for instance, uses the word "downscale" to discuss interventions that happen at a "local scale." See https://data.environment.sa.gov.au/Climate/SA-Climate-Ready/Pages/default.aspx, accessed 10 August 2017.

between the central government and the municipal bodies. In other words, while the SDGs provide an enabling conceptual and policy framework and incentivizes the implementation of RWH systems, the *Niti Aayog* can promote and support RWH by raising public awareness and developing an appreciation for conserving water resources among residents in the NCR of Delhi. They can additionally promote the uptake of rainwater harvesting by and through government institutions, including at the buildings they operate.

9.1.1 Methodological and Empirical Context

Drawing from a selection of documents, interviews, and case studies, this chapter demonstrates that several disincentives persist that deter people from taking up the clarion call of household-level rainwater harvesting. The data, evidence, and arguments highlighted in this chapter are supplemented by firsthand observations accrued over 12 years of short-term visits and long-term periods of residence in the city of New Delhi, which is the geographical region of focus. For 4 years total within that window of time, roughly from 2004 to 2016, the author served as an affiliate with several New Delhi based nongovernmental organizations and academic institutions working on issues of water rights and sustainable water resource management. Although never officially affiliated with the Centre for Science and Environment (CSE), the author's thinking on the ameliorative potential of rainwater harvesting for improving water access in Indian cities was greatly influenced by a week-long course on urban rainwater harvesting for water professionals that she took at CSE in the summer of 2009.

While exposure to a rainwater harvesting training, program was the author's "lightbulb moment," the empirical evidence serving as a focal point for the present discussion is a series of 30 interviews and conversations in New Delhi that took place over a period of 2 months in November of 2015 and October–November of 2016. The open-ended interviews specifically inquired into the promise and pitfalls of urban rainwater harvesting in India's capital city. The data from these interviews were supplemented by site visits to 25 locations mentioned by interviewees as well as by the informal and formal exchanges that took place with the people operating urban rainwater harvesting infrastructures within New Delhi, and the southern part of the city more specifically. In drawing from this content while using a political ecology analysis described below, this chapter argues that the successful uptake of urban rainwater harvesting requires enhanced centralized leadership and an increase in the construction and maintenance of municipality-operated rainwater harvesting infrastructures. At stake in this discussion is the fate of water self-sufficient Indian cities, and the viability of sustainable urban water management.

9.2 Thinking with and Researching Rainwater: Political Ecology and IWRM

The political ecology of water is a useful framework through which to examine the practice, promise, and pitfalls of urban rainwater harvesting because of the field's emphasis on the socioeconomic and scalar disparities of the water-power nexus. Numerous anthropologists and cultural geographers have made important contributions to this growing field with anthropologists such as John Donahue and Barbara Rose Johnston arguing in 1998 (1998) that power struggles are intricately intertwined with water resource management. Elaborations upon this insight have allowed several scholars to conclude that in many cases around the world the experience of water scarcity and insecurity is often manufactured by humans rather than being an outcome of inadequate water resources (Johnston 2005; Mehta 2005; Wutich and Brewis 2014).

Among the geographers operating in the political ecology of water, Erik Swynegedouw is widely cited for his efforts to string together the different economic, social, and political realms that impact water management outcomes. Explaining this approach, he writes, "Political-ecological perspectives on water suggest a close correlation between the transformations of, and in, the hydrological cycle at local, regional, and global levels on the one hand and relations of social, political, economic, and cultural power on the other…[2]"(2009, p. 56). In emphasizing the importance of this approach, he adds that "hydro-social research" usefully looks at the combined physical and social processes that impact the circulation of water in ways that attempt to transcend notions of a "modernist nature" propping up false nature-society boundaries (Swyngedouw 2009). In his perspective, and in the view of many political ecologists, nature and society are fused together in "inseparable manners" (Swyngedouw 2009; Swyngedouw 2006). This effort calls for water management studies to move beyond traditionally fragmented interdisciplinary approaches to the study of water (see also Bakker 2012).

The call for politically attuned interdisciplinarity is significant, and it is something that is increasingly demanded in studies of water resource management. As the late Ramaswamy Iyer once wrote, for instance, any "integrated" approach to water must move beyond disciplinary silos while focusing on basin-level planning rather than project-specific interventions (Iyer 2003a, p. 304). As Iyer once cautioned, while flagging the need to address power imbalances and politically motivated discourse, the language of integration should be more than a sounding board. As he explains:

> It is generally accepted that the basis for national water planning should be a *hydrological unit* (original emphasis) such as a basin or sub-basin. The national water policy says so. Thus, standard practice is to speak of *integrated* (original emphasis) planning. While these are unexceptionable ideas, our planning continues to be project-based. The basin merely provides a background for this, and the language of 'integration' is a gesture in the direction of current terminology, whereas the approach remains that of centralised planning of several large projects in certain selected locations with a sub-basin. The fact that an integrated multi-disciplinary approach is equally, if not more, consistent with micro watershed development is rarely recognised (Iyer 2003a).

In recognition of the importance of an integrated approach, the conclusion to this text merges the insights provided by the political ecologists of water with calls for integrated water resource management (also known as IWRM) to demonstrate that they are complementary and that they each point out different components of the urban rainwater harvesting challenge that needs to be comprehensively addressed. Despite the political-ecological complexities to be described in this chapter, in other words, the conclusion ultimately argues that urban rainwater harvesting is a significant component of an integrated approach to water management in Indian cities such as New Delhi. Urban rainwater harvesting initiatives are important, in other words, because they are part of the much-needed shift towards the kinds of integrated water resource management that scholars such as Bruce Mitchell (1990) argue is necessary. The IWRM quest is useful to keep in mind as we next look to the complex landscape of water management and its challenges as they manifest in contemporary New Delhi.

9.3 The Problem: A Landscape of Simultaneous Scarcity and Abundance

There are two different scenarios of New Delhi's hydrological landscape that are flagged in many discussions of the need for urban rainwater harvesting. The first scenario reflects the city's water balance as it stood in the latter years of the second decade of the twenty-first century; the second scenario offers an alternative future in which the urban hydrological imbalances have been rectified. To understand why rainwater harvesting is promoted in the urban topography of India's capital city, the following section outlines the scope of resource challenges alongside the ideas and imaginings of how New Delhi's hydrological balance might 1 day be more self-sufficient.

[2]In penning this statement, Swyngedouw self-cites a 2004 publication.

9.3.1 Scenario 1: Present Day Resource Inadequacy (Circa 2017–2018)

New Delhi is arid and densely populated, so the water that is in circulation is mostly sourced from interbasin water transfers. The result is that the city's water system is dependent on five rivers, and on the five states through which these rivers flow—Jammu and Kashmir, Himachal Pradesh, Punjab, and the state that is now known as Uttarakhand. Since the long-term availability of these waters is in question due to agricultural demands, rising populations, and the water stressing effects of warming temperatures, there are frequent political fights between the national capital and upstream states for their continued supply. The tensions between Delhi and Haryana alone can escalate to the point that some journalists are compelled to label the interstate dispute a "water war" in the making.[3] Even in the absence of a major altercation, the capital city is nonetheless perennially vulnerable to significant water stress. As Kalia (1997) observes, "Delhi's water supply system will work well if the river(s) on which it depends provide water; the state chief ministers remain docile to Delhi's rulers; and there is not sabotaging of the watercourse. Should any one thing go wrong anywhere, the whole water grid will go haywire" (78). These words are ominous given the political ecology of water resource management, which includes ongoing disputes and political battles over who has the right to the river waters upon which New Delhi depends.

At the time of this text's composition, New Delhi suffered from a freshwater supply gap that ranged between 200 and 250 million gallons per day, or MGD (Joshi 2016). Of the 900 MGD that it received, roughly 120 was sourced from the city's groundwater while 540–550 came from upstream diversions of the Yamuna River (and not the segment that flows through the city, which effectively serves as a sewage canal). Another 240 MGD came from transfers from the Upper Ganges Canal (Halder 2017). The gap in water supply versus demand is often met by private water tankers, who charge handsomely, as well as through "illegal groundwater extraction" (Joshi 2016). What is worse, the population is rapidly increasing and the rise in water demand has no end in sight. While in 2017 the population was estimated at between 18 and 25 million people, the United Nations contends that the city could reach as many as 36 million by 2030.[4]

When the city grew from 13 million to perhaps 18 million in the early years of the twenty-first century (Kumar 2014), projects, such as the Tehri dam of the Uttarakhand Himalaya, had to be completed to increase the supply from 650 MGD to 900 MGD. In the future however it is unclear what upstream projects can be added to the grid in order to increase the water supply by the same amounts (an additional 250 MGD); it is even more difficult to imagine what upstream projects might be created in order to double the supply of 900 MGD—which is ostensibly what would be needed assuming the population reaches 36 million residents, and assuming there are no major efforts to reduce demand by promoting water light lifestyles (which, incidentally, Ramaswamy Iyer widely argued is necessary since it is unrealistic to meet ever-expanding resource demands; Iyer 2003b, pp. 132–135).

---

[3]Press Trust of India. 2012. "Delhi-Haryana Water War Intensifies; Chief Ministers to Meet Today," 18 June, http://www.ndtv.com/india-news/delhi-haryana-water-war-intensifies-chief-ministers-to-meet-today-488630, accessed 27 March 2017. See also: Press Trust of India. 2016. "Delhi, Haryana Water War: Construct Your Own Canal To Carry Water, O.P. Dhankar Tells Arvind Kejriwal," 18 March, http://www.india.com/news/cities/delhi-haryana-water-war-construct-your-own-canal-to-carry-water-o-p-dhankar-tells-arvind-kejriwal-1039588/, accessed 27 March 2017.

[4]Media government reports in 2016–2017 often cited 18 million residents ("1.8 Crore," as per Joshi 2016) when discussing water demands. But reports such as the ones by the United Nations claim that perhaps as early as 2014 New Delhi had 25 million inhabitants. The same report estimated that New Delhi will have a population of 36 million by 2030. See: United Nations: Department of Economic and Social Affairs, Population Division, 2014, "Population Facts,"

http://www.un.org/en/development/desa/population/publications/pdf/popfacts/PopFacts_2014-2.pdf, accessed 2 August 2017.

The problem however is not just one of total supply; the efficient management of existing resources is also a critical challenge. Of the resources that flow through New Delhi's infrastructure, roughly 40% of the water is lost in transit through outdated leaky pipes and illegal connections (Rai 2011). What is left in the system is not enough to supply everyone with 24/7 tap water; at least 25% of the population makes do with 4 h or less of running water per day (Rai 2011). The per capita ratios of water access are also inequitable: a few wealthy urban residents receive as much as 500 L-per-day while a much larger percent of the population struggles to access the 20–50 L-per-day that is every person's right. To make up for the shortfall, and as earlier indicated, residents tap into scarce groundwater supplies that are not sufficiently regenerated and are in danger of drying up entirely leaving the capital without emergency reserves. In this scenario, a city essential to the smooth functioning of the world's tenth largest economy is vulnerable to a full-fledged "water crisis" (Chellaney 2013).

9.3.2 Scenario 2: Future Resource Self-Sufficiency

In the second scenario, which represents an alternative future, New Delhi is roughly water self-sufficient. Gone are the days when the majority of the city's freshwater was sourced from outside of the capital and its surrounding areas. Interbasin water transfers are moderate and interstate water conflicts emerge only when the annual monsoon rains are meager. Groundwater depletion rates have not only recovered but the groundwater system has been brought back into balance.

Driven by a policy turnaround, these feats have been achieved through the widespread implementation of urban rainwater harvesting. Rainwater harvesting involves catching rainwater where it falls, storing water in natural or concrete tanks, and distributing it to resource users. As the Centre for Science and Environment (CSE) states:

> In scientific terms, water harvesting refers to the collection and storage of rainwater and other activities aimed at harvesting surface and groundwater, prevention of losses through evaporation and seepage and all other hydrological studies and engineering interventions, aimed at conservation and efficient utilization of the limited water endowment of (a) physiographic unit such as a watershed. In general, water harvesting is the activity of direct collection of rainwater. The rainwater collected can be stored for direct use or can be recharged into the groundwater (CSE 2003, p. 3).

With confidence in the potentials of urban rainwater harvesting, organizations such as the CSE claimed at the turn of the twenty-first century that the city of New Delhi could be "entirely self-sufficient" if it were to capture, filter, and reuse every ounce of rainwater that falls in the city over the 3 months of the monsoon rains. This practice builds upon a time-honored approach that has allowed rural areas to thrive for centuries (Mishra 2001). Rainwater harvesting, in fact, is an engineering feature that enabled the flourishing of early Sultanate settlements (and subsequent Mughal settlements) set within the city's contemporary boundaries that date from the thirteenth Century onwards (CSE 2001; Lort 1995; Welch 2008; Westcoat 2014). It has however been neglected as a significant source of urban water supply in politically powerful cities such as New Delhi, which as earlier stated relies on diverted river waters to survive. This, according to CSE, reflects an unnecessary dependence on secondary as opposed to primary water supplies. As they state in their rainwater harvesting manual:

> Rain is the first form of water that we know in the hydrological cycle, hence is a primary source of water... Rivers, lakes and groundwater are all secondary sources of water. In present times, we depend entirely on such secondary sources of water. In the process, it is forgotten that rain is the ultimate source that feeds all these secondary sources and (we) remain ignorant of its value. Water harvesting (is a) means to understand the value of rain, and to make optimum use of rainwater at the place where it falls (CSE 2003, p. 3).

The CSE has put forward impressive calculations for harvesting this "first source" of water. Since the city of New Delhi has an annual average rainfall of 611.8 millimeters (mm), they argue that

Standard Rainwater Harvesting Equation	Sample Plot and Rainfall Volume		Sample Equation
(Area of Plot x Height of Rainfall) x .6 [60% success rate]	Area of Plot	100 square metres	(100 x 600mm) x .6 =36,000 Litres
	Height of Rainfall	.6 m (600mm) [or 24 inches]	

Fig. 9.1 Calculating Rainwater Harvesting Potential

the city's rainwater harvesting potential is 907 billion L/year. They add: "This is equal to about 270 days of water requirement for the entire city" (based on a demand 850MLD in 2003, CSE 2003, p. 4). These estimates are arrived at by using an equation that calculates the average annual rainfall with the total area of the city—which is 1486 km^2—and then applying a fractional extraction success rate of 0.6 (which equates to 60% success in the act of rainwater harvesting). The generic calculation, also used in their training programs, is as follows.

Note that the sample equation in Fig. 9.1 is for an individual household. Assuming a flat surface of 100 m^2, an average rainfall of 600 mm, and a 60% success rate, a household can expect to harvest 36,000 L/year. This volume, the CSE notes, "is about twice the annual drinking water requirement of a 5-member family" if you assume that the average daily drinking water requirement per person is 10 L.

In addition to work done by CSE, other scholars have also investigated the rainwater harvesting potential of residences in New Delhi. One study focusing on South Delhi calculated that urban water harvesting efforts can decrease residents' reliance on municipal supplies by 30–70% for those that adopt the technique (Said 2014).

9.3.3 Which Scenario Will Unfold and Perpetuate?

These two scenarios offer stark contrasts. The first scenario is one of water stress and conflict, whereas the second scenario promises something near to an urban water utopia. As bad as the first scenario seems based on current trends and challenges, it bears noting that if present trends continue the water reality may 1 day be far more dystopic; without a significant change in management practices, New Delhi's future hydrological challenges will be multiplied. With such high stakes, it is somewhat stupefying that urban rainwater harvesting is not more prevalent. The lack of uptake requires us to revisit the queries that began this text and to additionally ask: Why has not urban rainwater harvesting been more aggressively implemented in a city as important as New Delhi?

9.4 "Downscale" Challenges for Urban Rainwater Harvesting

To begin exploring why urban rainwater harvesting is not more prominent in New Delhi, a first point to note is that municipal and federal government agencies have long cited the benefits of rainwater harvesting and have repeatedly called for its expansion into urban zones (see, for instance, section 11.4 of the National Water Policy [GOI 2012]).[5] Over the last 15 years in

[5]According to one source, for instance, in 2009 the Central Ground Water Authority (CGWA), a body under the Ministry of Water Resources, "sounded an alarm" to call upon all federal states to adopt rooftop rainwater harvesting systems in government institutions. Yet by 2015, the implementation process in New Delhi was "still restricted" to only a few institutions. The article, cited below, notes that even in the "private sphere" only a few Resident Welfare Associations in New Delhi have made good use of the technology, "leaving most citizens unaware of the veracity of depleting surface water resource." See: Rooftop Rainwater Harvesting Can End Delhi's Water

New Delhi, in fact, the former and current Chief Ministers have made prominent declarations about the urgent need for immediate and large-scale interventions that would retrofit existing buildings and mandate substantial urban rainwater harvesting units be built into all future buildings. In 2014, the Chief Minister of New Delhi, Arvind Kejriwal, publicly encouraged its uptake as a means of water "self-reliance." Since the state was embroiled in yet another water sharing dispute with Haryana at the time, he indicated that urban rainwater harvesting might help alleviate such conflicts moving forward.[6] Notably, Kejriwal's statements in favor of rainwater harvesting built upon precedents set by Sheila Dikshit—the former, and longstanding, Chief Minister of New Delhi. Over the course of nearly 15 years in office, Dikshit spoke forcefully about the immediate need for the widespread uptake of rainwater harvesting. In 2007, she backed up her vocal support of the technology with the launch of a scheme called the "Best Rainwater Harvester Awards." This program awarded cash prizes to institutions and households deemed to have implemented exemplary rainwater harvesting structures.[7,8]

As a reflection of the political support for urban rainwater harvesting, policy changes have been made. Since at least 2014, buildings with a rooftop over 500 m^2 are required to have urban rainwater harvesting systems. In some instances, there are government subsidies to offset the costs of installation; in others, there is an added tax on the municipal water rate if there is noncompliance with rainwater harvesting guidelines. At earlier points in the policy development, the New Delhi municipality covered 50% of all water harvesting installment plans or, alternately, it offered a payment of 50,000 INR ($1047), whichever was less (Planning Department 2006). The residents that did not install the mandated water harvesting units were warned that they would be charged an additional tax of 1.5% on their water bills (Planning Department 2006).[9]

Despite the carrot-and-stick strategy, and despite the rhetoric of the policy moves, there remains widespread skepticism about the viability of urban rainwater harvesting in New Delhi. Indeed, stories abound about how new buildings are made to look compliant with rainwater harvesting regulations only for the initial inspections; after these compliance checks, the infrastructure is not maintained and is allowed to fall into disuse and disrepair.

To explain some of the complications around successful implementation of rainwater harvesting at the household or individual level, I defer to the perspective of a historian of Sultanate and Mughal India, and of Delhi's hydrological history. My interlocutor for this first insight is Mr. Sohail Hashmi (henceforth Hashmi), a man whose ancestry in the geogra-

Crisis. Zee News, 5 July 2015, http://zeenews.india.com/news/eco-news/rooftop-rain-water-harvesting-can-end-delhi%E2%80%99s-water-crisis_1608032.html, accessed 17 August 2017.

[6]"Delhi must harvest rainwater", *The Hindu*, 2 March 2015, http://www.thehindu.com/todays-paper/tp-national/tp-newdelhi/delhi-must-harvest-rainwater/article6950014.ece, accessed 15 March 2015.

[7]"Rainwater Harvester Awards Presented." *The Hindu*, 3 February 2007, http://www.thehindu.com/todays-paper/tp-national/tp-newdelhi/Rain-Harvester-Awards-presented/article14715361.ece, accessed 17 August 2017.

[8]The awardees became showcase venues to demonstrate how rainwater harvesting can be successfully implemented in a range of contexts. In 2009, and while enrolled in an urban rainwater harvesting training course through CSE, the author visited the two institutions that won the award in 2007. She also visited the home of one of the households that won the award in that same year. The institutions were awarded 100,000 INR each and the households were awarded 50,000 INR.

[9]This policy approach builds upon the precedent set by the city of Chennai in South India, a prominent example of centrally mandated rainwater harvesting about which readers may be aware. Work by Arabindoo (2011) and others however indicates that uptake and upscaling problems persist in cities such as Chennai—despite significant policy revisions and despite some initial improvements in the groundwater levels attributed to an increase of rainwater harvesting project. At the time of this text's composition, Chennai was undergoing yet another "water crisis" due to interstate water disputes, low reservoir levels, and the overextraction of groundwater (See, for instance, Priyanka Thirumurthy's online article, "Chennai Water Crisis: It is Time You Know What is Happening" from 4 July 2017, available at: https://www.thequint.com/environment/2017/07/04/chennai-water-crisis [accessed 24 August 2017]).

phy that is now called New Delhi goes back several generations. I first met Hashmi in 2016 when joining him and a group of 20 other people on a 5-h walking tour of Delhi's oldest living habitation: the enclave of Mehrauli. After our tour of Sultanate water features—which included large tanks and step wells dating back to the thirteenth century—I asked Hashmi what he thought of the discourse I was hearing about a renewed push towards water harvesting in the contemporary city. "Ah, urban rainwater harvesting," he remarked, "It works very well—on paper." This comment was met with a good-natured smirk that replaced additional explanation. For more of his thoughts on the topic, I arranged an interview at his home on October 28, 2016. It was during this interview that he would call urban rainwater harvesting, "Faff" and a bunch of "hot air."

Within minutes of sitting down at his living room table, Hashmi laid out a topographical map of New Delhi that predated the colonial era. His first task was to explain the water provisions of the seven cities that preceded the one set up by the British. These cities, he claimed, did not depend on river water. All but Shahjahanabad (with its seat of power at the Red Fort) were located at a distance from the flow of the Yamuna River. Their main source of water was springs and wells, along with a series of tanks that the early inhabitants of the Delhi plains created. In those days, the assessment of a good ruler was based on the number of water features that they created during their period of rule. Also known as "Raja dharma"—or royal responsibility—the emphasis on building water features aligns with a belief that one earns more religious merit when building a well than one does when building a mosque or a temple. As a result, the Delhi plains were populated by a vast number of wells, tanks, and step wells (*baolis*) (Fig. 9.2). This is the landscape that the British overtook when they moved the seat of power from Calcutta to New Delhi in 1931. In their refashioning of the city, and due to their concerns over malaria and other vector-borne diseases, they filled in and destroyed many of the water features that had served the land's residents for centuries.

Hashmi believes that if the current government of New Delhi was to revive the use of rainwater harvesting, and to expand the number of water features in a way that mirrors the city's pre-colonial past, then "Delhi will not need any water from outside." The increased blue cover (as opposed to green cover) could even help to bring down the highest summer temperatures. A study by INTACH, also known as the Indian National Trust for Art and Cultural Heritage, contends that the revival of Delhi's "traditional" water features would reduce the city's ambient temperature by one degree Centigrade (2008). In Hashmi's summary of this work, the result would mean that instead of "baking at 46 degrees" in the heat of the summer, "the maximum temperature at the peak of the monsoon would not cross 45 degrees." A return to past water management practices, following this perspective, is also a climate resilient move.

As water experts like Hashmi indicated, there is some disagreement over the extent to which individual households should be forced to engage in urban rainwater harvesting. While several interlocutors agreed that households could catch rainwater where it falls and use it for their gardens—and for the washing of pots, pans, and cars—there is a concern that households do not have the wherewithal to engage in the kind of upkeep for these structures that can substantially lead to the recharge of the groundwater table. Without substantial amounts of investment in maintenance, the rainwater captured might even mix with sewage as many of the existing storage drains function as open sewage lines in contemporary New Delhi. Thus, a major disincentive for urban residents is that they may have painstakingly invested time and money in a rainwater harvesting unit that might help save a modest amount of water that falls their property, but that will run into the storm drains (and sewers) as soon as their storage tanks overflow.

In New Delhi, the spatial limitations and infrastructural challenges mean that residents can realistically only "harvest" what they can store—and the storage potential of many households is relatively minor compared to the scale that is needed to bring the city's water supply into

Fig. 9.2 "Raja Ki Baoli" (King's Stepwell) in Mehrauli, New Delhi. (October 2016, Photo by the Author)

greater balance. Since a huge volume of water falls in New Delhi in a short duration of 3–4 months, and since the water must be stored throughout the year for the investment to pay off, individual efforts can improve household water access during the monsoon months, *but* they cannot improve water access throughout the year. It is because of this very short duration of rainfall that the CSE argues that cities such as New Delhi must focus on recharging groundwater aquifers (CSE 2003, p. 7). In other cities, where the rainfall is spread across the year however "…one can depend on a small domestic-sized water tank for storing rainwater, since the period between two spells of rain is short" (CSE 2003).[10]

Allow us to return to that point, and to restate it: The CSE, in its manual on rainwater harvesting for New Delhi residents, argues on page 7 that city, such as New Delhi must focus on recharging groundwater because individual plot sizes are often too small to accommodate the monsoonal deluge. This alone demonstrates that there is some slippage in the rationales upholding that individual should engage in household-level urban rainwater harvesting. In the same manual, there is also another element to point out that is a significant disincentive for residents—and that is the question of the proportional benefit to be received from engaging in urban rainwater harvesting efforts. They point out, for instance, that the average per person water consumption in New Delhi is estimated at 240 L, which at the time was the highest in India (CSE 2003, p. 1). According to the numbers arrived at in Fig. 9.1 of this text however a household lucky enough to have 100 m^2 for rainwater harvesting can expect

[10]Due to relatively consistent rains throughout the year, the 2003 report by CSE cites urban centres in Kerala and Mizoram as good candidates for year-round efforts to harvest water via modest storage tanks, as opposed to groundwater recharge (ibid.: 7).

to capture 36,000 L. If we assume that the household is a family of 5 (as the CSE assumes), then, that equals roughly 19.7 L per person per day because of urban rainwater harvesting. This is only 8.2% of the 240 L that an average New Delhi resident expects to receive. The other concern from a household perspective is that since it rains for only 3 or 4 months, the augmented water supply is seasonal and is unlikely to last throughout the year. So, even though it is impressive to capture 36,000 L per annum, it takes a mentality of thrift, and the spirit of environmental stewardship, for residents to be convinced of the need to implement *and maintain* the technology. The exception to this, of course, are the areas that are already water poor. In the southern parts of South Delhi, such as Mehrauli, some neighborhoods only get 35 L per capita per day (CSE 2003, p. 1). It is in these zones where the residential rainwater harvesting makes the most sense. The average household benefit of 19.7 L per person per day then goes from being 8.2% of the accustomed supply to 56% of the accustomed supply.[11]

9.5 "Upscale" Challenges for Rainwater Harvesting

It is because there is a pressing need for groundwater recharge—and for the proper centralized infrastructure needed to make it possible—that a representative of the World Wildlife Fund (WWF) in New Delhi contends that the real issue is good governance. As she argued during our October 2016 interview, it is impossible to implement rainwater harvesting effectively when there is still a substantial amount of pollution and sewage in the drains that fall under the purview of the municipality. In her words, "...you can't have a successful rainwater harvesting system unless you have solid waste management done successfully. So, they are all like dominos—steps which are linked one by one." Several other respondents concurred with the WWF representatives' perspectives while adding their own insights. As one interlocutor stated, based on several decades spent studying and critiquing government water policies while working at CSE, "Let me make one thing clear: If you talk about water, I promise that I will talk about sanitation. Because you cannot handle water problems without looking at sanitation." After making this statement, he clasped his hands together while saying, "They are like this." The gesture was meant to emphasize that the two issues were intricately intertwined.

Hashmi, in our conversations, agreed that the centralized system must be overhauled—and that addressing the use and maintenance of the stormwater drains was an important prerequisite for improving the rainwater harvesting metrics. He also pointed to what he called a "simple solution" that could provide the most benefit to the city's groundwater recharge while utilizing the city's storm water drains. As he explained, every road in the city has storm water drains on both sides. He suggested: "Now what you need to do is, every 200, 300, (or) 500 m, beneath the storm water drains, create a rainwater harvesting mechanism... Then your stormwater drains, which are all solid concrete now, will become your aquifers. They will gradually begin to replenish the subsoil water." So, he emphasized, "this is one thing that can be done." The problem, as Hashmi went on to note, is that such measures do not require the large-scale foreign expenses and investments of multinational corporations or MNCs. As he stated, "We have the technological know-how. We have the manpower. We can do this."

Hashmi also had suggestions for sewage management, which he preempted as an "even simpler" solution. In his opinion, the first point of action is to stop the flow of sewage water that moves through New Delhi's drains and canals. To do this, he argued that pipelines could be laid to "trap all the dirty water" and redirect them into solid waste treatment plants. These plants, or "STPs," can then remove all the "filth" that is flowing through the city's residential areas. As he

[11] Another implication is that extremely low-income residential areas, including what are known as 'slums', could greatly benefit from rainwater harvesting (assuming proper installation and regular maintenance). This is particularly true for slums that do not receive municipal water supplies.

explained: "Treat the water, and then release it into the drains. Eventually this water will get cleaned up before it reaches the Yamuna. And then you will not have to clean the Yamuna—the Yamuna will clean itself. Because 60% of the filth that is flowing into the Yamuna is generated by Delhi."

The problem with this solution, in Hashmi's opinion, is also its main benefit—which is that "...no one gets to make any money" off its implementation. Since he has spent years arguing for proactive and "simple" solutions, he reflected, "I am absolutely convinced that no one is interested in this—in simple solutions. Because simple solutions do not require massive financial outlays. And there are no kickbacks. The budgetary allocations would be so small that nobody would be interested in whatever kickbacks you can get." A major disincentive for the upscaling of urban rainwater harvesting, in other words, are the vested economic interests that stand to benefit from the building of large infrastructure projects rather than the patchwork of infrastructural modifications and improvements that the widespread implementation of urban rainwater harvesting entails.

Numerous interlocutors independently concluded that the management solutions most often enacted are ones that serve narrow economic and political interests. The author's conversation with water experts in New Delhi, in fact, were never far from veering towards the issue of corruption once initial questions were posed about the viability of urban rainwater harvesting. In the crosshairs of the critical commentary provided were the engineers, contractors, and politicians that are caught in a web of mutually beneficial economic exchanges. Also mentioned were the private water providers—the "water mafia"—whose service underserved parts of the city via water tankers. According to Yaffa Truelove (2011), some of these privately owned tankers get their water supplies from the municipal water tankers who have been paid off to redirect provisions that are often meant to go to slums and other economically struggling settlements. This example shows that profit seeking behavior can prevail even in times of water crisis. Since this behavior most benefits those who are in positions of socioeconomic and political power, it is no surprise that inefficient water management and delivery system is allowed to continue without significant overhaul. Such observations point to the struggles over capital capture that, counterproductively, make the effort to engage in sustainable efforts at urban water capture so difficult to achieve.

To stress the point that the municipality is not truly interested in the upscaling of urban rainwater harvesting, some interlocutors underscored the fact that very few of the city's municipal buildings have implemented this technology—even when government officials vocally express their support for its urgent uptake. An interview with an architect, who has studied various rainwater harvesting technologies across North India since the early twenty-first century, was particularly revealing on this point. After 25 min of discussion about the usefulness and longstanding histories of rainwater harvesting, he stated emphatically, "I mean, it is a *no brainer*—we *have* to do rainwater harvesting." Despite this need however he claimed that the city of New Delhi was looking to build new dams upstream (at places such as Renuka) while "doing *zero* rainwater harvesting" on its own buildings. This, he stated, includes the buildings of the municipal water supplier, also known as the Delhi Jal Board. He explained:

> Somebody was telling me that they live in some posh neighbourhood (and) they got a notice from the Delhi Jal Board saying that 'We are going to charge you four times the amount of money that we used to charge because you don't have rainwater harvesting.' So..., water supply billing is being increased manifold if you do not have rainwater harvesting. But the Delhi Jal Board itself is doing hardly any rainwater harvesting; there are only one or two of their campuses where they are doing rainwater harvesting. So, I think (they are) just putting it out and saying, 'Everybody else should do it'...[12]

[12]For the time being, let us put aside the fact that the interlocutor may have slightly embellished his previous statement as his subsequent comments do show that there is *some* implementation of rainwater harvesting on select municipal buildings (as opposed to his earlier estimate of *zero* uptake at the level of municipal infrastructure).

A main critique provided by this architect was the fact that the municipal water supplier was holding residents of housing enclaves in New Delhi to new rainwater harvesting requirements while their own buildings did not feature this technology. This is telling of the government's priorities, he felt, since the city's "biggest landowner" was not, to paraphrase a saying, "walking the talk." What this interlocutor ultimately wanted to underscore is that, as he later said, "It doesn't make sense," that the Delhi Jal Board has devised a strategy to penalize residents for not installing rainwater harvesting units when many of their own buildings have failed to follow their own advice. In the architect's opinion, this further underscored the odd financial equations that drive how, and when, strategies for promoting urban rainwater harvesting are pursued. His statements also bolster the overarching point that even modest efforts to promote urban rainwater harvesting are mired in political economic factors that tie into the overarching political ecologies of water management in New Delhi.

9.6 Harnessing the Raindrop: Concluding Remarks

The issues and perspectives itemized demonstrate that the problem of water stress in a city of over 18 million is as much an issue of economic and political power as it is one of actual resource scarcity. At stake is the power to decide how water is managed and how water is valued as a commodity rather than a public good (Maria 2008). Also at stake is the power to determine who receives more water than others, and at what cost (Swyngedouw 2009). At the top of the resource use power structure—and in descending order—are the politicians, municipal employees, engineers, contractors, and private water suppliers. Each of these groups plays a role in stalling, derailing, or misguiding the efficient and equitable management of water resources based on the narrow institutional, and sometimes even personal, interests. It is in this way that the complicated, unruly, and unjust water landscape of New Delhi is intricately linked to the political ecology of water resource management. As Nikhil Anand (2017) cautions, "...infrastructures are made by and constitutive of diverse political rationalities, past and present" (13).

Following post-structural analyses of landscape management, such as the one on Delhi's urban governmentalities offered by Stephen Legg (2007), urban rainwater harvesting may, at present, be little more than a discursive emphasis positing the benefits of a knowledge practice that—while promoted under the auspices of good intentions—enables a set of power configurations to remain in place without substantial change to their fundamental structure. In this political-ecological landscape, rainwater harvesting units (and their absence) serve as infrastructural sites in which water management is a visible hallmark of power relations (Hannah 1997). That said, the presented critiques of the failure to successfully implement rainwater harvesting serve as a significant counter discourse. Through such counter discourses, alternatives are suggested and imagined even as prevailing inequities and power imbalances are critiqued. Although it seems a far cry to suggest that these counter discourses will lead to the hopeful second scenario presented at the beginning of this discussion, it is through these critiques that the momentum for systems change is built.

While several challenges persist, the observations made in this text indicate that, given the *political will* for change, there is still scoped to persevere in the aim of integrating urban rainwater harvesting into the water management schemes of large cities such as New Delhi. An integrated approach is needed because, as Saif Said notes based on an assessment study from 2014, rainwater harvesting, "...can prove to be an enormous supplement to our existing resources if it is utilised wisely and judiciously" (145). This is particularly true, he goes on to note, if every individual were to engage in demand reduction by curtailing their water usage by "at least 20%" per day (Hannah 1997). Other experts agree that rainwater harvesting is an important part of an efficient water management approach. As Aditya Sharma, cofounder of an initiative known as RAIN Water, states: "Rainwater harvesting is not a remedy to the problem, but (it is) a kind of insurance. We need to insure against some degree of water scarcity."[13] According to such perspectives, urban rainwater harvesting is a useful and necessary means to improving water supply and water resource access in cities such as New Delhi. As such, it deserves to be integrated into government approaches to water management to a greater degree.

The assertions made herein are bolstered by the significant work of Bruce Mitchell, who has argued for the importance of an integrated water resource management for several decades. According to Mitchell and his collaborators (Mitchell et al. 2014, Mitchell et al. 2015), an integrated approach to water management embraces the importance of understanding interconnections among variables and relationships in a relevant socio-ecological system. An integrated approach, furthermore, focuses "only on those (variables) which are key drivers for change and (which are) amenable to being managed)" (Mitchell et al. 2015, p. 719). This, Mitchell and his colleagues argue is different from a "holistic" or "comprehensive" approach to water resource management; these later approaches take into consideration every variable or relationship associated with a given socio-ecological water management system. While admirable, the downside of a holistic or comprehensive approach is that the "all-inclusive perspective" may become overwhelming in terms of one's ability to examine all the interconnections (Mitchell et al. 2015). Or, as Mitchell self-quotes, these approaches may lead to "inordinately lengthy periods of time for planning" as well as the drafting of plans that are not "sufficiently focused" enough to help water managers (Mitchell et al. 2015, p. 4). It is for reasons of efficiency that Mitchell and his colleague argue for simplified yet integrated solutions (Mitchell et al. 2015, p. 719).

While this chapter has demonstrated the complexities of "downscaling" and "upscaling" urban rainwater harvesting in New Delhi, the perspectives of key interlocutors featured indicate that it could nonetheless provide a necessary intervention that can improve the overall integration of New Delhi's water system. The data however also indicates that the simplest means of doing this is to focus the municipality and the central government on improving the centralized infrastructure. The conversion of storm water drains into groundwater recharge infrastructure, for instance, could do more to improve the aggregate amount of water available in the ground—and it can do it faster and more efficiently than getting individuals and households to retrofit their residences for the same purpose. This means that the urban "(rain)water revolution" has a better chance of receiving improved centralized treatment than it does undergoing successful downscaling and decentralization. The challenges of downscaling are particularly robust given that small household rainwater harvesting systems lack sufficient storage and they require consistent maintenance to stay effective. The centralized infrastructure, by contrast, covers a vast amount of area within New Delhi and it can be maintained by paid staff members who receive the training needed to keep the systems running properly.

In closing, and to restate the main point, it appears that urban rainwater harvesting in a city such as New Delhi needs to be upscaled at the municipal level—and hence centralized to a much greater degree. This would be an important step towards leading by example. It would also help transform the water imbalances that have characterized New Delhi's hydrological flows for the past several decades. Such a transition would allow India's capital to supplement the water that it draws in from other regions and water basins. It would create, in short, a more integrated approach to the city's water resource management by focusing on a basin-level approach rather than a project-based approach.

Acknowledgments The author would like to thank Mr. Sohail Hashmi for his insights and comments on an early version of this text. She also gratefully acknowledges the funding support of the Australian Research Council (DE160101178). Additional thanks go to Professor Baleshwar Thakur for his kind invitation to contribute to this edited volume.

[13] Rooftop Rainwater Harvesting Can End Delhi's Water Crisis. Zee News, 5 July 2015, http://zeenews.india.com/news/eco-news/rooftop-rain-water-harvesting-can-end-delhi%E2%80%99s-water-crisis_1608032.html, accessed 17 August 2017.

References

Agrawal A, Ostrom E (2001) Collective action, property rights, and decentralization in resource use in India and Nepal. Polit Soc 29(4):485–514

Anand N (2017) Hydraulic City: water and the infrastructures of citizenship in Mumbai. Duke University Press, Durham

Arabindoo P (2011) Mobilising for water: hydro-politics of rainwater harvesting in Chennai. Int J Urban Sustain Dev 3(1):106–126

Bakker K (2012) Water security: research challenges and opportunities. Science 337(6097):914–915

CSE [Centre for Science and Environment] (2001) Making water Everyone's business: practice and policy of rainwater harvesting. CSE, New Delhi

CSE [Centre for Science and Environment] (2003) A water harvesting manual for urban areas: case studies from CSE. CSE, New Delhi

Chellaney B (2013) Water, peace, and war: confronting the global water crisis Rowman and Littlefield, New Delhi

Donahue JM, Johnston BR (eds) (1998) Water, culture, and power: local struggles in a global context. Island Press, Washington, DC

Government of India [GOI] (2012) National Water Policy, Weblink: http://wrmin.nic.in/forms/list.aspx?lid=1190

Halder R (2017) Haryana sending less water to capital: Delhi may face major supply woes. Hindustan Times, 12 July, http://www.hindustantimes.com/delhi-news/haryana-sending-less-water-to-capital-delhi-may-face-major-supply-woes/story-sCQR6zYiPTpiYksXfzyLzJ.html. (Accessed 3 August 2017)

Hannah M (1997) Space and the structuring of disciplinary power: an interpretative review. Geogr Ann 79B(3):171–180

INTACH [Indian National Trust for Art and Cultural Heritage] (2008) Conservation of waterbodies in Delhi. INTACH, New Delhi

Iyer R (2003a) Wanted: fresh ideas; effective water resources planning needs a shift of fresh air infused into stale minds. In: Agarwal A, Narain S, Khurana I (eds) Making water Everybody's business: practice and policy of water harvesting. Centre for Science and Environment, New Delhi, pp 303–304

Iyer R (2003b) Water: perspectives, issues, concerns. Sage Publications, New Delhi

Johnston BR (2005) The commodification of water and the human dimensions of manufactured scarcity. In: Whiteford L, Whiteford S (eds) Globalization, water, and health: resource management in times of scarcity. School of American Research Advanced Seminar Series, Santa Fe, pp 133–152

Joshi M (2016) Why New Delhi can never be self-reliant in water. Hindustan Times, 27 February, http://www.hindustantimes.com/delhi-news/why-delhi-can-never-be-self-reliant-in-water/story-m0GjHufZsb7ptYRmrW12cI.html. (Accessed 3 August 2017)

Kalia D (1997) Delhi: the murder of a river. In: Agarwal A, Narain S (eds) Dying wisdom: rise, fall, and potential of India's traditional water harvesting systems. Centre for Science and Environment, New Delhi, p 78

Kumar DM (2014) Thirsty cities: how Indian cities can meet their water needs. Oxford University Press, New Delhi

Lort JE (1995) Curious seen: Baolis of the Delhi sultanate. University of Victoria, Master of Arts Thesis for the Department of History in Art

Maria A (2008) Urban water crisis in Delhi: stakeholders' responses and potential scenarios of evolution. Paris, Iddri–Idées pour le débat N° 06, http://www.iddri.org/Publications/Collections/Idees-pour-le-debat/Id_0806_Maria_Urban-Crisis-Water-Delhi.pdf. (Accessed 20 August 2017)

Mehta L (2005) The politics and poetics of water: Naturalising scarcity in Western India. Orient Longman, New Delhi

Mishra A (2001) The radiant raindrops of Rajasthan. Research Foundation for Science, Technology, and Ecology, New Delhi

Mitchell B (1990) Integrated water management. In: Mitchell B (ed) Integrated water management: international experiences and perspectives. Belhaven Press, London and New York, pp 1–21

Mitchell B, Charles P, Shrubsole D, Veale B, Walters D (2014) Integrated water resource management: lessons from conservation authorities in Ontario, Canada. Int J Water Resour Dev 30(3):460–474

Mitchell B, Kathryn B, Richardson S (2015) Integrated' approaches to water and natural resources Management in South Australia. Int J Water Resour Dev 31(4):718–731

Planning Department (2006) Evaluation study of the scheme of rainwater harvesting. New Delhi: Delhi Secretariat., http://delhiplanning.nic.in/Reports/RWH.pdf. (Accessed 8 March 2015)

Rai SC (2011) Water Management for a Megacity: National Capital Territory of Delhi. Water Resour Manag 25(9):2267–2278

Said S (2014) Assessment of roof-top water harvesting potential in South Delhi, India: a case study. Int J Environ Res Dev 4(2):141–146

Swyngedouw E (2006) Circulations and metabolism: (hybrid) natures and (cyborg) cities. Sci Cult 15(2):105–122

Swyngedouw E (2009) The political economy and political ecology of the hydro-social cycle. J Contemp Water Res Educ 142(1):56–60

Truelove Y (2011) (Re-) conceptualizing water inequality in Delhi, India through a feminist political ecology framework. Geoforum 42(2):143–152

Welch A (2008) The emperor's grief: two Mughal tombs, in frontiers of Islamic art and architecture: essays in celebration of Oleg Grabar's eightieth birthday (Muqarnas volume 25). Leidon, Brill, pp 255–274

Westcoat JL (2014) Conserving urban water heritage in multicentred regions: an historical-geographic approach to early modern Delhi. Change Over Time 4(1):142–166

Wutich A, Brewis A (2014) Food, water, and scarcity: toward a broader anthropology of resource insecurity. Curr Anthropol 55(44):444–468

10 Rethinking Capacity Building in Water Governance: Factors Influencing Risk Interpretation and Decision-Making in Delhi

Shabana Khan

Abstract

The presence of multiple stakeholders, and their varied perceptions and unclear accountability, within the complicated institutional arrangements increase the complexity of water governance in Delhi. In such context, it is highly crucial for stakeholders to well-interpret risks for effective decision-making to tackle water-related hazards. This study aims to understand the interplay of risk interpretation and decision-making in the current water governance system in Delhi by using the Risk Interpretation and Action (RIA) framework. It explores various factors suggested in the RIA framework, such as uncertainty, experience, learning, trust, complexity, scale and social context, and the influence of these factors on risk interpretation and subsequent decision-making. In this study, 30 in-depth interviews were conducted with key stakeholders including members of the private sector, government, non-government, educational, and research organizations. The results highlight a need to rethink capacity building in terms of preparing varied stakeholders for their greater engagements and participation in the development of effective water governance that addresses various implied risks.

Keywords

Capacity Building · Delhi · Risk · Risk Interpretation and Action (RIA) · Water Governance

10.1 Introduction

Water is essential for life! It is an incontestable fact, yet the ignorance towards water in Delhi is not just widespread but threatening. Despite being the capital of India and having a close access to the political and economic power, the city has a weak profile of urban water risks governance. Delhi has been ranked as one of the worst performing metro cities in Asia in terms of providing safe and sufficient water to its residents (ICNP 2001). The maladministration of sewage in Delhi has not only led to the degradation of water quality in the Yamuna but has also claimed it the status of a "dead" river (Sharma and Kansal 2011; Chauhan 2015). Groundwater is overexploited in seven out of nine districts of Delhi (Chatterjee et al. 2009; Shekhar et al. 2009). Besides, the city is exposed to multiple water-related risks, such as water scarcity, flooding, groundwater pollution, and water insecurity in the face of climate change. Despite significant resource investments, a little progress has been achieved in reducing its water-related risks in Delhi. The presence of multiple stakeholders,

S. Khan (✉)
Indian Research Academy, New Delhi, India

B. Thakur et al. (eds.), *Resource Management, Sustainable Development and Governance*, Sustainable Development Goals Series, https://doi.org/10.1007/978-3-030-85839-1_10

their varied perceptions, and unclear accountability within complicated formal and informal arrangements increase the complexity of the water governance in the city. While the risks related to water are portrayed at different levels, various subtle factors, such as how varied risk interpretations and decision-making are influencing water governance at the local level have not been studied in-depth. In a highly complex system of water governance in Delhi, it is critical that all stakeholders interpret risks accurately for effective decision-making to tackle various water-related hazards. This study aims to understand the interplay of multiple factors governing risk interpretation and decision-making in the current water governance system in Delhi by using the Risk Interpretation and Action (RIA) framework developed by Eiser et al. (2012). Various factors discussed in the model and their implications are discussed to highlight the gap in the capacity building that needs to be addressed for effective water governance at the city scale.

10.2 Conceptual Framework

Water risks in the megacities remain one of the most poorly articulated issues (Vörösmarty et al. 2000). The rising threats in the face of climate change have the potential to destabilize the existing urban water management practices either by creating a need for change or by reinforcing the traditional methods (Keath and Brown 2009). Sustainable Development Goals (2015–2030) point explicitly towards the vulnerability of cities to water disasters that need to be managed in a holistic manner (United Nations 2016). Multiple approaches and discourses applied for water resource management at the city scale continue to fail for various reasons.

Box 10.1 Sustainable Development Goals

Sustainable water governance in India's capital Delhi is crucial given this metropolis is the second largest populated in the world, and water as a resource is scarce. Good water governance is also critical to sustainable development as it is pivotal for economic growth, social inclusion, and environmental sustainability. Water governance in Delhi is challenged by the fact that there is too little and too polluted water to sustain the demand for domestic and industrial water needs and sanitation services. The effects of climate change, urbanization, and growing population among others continue to drive water resources demand, availability, and quality, now and in the future. In this context, the role of water governance for improved water policy design and implementation is now undisputed and will remain key in addressing water challenges. To that end, water runs across all the 17 sustainable development goals (SDGs). The author recognizes water governance challenges in Delhi and identifies "capacity building" across stakeholders, as crucial to address issues of efficiency, effectiveness, and inclusiveness. Using a risk interpretation and action (RIA) framework the author observes that capacity building is still catching up to implementation needs. The research and conclusion drive home the fact that capacity building (including customized training, relevance of auditing for evaluating change over time) will remain crucial for effective to deliver water governance in order to maintain healthy ecosystems, and in the mitigation and adaptation to climate change.

Wong and Brown (2009) note an overhaul of the hydro-social contract that underpins the conventional approaches of water management and associated investments that undermine the proposition of sustainable cities. It is observed that there is a minimal impact on the large-scale infrastructural projects in the face of ongoing sustainability agenda despite their overwhelming impacts on the environment (Crow-Miller et al. 2017). The query for development and addressing the needs of the rising population often keep the focus away from the sustainable use of resources. Further, watershed management has become an arena for social dilemmas exacerbating the conflicts as well as hazards that are internalized within watershed (Navarro-Navarro et al. 2017). Specific insights derived from an empirical analysis show that water management (WM) is socially embedded in dense networks of family, friends, farmers, and the local government. These stakeholders share varying degrees of information about local water crises. It is found that while irrigation water user representatives (WUR) are connected across communities within their municipalities, but inter-watershed social links with other WUR were virtually nonexistent, despite high levels of awareness of cross-municipality WM problems (Navarro-Navarro et al. 2017).

Many studies look at water management from the economic and efficiency point of view (Briscoe 1997; Briscoe and Malik, 2006; Toteng 2008; Yuling and Lein 2010; Raul et al. 2011). Molle and Berkoff (2007) have documented the history of the idea of "water pricing" and found that there are, in the South, virtually no examples in which pricing does the allocative- and efficiency-enhancing work that mainstream economics want it to do. Privatization of water is also sought for improved services, but it is more closely linked with governance failure rather than better resource management. The social distribution of the cost is highly unequal, and the poor are affected more consistently negatively than the affluent groups. The mainstream perspectives tend to give a commodity status to water by the name of providing value to water, its services, and infrastructure (Obertreis et al. 2016).

Other studies that focus on a closely related issue are the ones looking into the public behaviors of water consumption and its uses in cities (McMohan and Weeks 1973; Maidment et al. 1984; Dube and Zaag 2003; Van Rooijen et al. 2005; Chu et al. 2009; Sohn 2011; Zhi et al. 2015). In this context, the studies focusing on the population water use behavior or consumption model further notify the gaps that administration is forced to fulfill by enhancing access to more water. The focus on distribution, equity, and quality of the water keeps the mind on the need for more water that results in water conflicts, a common feature particularly noted in the cities of developing counties like India (Janakarajan et al. 2006).

One of the most advocated approaches is Integrated Water Resources Management (IWRM). It is defined as a process that promotes coordinated development and management of water, land, and other related resources, for maximizing the subsequent economic and social welfare in an equitable manner without compromising the sustainability of vital ecosystems (GWP-TEC 2000). It is a step forward towards integrated solutions at the local, regional, national, and transnational scales. Megacities though have specified boundaries and fall within a national boundary, they tend to have very complicated water management systems that defy the general principles of water use applied across the country (Khan 2017). Various hazards relating to the unequal distribution of water, poor water quality, wastewater, leakages, overflow, or cities' overdependence on external resources for their water supply enhance local vulnerability to extreme events. As the increasing struggle to meet the gap between rising water demand and failing resources dominates the focus of urban water governance, the responses to extreme events are frequently ad -hoc and short-term. Cossío and Wilk (2017) note that in contrast to the river basin approach applied in the IWRM framework by the water professional based on a perceived dimension of space for water management, lived dimension of space is more important as they are more flexible, and the local people and organizations relate to them personally for managing water resource than that of the river basin space. The lived spaces also include the local people's perception, experiences, and practices of water use and management.

A shift can be seen in the domain of water management from seeing it as purely a technical domain to one that involves multi-stakeholders for inputs and cooperation (Nguyen and Ross 2017). There is an increasing consensus in the literature which notes that the current water crises are not due to the lack of water supply and technology but due to failure in water governance (Miranda et al. 2011). A surge can also be seen in the literature focusing on water governance and its various characteristics, including participatory, interactive, or reflexive governance to address the need for democratic decision-making in the scenario of multiple stakeholders (Shove and Walker 2007; Torfing et al. 2012; Pfeffer et al. 2013). Megacities, particularly those in the developing countries with inadequate infrastructure and facilities are served by multiple actors and their networks, including government bodies, non-government organizations, private industries, small water enterprises, and many other units from organized and unorganized sectors (McGranahan et al. 2006). Studies note that diverse interests, rights, and knowledge of multiple actors, stakeholders, and institutions operating at various scales are difficult to reconcile (Sansom 2006), which goes along with the fact that very few studies have attempted to do such a comprehensive analysis. Often the policies adopted at the national or international level fail due to multiple reasons. Studies note that policy measures are weak, particularly in developing countries like India and China due to a lack of robust legal mechanisms to control pollution at different levels, lack of coordination, and gaps in policy implementation (Wang et al. 2016). A part of it can also be attributed to varied perceptions and awareness about the issues relating to water resources and risks that create conflicts in response measures supported or implemented by different stakeholders.

The studies focusing on the perception of water practitioners and decision-makers are rather few and fragmented. Such studies tend to assess perception from the water resource management or policy perspective. Baggett et al. (2006) note that dissonance in the perception of key stakeholders can be useful in building knowledge and participatory planning. Dobbie and Brown (2014) also stressed on the need to acknowledge diversified and subjective risk perception of all stakeholders and water practitioners for a sustainable water future. However, there is a lack of studies focusing on the perception of various water practitioners and stakeholders for their understanding of various water risks and responses in a city, and how they impact sustainability. While the stakeholders' perception of water risks and vulnerabilities in Delhi has been studied (Khan 2014, 2015; Khan et al. 2015), there is no study on the variations in the risk interpretation and decision-making for their influence on water governance.

Eiser et al. (2012) developed the Risk Interpretation and Action [RIA] framework to explain the factors that influence the interpretation of risk and subsequent decision-making and actions. The framework is based on the premise that the judgments underlying risk interpretation and action are not merely personal but interpersonal, which are influenced by several factors, such as uncertainty, experience, learning, trust, heuristics, complexity, scale, and sociocultural contexts. While the RIA framework is developed in the broader context of disaster management, it brings forth multiple factors that influence the response to risks applicable to water as well. Although many studies have addressed these factors in isolation, there is a gap of research focusing on how these multiple factors influence risk interpretation and decision-making for water within an urban context. While RIA framework is not comprehensive in listing all the factors that affect risk interpretation and action, it highlights dominant elements that can be assessed for an overall outlook (Khan et al. 2017). Addressing these factors in the context of water management can highlight gaps in the understanding of the water risks and effective response.

10.3 Methodology

The study was designed to be explorative and qualitative. Both primary and secondary data have been used to reflect on various aspects of water management, hazards, and other related characteristics. Secondary data was collected from the literature, newspaper articles focusing on water issues, websites of different water-related institutions, and published data from the Census of India. For the primary data, scheduled structured interviews were conducted with the key stakeholders. Samples were selected by using non-probability purposive sampling method. Informed and available interviewees from various government, nongovernment, private, education, research, and local bodies were chosen to understand water issues from different perspectives. In total, 30 in-depth interviews were conducted with varied actors across the city. Eleven interviews conducted for another project chance2sustain on the same issue is also used in this sample, due to the non-availability of many senior officials for the second interview. A larger proportion of interviewees were government officials as water is a state issue in India, which places government as the leading stakeholder in this sector. Out of ten interviews conducted with public institutes, seven were national institutes, one regional, and two were the local governing bodies. This variation helped to understand various complexities in addressing multiple challenges associated with water at different levels. Seven interviews were conducted with nongovernment organizations that deal with water in general and issues relating to water in Delhi in particular. Many of these institutions work closely with government institutions and play a critical role in bringing changes in water management and addressing challenges faced at the local level. Academics and researchers formed the third largest sample. Three universities and three national research institutes were also included in the sample to bring forth different perspectives. Interviews were also conducted with media professionals including one from the leading newspaper of the city and another from a magazine explicitly focusing on water-related issues in India. They helped to bring critical local variations and innovations to address water-related challenges. To include the community perspective, three resident welfare associations were also interviewed. They helped to clarify the details of sustainability challenges faced at the local level. Two private organizations were also interviewed to bring forth their perspective in dealing with water risks.

10.4 Delhi: The Context of Water Governance and Stakeholders

Delhi is the capital of India, and, therefore, is the home of both central and state water governance systems. The responsibilities relating to water in India are spread across multiple ministries and departments that investigate various aspects of water, and they are interlinked through a complex network. At the national level, the Ministry of Water Resource, Government of India is the apex body to formulate water policy and distribution of roles and responsibilities relating to water. It works through a network of organizations some of which have a direct role to play in the water availability, distribution, and quality in India. These include various research departments, river control boards, and major river projects along with public and autonomous bodies of national significance. Other ministries that deal with water include Ministry of Drinking Water and Sanitation, Ministry of Urban Development, Ministry of Environment and Forests, Ministry of Home Affairs, Ministry of Earth Sciences, Ministry of Science and Technology, Ministry of Agriculture, and Ministry of Power. They deal with water for its varied uses, state, hazards, and research.

Apart from ministries, there are several central institutes, which do not fall under any ministry but are formed as critical players through national acts. These include Niti Ayog (earlier Planning Commission), National Disaster Management Authority, and National Institute of Disaster Management. Besides, there are several departments, which work at the national level, but not all of them work in Delhi. Some of the departments that play an essential role in water governance of Delhi, include Central Water Commission, Central Ground Water Authority (CGWA), Central Ground Water Board (CGWB), Central Public Works Department (CPWD), and Water Quality and Assessment Authority. The Central Water Commission is the technical organization attached to the Ministry of Water

Resources. Central Ground Water Board is engaged in research relating to groundwater in India, while the Water Quality and Assessment Authority investigates issues relating to water quality. While these structures are created to have integration at different levels, the multiplicity of organizations and their different focus hinders holistic thinking. Different departments tend to follow diverse approaches as per their mandate, which creates barriers in communication. For example, an engineering approach is found more dominant in the Central Water Commission than the National Institute of Disaster Management, which looks at the ecosystem approach for a sustainable future. Such variations lead to different understanding of the issue, and hence a different response that may or may not be compatible with each other.

Delhi's most important drinking water source is the river Yamuna, which originates from the glaciers of the Mussoorie range of the lower Himalaya near Yamunotri in Uttarakhand state. Before entering the NCT of Delhi at Palla, the Yamuna flows through three riparian states of Uttarakhand, Haryana, and Uttar Pradesh. At this river basin level, the Upper Yamuna River Board (UYRB) is responsible for the water distribution among states, and to maintain optimal flow in the Yamuna River. A member of the Water Planning and Projects Wing of the Central Water Commission chairs the UYRB. In addition to the representatives from the riparian states (irrigation departments of Uttar Pradesh, Uttarakhand, Rajasthan, and Himachal Pradesh, the chief engineer from Yamuna Water Services Haryana, a member of the Delhi Jal Board) the Central Electricity Authority (Ministry of Power), the Central Ground Water Board and the Central Pollution Control Board are members of the UYRB. Despite this arrangement, water at the basin level is subject to frequent interstate conflicts for both quality and quantity. Apart from the usable water, Yamuna is also the source of flooding in Delhi. The Irrigation and Flood Control Department of Delhi is singularly in-charge of execution, repair, and maintenance of flood control work on river Yamuna and the Najafgarh drain system. The primary functions of the department are to protect the city of Delhi from floods in the Yamuna River, and planning, execution, and providing support to flood protection and river training works through construction, strengthening, and maintenance of marginal embankments.

Delhi Jal Board (DJB) is the sole government agency responsible for meeting water demands of the city. It was created in April 1998 through an Act of the Delhi Legislative Assembly integrating the Delhi Water Supply and Sewage Disposal Undertaking under one body. Apart from being responsible for the treatment and distribution of drinking water from surface water received from river Yamuna, Bhakra dam, and Ganges along with groundwater, the Delhi Jal Board is also in-charge of sewage treatment and disposal of wastewater. The process of water distribution is however not so straightforward. DJB supplies water directly to the households located within the Municipal Corporation of Delhi (MCD), provides water in bulk to the New Delhi Municipal Corporation (NDMC) and the Delhi Cantonment Board (DCB) for further distribution. Besides, the Delhi Development Authority (DDA), MCD, DCB, and NDMC are also engaged in the provision, development, and maintenance of water supply, mainly in newly developed areas. This creates inequity in the amount and quality of water received by different areas. Presently, nearly 83% of households meet their water requirements through piped water supply system (DJB 2016a). The rest depend on tube wells, deep bore wells, hand pumps, public hydrants, and other sources, which are mainly private and informal. The issues of inequity create resentments in the residents. While, at the one hand, some areas receive 24 h of water supply, on the other hand, there are areas that are yet to be connected to piped water supply systems, and others which are connected but receive water for less than an hour in a day.

Equipped with the powers of the capital city, the Delhi Government, mainly through the Delhi Jal Board, has been able to shape the discourse in such a way that Delhi's water crisis is predominantly understood as a demand-supply gap. Today, besides the Yamuna water, Delhi gets its water through inter-basin transfers from the Beas River and the Ganges. Therefore, Delhi's water hinterland exceeds far beyond its local watershed and has a substantial impact on the water governance of distant regions and communities. It has brought a greater emphasis on the raw water supply augmentation through the construction of new dams in the Himalayas, and yet the shortage continues (Rohilla, 2012; DJB 2016b). There is enough data to prove the water scarcity in the city.

In 2010, Delhi's installed capacity for water treatment was about 745 million gallons per day (MGD) from the surface water. With an additional groundwater augmentation of approximately 100 MGD, the Delhi Jal Board supplies about 845 MGD water to the residents of the city. This supply is more than 100 MGD short of the actual demand (DDA 2010). This gap increases further during the summer season. Besides, as per the Delhi Master Plan 2021, the demand for water increases by approximately 20 MGD per year that add to the existing gap (ibid). Wastage of water is another primary cause of the shortage of water in Delhi. The unaccounted water losses in transmission and distribution of the water supply are noted to be about 42%, indicating further deficiency (Department of Urban Development 2006). Besides, nearly 40% (90 lpcd) of the treated water supplied by the DJB is used for domestic chores, flushing of toilets, and not for drinking purpose. This further reduces the quantity of clean water available for drinking purpose, and, therefore, people depend on sources other than DJB for water. Interestingly, the Delhi urban area has the highest per capita supply of water in India, which indicates further worse situation in other cities.

Consistently rising demand causes a greater dependence on groundwater, which results in groundwater depletion and pollution that affect both human health and environmental carrying capacity. Groundwater is a significant source of water in Delhi, particularly in unplanned or newly developed areas with inadequate water supply. A few studies suggest that in contrast to the official figure of 11% of Delhi's total water use, nearly 50% of Delhi's water comes from groundwater (Ruet et al. 2002). According to the Central Ground Water Board, the total groundwater potential for Delhi has declined from 428.07 million m^3 in 1983 to 292 million m^3 in 2003 (DDA 2010). Groundwater is confined to semi-confined state at the depth varying from 1 to 10 m below ground level in alluvial terrain and up to 70 m in sandy aquifers (DDA 2010). The maximum decline was noted in the Najafgarh and Mehrauli blocks, i.e., areas experiencing rapid urbanization and population growth (DDA 2010).

Although Delhi uses minimum water from the Yamuna, it discharges nearly 80% of the total pollution load in the river (CPCB 2006). As a result, the 22 km stretch of the Yamuna in Delhi is the most polluted stretch of the entire river. Yamuna is comparatively clean in the upstream of Delhi, despite a deterioration noted in the water quality from 1999 to 2005 due to an increasing number of coliform bacteria and concentration of ammonia (CPCB 2006). However, a threat exists with increasing population and industrial development in the cities upstream, e.g., Yamuna Nagar, Karnal, Sonipat, and Panipat, for a rise in pollution levels if treatment capacities for domestic and industrial wastewater continue to be inadequate. Due to the abrupt release of high qualities of wastewater from industries (e.g., distilleries), incidents of higher pollution levels have led to massive problems and the temporary shutdown of water treatment plants in Delhi. The Delhi Jal Board is now facing major difficulties in demanding unpolluted freshwater from Haryana when it is responsible for the same problem for downstream states, which receive polluted water generated by Delhi. Unable to resolve Delhi's discrepancy of sewage generation (about 700 MGD) and existing sewage treatment capacity (about 514 MGD), Delhi is responsible for the severe pollution of the river Yamuna. The situation gets further complex since the neighboring states Haryana and Uttar Pradesh represent the interests of both upstream and downstream water simultaneously. In such case of upstream-downstream relationship, the role of state governments in the regional water governance setup of UYRB is crucial.

In Delhi's water supply, the private sector also plays a significant role. A substantial number of private tanker companies are employed in Delhi to supply water to areas facing water shortage. The DJB has awarded contracts to three private companies to run 385 water tankers. It is in conjunction with the Delhi Government's plans to privatize essential water services in the wake of DJB's inability to successfully supply water to the unserved or underserved areas throughout Delhi. In July 2012, three companies—City Lifeline Travel, VSK Technologies, and Ramkey Enviro Engineers were awarded contracts for 10 years to run the 385 water tankers in five zones. Besides, the private sector also plays a dominant role in water purification either through supplying the bottled water or through reverse osmosis systems in Delhi.

The local and national non-governmental organizations (NGOs) working on water, sanitation, and environment are other essential stakeholders in the region. Environmental NGOs play an important role in improving public access to the proper information about the local problems, including drinking water quality and the water-related risks for human health. Many of these NGOs organize seminars and workshops on water problems for the communities and local authorities, multi-stakeholders' debates on water supply development, technical solutions, public hearings, and sanitation plans at the local and national levels. They also publish and disseminate many educational and information materials on different water and health problems.

The Supreme Court and the High Court also emerge as key stakeholders in case of state government failure and public cry for water. Based on a Public Interest Litigation filed by the environmentalists in 1992, the Supreme Court of India forced the riparian states to maintain a minimum flow of 10 cumecs (353 cusecs) of water downstream of Tajewalla to Okhla for ecological reasons. Again, in February 1996, the Supreme Court of India by giving priority to drinking water supply also forced Haryana to maintain the required pond level at the Wazirabad Barrage to ensure uninterrupted functioning of Delhi's water treatment plants. The Supreme Court thereby suspended the existing MoU in favor of drinking water supply in Delhi. Besides, the CPCB has been given the mandate of monitoring the river water quality by the Supreme Court. Apart from the Supreme Court, the High Court has also passed a few orders to Delhi Jal Board to supply regular water to the residents of Delhi.

Another important water stakeholders in Delhi include media and press. Apart from the multiple water problems of Delhi, the social and physiological impacts of policymaking interest the media groups. Journalists and reporters cover the water supply and management sector extensively. Media groups help to understand how politics, demand, supply, and other factors affect the actual availability of the resource for the public. Several print and electronic media reports have extensively covered the water supply and management situation across the city. The role of print and electronic media is vital in generating awareness among the general population. They also help to create a link between the government and the people by proving information on policy and their implications to society.

Besides, Delhi has a population of over 16.75 million, a key stakeholder that manages water at the very local level (Census of India 2011). The State of The World's Cities 2012–2013 suggests that the Delhi urban agglomeration is likely to have a population of 28.6 million by 2025. Its increasing population contributes to the rising demands and additional pressure on the limited water resource of the city. Due to the regular demand-supply gap, incomplete coverage, unreliable supply, the residents of Delhi make their own private water provisions. The residents are paying a high price to augment water and the poor suffer most from such a situation. These informal and alternative arrangements of water in Delhi are (1) private-owned bore wells and tube wells, (2) private small-scale piped water provision, (3) private water tanker, (4) private packaged water or bottled water, (5) informal reselling of water through pushcarts and bicycle operations, (6) DJBs' water tanker, and (7) DJBs' packaged water (Biswas 2011). The willingness-to-pay survey carried out under a study project estimates that around 23% of the households use such sources for at least part of their water requirement (Economic Survey of Delhi 2005–2006). Most of these alternatives are grossly unsustainable and available to people at high environmental, economic, and health costs. Illegal private water tankers and packaged water vendors charge a very high price, which is unaffordable to the poor, but they are flourishing due to poor water governance in the city.

10.5 RIA Framework: Influences on Decision-Making and Response to Water Risks

This section assesses various factors discussed in the Risk Interpretation and Action framework to identify their influences on decision-making and response to water risks by multiple stakeholders in Delhi.

10.5.1 Uncertainty

Risk emerges from uncertainty, but the meaning of uncertainty would differ from people to people, from the one which is associated with the likelihood of the event to the one related to the value of the consequences which may vary further both objectively and subjectively (Eiser et al. 2012). The studies on urban water management have modeled uncertainty about water demand and supply along with associated hazards, i.e., of scarcity, flooding, desalinization, and so on (Zelazinski 1998; Singh et al. 2010; Ray and Shaw 2015). Some studies have also explored the influences of uncertainty on decision-making (Bender and Simonovic 2000; Zelazinski 1998). However, less attention has been paid to the qualitative differences in the meaning of uncertainty to which people respond. In case of Delhi, uncertainty did not emerge as a leading factor in water management. The respondents however noted various sources of uncertainty that threaten water supply throughout the city, i.e., of erratic rainfall, lack of infrastructure, inadequate capacity to manage existing water demands, changing climate, and the complexity of water management systems. While most respondents admitted uncertainty in the water supply, only a few noted it to be an essential factor that influences decision-making, mainly the respondents from the government and research organizations. They found that varied uncertainty of hazards determines their priority for management. For example, a hazard of high frequency and spatial certainty receives higher priority as compared to a rare event with diffused locus and undefined spatial boundaries. These respondents also found themselves to be well equipped to deal with uncertainties because of their access to both science and technology. They mentioned that uncertainty has also played a key role in driving investments, e.g., towards prediction methods, early warning systems, or developing new seeds with a greater threshold of drought and flood resistance. Contrary to this, uncertainty is found to be less useful by the respondents from the non-government organizations and resident welfare associations. These respondents resolve conflicts based on evidence, and do not necessarily deal with uncertainty. They noted that their decision-making depends on the culture of responsibility where they must act despite having varying degrees of uncertainties being attached to a hazard. It causes continuation of various practices and unsustainable designs that are built to manage the current environmental challenges without necessarily focusing on sustainability in the face of future risks.

10.5.2 Trust

RIA Framework identifies trust in others as of central importance to any hazard response. It notes that trust is ingrained in the prior belief systems which vary individually as well as socially (Eiser et al. 2012). The role of trust has been studied mainly on the periphery of water management practices. There are some theoretical accounts of how trust can enhance cooperation between multiple stakeholders engaged in the process of water management (Richter et al. 2003; Ogden and Watson 1999; Wade 1988). However, the evidence of its influences on decision-making associated with water risks and security is limited. Most of the respondents in Delhi were found to be divided over the influence of trust in their risk perception and response. While respondents from the government organizations found that trust has little to do with their work as they follow the intent, procedure, and guidelines of the organization, it was noted to be a dominant factor by the respondents from the non-government organizations, education sector, and research institutes. Lack of trust between different stakeholders was evident in interviews. There is a lack of trust among communities in getting any assistance from the state authorities. It thus made it difficult to bring both people and members of the Delhi Jal Board on one platform as noted in Dwarka. One respondent from Shahadra stated that while one department plans by trusting other, but different departments do not respond as expected, which creates a discrepancy in response. Respondents also found a need for the government to start conversations to make people aware of risks as well as new knowledge and technologies. They note that government should provide subsidies, insurance, and required knowledge to gain farmers' trust, who are critical stakeholders in Delhi's water. Low trust in the system affects the type of solutions one chooses to deal with water-related issues, e.g., political, technological, or expressing grievances with municipal authorities. One respondent also mentioned that foreign organizations are trusted more for the local solutions as noted in the case of Public-Private Partnerships [PPP]. Another respondent from a non-government organization found that the government cannot be trusted for the pro-poor solutions, and, therefore, they must intervene and promote local interests. While all stakeholders are essentially working towards solving water issues for the state, such discontinuities in trust, not only could hamper the success of existing policies but may also affect future response in case of a major hazard.

10.5.3 Experience

Past experiences not only influence risk interpretation regarding severity and magnitude of the potential consequences, but also the perception of one's ability to deal with the uncertainty and other implications (Eiser et al. 2012). Experience has rather been studied more frequently in the domain of water management than any other factor (Lange 1998; Mitchell 2006; Woltjer and Al 2007; Domenech and Sauri 2011). Experiences tend to affect the decision-making in this sector directly. The success and failures of different approaches, methodologies, and technologies are of particular interest in this field, and their applications are regularly shared so that the acquired knowledge can be readily applied with slight or no modifications. In case of Delhi, most of the respondents had experienced water hazards including scarcity, contamination, flooding mainly in the form of waterlogging, pollution, and land subsidence. The role of experience was not easy to articulate in research but was clearer in practice. Most of the respondents from the organizations which directly respond to hazard such as government, non-government, or Resident Welfare Associations found that their past experiences immensely affected their decision-making. They found that experience helped them to prepare better for the contingencies. On the other hand, those engaged in the works that address hazards on the periphery found it less significant regarding its influence on decision-making. In general, researchers found that personal experience should not affect their writings on larger issues that affect a wider population. On the other hand, those engaged in practices of water management explained that experience adds to their knowledge but does not affect their decision-making. It is also found that in some cases, despite having a personal experience of a negative outcome, respondents continue to do what is asked of them. For example, one engineer mentioned that "by my experience, I wouldn't make rainwater harvesting structure on floodplain but because it is a government mandate, I have to make it," and in such case, experience has a little or no role to play in decision-making.

10.5.4 Learning

Learning plays an essential role in risk interpretation and action as it has the power to transform existing social belief. RIA finds that all learning is dynamic (Eiser et al. 2012). In the domain of water resource management and sustainability, there is a growing emphasis on learning, particularly, social and policy learning (Pahl-Wostl et al. 2007; Pahl-Wostl et al. 2008; Huntjens et al. 2011). In Delhi, learning emerged as a leading factor, and it is found to be extremely important in government, non-government, and research organizations. Here, learning has been mainly associated with water management, hazard characterization, and mitigation techniques. The most frequent source of learning is found to be research, literature, personal experience, newspapers, and other media channels. Various national government and research organizations mentioned pursuing research and expert opinion for making guidelines and suggestions, while the non-governmental organizations used their learning for advocacy and policy recommendations. The agencies which did not find learning to be important were those either dealing with a particular aspect of risks, such as running campaigns, writing magazine articles, or were Resident Welfare Associations, who must deal with water challenges on a day-to-day basis, and they address it accordingly.

10.5.5 Complexity

The complexity in risk interpretation and action emerges from various reasons ranging from that of a scale of hazard to its understanding, multiple interactions, learning, and sometimes diverse and opposing interests of varied stakeholders (Eiser et al. 2012). The complexity is widely recognized in water governance and boundary issues along with its multiple uses and users (Berger et al. 2007; Bressers and Lulofs 2010). In Delhi, while complexities prevail at different levels and scales of water management, not every respondent could relate to it, and, therefore, they found it less important. The respondents from the government, non-government organizations, and universities could clearly see the penetrative role of complexities in water management. However, the agencies involved in practice of hazard management and Resident Welfare Associations found it to be of low value in affecting their decisions because they deal with the issues at hand, and not necessarily have to work on the complexities associated with them. The sources of complexities in the water management in the city are found to be varied and dispersed ranging from a lack of knowledge and awareness to poor political will, social structure, lack of sectoral coordination, and uncertainty. Respondents noted that people tend to order water tankers in case of shortages and do not realize that a tanker also gets its water from the same aquifer. "Out-sourcing" the extraction of groundwater to another geographical location though solves the immediate water requirement; it creates deficiency with long-term negative environmental impacts. Similarly, the construction on the riverbed affects the ability of soil to absorb water, which leads to the problem of waterlogging as seen in Budh Vihar. Lack of coordination and communication between different departments is also a source of complexity in Delhi. A respondent noted that dealing with urban flooding should be the responsibility of the Ministry of Urban Development, but this is not the case. A multitude of civic bodies is interfering with water; the vastness of the network system along with the new trend of concealment is the major source of complexity. Earlier buildings had exposed water systems with uncovered pipes, open drains, canals for the supply of water, and as they were visible, people had a greater sense of belonging and responsibility for the water system. Now, since most pipes are hidden either behind walls or underground, it is difficult to detect any leakage, break or collapse instantly. As people must wait for external agencies to spot the fault and restore the system, it causes greater loss at times depending on the nature of the problem and complexities involved.

10.5.6 Scale

The importance of scale is recognized widely and applied in the water management sector in different domains be it water supply or issues relating to water use or users (Kurian 2004; Faysse 2004). In Delhi, respondents found water hazards at all levels, starting from the local neighborhood to sub zonal, zonal, city, and regional levels. Water scarcity is found to be spread around the city mainly in slums and unauthorized areas along with some planned areas, particularly during summers, while flooding is found concentrated in the Yamuna floodplain. Groundwater depletion and pollution are found across the city whereas surface water pollution is found in the river Yamuna and along the major drains that carry sewage. Respondents also noted that not just the character of hazards, but also the vulnerability varies across scales. Some sections of society are affected by many hazards. Apart from the poor sections of the society, the areas of very high density of population are also affected greatly due to water scarcity. Although the risk of water-related hazards, such as scarcity, groundwater depletion, pollution, and related health hazards exist at the city level, it is often not perceived like that. It is mainly because it is addressed at the local level. Resource conflict is also a major concern for the related government bodies. Scale of an issue directly influences decision-making. The scale of hazards determines its priority and the level of resource mobilization along with institutional and political attention that can be pulled for certain hazards. Also, the nature of response changes at different scales as noted by the research organization. For example, at the urban scale response include the introduction of water recharge methods and interactions with policymakers and government bodies, while at the social scale, it is important to speak with residents, use rainwater harvesting, and engage in greywater recycling, and at an individual scale, talking to individuals including children and residents through workshops to bring down water consumption gains prominence.

10.5.7 Social Context

The role of communities is found to be very important in the RIA Framework as they communicate and create the social system to deal with any risks. The water uses and associated practices not only emerge from social and cultural contexts but also impact them in multiple ways (Kley and Reijerkerk 2009; Donahue and Johnston 1998). In Delhi, the social context is found to be relevant by most respondents. They explicitly noted that rich people can adapt to hazards, while the poor are highly vulnerable and lack adaptive capacity. Slum population lacks a voice on the policy front. Delhi as a city is attracting migrants from different states with varied cultural backgrounds, which means that the people perceive risks differently. Most of the respondents engaged in research, teaching, and non-government organizations found the social context to be a significant factor behind decision-making, while government organizations and reporting agencies found it less important as they tend to respond to all groups. Non-government agencies mentioned that they respond to the needs of the society, so the social context is more important for them than hazards. Equity, poverty, affordability, access, and awareness of water rights are essential criteria for this. Also, what kind of solutions are suggested to a community is primarily dependent on the social context. It is noted by a respondent that the better aware public is likely to take better decisions, and it can also influence political will for specific issues and the type of solutions to be adopted.

10.6 Discussion

Water management stands as a distinct field with its explicit theories and specific practices for managing this vital resource as well as its risks. RIA framework was developed to understand human decision-making and actions in the face of disaster risks. Integrating the theory and practices from these two fields was though not obvious, the results of this exploratory study brought forth some interesting findings that need attention. While many of the factors identified in the RIA framework have been studied in the water management sector, they have not been assessed for their influences on decision-making at various levels. In the case study of Delhi, all respondents were asked to rank multiple factors of RIA framework for their influences on decision-making, which showed some interesting trends when averaged at the city scale (Fig. 10.1).

Among various RIA factors that influence decision-making, learning and experience turned out to be the leading factors, while complexity and occurrence of hazards had minimum influences on the decision-making. A significant finding is that the importance of the selected factors varied for different stakeholders. Various stakeholders were divided over the influences of trust and socio-cultural context. It is evident from the results that not only the understanding of different factors varied among various stakeholders but also their inclusion in the process of decision-making. However, most stakeholders could see some influence of most factors suggested in the RIA framework on water management, even though there were very few factors for which all agreed to be of high importance.

The importance given to a factor for its influence on decision-making also depended on the role of the respondents and their respective sphere of influence in decision-making. Trust is found to have minimal impact in the environment of protocols followed by the government sector. It is important to note that many civil authorities were aware of the lack of trust in public for them, but less effort has been put on the ground to bridge this gap. Trust is rather found significant by the non-government sector in dealing with water risks, as they must engage people in various activities. Such deviations are often not planned. Similarly, the variations in the significance given to complexity in decision-making often go unnoticed and unaddressed.

It is also interesting to note that there is a poor understanding of many factors mentioned in RIA and how they influence decision-making by different stakeholders. Many participants found it difficult to understand the influence of heuristics. The use of heuristics in the water sector is limited to the engineering and technical domain (Yang et al. 2012; Cabrera and Cabrera Jr 2003). In this context, heuristics refers to the process of learning from past anecdotes in contrast to experience that refers to enhanced professional knowledge concerning water management practices. The lack of its clear understanding and application in the Delhi water management highlights a gap that can be addressed for a better response.

The participants could also identify other factors that were not included in the RIA Framework, but they influence the way decision-makers perceive and interpret water risks, such as the political will or lack of coordination. Political will is found to be an important factor for its influence on water management decisions, which affects the way water issues are addressed. The solutions to enhance the quality and quantity of water along with reforms in decision-making often fail to be applied or produce the result in Delhi attributed to ideological contestation, fear of price hike, unequal access, and politics (Janakarajan et al. 2006).

Water in Delhi is mainly found to be a management issue, where there are too many stakeholders with minimum coordination. Delhi Jal Board is the nodal agency for the water management within the city however multiple stakeholders are making decisions around water every day

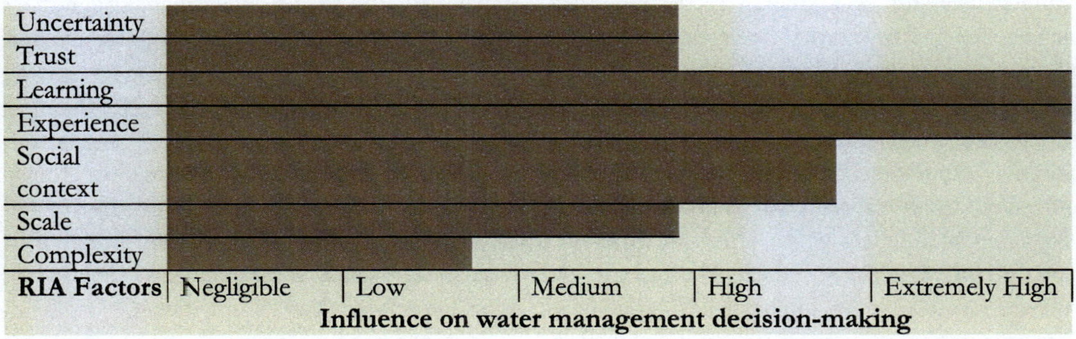

Fig. 10.1 Influence of RIA Factors on the Water Management Decision-Making in Delhi

at different levels with a completely different set of criteria, and at times with opposing interests. The conflicts of interest and power create more of water mismanagement than management. It is found that frequently, the cause of conflict is not resource scarcity but a human-made and entangled with ideology (Janakarajan et al. 2006). Lack of coordination, contestation, and political conflicts noted in the water management situation in Delhi, that often render the existing services less successful.

The existing capacity building programs revolve around the components of technical performance, economic efficiency, management issues, and policy applications, which fail to deal with the local conflicts based on varied risk interpretations and actions taken by the multiple stakeholders. According to Frantzeskaki and Loorbach (2012, p. 21), dealing with complex risks and uncertainties requires "a different set of guiding principles in the context of sustainability transitions. Transitions cannot be governed linearly by simple objectives and targets following regular implementations models." Within a centralized water management system, the participatory or reflexive water governance fails to be applied. Besides, it is essential that water-related risk management must find at the place within currently institutionalized practices of governing water (Conca 2015). Currently, drought and floods are managed by separate departments, which have little to do with the water governance system of the city.

Werbeloff and Brown (2011) highlight the vulnerability associated with the centralized institutional infrastructure and propose "security through diversity" as a strategy for urban water management. The current water management practices are based on engineering, economic, and ecological principles, and there are no distinctive variations based on the heterogeneity present within the community. It can also be attributed to the absence of traditional water management practices within cities. However, as noted in this case study, the decision-making by different stakeholders is influenced by varied factors, it is important to consider variations in risk interpretation and action. It can be particularly significant in managing major risks associated with water including, scarcity, flooding, or water insecurity.

It is noted that in a situation of a diverse group of stakeholders, dialog is an effective tool for better resource management than implementing more laws or policy reforms (Janakarajan 2003). The current water governance system puts a little emphasis on the complexity of multi-stakeholders dealing with the issues. While the aspirations of IWRM also include making it more participatory, knowledge-driven, and rational in economic, ecological, and hydrological terms, there are documented cases of IWRM-inspired "reforms" that create institutional rigidity without the benefit of better participatory decision-making (Conca 2015). Rules and guidelines are so rigid that despite having a pertinent knowledge, the officers and experts prefer to follow the rules than making a request for change or introducing flexibility to meet the changing situation.

This is further critical in the face of climate change. Mall et al. (2006) find that climate change brings uncertain future at the global, national, regional boundary conditions along with technology, law, socioeconomic development, politics, value judgments, and consumer habits that are likely to affect the water demand and supply in the Indian context. Climate change though addressed at the national level within the national water policy, it is found missing as an important factor for decision-making and planning response. Climate change adaptation in Indian cities finds little priority as most municipal authorities are challenged with a significant deficit in infrastructure and services (Sharma and Tomar 2010). In such a situation, recurrent messages of integrating climate change adaptation with that of disaster management and resource planning fail to impact the local reality due to a lack of coordination and support across the departments (Aromar 2008; Sharma and Tomar 2010). It is however not merely because of the differentiated decision-making power but also because of varied understandings of uncertainty, risks, and services that go unnoticed.

Disconnect between people and administration continues to exist despite several efforts on the political front in case of Delhi. This is primarily due to the lack of training of various engineers or officers who perceive water management as their job and people as the receivers. Less emphasis is being paid to social learning in the existing water governance. In the era of rapid technological change and public awareness, the dependence of people is likely to decline further creating an even more significant gap between the people and administration causing more water misuse and ecological degradation than planning and preparing for a sustainable future. It is crucial that capacity building is not just limited to the training of a few officials, but it also focuses on building responsible communities by promoting participatory water governance.

10.7 Conclusion

Water management and associated hazards are assessed and responded by different departments than that of the general disaster management. While it does bring specialized attention to water, it also creates some gaps due to a lack of communication across these departments. RIA model although developed in the context of disaster management, it highlights many critical aspects of decision-making, which are not understood and strategically used in water management practices. The gaps in understanding and knowledge of these factors result in a fragmented and ad-hoc response to water hazards and other management issues. These differences can be minimised by meaningful engagements of different stakeholders to bring a common understanding for sustainable water management practices. Water problems in Delhi are not merely the result of inadequate water provision, but emerge from a complex water governance system where stakeholders are aware of different realities of water and associated issues that they addresses in their own way. It is essential that all the stakeholders realize the risk that the city faces and respond accordingly within their domain of understanding and power. More officials found it better to follow guidelines than understanding the risk context. It implies that existing practices and capacity building at the city scale need a rethinking and research for a more holistic water risks management.

Acknowledgment The author is grateful for the generous financial support from IRDR International Center of Excellence, Taipei and START (The Global Change Systems for Analysis, Research and Training), and acknowledges the support provided by the World Social Science Council to promote young researchers.

References

Aromar R (2008) Climate change risk: an adaptation and mitigation agenda for India cities. Environ Urban 20(1):207–229

Baggett S, Jeffrey P, Jefferson B (2006) Risk perception in participatory planning for water reuse. Desalination 187:149–158

Bender MJ, Simonovic SP (2000) A fuzzy compromise approach to water resource systems planning under uncertainty. Fuzzy Sets Syst 115(1):35–44

Berger T, Birner R, Díaz J, McCarthy N, Wittmer H (2007) Capturing the complexity of water uses and water users within a multi-agent framework. In: Craswell E, Bonnell M, Bossio D (eds) Integrated assessment of water resources and global change: a north-south analysis. Springer, New York

Biswas R (2011) Water management approaches in group housing development in Dwarka. Spatio-economic development record 18(5):125–129

Bressers H, Lulofs K (2010) Governance and complexity in water management: creating cooperation through boundary spanning strategies. Edward Elgar, Cheltenham

Briscoe J (1997) Managing water as an economic good: rules for reformers. Water Supply 15(4):153–172

Briscoe J, Malik RPS (2006) India's water economy: bracing for a turbulent future. Oxford University Press, New Delhi

Cabrera E, Cabrera E Jr (2003) Pumps, electromechanical devices and systems applied to urban water management. A.A. Balkema Publishers, Tokyo

Chauhan C (2015) Yamuna a Dead River, says report, even as focus on clean ganga. Hindustan times. New Delhi. April 18. Accessed on 20th September 2017 from http://www.hindustantimes.com/delhi-news/yamuna-a-dead-river-says-report-even-as-focus-on-clean-ganga/story-4R6VXEcjNOlLSelnREqrxN.html?isab=trueandmeta-geo=%2D%2D%2D%2D

Chatterjee R, Gupta BK, Mohiddin SK, Singh PN, Shekhar S, Purohit RR (2009) Dynamic groundwater resources of National Capital Territory, Delhi: assessment, development, and management options. Environ Earth Sci 59(3):669–687

Chu J et al (2009) Agent-based residential water use behavior simulation and policy implications: a case-study in Beijing City. Water Resour Manag 23:3267–3295

Conca K (2015) Which risks get managed? Addressing climate effects in the context of evolving water-governance institutions. Water Altern 8(3):301–316

Cossío V, Wilk J (2017) A paradigm confronting reality: the River Basin approach and local water management spaces in the Pucara Basin, Bolivia. Water Altern 10(1):181–194

CPCB [Central Pollution Control Board] (2006) Water Quality Status of Yamuna River (1999-2005). Accessed on 16 October 2012 from http://www.cpcb.nic.in/newitems/11.pdf

Crow-Miller B, Webber M, Molle F (2017) The (re)turn to infrastructure for water management? Water Altern 10(2):195–207

Department of Urban Development (2006) City development plan Delhi. JNNURM, Government of Delhi, Delhi

DDA (2010) Delhi master plan 2021 with perspective for the year 2021. Delhi Development Authority, New Delhi

DJB (2016a) About Us. Accessed 7 January 2018 from http://delhi.gov.in/wps/wcm/connect/doit_djb/DJB/Home/About+Us

DJB (2016b) Delhi Water Policy. Delhi Jal Board, Delhi. Available at http://delhijalboard.nic.in/sites/default/files/All-PDF/water%2Bpolicy_21112016_0.pdf

Dobbie MF, Brown RB (2014) Transition to a water-Cycle City: sociodemographic influences on Australian urban water practitioners' risk perceptions towards alternative water systems. Urban Water J 1(6):444–460

Domenech L, Sauri D (2011) A comparative appraisal of the use of rainwater harvesting in single and multi-family buildings of the metropolitan area of Barcelona (Spain): social experience, drinking water savings and economic costs. J Cleaner Prod 19(6–7):598–608

Donahue JM, Johnston BR (eds) (1998) Water, culture, and power: local struggles in a global context. Island Press, Washington, DC

Dube E, Zaag P (2003) Analysing water use patterns for demand management: the case of the City of Masvingo, Zimbabwe. Phys Chem Earth 28:805–815

Eiser JR, Bostrom A, Burton I, Johnston DM, McClure J, Paton D, Van der Pligt J, White MP (2012) Risk interpretation and action: a conceptual framework for responses to natural hazards. Int J Disaster Risk Reduct 1:5–16

Economic Survey of Delhi (2005–2006) Water supply and sanitation, Chapter – 13, p. 157–168. Planning Department, Government of Delhi

Faysse N (2004) Challenges for fruitful participation of smallholders in large-scale water resource management organizations: selected case studies in South Africa. Agrekon 43(1):52–73

Frantzeskaki N, Loorbach D, Meadowcrof J (2012) Governing transitions to sustainability. Int J Sustain Dev 15(1–2):19–36

GWP-TEC (Global Water Partnership – Technical Advisory Committee) (2000) Integrated Water Resources Management. TAC Background Paper No. 4. GWP, Stockholm, Sweden

Huntjens P et al (2011) Adaptive water management and policy learning in a changing climate: a formal comparative analysis of eight water management regimes in Europe, Africa, and Asia. Environ Policy Gov 21(3):145–163

ICNP (2001) Rainwater harvesting. International conference on new perspectives on water for urban and rural India, 18–19 September, New Delhi

Janakarajan (2003) Multi-Stakeholders' Dialogue as A Tool for Natural Resource Management, Paper presented in a National Seminar on Water, organized by CESS, Hyderabad, 30–31 July 2003. Accessed on 25 September 2017 from http://www.cess.ac.in/cess-home/wp/VUMURHJ1.pdf

Janakarajan S, Zérah MH, Llorente M (2006) Urban Water Conflicts in Indian Cities: Man-Made Scarcity as A Critical Factor in Urban Water Conflicts: An Analysis of The Origins and Nature of Water Related Unrest and Conflicts in The Urban Context. UNESCO Working Series SC-2006/WS/19. International Hydrological Programme (IHP) of the United Nations Educational, Scientific and Cultural Organization (UNESCO), France. Accessed on 22 September 2017 from http://www.chs.ubc.ca/archives/files/Urban%20Water%20Conflicts%20an%20Analysis %20of%20the%20origins%20and%20....pdf

Keath NA, Brown RR (2009) Extreme events: being prepared for the pitfalls with progressing sustainable urban water management. Water Sci Technol 59(7):1271–1280

Khan S (2014) Towards sustainability: managing integrated issues of the Brown and the Green agenda in water governance and Hazard mitigation. Policy Brief. http://chance2sustain.eu/67.0.html

Khan S (2015) Integrating brown and green agenda for sustainable and resilient cities through participatory water governance: a case study of Dwarka, Delhi. In: Design to sustain - towards effective water management through habitat development. Agha Khan Planning and Building Services, Mumbai, pp 154–171

Khan S, Sara LM, Sydenstricker-Neto J, Sutherland C, Hordijk M (2015) Scenario building as a process and tool for urban governance. In: Gupta J, Pfeffer K, Verrest H, Ros-Tonen MAF (eds) Geographies of urban governance: advanced theories, methods, and practices. Springer International Publishing, Cham

Khan S (2017) Water in megacities: risks and research gaps. GYA Connections: Magazine of the Global Young Academy, Issue 5. pp. 5–7

Khan S, Mishra J, Doyle E, Kuna-hui EL (2017) Rethinking communication in risk interpretation and action. Nat Hazards 88(3):1709–1726

Kley LS, Reijerkerk L (2009) Water: a way of life: sustainable water management in a cultural context. CRC Press, London

Kurian M (2004) Institutional analysis of integrated water resources Management in River Basins: a methodology paper. Working paper 79. International Water Management Institute, Colombo

Lange G (1998) An approach to sustainable water Management in Southern Africa Using Natural Resource Accounts: The Experience in Namibia
Maidment DR et al (1984) Time patterns of water use in six Texas cities. J Water Resour Plann Manag 110:90–106
McMohan TA, Weeks CR (1973) Climate and water use in Australian cities. Aust Geogr Stud 11:99–1088
Mall RK et al (2006) Water resources and climate change: an Indian perspective. Curr Sci 90(12):1610–1626
McGranahan G, Njiru C, Albu M, Smith M, Mitlin D (2006) How small water enterprises can contribute to the millennium development goals: evidence from Dar Es Salaam, Nairobi, Khartoum, and Accra. WEDC, Loughborough University. Accessed on 20th January 2017 from http://pubs.iied.org/pdfs/G00503.pdf
Miranda L, Hordijk M, Molina RKT (2011) Water governance key approaches: an analytical framework. Literature review. Chance2Sustain, University of Amsterdam, Amsterdam
Mitchell VG (2006) Applying integrated urban water management concepts: a review of Australian experience. Environ Manag 37(5):589–605
Molle F, Berkoff J (2007) Water pricing in irrigation: the lifetime of an idea. In: Molle F, Berkoff J (eds) Irrigation water pricing: the gap between theory and practice. CABI Publishers, Oxfordshire
Navarro-Navarro LA, Moreno-Vazquez JL, Scott CA (2017) Social networks for Management of Water Scarcity: evidence from the San Miguel watershed, Sonora, Mexico. Water Altern 10(1):41–64
Nguyen TH, Ross A (2017) Barriers and opportunities for the involvement of indigenous knowledge in water resources Management in the gam River Basin in North-East Vietnam. Water Altern 10(1):110–159
Obertreis J, Moss T, Mollinga P, Bichsel C (2016) Water, infrastructure and political rule: introduction to the special issue. Water Altern 9(2):168–181
Ogden S, Watson R (1999) Corporate performance and stakeholder management: balancing shareholder and customer interests in the U.K. privatized water industry. Acad Manag J 42(5):526–538
Pahl-Wostl C, Crabs M, Dewulf A, Mostert E, Tabara D, Taillieu T (2007) Social learning and water resources management. Ecol Soc 12(2):5
Pahl-Wostl C, Mostert E, Tàbara D (2008) The growing importance of social learning in water resources management and sustainability science. Ecol Soc 13(1):24
Pfeffer K, Baud ISA, Denis E, Scott D, Sydenstricker-Neto J (2013) Participatory spatial knowledge management tools: empowerment and upscaling or exclusion? Inf Commun Soc 16(2):258–285
Raul SK et al (2011) Integrated water resource Management in a Major Canal Command in Eastern India. Hydrol Process 25:2551–2562
Ray B and Shaw R (2015) Community response to water stress in the megacity of Kolkata and implications to urban resilience. New Technologies for Urban Safety of Megacities of Asia. Kathmandu, Nepal
Richter BD, Mathews R, Harrison DL, Wigington R (2003) Ecologically sustainable water management: Managing River flows for ecological integrity. Ecol Appl 13(1):206–224
Rohilla SK (2012) Water, city and urban planning: Assessing the role of groundwater in urban development and planning in Delhi. CSH Occasional Paper No. 31/2012. CNRS, Delhi.
Ruet J, Saravanan VS, Zerah M (2002) The water and sanitation scenario in Indian metropolitan cities: resource and Management in Delhi, Calcutta, Chennai, Mumbai. French Res Inst in India, Delhi
Sansom K (2006) Government engagement with non-state providers of water and sanitation services. Public Admin Dev 26(3):207–217
Shekhar S, Purohit RR, Kaushik YB (2009) Groundwater Management in NCT Delhi. Accessed on 20 September 2017 from http://www.cgwb.gov.in/documents/papers/incidpapers/Paper%2015-%20Kaushik.pdf
Singh A, et al (2010) Analyzing uncertainty and risk in the Management of Water Resources for the state of Texas. Texas Water Development Board. Retrieved on 14 September 2017 from http://www.twdb.texas.gov/publications/reports/contracted_reports/doc/0904830857_U ncertainty_waterResourcemgmt.pdf
Sharma D, Kansal A (2011) Water quality analysis of river Yamuna using water quality index in the National Capital Territory, India (2000–2009). Appl Water Sci 1(3–4):147–157
Sharma D, Tomar S (2010) Mainstreaming climate change adaptation in Indian cities. Environ Urban 22:451–465
Shove E, Walker G (2007) CAUTION! Transition ahead: politics, practice, and sustainable transition management. Environ Plan A 39(4):763–770
Sohn J (2011) Watering cities: spatial analysis of urban water use in the South-Eastern United States. J Environ Plan Manag 54(10):1351–1371
Torfing J, Peters G, Pierre J, Sorensen E (2012) Interactive governance, advancing the paradigm. Oxford University Press, Oxford
Toteng EN (2008) The effects of the water management framework and the role of domestic consumers on urban water conservation in Botswana. Water International 33(4):475–487
United Nations (2016) Transforming Our World: The 2030 Agenda for Sustainability Development. A/Res/70/1. Accessed on 17 January 2017 from https://sustainabledevelopment.un.org/content/documents/21252030%20Agenda%20for%20Sustainable%20Development%20web.pdf
Van Rooijen DJ, Turral H, Biggs TW (2005) Sponge City: water balance of megacity water use and wastewater use in Hyderabad, India. Irrig Drain 54:S81–S91
Vörösmarty CJ, Green P, Salisbury J, Lammers RB (2000) Global water resources: vulnerability from climate change and population growth. Science 289(5477):284–288

Wade R (1988) The management of irrigation systems: how to evoke trust and avoid prisoner's dilemma. World Dev 16(4):489–500

Wang Y, Mukherjee M, Wu D, Wu X (2016) Combating River pollution in China and India: policy measures and governance challenges. Water Policy 18(S1):122–137

Werbeloff L, Brown R (2011) Working towards sustainable urban water management: the vulnerability blind spot. Water Sci Technol 64(12):2362–2369

Woltjer J, Al N (2007) Integrating water management and spatial planning: strategies based on the Dutch experience. J Am Plann 73(2):211–222

Wong THF, Brown RR (2009) The water Sensitive City: principles for practice. Water Sci Technol 60(3):673–682

Yang X, Gandomi AH, Talatahari S (eds) (2012) Metaheuristics in water, geotechnical and transport engineering. Elsevier, Amsterdam

Yuling S, Lein H (2010) Treating water as an economic good: policies and practices in irrigation agriculture in Xinjiang, China. The Geographical Journal, 176(2):124–137. https://doi.org/10.1111/j.1475-4959.2009.00345.x

Zelazinski J (1998) Uncertainty in water management. Geographia Polonica 71:109–119

Zhi Y et al (2015) Evaluating and forecasting the drivers of water use in a city: model development and a case from Beijing. J Water Resour Plann Manag. https://doi.org/10.1061/(ASCE)WR.1943-5452.0000548

Part IV

How Resource Management, Sustainable Development and Governance Work: Case Studies

Check for updates

11 Regional Environmental Governance: Analytical Framework for Resource Conservation in the Eastern Ghats, India

Kapil Kumar Gavsker

Abstract

Regional environmental challenges arising from contemporary development practices are unavoidable. They cause irreversible impacts from local to global scale. Development in terms of industrialization, urbanization, and exploitation of natural resources produce negative environmental consequences, which pose a serious threat to the local ecosystem. Much of natural resources are in the areas where aboriginals live and depend on them. The development practices have affected the ecological settings as they aim at unscrupulous exploitation of resource base. Understanding this requires a strategic framework and institutional responsibility in the region, such as regional environmental governance. The Eastern Ghats, a less explored ecological settings and sensitive area, in Eastern India, requires integrated and collective regional approach for their conservation and sustainable management. This chapter deals with role of Space, stakeholders, institutional structure, and socio-ethnic elements in dealing with regional environmental challenges in the Eastern Ghats region. It also examines the contemporary development process and practices and offers an analytical framework for the conservation of the same.

Keywords

Conservation · Eastern Ghats · Ecosystem · Regional Environmental Governance · Resource Conservation · Sustainable Development

11.1 Introduction

Unique physical features and climatic situations have formed ecological habitats like forests, grasslands, wetlands, coastal, marine ecosystems, and desert ecosystems, which harbor and sustain immense biodiversity on the surface of the Earth. In a country struggling with scarce resources, growing population, unprecedented urbanization, and the consequences of climate change, the mountains as ecological regions can contribute more to the sustainable future. On a broader scale, mountains are home to one-tenth of the world's population and cover 25% of the Earth's surface. However, the challenges of development in mountains are rarely reflected in national policies. India is endowed with different types of natural resources, such as fertile soil, forests, minerals, and water. Nature provides humans with all resources necessary for life: energy for heat, electricity, and mobility; wood

K. K. Gavsker (✉)
Department of Geography, Institute of Science, Banaras Hindu University, Varanasi, India

B. Thakur et al. (eds.), *Resource Management, Sustainable Development and Governance*, Sustainable Development Goals Series, https://doi.org/10.1007/978-3-030-85839-1_11

for furniture and paper products; cotton for clothing; construction materials for our roads and houses; and food and pure water for a healthy diet. Upto the middle of the twentieth century, natural resources seemed to be inexhaustible. But, during the past decades, it is getting more and more clear that nature is not infinite. In a way, accessible types of reserves, resources, and raw materials depend upon the progression level, structure, and organization of society.

It is true natural materials become resources when humans value them for the satisfaction of wants. The use and value of resources change between cultures and over time. There is also no direct connection between the availability of rich natural resources and level of economic development of the concerned region. However, many countries have developed their economies by using their natural resources. Some also get income from their resources in the form of tourism and recreation. Brazil and Peru, for example, make a lot of money from the Amazon forests, which is diverse in trees and animals. Growth of the human population is a major factor affecting the environment. Simply put, overpopulation means that there are more people than there are resources to meet their needs. The human population is at 6 billion; with an annual global growth rate of 1.8%. Simultaneously, the World has experienced an annual economic growth rate of 2.7% over the past 3 years.

Box 11.1 Sustainable Development Goals

In the context of Eastern Ghats of India, regional environmental governance is a key driver for the achievement of sustainable development. In this chapter, the author deconstructs the inextricable links between the environment and the social and economic dimensions of sustainable development which relies on decision-making processes, effective institutions, policies, laws, standards, and norms. The UN agenda 2030 and its goals, targets, and indicators have identified several practical steps that can foster more efficient regional environmental governance, making better use of the resources available and designed in a way that will be more helpful to the implementation of regional environmental needs. The author in his research recognizes the need for collective approaches to regional environmental issues in the Eastern Ghats, with local stakeholders having the bulk of implantation responsibilities, influencing decision-making for a wide variety of biodiversity and ecosystem dynamics. The authors' research raises several questions about the increasingly complex nature of recent scholarship and practice around environmental governance. Indeed, political ecology has emerged as a critical lens for the evaluation of regional governance practice. The author utilizes the broader literature on political ecology to situate theoretical arguments linked to regional environmental governance in the context of the Eastern Ghats. That the challenge faced by Eastern Ghat's sustainable development is the development and adoption of a solid environmental governance process.

11.2 Conceptual Framework

This chapter adopts a regional environmental governance framework to understand unequal power relations, conflict, and cultural "modernization" under a global capitalist political economy as key forces in reshaping and destabilizing human interactions with the physical environment. Thus, the framework focuses to explain how regional and local resources are under threat the way development is perpetuating in the Eastern Ghats. Another one looks at rescaling of governance and a cohesive regional strategy for conservation of biodiversity and natural resources. This chapter discusses how regional environmental governance can be a reality and an effective strategy in addressing the crucial issues towards biodiversity and natural resources, conservation, and their sustainable use. It presents the role of the state, stakeholders, institutional structure, and socio-ethnic elements in dealing with regional environmental challenges. Further, the study offers a general framework to deal with regional environmental challenge with reference to India (Buck 2006).

The analytical framework used here is drawn from Buck's (2006) study of *The Global Commons*. It enables one to look at how certain resources remain as exceptions prior to wider use of modern technology and the value they contain is recognized by justifying the effort of acquiring them (Buck 2006). This interdisciplinary perspective deals with the commons and common pool resources and the way they are used and consumed. Buck (2006) argues that "to understand commons and common pool resources, we must understand the differences and connections between resources, resource domains, property and property rights" (Buck 2006).

Natural resources, defines Philip's Geography Dictionary (2000), are "those materials used by man which are found naturally, for example, mineral deposits, energy, timber, fish, wildlife, landscape." Most of these are exploited due to human activities. There are other *spatial-extension resources*, which have value because of their location, such as stationary orbits as they cannot be extracted or converted from their natural state. Buck suggests that resources located in fixed spatial dimension are known as *resource domains*. In a similar way, wildlife and biodiversity resources are found in the ecological resource domains such as the Western Ghats and the Eastern Ghats. So far *property* is concerned, Buck notes it is "aggregate of rights which are guaranteed and protected by the government." The point is that property rights and access to and control over land and natural resources can generate critical incentives for conservation and sustainable use, management, and governance of natural resources. Property rights may be held by individuals or by groups of individuals, such as communities, corporations, or nation-states. She notes that "in Roman law, the property was thought to exist in one of the four property regimes: *res publica*, res *communes*, *res nullius*, and *res privatae*" (Buck 2006).

Further, Buck develops a helpful approach to consider two attributes of resources: the difficulty or feasibility of excluding others from using the resources (known as *exclusion*) and the degree to which one appropriator's use of the resources diminishes the number of resources left for the other (termed as *subtractability*). The common pool goods have high subtractability and are difficult to exclude others from using. The timber is a common pool resource. At the same time, sets of rules (laws, treaties, regulations, customs) that define property rights are *property regimes*. The definition of common pool resources is that it is managed under a property regime. So far *commons* are concerned; they are resource domains in which common pool resources are found.

11.3 Changing Nature-Society Relations

Affluence is a problem because with increasing affluence comes an increase in the per capita resource utilization. The past two centuries have seen unprecedented growth in human population and economic well-being for a part of the World. This growth has been fed by equally natural resource consumption and environmental impacts. The main cause of worldwide natural resources, as well as biodiversity loss, is the conversion of natural habitats to other land uses. Developing countries are more dependent on natural resources as their primary sources of income, and many others depend on them for their livelihoods. According to the United States Institute of Peace (2007) "Some natural resources play a central role in the well-being of local community, and some are used for trade purposes. They are controlled by the state and are used as exports by the government to attain profit and power. Developed countries rely on imports of natural resources, and mineral-rich countries are positioned to supply the demand. Many of these resources have great values in the global market and participate in the international economic system."

Studies reveal that habitat loss and degradation create the biggest source of pressure on biodiversity worldwide. In some areas, it has recently been partly driven by the demand for biofuels. Most of the natural resources are in areas where aboriginals live and depend on the land-products. The development practices have affected the ecological settings of such areas. This requires a strategic framework and institutional responsibility across the region, such as the "regional environmental governance." Sharma (1989) argues that "to a considerable degree, the interest of our (own) society in the availability, renewability and exploitability of natural resources springs from the technological innovations." Therefore, cultural requirements and societal progress depend on the nature of technological developments leading to increasing exploitation and consumption of resources. Therefore, collaborative efforts between institutions along with active involvement of the communities can lead towards a productive and sustainable approach in this direction. Geographical regions of India, viz., the Eastern Ghats—a less explored ecological setting, and other ecologically sensitive areas require an integrated and collective regional approach for their conservation and sustainable management. India's biodiversity hotspots cover a significant proportion (16%) of the total global area under biodiversity hotspots.

India has extensive laws, and policies to promote environmental conservation and sustainable use of natural resources. The country has provided for the protection and improvement of the environment in its Constitution and has taken several steps in this direction through planning and policies to overcome the environmental problems. In all its Five-Year Plans, India has focused on the protection of environmental and conservation of ecologically sensitive regions. Environmental regulation is seen to play different roles during different phases of the country's development.

There are two significant phases through which environmental regulation can be understood in the case of India. Initially, environmental regulation served a reactive purpose as a means of cleaning up the (polluted) environment after the use of new technologies and new industrial establishments. During this initial phase of development, environmental degradation seems unavoidable from a process of economic growth. In this context, it is essential to focus on *environmental governance*—conservation of natural resources areas through collective goals.

11.4 Biodiversity and Ecological Degradation

India with a total geographical area of 329 million hectares is the second largest nation in Asia and seventh in the world. The diversity of habitats owing to different climates and altitudes endorsed the country's rich and diverse flora. India is fortunately endowed with a wide range of agroclimatic conditions that support the growth of an equally diverse range of plant and animal species. Biodiversity, as measured by the number of plant and vertebrate species, is greatest in the Western Ghats and the Northeast. This is because of the presence of tropical rainforests that are the richest habitats for species diversity. According to the Ministry of Environment and Forests Report (1996), India had over 45,000 plant species and 81,000 animal species representing 7% of the world's flora and 6.5% of its fauna.

India contains a great wealth of biological diversity in its forests, its wetlands, and in its marine areas. India has many endemic plant and vertebrate species. Among plants, species endemism is estimated at 33% (Botanical Survey of India 1983). Areas rich in endemism are northeast India, the Western Ghats, and the northwestern and eastern Himalayas. A small pocket of local endemism also occurs in the Eastern Ghats. India contains 172 species of animals considered globally threatened by the International Union of Conservation of Nature (IUCN). These include 53 species of mammals, 69 birds, 23 reptiles, and 3 amphibians. Four of the 34 globally identified biodiversity hotspots, namely, the Himalayas, Indo-Burma, the Western Ghats-Sri Lanka, and Sunderland are represented in India. This richness is shown in Table 11.1 in absolute numbers of species in India and the proportion they represent of the world total including Mammals, Birds, Reptiles, Amphibians, Fishes, and Flowering plants.

Table 11.1 Comparison between the Number of Species in India and the World

Groups	Number of Species in India (SI)	Number of Species in the World (SW)	SI/SW (%)
Mammals	350	4629	7.6
Birds	1224	9702	12.6
Reptiles	408	6550	6.2
Amphibians	197	4522	4.4
Fishes	2546	21,730	11.7
Flowering plants	15,000	250,000	6

Source: http://ces.iisc.ernet.in/hpg/cesmg/indiabio.html

Chitale et al. (2015) note that "the biodiversity hotspots situated in densely populated tropical countries are experiencing dynamics due to urbanization, agricultural expansion and rapid economic development." The loss of biodiversity is a very serious problem for the country. In addition, climate change and over increasing human disturbance are major causes for forest destruction and species extinction. Several species of living organisms are disappearing, and biodiversity is more threatened now than at any time in the past. Basha et al. (2014) observe that "effect of anthropogenic activities not only threatens the biodiversity, but also affects the socio-economic condition of the indigenous people of the forest. Various activities like habitat loss, deforestation, clear felling and overexploitation amplify the impact of climate change on biodiversity." A related issue with declining biodiversity is environmental degradation.

The UNDP India Report (2009) elaborates that "environmental degradation is a major factor in enhancing and perpetuating poverty, particularly among the rural poor, and such degradation factors impact soil fertility, quantity and quality of freshwater, air quality, forests and fisheries." In fact, the high standard of living that accompanies the increased production and consumption of goods is the major cause of pollution and environmental degradation. The growing trends of population and consequent demand for food, energy, and housing have considerably altered land use practices and severely degraded India's forest, viz-à-viz, environment also. The growing population put immense pressure on land extensification at cost of forests and grazing lands because the demand of food could not increase substantially to population. Major threats to biodiversity include habitat disturbance, habitat fragmentation, overexploitation, exotic species, climate change and pollution, and natural calamities.

11.5 Development Effect on Natural Resources

Sustenance and welfare of mankind depend upon the exploitation of different natural resources. The role of natural resources in economic development has been recognized by classical economists. Utilization of soil, water, minerals, coal, electricity, oil, gas, and nuclear energy is very important for the development of nation. These resources have changed the level of living standard of man. Natural resources are highly valued because human beings are dependent on them to fulfill their fundamental needs. Natural resources are materials and constituents formed within environment or any matter or energy that are resulting from environment, used by living things that humans use for food, fuel, clothing, and shelter. These comprise of water, soil, minerals, vegetation, animals, air, and sunlight. People require resources to survive. Everything which happens naturally on Earth are natural resources, that is minerals, land, water, soil, wind that can be used in many ways by human beings. In India, as argues Vandana Shiva that the colonial rule introduced dramatic breaks in the way in which forests in India were perceived and used. The perception of forest ecosystems with multiple functions for satisfying diverse human needs for air, water, and food was superseded by the growth of one-dimensional scientific forestry during the colonial period. Its objective was maximization of the production of commercially valuable timber and wood while ignoring the other ecological and economic objectives for the utilization of forest resources. Structural changes in a society also lead to transformations in consumption patterns and lifestyles, which, then, impact resource use. India is witnessing transformations due to its rapid economic growth which is influenced by interlinked factors, such as growing population, expanding industrial and service-related production, investment in various sectors, growing middle class, and increasing urbanization.

Resources are of two types, renewable and nonrenewable resources as shown in Fig. 11.1. Renewable resources can be described by scientists as a resource that can be replenished or reformed either naturally or by systemic recycling of used resources. Renewable is the source of energy that is replaced naturally or controlled carefully and can therefore be used without the risk of finishing. Nonrenewable resources cannot be reproduced or regrown and are therefore they are available in limited supply. Nonrenewable resources are resources for which there is a limited supply. Their supply comes from the Earth itself and, it takes millions of years to develop, is finite. Bhat (2010) remarks that "during the last century, with the growth of population, the environment has been subject to a harsh treatment by various human activities based on the development of scientific knowledge. The result is that mankind has altered the characteristics features of the Earth and its surroundings."

Forests play an important role in enhancing the quality of environment by maintaining the ecological balance and life support system. India with less than 2% of the world's total forest area, supports 18% of its population. The total area under forests was 675.54 thousand square kilometers in 2001, which was 21% of the total geographical area, as against the National Forest Policy (1988) stipulation of a target of 33%. The states which have shown significant decline in the forest cover are Andhra Pradesh and Madhya Pradesh. In India, the spread of grassland and shrubland is put at 12%. The wetland covers 3% of the Indian landmass or nearly 100,000 km^2. It is estimated that after independence, India has lost 4696 million hectares of forest land, while 0.07 million hectares of forest land have been illegally encroached upon, 4.37 million hectares has been subjected to cultivation, 0.52 million hectares given to river valley projects, 0.14 million hectares to industries and townships, 0.16 million hectares for transmission lines and roads, and rest for miscellaneous purposes (India: State of the Environment 2001). Many rivers have headwaters in forests. The forest sector contribution to GDP, though low (at 1.7% during 2011), could be high for the livelihood of forest-dependent communities or poor in general. Currently, 20,000 known deposits with as many as 87 minerals (including 4 fuels, 11 metallic, 50 nonmetallic, and 22 minor minerals) are exploited in the country. They necessarily lead to land degradation.

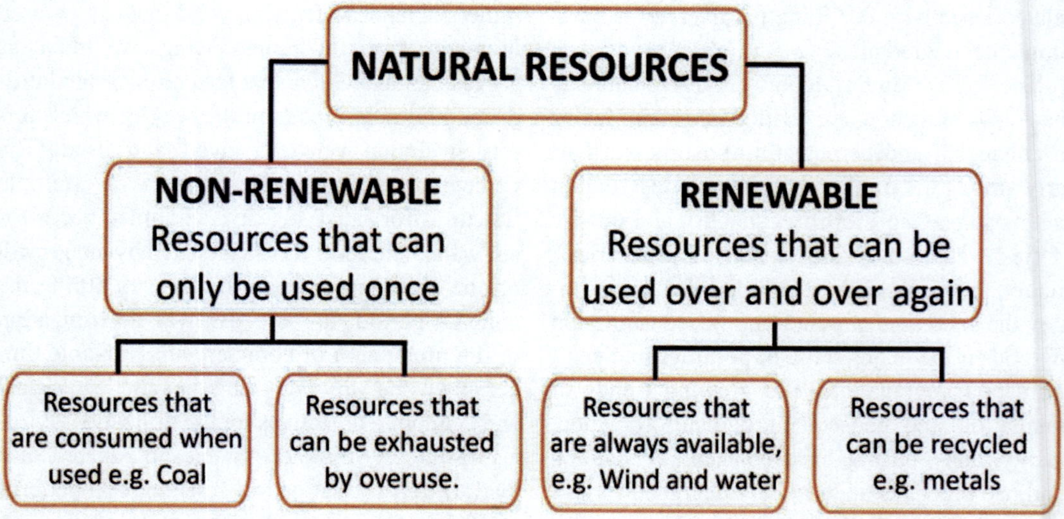

Fig. 11.1 Types of Natural Resources
Source: Based on Haggett, Peter (1979) *Geography: A Modern Synthesis*

A report on the environmental scam observes that according to the Greenpeace India Report over 1.1 million hectares of forest are at risk from coal mining in 13 coalfields in central India (see http://www.thehindubusinessline.com). Further, it says, when colonization ended, resource use policies continued along colonial lines. Today, those affected are the politically weak and socially disorganized whose survival is primarily dependent on products of nature outside the market system. Recent changes in resource utilization have almost wholly bypassed the survival needs of these groups. The Indo-German Environment Partnership Report on *India's Future Needs for Resources* (2013) elucidates that "industrialized countries already have high levels of resource consumption, while emerging countries need resources to provide appropriate living standards for their populations. Coordinated and collaborative efforts are required to ensure both availability and conservation of natural resources."

However, there has not been a uniform approach across the states of India towards the use of resources. Conservation of biodiversity has been a two-way process—in situ and ex situ—ignoring the integrated approach. Kale and Mazaheri (2014) indicate that despite vast endowments of natural resources and similar socioeconomic profiles, the Indian states of Bihar and Odisha pursued markedly different development strategies during India's first decade of economic liberalization. Whereas Bihar turned away from its natural resource sector and adopted policies of social empowerment, Odisha courted private investment in extractive industries and aggressively pursued market reforms.

11.6 The Eastern Ghats: Ecologically Underexplored Region

The Eastern Ghats, also known as Purvaghat or Mahendra Parvat, are a discontinuous range of mountains. This series of ranges runs parallel to the Bay of Bengal along the eastern coast from West Bengal in the north, through Odisha and Andhra Pradesh to Tamil Nadu in the south. Some parts of eastern Karnataka have also one of the important physiographic units with great environmental, socioeconomic, cultural, and spiritual significance in the Peninsular region. The Eastern Ghats are located between 11°30′ and 22°N Latitude and 76°50′ and 86°30′ E longitude in a North-East to South-West direction (Fig. 11.2). Much older than the Western Ghats, the Eastern Ghats are located at a height lower than the former. They are eroded and cut through by the four major rivers of southern India, the Godavari, Mahanadi, Krishna, and Kaveri. According to the EPTRI-ENVIS Newsletter (2015) "they (The Eastern Ghats) extend over a length of 1700 km in a north-east and south-west strike in the Indian Peninsula covering an area of 2,50,000 km^2 with an average width of 220 km in the north and 100 km in the south." Odisha has 25%, Andhra Pradesh 45%, Tamil Nadu 25%, Telangana 3%, and Karnataka has 2% of the hill ranges. The Eastern Ghats in the south are a rugged, long chain of broken hills and mountain ranges running almost parallel to the eastern coast of India. It constitutes the watershed of several important rivers. The seaside flatlands can be found between the Eastern Ghats and the Bay of Bengal. The hills are rich provinces for bauxite, iron ore deposits, and other earth resources.

The Eastern Ghats are an invaluable ecological treasure and natural heritage of our country. The hills are nestled with rich biodiversity with endemic and endangered species of plants and animals. The northernmost section of the Eastern Ghats is found in the mountains of Odisha. Between Mahanadi and Godavari, the average elevation of the Eastern Ghats is about 1100 m. The Singaraju Peak (Odisha) with an elevation of 1516 m is the highest peak of the entire range. Among other peaks, Nimagiri (1515 m) in the Koraput district and Mahendragiri (1501 m) in the Ganjam district are important hills. Between the Krishna and Chennai are the Kondavida, Nallamalai, Velikonds, Palkonda, and Erramala ranges. Sirumalai and Karanthamalai hills of Tamil Nadu lie in the southernmost part of the Eastern Ghats. North of the river Kaveri in Tamil Nadu are higher mountain ranges of the Eastern

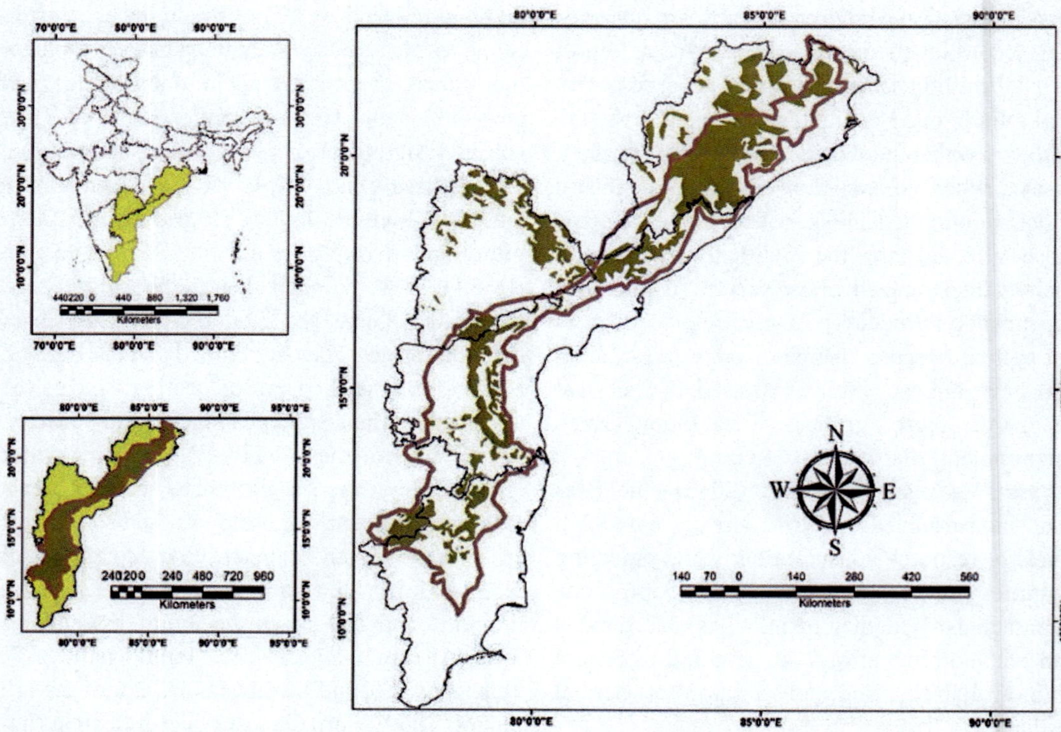

Fig. 11.2 Geographical Extension of Eastern Ghats, India
Source: Based on Murthy, K. S. Rama et al. (2013)

Ghats, namely, Kollimalai, Shevaroy, Kalrayan, Chitteri, Palamalai, and Mettur hills. The climate of the higher hill ranges is generally cooler and wetter. The Biligiri hills, which run east from the Western Ghats to the river Kaveri, form a forested ecological corridor that connects the Eastern and the Western Ghats. The Eastern Ghats join the Western Ghats at the Nilgiri hills, and the highest point is Doda Beta which is about 2637 m above sea level. The Indian Institute of Science reports that "the Eastern Ghats region falls under tropical monsoon climate receiving rainfall from both South-West monsoon and North-East retreating monsoon. The Eastern Ghats are rich in floristic diversity. Vegetation varies considerably with altitudes and shows a distinct zonation of forest types on the Eastern Ghats" (http://florakarnataka.ces.iisc.ac.in/hjcb2/easternghats.php).

11.7 Biodiversity, Flora, and Fauna of the Eastern Ghats

One of the widely recognized works dealing with diverse aspects of biodiversity, and flora is Pullaiah and Rao's work on the *Flora of Eastern Ghats*. This piece of research provides a vivid profile and comprehensive information on the physiography, vegetation, and natural habitats of the region. The Eastern Ghats are enriched with diverse flora and fauna. They note that North Coastal districts—Srikakulam, Vizianagaram, Visakhapatnam, and East Godavari are one of the biodiversity-rich areas of the Eastern Ghats. About 30% of Andhra Pradesh Forest are in the North Coastal districts. In the Eastern Ghats, besides natural vegetation, there are numerous exotic species widely spread throughout the

Eastern Ghats whose floristic composition differs from the natural vegetation. On the basis of dryness, Pullaiah and Rao, Ahmedullah, and Nayar divided Eastern Ghats into two types (1) Northern Zone of moist deciduous type with S. robusta, and (2) Southern dry deciduous forest region with Shorea tumbuggaia, Hardwickia binata type. Further, they cite Champion and Seth's classification which divides vegetation into nine categories: (1) Evergreen forests, (2) Tropical semi-evergreen forests, (3) Tropical moist deciduous forests, (4) Southern tropical dry deciduous forests, (5) Northern mixed dry deciduous forests, (6) dry Savannah forests (7) Scrub forests, (8) Tropical dry evergreen forests, and (9) Tropical dry evergreen scrub. The vegetation in the Eastern Ghats ranges from moist deciduous type in the north to dry deciduous type in the south. These forests are composed of tropical, subtropical, and temperate elements along with evergreen types which occur at high elevations.

There are five ecological hotspots with endemic and endangered species in India, out of which two of them are found in the Eastern Ghats. For millions of years, the Eastern Ghats have been cradles of life and civilization. The mountain ranges are rich in biodiversity. Asia's biggest tiger reserve, Nagarjunasagar-Srisailam Tiger Sanctuary is in the Nallamala range of the Eastern Ghats. As per the Ministry of Environment and Forests, some 15 wetlands have been identified in Andhra Pradesh, Odisha, Tamil Nadu, and West Bengal for management and conservation. Sixteen mangroves sites have been identified in Andhra Pradesh, Odisha, Tamil Nadu, and West Bengal for the protection of mangroves. The entire region of the Eastern Ghats is enriched with agri-horticultural diversity including valuable diversity in paddy, millets, pigeon pea, hyacinth bean, rice bean, niger, brinjal, cucurbits, yams, banana, mango, custard, apple, ginger, turmeric, etc. The whole region is inhabited by nearly 54 tribal communities, which constitute nearly 30% of the total population. Most of the tribal inhabitants are small and marginal farmers and are engaged in slash and burn and shifting cultivation. The land is also occupied by quite a few tribes which include Savara, Jatapu, Konda Dora, Gadaba, Khond, Manne Dora, and Mukha Dora. These indigenous people have their own unique cultural heritage. They follow age-old customs and traditions. Majority of them are still dependent on the forest produce and hunting for their livelihood and survival.

11.8 Unsustainable Exploitation of Resources and Threats to Biodiversity

The development-induced activities in bioresource-rich areas, unsustainable exploitation of resources, and increasing economic needs have affected biodiversity and caused an ecological imbalance. The ecosystems in the Eastern Ghats are fragile to degradation owing to fragmented and narrow distribution and heavy anthropogenic pressure. GrACE (Green's Alliance for Conservation of Eastern Ghats 2016) comments that rampant deforestation and devouring land use changes and encroachments in the hill causing a serious threat to the very existence of ecosystems, indigenous communities, and others in the region. It further complains that forest fires, grazing, mining, dams, roads, tourism, hunting, poaching, invasive species, etc., are some of the exhaustive list of instances undesirably altering the rich texture and mosaic of ecosystems. Mishra (2013) argues that "the depletion of forest cover in the Eastern Ghats range due to population explosion and developmental activities, including mining, have further accentuated the ecological imbalance and loss of rich biodiversity in that area raising a grave concern among the environmentalists and concerned citizens." In a sense, there is an imminent need for concerted and macrolevel conservation movement by collectivizing individuals, communities, and civil society groups.

Several laws have been enforced to safeguard both the environment and tribal communities, which include the Wildlife Protection Act 1972, Environmental Protection Act 1986, Forest Conservation Act 1981, Environmental Impact Assessment Notification 1996, Biodiversity Act 2002, National Green Tribunal Act 2010,

Scheduled Five of the Indian Constitution, Forest Rights Act 2005, and the Panchayat Extension to Scheduled Areas 1996. While these laws in principle could be effective in providing necessary checks for ecological disturbances and damages, in practice they have been found inadequate in many cases.

The more recent controversies of Niyamgiri and POSCO, where mineral extraction was proposed, have proven the complexities of political and legal battles fought by local communities and environmental and human rights groups. They are fighting for justice for nature and people. A similar situation will arise out of the decision of the Andhra Pradesh government's proposal to mine bauxite in Visakha and East Godavari region, as also the decision to go ahead with the Polavaram project, which would displace more people than the Sardar Sarovar Project on Narmada River. The effect of globalization has far-reaching consequences in the Eastern Ghats. With the advent of liberalization, the Eastern Ghats witnessed a sudden invasion of macroeconomic forces for the exploitation of its rich natural resources. Virgin forests, with their enormous wealth of timber, mineral, and non-timber resources have been ruthlessly, unscientifically, and irresponsibly tapped first by the public sector industries and from the 80s onwards, by the private and multinational industries. A report highlights that since long, Andhra Pradesh government has been trying hard to unearth the vast bauxite reserves in the Eastern Ghats despite campaigns and protests from the tribal community. The tribal communities believe that bauxite mining would not only render thousands of tribal people homeless, it would also sound the death-knell for the cultural diversity of the community and the endemic biodiversity of the Eastern Ghats (refer http://www.sspconline.org). The Polavaram Project will affect the biodiversity of the Eastern Ghats. Polavaram, when completed will submerge forests of the Papikonda National Park (in East Godavari district, West Godavari district, and Khammam district), which is close to the dam site.

Though the Eastern Ghats are not neglected in the biodiversity and natural resource conservation process, it is not also of extreme importance if one compares with the Western Ghats. Only a few areas are paid rigorous attention from the conservation viewpoints. Otherwise, with development race, states have adopted neoliberal policy which allows the private sector and transnational corporations to enter in mining and exploitation of other crucial resources in the region. Even there have been rampant cases of land acquisition in some places where communities reacted with resistance to a certain extent. *Dhaatri—Resource Centre for Women and Children* reports that however the adivasis today are far from being protected against exploitation and violence. Control over natural resources and freedom of rights is so threatened by the state today that there is an entirely new phenomenon of political conflict emerging from the *adivasi* regions. The story of Anantagiri and Kasipur: Anantagiri mandal of Vizag district in Andhra Pradesh and Kasipur taluk in Raigada district of Odisha are two evocative examples of the combined might of state and industrial oppression on adivasi communities. Anantagiri mandal has a history of erratic and unscientific mining on a small scale. With the liberalization of the economy, the state government opened the tribal area since 1990 to large-scale mining by giving leases to 17 companies, the biggest of them being Birla Periclase, a company of Indian Rayon and Industries. In Kasipur block of Raigada district, the tribal women opposed the entry of big corporations like Norsk Hydro, L and T, ALCAN, INDAL, etc. Here their struggle has been against much stronger lobbies of multinational and national industries controlling state power.

11.9 Regional Environmental Governance Framework

Natural resources are deeply tied to local communities through income, culture, and identity and thus often hold more than purely economic value for individuals and communities. With formation of State, resources fall under direct control of the government. Governments provide oversight for resource management, regulate trade and development, establish licensing protocols, levy taxes on resources industries, and engage in natural resource extraction directly or by selling extraction rights. The number of natural resources extracted to produce goods and services is steadily increasing. The fact is that governments that receive substantial income from natural resources may have less incentive to enter power-sharing arrangements or promote democracy-building efforts because they have the means to intimidate their opponents. However, indicators of successful management of natural resources that have contributed to peace and protection of interests of indigenous people include establishment of standards and agreements and efforts at cooperation, comanagement, and conservation.

Kurup (2008) indicates that though an overwhelming majority of India's tribal people inhabit the Peninsular region and the Eastern Ghats for that matter, they were recently introduced to decentralization when the Indian Parliament legislated the *Panchayat (Extension to Scheduled Areas) Act, 1996* (or PESA) exclusively for these areas. PESA mandated the states to devolve certain political, administrative, and fiscal powers to local governments elected by the communities (whether tribal or non-tribal). What PESA has done was, it did not amend the Fifth Schedule of the Constitution. The Fifth Schedule was, until PESA was legislated, an entirely centralized system where the communities—the majority being tribal—were directed in their affairs by the provincial governors. The Schedule permitted the states to extend their executive powers to these areas. Instead, PESA sought to secure the participation of the tribal communities through limited self-government, expecting this arrangement to be better suited to their "level of advancement." The ground situation is that the PESA has not been successful in achieving the stated objectives because of its several drawbacks. There has been rampant violation of tribal rights and their interests. These ongoing practices must be curbed so that resources and rights of the tribal and other communities can be saved, and development process is made sustainable. Thus, by and large, there is ecological interdependence in nature, as shown in Fig. 11.3, and it is largely determined by geographical proximity, climate contiguity, and ecological interconnections at regional scale (Sangmin 2016). What is known as common pool resources and shared environmental resources are very much driven by the elements and components of ecological interdependence. Uncontrolled exploitation of natural resources leads to make them vulnerable. Well-established environmental governance at regional scale can play a vital part in balancing relations between society and natural resources. Vulnerability and challenges can be addressed through a regional cooperation and integration of governing strategies adopted by the states towards common pool resources and sustainable management of the region.

Environmental governance is about how societies deal with environmental problems. It is concerned with interactions among formal and informal institutions and the actors within society. These interactions influence how environmental problems are identified and addressed in a collaborative manner. Regional environmental governance in the context of the Eastern Ghats among the states must take a lead towards access to information, transparency, accountability and political responsibility, application of subsidiary principle, acceptance of pluralism, discursive forums, and regional agreements. Protecting and conserving biodiversity—the variety of life in all its forms, including genetic, species, and ecosystem diversity—and its ability to change and evolve is fundamental to sustainable development. Integrated environmental governance at regional level may lead to sustainable management of both biodiversity and natural resources, such as the Eastern Ghats. Taking a dig at sustainable development as a political process, one may identify that this process has the potential to generate multiple sustainabilities. Therefore, "sustainable development" may refer to a procedural norm or principle for reconciling actual and potential conflicts between environmental, economic, and social goals (Balsiger 2012).

Drawing from the European experience, the Alpine Convention, an international agreement among eight European states and the European

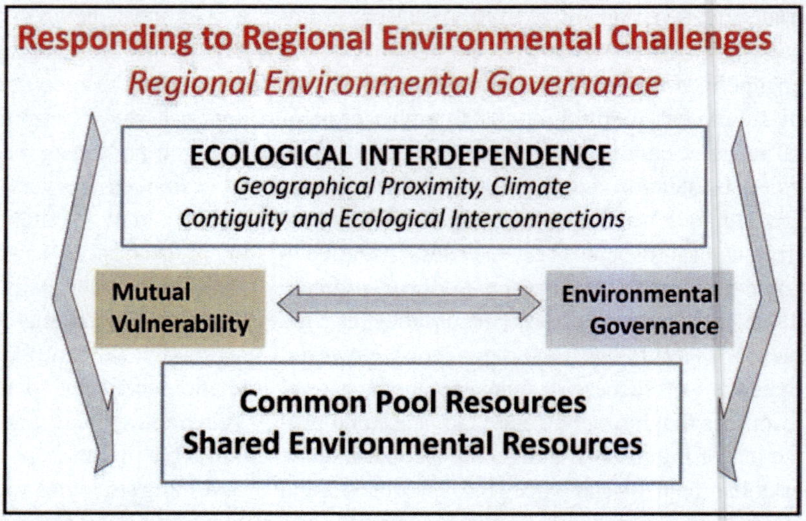

Fig. 11.3 Showing Inter-Linkages between Ecological Dependence and Common Pool Resources
Source: Nam, Sangmin (2016)

Union signed in 1991, seeks the protection and sustainable development of a globally significant mountain region. In geographical terms, the Alps as defined for the Alpine Convention have an area of 190,568 km^2, extending more than 1000 km from east to west and, at the widest, almost 300 km north to south. The highest peak is Mont Blanc (4810 m), on the French/Italian border. The total population of the Alps is about 14 million people, giving an average population density of 73 inhabitants/km^2. The Convention fostered the establishment of many trans alpine organizations as well as a nascent alpine identity. Balsiger (2012) notes that "the Alpine Convention recognizes the Alps as 'an economic, cultural, recreational, and living environment, accentuates the need for economic interests to be reconciled with ecological requirements' and portrays itself as a blueprint for sustainable development." While India's transition from a dirigisme to a modern, incentive-based model of economic management has fuelled rapid economic expansion, environmental management practices have failed to keep pace with these changes. Environmental governance still adheres to a command-and-control approach recognized as increasingly inadequate for managing the country's diverse set of environmental challenges. The instruments available to the environmental authorities no longer match the complexity of the Indian economy and the multiple sources of environmental pollution, including a heterogeneous industrial sector, unmanaged urban development, and overstretched infrastructure.

11.10 Conclusion

In a country struggling with scarcer resources, growing population, unprecedented urbanization, and the consequences of climate change, the mountains as the ecological regions can contribute more to the sustainable future. India is endowed with different types of natural resources such as fertile soil, forests, minerals, and water. Nature provides humans with all resources necessary for life: energy for heat, electricity, and mobility; wood for furniture and paper products; cotton for clothing; construction materials for our roads and houses; and food and pure water for a healthy diet. Growth of the human population is a major factor affecting the environment. Simply put, overpopulation means that there are more people than there are resources to meet their needs. Almost all the environmental problems we face today can be traced back to the increase in population in the world. The main cause of worldwide natural resources, as well as biodiversity loss, is the conversion of natural habitats to other land uses and increasing level of exploitation. Developing countries tend to be more dependent on natural resources as their primary sources of income, and many other individuals depend on these resources for their livelihoods. Major threats to biodiversity include habitat disturbance, habitat fragmentation, overexploitation, exotic species, climate change and pollution, and natural calamities. For example, the ecosystems in the Eastern Ghats are fragile to degradation owing to fragmented and narrow distribution and heavy anthropogenic pressure. The indicators of successful management of natural resources have contributed to peace and protection of interests of indigenous people include establishment of standards and agreements and efforts at cooperation, comanagement, and conservation.

References

Balsiger J (2012) New environmental regionalism and sustainable development in the European Alps. Global Environ Polit 12(3):58–78

Basha SK et al (2014) Impacts of climate change on Yerramalais forest of eastern Ghats of Kurnool District, Andhra Pradesh, India, and options for adaptation. Int J Biodivers Conserv 6(3):210–216

Bhat S (2010) Natural resources conservation law. Sage Publications India Pvt. Ltd, New Delhi

Buck SJ (2006) The global commons: an introduction. Earthscan Publications Ltd., London

Chitale VS, Behera MD, Roy PS (2015) Global biodiversity hotspots in India: significant yet under studied. Curr Sci 108(2):149

EPTRI-ENVIS Newsletter (2015) The. Eastern Ghats 21(2):1–7

GrACE (2016) Save eastern Ghats. Council for Green Revolution, Hyderabad

Peter H (1979) Geography: a modern synthesis. Harper & Row, Publishers, New York

Kale SS, Mazaheri N (2014) Natural resources, development strategies, and lower caste empowerment in India's mineral belt: Bihar and Odisha during the 1990s. Stud Comp Int Dev 49(3):343–369

Kurup A (2008) Tribal law in India: how decentralized administration is extinguishing tribal rights and why autonomous tribal governments are better. Ind Law J 7(1):87–126

Mishra S (2013) Save eastern Ghats ecosystem launched. The Industry and Mines Observer, June 16–30: 22

Philip's Geography Dictionary (2000) George Philip Limited, London

Pullaiah T, Rao DM (2002) Flora of eastern Ghats: hill ranges of Southeast India. Regency Publications, New Delhi

Sharma SK (1989) Resource development in tribal India: an example of the Baghelkhand plateau, Madhya Pradesh. Northern Book Centre, New Delhi

Shiva V (1991) Ecology and the politics of survival: conflicts over natural resources in India. Sage Publication India Pvt Ltd., New Delhi

India UNDP (2009) Analysis of existing environmental instruments in India. Administrative Staff College, Hyderabad

Web sources

Global Biodiversity Outlook 3 (n.d.). accessed at https://www.cbd.int/gbo3/?pub=6667&section=6711

Geography Standard 16- The Changes that Occur in the Meaning, Use, Distribution, and Importance of Resources (n.d.). accessed at http://nationalgeographic.org/standards/national-geography-standards/16/

India: State of the Environment 2001. accessed at http://envfor.nic.in/soer/2001/ind_bio.pdf

Sangmin N (2016) Regional Environmental Governance and North-East-Asia accessed at https://slideplayer.com/slide/10835116/

The Great Environmental Scam (n.d.). accessed at http://www.thehindubusinessline.com/opinion/the-great-environment-scam/article5075991.ece

United States Institute of Peace- Natural Resources, Conflict and Conflict Resolution (n.d.). accessed at http://www.usip.org/sites/default/files/file/08sg.pdf

India's Future Needs for Resources- Dimensions, Challenges and Possible Solutions (n.d.). accessed at http://www.igep.in/live/hrdpmp/hrdpmaster/igep/content/e48745/e50194/e58089/ResourceEffic iency_Report_Final.pdf

Eastern Ghats (n.d.). accessed at http://florakarnataka.ces.iisc.ac.in/hjcb2/easternghats.php

Bauxite Mining: Threatened Eastern Ghats of India (n.d.). accessed at http://www.sspconline.org/opinion/BauxiteMining-ThreatenedEasternGhatsofIndia_SrinivasGanjivarapu_080307

Tribal Women's Struggles against Political Violence in the Eastern Ghats by Dhaatri-Resource Centre for Women and Children (n.d.). accessed at sjjhttp://www.dhaatri.org/dhaatri/index.php?option=com_content&view=article&id=105:kbhanumathi-tribal-womens-struggles-against-political-violence-in-the-eastern-ghats&catid=54:resources&Itemid=69

12 Groundwater Sustainability in Haryana: Challenges to Governance

Inder Jeet

Abstract

In the alluvial heartlands of the Green Revolution, tube well technology played a significant role in the development of agriculture. Agriculture and particularly irrigation systems in this part of the country strongly depend on groundwater resources. Due to appropriateness of this technology, Haryana state is moving into crisis zone where the groundwater development has reached 133%. Pressure on groundwater resource has become linked to a few major developments in the cropping system. This change in the cropping pattern and certain governmental policies has negative effects on groundwater resources. Resultantly, Haryana has reached a stage where even their current level of groundwater extraction exceeding recharge and is therefore unsustainable. This research inquires the evolution, trends, present state of groundwater development, management, and governance. Some indicators like groundwater level and groundwater quality have been adopted to measure groundwater sustainability. This chapter argues that small landholdings, intensive agricultural and government policies are the main causative factors of groundwater exploitation in Haryana.

I. Jeet (✉)
Department of Geography, Maharshi Dayanand University, Rohtak, India

Keywords

Governmentality · Groundwater · Haryana · Sustainability

12.1 Introduction

India's development has become strongly dependent on its groundwater resources. The rise in absolute and per capita water demands that is linked to population growth and changing consumption patterns is increasingly met from groundwater sources. The agricultural sector and the industrial both thrive from free access to water from wells. Governing the groundwater is simultaneously a growing challenge in large parts of the country, where the water table is steadily sinking. Overexploitation and quality deterioration are deterred by different policy and reform choices at federal, state, and local levels. The widening availability and demand gap is often held to be a governance problem, commonly interpreted as due to misguided policies, unenforceable legislation, inefficient bureaucracy, institutional fragmentation, low capacity, outdated knowledge, poor accountability, corruption, "vote-bank" politics, lack of stakeholder involvement, and so on. Quality deterioration suffers from partly the same conditions. The pressure on available groundwater resources in India necessitates sound, scientifically based reg-

B. Thakur et al. (eds.), *Resource Management, Sustainable Development and Governance*, Sustainable Development Goals Series, https://doi.org/10.1007/978-3-030-85839-1_12

ulations to prescribe behavior relating to use and abuse. The situation with groundwater use is well described in several recent publications, such as Deep Wells and Prudence (World Bank 2010) and Taming the Anarchy (Shah 2009a, b).

In 2016, India had more than 19 million tube wells, compared to less than a million in 1960. This technological revolution has played an important role in the country's efforts to combat poverty, but the ensuing development of irrigation has, in turn, resulted in significant water stress in some regions of the country. Studies point to the large variation in future groundwater levels across India. Under future climate change, notably, some districts will do better and may even be able to rely solely on sustainable water supplies allowing groundwater levels to recover. Others will see slower rates of groundwater decline and yet others will experience declines for the first time. But most of Punjab and Haryana, northern areas of Rajasthan and Gujarat, and parts of Uttar Pradesh and Tamil Nadu will face a continued decline in groundwater level. As the levels become deeper, rising pumping costs can make extraction prohibitive and directly impact welfare.

It has been established that a large part of the Haryana plains constitutes a widely spaced topographic depression between the Shiwalik hills and the Aravalli hills which have created the typical internal drainage conditions. Resultantly, there is a falling groundwater table zone in eastern and southern parts (Yamunanagar, Karnal, Panipat, Sonepat (part), Faridabad, and Gurugram districts) of the state there is a rising water table zone, leading to soil salinization and degradation, in the central and western parts (Rohtak, Jhajjar, Jind, Bhiwani, Hisar, Sirsa, and part of Sonepat districts). The state can thus be broadly divided into two distinct zones: the rising water table zone (52% of the state) and the falling water table zone (eastern and southern parts) (Jeet 2001, 2005).

Haryana, from being a food deficit state in 1966 at the time of its inception, has now emerged as a major contributor to the national pool of food grains. Agriculture accounts for 31% of the state GDP and, along with Punjab, Haryana led India's Green Revolution. Grain yields are some 30–40% above the national average and with just 1.4% of India's area, this small state provides 30% of the national procurement of wheat and 10% of its rice (Hellegers 2007). Development of water for irrigation can be cited as one of the major contributors to Haryana's agricultural success. The Western Yamuna Canal (WYC) with a majority of its command area falling in Haryana and liberal use of groundwater can be the most significant influence on the agricultural turnaround in the state of Haryana. The state has total availability of surface and groundwater is 13.43 MAF against the requirement of about 32MAF at present of water. The Yamuna, Sutlej, Ravi-Beas, and Ghaggar rivers have water availability of 9.24 MAF. The available 4.21 MAF groundwater is used in irrigation through 607,098 tube wells.

However, this success resulted in second-generation problems, such as declining resource base, hydrological imbalance, decline in underground and above ground biodiversity, and pollution of soil, water, and environment. Also, there had been a gradual decline in water table in areas having good quality groundwater due to cultivation of high-water requiring crops like rice and sugarcane. On the other hand, inland basin with underground brackish water, introduction of canal irrigation with poor on-farm water management, in the absence of effective drainage, has resulted in the rise of water table and soil degradation (salinization, solidification, and waterlogging). In the southwestern region, having poor-quality groundwater and low rainfall, drastic decline in water table has taken place due to dominance of sprinkler system of irrigation.

This scenario of groundwater use and development in the state is indicating an unchallenged threat to groundwater sustainability. It is defined as "the maintenance and protection of groundwater and related ecosystems to balance current and future environmental, economic and human

Box 12.1 Sustainable Development Goals

Acknowledging the synergy and trade-offs between groundwater and sustainable development is paramount to successfully implementing the United Nations SDGs. The author through his research displays both the geophysical characteristics and significance of groundwater as well as its impact on human development broadly in the state of Haryana, India. In his understanding, groundwater here has untapped potentials which includes climate change adaptation, hydrological resilience, hydrogeological storage of carbon emission, and access to renewable energy, among others. However, as Haryana continues to use groundwater for domestic use, enabling food production and sustaining critical ecosystem's function, as it relates to various aspects of human development including poverty eradication, human dignity and wellbeing, sustainable groundwater use, and development of policies face a paradoxical challenge. In Haryana, this challenge has reached a stage wherein the current level of groundwater extraction exceeds recharge and is therefore unsustainable. As this chapter recognizes, groundwater is a key resource for the achievement of SDGs but is unfortunately weakly conceptualized in the sustainable development goals, targets, and indicators. There is tremendous potential for revising, revisiting, and redefining SDG targets 6.3, 6.4, and 6.6. Meanwhile, the planning wing of the Government of India—*Niti Aayog* has developed a "composite water management index" (CWMI) as a useful tool to assess and further improve the performance in the efficient management of groundwater resources. Together, between the CWMI and the SDGs there is tremendous potential for a dent in understanding and articulating the narrative around the meaning and value of sustainable groundwater use and management with an evolving framework of governance taking into consideration the interrelationship between various sectors, and what it will take to achieve them.

(social) requirements" (Gordon Report 2011). Once the groundwater sustainability issues have been identified, the indicators that best allow the assessment and monitoring of the issue should be selected and developed. Indicators help describe relevant information on *trends in groundwater systems* in a clear and simplified way. Without indicators, it would be difficult to organize and present the information in an accessible manner. The criteria needed to develop and design groundwater sustainability indicators need to be sensitive, scientifically robust, measurable, and representative. Indicators need to be created from available or obtainable information, be consistent and reproducible. Ideally, groundwater indicators should be reliable and appropriate at different scales, plus comparable between different hydrogeological regimes (Li 2013). Strong consideration should be given to selecting already established indicators from internationally recognized organizations. In the present study, indicators recommended by UNESCO (2007), namely, *quantity, quality, ecosystem, socioeconomic, and governance* are adopted to evaluate the sustainability in the state. However, established indicators from recognized organizations can be used as a starting point and as references, but the most useful indicators are often those designed and customized based on local issues and the historical indicators used in the study area.

12.2 Trends in Groundwater Sustainability

12.2.1 Groundwater Availability

Availability of surface water in the state is not sufficient to cater to the demand of water for irrigation and other uses. Over the past decades, the state has witnessed extensive development of groundwater through tube well irrigation. The number of tube wells in the state has increased from less than 40,000 in 1960 to 6.97 lacs (0.697 million) during the year 2011. Resultantly, there is unabated exploitation of groundwater. The unplanned and unchecked growth of groundwater uses has resulted in depleting groundwater and the state is quite concerned about the issue. The climate change scenario is likely to further increase the unabated usage of groundwater. Haryana has among the highest rates of groundwater extraction in the country. Based on information published in India's annual groundwater report, and by comparing the data in the 2009 report to the 2014 report, one can see the net level of groundwater availability has declined. A satellite-based study from NASA observed that, over the Indian states of Rajasthan, Punjab, and Haryana, groundwater is being depleted at a mean rate of 4.0 ± 1.0 cm/year equivalent height of water (i.e., 17.7 ± 4.5 km^3) between August 2002 and October 2008. During this period, groundwater depletion was equivalent to a net loss of 109 km^3 of water, which is double the capacity of India's largest surface water reservoir (Rodell et al. 2009). Mall et al. (2006) further suggest that changes in the cropping pattern and land use pattern, overexploitation of water storage, and changes in irrigation and drainage in the Gangetic basin show a reduction in the Ganges discharge by 60% over 25 years. This has led to about 50% drop in water availability in surface water resources (Adel 2002).

In February 2016, Haryana was reported to be the first state to complete the mapping of underground aquifers. Haryana was among the eight States for scientific mapping in the first phase. The State Agriculture Department in July 2016 claimed that 18 districts out of 21 had witnessed an alarming decline in the water level in the state since June 1974. There was a total of 71 overexploited blocks across the state where groundwater had been exploited above 100%, while 15 blocks were listed in critical category and 7 blocks were in semi-critical category.

The annual replenishable groundwater resource of the state has been estimated as 10.78 billion cubic meter (bcm) and net annual groundwater availability is 9.79 bcm. The annual groundwater draft is 13.05 bcm and the stage of groundwater development is 133%. Out of the total 116 assessed blocks taken for study, 71 blocks (61%) are overexploited, 15 blocks (13%) are critical, 07 blocks (6%) are semi-critical, and 23 blocks (20%) are in safe category. The analysis of present groundwater resource assessment indicates that there is a marginal decrease in net availability of groundwater resources as compared to the previous assessment carried out for the period 2004–2008. The groundwater draft has increased by about 5%. It has been observed that nine blocks have shown change to higher category owing to increased groundwater draft for irrigation and other uses.

12.2.2 Quality

Continued decline in the groundwater table accompanied by signs of declining quality is the single most important factor contributing to unsustainability. Several contributing factors are recognized. Excessive withdrawal over annual recharge is obviously the main issue. Most of the area receives a mean annual rainfall of about 600–700 mm. Puddling soils using tractor power prior to transplanting rice is an important energy intensive agronomic practice aimed at reducing infiltration to help maintain water on soil surface which is considered important for obtaining high rice yields.

As per Natural Resource Management Working Group Report (NRM Report 2015), there has been deterioration in the groundwater quality due to overexploitation of groundwater, and, consequently, mixing of brackish water from adjoining/deeper aquifers. Deterioration in the physical properties, as well as decline in water table and quality due to overexploitation of groundwater with a tendency to extend rice planting increasingly to rainless months, is the principal factor in this rice-wheat system zone. There are raving concerns over water-level declines, in several districts in Haryana, coupled with multiple contaminants occurring above their MPLs that challenges safe and sustainable drinking water supply. Out of a total of 119 blocks in the state, elevated fluoride and nitrate have been reported from groundwater from 53 and 63 blocks, respectively.

According to a report by the Central Groundwater Board (CGWB) 2013, groundwater in most of the area in 11 districts of southern and western Haryana is unfit for consumption. The reason for this is salinity or high concentration of nitrate or fluoride. According to the CGWB report, the worst hit districts are Bhiwani, Fatehabad, Jhajjar, Mewat, and Sirsa. As many as 70% of samples, drawn from wells and hand pumps, in these districts, failed the test as chemical parameters were higher than permissible limits. In five other districts, Faridabad, Gurugram, Hisar, Mahendergarh, and Rewari, 30–50% of water has potable quality as per the Bureau of Indian Standards (BIS) 2012. The report was finalized in September 2016 for the years 2015–2016 and is based on BIS 2012 norms. Parameters to evaluate suitability of drinking water were salinity, nitrate, sulfate, fluoride, hardness, and alkalinity.

According to CGWB report findings, groundwater in these districts is not only unsuitable for drinking but also for irrigation. Prepared by the north-western region of CGWB, the report has stated that Ambala, Jind, Kaithal, Karnal, Kurukshetra, Palwal, Panipat, Panchkula, Rohtak, Sonepat, and Yamunanagar have more than 50% water fit for human use. The salinity of water is checked through its electric conductivity, where low conduction (measured in microsiemen per centimeter or S/cm) shows low salinity. The Report said low salinity (<750 S/cm) was found mostly in Ambala, Gurugram, Panchkula, Panipat, Karnal, Kaithal, Kurukshetra, Sonepat, and Yamunanagar districts, water with intermediate salinity (750–3000 S/cm) was found mostly in all districts and samples with high (>3000 S/cm) salinity were found scattered in Bhiwani, Faridabad, Gurugram, Hisar, Jhajjar, Kaithal, Mahendergarh, Mewat, Palwal, Rewari, Rohtak, Sirsa, and Sonepat districts.

Declining quality of groundwater is a related and important aspect. Two main contributing factors are seen. First, the region is among the highest per ha use of chemical fertilizers. With most studies showing that efficiency of fertilizer uses rarely exceeding 40–45%, a significant fraction of applied fertilizers is likely to join water bodies including groundwater. Second, with declining water table in the region, there are fears that subsurface flows from adjoining high salinity groundwater in some areas are already causing water quality decline.

Similarly, there are about 8804 medium and large industrial units working in Haryana, bulk of them are concentrated in six cities, namely, Ambala, Yamunanagar, Panipat, Sonepat, Gurugram, and Faridabad. The major industries in Ambala are metal (127) and food processing (46). It is observed that the groundwater occurring at shallow depths is alkaline and is of NaHCO3 type. Some of the well waters have been found with high NO3 due to contamination with domestic sewage. Thermal power plant, Sugar mill, National fertilizer, and Panipat Oil refinery are some important units located in and around Panipat City. There are more than 175 handloom and textile units that use large amounts of chemicals for processing and dyeing of the textile. The groundwater in most parts of the nearby area has been polluted due to discharge of effluent either in ponds or cesspools or in the *Ganda nala* flowing through the city. Gurugram has several mechanical, electrical, textile, electroplating, and chemical industries. The waste generated by these units is dumped untreated either on land or into the city sewage drains. Faridabad and Ballabgarh are major industrial towns and there are about 1500 registered factories housed in these towns, some of which generate hazardous wastes. The waste from industries engaged in electroplating works, manufacture of textile, fertilizer, plastic, etc., are normally rich in toxic trace metals. Besides, industrial effluent, discharge of untreated sewage in the roadside unlined channels may pollute the groundwater due to seepage. Soils in the nearby areas have deteriorated, turned acidic, due to continuous release of spent wash on the soil during irrigation and mud in the fields by sugar mills with attached distilleries at Yamunanagar, Panipat, and Rohtak. Though waste from hospital and nursing homes are required to be collected separately, in most cities and towns in the state, such waste, form a part of municipal solid waste. The waste is normally dumped in the low-lying areas for natural decomposition. During rainy season, the waste emits foul smell and becomes a potential breeding ground for flies, mosquitoes, and other insects.

12.2.3 Socioeconomic Aspect

Depleting groundwater resources not only disrupt ecological balance but also put heavy financial burden on farmers and give rise to socioeconomic inequality in its distribution. For Haryana, water is the key issue for sustainable growth of agriculture. The General Circulation models predicted that the Indian Subcontinent will be warmer by about 1.5 °C during the middle of the current century. It is also the fact that each

1 °C rise in temperature will increase the demand for irrigation water by 2–3% to sustain production at the current level, and the competing demands of freshwater for drinking and industrial purposes will further reduce the availability of freshwater supplies for agriculture in the state. It implies that agriculture will be the major user of poor-quality waters; hence, unproductive loss of water through evaporation and other processes must be reduced to sustain food production. The productive and economic efficiency of water and other inputs are interlinked and could be increased by maintaining proper soil health, resource conservation and augmentation, selection of location-specific water management technology, and crop diversification as well as by shifting the focus from purely crop commodity approach to integrated farming system approach to help the resource-poor farmers of the state. Thus, the complex and interlinked issues concerning land use, soil and water resources, biodiversity, climate change, and environment need to be critically addressed in a holistic way for their critical monitoring, conservation, augmentation, and utilization for sustainable progress of agriculture in the state.

12.2.4 Irrigation

Irrigation in the region has largely involved conjunctive use of water (i.e., a combination of surface and groundwater). However, with an increasing variability, lower controls on groundwater extraction, subsidies in electricity needed for pumping groundwater, and improvements in groundwater extraction technologies, the trend is a significant increase in groundwater for crop production in Haryana. Around 2009, groundwater served 60% or more of irrigated lands (Shah 2009a, b). A study reposted by Erenstein (2009) estimates the water productivity indicator for paddy and wheat in the region. The Report indicates that, on average, farmers irrigated wheat 3.4 times, while they irrigated paddy 34.5 times. Estimated irrigation volumes for paddy are also a multiple of those for wheat: a factor of 8.4 in Haryana. Physical productivity markers (crop yield per volume of physical inputs) for paddy are therefore markedly lower than those for wheat, reflecting significantly higher water inputs in paddy cultivation with relatively similar yields (Table 12.1). Compared to wheat, financial water productivity is also lower for paddy in each site, as the higher net revenues for paddy are offset by the higher water inputs. Estimated physical water productivity indicators for wheat are about 1.5 kg/m^3 and about 0.2 kg/m^3.

Further, the latest published groundwater report (Government of India 2014) provides data on annual groundwater at the state level—as shown in the synthesis in Table 12.2. Total replenishable groundwater provides the overall level of inflow or recharge of groundwater. Total annual groundwater draft is the total level of extraction for different uses, primarily irrigation and domestic/industrial uses. The difference between these two gives the level of availability and, in the case of Haryana, this is a negative amount for the latest reported data in 2009. With a 127% overall groundwater development, Haryana is at the third highest level of groundwater development in a country that ranks highest in the world for groundwater use for irrigation.

In Haryana State, the number of tube wells have increased linearly since the mid-1960s (Sharma et al 2008). For a given land holding the increase in crop productivity on irrigated land leads to greater number and deeper tube wells

Table 12.1 Selected Crops and Irrigation Information for Haryana

	Paddy	Wheat	Sugarcane
Irrigation costs (INR/ha)	6820.18	4066.90	2680.43
Derived yield (quintal/ha)	44.14	50.78	654.52
Water productivity indicator (kg/m^3)	0.2		1.5
Area under principal crops (2007)	1041	2365	140

Source: IISD Analysis

Table 12.2 Annual Groundwater Status and Stage of Development for Selected States (billion cubic meter)

State	Annual replenishable groundwater resources			Natural discharge (non-monsoon)	Net annual groundwater availability	Irrigation	Domestic and industrial use	Total	Projected demands upto 2025	Groundwater availability for future irrigation use	Stage of groundwater development (%)
	Monsoon	Non-monsoon	Total								
Punjab	16.43	6.12	22.55	2.21	20.35	33.97	0.69	34.66	0.95	−14.5	170
Rajasthan	9.43	2.43	11.86	1.07	10.79	12.86	1.65	14.52	1.84	0.75	135
Haryana	6.22	4.26	20.48	0.68	9.80	11.71	0.72	12.43	0.79	−2.70	127

Source: Central Ground Water Board (2014a, b)

which consequently decline water table (Foster and Rosenzweig 2005). Irrigation through groundwater uses high-capacity electric or diesel pump sets. In the case of electric agricultural pump (AP) sets, the electricity tariff applicable falls under two categories: AP metered consumers billed on an energy-consumption basis and AP unmetered consumers who are currently paying a flat rate based on pump rating per month. These categories are used to determine the quantum of subsidies to be set aside each year.

Haryana Electricity Regulatory Commission in its annual Tariff Order based on Annual Revenue Requirement filings by the state's distribution utilities. AP users must pay only a small fraction of the actual tariff, with the result that, each year, subsidies run into thousands of crores (1 crore = 10 million) for the agriculture sector. It must be noted here that the entire revenue gap in the AP consumer category is bridged by way of the AP subsidy from the state government, and no consumer category is cross subsidizing the AP consumers. However, the subsidy from the state government is not always reimbursed, which has invariably resulted in state DISCOMs operating in a state of perpetual loss and poor financial health. With such substantial levels of subsidies being afforded to farmers, the number of electric pump sets has risen steadily over the years in Haryana, as shown in Table 12.3. As per the latest figures, the total number stands at 772,310 pump sets with 556,664 (72%) being electric. A direct consequence of this surge in electric pumps has been mounting financial burden on the state's DISCOMs due to excessive use of electricity, nonpayment of bills, and drastic reductions in groundwater levels across the state.

Another interesting observation was the district-level distribution of pump sets, which perhaps could indicate the energy intensity of crops being sown. Sugarcane and paddy are water-intensive crops, and, hence, require more pumps per hectare than wheat, as can be seen from Table 12.4. The high density of pumps in their respective districts is indicative of this fact and represents the trend of opting for groundwater for irrigation due to very low electricity tariffs. This has been the major reason for the depletion of groundwater resources in these districts.

Over the last two and a half decades a shift has been observed in the cropping pattern for irrigated crops in Haryana. The area devoted to wheat cultivation has increased by more than 150%, and the area used for the cultivation of

Table 12.3 Agriculture Pump Sets in Haryana

Year	1970–1971	1980–1981	1990–1991	2000–2001	2010–2011
Diesel Pump sets	17,903	109,353	155,842	255,302	231,146
Electric Pump sets	86,455	222,674	341,729	334,171	492,311
Total	104,358	204,736	497,571	589,473	723,457

Source: Department of Agriculture and Statistical Analysis, Haryana (2012, 2013, 2014, 2015)

Table 12.4 District-Level Distribution of Pump Sets in Haryana

Crop	District	Types of pumps			Pump density (per 1000 ha)
		Diesel	Electric	Total	
Paddy	Karnal	184	43,416	43,600	382
	Kaithal	18,935	44,203	63,138	380
	Kurukshetra	8915	67,627	76,542	276
Wheat	Sirsa	19,062	39,147	58,209	688
	Hisar	19,556	11,416	30,972	566
	Fatehabad	8750	31,163	39,913	415
Sugarcane	Yamunanagar	5802	26,782	32,584	204
	Ambala	4873	22,919	27,792	189

Source: Department of Agriculture, Haryana (2014)

paddy, a highly water-intensive crop, has increased threefold (Gangwar and van den Toor 1987). After remunerative price policies for paddy rice were initiated in the late 1970s, the area devoted to rice cultivation expanded, resulting in the exploitation of groundwater resources. Subsidies for the use of electricity in the rural sector and the lack of regulatory measures for the use of groundwater only exacerbated the exploitation and encouraged inefficient use of groundwater. As a result, the water table declined rapidly, especially in the semiarid region.

The study shows that there is an immense increase in groundwater draft since 1974, causing enormous burden on groundwater reservoir in the study area. Private-owned shallow tube wells have increased many folds during the last three decades, which are extracting huge quantity of groundwater in an injudicious and unplanned manner. It is observed that majority of districts have experienced more than 90% increase in groundwater development in this span of time. Therefore, according to Groundwater Estimation Committee (1997), these districts are overexploited and need thorough managerial attention from government agencies. Furthermore, these districts require exhaustive monitoring and evaluation for future groundwater development.

As far as countermeasures and possible means of controlling the groundwater economies of these regions are concerned, it is emphasized that current experiences derive from the development of groundwater and not from direct and proactive formal and institutionalized control of the resource users. The stakes are high and contradictory goals are at hand, one of maximizing poverty alleviation using groundwater and the other, the concern for sustainability of the endeavor. The impending and increasing dilemma as the resource gets scarcer in many areas is how to secure access to the resource of the poorest and generally deprived farmers. In water-rich, but poor regions a still unclear strategy needs to be formulated of how to better increase access to energy, as well as other basic inputs and requirements for production increases, for the millions of small-scale farmers with inadequate livelihood opportunities to escape poverty.

12.3 Need of Governance

Groundwater governance is a major concern in India, particularly in Haryana where irrigated agriculture heavily depends upon groundwater. The groundwater crisis in the country was ignored until recently because governments were under pressure to produce more food for the growing population and groundwater generated prosperity. Groundwater exploitation was, in any case, mainly privately financed. Government is only now beginning to develop a management policy to control groundwater issues.

Groundwater has become a major contributor to GDP in the state. It is the foundation on which agriculture, urban development, rural jobs, and safe drinking water supply systems depend. Indeed, access to groundwater through private tube wells was a key factor in the Green Revolution in this part of the country. This explosion of groundwater use has occurred in a largely unplanned and uncontrolled way, taking place almost unnoticed in many parts. Consequently, in many places, the unplanned and massive use of groundwater has resulted in serious and growing problems of depletion and quality deterioration. Because of its local availability and generally good quality, limiting treatment costs, groundwater is often cheap compared to alternative sources of supply. When nearing depletion, supply from groundwater will have to be replaced by more expensive alternatives, claiming valuable economic resources that are not available for other investments. Increased water costs will translate in higher water bills, impacting the urban poor and middle classes most, or in higher fiscal cost.

The exploitation of fresh groundwater resources provided an opportunity for farmers to supplement their irrigation requirements and cope with the vagaries of the surface supplies thus increasing accessibility and reliability of water, while increasing crop production. However, due to uncontrolled and unregulated use of groundwater, the problems of over-draft of the aquifer and saline water intrusion have emerged in many areas of the Indus Basin (Kijne 1999). In Haryana, more groundwater is being pumped out than is being recharged leading to declining water table in many areas (Meinzen Dick et al. 1997). The increasing water table depth and high diesel cost is making groundwater use quite uneconomical. Likewise, salinization associated with the use of poor-quality groundwater for irrigation has further compounded the problem(Foster and Chilton, 2003). Therefore, salt-affected soils are becoming an important ecological entity in the study area. Generally, the major reason for emerging groundwater problems is that the management of groundwater resources could not keep pace with its development. The major issues pertaining to groundwater governance in Haryana are high population density, exceedingly large number of groundwater users, low levels of resource management capacity, high share of agriculture in GDP, and dependence of rural livelihoods on tube well irrigation, poor institutional arrangements, and lack of information on groundwater use.

Governance is understood as the operation of rules, instruments, and organizations that can align stakeholder behavior and actual outcomes with policy objectives (Marcus et al. 2012). Essentially, there has been a surge in the uncontrolled private exploitation of resource, and governance frameworks have been ill-adapted to control it. The result has been depletion and quality deterioration and, in some cases, the misallocation of the resource to uses on which society places a lower value. Governance today has to take account of the reality that in many locations "the cat is out of the bag." Once groundwater rights also have been asserted ahead of any governance systems that might have contained them, it is incredibly difficult to recover control. This is especially true in countries, where all the incentives are in favor of development and abstraction, particularly where agricultural policy coincides with farmers' own motives to produce evermore. These external incentives are compounded by the powerful incentives inherent in the resource itself that lead farmers to prefer groundwater to all other water sources.

Governance frameworks have proved very frail in the past to resist such powerful motives. Traditional and local governance developed to manage springs or oasis have rarely been able to adapt to the new tube well technology. Very few governments have been able to align agricultural policy with good water resource management, and even fewer have been able to recover control over groundwater once that control has been lost.

This chapter provides an overview of established practices as well as recent trends in groundwater legislation. It discusses key challenges that legislation must address to manage and protect groundwater and to effectively counter the unsettling effects of groundwater depletion and pollution. The right to water entitles everyone to sufficient, safe, acceptable, physically accessible, and affordable water for personal and domestic uses. The right to water poses an obligation upon the state to ensure such access in a nondiscriminatory manner. It has implications for water resources legislation and for the regulation of the water industry.

The existing legal framework governing groundwater is based largely on principles developed during the second part of nineteenth century and applied consistently until today. Basic rules governing access to and use of groundwater in Haryana were laid down in English decisions. Since judges developed this law, it should have given it ample scope for changing over time in line with changing circumstances and understanding of the science underlying the uniqueness of place. The system of land-based groundwater rights was developed at a time when groundwater was not a major source of freshwater, and the technology was not developed enough to facilitate the unsustainable extraction of groundwater. As such, groundwater was not a serious concern.

More recently, the calls for more effective interventions to halt aquifer depletion and deal with quality issues have grown. In response, many experts have shed light on the specific problems and possible solutions of groundwater in India. Not the least, this has happened within Steering Committees and Working Groups set up in connection with the Government's Planning Commission for preparation of its Twelfth Five-Year Plan (for 2012–2017). Some of these were commissioned to analyze the scope for reform of the law on groundwater.

The Central Government clearly wishes to promote institutional reforms at State level with more room for local adaptation and implementation. Part of this shift includes greater involvement and participation from local governments and communities (such as the elected village councils—Gram Panchayats—and Water Users Associations for irrigation).

However, to regulate and manage the groundwater development in Haryana, a draft bill has been adopted which is read as "Haryana State Groundwater Management and Regulation Act, 2013." This Act explicitly calls for a number of measures to protect and enhance recharge, namely, water harvesting, including rooftop rainwater harvesting, catchment conservation using appropriate groundwater structures or pits, the creation of protection zones in natural recharge areas and in areas that require special attention with regard to the artificial recharge of groundwater, programs for the recharge of aquifers, setting up artificial recharge structures, afforestation, and reforestation. This Act explicitly enables the relevant authority to issue guidelines for constructing appropriate rainwater harvesting structures in all residential, commercial, and other premises and larger open spaces.

The overall objectives of the Act, 2013 are to: (1) Regulate and control iniquitous groundwater use and distribution, based on priority of allocation to ensure that the drinking water/domestic needs of every person and irrigation needs of small and landless farmers can be met, (2) Ensure safe and secure drinking/domestic water for all people, particularly in groundwater-dependent regions, (3) Regulate the over-extraction of groundwater in order to ensure the sustainability of groundwater resources, equity of their use and distribution, and to ensure fulfillment of ecosystem needs, (4) Promote and protect community-based, participatory mechanisms of groundwater management, that is, adapted to specific locations considering resource, enhancement, and socio-economic set up, (5) Prevent and mitigate contamination of groundwater resources, (6) Promote and protect good conservation, augmentation (recharge) and management practices, and (7) Protect areas of land that are crucial for the sustainable management of groundwater resources and ensure that high groundwater consuming industries are not located in areas unable to support them.

12.4 Conclusion

There are several central and state groundwater authorities functioning in the state. These agencies have a wide network of monitoring wells from which the level of groundwater exploitation in a block or district can be calculated. The Haryana State does not restrict the number of connections awarded in a year based on these reports. The state water policies formulated by these agencies have started to advocate the use of artificial recharge methods to replenish the aquifers. Micro-irrigation techniques, which promote sustainable use of water, were not propagated much by the state government though some subsidies are now available. However, the villagers of the study area lack awareness on these technologies. A new proposal by the Soil Conservation Department allows a farmer with a certificate from the Horticulture/Soil Conservation Department showing that he is adopting sprinkler/drip irrigation for his plantation or vegetable crops, will provide him with priority access for a tube well connection by the Haryana State Electricity Board.

The government policy of appeasing the farmers lobby with zero electricity tariff can be a double edge to the sword as it can lead to unsustainable groundwater exploitation. The current need is the promotion of artificial recharge methods, which can replenish the aquifers, and adoption of micro-irrigation techniques, especially for the plantation crops. Under these circumstances, some stringent regulations are needed for better sustainability and governability.

References

Adel MM (2002) Man-made climactic changes in the Ganges Bason. Int J Climatol 22(8):993–1016

Central Groundwater Board (2013) Groundwater Yearbook India 2012–13. Weblink: http://www.indiaenvironmentportal.org.in/files/file/Groundwater, 2012-13.pdf

Central Groundwater Board (2014a) Groundwater scenario in India Premonsoon. Weblink: http://www.cgwb.gov.in/Ground-Water/GW.2014.pdf

Central Groundwater Board (2014b) Groundwater Yearbook, 2013–14. Weblink: http://www.cgwb.gov.in/documents/2013-14.pdf

Department of Agriculture and Statistical Analysis (2012, 2013, 2014, 2015) statistical abstract. Haryana

Erenstein O (2009) Comparing water Management in Rice–Wheat Production Systems in Haryana, India and Punjab, Pakistan. Agric Water Manag 96:1799–1806

Foster SSD, Chilton PJ (2003) Groundwater: the process and global significance of aquifer degradation. Philos Trans R Soc B Biol Sci 358:1957–1972

Foster,A.D., Rosenzweig (2007) Economic Development and the Decline of Agricultural Employment.https://doi.org/10.1016/S1573-4471(07)04047-8

Gordon Report (2011) Sustainable groundwater management: preliminary approach for assessing the sustainability of groundwater. CCME Water Management Development Committee. pp. 48

Jeet I (2001) Water resources and management. Manisha Publications, Delhi

Jeet I (2005) Groundwater resources of India. Mittal Publications, Delhi

Kijne JW (1999) Improving the productivity of Pakistan's irrigation: the importance of management choices. International Water Management Institute, Colombo

Li Q (2013) Groundwater sustainability in Prince Edward Island. The Department of Environment, Labor and Justice PEI, Charlottetown, p 32

Mall RK, Gupta A, Singh R, Singh RS, Rathore LS (2006) Water resources and climate change: an Indian perspective. Weblink: http://www.iisc.ernet.in/currsci/jun252006/1610.pdf

Meinzen-Dick RS, Brown LR, Feldstein HS, Quisumbing AR (1997) Gender, property rights, and natural resources. FCND discussion papers 29, International Food Policy Research Institute (IFPRI)

Marcus LJ, C.D. Bery, Eric. J .McNulthy (2012) The Walk in the Woods: A Step-by-Step Method for Facilitating Interest-Based Negotiation and Conflict Resolution. https://doi.org/10.1111/j.1571-9979.2012.00343.x

NRM Report [Natural Resource Management Working Group Report] (2015) Working group Report on natural resource management, Haryana Kishan Ayog, pp. 30

Rodell. M, Isabell.V, Famiglietti (2009) Satellite-based estimates of groundwater depletion in India.https://doi.org/10.1038/nature08238.

Shah T (2009a) Indian irrigation in transition: growing disconnect between public policy and private enterprise. Water Resource Management-Economic Instruments, Indira Gandhi Institute of Development Research, Mumbai

Shah T (2009b) Taming the anarchy? Groundwater governance in South Asia. RFF Press, Washington, DC

UNESCO (2007) Groundwater resources sustainability indicators. IHP-VI series on groundwater no. 14. Weblink: http://unesdoc.unesco.org/images/0014/001497/149754e.pdf

World Bank (2010) Deep Wells and Prudence : Towards Pragmatic Action for Addressing Groundwater Overexploitation in India. World Bank. © World Bank. https://openknowledge.worldbank.org/handle/10986/2835 License: CC BY 3.0 IGO."

13

Wetland Resources in the Brahmaputra Valley, Assam: Characteristics, Use, and Sustainable Development

A. K. Bhagabati and Nityananda Deka

Abstract

The Brahmaputra Valley in Assam is dotted with more than 2500 wetlands of various sizes, shapes, and depths. While most of them are fluvial in origin, some are proved to be of tectonic origin. They play a significant role in regulating surface and subsurface water flow in the valley and serve as repositories of diverse bio-resources. However, due to the growing pressure of population and uncontrolled exploitation of resources, on the one hand, and the increasing transfer of land in the wetland fringes to certain unsustainable uses, on the other hand, the wetlands are rapidly losing their areal extent and rich biological contents. This study provides an outline of the distribution of wetlands, their status as natural water bodies in the valley in general and a village, and presents an inventory of the water and biological resources available in the wetland environments. It assesses the threats and pressures on the wetlands and, finally, evolves some workable strategies and action plans for their sustainable development. The current relevance of the traditional knowledge systems associated with the wetland ecosystems among different tribal and non-tribal communities has also been examined in the changing environmental context of the area.

Keywords

Brahmaputra Valley · Degradation · Floodplain · Ramsar Convention · Wetland

13.1 Introduction

Wetlands of different types and sizes occur almost in all parts of the world. They cover an area of about 800 million hectares accounting for approximately 6% of the earth's land surface (Reddy and DeLaune 2008). Wetlands are the most sensitive and fragile habitats for diverse species of plants and animals which provide a variety of goods, services, and functions to the concerned ecosystems and human society (Maltby 1991; Moore 2006; Aber et al. 2012). Having transitional location between terrestrial and aquatic ecosystems and possessing some characteristics of both, these complex and biologically most productive ecosystems represent interfaces of natural and human systems. They act as ecotones across which one natural system gives way to another (Maltby 2009; Keddy 2010).

Drained intensively by the Brahmaputra and its numerous tributaries, the Brahmaputra Valley (BV hereafter) of Assam is dotted with numerous wetlands of different shapes, sizes, and character-

A. K. Bhagabati (✉) · N. Deka
Department of Geography, Gauhati University, Guwahati, India

B. Thakur et al. (eds.), *Resource Management, Sustainable Development and Governance*, Sustainable Development Goals Series, https://doi.org/10.1007/978-3-030-85839-1_13

istics. With as many as 2930 prominent wetlands as recorded in 1997 and thousands of unaccounted natural and man-made smaller wetlands scattered throughout, the valley is one of the richest in India so far, the number and diversity of wetlands are concerned. It is noteworthy that the biological diversity and productivity of the BV are greatly influenced by the wetlands just like the dense reserve forests, wildlife sanctuaries, and national parks. The wetlands constitute an important element of the floodplain landscapes which are intrinsically associated with other ecological settings of the valley. As the BV experiences heavy and continuous rainfall during the summer and occasional winter droughts, the wetlands play a crucial role by regulating the local and regional hydrological and ecological systems.

Importantly, the wetland ecosystems of the BV and the associated activities performed traditionally by the local people have contributed to the development of some peculiar landscapes in the region. The early settlers, who behaved as ecosystem people, earned their livelihoods depending mainly on the wetlands. Thus, the local people have been traditionally following an ecofriendly attitude towards using, protecting, restoring, and reviving the wetland systems. The rural settlements in the floodplain environment had originated and got spatially organized around the wetlands as they provided a variety of services to the nearby communities (Deka and Bhagabati 2015). Interestingly, most of the villages once had community ponds for procuring drinking water. It is also important to note that the rural systems in the valley are basically governed by subsistence mode of farming, livestock rearing, and traditional fishing which are closely connected with the neighboring wetlands (Deka et al. 2011).

Interestingly, during the British rule in Assam, the wetlands were treated as wastelands and the habitats of mosquitos were carrying malaria parasites. Subsequently, many of the prominent wetlands in and around some urban areas of the valley were filled up to transform them into human habitats. Thus, with the modernization of society, particularly during the last few decades, the traditional use of wetlands has experienced remarkable change. Moreover, the overexploitation of wetland resources, irrational mode of fishing, invasion by weeds, siltation, pollution caused by agrochemicals, adoption of flood protection measures (embankments and switch gates), construction of roads and railways have posed serious threats to the wetland ecosystems leading to their partial or complete disappearance (Bhagabati 2012). In addition, the continued poverty among the villagers, their ignorance about and indifferent attitude towards the wetlands have adversely affected the wetland ecosystems. Lack of proper attention to and understanding of the rural situations, defective development plans, particularly those guided by macro-regional considerations and principles of profit-maximization, have significantly impaired the natural resource base provided by the wetlands (Bhagabati and Deka 2017).

The BV has witnessed a remarkable change in land use/land cover pattern during the post-independence period as large number of people from different parts of India as well as Bangladesh and Nepal started migrating into the valley raising the density level to higher than many other states of the country. The total population in the valley in 1872 was only 18.84 lakhs (1.884 million) which increased to 251.22 lakhs (25.122 million) in 2011 putting tremendous pressure on the floodplain environment including the wetlands. Keeping all these in view the present study attempts to investigate the use and sustainable development of the wetlands in riverine environment of the BV, Assam, and the associated issues.

13.2 Relation Between Rivers and Wetlands

As elsewhere, the relation between rivers and wetlands in floodplains of the Brahmaputra is also very intrinsic and dynamic. Due to the sharp decline of slopes from highlands around to floodplains, rivers are found to change their morphological patterns and hydrological characteristics creating a range of wetlands in floodplains within the valley. Many floodplain wetlands, which are

Box 13.1 Sustainable Development Goals

Wetlands are a vital part of any ecosystem related to water, food, and climate and have a significant role to play in the regulation of surface and subsurface water flow, providing ecosystem services, and managing biodiversity. In the context of Brahmaputra valley of Assam, wetlands have become vulnerable to population pressure, incessant resource exploitation, and transfer of fringe land in the wetland to unsustainable use. The authors present through their research issues of risk and water scarcity to which wetlands are exposed in the Brahmaputra Valley region. As a result, policy imperative suggests that wetland management has of late shifted towards a focus on ensuring the sustainable development of individual sites and suites of wetlands across the Brahmaputra valley landscape. The Ramsar convention on wetlands recognizes conservation and wise use of wetlands in the valley. According to at least, seven out of 17 sustainable development goals, restoring wetlands is crucial to meet SDG targets (Ramsar Convention Secretariat 2007a, 2007b). The authors study and investigate the status and development prospects of wetlands in the Brahmaputra valley but also reiterate the broad objectives and vision of both the Ramsar Convention and the United Nations SDGs 2030 targets and indicators in the conservation and integrated sustainable management of the same. Their analysis highlights the need for synergy between integrated management, policies, investments, and practices.

formed in cut-off meandering courses, occur along the middle and lower reaches of rivers (Rogers and Ralph 2011). Besides, natural levees that stretch parallel to the northern bank of the Brahmaputra prevent the run-offs from directly reaching the main channel and thereby help in the formation of wetlands in the floodplain landscape.

It is observed in floodplain landscapes of the BV that almost all the natural wetlands serve either as the source or mouth of many small rivers and rivulets. During flooding, wetlands trap sediments, build floodplains, alter the hydraulic gradient, and promote the diversion of floodwater to alternative sites (Saintilan 2011). Interestingly, the diversity of flood-reliant and flood-tolerant flora and fauna and microorganisms including endemic and threatened species in wetland habitats are sustained by the flows and flood regimes of rivers (Rogers and Ralph 2011). The floodplain wetlands tend to experience geomorphic and ecological adjustment during floods and can respond rapidly to the processes of new channel formation, channel abandonment, and associated changes in flood patterns (Semeniuk and Semeniuk 1995; Thoms 2003; Ward et al. 2002).

The hydrological relation between rivers and wetlands helps in the growth and diffusion of fish population. The local fish species usually migrate to wetlands for spawning through feeder channels. Linking together all other habitats, wetlands become the last refuge of certain animals forced out of other habitats by anthropogenic disturbances. Thus, floodplain wetland biota maintains an intrinsic relationship with water dynamics of rivers, and flooding serves as "principal driving force responsible for the existence, productivity and interactions of major biota" (Junk et al. 1989; Rogers and Ralph 2011). As many of the wetlands are directly or indirectly linked with rivers, restoring riparian corridors may be a natural means for maintaining their hydrological dynamics and for reviving the ecological community. The natural processes and the environmental conditions responsible for wetland formation should be kept uninterrupted for the sustenance of wetland habitats (Jordan et al. 1987; Dahm et al. 1995; Mitsch and Gosselink 2007). As a matter of fact, rivers that are naturally connected to wetlands should be considered as part of the wetland ecosystem and such rivers must be protected, maintained, and restored to conserve the related broader ecosystem.

Although low-lying rice fields, especially *bao* paddy fields, are not normally considered as typical wetlands, they provide immense ecological services and habitats for faunal species, especially fish during summer season. During flood waves, many small fish, insects, and other animals take shelter in rice fields and with the decline of water level in winter, they move to nearby rivers and wetlands. As villagers find it difficult to catch fish in the summer rivers and wetlands due to very high water level, rice fields provide grounds for fishing for domestic consumption. The water depth in rice fields during summer (June–August) remains up to 2–3 ft which can temporarily support the presence of many aquatic species. Although rice fields reserve a smaller volume of water than the perennial wetlands, they attract a variety of birds. This is due to the presence of a variety of insects and some grains left out in fields after harvesting. Thus, summer rice fields can also be considered as a kind of wetland in the ecological context of the BV (Fig. 13.1).

13.3 Wetlands in the Brahmaputra Valley

The BV in Assam covering an area of 56,194 km^2 is a unique river valley with innumerable tributaries, streams, and wetlands (Fig. 13.2). As the valley is surrounded by hills, mountains, and plateaus on its three sides, the tributaries coming down from highlands onto flat plains to join main streams naturally follow meandering courses (Fig. 13.3). These courses are subject to frequent change, and, thus, create a series of wetlands of different shapes and sizes. Some of the wetlands

Fig. 13.1 (**a**) Transplanting of paddy, (**b**) women catching fish in the paddy field, (**c**) an Adjutant stork in a deep water paddy field

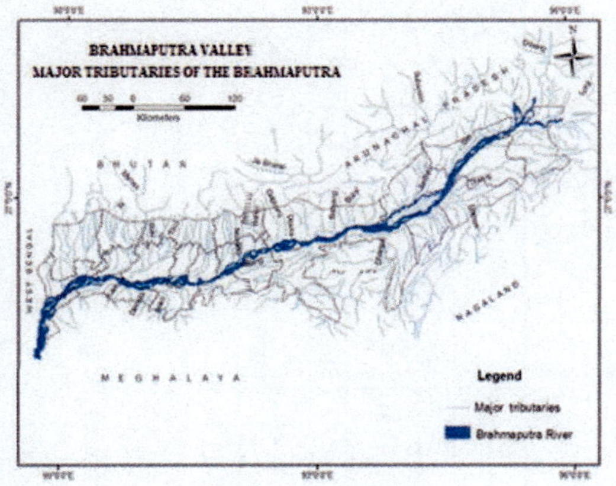

Fig. 13.2 The Brahmaputra and its tributaries in Assam

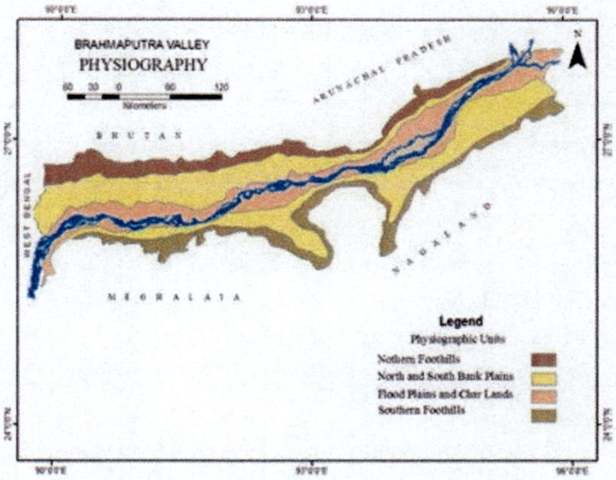

Fig. 13.3 Physiography of the Brahmaputra Valley

in the valley also originated due to subsidence of areas caused by tectonic events.

According to a survey conducted by Assam Remote Sensing Application Centre in 1997, there were as many as 2930 wetlands in the valley which covered an area of 84032.4 ha, accounting for only 1.43% of the valley's total area (Table 13.1). Various types of wetlands in the valley such as ox-bow lakes, swamps, marshes, abandoned channels (dead channels), reservoirs, tanks, seasonally waterlogged areas, and ponds make the valley's aquatic ecosystems more diverse in character (Fig. 13.4). Of these, ox-bow lakes registered the highest number (790) followed by the seasonal waterlogged areas (739), swamps and marshes (680), lakes and ponds (614), and reservoirs and tanks (107).

13.4 Wetland Degradation

Wetlands are extremely fragile and vulnerable habitats. They are very easily damaged, possibly beyond repair and they continue to remain under threat because of their potentiality for alternative uses (Moore 2006). The industrial revolution and subsequent urbanization affected wetlands all over the world and made them degraded. The wetland habitats were lost because of many physical, economic, political, and sociocultural factors, such as floodplain reclamation, construction of flood control structures, dams, draining and filling of wetlands for agriculture, habitat fragmentation, unscientific aquaculture, overharvesting of wetland products, clearing of aquatic vegetation, contamination through agricultural and industrial runoff, public and private attitudes and perception towards wetlands, excavation of peat for fuel and modification, and straightening of river channels in favor of navigation. (Bobbink et al. 2006; Garone 2011; Parisopoulos 2011).

The process of degradation of wetlands attained a new dimension after the great earthquake of 1950. This got accelerated recently due to growing population pressure and demand for food and other agricultural and industrial resources. In the rural areas, many wetlands are getting converted into agricultural lands and settlement sites, while in urban and suburban areas they are used for raising massive apartments, hospitals, hotels, restaurants, roads, and so on. Wetlands are also used as grounds for dumping urban wastes and garbage. There is evidence that many small and shallow wetlands have been already filled by sediments drained from the cultivated adjacent fields and degraded hills. It is also found that wetlands in the valley seem to remain intact, but their biological diversity is declining.

The frequency of flooding is essential for species composition, diversity, and functioning of wetlands (Keddy 2010). This study finds that due to the irrational human interferences, many streams in the BV got disconnected from the associated wetlands. This also suffered from deg-

Table 13.1 Distribution of wetlands in the Brahmaputra Valley, 1997

District	Number and area of different wetlands						
	Lake/ponds	Ox-bow lake/meander cut-off	Waterlogged (seasonal)	Swamp/marsh	Reservoir and tank	Total	Total area (ha)
Dhubri	73	68	25	65	2	233	6459.5
Goalpara	31	32	31	68	3	165	3832.5
Bongaigaon	25	36	17	21	1	100	3158.5
Kokrajhar	12	38	20	14	1	85	1578.4
Barpeta	17	30	7	42	1	97	3301.0
Nalbari	10	29	–	24	5	68	1987.0
Kamrup	49	65	64	155	19	352	11407.0
Darrang	11	45	22	18	7	103	3516.0
Sonitpur	23	35	110	22	16	206	3651.0
Morigaon	37	41	4	59	3	144	11658.0
Nagaon	68	71	138	92	10	379	11295.0
Golaghat	113	104	80	29	4	330	5466.5
Jorhat	39	24	34	9	3	109	2108.5
Lakhimpur	38	43	56	7	7	151	3033.5
Dhemaji	36	22	61	17	3	139	3960.0
Sibsagar	16	49	16	8	20	109	2135.0
Dibrugarh	13	36	23	12	2	86	2752.5
Tinsukia	3	22	31	18	–	74	2732.5
Brahmaputra valley	614	790	739	680	107	2930	84032.4

Source: Bahuah et al. (1997)

A B

C D

Fig. 13.4 Human pressure on wetlands. (**a**) Brick factory encroaching wetland, (**b**) rice cultivation in wetland fringe, (**c**) brick wall constructed in wetland fringe, (**d**) traditional fishing in wetland

radation in respect of species diversity (Fig. 13.4). Interestingly, the agricultural fields are often linked with the neighboring wetlands. The overexploitation of water for agricultural fields and unscientific application of pesticides and fertilizers in agriculture affect the floral and faunal diversity of wetlands. Another factor causing wetland degradation is the arrival of water hyacinth from Central America. This fast-growing weed deprives the microflora and fauna from sunlight and produces sedimentation and eutrophication.

Another reflection on this study, especially in Guwahati City area; the Deepar Beel, a Ramsar site and one of the largest wetlands of Assam, has been seriously affected by wide-ranging factors, such as heavy siltation, overfishing, hunting of water birds, pollution from pesticides, human encroachment, dumping of municipal solid wastes, and sewage from the city. Surprisingly, the Beel, which once covered an area of 10 km^2 and up to 40 km^2 during flood periods, at present remains confined to only 5.89 km^2. Another important wetland, called Silsako Beel in the heart of Guwahati City, which covered an area of 2.0 km^2 till 1980s, has now shrunk to almost half due to continued human encroachment. In other towns and cities of BV wetlands have faced similar problems.

13.5 Conservation Efforts

Up to the first half of the twentieth century, the multidimensional role of wetlands in ecosystems and livelihoods of associated communities were largely neglected. However, during the second half of the century, the ecological services rendered by the wetlands were given importance and conservation of them became a matter of concern (Aber et al. 2012). With the rise of contemporary environmentalism during the 1960s and 1970s, ecologists had given much importance to understanding the physical and biological processes of wetlands and their ecological and cultural roles. Worldwide wetlands have drawn attention of the environmentalists, ecologists, and geographers over the past few decades especially because of their ever-increasing degradation caused by irrational human activities. The international convention on wetlands held in Ramsar of Iran in 1971 had provided the framework for national action and international cooperation towards conservation and wise use of wetlands and their resources (Reddy and DeLaune 2008; Garone 2011).

The development of science of wetland ecology has strengthened legislation for wetland conservation across the world. On the other hand, the Global Convention on Biological Diversity, 2003 (CBD) considers "ecosystem approach" (EA) as a strategy for the integrated management of land, water, and living resources. Predictably then, the ecological principles need to be integrated into assessment and management for the conservation of wetlands.

Like many parts of the world in the BV also measures, such as wetland restoration, reclamation, enhancement, creation, and mitigation, have been suggested by the ecologists, hydrologists, and geographers conserving wetlands. Restoration can be made by repairing the human-induced damages to the structure and functions of natural ecosystems. Reclamation emphasizes on the restoration of functions and processes within a wetland ecosystem. Recently, emphasis has been placed on wetland enhancement including functions performed by an existing wetland. Wetland creation involves converting a non-wetland (either dry land or non-vegetated water) to a wetland (van der Valk 2009). Mitigation mandates that development or conversion activity impacting wetlands must be balanced with restoration, enhancement, or creation of wetland (Lewis 1989).

It is also important to note that any plan and policy for conservation of wetlands must incorporate related social and physical elements of concerned landscape or watershed. Strong legislative framework is also required for conservation and management of wetlands. Lack of awareness about multidimensional functions of wetlands on the part of people and the tendency to classify them as wastelands show indifference towards wetland conservation. Overall, wetland management in the BV, falls under the jurisdiction of more than one government department, e.g., while primary wetland management falls under the jurisdiction of the Ministry of Environment and Forests, Departments of Agriculture, Fisheries, Revenue, and Water also exercise some jurisdictional controls (Prasad et al. 2002). Therefore, it is difficult for the concerned people and organizations to design a concrete wetland management plan. A combined effort of the stakeholders including the government departments, technocrats, and nongovernmental organizations along with local communities can only conserve the wetland resources, meaningfully.

It is disheartening that the National Wetland Conservation and Management Rules, 2010 have not been adequately implemented in Assam to restore, classify, and update documentation of important wetlands. These rules under the Environment Protection Act, 1986 focused only on economic aspects of wetlands neglecting their values and services. The "Assam Hill Land and Ecological Sites Protection and Management Act, 2006," "Guwahati Water Bodies Preservation and Conservation Act, 2008" have not been successful due to their poor implementation. Therefore, it is necessary to formulate a separate wetland policy and form a separate wetland authority for sustainable management of wetland resources.

Subsequently, it is realized now that the indigenous resource-use strategies, traditional knowledge systems, experiences of the wetland-dependent communities, and cultural values attached to wetlands by the people should be considered in formulating wetland conservation and management plans and policies (Hammer 1996; Hay and Philippi 1999; Mitsch 2006; Wood 2006; Ramseier et al. 2009; Malacou 2011). Focusing on the past and present cultural values, as well as indigenous knowledge on wetlands—be they historical, traditional, or contemporary—can lead to a more flexible and site-specific management of wetlands (Papayannis and Pritchard 2011).

13.6 Development Prospects

In this study, it is now worth noting that important factors responsible for increase in intensity of flood hazard in the BV is degradation of its wetlands. As the wetlands serve as natural reservoirs, they have the potential to minimize flood intensity by storing excess surface runoff during the monsoon season. Therefore, there is an urgent need to rejuvenate the degraded wetlands and to create some new wetlands wherever feasible for flood cushioning, especially in flood-prone areas.

The floodplain wetlands of the BV provide livelihood to a large number of people. These wetlands have immense potentialities for fulfilling the State's domestic demand for fish as well as boosting up the rural economy of Assam. It is estimated that the demand for fish in Assam is about 2.21 lakh tons (0.221 million) per year, while its own production is only 1.55 lakh tons (0.155 million). Although the average productivity of wetlands of Assam was only 170 kg/hectare/year, the productivity potential of these wetlands can go up to 450 kg/ha/year, if they are cultured for commercial purpose. The pen culture technology developed by CIFRI is found to be an effective tool to meet the growing demand for quality fish seed (Chandra 2010).

The beautiful wetlands of the valley have immense potentiality for the promotion of ecotourism in the area. Ecotourism developed in and around the wetlands in a planned way may offer a variety of opportunities to the local people and conserve the dying wetlands. Many tourist attractions, like angling, water sports, boating, and aquatic gardening may be promoted in wetlands without disturbing their ecological settings. Involving the local people in tourism development may ensure dual benefits—generating employment opportunities and promoting wetland conservation goals (Scheyvens 1999).

The wetlands of the BV play an important role in agricultural development by providing water for irrigating the fields during the dry season. The wetlands and the neighboring low-lying paddy fields have also provided a natural base for integrated rice and fish farming in the valley. This practice may be renovated by introducing floating or hydroponic agriculture in the wetlands, wherever it is feasible and favorable.

Additionally, wetlands are the treasure house of many plant species having tremendous ecological and medicinal values. Many of them are however disappearing fast because of human interferences including human-induced water pollution. To conserve these plant species, some wetlands with greater potentiality may be upgraded as aquatic botanical gardens. Such gardens may attract tourists and provide scope for further biological research.

At present, there has been growing demand for flowers mainly in the urban markets which is met mostly with flowers imported from other States of the country. As several aquatic flowering plants grow in the wetlands of the BV which are traditionally in use in religious and other festivals and ceremonies, ecologically feasible floricultural practices may be introduced in wetlands involving poor farmers from the local communities (Table 13.2). It may help them raise their income level and ensure the conservation of the concerned wetlands. This may be feasible in some of the wetland ecosystems, and it may prove useful (Tables 13.3 and 13.4).

13.7 Wetlands in a Floodplain Village

This is a study on the nature and type of wetlands, their change in area, and utility within a village called Muktapur. It is in the lower Brahmaputra floodplain, Assam. The wetlands in the village cover an area of 21.44 ha accounting

Table 13.2 Checklist of some common plants found in wetlands

Local name	Scientific name	Local name	Scientific name
Kolmou	*Ipomoea aquatic*	Keheraj	*Eclipta prostrate*
Pani khutura	*Ludwigia adscendens*	Nikori	*Euryale ferox*
Helochi	*Enhydra fluctuans*	Pani long	*Ludwigia octovalvis*
Tita helochi	*A. philoxeroides*	Sorupuni	*L. perpusilla jorney*
Dal ghanh	*Nymeneche assamicanith*	Guri puni	*A. pinnata*
Petuli dol	*Hygroryza aristata*	Borpuni	*Salvania natans*
Boss	*A. calamus*	Pani meteka	*Monocharia vaginalis*
Kuhila	*A. aspera*	Bih meteka	*E. crassipes*
Pati doi	*Schumannianthus dichotoma*	Pani likosi	*Najas indica(willd) Cham*
Jora	*Alpinia galanya*	Pani tengesi	*M. quadrifolia*
Mati kanduri	*Alternanthera sessile*	Kona simalu	*C. benghalensis*
Bhet	*Nymphaea stellata willd*	Amarlota	*I. fistulosa*
Boga bhet	*Nymphea lotus*	Eralibon	*Leersia haxadra*
Ronga bhet	*Nymphaea rubra*	Pani chuli	*Nymphoides cristatum*
Suhoni	*Spilenthes clava*	Bon paleng	*Rumex nepalensis*
Padum	*Nelumbo nucifera*	Bihlongoni	*Polygonaceae*
Biyani bon	*C. muricata*	Leheti	*Renunculus sceleratus*
Sial bhobora	*Cerratophyllum demarsum*	Tora	*Alpinia jalanga*
Uria bon	*Cyperus corymbosus*	Sereka bon	*C. pilosus*

for 7.33% of the total operational area of the village. As a part of the livelihood support system, wetlands provide scope for fishing, collecting edible plants, flowers, facilities for bathing, washing, and irrigating the agricultural fields, and fodder for the cattle. They also function as microhabitats for a variety of aquatic flora and fauna within the agroecological setting of the village.

A comprehensive field survey covering 408 households of the village was conducted during 2006–2008 through a purposively designed schedule to acquire data. Moreover, through participatory rural appraisal (PRA) and interview with the villagers individually and in groups, wetlands have been classified and mapped according to their characteristics and use. Besides, oral discourses with some elderly and experienced persons were made to understand the pattern of change in the availability and use of wetland resources in the village.

The updated cadastral map (*dag* map) of the village obtained from the Revenue Circle Office at Goreswar, Assam is taken as the base to map the distribution of different types of wetlands. *Dags* are the smallest land revenue units of the village landscape which are generally demarcated by the village surveyors (*Mandals*) using some numbers systematically. The location of wetlands has been determined with reference to *dags* in consultation with owners in the field. A toposequence has been prepared following the PRA method to understand the topographical characteristics of wetlands in the local agroecological context. Tools and software, such as GPS, ArcGIS 9.2 were used in preparing the land use map and distribution of wetlands in the village.

13.8 Background of the Village

Muktapur village had originated and flourished by the side of a dead river channel called Punai. The term "Muktapur" stems from two local words: *mukta* (pearl) and *pur* (abundance). The plethora of resources that endowed the village is mainly the gift of the Punai dead channel. This channel served as the axis of life of the villagers as it provided abundant fish, water, aquatic vegetables, and fertile alluvium for growing crops. Interestingly, before the 1980s, there was no tube well or any other modern sources of drinking water and villagers had to depend solely on the natural or man-made water bodies.

Table 13.3 Checklist of some common wetland fish

Local name	Scientific name	Ideal habitats
Aarie	*Aorichthys seenghala*	River, dead channel
Bami	*M. armatus*	Dead channel, *beel*, *khal*
Bogi bhangon	*L. boga*	River, pond
Bardaia	*A. coila*	River, dead channel
Barali	*W. attu*	River, *beel*, *khal*
Besa	*Setipina phasa*	*Khal*, *beel*
Boriyola	*B. barila*	River, dead channel
Bhokua	*C. catla*	Pond, river
Bhangon/Nara	*L. bata*	River, pond
Botia	*Botia Dario*	*Khal*, *beel*, dead channel
Balisonda	*B. barna*	River, *beel*, dead channel
Bhasaylee	*Colisa colisa*	*Beel*, dead channel
Chanda	*C. nama*	River, *khal*, *beel*
Chengeli	*C. gachua*	*Khal*, marsh,
Cheniputhi	*P. sarana*	River, *beel*
Common carp	*C. carpio*	Pond, river
Dum vacheli	*B. badis*	*Khal*, *beel*
Darikana	*E. danricus*	Dead channel, *khal*
Gangatope	*Tetradon cutcutia*	River, *beel*
Gagal/Borsingora	*M. cavasius*	River, *beel*
Goroi	*C. punctatus*	*Khal*, dead channel, marsh
Gedgedi/Bhetki	*N. nandus*	*Beel*, *khal*, dead channel
Garua	*B. bagarius*	River
Grass carp	*Ctenopharyngodon mola*	Pond, *beel*
Kanduli	*N. notopterus*	Pond, dead channel
Kawoi	*A. testudineus*	Marsh, swamps, *khal*
Kholihona	*C. fasciata*	*Khal*, *beel*
Kokila	*X. cancila*	River, dead channel, *beel*
Karoti	*Gudusia variegate*	*Khal*, *beel*, river
Kuchia	*M. cuchia*	Swamps, marsh, dead channel
Lasi bhangun	*C. reba*	River, *beel*, pond
Laupati	*C. cachius*	*Khal*, dead channel
Magur	*Clarias Magur*	Marsh, swamps, *khal*
Moa	*Ambiypharyngodon mola*	River, *beel*
Mirika	*C. mrigala*	River, dead channel
Panimutura	*Glossogobius gutum*	River, *khal*, *beel*
Puthi	*P. sophore*	Dead channel, *khal*, *beel*
Pabhoh	*O. bimaculatus*	River, dead channel, *beel*
Rau	*L. rohita*	River, *beel*, pond
Selkona	*Chela atpar*	Dead channel, *beel*, *khal*
Singi	*H. fossilis*	Swamps, marsh, *khal*
Sol	*Channa striatus*	Dead channel, *khal*, *beel*
Silgharia	*L. dyocheilus*	River
Silver carp	*Hypothalmichthys molitrix*	Pond
Haal	*C. marulius*	*Beel*, dead channel
Tingara	*M. tengara*	Dead channel, *khal*, *beel*
Tora	*M. aral*	*Khal*, *beel*
Uporchakuamas	*G. giuris*	Dead channel, *khal*

Table 13.4 Checklist of some common wetland birds

Local name	English name	Scientific name
Masroka nila	Common kingfisher	*A. atthis*
Masroka tikonithoka	Pied kingfisher	*C. rudis*
Bogoli	Little egret	*E. garzetta*
Bortokola	Greater adjutant stork	*L. dubius*
Dawok	White breasted Waterhen	*A. phoenicurus*
Pani kauri	Little cormorant	*P. niger*
Konamusori	Indian pond-heron	*A. grayii*
Sorali	Whistling duck	*D. javanica*
Balimahi	White wagtail	*M. alba*
Chiloni	Black kite	*M. migrans*
Haargilla	Lesser adjutant stork	*L. dubius*
Samukbhonga	Asian Openbill stork	*A. oscitans*
Kaamsorai	Whiskered tern	*Chlidonias hybrid*
Dolmora	Red wattled lapwing	*V. indicus*
Kukuha	Greater Coucal	*C. sinensis*
Moniwori	Oriental darter	*A. melanogaster*
Hudu	Great horned owl	*B. bubo bangalensis*
Loriyuli	Eurasian Golden plover	*P. apricaria*
Dolpunga	Bronze winged Jacana	*M. indicus*
Fechu	Fork-tailed Drongo	*D. adsimilis*

Muktapur village is in Goreswar Revenue Circle of Rangia Subdivision, Kamrup district, Assam (Fig. 13.5). This is a typical Assamese village inhabited by indigenous non-tribal people. It is located on the north bank of the Lower Brahmaputra at nearly 35 km from Guwahati city and 40 km south from Bhutan Himalayan foothills. The village covers an area of 3.67 km^2 with a population of 2080 as per the household survey carried out in 2006. Distributed in 11 hamlets (*chuburi*) the total number of households in the village is 408.

13.9 Wetlands in the Village

Generally, called *khal-beel* by the villagers, wetlands of the village include man-made ponds (*pukhuri*), natural ponds (*khal*), dead channel (*mora nadi*), marshes (*pitoni*), water loggings (*hola*), and canal (*khawai*). Except for the man-made ponds, all other wetlands, such as *khal*, *pitoni*, *hola*, and *khawai* are natural in origin. There are as many as 471 wetlands of different types and sizes in the village with a total area of 21.44 ha accounting for 7.33% of the total operational area of the village (Table 13.5).

Out of the total 408 households of the village, 63% possess wetlands, the number of wetlands per household being 1.83 as high as 72.37% of the households possessing wetlands, have man-made ponds (*pukhuri*), 75.49% have natural ponds (*khal*), 5.06% have ownership over the marshes (*pitoni*) and 7.39% have small canals (*khawai*). It is evident from Table 13.5 that the man-made ponds with an area of 14 ha constitute the highest proportion of wetland area in the village.

The number and area of man-made ponds have increased significantly during the period 1980–2006 (Table 13.6). This increase is mainly due to the growth of some new households in the village. The number and area of the natural wetlands, on the other hand, have gradually declined. The overdependence on wetland resources, excessive siltation, alarming growth of weeds, and transformation of wetland margins into paddy fields are the main causes behind the rapid degradation of the natural wetlands within the village. Again, the process of fragmentation of the natural wetlands to bring them under different uses and the provision created according to individual ownership on wetlands have rendered them increasingly vulnerable to encroachment and degradation.

13.10 Indigenous Typology of Wetlands

The rural landscape of the BV is endowed with wetlands of various types. These may be identified on the basis of their micro-geomorphic char-

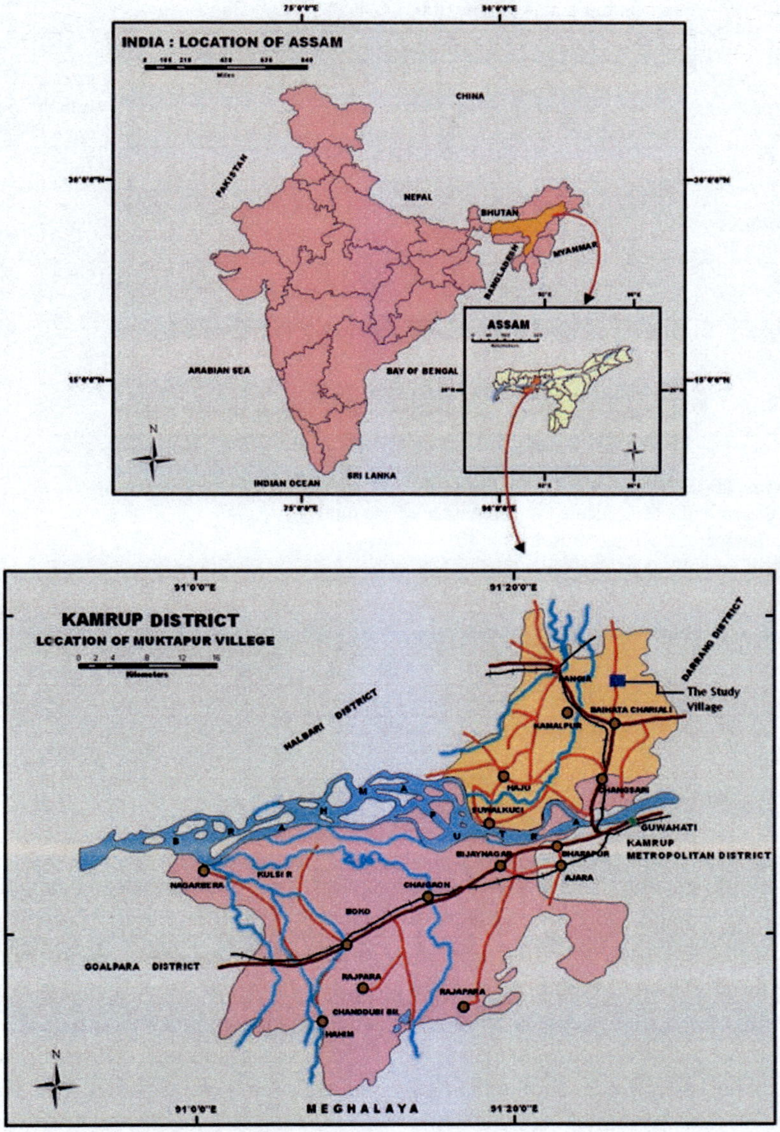

Fig. 13.5 Location of Muktapur village. *Tora (Mastacembelus)*, *Chanda (C. nama)*

acter and the nature of response as well as perception of the local people. Although the genesis of wetlands appears to be similar throughout the BV, some variations are observed in terms of their location, size and shape, depth, seasonal fluctuation of water area and aquatic resources. Their utility pattern also varies according to the social background, tradition, and economic condition of the people around. The rural people, whose livelihoods are intimately associated with wetlands around, have traditionally developed the practice of classifying the wetlands based on their perception, experience, and benefits derived from them through generations.

The people of Muktapur have evolved a classification system of water bodies present in and around their village (Fig. 13.6). Accordingly, seven categories of wetlands are identified in the village: (1) man-made pond (*pukhuri*), (2) natural pond (*khal*), (3) dead channel (*mora nadi*), (4) marshes (*pitoni*), (5) waterlogged area (*hola*), and (6) canal (*khawai*). These wetlands are per-

Table 13.5 Status of wetlands in the village, 2006–2007

	Number and area of wetlands possessed by households			
Type	Number	Area (in ha)	No. of households possessing wetland	Number of wetland per household (average)
Man-made Pond (*Pukhuri*)	218 (46.28)	14 (65.29)	186 (45.58)[a]	1.17
Natural pond (*Khal*)	207 (43.95)	4.84 (22.57)	194 (47.55)[a]	1.07
Marshes (*Pitoni*)	15 (3.18)	1.24 (5.78)	13 (3.19)[a]	1.15
Waterlogging (*Hola*)	19 (4.03)	0.98 (4.57)	19 (4.66)[a]	1
Canal (*Khawai*)	12 (2.55)	0.38 (1.77)	10 (2.45)[a]	1.2
Total	471	21.44	257[b]	1.83

Source: Field survey, 2007–2009
Note: Figures in the parentheses indicate the percentage of the total
[a]Indicates the percentage of the total households (408)
[b]Total households of the village having same type of wetlands

Table 13.6 Change in number and area of wetlands

	Number			Area (in ha)		
Type of wetland	1980	2006	Change in %	1980	2006	Change in %
Man-made pond (*pukhuri*)	163	218	+33.74	9.46	14.0	+48
Natural pond (*khal*)	217	207	−4.61	5.07	4.84	−4.54

Source: Field survey, 2007–2009

ceived to have their own characteristics and roles in providing ecological and economic services to the people.

Man-made pond: This is a type of artificially created freshwater body usually located within or near the homesteads, their average distance being 75 m from the owner's residence. Unlike the natural ponds (*khals*), fish are reared here for home consumption as well as for selling. On the banks of the ponds, a variety of economic plants like coconut, banana, and bamboo is grown. Interestingly, the diversity of aquatic flora and fauna in these man-made ponds is relatively less. Besides pisciculture, these are used for bathing, washing of clothes and utensils, soaking seeds and jutes. Woods and bamboo for domestic use are also soaked in water to enhance their durability. The average depth of ponds in the village remains around 2 m and 1.5 m during summer and winter, respectively.

Natural pond: Locally called *khal*, these are naturally formed water bodies. In the village under study, the Punai dead channel has provided scope for the formation of several natural ponds in its long bed. These natural ponds are rich in aquatic flora and fauna. These are regularly inundated and get connected with open sheet of floodwater during summer. However, most of these ponds dry up during winter. The *khals* in the village are usually located relatively away from the homesteads, average distance being 400 m.

Dead channel: A segment of the abandoned course of a stream called Punai is there in the village. It lies along the boundary of the village measuring a length of about 700 m. The dead channel (*mora nadi*) usually gets flooded during the summer. In winter, the water level of the channel goes down considerably. This dead channel forms a good habitat for a variety of fish and other animals and plants.

Marshes: Marshes (*pitonis*) are the permanently low-lying wet area that favors the growth of many herbaceous vegetations like reeds, sedges, and grasses. They constitute the biologi-

Fig. 13.6 (**a**) Man-made pond, (**b**) *khal*, (**c**) marsh, (**d**) dead channel

cally most diverse and productive aquatic ecosystems providing habitats for insects, reptiles, and birds. The villagers collect necessary fodders for the cattle mainly from these marshes.

Waterlogged area: These are small depressions inundated in summer and dried up in winter. They remain covered by stagnant or slow-moving water usually and get connected with the dead channel. When the water goes down in winter only the deeper parts of the depressions contain some water. They act as the last winter resort for the small fish and other minor aquatic fauna which normally move around the nearby paddy fields during the summer. Locally called *holas,* these are used by the villagers as their common fishing grounds.

Canal: These are the shallow elongated depressions saturated by surface run-off or seepage of water from the adjacent interfluves and remain wet during summer and become dry in winter. The villagers use them as a source of irrigation by diverting their flow to paddy fields. These wetlands constitute good habitats for small fish species like *Darikona (Esomus dandricus), Mowa (Amblypharyangodon mola), Cheng (C. orientalis), Garai (Chenna punctata), Magur (C. batrachus), and Singi (Heteropneustes).*

13.11 Size and Distribution

The wetlands of the village are mostly of small size and scattered throughout the village area (Fig. 13.7). The fragmentation of landholding resulting from the law of inheritance and transfer of ownership affects the size and spatial distribu-

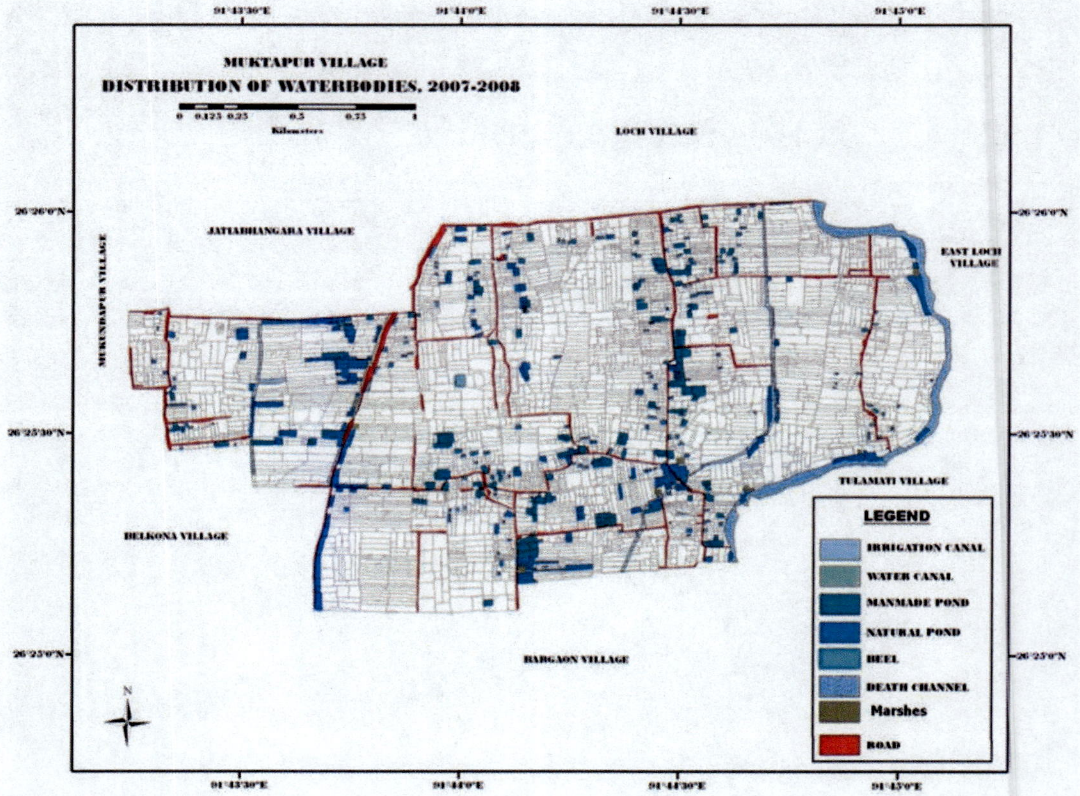

Fig. 13.7 Distribution of wetlands in Muktapur village

Table 13.7 Number of *khal*s and ponds in different size classes

Size class (in ha)	Natural wetland (*khal, pitoni, hola, khawai*)		Man-made pond (*pukhuri*)	
	Number	Area (in ha)	Number	Area (in ha)
<0.04	202 (79.84)	4.15 (55.78)	140 (64.22)	7.05 (50.36)
0.04–0.08	46 (18.18)	2.70 (36.29)	49 (22.48)	2.93 (20.93)
0.08–0.12	3 (1.19)	0.31 (4.17)	14 (6.42)	1.44 (10.29)
0.12–0.16	2 (0.79)	0.28 (3.76)	8 (3.67)	1.11 (7.93)
0.16–0.20	0	0	4 (1.83)	0.72 (5.14)
0.20–0.24	0	0	1 (0.46)	0.21 (1.5)
>0.24	0	0	2 (0.92)	0.54 (3.86)
Total	253	7.44	218	14.00

Source: Field survey, 2007–2009

Note: Figures in the parentheses indicate the percentage of the total

tion of wetlands. The wetlands, particularly the natural ones, show remarkable variation in terms of their size and scattering (Table 13.7). The small size and scattering have a negative impact on the wetland biota as most of the aquatic species require a minimum path size as well as composition of wetlands suitable for their sustenance (Jacques et al. 2007). Most of the natural wetlands (79.84%) are of less than 0.04 ha size and cover a total area of 4.15 ha. On the other hand, 64.22% of man-made ponds are below 0.04 ha size and cover 50.36% of the total pond area. It has been found that the size of natural wetlands lies within the range of 0.04–0.16 ha, while it is above 0.24 ha in the case of man-made ones. Only two ponds have the size of 0.54 ha each.

The distribution of wetlands in the village indicates a peculiar relationship between the people and topographic characteristics, which they have evolved through modification of natural landscape as per convenience. The man-made ponds are mostly located near the homesteads so that they may be used frequently and maintained with care, while the natural ones are in the low-lying areas at some distance from the homesteads. Thus, the size and distribution of the wetlands in the traditionally evolved agroecological landscape of the village reflect the careful adaptation of inhabitants to microecological settings of the area.

13.12 Use of Wetland Resources

Wetlands serve as an important component of livelihood support system for rural communities. As has been already mentioned, they play a remarkable role in the life and living of rural people by providing scope for fishing, collecting drinking water, edible plants and flowers, fodder for the cattle, bathing and washing, and irrigating agricultural fields (Fig. 13.8). People have thus traditionally developed a spontaneous relation with wetlands and reliable knowledge system on resources made available to them by nature.

Finally, we attempt to look more closely at the use of wetlands. Seventy six percent of households use them for fishing, 73% for irrigation purpose, 24% for collecting aquatic vegetables, 74% for providing drinking water to their cattle, and 11% for bathing and washing purposes (Tables 13.8 and 13.9).

The villagers, generally, collect resources for domestic use. However, a small proportion of them is also sold by some of the villagers for cash. There is a peculiar system in practice traditionally in the village. The natural wetlands owned in some cases by individuals are open to all for fishing during the summer season. This provides scope for the poorer section of the people, who do not have their own wetlands, to collect fish and other aquatic resources for home consumption and for sale as well. However, the poor people nowadays suffer greatly from dearth of natural fish supply during the winter. Just like Muktapur, all other villages in floodplain of the BV possess wetlands of different shapes and sizes. However, there is no systematic effort on the part of people and the government yet to use and scientifically conserve wetlands resources for further development of floodplain sustainably.

13.13 Conclusion

Wetlands are biological systems with internal biochemical cycling and metabolic productivity (Wetzel 2001). Sustainability is the goal of an effective management regime. The physical, chemical, physicochemical, biochemical, and biological processes operating in wetlands are often particularly complex and highly dynamic, and, therefore, understanding and managing such ecosystems become a real challenge. As a result, wetland hydrological studies and water management plans can neither be conducted in isolation nor implemented independently from the overall basin context (Hollis 1992). This is true in the case of the BV also, which is a constituent of the great Brahmaputra basin measuring a total area of about 5.80 lakh km^2.

The preceding sections have presented aspects of characteristics, distribution, and use of various types of wetlands in the BV, in general and a floodplain village. This study has provided analysis of wetlands scattered in the Brahmaputra Valley in Assam which are significant so far as the conservation of aquatic biodiversity is concerned. As microecosystems, wetlands contribute directly to the local environment and the associated rural economy. It has been observed that the rural development plans and programs followed in the country and the State as well do not pay due attention to the status and problems of wetland ecosystems. Again, most of the programs adopted for conservation and development of wetlands do not bear expected fruits because of the indifferent attitudes on the part of implementing agencies towards the need and aspiration of the people directly sharing the concerned microecological settings.

Further research needs to be carried on in the following directions. The traditional management practices evolved by the communities through their long-continued trial and error methods need to be examined and integrated into the

Fig. 13.8 (**a**) Traditional fishing, (**b**) spraying water over rice seedbeds, (**c**) bathing of cows, (**d**) soaking of wood and bamboo, (**e**) cows drinking water, (**f**) catching fish in a drained area

development processes chalked out for the rural areas. Scientific measures for protecting the wetlands must be best integrated into the ecological needs and economic benefits of the local people (Junk 2002). Proper attention should be paid to the problem of inequality in the ownership pattern of wetlands in the villages and to eradicate rural poverty by providing them alternative opportunity before implementing the wetland conservation plans and programs. Again, the positive perception and traditional knowledge of the villagers on wetlands, their preference for resource use locally should be given due weightage and, accordingly, separate strategy to con-

Table 13.8 Use of natural wetlands

Type of use	No. of households	Purpose	
		Domestic use (no. of households)	For sell (no. of households)
Fishing	312 (76)	283 (91)	29 (9)
Irrigating fields	298 (73)	298 (0)	–
Collecting edible plants	97 (24)	90 (93)	7 (7)
Drinking water for cattle	302 (74)	–	–
Bathing and washing	47 (11)	-	–

Source: Field survey, 2007–2009
Note: Figures in the parentheses indicate the percentage of the total households

Table 13.9 Income generated from ponds and fish consumption pattern

Occupational category	From man-made ponds		From natural ponds	
	Annual income per household (Rs)	Annual fish consumption in money value per household (Rs)	Annual income per household (Rs)	Annual fish consumption in money value per household (Rs)
Farmers	572.16	1200.00	950	3100
Wage laborers	0	350.00	746	2700
Businessmen	954.55	1336.36	1150	454
Service holders	1020.00	1848.00	1300	640

Source: Field survey, 2007–2009

serve and manage each type of wetlands should be evolved and implemented with the active participation of local people.

References

Aber JS, Pavri F, Aber SW (2012) Wetland environments: a global perspective. Wiley-Blackwell, West Sussex

Bahuah P et al (1997) Wetlands of Assam. Assam Remote Sensing Application Centre, Guwahati and Space Application Centre, Ahmedabad

Bhagabati AK (2012) Wetlands in Assam. In: Baroowa et al (eds) Aspects of land policy in Assam: continuity and change. Vivekananda Kendra Institute of Culture, Guwahati, pp 197–213

Bhagabati AK, Deka N (2017) Natural resource in a village environment of the Brahmaputra floodplain. In: Thakur B et al (eds) Regional development theory and practices. Concept Publishing Company Pvt. Ltd, New Delhi, pp 356–379

Bobbink R, Whigham DF, Beltman B, Verhoeven JTA (2006) Wetland functioning in relation to biodiversity conservation and restoration. In: Bobbink R, Beltman B, Verhoeven JTA, Whigham DF (eds) Wetlands: functioning, biodiversity conservation and restoration. Springer, Berlin, pp 1–12

Chandra G (2010) Impact of adoption of pen culture technology on Well-being of fishers of Haribhanga wetland in Assam. Indian Res J Ext Educ 10:61–65

Dahm CN, Cummins KW, Valett HM, Coleman RL (1995) An ecosystem view of the restoration of the Kissimmee River. Restor Ecol 1995:225–238

Deka N, Bhagabati AK, Ando K (2011) Rural land use in Brahmaputra floodplain environment, Assam: a case of Muktapur Village. Contemp India 1:177–193

Deka N, Bhagabati AK (2015) Wetlands in the village environment: a case from Brahmaputra floodplain, Assam. Trans Inst Indian Geogr 37:35–45

Garone P (2011) The fall and rise of the wetlands of California's great Central Valley. University of California Press, Berkeley

Hammer DA (1996) Creating freshwater wetlands, 2nd edn. CRC Press, Boca Raton, FL

Hay DL, Philippi NS (1999) A case for wetland restoration. Wiley and Sons, New York

Hollis GE (1992) The hydrological functions of wetlands and their management. In: Gerakis PA (ed) Conservation and Management of Greek Wetlands. Proceedings of Workshop on Greek Wetlands, Greece

Jacques K, Congalton RG, Babbitt K (2007) Effects of urbanization on the spatial distribution and size of wetlands in new hampshire, ASPRS, Annual Conference, Florida

Jordan WRII, Gilpin ME, Aber JD (1987) Restoration ecology: synthetic approach to ecological research. Cambridge University Press, Cambridge

Junk WJ, Bayley PB, Sparks RE (1989) The flood pulse concept in river-floodplain system. Canadian Fisheries and Aquatic Sciences, Special Publication, pp. 110–127

Junk WJ (2002) Long-term environmental trend and the future of tropical wetlands. Environ Conserv 29(4):414–435

Keddy PA (2010) Wetland ecology principles and conservation. Cambridge University Press, Cambridge

Lewis RR (1989) Wetland restoration/creation/enhancement terminology: suggestions for standardization.

In: Kusler JA, Kentula ME (eds) Wetland creation and restoration: the status of the science, vol 2. U.S. Environmental Protection Agency, Washington, DC

Malacou M (2011) Traditional ecological knowledge, conservation and socio-ecological perspectives in the Prespa Transboundary Park. In: Papayannis T, Pritchard D (eds) Culture and wetlands in the Mediterranean: an evolving story. Mediterranean Institute for Nature and Anthropos, Athina, pp 329–340

Maltby E (1991) The World's wetlands under threat – developing wise use and international stewardship. In: Hansen JA (ed) Environmental concerns: an interdisciplinary exercise. Elsevier, London, pp 109–136

Maltby E (ed) (2009) Functional assessment of wetlands towards evaluation of ecosystem services. Woodhead Publishing Limited, Cambridge

Mitsch WJ (ed) (2006) Wetland creation, restoration and conservation: the state of the science. Elsevier, Amsterdam

Mitsch WJ, Gosselink JG (2007) Wetlands. John Wiley & Sons, Hoboken

Moore PD (2006) Wetlands. Chelsea House Publishers, New York

Papayannis T, Pritchard D (eds) (2011) Culture and wetlands in the Mediterranean: an evolving story. Mediterranean Institute for Nature and Anthropos, Athina

Parisopoulos G (2011) Water management challenges in the Mediterranean: past, present and future. In: Papayannis T, Pritchard D (eds) Culture and wetlands in the Mediterranean: an evolving story. Mediterranean Institute for Nature and Anthropos, Athina, pp 237–245

Prasad SN, Ramachandra TV, Ahalya N, Sengupta T, Kumar A, Tiwari AK, Vijayan VS, Vijayan L (2002) Conservation of wetlands of India: a review. Trop Ecol Int Soc Trop Ecol 43:173–186

Ramsar Convention Secretariat (2007a) Managing wetlands: frameworks for managing wetlands of international importance and other wetland sites. In: Ramsar handbooks for the wise use of wetlands, vol 16. Ramsar Convention Secretariat, Gland

Ramsar Convention Secretariat (2007b) Wise use of wetlands: a conceptual framework for the wise use of wetlands. In: Ramsar handbooks for the wise use of wetlands, vol 1. Ramsar Convention Secretariat (RCS), Gland

Ramseier D, Klotzli F, Bollens U, Pfadenhauer J (2009) Restoring wetlands for wildlife habitat. In: Maltby E, Barker T (eds) The wetland handbook. Blackwell Publishing, Oxford, pp 780–801

Reddy KR, DeLaune RD (2008) Biogeochemistry of wetlands: science and applications. Taylor & Francis Group, Milton Park

Rogers K, Ralph TJ (2011) Impacts of hydrological changes on floodplain wetland biota. In: Rogers K, Ralph TJ (eds) Floodplain wetland biota in the Murray-Darling basin. CSIRO Publishing, Australia, pp 311–313

Saintilan N (2011) Management of Water for floodplain wetland biota. In: Rogers K, Ralph TJ (eds) Floodplain wetland biota in the Murray-Darling basin. CSIRO Publishing, Clayton, pp 329–334

Scheyvens R (1999) Ecotourism and the empowerment of local communities. Tour Manag 20:245–249

Semeniuk CA, Semeniuk V (1995) A geomorphic approach to global classification for inland wetlands. Vegetatio 118(1–2):103–124

Thoms MC (2003) Floodplain-River ecosystems: lateral connections and the implications of human interference. Geomorphology 56:335–349

van der Valk AG (2009) Restoration of wetland environments: lessons and successes. In: Maltby E, Barker T (eds) The wetland handbook. Blackwell Publishing, Oxford, pp 729–754

Ward JV, Tockner K, Arscott DB, Claret C (2002) Riverine landscape diversity. Freshw Biol 47:517–539

Wetzel RG (2001) Limnology: Lake and river ecosystems. Academic Press, San Diego, CA

Wood A (2006) Economic contribution of headwater wetlands: experience from Western Ethiopia. In: Krecek O, Haigh M (eds) Environmental role of wetlands in headwaters. Springer, Dordrecht, pp 221–232

14 Impact of Physical Factors on Transboundary Water Management and Governance in the Kosi Basin

Ramashray Prasad

Abstract

The Kosi River Basin which spreads over Tibet, Nepal, and India has complex riparian issues and challenges of transboundary water management with some of the highest peaks of the world located therein. Despite huge water availability and very high demand of the same, the economy of the basin is very poor and sharing this river's water is fraught with challenges. In this context, this chapter (1) assesses the water resources of the Kosi River, (2) evaluates the water management strategies, and (3) examines the mechanism of water governance in the basin. The entire Kosi Basin is selected as the study area. The study is based on secondary information collected by different organizations concerning land, climate, and water availability, need, utilization, and supply in present-day context as well as in future, too. A brief history on Kosi River related to recommendations or actions taken is traced to see the chronological development about monitoring and governing the water between India and Nepal. This study sets out to discuss Kosi rivers transboundary water management and sustainable development for eye opening towards better living. It is found that there is insufficient attention to the concurrent solution of diverse problems. Findings indicate that the Kosi River water is drained to the Bay Bengal without giving proper benefits to the region. In addition, results call attention not only for its own development, but also for the betterment of the region in general and both nations, in particular.

Keywords

Data Inventory · Flood Hazard · Kosi Basin · Sustainable Development · Transboundary · Water Management

14.1 Introduction

The term "transboundary" is made up of two words—"trans" and "boundary." "Trans" is a Latin word used as a preposition meaning thereby "across" or "on the other side" and "boundary" means "a demarcating line of one's authority." Thus, "transboundary" means beyond the territorial limit of a state (nation). When two or more than two nations are associated with any issue, it is said to be a transboundary concern. "Transboundary water" is related to the concern of two or more than two nations for the fulfillment of their needs. It also refers to the freshwater shared among different nations according to

R. Prasad (✉)
Department of Geography, B.R. Ambedkar College, University of Delhi, Delhi, India

B. Thakur et al. (eds.), *Resource Management, Sustainable Development and Governance*, Sustainable Development Goals Series, https://doi.org/10.1007/978-3-030-85839-1_14

their needs and requirements. Here, water across the boundary is used collectively in an economical way taking the judicious need without being "big-brother" role by anyone. When multiple nations have common rivers and their waters are managed, allotted, and utilized in a judicious manner by the concerned nations is known as "appropriate transboundary water management."

By governance, we mean the way something is managed at the top level and its regulation percolating to the down level. Hence, governance is how the rules, norms, and arrangements are planned, regulated, and executed regarding any aspect. When the transboundary water is considered for governance, it is delimited by several concerned nations which are beneficiaries. With the growing need of the people as well as even the rise of population has created stress on limited available water for the use by masses. The per capita availability of usable water is decreasing due to rise in population/users. The pollution of surface and groundwater is also creating pressure on further usable water as it becomes unsuitable. It is of concern because its availability is further reduced and so is the per capita of water. Therefore, governing the water in a judicious manner is big and this challenge is bound to be difficult when it comes to transboundary water management.

All resources on the Earth are getting reduced if we calculate per head availability. Their qualitative deterioration further puts discomfort on the resources. For better quality of life, everyone wants resources to meet their needs. Water is one of the most required substances even for our survival. Hence, sustainable development is the need of the hour today. Sustainable development is defined as "development that meets the needs of the present without compromising the ability of future generations to meet their own needs" (World Commission on Environment and Development, WCED 1987). A competition between demand and supply is created due to the tendency to utilize more transboundary water by the upper riparian which overlooks the need of the lower one. It is critiqued for creating countless hardships. The need of water and sustainability becomes an issue of bargaining power between/among the nations. The reduction of void creates several complexities in implementing and maintaining the balance. Hence, it is an issue to be resolved mutually by sitting on the discussion table by the representatives of con-

Box 14.1 Sustainable Development Goals

Transboundary water cooperation is critical from the perspective of the UN 2030 Agenda for Sustainable Development Goals (SDG6: water). This is largely because it concerns issues across multiple states, SDGs, and targets regarding agriculture, energy, ecosystems, climate adaptation, health, and peace and security. In this context, the author argues that sustainable development of Kosi River basin is inherently linked to, and requires transboundary water cooperation. The transboundary water cooperation and governance of Kosi River basin spread between Nepal and India is not only key for integrated water management but also for sustainable development, regional integration, and peace. According to the author who has been researching the Kosi River basin for more than two decades, it is important to understand the power of the Kosi basin given the role it plays in the irrigation of downstream areas and has a large potential for hydropower development and hence has riparian issues. On the other hand, the basin presents challenges as it is highly prone to erosion, sedimentation, and natural hazards, which may further worsen with climate change and its varying impacts. All of this has added pressure on the basin's freshwater ecosystem and the fragile livelihood of men and women in the region. Agenda 2030's sustainable development goals become more significant as there is a need to promote inclusive, community-centered river basin management to promote the sustainable use of transboundary water resources across both Nepal and India. The author recommends the imperative need to prioritize issues concerning gender and inequality and their linkages to drivers of change and river basin management.

cerned nations. Thus, sharing of international water in a river basin has a long way to go and its implication will decide the level and intensity of sustainability in the region.

14.2 Purpose of Study

The purpose of this study is to: (1) assess the water resources in the basin, (2) evaluate the water management strategies between/among nations, and (3) examine the mechanism of water governance in the region. This study is based on secondary source of data collected by concerned body of organizations related to land, climate, water availability, its need, utilization, demand, and supply with respect to not only in the present day, but, in the future as well. For this purpose, numerous works of researchers, authors, and concerned suitable reports were consulted and referred. Some important reports are prepared by International Centre for Integrated Mountain Development (ICIMOD 2009), NITI Aayog (2021), Flood Management Improvement Support Centre (FMISC 2014), Asian Development Bank (ADB 2019), Water and Energy Commission Secretariat (WECS 2011), United Nations Economic and Social Commission for Asia and the Pacific (UNESCAP 2016), Inland Waterways Authority of India (IWAI 2019), Ministry of Water Resources (1985) Kosi Project Agreement (1954, revised in 1966 and 1975).

14.3 Review of Literature

Mishra (1997) narrates the story of people living in the Kosi basin especially in Bihar (India). He deals with the geography of the basin, history of floods, measures to control floods, its failures, people's response, their plight, and relief measures by the government. He has traced different actions taken on the Kosi River, its flood mitigation measures, etc., in a chronological order. This book has attempted to trace the genesis of floods, flood control measures adopted, and their aftereffects. He is of the opinion that "living with floods" is the best option to manage the flood in the basin. Shifting of the Kosi channels in the plain, in India, is the greatest problem of Kosi water management (Das 1968; Mishra 1997; Wells and Dorr 1987; Prasad 2000). Hindu Kush Himalayan (HKH) region constitutes eight countries, namely, Afghanistan, Bangladesh, Bhutan, China, India, Myanmar, Nepal, and Pakistan have huge water potential in the form of yearly replenishment through heavy precipitation and enormous ice accumulation on the high altitudes. This region has really several conflicts of varied nature, but those conflicts could be resolved by bringing nations together through transboundary water cooperation at national or regional levels (Molden et al. 2017).

Two close neighbors—India and Nepal—are very close in terms of history, culture, and people-to-people interaction, but cooperation to manage transboundary water is not to the satisfactory mark. For better Indo-Nepal water cooperation, it is essential to have a glance at past misunderstandings and hindrances, try to find the way to resolve them, identify the mutual interest, trust, and risk working on them for a better future (Upreti and Acharya 2017). Nepal has a huge theoretical hydropower potential of 90,000 MW out of which 42,000 MW is technically and economically feasible to generate, but so far, Nepal is utilizing only 847 MW (only 2% of the feasible capacity). There is huge demand for power and it is increasing by 10% annually when there is no power connection to 60% of the people of the country (Alam et al. 2017). The upper Kosi River basin has a high potential for investments in hydropower development and irrigation in the downstream areas (Wahid et al. 2017). Due to underutilized water resources in the Kosi basin, it is subject to water-related hazards like flood, drought, landslide, etc., they could very well be minimized by proper and judicious management of water resources in the basin.

Despite the great potential for regional cooperation, the Ganga-Brahmaputra-Meghna River basin development and management have several geopolitical constraints (Brichieri-Colombi and Bradnock 2003). Individual country-wise water management steps taken have created disputes which are ranked among the most well-known transboundary water conflict in the world (Salehin et al. 2011). The goodwill shown both by India and Bhutan has resulted in a very good result in water management but the same is not with India, Nepal, and Bangladesh due to lack of trust. It has compounded the deprivation not only for a country but for whole of the region leading to underdevelopment (Biswas 2011). Hanasz (2017) is of the view that there is significant distrust between nations and resentment about India's hydro-hegemony; and bilateral, rather than multilateral arrangement. These factors make collective action more urgent and more difficult. Once, the water cooperation comes into being, it will resolve historical grievances, strengthen water sharing, and build trust between riparian states and work towards outcomes based on principles of water justice in the Ganga-Brahmaputra *problemshed.*

Adhikari (2014) states about conflict and cooperation on water resources of South Asian countries, primarily with respect to transboundary rivers. According to him, the primary cause of conflict is greater national interest and nationalism by smaller countries and the "big-brother" role of large and strong country. In this condition, cooperation could be possible when the reality of the water resources is understood well to all and come closer by leaving their "historical baggage" behind and move forward with a sense of trust and understanding. This will pave the way for cooperation for the better well-being of the people of the region. Chinnasamy et al. (2015) are of the view that the political instability and changing governments have delayed the implementation of water resource development projects which can be avoided by examining the pros and cons of the proposed projects from the perspective of each nation. Compensation for land submergence in the upstream nation should be considered while estimating hydropower benefit for downstream nations. For that purpose, there must be a balance for both nations involved.

Different aspects and levels of water governance like—international, national, regional, and local have been dealt by Grumbine et al. (2015). They have very meticulously elaborated the data sharing, transparency, accountability, participation, equitability, and rule of law regarding water governance at various levels. A detailed analysis of Nepalese water resources has been taken up by Upreti (2006). Assessment of potential water resources in Nepal, downstream benefits like hydropower, irrigation, flood control, navigation, and overall regional cooperation have been highlighted. Excess water during monsoon and scanty during lean season poses a hardship for water and it could be managed by proper cooperation and appropriate water management in the concerned basin. He suggests that cost and benefit can very well be shared between India and Nepal according to the Columbia River Treaty Agreement between the USA and Canada.

Under the climate change impact study on water resources in the Kosi River basin, Bharati et al. (2019) found that the availability of water is very high in the Himalayan region while its maximum utilization is in the plain and experiences extreme drought. They have estimated water availability based on 8 years of hydrological, climatic, soil, and earth surface data by using Soil and Water Assessment Tool (SWAT). They found that monsoon water yield will increase and river flow in winter will decrease leading to severe flood and extreme drought which could be mitigated by judicious water management through unbiased cooperation between/among the nations. International water law works as a platform to identify and integrate the relevant legal, scientific, and policy issues and aspects related to transboundary water utilization (Wouters 2013). International law provides a variety of options in the implementation and resolution of problems of transboundary watercourses by explicit rights and duties as well as procedures to be followed to act upon.

Water is the basic need for the existence of biotic life including human beings. Water is the core of the 2030 agenda for sustainable development. Availability and sustainable management of water and sanitation for all is the sixth goal of Sustainable Development Goal (SDG). In this perspective, several progress reports have been prepared under the leadership of United Nations Organizations and Progress on Transboundary Water Cooperation, global baseline for SDG indicator 6.5.2, is one of them (United Nations 2018). United Nations Environment Programme (2016) is of the view that more efforts are needed on transboundary agreements; expansion and construction of water infrastructure need cooperation and capacity building is required within concerned countries to meet the transboundary objectives.

Uitto and Duda (2002) and Siwakoti (2011) have studied the transboundary river water management as a way of resolving conflict between/among concerned nations. Uitto and Duda (2002) emphasized lessons learnt from international cooperation for resolving conflict through transboundary water resource management particularly from Africa, Central Asia, and Latin America. This study highlights the promotion of collaborative environmental management, sharing of benefits, and justifiable proper utilization of transboundary waters, which ultimately lead to harmonious relation among the concerned nations and better life led by the people. This sort of cooperation enhances the peace with meeting the needs and overall better condition for everyone. Siwakoti (2011) has taken up the study of South Asia. Water has a great attractive character, and it energizes all sectors of society (Jagerskog and Zeitoun 2009). Appropriate water availability is at the root of every sort of developmental activity. Hence, its growing demand from several fronts places the earth's water resources under increased strain but managing the demand and meeting the supply is not that straightforward. Water availability varies very significantly in volume over space and time, and climate change makes these shifts even less predictable.

14.4 Study Area: The Kosi River Basin

The location of the Kosi basin in South Asia under the Himalayan region can very well be identified with the inset map accompanied with Fig. 14.1. Since it is passing through more than one nation, it is called transboundary river. This river occupies across three countries—China, Nepal, and India. Its basin is bounded by the water divides of rivers—the Tsangpo/Brahmaputra in the north, the Gandak in the west, the Mahananda in the east, and the Ganges in the south. It originates from the autonomous region of Tibet in China, south of the water divide of the Tsangpo River. It drains the northern slopes of the Nepal Himalaya in Tibet and the southern slopes of Himalaya in Nepal, while in India, almost entire area is plain. There are several tributaries in the Himalayan areas most important among them are three—the Sun, the Arun, and the Tamur, from west to east, respectively. All these three join at *Tribeni* (confluence of three rivers) in eastern Nepal, near Dhankuta, upstream to Chatra gorge. Apart from these three, other important tributaries are—Indrawati, Tama, Likhu, and Dudh. All these four join with the Sun Kosi. Altogether, there are seven important rivers. After *Tribeni*, the river is known as *Sapt* (seven) Kosi, named after the draining of water of seven rivers (Kosi).

The Kosi basin is situated between 25°24′43″N to 29°08′47″N latitudes and 85°09′00″E to 88°55′00″E longitudes. The total Kosi basin area occupies 87,311 km^2 (Bharati et al. 2019) of which 28,300 km^2 (32.4%) in Tibet (China), 39,407 km^2 (45.1%) in Nepal, and 19,604 km^2 (22.5%) in India. Altogether, this basin spans over

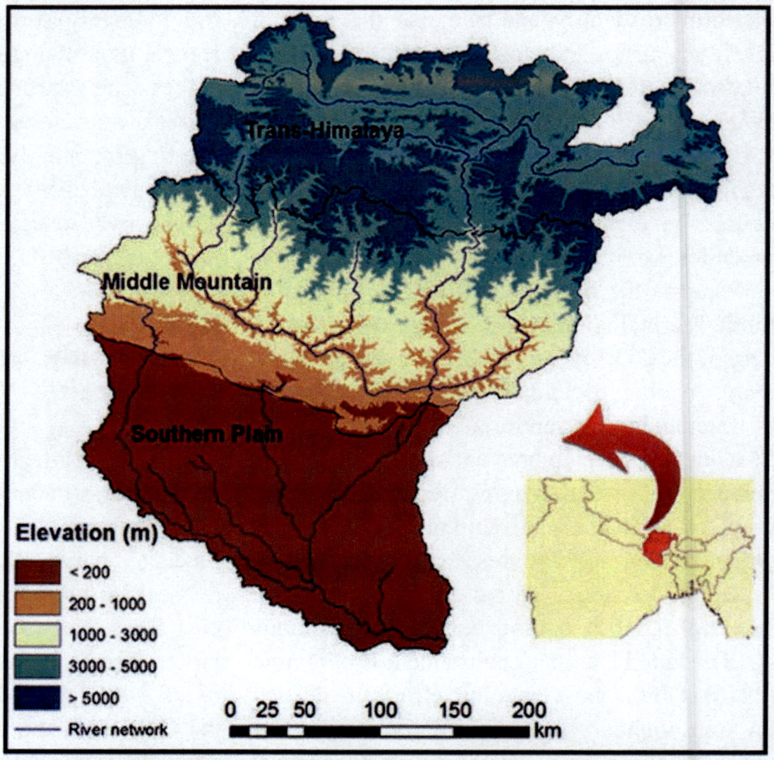

Fig. 14.1 Physiography of the Study Area: Kosi Basin

Source: https://html.scirp.org/file/5-4700452x7.png

five districts of China, 27 districts of Nepal, and 16 districts of India (Bharati et al. 2019). The Tibetan region of the Kosi basin is characterized by very high altitudes of the Tibetan Plateau constituting large number of glaciers and glacial lakes. The area of the basin falling within the territory of Nepal is lofty Himalayan slopes covered with variety of vegetation with wide range from tropical in the southern foothills to alpine in the northern region. The Indian territory of the Kosi basin is marked by plain with an elevation variation from 30 m in the southern limit to around 200 m in the northern side with the Nepal boundary. The river originates from trans-Himalayan highlands of Tibet, passes through Nepal, and joins the Ganges River at Kursela in Bihar State of India. Its length is approximately 730 km. Its height varies from 30 m in the southern tip at confluence with Ganges near Kursela to the highest peak of the world Mt. Everest, 8848 m in Nepal.

There are several geographical factors making the Kosi Basin a unique one. Important among them are relief, climate, precipitation, drainage, people and agriculture, changing climate and change in water availability, etc. Before the water is managed, these factors are needed to be deliberate upon. They are discussed in brief below.

14.4.1 Relief

The Kosi basin has very high relief because of lower value of altitude in the southern limit of the plain to the world's tallest Mt. Everest falling within north central part of the basin. The difference between the highest and the lowest is very high (8848–30 m = 8818 m). This much drop in height is observed in less than 300 km of crow-fly distance, while along the water path, it is less than 450 km. Kanchenjunga (8586 m), the third highest peak of the world also falls in its eastern part of the basin. Apart from these two, six more peaks with more than 8000 m, 36 glaciers, and 296 glacial lakes (Bajracharya et al. 2007) are also situated in its basin.

The local relief in the mountainous region is high because of its topography, whereas in the plain, it becomes very low. Based on the altitude and geology of the basin, it can very well be divided into five groups. From north to south, they are: (1) Trans-Himalayan Highlands, (2) Great Himalayas, (3) Middle or Lesser Himalayas, (4) Shiwalik or Outer or sub-Himalayas, and (5) Indian Plains. These variations are seen from Fig. 14.1. Trans-Himalaya is characterized by high altitude. It ranges between 4000 m and 6000 m. The trans-Himalaya's highest altitude is observed to the north (to the left bank side) of the Arun River. This area is marked grasslands and very low temperature in winter. In this region, the right bank tributaries of the Arun River are flowing from the northern slope of the Great Himalayas.

The Great Himalayas is characterized by steep-sided high peaks, eroded deep valleys and gorges, large and fast-moving glaciers, and overall great source of water from where many rivers are starting their journeys. Due to great height (Mt Everest in the basin) with large and numerous glaciers and very low temperature, it is also termed as Third Pole of the Earth. Middle or Lesser Himalayas is noted for lush green and dense varied vegetation zones based on its altitudes and prevailing temperature condition. It is also the source of many tributaries of the Kosi River. Most of the rivers originating from this region are from the southern slopes of the Great Himalayas and flowing through the Lesser Himalayas. Shiwalik or Outer Himalayas is the newest in terms of its evolution and is composed of detritus eroded materials. The plain of the Kosi basin in India is known as one of the biggest and largest alluvial fans sometimes termed as inland delta (Das 1968, Mishra 1997; Wells and Dorr 1987). The Kosi River is notorious for shifting its courses very frequently. It is said to be the "sorrow of Bihar" for its channel shifting and flood havoc.

Trans-Himalayan Highlands are known as Tibetan Plateau from where the Kosi River originates. The origin is marked by glaciers and permafrost areas whose meltwater gives rise to this river. About one-third (32%) of the entire Kosi basin (28,300 km^2) spreads over a plateau whose average height is about 5000 m. It is also known as tableland with less variation in altitude in comparison to the Himalayan region. The entire Himalayan mountain region is confined to Nepal. Little less than half (45%) of the basin area falls in Nepal (39,407 km^2). It is highly dissected by numerous flowing rivers. The area within the Indian domain is composed of plain. The local relief is very negligible which covers little less than one-fourth (23%) of the total area. If one moves from the foothills to the Ganges River, the slope keeps on declining.

14.4.2 Climate and Precipitation

There is a great variation in climate in the Kosi basin. In the southern part of the Indian plain, the climate is humid tropical but in the northern part, it is alpine in nature (Agarwal et al. 2014; Bharati et al. 2014, 2016). The northern and central region of the basin is very cold, and the temperature is sub-zero where the ice is accumulated.

By and large, the climate is controlled by the South Asian monsoon in the southern and central regions of the basin, while the climate in the trans-Himalayan region is influenced by its location and physiographic situation. In India and Nepal, there are four distinct climatic seasons. They are (1) pre-monsoon (March to May), (2) monsoon (June to September), (3) post-monsoon (October to November), and (4) winter (December to February). Depending upon the variation in climate, the requirement of water is also different during different seasons. At one point of time during monsoon, water is so abundant that there is difficulty in managing it and it causes widespread flooding in the low-lying areas. During winter days, it is very scantily available. In a cycle of 1 year, the same area is observing both flood and drought. It is due to non-planned or ill-planned approach towards water management. This problem can very well be addressed by multilevel governments of the concerned nations.

The Kosi basin receives handsome amount of precipitation in its catchment because a large

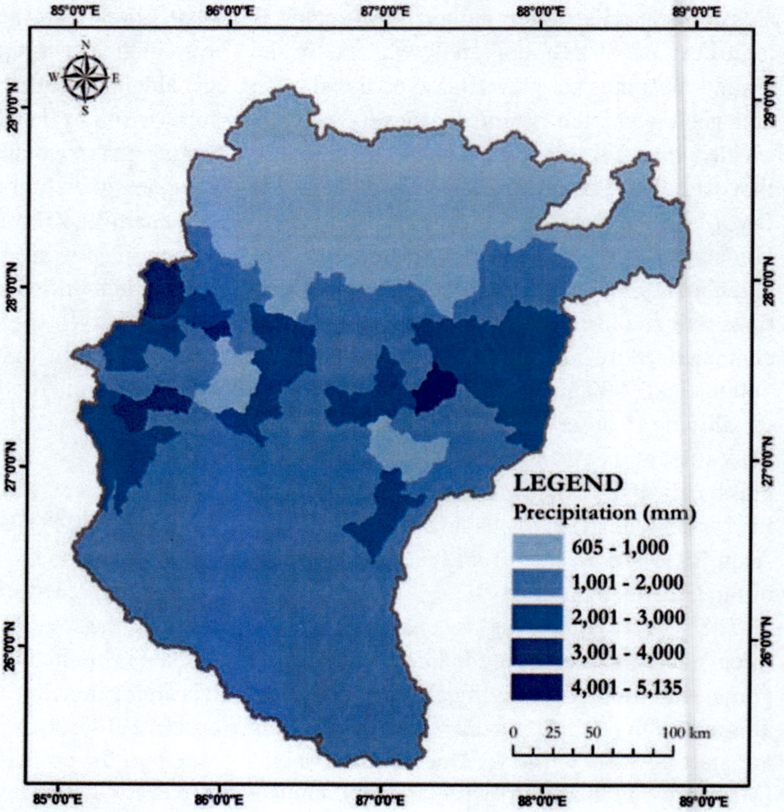

Fig. 14.2 Average Distribution of Precipitation: Kosi Basin

Source: After Bharati et al. 2019

part of its basin is lying in the monsoonal wind direction. The distribution of precipitation is governed by prevailing climatic conditions and obstructions in the path of monsoonal winds. The northernmost trans-Himalayan region receives the least amount of precipitation due to leeward side to the monsoonal rain-bearing winds. In the areas of high altitude, both rainfall and snowfall are observed. The precipitation in these areas varies between 300 cm and 500 cm in certain pockets. A large part receives rainfall between 200 cm and 300 cm. Considerable part of mountainous Nepal Himalaya as well as entire Indian plain receives rainfall between 100 cm and 200 cm (Fig. 14.2). Based on the analysis of basin temperature data of about 50 years by Nepal (2016), the average rise in maximum temperature is 0.058 °C/year while that of minimum is 0.014 °C/year for basin.

14.4.3 Drainage System in the Kosi Basin

The main river (Arun Kosi) is an antecedent river, i.e., it has been flowing before the evolution of the Himalayas about 70 million years ago. The erosive power of this river was greater than the rate of upliftment of the Himalayas. Thus, it has maintained its path and direction and crossed the Himalayas. In general, the drainage network is very high particularly in the mountainous region because of high erosion, dissections, and greater slope. It is an ideal condition for more dense drainage, but when it coincides with more precipitation and permanent water feeding glaciers, it translates into perennial rivers. There are seven major tributaries in the upper Himalayan catchment. Upstream to the Barahkshetra gorge, it is known as Sapt-Kosi, water of seven rivers. Some

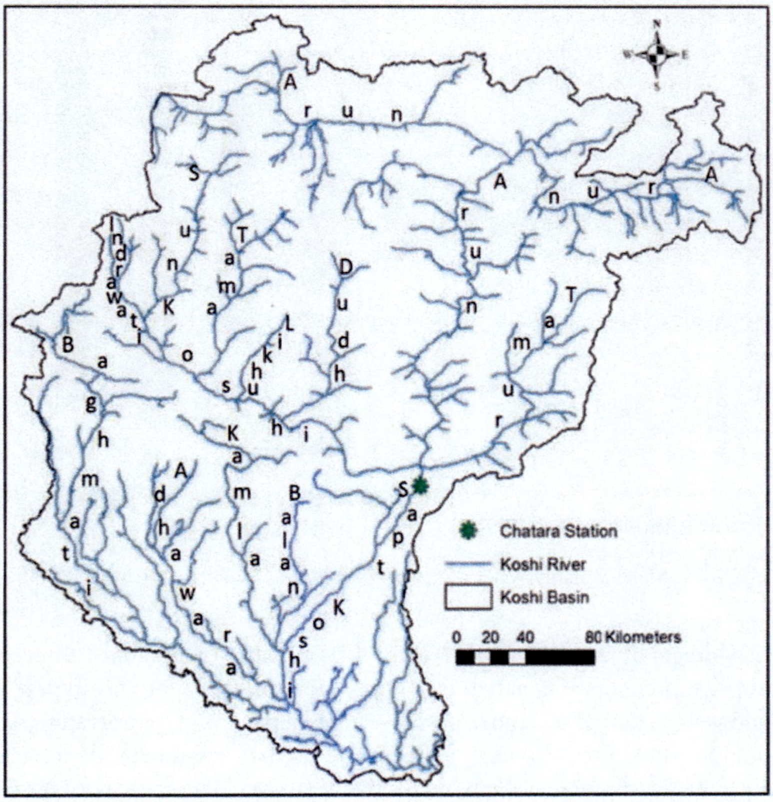

Fig. 14.3 The Kosi River System

of its tributaries originate in the foothills of the Himalayas and they join Kosi in the plain of Indian territories. These tributaries are Baghmati, Adhawara, Kamla, and Balan from west to east, respectively (Fig. 14.3).

14.4.4 Temporal Water Availability

About 80% of total rainfall in the Kosi catchment is received by south-westerly monsoon from May to October (Shrestha 2015). As mentioned above, about 75% of rainfall is concentrated in 4 months—June to September. This period is associated with high temperature and a large amount of snow/ice-melt of Himalayan glaciers as well as torrential rain. This leads to a high surge of water discharge. Based on the surge in water level of Kosi at Barahkshetra, Kosi water rises by 9 m in a day (Leopold et al. 1964, Spate and Learmonth 1967; Wells and Dorr 1987). When this much rise in water level in the gorge is increased, that much of rushing and gushing water from very steep slope creates havoc in the plain. It results in a widespread large-scale flood in the foothills and plains. That floodwater may be moderated by large dams in the Himalayan region and slowly that water may be released and at the end of monsoon, sufficient water may be stored behind the dams for utilization in the non-monsoon period by releasing based on the need in the downstream areas. The maximum peak discharge of Kosi recorded at Chatra is 25879 cumecs (cubic meter per second) in 1968 followed by 24,241 cumecs in 1954 and 24,000 cumecs in 1980 (Devkota et al. 2012).

Luna Bharati et al. (2019) have done a commendable job by studying the impact of climate change on water resources of Kosi River basin from mountain to the plains based on the data for 11 years (1998–2008). They have prepared an illustration representing precipitation, actual evapotranspiration, net water yield, and change

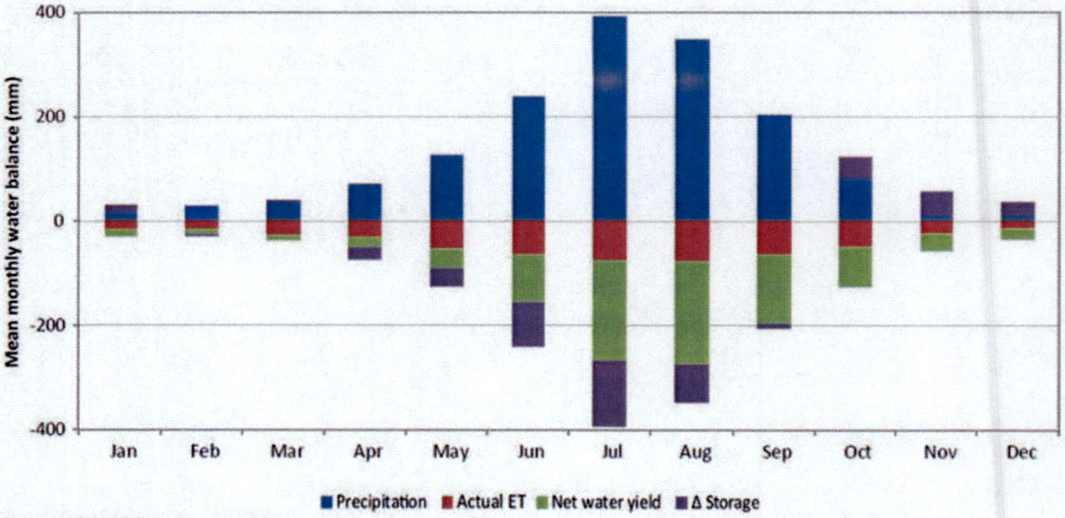

Source: After Bharati et al. 2019

Fig. 14.4 Mean Monthly Water Balance (1998–2008)

in storage of available water (Fig. 14.4). The average precipitation, actual evapotranspiration, and water yield of the entire basin based on simulation study of 11 years were found to be 172 cm, 52 cm, and 112 cm, respectively (Luna Bharati et al. 2019). According to the characteristics of the rainfall during monsoon months, the water balance components are directly affected. Greater availability of water through precipitation, more and more are evapotranspirated during this period. Large storage of water is attributed to recharge of groundwater table. This is reflected in groundwater base flow during lean season and feeding the streams in the basin.

The mean seasonal precipitation distribution during winter (December–January) is 4.9 cm, pre-monsoon (February–May) is 24.1 cm, Monsoon (June–August) is 134.5 cm, and post-monsoon is 8.4 cm. Likewise, actual evapotranspiration is 4.9 cm, 13.1 cm, 29.0 cm, and 8.0 cm during winter, pre-monsoon, monsoon, and post-monsoon, respectively. Net water yield is the highest in the mountains, followed by the hills, Kosi plains, and trans-mountain (Tibet) region (Bharati et al. 2019). Figure 14.4 shows varying behavior of these components over different months. The precipitation and net water yield are the lowest in Tibetan region and highest in mountains followed by hills in Nepal. Actual evapotranspiration is the product of precipitation, temperature, land cover, nature of soil, and mountain slope. Therefore, it is high on south-facing slope of the forested area of the Himalayas as well as the plain region of India where different crops are grown.

14.4.5 Population and Major Economic Activities

The gradient (low to high) of population density is from north to south, i.e., Tibet to India via Nepal. The total population in the Tibetan part of Kosi basin is about 1 lakh, while it is about 6 million in Nepal. The Indian part of the basin gives shelters to about 20 million people (Chen et al. 2013). Altogether, the Kosi basin is occupied by more than 26 million people. The population density in the Tibetan plateau is less than 5 persons per km^2 (Wahid et al. 2017), it increases to 100–200 in the middle hills of Nepal but in Kathmandu, it exceeds 500 persons per km^2. For Nepal section of the Kosi basin, the average density is 176 persons per km^2 (Dixit et al. 2009; Shrestha 2015). In Indian part of the basin, the density of population is more than 1000 persons per km^2 (Chen et al. 2013). The need of the people is to be fulfilled from the area where they are residing. Water is the most precious resource for people. Water is also found in abundance, but it is draining to the ocean in its journey.

The Kosi basin is primarily characterized by rural economy starting from the origin to its confluence with the Ganga River in India. People are mainly involved in primary activities—prominently in agriculture. The rainfall is mainly confined to monsoon months. Even during monsoon months water requirement of crops is not fully met as monsoon is erratic. Irrigation is practiced through surface water or groundwater. Proper water management and water storing is the important aspect during less rainfall receiving months. Water storing capacity in the upper catchment in Nepal is poor although several plans and programs have been studied through joint consultation between India and Nepal. This focused on dam construction, hydropower generation, silt trapping check dams, and canal irrigation but they are not yet implemented to a satisfactory limit. It is political in nature between two countries where general mass population is affected by floods and droughts.

14.4.6 Climate Change and Water Demand

Climate change is now an established reality but rise in temperature over the globe is not uniform. According to National Oceanic and Atmospheric Administration (NOAA) 2019, average temperature increase of the entire globe is 0.07 °C per decade since 1880. But in recent times it has increased to 0.18 °C per decade since 1981 which is more than double in comparison to the previous period. The ten warmest years have occurred since 1998 and nine of them have been observed since 2005 (Lindsey and Dahlman 2020). Therefore, the rise in recent years is very rapid and is expected to be, probably, in the same trend or even the trend may be intensified.

The upper catchment of the basin is characterized by high altitude with snow cover and glaciers. Ices and glaciers are the sources of water in the river when the precipitation is low. However, impacts of changing climate in the basin are the most important aspects for precise, accurate, and suitable data. Since 1940s, hydrological and climatological data collections have relatively enhanced. But, still, that data need to be continuous, precise, detailed, and well-distributed. Whatever the data is available, Nepal Climate Vulnerability Study Team (NCVST 2009) has studied the climate change in Nepal by Global Climate Model (GCM) and has come to

Table 14.1 Anticipated Climate Change Impacts in Nepal

Temperature	Significant rise in temperature: • 0.5 °C to 2.0 °C by 2030: With a mean of 1.4 °C • 1.7 °C to 4.1 °C by 2060: With a mean of 2.8 °C • 3.0 °C to 6.3 °C by 2090: With a mean of 4.7 °C • Increase in the number of days and nights considered hot by current climate standards. • Highest temperature increases during the months of June to August and at higher elevations.
Precipitation	Wide range of mean annual precipitation changes: • −34% to +22% by the 2030s: With a mean of 0% • −36% to +67% by the 2060s: With a mean of +4% • −43% to +80% by the 2090s: With a mean of +8%. Increase in monsoon rainfall towards the end of the century: • −40% to +143% by the 2030s: With a mean of +2% • −40% to +143% by the 2060s: With a mean of +7% • −52% to +135% by the 2090s: With a mean of +16%.
Runoff	• Higher downstream flows in the short term, but lower downstream flows in the long term due to retreating glaciers and snowmelt and ice-melt. • Shift from snow to rain in winter months. • Increased extreme events, including floods, droughts, and glacial lake outburst floods (GLOFs).

Source: From NCVST 2009 (also explained by Bartlett et al. 2010, Bharati et al. 2014 and Patra and Terton 2017)

the following conclusion which is compiled (Table 14.1) by Ryan Bartlett et al. (2010).

14.4.6.1 Temperature in Changing Climate

Since 1880, the global annual temperature is rising by an average rate of 0.07 °C per decade (Lindsey and Dahlman 2020) which is equivalent to 0.7 °C per hundred years. Lindsey and Dahlman found that five warmest years from 1880 to 2019 have occurred since 2015, all in a row and ten warmest years happen to be after 2005. Based on this, it can be stated that the rise in temperature is very high in recent years. This analysis provides a new perspective on comparison between the rise in global temperature and the rise in temperature in Nepal (Table 14.1). The minimum temperature rise in Nepal could be 0.5 °C and the maximum could be 2.0 °C with an average of 1.4 °C by 2030s. It is much beyond the world limit to around 1.0 °C. The average rise in temperature is expected to be 2.8 °C by 2060s and 4.7 °C by 2090s (NCVST 2009). The temperature increase in the Nepal Himalayas is more than double in comparison to average global temperature rise (Hijioka et al. 2014). Depending on the direction of slope, altitude, and mountain ranges in certain region, the projected temperature by 2040 to 2060 is estimated to be 1.5 °C and 3.0 °C, respectively (Yu et al. 2010, Gautam et al. 2013). The greater changes are expected to happen by the end of this century (Zomer et al. 2014).

The increasing temperature will have chain effects on pattern of water availability or scarcity. It is expected that the winter will be smaller and warmer than the present. The summer is supposed to be much warmer. Due to rise in temperature, high-altitude glaciers will shrink, and the store of water will decrease. The "Third Pole Ice" will lose its extreme character. The tree line will shift to upper slope. Xu and Grumbine (2014) have estimated that by the mid-twenty-first century, climate change will lead to a major regional "ecosystem shift" where upslope shrublands will replace grassland and forest cover will move higher elevation. All types of mountain vegetation will shift upslope due to warming. Many local people of Nepal and Tibet residing on the mountain and highlands observe such changes (Biggs and Watmough 2012, Haynes and Yang 2013). World Bank Report (Gautam et al. 2013) also estimated that the extreme climatic events are bound to increase and mountain ecosystem will deteriorate.

14.4.6.2 Precipitation in Changing Climate

Climate change or global warming is a reality. The amount of water vapor in the atmosphere is the direct outcome of the temperature of the air. Since the air temperature is rising, its effect will be linked to water vapor quantity in the air, condensation and precipitation will directly be affected. The current trend of precipitation is likely to change. It should be noted that wet areas will become wetter and dry and arid areas will become drier (Barnett et al. 2005, Held and Soden 2006, Rees and Collins 2006, Trenberth 2011, NAS 2012, Seneviratne et al. 2012). It is also important to note that the pattern of precipitation change is also showing the wetter conditions with higher latitude and higher altitude. In terms of seasonal variation, more precipitation is expected to be during monsoon season in the Himalayan region, while the lean season water supply will reduce drastically.

The precipitation departure between rainy season and lean season will further be widened and it will affect the livelihood very hard. The flood in the plain is very terrifying today and it will further be intensified due to more rain-producing conditions due to warming in the coming decades. In the same way, non-monsoon months will additionally be drier. The existing drought condition will more be exaggerated due to less precipitation. Table 14.1 shows the increasing precipitation is, on an average, 4% and 8% higher by 2060 and 2090, respectively. In the same way, the average increase in monsoon rain is likely to be 7% and 16% more by 2060 and 2090, respectively. It is quite evident that water management is going to be a major challenge in the coming decades in the Kosi basin. Therefore, it is high time that concerned countries try to resolve the water problem, both excess as well as scanty, amicably by showing coexistence with cooperation. This approach must be in the mind when talking about negotiation for water management.

14.4.6.3 Runoff in Changing Climate

The runoff of not only the Kosi River but of all the rivers over the globe will be impacted due to the rise in temperature. The warmer condition will lead to retreat/melt of the high-altitude glaciers of the Himalayas. Higher temperature is also associated with more moisture in the air, and, hence, more rainfall accompanied, and more severe flood situation in the downstream in a shorter duration. Therefore, the severity of the flood will be higher. Other associated problems like more landslides, sediment-loaded water from the hills and clogging of river channels in downstream, spreading of coarse sediment in the plain and soil infertility and several related problems will be the outcomes. It is also expected that where snowfall occurs in the hills during winter, will receive rain during incoming decades or by the end of this century. It is also estimated that the precipitation in the non-monsoon months will decrease sharply. The base flow of the river will be lessened as the supply will be reduced. It will be the cause of still more severe drought. All these discussions necessitate the proper management of water resources. Since Kosi is a transboundary river, its water management will depend on the cooperation of concerned nations.

14.4.6.4 Water-Related Disaster

Water-related hazards can be grouped into two categories: (1) excess and (2) deficit of water. Excess of water causes widespread damage to land, life, crops, and infrastructures. Fatal causalities, injuries, missing of people, damages of properties, human sufferings, disruption of socioeconomic fabric, and overall environmental degradation (Baral 2009) are the results of floods. Deficit of water equally impacts land, life, and people. Depending upon the severity of drought, decrease in the farm yield, loss of employment, decrease of income are bound to happen as direct impact (Udmale et al. 2014). Apart from them, some social impacts are people migration, health deterioration, hopelessness, conflict in society, and malnutrition. Pasture/plant/forest degradation, lowering water quality, damage to water-life habitat, groundwater depletion (Udmale et al. 2014), etc., are some of the environmental impacts of drought. Flood disaster in the lower Kosi basin forces the displacement of people and it is a big challenge for the government to rehabilitate (Gangwar and Thakur 2018). Both excess and deficit of water are not in congruence with good lives. Both conditions create hardship to the entire society. It is very contrasting because both flood at one time and drought at another is observed in the same area. To overcome this problem of excess as well as deficit, a justified and proper balance between the two is the right perspective to work on. These could be stated to be right mitigation measures which could be possible by the concerned governments (national and international), nongovernmental organizations (NGO), local governments, representatives of the society as well as affected common people.

14.5 Data Inventory of the Kosi Transboundary Water

Data inventory refers to accurate collection and dissemination of transboundary water updated status and information to concerned persons, researchers, organization, or authorities. For any research or management status of the subject concerned, its availability and need are to be identified and quantified. Therefore, data inventory is the prime concern to know the items to be managed. The main purpose is to provide a sound scientific basis for informing discussion and fostering dialog on the precious water resources that have become increasingly important to sustain development in an era of growing demand and dwindling supply (UN-ESCWA [United Nations Economic and Social Commission for Western Asia] 2013). Hence, all shared water resources within the Kosi basin should be identified and collected. It is also important to know how much annual water is available and how it is used.

Political willingness with a high commitment is a key element to develop interinstitutional cooperation and data management which pave the way for water data policy and its management. It includes a combination of legislative laws, decree, documents, strategic procedures, etc. (IWRM n.d. undated). Acquisition of data requires huge expenses in terms of developing data, collecting infrastructure and facilities. This also requires sharing if the area in which data is collected falls under transboundary. For data governance, it must be collected rationally and accurately pertaining to any aspect of study. Thus, data is processed and analyzed by various individuals, bodies, institutions, and organizations. Therefore, proper and appropriate data is the base for integrated water resource management.

14.6 Upstream/Downstream Conflict for Water Between India and Nepal

Nepal and the water draining off its soil is of prime importance to both Nepal itself and India. On the one hand, the excess water causes flood hazards in Indian plains, while, on the other hand, it can be a boon to meet the water need to the people of both countries, provided it is managed well. Kosi is known as "sorrow of Bihar" due to flood and changes in river course. Entire Nepal lies under the Ganga basin with only 13% area, and it contributes 45% of average annual flow measure at Farraka Barrage. Its lean season runoff contribution is 70% of the total. Thus, we can conclude that proper management of water will reduce the peak flow of river and lower the flood intensity in the plain and enhance the water supply during dry season.

The Kosi has already been notorious for changing its courses (Das 1968, Mishra 1997; Wells and Dorr 1987; Prasad 2000). Recently, the challenging problem has been flood occurrence several times with wet spell during monsoon period. For handling flood and other related problems, survey, investigation, and data collection are the primary needs to be carried out for any concrete step to be taken. To prepare an intensive project, these steps were taken up in 1946. As a result of the survey, a multipurpose

Table 14.2 Chronological Development with Respect to the Kosi River

Duration	Events/recommendations/ actions
Beginning of the sixteenth century	Flowing east of Purnea known as Kamla
1556–1605	Flowing west of Purnea known as Kajri or Kali Kosi
1731	Further westward new channel known as Kamla Kosi, to southward, joined Kali Kosi
1770–1807	A new channel created further westward in 1770 and existed till 1807, known as Libri Kosi
1807–1839	Westward a new channel was created, known as Dhamdaha Kosi
1839–1873	Westward shifted channel known as Hiran Kosi
1869 and 1870	Serious flood in the Indian plain, but very poor coordination between Nepal and British government, no efforts were possible for flood control measures
1873–1893	Further shift of channel known as Dhaus Kosi
1891	Slowly the relation between the two countries improved and an effort was made to construct embankment along the river. In this respect, Suguali pact was signed
1893	The channel shift was expected. The then chief engineer of Bengal Province undertook a detailed tour of the Kosi area but he refrained from suggesting any measure to tame the Kosi
1893–1921	Loran Kosi took south-westerly direction from upstream of Chatra in Nepal

(continued)

Table 14.2 (continued)

Duration	Events/recommendations/ actions
February 24, 1897	Calcutta flood conference, flood control measures with respect to Kosi river was discussed but no conclusive measures were finalized and decided not to interfere with the Kosi. "No steps are feasible to control the course of this tremendous river with its numerous channels and their wide and elevated beds, beyond protecting by short length of embankments isolated tracts exposed to floods"
1922	Dhusan Kosi joined at Bagian Ghat in Indian territory
1926	Dhusan Kosi again turned towards west downstream to Chatra and it was known as Tilawe Kosi
1936	Kosi again changed its course below Hanuman Nagar, termed as Soharain Kosi
November 10–12, 1937	a. Three-day conference to deliberate Bihar flood problem and its mitigation b. The then governor of Bihar, Hallet was of the view that similar problems in other countries like China and America had no success c. Dr. Rajendra Prasad was of the opinion that private embankments railways and roads were creating drainage problems and waterlogging d. Then chief engineer captain G.F. hall was of the firm opinion that flood prevention was undesirable and embankments increase flooding e. The, then, public works department (PWD) secretary Jimut Bahan Sen agreed for dam across Kosi in Nepal

(continued)

Table 14.2 (continued)

Duration	Events/recommendations/ actions
1937–38	With a view to erect embankments, a survey of all old and new channels was carried out by newly elected government headed by Srikrishna Sinha
1941	Claude Inglis suggested that the higher land level west of Tamuria (Madhubani district) would prevent further westward shifting
1942	After detailed study, PC Ghosh proposed afforestation and construction of dams on the tributaries of the Kosi River in Nepal for velocity reduction
1945	A plan was suggested to restrict the Kosi within two embankments with a span of about 16 km, though not approved by the central government.
1946	The central water irrigation and navigation commission (CWINC) prepared a preliminary report on Barahkshetra dam with the assistance of geological survey of India and the United States Bureau of Reclamation
April 6, 1947	A conference on Kosi sufferers was held and attended by top leaders including CH Bhaba, Rajendra Prasad, Binodanand Jha, Srikrishna Sinhaetc and deliberated on 229 m high dam at Barahkshetra in Nepal at a cost of Rs 100 million and its completion in 10 years duration, with multiple benefits
1946–1951	Widespread media coverage about the proposed dam however no work was undertaken and scheme got delayed and the cost escalated to Rs. 177 million

(continued)

Table 14.2 (continued)

Duration	Events/recommendations/ actions
1951	A committee was constituted to give an expert opinion on the dam project. In principle, it approved the project but said that huge money was needed and hydropower generated would not be utilized and huge capital would be blocked. Hence, it came out with Belka reservoir scheme which highlighted: • Construction of 26 km earthen dam, with a concrete spillway downstream Belka Hill. • Hydropower unit to produce 68 MW, for use in India and Nepal. • Eastern Kosi canal system to irrigate 615,000 hectare land in Bihar. • Western Kosi canal system to irrigate 47,000 hectare land in Nepal and Bihar. • A 56 km embankment to arrest westward movement of Kosi.
Sept 11, 1953	Belka dam project was debated in Bihar assembly and dissatisfaction was expressed by many members
Nov 1953	A committee was constituted to provide expert opinion on the status of Kosi project The recommendation was accepted by CWPC (Central Water and Power Commission), it was known as Kosi Project 1953 which include: (a) Construction of a barrage at Bhimnagar, 48 km below Chatra to serve as a control structure to divert water to canals. (b) Construction of embankments on both banks of the river to prevent flooding and to arrest westward channel shifting. (c) Annual irrigation of 14.36 lakh acres of land in Purnea and Saharsa districts through eastern Kosi canals.

(continued)

Table 14.2 (continued)

Duration	Events/recommendations/ actions
1954	Agreement between India and Nepal for Bhimnagar barrage and Chatra irrigation project. The administrator of Kosi project dumped Barahkshetra dam project and Belka proposal as unviable and sought a scheme that suited the financial resources and could be implemented After the visit of experts to China to study a similar project, they approved the technical soundness of Kosi project In September, the government of India unveiled first flood control policy—With short- and long-term plans For the first time, heavily debated and long-term solution of flood control by construction of dam was put to rest by that policy
1955	Bihar government published detailed map of proposed embankments Foundation was laid for Kosi project Construction of eastern and western embankments started Spurs were also erected to keep Kosi channel away from the embankment
1957	Commencement of the construction of eastern Kosi Canal
1958	Construction of Bhimnagar barrage started
1959	Construction of Western Kosi Canal started, but it was stopped due to non-clearance from Nepal government, which later started in 1965 As per the plan embankments were completed: • 125 km long eastern embankment from Birpur to Koparia • 126 km long western embankment from Bhardah to Ghonghepur.

(continued)

Table 14.2 (continued)

Duration	Events/recommendations/ actions
March 1963	Construction of Bhimnagar barrage completed
1966	1954 agreement was modified to remove objections raised by Nepal for its benefits
1965–1975	Construction of barrage at Chatra from where canals were constructed to irrigate Nepalese land
Major embankment breaches Immediately after completion of the Kosi project, breach in embankment was observed. Several times the breaches were very severe and huge life and property were lost. Some of the major breaches were: • 1963: On 20th August, first breach in embankment at Banarjhoola near Dalwa in Nepal, western side • 1968: On 6th October breach occurred at Jamalpur and other four places. It was associated with torrential rain in the upper Himalayan catchment in Nepal with a peak discharge 25,900 cumecs at Bhimnagar barrage • 1971: On 12th August breach occurred at Bhatania, near Supaul • 1980: On 8th August breach occurred at Bahuarawa • 1984: On 5th September breach occurred at Navhatta, 75 km from Bhimnagar barrage along Eastern Embankment • 1991: On 17th July breach at Joginia along Western embankment • 2008: On 18th August, first time breach occurred, upstream to Bhimnagar Barrage at Kusaha (12 km from barrage). 2008 breach completely changed the course of the river to about 120 km towards east from its existing flow. Kosi River started to flow southward through its abandoned channel. For a natural flow of a river, it was the best adjusting course, but by the beginning of the monsoon 2009, embankment breach was repaired and the river was forced to flow between the two embankments to utilize irrigation infrastructure facilities already developed.	

Source: Compiled from different sources but mostly are from D.K. Mishra (2008)

scheme was proposed, like construction of a high dam at Barahkshetra with a height of 235 m was proposed, but immediate emphasis was flood control and reduction in flood losses for which 1150-m-long barrage was taken up on priority basis at Bhimnagar in India (Bihar), about 5 km upstream of Hanuman Nagar (in Nepal). The treaty between the two countries was signed in 1954. The barrage construction was completed in 1963, while the embankments along the two banks of Kosi were completed by 1958.

Soon after signing the agreement of Kosi Treaty between India and Nepal in 1954, it was criticized by the Nepalese people. Subsequently, it was argued that this Treaty was favoring India and not Nepal. It was also mentioned that Nepal had given "extraterritorial right to India for an indefinite period, without getting adequate compensation and benefits from the project. It was also alleged that Nepal had to lose its fertile land without equivalent gains in exchanges of it." (Pahl-Wostle et al. 2016). A very brief description of the events or the actions taken with respect to the Kosi River is given in Table 14.2:

14.6.1 Kosi Water Discharge and Irrigation in the Plain

It is already mentioned that Kosi River drains huge amount of water. The, then, Deputy Director of Flood Management Improvement Support Centre (FMISC) Dilip Kumar computed the annual mean discharge of Kosi River which is equivalent to 1557 cumecs (Kumar 2015). It is equivalent to 4,910,155-hectare meter or the total quantity is 49.10 km^3. The discharge during monsoon months is excessively very high. It is 3860 cumecs (Prasad 1999; Prasad and Thakur 2005). Only scanty amount of water is utilized by people to irrigate the land. Initially, the target of land irrigation was set to be 7.34 lakh hectares as per the Kosi Barrage Project, but it was slashed to 3.74 lakh hectares by Eastern Kosi Canal (Mishra 1997). The target to irrigate by Western Kosi Canal is 1.9 lakh hectares. The total target of 5.64 lakh hectares was never realized. The biggest challenge is trapping the sediment of the Kosi water. Very frequently, the canals are filled with sediment and the water carrying capacity of the canal is reduced. It has become almost annual affair to desilt the canal, which is not done on regular basis. This sort of annual maintenance is also very much there for water governance. A schematic/diagrammatic

Fig. 14.5 Canal Irrigation in Lower Kosi Basin

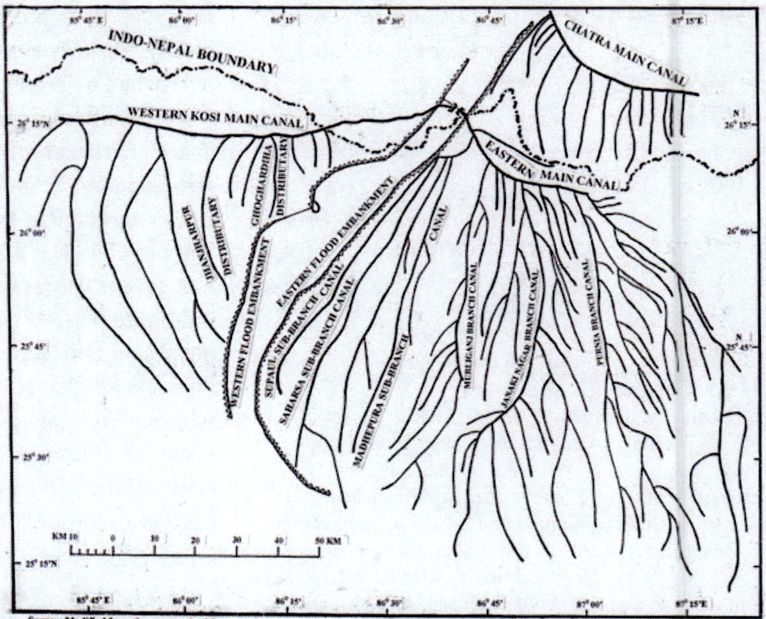

Source: Modified from the map received from the Research and Investigation Division, Kosi Project, Birpur, Bihar.

illustration of the canal system is presented in Fig. 14.5.

14.6.2 Water Governance in India and Nepal

Since time immemorial, water has been a central theme in Indian society and has always been advocated to use it judiciously without any waste. Increase in population with every passing day, the utilizable water per head kept on reducing, and hence National Water Policy was prepared by the government, for the first time in 1987 and reformulated with improvement in 2002. It was further improvised in 2012. In short, it focused on developing data bank, estimating available water, prioritizing water, meeting water needs, encouraging participation of stakeholders, monitoring water quality, promoting conservation, developing flood control and management system (Cullet and Gupta 2009).

Nepal's water policy was announced and termed as Water Resource Act 1992 and slightly modified as Water Resource Regulation in 1993. They emphasized on governance of water resources, procedural registration for water allocation, rights, and obligations of water use associations and ensuring water quality by controlling pollution (Adhikary 2005), etc.

Despite being no pronounced water policy prior to notification, practically there have been several mechanisms to address the concerned issues. Important among them are bilateral cooperation regarding flood control measures and irrigation schemes of 1954 and 1966. These documents are the guiding force in water cooperation between two countries. Under these agreements, the mechanism of joint ministerial commission (both countries) for water resources is there to discuss and decide plans for maximizing benefits. It is supposed to meet at least once in a year. Indo-Nepal Joint Commission on water resources was established in 2000 to discuss and decide on issues of cooperation in water sector. It is headed by the Secretaries of Ministry of Energy from Nepal and Ministry of Water Resources from India. This commission ensures the implementation of existing agreements. This commission used to meet on regular basis. Joint Standing Technical Committee was created to rationalize technical committees and subcommittees between the two countries. It is related to flood control management, flood forecasting, and hydropower. For proper and smooth functioning, there are other several bilateral committees like joint committee on inundation and flood management, power exchange committee, joint committee on Kosi and Gandak projects, joint team of experts, and joint Kosi high-level committee (The Asia Foundation 2015).

14.7 Transboundary Kosi Water Management: Sustainable Development

If water is made available to use, there would be overall development in every sector of living. But if we fall short, the life would be miserable and sustainable development would be a distant reality. It is also now widely perceived that the economic condition of the basin is one of the worst in the world, despite very heavy precipitation and the best soil available in terms of fertility. Nature has gifted the region with ample resources, but still, it is one of the least developed areas in the world. The hard reality is that lesser resource available regions are very well developed in the world, but it is not the Kosi basin. Two prime reasons can be attributed to it. First, there is excess precipitation, high and precipitous slope with weak geological structure leading to huge sediment yield and new plain formation in terms of geomorphological processes. Culmination of all of them is uncontrolled rushing water and flood havoc. Its management is not so easy. Second, there is very poor cooperation between the two concerned nations. It is the immaturity in terms of finding out the cooperative and mutual benefits and careless attitude towards water management. Insensitive approach at bureaucratic and government level and hostile attitude with neighbor is the prime concern. It is more problematic when some other nations come in between the two to regulate as per whims and fancy of its own.

14.8 Conclusion

This chapter has deliberated the different geographical factors which makes Kosi basin a unique one, assessed the water resources, evaluated the water management strategies, and examined the mechanism of water governance in the transboundary Kosi River basin. The river originates from the Tibetan Plateau and passes through Nepal encompassing Mt. Everest and Kanchenjunga in its catchment. Enormous amount of precipitation and snowmelt during monsoon period produce high discharge causing widespread flood in Nepal and Bihar (India) plains. The ample natural resources given by nature in the Kosi basin are not utilized properly and it has led to poor conditions of the people in the region. Enormous water is directly drained into the ocean without serving the humanity. Before meeting the Bay of Bengal, it terminates the crop produce by flooding, kills cattle and human lives, annihilates everything in the plain, and creates miseries in the plain in general and the people.

The findings elucidate that despite various resources in good quantity and quality, it has not been a boon to the inhabitants. People-to-people contact and relation of two nations are very intense because of porous boundary intermingling, the congenial natural cooperation and understanding are lacking. This study highlighted the weak points of coordination and cooperation and this would prove to be heart changing of the government officials for the betterment of the people of the two nations. The transboundary water between India and Nepal has not yet worked as per the desire and need of the people. Great cooperation and understanding between the two nations need to be established. For that, people to people interaction and viewpoints should also be prioritized. Two of the nations are very close to each other at people level. Complete porous boundary and intermingling of the people are quite common but the political parties and governments sometimes become very hostile and it is leading to overall neglect of the poor cooperation. Common people of both countries are sufferers. Though the Himalayan region falls under severe earthquake-prone zone, now the technology is well developed to construct a dam on the Kosi River and its tributaries. Smaller and multiple check dams would trap sediment yield locally, reduction in sediment would be easier to check the channel shift, to some extent, in the plain, prosperity of the plain people would be on the card. On the other hand, India has a huge demand for power. If clean energy, hydropower is degenerated at dam site, Nepal will benefit prodigiously. The economic condition of Nepal will automatically improve, but nothing is visible like this. Hence, the development of this basin lies in the hands of two governments for their mutual benefits and sustainable development.

It was found that the data inventory of the Kosi basin is very poor. Data inventory is the backbone to quantify the available resources. Until it is known completely, real scientific research cannot be conducted. The data pertaining to climatology, hydrology, forest, land use and land cover, physiography, soil types, groundwater inventory, etc., are not that easy to extract correctly. Unfortunately, due to internal character of the study area, the author faced tremendous difficulties in secondary data collection. On the other hand, flood hazards, shifting channels, and relatively undeveloped stretches of the upper basin posed numerous challenges. Therefore, it is the need of the hour to first cooperate in the collection of data pertaining to all types of resources, make it available to researchers and scientific organizations so that proper study is conducted, and people are made aware of the results. For this purpose, huge investment is also required to create infrastructure for data collection. Therefore, it is the responsibility of booth concerned to take wise decision in this regard.

Recent advances in governance theory are influencing the field of natural resource management. This has attracted attention as a policy concern both in developed and developing countries. Understanding the dynamics of water management and governance is essential because governance is related to water scarcity. Therefore, further research direction needs to be carried on in the field of water governance between India and Nepal. Such study would address and examine: (1) water policy and its application, (2) role of governmental, nongovernmental, and civil society, (3) which specific water challenges require emphasis, (4) strengths and weaknesses of policymakers, (5) conflicts of values and science, (6) information generation and sharing, and (7) finding out missing links in funding and information. Transboundary water problems may be solved by questionnaire information served to the lay public, technical experts, and government officials. Future research should always be such which help humanity in improving its way of life. Hence, intensive cooperation at heart-to-heart level is the foundation of success which should be the ultimate aim of people and government. This would pave the way for better governance of water and ultimately lead to better living conditions.

References

ADB (2019) Managing Nepal's Dudh Koshi River system for a fair and sustainable future. Asian Development Bank, Mandaluyong

Adhikari KN (2014) Conflict and cooperation on south Asian water resources. Islamabad Policy Res Inst 14(2):45–62

Adhikary S (2005) Water Laws in Nepal: Laws relating to drinking water, sanitation, irrigation, hydropower and water pollution. Water Aid Nepal, Lalitpur

Agarwal A, Babel MS, Maskey S (2014) Analysis of future precipitation in the Kosi River basin, Nepal. J Hydrol 513:422–434. https://doi.org/10.1016/j.jhydrol.2014.03.047

Alam F, Alam Q, Reza S, Khurshid-ul-Alam SM, Saleque K, Chowdhury H (2017) A review of hydropower projects in nepal energy. Procedia 110:581–585. https://doi.org/10.1016/j.egypro.2017.03.188

Bajracharya SR, Mool PK, Shrestha BR (2007) Impact of climate change on Himalayan glaciers and Glacial Lakes: case studies on GLOF and associated hazards in Nepal and Bhutan. ICIMOD and UNEP, Kathmandu

Baral M (2009) A report of east and Southeast Asia. Regional Seminar on Flood Hazard Mapping, Manila, Philippines, 17–19 February. Accessed on August 28, 2020. https://www.preventionweb.net/files/9333_progressreport2009nepal1.pdf

Barnett TP, Adam JC, Lettenmaier DP (2005) Potential impacts of a warming climate on water availability in snow dominated regions. Rev Nat 438(17):303–308

Bartlett R, Bharati L, Pant D, Hosterman H, McCornick P (2010) Climate change impacts and adaptation in

Nepal. IWMI working paper 139, International Water Management Institute, Colombo, Sri Lanka

Bharati L, Bhattarai U, Khadaka A, Gurung P, Neumann LE, Penton DJ, Dhaubanjar S, Nepal S (2019) From the mountains to the plains: impact of climate change on water resources in the Kosi River basin. Working paper 187, international water http://www.iwmi.cgiar.org/Publications/Working_Papers/working/wor187.pdf

Bharati L, Gurung P, Jayakody P, Smakhtin V, Bhattarai U (2014) The projected impact of climate change on water availability and development in the Kosi Basin, Nepal. Mt Res Dev 34(2):118–131

Bharati L, Gurung P, Maharjan L, Bhattarai U (2016) Past and future variability in the hydrological regime of the Kosi Basin, Nepal. Hydrol Sci J 61(1):79–93

Biggs E, Watmough G (2012) A community-level assessment of factors affecting livelihoods in Nawalparasi District, Nepal. J Int Dev 24:255–263

Biswas AK (2011) Cooperation or conflict in transboundary water management: case study of South Asia. Hydrol Sci J 56(4):662–670

Brichieri-Colombi S, Bradnock RW (2003) Geopolitics, water and development in South Asia: cooperative development in the Ganges–Brahmaputra Delta. Geogr J 169(1):43–64

Chen NS, Hu GS, Deng W, Khanal N, Zhu YH, Han D (2013) On the water hazards in the transboundary Kosi River basin. Nat Hazards Earth Syst Sci 13:795–808

Chinnasamy P, Bharati L, Bhattarai U, Khadka A, Dahal V, Wahid S (2015) Impact of planned water resource development on current and future water demand in the Kosi River basin, Nepal. Water Int 40(7):1004–1020

Cullet P, Gupta J (2009) Evolution of water law and policy in India. In: Dellapenna JW, Gupta J (eds) The evolution of the law and politics of water. Springer Academic Publishers, Dordrecht, pp 159–174

Das KN (1968) Westward shift in the courses of the Kosi. Natl Geogr J India 14(1):23–31

Devkota L, Crosato A, Giri S (2012) Effects of the barrage and embankment on flooding and channel avulsion: case study of Kosi River, Nepal. Rural Infrastruct J Infrastruct Dev 3:124–132

Dixit A, Upadhya M, Dixit K, Pokhrel A, Raj DR (2009) Living with water stress in the hills of the Kosi Basin, Nepal. International Centre for Integrated Mountain Development (ICIMOD), Kathmandu

FMISC (2014) Environmental impact assessment for protection and restoration of Kosi River embankments in Bihar prepared by Environmental Infrastructure & Services Ltd.

Gangwar S, Thakur B (2018) Disaster, displacement and rehabilitation: a case study of Kosi floods in North Bihar. Natl Geogr J India 66(2):76–92

Gautam M, Govinda R, Timilsina R, Acharya K (2013) Climate change in the Himalayas: current state of knowledge. Policy research working paper 6516. World Bank, Washington DC

Government of Nepal (1954) Agreement Between the Government of The Kingdom of Nepal and The Government of India on The Kosi Project, 1954

Grumbine RE, Nizami A, Rana Tharu B, Niraula R, Su Y, Xu JC (2015) Water governance in the Asian highlands. ICRAF working paper 198. World Agroforestry Centre East and Central Asia, Kunming, China

Hanasz P (2017) Muddy waters: international actors and transboundary water cooperation in the Ganges-Brahmaputra problem shed. Water Altern 10(2):459–474

Haynes M, Yang Y (2013) Adapting to change: transitions in traditional range Management of Tibetan Yak Herders in Northwest Yunnan. Environ Dev Sustain 15:1065–1077

Held IM, Soden BJ (2006) Robust responses of the hydrological cycle to global warming. J Clim 19(21):5686–5699

Hijioka Y, Lin E, Pereira JJ, Corlett RT, Cui X, Insarov GE, Lasco RD, Lindgren E, Surjan A (2014) Climate change in Asia: impacts, adaptation, and vulnerability. In: Barros VR, Field CB, Dokken DJ, Mastrandrea MD, Mach KJ, Bilir TE, Chatterjee M, Ebi KL, Estrada YO, Genova RC, Girma B, Kissel ES, Levy AN, MacCracken S, Mastrandrea PR, White LL (Eds) Contribution of Working Group II to the Fifth Assessment Report of the Intergovernmental Panel on Climate Change. Cambridge University Press, Cambridge, United Kingdom and New York, NY, USA, pp. 1327–1370 https://www.ipcc.ch/site/assets/uploads/2018/02/WGIIAR5-Chap24_FINAL.pdf

HMGN (1966) Revised agreement between his Majesty's government of Nepal and the government of India on the Kosi project. Ministry of Water and Power. His Majesty's Government of Nepal, Kathmandu

HMGN (1975) Revised agreement between his Majesty's government of Nepal and the government of India on the Kosi project. Ministry of Water and Power. His Majesty's Government of Nepal, Kathmandu

ICIMOD (2009) Living with water stress in the hills of the Kosi Basin, Nepal; International Centre for Integrated Mountain Development, Kathmandu

IWAI (2019) Detailed hydrographic survey of Kosi River, Bihar, submitted by precision survey consultancy

IWRM (n.d.) (undated) Integrated Water Resource Management. https://www.aquacoope.org/data4iwrm2/en/about

Jagerskog A, Zeitoun M (2009) Getting transboundary water right: theory and practice for effective cooperation. Report no. 25. Stockholm International Water Institute, Stockholm

Kumar D (2015) Flood and sediment Management in Kosi River. In: Proceeding of 15th world water congress held at Edinburgh, May 25-29. Accessed on 15 August 2020. https://www.iwra.org/member/congress/resource/2989645.pdf

Leopold LB, Wolman MG, Miller JP (1964) Fluvial processes in geomorphology. Freeman, San Francisco

Lindsey R, Dahlman L (2020) Climate Change: Global Temperature https://www.climate.gov/news-features/understanding-climate/climate-change-global-temperature

Mishra DK (1997) The Bihar flood story. Econ Polit Wkly 30:2206–2217

Mishra DK (2008) Trapped! Between the devil and deep waters: the story of Bihar's Kosi River. Peoples' Science Institute and South Asia Network on Dams, Rivers and People (SANDRP)

Molden D, Sharma E, Shrestha AK, Chettri N, Pradhan NS, Kotru R (2017) Advancing regional and transboundary cooperation in the conflict-prone Hindukush Himalaya. Mt Res Dev 37(4):502–508. https://doi.org/10.1659/MRD-JOURNAL-D-17-00108.1

NAS [National Academy of Sciences] (2012) Himalayan Glaciers: Climate Change, Water Resources and Water Securities. Report of the committee on himalayan glaciers, hydrology, climate change, and implications for water securities. The National Academies Press, Washington. http://www.nap.edu/catalog.php?record_id=13449

NCVST (2009) Vulnerability through the eyes of vulnerable: climate change induced uncertainties and Nepal's development predicaments. Institute for Social and Environmental Transition-Nepal (ISET-N, Kathmandu) and Institute for Social and Environmental Transition (ISET, Boulder, Colorado) for Nepal Climate Vulnerability Study Team (NCVST), Kathmandu

NITI Aayog (2021) Report of the committee constituted for the formulation of strategy for flood management works in entire country and river management activities and works related to border areas (2021–26), National Institution for Transforming India, New Delhi

Pahl-Wostle CA, Bhaduri A, Gupta J (2016) Handbook on water security. Edward Elgar Publishing Ltd, Cheltenham

Udmale P, Ichikawa Y, Manandhar S, Ishidaira H, Kiem AS (2014) Farmers' perception of drought impacts, local adaptation and administrative mitigation measures in Maharashtra state, India. Int J Disaster Risk Reduct 10:250–269

Patra J, Terton A (2017) Review of current and planned adaptation action in Nepal. CARIAA working paper no. 20, International Development Research Centre, Ottawa, Canada and UK aid, London, United Kingdom

Prasad R (1999) Dynamics of River Environment: A Case Study of Kosi Embankment Area. Unpublished Ph.D. Thesis, Department of Geography, University of Delhi, Delhi

Prasad R (2000) Progression of the Kosi channels. Annal Natl Assoc Geogr India 20(2):47–65

Prasad R, Thakur B (2005) Floods, embankment and development in the lower Kosi plain. In: Thakur B, Sinha VNP, Prasad M, Sharma N, Pratap R, Mandal RB, Singh RBP (Eds) Urban and regional development in India, Vol. 2, pp. 632–657

Rees HG, Collins DN (2006) Regional differences in response of flow in glacier-fed Himalayan Rivers to climatic warming. Hydrol Process 20(10):2157–2169

Salehin MS, Khan MSA, Prakash A, Goodrich CG (2011) Opportunities for transboundary water sharing in the Ganges, the Brahmaputra, and the Meghna basins. India infrastructure report 2011, water: policy and performance for sustainable development. Oxford University Press, New Delhi, pp. 29–43

Seneviratne SI, Nicholls N, Easterling D, Goodess CM, Kanae S, Kossin J, Luo Y, Marengo J, McInnes K, Rahimi M, Reichstein M, Sorteberg A, Vera C, Zhang X (2012) Changes in climate extremes and their impacts on the natural physical environment. In: Field CB, Barros V, Stocker TF, Qin D, Dokken DJ, Ebi KB, Mastrandrea MD, Mach KJ, Plattner GK, Allen SK, Tignor M, Midgley PM (Eds) Managing the Risks of Extreme Events and Disasters to Advance Climate Change Adaptation. A Special Report of Working Groups I and II of the Intergovernmental Panel on Climate Change (IPCC). Cambridge University Press, Cambridge, UK, and New York, USA, pp. 109–230

Shrestha R (2015) Kosi River Basin Inventory, Nepal. https://www.researchgate.net/publication/336278788

Siwakoti CG (2011) Transboundary River Basi in South Asia: option for conflict resolution. International Rivers Focus on People – Water - Life. https://www.internationalrivers.org/sites/default/files/attached-files/transboundaryriverbasins.pdf

Spate OHK, Learmonth ATA (1967) India and Pakistan: a general and regional geography. Methuen, London

The Asia Foundation (2015) Status and Implementation of Transboundary River Agreements on Kosi River in Nepal: An Executive Summary. Issue Brief No.1, The Asia Foundation, New Delhi

Trenberth KE (2011) Changes in precipitation with climate change. Clim Res 47:123–138. https://www.int-res.com/articles/cr_oa/c047p123.pdf

Uitto JI, Duda AM (2002) Management of Transboundary Water Resources: lessons from international cooperation for conflict prevention. Geogr J 168(4):365–378

UN-ESCWA [United Nations Economic and Social Commission for Western Asia] (2013) Inventory of shared water resources in Western Asia. Beirut

United Nations (2018) Progress on transboundary water cooperation. UNECE and UNESCO, Paris

United Nations Environment Programme (2016) Transboundary River basin: status and trends, UNEP

Upreti BR, Acharya G (2017) Shared water resources: indo-Nepal conflict and cooperation. In: Aneel SS, Haroon UT, Niazi I (eds) Securing peace and prosperity. Sustainable Development Policy Institute and Sang-E-Meel Publications, Lahore

Upreti T (2006) Prospects and problems of Nepalese water resources. In: Upreti T (ed) International watercourses law and its application in South Asia. Pairavi Prakashan, Kathmandu

UNESCAP (2016) Flood forecasting and early warning in Transboundary River basins: a toolkit. The United Nations Building, Bangkok

Wahid SM, Kilroy G, Shrestha AB, Bajracherya SR, Hunzai K (2017) Opportunities and challenges in the transboundary Koshi River basin. In: Sharma N (ed) River system analysis and management. Springer, Singapore, pp 341–352

WCED (1987) Our common future: report of the world commission on environment and development. Oxford University Press, Oxford

WECS (2011) Water resources of Nepal in the context of climate change. Water and Energy Commission Secretariat, Kathmandu

Wells NA, Dorr JA (1987) Shifting of the Kosi River, Northern India. Geology 15:204–207

Wouters P (2013) International law-facilitating transboundary water cooperation. Global water partnership technical committee background paper no 17

Xu J, Grumbine RE (2014) Building ecosystem resilience for climate change adaption in the Asian Highlands, in WIREs Climate Change, 5:709–718

Yu H, Luedeling E, Xu J (2010) Winter and spring warming result in delayed spring phenology on the Tibetan plateau. Proc Natl Acad Sci U S A 107:22151–22156

Zomer RJ, Trabucco A, Metzger M, Metzer MJ, Wang M, Oli KP, Xu J (2014) Projected climate change impacts on spatial distribution of bioclimatic zones and ecoregions within the Kailash sacred landscape of China, India, and Nepal. Clim Chang 25:445–460

15 Governance and Management of Teesta River Water Resources: A Geopolitical Appraisal

Sudeepta Adhikari and Subinita Kamle

Abstract

Teesta, a *transnational river* between India and Bangladesh faces the crisis of droughts in the lean period and floods in the rainy season. This chapter is concerned with the Teesta River conflict which has been continuing for more than three decades between the two nations based on their respective claims and still lacks a proper framework for effective governance and cooperative management. Due to this long-standing conflict, sustainable management of Teesta water has been neglected, which has further fueled ecological, socioeconomic, and geopolitical conflicts in this river's basin. In this context, this chapter aims to analyze the multidimensional facets of the Teesta River water conflict by identifying the intergovernmental steps to address the problems arising out of the contrasting claims and to find the peaceful resolution of the same. The study concludes that there is an urgent need to convert the *Interim Deal* to a permanent Teesta River Water Treaty, as a possible resolution to the conflict. Also, to resolve the long-standing conflict, initiating an Integrated Teesta River Water Management (ITRWM) mechanisms for the whole basin of the river will be most fruitful.

Keywords

Bangladesh · Geopolitics · Governance · Hydropower · India · Teesta River · Teesta River Water Treaty · Water resource

S. Adhikari · S. Kamle (✉)
Department of Geography, Patna University, Patna, India

15.1 Introduction

Governance of a resource (natural) is often defined as the interaction among structures, processes, and traditions that determine how power and responsibilities are exercised, how decisions are taken, and how the stakeholders have their say? (Lockwood et al. 2010). Good governance of available recourses requires mutual and inclusive consensus among the stakeholders to ward off any kind of conflictual among them. But it is highly *normative*.

Management of resources is integral and inseparable to the concept of good governance of a resource, as the latter facilitates utilization, regulation, and conservation of available resources, while the former plays a fundamental role in the governance of resources. The management of a resource has a different nuance, from the man-

B. Thakur et al. (eds), *Resource Management, Sustainable Development and Governance*, Sustainable Development Goals Series, https://doi.org/10.1007/978-3-030-85839-1_15

agement of a social, political, or economic affair. So far as, the management of a resource is concerned, it speaks of *sustainability*, i.e., the resource must be utilized and managed in such a way that the future generations should not have to starve for its extinction or *loss*. Management is also normative, linked with the *norms*, in terms of *practices*, and about uses of resource(s).

Geopolitical (adj.) pertaining to *geopolitics* that is concerned with the rivalry relations between the nations for domination and/or control over natural resources, particularly natural resources, like water, natural gas, petroleum, and other such mineral resources, which have *transboundary* continuities and occurrences, beyond political frontiers of two or more nations. Geopolitics is not normative as there are no norms as such to regulate the political behavior of the nations, about the governance and management of such natural resources, with transboundary characteristics. Geopolitics is therefore *geographically dimensioned* political aims, with political, economic, social, and military/strategic overtones, in commensurate with *national objectives* and *interests*.

At present, there are more than 263 international rivers covering 45.6% of earth surface that affects more than 40% of global population and account for nearly 80% of global water flows. Population of various nations of the world are necessarily linked by transboundary basins and aquifers, making them interdependent, not just hydrologically, but also socially, economically, politically as well as culturally.

However, about 158 (60%) of the world's 263 transboundary river basins still lack a framework for governance and cooperative management thus experiencing not only continued dereliction of water resources but also socioeconomic and geopolitical conflicts in these areas. This chapter deals with such a transboundary river of South Asia, i.e., the Teesta that flows across the two nations: India and Bangladesh. Its origin lies in the higher reaches of Sikkim, and flowing through the plains of West Bengal, the *river* drains into and/or merges with the Brahmaputra in Bangladesh. Sikkim and West Bengal are federating political units (called *states*) of the *Indian Union*. The Teesta is a *historical* river because for long rather over centuries, its water resources were used and shared by the people of Sikkim

Box 15.1 Sustainable Development Goals

The sustainable governance and management of Teesta River between India and Bangladesh has been a bone of contention for numerous issues including sharing energy from hydroelectric projects and other dams, usage conflicts, seismic concerns as well as climate change impacts. The 2030 Agenda for Sustainable Development clearly outlines through SDG 6 (more specifically target 6.5) as well as through 21 other related links between the SDGs and 31 previously established links between water and sanitation and the other SDGs how transboundary water cooperation is crucial for sustainable management of water resources. In this background, the authors argue that sustainable development is inherently linked to and requires transboundary water cooperation between India and Bangladesh. Unfortunately, achievement of sustainable development without transboundary water cooperation is challenged by existing and emerging rules and institutions that seeks to halt and minimize the negative consequences of poor transboundary water management. As such, the authors observe that transboundary water cooperation in the context of Teesta River water resource governance is critical in meeting the larger objectives of SDGs. Stimulation of transboundary water cooperation in an environment of shared transboundary freshwater and ecosystems cut across myriad sectoral needs and political boundaries. Conclusively, these would set effective policy goals coupled with the need for investments. This means there is a need for governance to work at multiple scales with a range of public and private stakeholders throughout the river basin.

(then a semi-independent kingdom under British Protectorate) and Bengal Province until the division or partition of the Bengal Province in 1947, with the creation of an independent East Pakistan, together with West Pakistan in North-West India. East Pakistan, later emerged as a sovereign nation-state, with the name of Bangladesh. Though Sikkim was a *dependent territory* of British India, the British enjoyed the water resources of the Teesta River. Nevertheless, even during British India, the Teesta River used to be a transboundary river, because of its origin in Sikkim, which was a *quasi-sovereign kingdom* in the Himalayan region, despite being a dependent territory of British India. Sikkim maintained its quasi-sovereign status until its merger with the Indian Union in 1975. Before Sikkim's merger with the Indian Union, the Teesta River used to flow through more than two sovereign political units: Sikkim, India, and Bangladesh. The transboundary character of the river is thus historic, and political as well. From its source in Sikkim to its mouth, i.e., the point/site where the Teesta River empties into the Brahmaputra in Bangladesh, it is approximately 414 km long with a greater part of it—151 km in Sikkim and 142 km in West Bengal within India and the remaining 121 km in Bangladesh. The Teesta River is a *contested* transboundary river within India between Sikkim and West Bengal on the one hand, and between India and Bangladesh, on the other hand.

The present study is about the governance and management of water resources of the Teesta River and *involves* geopolitics between India and Bangladesh, about the problem of *sharing* and *uses* of water resources of the *river*.

15.2 About the Teesta River

The Teesta River is the fourth largest of the 54 transboundary rivers, between India and Bangladesh. It has its origin in northern Sikkim, close to the Sino-Indian border, at an altitude of 5280 M, in the Tso Lamo Lake, and the *river* is fed by the Teesta Khangse Glacier that sustains its turbulent continuity in the state. It abruptly descends from the Pahaunri peak, to 213 M—a steep drop from the higher reaches. The Teesta River, during its 151 km flow in Sikkim, has cut deep gorges, and narrow valleys, throughout its course. Several rapids occur across its course, which form ideal sites for hydropower generating locations in Sikkim. The Teesta River defines eastern boundary between Sikkim and West Bengal. The *river* enters West Bengal through the districts of North Bengal, such as Darjeeling, Kalimpong, Jalpaiguri, and Cooch Behar, and traverses for about 142 km in the state, before it enters Bangladesh near the town of Mekhliganj in Cooch Behar. In West Bengal, much of its course flows through the plains. In Bangladesh, the Teesta River flows across the Rangpur Division through the districts of Gaibhande, Kurigram, Lalmunirhat, Nilphamari, and Rangpur, for 121 km before it merges with the Brahmaputra River near the town of Chilmari. The *river's* braided course is further widened between 500 and 550 M (Islam 2016; Noolker-oak 2017) (Fig. 15.1).

The total area of the Teesta River basin that spreads across India and Bangladesh is 12,159 km^2, of which 2004 km^2, or about 16.48% lies in Bangladesh, and the rest is in India, i.e., 10,155 km^2, or about 83.51%. Of 10,155 km^2, the share of Sikkim is 6930 km^2 (56, 99%), and the remaining 3225 km^2 (26.52%) lies in West Bengal. The flow of the *river* is highly variable. The average maximum flow of the *river* at the Dalia Barrage in Bangladesh has been found to be recorded as high as7900 m^3/s, while the average minimum flow was found to be as low as 280 m^3/s. The average annual flow of the Teesta River is about 60 billion cubic meters (BCM), and the average seasonal variation rate is about 1:10, i.e., 90% of its water that is roughly 54 BCM flows in the rainy season (from June to September). However, the flow for the greater part of the year, or during the remaining part of the year is mere 6 BCM. It has been found rather recorded that construction of large number of dams, and control over the flow of the *river,* without measuring the ecological impact of construction over the biodiversity in the basin area, has

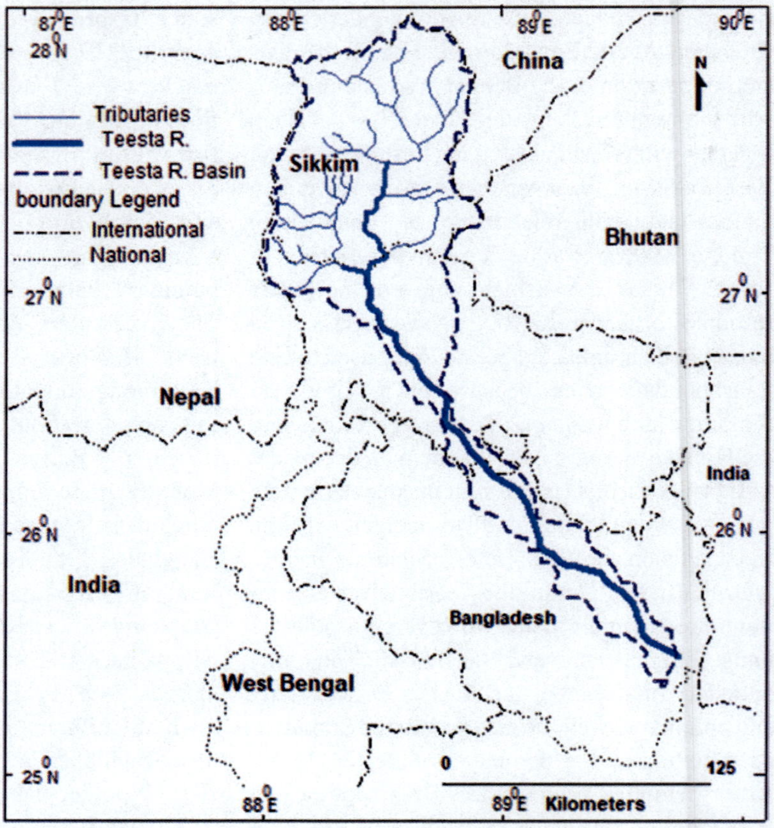

Fig. 15.1 Location of Teesta river basin

led to a sharp reduction in the flow up to 28 BCM or even 14 m^3/s, particularly during the drought or the lean period (Mondal Md and Islam Md 2017).

The Teesta River basin comprises a linguistic region as the population of it speaks a homogeneous Bengali language, of course, with minor variations in dialects and speeches, but in terms of religion, the basin is *geo*politically divided as the Buddhist-Hindu Teesta basin and the Muslim Teesta basin, the former lies in India, and the latter is in Bangladesh., with both the parts of basin having different political structures and mindsets. The basin has a total population of 30 million, out of which 29% live in India and the remaining 71% in Bangladesh. However, the *genre de vie* of the population of the basin is the same, with the Indian part of it is a little bit *developed* economically.

15.3 Literature Review

There are large numbers of literature available on the Teesta River, and most of them are concerned with India-Bangladesh dispute over the sharing of water resources of the *river*. Lot of these studies have been conducted in different disciplines on river water management by the authorities and agencies within appropriate frame of spatiotemporal reference. Most of the studies completed so far deal with the problem of river water management and their consequential effects from the perspectives of basin ecology and environment, hydro-politics, socioeconomic, and regional geopolitics.

Since space does not permit for an elaborate literature survey in this chapter therefore an attempt has been made to survey a few literatures regarding the governance and management of transboundary rivers with particular focus on the Teesta River. Islam (2016) investigates how India-Bangladesh conflictual about the sharing of the Teesta River badly affects the socioeconomic conditions of inhabitants of the Teesta valley region in Bangladesh. This chapter also focuses on the poverty scenario of the region. Mondal Md and Islam Md (2017) study the hydrological aspects related to the maximum and the minimum flows of the Teesta River in Bangladesh in historical perspectives and their geostrategic implication in the water resource management.

Singh (2010) analyses the pattern of hydropolitics in South Asia with reference to India, Pakistan, and India-Bangladesh river water dispute and points out how much of peace in the region depends on equitable joint governance and management of water resource management of transboundary rivers. Chan et al. (2016) made a SWOT analysis to assess the role of institution and political bodies in water resource management in Bangladesh and point out the local challenges before the IWRM (Integrated Water Resource Management) in its pursuit of sustainable management.

Ahmed (2015) studies the interstate relations about the sharing of the Nile River water resources in the Nile basin, with focus upon the equitable water resource management. Afroz and Rahman Md (2013) make a critical assessment of water management of the Ganga and Teesta River in Bangladesh. Banik (2016) measures the impact of the Gazaldoba Teesta valley multipurpose project upon its neighborhoods' socioeconomic landscape. Moreover, it also studies how the project has caused landscape change in the areas around the Gazaldoba. Bauer and Quiroz (2013) discuss various approaches to resource governance and management including water resource management with global examples. Mirchandani (2016) provides a structural interpretation of the nature of India-Bangladesh conflicts over the sharing of Teesta water resources and its impact on regional geopolitics and economy.

Mukerjee and Saha (2016) attempted to investigate the cause of unattained goals of the Teesta Barrage Project in India and the changes that occurred in the socioeconomic landscape of basin area due to the failure to reach the goals. Noolkeroak (2017) discusses India-Bangladesh conflictual relations over the sharing of the Teesta River water resources from geopolitical perspectives. Prasai and Surie (2013) have discussed the apportionment of Teesta River water resources between the involved political entities of the river basin. They have also examined the feasibility of the distribution of surplus water resources, particularly during the wet season. Rudra (2003) suggests the means of controlling the wasteful flows of the Teesta River and discusses some concrete measures so that the flows should not go waste. Syed et al. (2017) measures the potentials of the Teesta River water resources for hydel-power generation and agricultural sustainability for all the stakeholders of the Teesta basin.

All the above-reviewed literature surveys are discursive as they have attempted to conclude with reasons. This chapter also aims to be a discursive one, but in a more holistic way as it proposes to take up all the issues of governance and management of the Teesta River, with a particular attention to regional geopolitics. Therefore, an attempt has been made in this chapter to integrate all the different outlooks of the Teesta River water resources to create a more holistic approach to represent a comprehensive geopolitical analysis of the problems of transboundary governance and management of water resources of the Teesta River.

15.4 Research Problem

As mentioned above, the Teesta River basin spreads across India and Bangladesh, and each nation has its own perception, based on individualistic claims, about the use of the water resources of the *river*, in a way that has affected the interests of the *involved* nations, which has resulted in *conflictual* between them, despite both the nations being good neighbors in South Asia. The above account about the *river* reveals that India, being the upper riparian nation, with approximately 83.51% of the basin area under its territorial jurisdiction, has *unilateral* physical control over water resources of the Teesta River, and

Bangladesh, being the lower riparian nation, with just 16.48% of the basin area under its sovereign jurisdiction, faces the crisis of droughts in the lean period, and floods in the rainy season. Bangladesh depends on India for its water requirements, and India says it has its own compulsion, about the sharing of water resources of the *river*. Though the Teesta River basin is geographically a single unified unit, given its spatial characteristics, it is *geo*politically a *split* political basin, with a substantial part of the *river*, falling in India, and the rest in Bangladesh.

The lack of any concerted effort on the part of *contested nations*, India and Bangladesh, for equitable uses of water resources in the basin area of the Teesta has caused ecological damages not only in the lower reaches (Bangladesh) but also in the middle reaches (West Bengal), with perceptible impacts over the *genre de vie* of people on either side. Each nation therefore has its own methodology of governance and management of water resources of the *river*, which has affected the normal flows of the river to the extent as to have caused *distress* in one unit, or the other of the basin. Individualistic approach to governance and management of the riparian nations has stood in the way of a meaningful resolution to the Teesta problem, and all attempts since 1983 have proved to a futile exercise until 2011 when an interim deal between India and Bangladesh was put on the record, but it is yet to be finalized.

15.5 Objectives of the Study

This chapter addresses the following pertinent objectives: (1) to critically assess the total volume of water resources of the Teesta basin, i.e., to appraise the total availability of year-round water resources in the *river*, (2) to study the seasonal nature of flows in the Teesta valley, particularly in the dry season and, in the rainy season, with special reference to flow in the upper and lower reaches, (3) to identify and evaluate the problems of water-sharing of the river, between India and Bangladesh, in terms of seasonal requirement, and the problems inherent in the water-sharing between the two nations, (4) to discuss and analyze the problems, involved in the governance and management of water resources of the Teesta basin, (5) to assess the impact of damming and control of the flow of the river in the upper reaches of the valley on the *genre de vie* of people of the lower reaches of the valley, together with the ecological impact of the concerned areas, (6) to study the status of the *2011 Interim Deal*, with regard to the problems of sharing of water resources of the *river*, between India and Bangladesh, with special reference to West Bengal's opposition to any kind of understanding and/or treaty with Bangladesh over the allocation of the Teesta water resources, (7) to assess the nature conflict, and the involved geopolitics between the two nations, about the governance and management of water resources of the Teesta River, and (8) to suggest comprehensive and acceptable mechanisms of governance and management, about the equitable and just uses of water resources, between the contested nations India and Bangladesh.

15.6 Methodology

The present study is based on the published documents, records, literatures, data set, etc., pertaining to the Teesta River. The study necessarily relies upon the analysis of (1) hydrological data collected from government sites and publications related to Teesta from the same sites. An utmost attempt has been made to collect various data regarding flow, discharge, and volume of water of the river from the source of Bangladesh, though the volume of data collected was not found to be adequate. (2) Relevant information, papers, and documents concerning environmental and socio-economic conditions of the Teesta River basin and the people of the involved nations. (3) Besides, political literatures related to Teesta River dispute, case studies conducted by different government organizations and agencies as well as Ministries regarding impacts of various constructions across the river Teesta has been considered here.

However, not much case studies done in Bangladesh were considered in the study because of nonavailability of such literatures there. Content analysis of the literatures, documents, records, and various published papers pertaining to the governance and management of the Teesta River water resources, has been done for explanation of the objectives of this chapter. Most of the data and information used in the present analysis are quite old, because of the restriction on the free availability of data for security reasons. Moreover, there is a paucity of data and required information in Bangladesh, and whatever data are available, they are too old, also.

15.7 Analysis and Discussion

Transboundary water governance generally refers to "equitable and reasonable utilization" of shared water sources by two or more co-riparian countries "without causing of significant harm to other watercourse states" (Article 2, 5, 6, and 7 of 1997 NY UN Convention). Therefore, transboundary water governance is a concept envisaged upon the ideals of mutual trust, coordination, collaboration, information sharing, and joint action for the benefit of all riparian countries. But, due to lack of transparency, understanding, and effective policy planning among the institutional stakeholders of upper riparian and lower riparian countries, negotiation has become a formidable challenge.

India-Bangladesh share one of the longest borders in the world that stretch across 4156 km and is traversed by 54 transboundary rivers. These rivers contribute to more than 96% or 1140 BCM of total river runoff in Bangladesh. Most of these transboundary rivers are part of Ganga-Brahmaputra-Meghna River system which is also the most important river system in South Asia. Out of all 54 transboundary rivers, understanding between India and Bangladesh was reached for only the Ganga River and that too after more than two decades of negotiation. As for Teesta River water-sharing, the negotiation has been going on for more than 22 years with no conclusive settlement. The prospect of Teesta treaty hinges entirely on the negotiation and collaboration of three main sociopolitical stakeholders, i.e., the states of Sikkim and West Bengal of India and Bangladesh. So, the following discussion attempts to uphold the different dimensions of Teesta water resource governance and management from the perspectives of the stakeholders of Sikkim, West Bengal, and Bangladesh.

15.7.1 Total Available Resources across Space and Time

To establish effective management of the Teesta water resources, assessment of total available discharge is very significant. But due to climatic as well as human interventions, the availability of the discharge and/or volume of water varies according to space and time. The mean annual discharge of the Teesta studied at different stations of the river basin highlights this fact; for example, at the coronation bridge in Darjeeling district of West Bengal, the annual discharge is about 580 m^3/s (2003), near Gajoldobha barrage the flow was recorded to be 300 m^3/s (in 2010), whereas in Bangladesh at the Dalia barrage it was only about 200 m^3/s (1998–1999) (Table 15.1).

Also, various chronological studies reveal that the flow of the river varies intensely from wet to dry seasons. At the upper stream point, i.e., near Darjeeling and Coronation Bridge, the peak discharge in the monsoon season is about 4000–5000 m^3/s, which gets diminished to 90 m^3/s in the dry season. While water flow of the Teesta at Dalia Barrage in Bangladesh varies from average maximum flow of 700–600 m^3/s in the wet season to a minimum of 30–40 m^3/s in the dry season (October–April). The months of March and April are the leanest period when sometimes it gets reduced to as low as average of 1000 cusec (28.31 m^3/s).

15.7.2 Nature of Conflict

One of the inherent problems that are fueling the bitterness among the two nations regarding the Teesta water-sharing is the lack of sufficient water for irrigation in the lower stream areas of northern Bangladesh, mainly due to the divergence of Teesta waters in the upstream areas of West Bengal. The Teesta basin is home to about 30 mil-

Table 15.1 Discharge characteristic at Teesta barrage

Discharge in Cumec (m^3/s)(during pre and post Gozoldoba barrage operation period)							
Year	Max	Mean	Min	Year	Max	Mean	Min
1978–1979	721	541	361	1988–1989	667	397	127
1979–1980	670	432	195	1989–1990	645	409	173
1980–1981	522	330	137	1990–1991	729	427	125
1981–1982	666	400	135	1991–1992	653	391	135
1982–1983	652	415	177	1993–1994	453	296	138
1983–1984	883	523	164	1994–1995	633	412	190
1984–1985	795	488	182	1995–1996	459	252	44
1985–1986	760	487	214	1996–1997	478	259	39
1986–1987	660	425	190	1997–1998	672	353	34
1987–1988	527	301	76	1998–1999	364	200	36

Source-BWDB, retrieved from Islam Md and Higano, 1999

lion populations, 2% of which belong to Sikkim (about 0.61 million), 27% to West Bengal (i.e., 8.5 million, the population of North Bengal districts drained by Teesta River), and the rest 71% which is approximately 20 million population belong to the northwestern part of Bangladesh (Syed et al. 2017). Majority of the population across borders are engaged in subsistence agriculture with rice as their main crop for three seasonal harvests per year. But, due to uncertainty of monsoon, lack of availability of water during the lean period, frequent drought and steadily depleting groundwater level intrinsically make agriculture here highly dependent on irrigated waters of Teesta. Also, as the northwestern Bangladesh experiences a much longer dry season persisting for more than 8 months (October–May); therefore, cultivation especially of rice three times a year without irrigation is impossible. Lack of irrigable water in Teesta and higher water requirement in drought prone areas of north Bengal and northwestern Bangladesh has propelled the adversarial relations between India and Bangladesh.

15.8 The Nature of Governance and Management of Water Resources across the Teesta Basin

The Teesta, being a transboundary river, traverses across two politically independent states, i.e., India and Bangladesh, and on account of its being a transboundary river, it is rather difficult for proper governance and management, particularly its water and other resources. Under Indian Constitution, water is a *state subject*, and the *Center* cannot impose its will over the state rather it has to *respect* the state, even if the state's view is not in the line of the national interest. Both, Sikkim, and West Bengal have different interests, about their respective requirements, as such Sikkim is more interested in generating hydropower, whereas West Bengal has its interest in the irrigational value of the Teesta water for its rather *poor* north Bengal. The center cannot simply interfere in the domestic interests of the respective states of the Teesta Basin of the Indian part. Because of the respective state governments' diverse domestic interests, and the states' growing rights sentiments, the center is unable to draw up a comprehensive strategy, regarding the uses of the Teesta water resources. Since Sikkim and West Bengal have diverse interests, it is difficult for the center to draw up a comprehensive treaty with Bangladesh over the sharing of water resources of the river. Since each stakeholder, Sikkim, West Bengal, and Bangladesh have their own approach of governance and management of the Teesta water resources, it is necessary to investigate the mechanism that each had adopted for governance and management of the Teesta water resources, and, hence, separate discussion is made of each stakeholder in this respect.

15.8.1 Teesta under the Governance and Management of Sikkim State Government

The Teesta River, originating in the snow-covered mountains of northeastern Sikkim, drains about 97.66% of the total geographical area of the state. Because of the mountainous profile, steep slopes, and swift-flowing snow-fed channels, Sikkim has been identified as the state with much innate potential for hydroelectric generation by Central Electricity Authority. And the Teesta, being its prime river, has been recognized as one of the main sources and medium of fulfilling the potentiality of producing 8000 MW electricity.

Therefore, the governance of Teesta waters mostly centers around producing hydroelectricity in Sikkim. By the year 2007, the total hydro projects planned and initiated in Sikkim accounted to 27, including projects under both Government and privately owned companies. This excess energy generated was sold to different Indian states like Punjab, Uttar Pradesh, Haryana, etc., and that caused economic boost to the state, so it went for dam building spree. Consequently, by 2011, the government of Sikkim sanctioned 38 new hydropower schemes (Noolker-oak 2017). At present, there are more than 25 governmental and private hydropower projects in Sikkim with an installed capacity of 5284 MW.

However, building so many dams in tectonically fragile and seismically active region have raised many concerns; particularly among the environmentalist, who warned about the impending consequences of restricting the mighty river in its course, as it might increase the chances of landslides, earthquake, flash flood, and ecological displacement of species. with cumulative impact on the landscape and the life changes of people of the basin. This triggered numerous protests, petitions, hunger strikes, and sometimes angry outrages against the dam construction over the Teesta in Sikkim that often sustained continuous struggle from environmentalists, public, monks, activists under various groups and organizations, such as Affected Citizens of Teesta (ACT), Sangha of Dzongu (SOD), Concerned Lepchas of Sikkim (CLOS), Citizens Forum of Sikkim (CFS), and the Sikkim Association for Environment (Table 15.2).

Recently, because of continuous demonstrations and protests by various organizations, 12 newly commissioned dam constructions were canceled by the Sikkim state government. Therefore, by ignoring the environmentalist's concerns, making the unsustainable choices, and forsaking the interests of lower riparian states, Sikkim government has failed in proper management of the Teesta water (Fig. 15.2).

15.8.2 Teesta under the Governance and Management of West Bengal State Government

The Teesta Barrage project at Gozoldoba in Jalpaiguri district is a very significant project for north Bengal. Not only it is one of the largest irrigation projects in West Bengal but also the whole of eastern India. The project completed in 1993 is also known as Gozoldoba Barrage, it is a multipurpose project aimed to use water resources of the Teesta River for irrigation, hydropower generation, flood mitigation, navigation, aquaculture, recreation, and tourism development. This 210 km long intricate canal system envisions irrigating approximately 9.22 lakh hectares of land, but at present, it can irrigate only 66,000 hectares of area under six districts of north Bengal (Banik 2016). The main canals are Teesta-Mahananda Link Canal, Mahananda Main Canal, Dauk Nagar Main Canal, Nagar Tangon Main Canal, Teesta-Jaldhaka Main Canal, etc. Three more minor projects are under construction which will increase the total irrigable area soon. It has actively reduced the flooding in the districts of Jalpaiguri, Uttar Dinajpur, Maldah, KochBihar, Darjeeling, and Dakshin Dinajpur.

Due to frequent shifting of courses of Teesta, waterlogging and flooding are common problems in the north Bengal plains, and to combat this problem cross drainage structures and regulators have been made that have successfully diverted the excess rainwater in the canals for irrigation. The project has also a significant contribution to the transportation and communication in this

Table 15.2 Government and Private Hydroelectricity projects sanctioned by State Government of Sikkim

S. No.	Name of Project	Installed Capacity MW	Latest Status
1	Teesta I (Teesta stage I HEP)	280	MOU/IA canceled as these areas fell within the vicinity of Kangchendzonga National Park
2	Teesta II (Teesta stage II HEP)	330	MOU/IA canceled due to nonperformance of the developer.
3	Teesta III (Teesta stage III HEP)	1200	Commissioned
4	Teesta IV (Teesta stage IV HEP)	520	Major works still not started
5	Teesta V (Teesta stage V HEP)	510	Project commissioned
6	Teesta VI (Teesta stage VI HEP)	500	Under construction (facing financial problems)
7	Lachen HEP	210	–
8	Panan HEP	300	Pre-construction works started
9	Rangyong HEP	117	Projects canceled/ not taken up as these areas fell within Dzongu area and in the vicinity of Kangchendzonga National Park.
10	Rongnichu HEP	96	Under construction
11	Sada Mangder HEP	71	MOU/IA terminated as no activities started at the site.
12	Chuzachen HEP	99	Project commissioned
13	Bhasmey HEP	51	Engineering and most items completed
14	Rolep	36	MOU/IA canceled due to nonperformance of the developer.
15	Chakhungchu	50	MOU/IA canceled due to nonperformance of the developer.

(continued)

Table 15.2 (continued)

S. No.	Name of Project	Installed Capacity MW	Latest Status
16	Ralong	40	MOU/IA canceled due to nonperformance of the developer.
17	Rangit II HEP	66	Under construction since 2005 (facing financial problems)
18	Rangit IV HEP	120	Under construction since 2005 (facing financial problems)
19	Dikchu HEP	96	Under construction since 2006
20	Jorethang loop (HEP)	96	Commissioned
21	Lingza	120	Projects canceled/ not taken up as these areas fell within Dzongu area and in the vicinity of Kangchendzonga National Park.
22	Thankgchi	40	-
23	Bimkyong HEP	99	Revised timeline awarded
24	Bop HEP	99	Revised timeline awarded
25	Ting ting HEP	99	Project canceled vide Govt. notification no. 12/ Home/2012 as milestones as per MOU not achieved.
26	Rateychu Bakchachu	40	LOI issued
27	Tashiding HEP	97	98% completed
28	Lachung HEP	99	Revised timeline awarded
29	Lethang HEP	96	Project was not granted clearance by National Wild Life Board, GoI. Project canceled vide notification no. 12/Home/2012.
30	Suntaleytar HEP	30	MOU/IA canceled due to nonperformance of the developer.
31	Rangit III HEP	60	Project commissioned

(continued)

Table 15.2 (continued)

S. No.	Name of Project	Installed Capacity MW	Latest Status
32	Kalez Khola-I HEP	27.5	MOU/IA canceled due to nonperformance of the developer.
33	Kalez Khola-II HEP	60	MOU/IA canceled due to nonperformance of the developer.
34	Rechu HEP	26	–
35	Rahikyoung HEP	25	Yet to start
36	Rukel HEP	33	Projects canceled/ not taken up as these areas fell within Dzongu area and in the vicinity of Kangchendzonga National Park.
37	Ringpi HEP	30	Projects canceled/ not taken up as these areas fell within Dzongu area and in the vicinity of Kangchendzonga National Park.
38	Rathangchu HEP	320	Project scraped due to religious sentiments.

Source: Retrieved from Noolker-Oak (2017)

region. It was proposed that 166 bridges, crossings on Railway Lines, National Highway, State Highway, and Major District Roads covering a length of 4200 km (approx.) will be constructed in six districts of north Bengal (Mukerjee and Saha 2016; Banik 2016).

The Chief Minister of West Bengal also recently inaugurated the Gozoldoba eco park, known as "*Bhorer Alo*" (Morning Light) tourism hub, which aims to incorporate forest safari, boating, cycling trail, orchid park, angling, elephant rides, etc., and to promote the barrage as a tourist point. (Times of India, Jul 15, 2016).

However, the implementation of the project has left a deep positive impact on the socioeconomic life of the people and has created growth and development in this region. But the critics have cited different instances and environmental warnings against the project. Excessive control of Teesta water at Gozoldoba Barrage has lasting effect on the flow pattern of downstream river. Thus, the governance and management of the Teesta by West Bengal state Government does not seem to be sustainable either.

15.8.3 Teesta under the Governance and Management of Bangladesh Government

To overcome the irregularity of flows of the Teesta waters and insecurity that come with them, the people of northern Bangladesh constructed extensive surface water irrigation system with a large capacity barrage 16 km downstream from the Indian border at Duani and Dalia in the Lalmonirhat district of Bangladesh. This Teesta Barrage (TBPB) also known as Dalia Barrage was developed to control frequent flooding and droughts affecting the area. The planning began as early as *1979 but it became fully operational in 1993*) Islam 2016). It spans across 154,250 ha of land covering seven districts, of which 111,406 *hectares* are accessible for irrigation.

The projects have helped increase not only the total agricultural production but have also increased the cropping intensity of the region through well-planned canal irrigation system. It has played a significant role in flood control as well as increasing the lean period flow of Teesta water by 2.5 times (Afroz and Rahman Md 2013). But the construction of Gozoldeba Barrage in West Bengal put serious constrains on the Dalia Barrage during the lean period, as it faces a serious shortage of water, which ultimately leads to the shrinkage of the target area of irrigation; accordingly, the amount of production also receives quite a setback. And then, again during the wet season, the surplus water released from the Gozoldoba Barrage causes flooding and destruction of the paddy fields. It destroys not only the human habitats, infrastructures, crops, livestock, and soil fertility thus failure to create transparency and understanding among the stakeholders does not only result in physical and/or environmental loss but its effects impinge upon every fiber of society and economy.

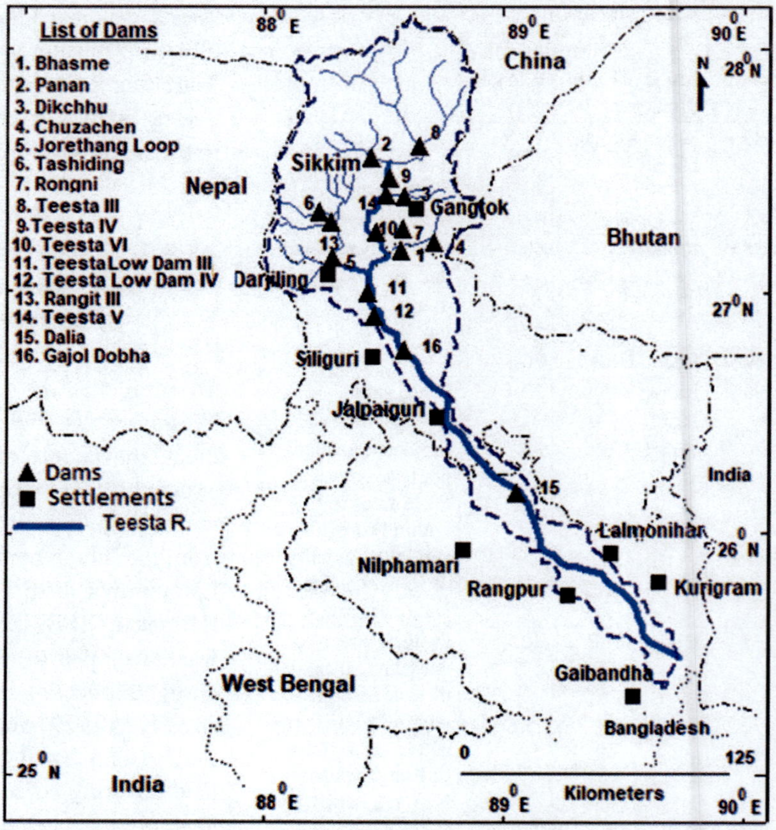

Fig. 15.2 Maor dams in Teesta basin

15.9 Socioeconomic Impacts of Governance and Management of Teesta Water Resources

The Teesta River descends from the Himalayas to the plains harboring immense hydroelectric potential; thus, to harness its capacities to optimum level, many dams and barrages were built along its course. Building numerous dams in seismically active zone and storing water in particular areas, while depriving other regions do have severe consequences not only on environment but also on every sphere of human life. In the following part these consequences have been discussed.

Most of the hydel projects sanctioned in the mountainous region of Sikkim do not follow proper geo-hydrological assessment before, constructing dams; thus, increasing the propensity of hazards like landslide, flash flood, soil erosion, etc. Frequent disruption to the natural flow of river, surface pollution of the dammed water, unauthorized quarrying, etc., is responsible for rapidly decreasing physical and biological qualities of water in these areas. Furthermore, these hydel projects need large open areas and for that they have destroyed vast forested areas and agricultural lands. They have also displaced thousands of ecological species, and indigenous people without proper rehabilitation program. According to an estimate, over 2000 hectares of land have been acquired so far which directly affects the life, livelihood, and employment of more than 4000 rural people of Sikkim. Also, due to rude disruptions in native lifestyle, occupations, traditions, and culture, together with sociocultural degradation have become apparent in the region with unfortunate outcomes, such as alcoholism, unemployment, poverty, and increased rate of venereal diseases (STDs).

In the lower reaches of the river two large Teesta Barrages have been constructed by the respective governments of West Bengal and

Bangladesh within the distance of 100 km from each other. The statistical analysis of the average flow characteristics of Teesta River at the Gozoldoba Barrage (West Bengal's largest Teesta project) and Dalia Barrage (Bangladesh's biggest Teesta project) reveal that the average water flows have been significantly reduced due to the upstream barrage in West Bengal which actively affects the productivity of the area (Mondal Md and Islam Md 2017; Afroz and Rahman Md 2013; Khan Md et al. 2015). According to another research by Afroz and Rahman Md (2013) in the months of November to May (dry season), the discharge of the Teesta decreased at a rate of about 88% after the operation of Gozoldoba Barrage, while the lowest annual discharged decreased by 85%.

This is indeed of much concern for the lower riparian state of Bangladesh. Due to constant lack of water in Dalia Barrage, the vast irrigation facilities are rendered useless in the dry season resulting in consistent droughts, regular crop failures, unemployment, poverty, hunger, and food insecurity manifested in the form of socio-economic deterioration of the region. Furthermore, lack of water in the Teesta in dry season causes a sharp decline in fishing species and their populations, rendering many fishermen jobless and vulnerable during these months.

Due to excessive control of water at upper stream and sluggish flow of Teesta during the dry season, massive sedimentation takes place each year at the lower riparian region, which on the other hand increases the chances of flooding during the wet season when excess water is released from Gozoldoba Barrage. Thus, the people of this region remain vulnerable to the whims of the upper stream region in both dry and wet seasons.

Local stakeholders have also highlighted other concerns, such as waterlogging in areas of northern Bangladesh due to erratic flow and frequent course change of the river, development of "char"/sandbars due to sedimentation, siltation, accelerated rate of riverbank erosion, development of braided river hindering navigability in the lower course due to decreased flow and precarious condition of several aquatic flora and fauna, attributed to declined flows of the Teesta in lean period (SFG Report 2013). Thus, the lack of water does not only directly and indirectly affect the rural agriculture and allied occupations, but also raises "the specter of future food deficit, unemployment and poverty" in the region (Islam Md and Higano 1999).

15.10 The Status of Teesta River Treaty

During the early 1950s and 1960s, when the dams were declared the "temples of modern India" and rivers were considered "national property"; at that time to fund India's rapid industrial growth and to meet their increasing energy requirements, several projects were taken up. The idea of the Farraka barrage and the Teesta Barrage project also started to take shape from this period and although the governments of both sides met to discuss several things, no serious dialogs took place on transboundary water issues between India and then, East Pakistan (now Bangladesh).

In 1971, after the emergence of a sovereign Bangladesh, the actual discussion began regarding the transboundary water issues. To examine and deliberate on the issues of transboundary water governance and management, a Joint Rivers Commission (JRC) was established in 1972. But it is mainly after the completion of the Farraka Barrage in 1974; Bangladesh took steps to negotiate for greater share of the Ganga water. At the same time, Bangladesh also started to deliberate on Teesta water-sharing issue, but no concrete steps were taken. In July 1983, an ad hoc arrangement was contrived for rationalized sharing of the Teesta waters, according to which, Bangladesh to get 36% and India 39% of shared waters, while rest of 25% to remain unallocated. However, this arrangement was short-lived, because of the construction of Gozoldoba Barrage that affected the flows of the Teesta that decreased dramatically; therefore, Bangladesh started demanding 50% share of the Teesta waters during the lean periods which once again renewed the disagreement as both the governments remained uncompromising on their demand.

Since then, 37 meetings were held at the ministerial level of the Joint Rivers Commission. Finally, in 2010, the Joint Commission came up with a new draft agreement, that proposed India and Bangladesh each country would get 40% of the actual flow available at Gazaldoba Barrage in West Bengal, while 20% of the remaining flow available at Gazaldoba would be reserved as environmental flow. The agreement included the provision of biennial review of the said agreement; and proposed to establish a Joint Committee consisted of specialists from both the countries to study the changes in the flow discharge and, accordingly, to take steps toward conflict resolution for the mutual benefit.

In September 2011, the then Indian Prime Minister Manmohan Singh was to visit Dhaka to sign the much-awaited Teesta River Water Treaty. The West Bengal's Chief Minister to accompany him to Dhaka to witness the ceremony on the signing of the Treaty, but at opportune moment, she refused to accompany the Prime Minister to Dhaka for the purpose of the visit was proposed. Her refusal to go to Dhaka spoiled the spirit of the Teesta Water Treaty. The final draft of the Treaty was agreed upon at the 37th meeting of the Joint Rivers Commission (JRC). She wanted more share of the Teesta water than specified in the final draft proposal, and taking the Constitutional Provision of the water being a states' subject, West Bengal succeeded in stalling the signing of the Teesta Water Treaty. The NDA government in New Delhi that came in power in 2014 was sympathetic to the Treaty because of geopolitical reasons, and sought for West Bengal's cooperation in this regard that the Treaty should be signed so as to avoid unnecessary delay. But the Chief Minister was reluctant to accept signing of the Treaty, while Bangladesh was anxiously waiting for the Teesta Water Treaty for domestic reasons and/or local geopolitical compulsion. The Government of West Bengal argued that allowing more water to Bangladesh would lead to water insufficiency in north Bengal with low crop productivity in the region, and if the draft proposals of the Treaty were accepted, then it would be an injustice to the farmers of north Bengal.

The NDA Government at the center could not proceed further to apprise the Government of West Bengal of the merits of the proposed Teesta Water Treaty, besides the regional geopolitical compulsion of the agreement for both the nations. The inflexible attitude of West Bengal stands in the way of cooperative bilateral relationships between the nations. Bangladesh would be going to the general election shortly, while India will have its general election in a sixth months' time thus the Teesta seems to be a very sensitive issue for both the ruling parties of Bangladesh and India, respectively.

Moreover, the failure to achieve the Teesta agreement further complicated the desired initiatives on the governance of 54 other transboundary rivers, such as the Feni, the Manu, and the Muhuri, whose water-sharing formula remains undecided. However, West Bengal Chief Minister during her visit to Dhaka in 2017 reassured Bangladesh that a fair solution would be agreed upon which would be beneficial for both the nations particularly North Bengal, on the one hand, and Northwestern Bangladesh, on the other hand; she also proposed to share Torsa river waters instead of Teesta to resolve Bangladesh's water scarcity problems.

15.11 Geopolitical Implications of Teesta River Treaty

Though, it is a subject matter of national and international politics but the local as well as domestic political issues also play a very significant role in determining the course of the transboundary resource governance and management. The case of the Teesta River Treaty is colored more by internal domestic politics of the involved nations, and the stakeholders therein, together with the changing socio-*geo*political circumstances rather than actual environmental availability of the Teesta water resources.

The bilateral discussion over the sharing of the Teesta waters and its failure to reach a conclusive agreement cannot be put under one state or nation as in different points in time, different agenda, and issues became more important to the representative governments. In 1972, a Joint River Commission was established to investigate the interests of both the nations, but not much progress was made regarding the Treaty.

During the reign of Begam Khaleda Zia (2001–2006) Bangladesh took up an aggressive and often termed as "anti-India" policies. During this period, India witnessed greater insurgencies and emergence of unlawful activities, with the sources, sustaining these activities, being in Bangladesh, but the Government of Bangladeshi took no step to stop such activities of insurgencies against India. Failure on the part of Bangladesh to contain the anti-India activities on its territory resulted in lack of initiative on Indian part, about the transboundary negotiations over the Teesta River.

Later, India-Bangladesh relations improved with the coming of Awami league to power and a new era of bilateral ties began. Sheik Hasina as the Prime Minister of Bangladesh took positive steps to check the insurgencies, boundary disputes, strengthening national and local security against extremist and insurgent groups, and other militant groups, which were spearheading anti-India activities from there. This mutual understanding that arose, finally, resulted in the 2011 *Interim Deal*, which was however eventually rendered unsuccessful due to the Center and West Bengal conflict in India. The opposition parties, such as Bangladesh Nationalist Party (BNP) and National People's Party (NPP) criticized the Bangladesh's ruling Awami League government for their inability to sign the Teesta River Water Treaty. The opposition parties, including the BNP wanted the Government of Bangladesh to raise the issue of 54 transboundary rivers at the General Assembly of the United Nations for fair share of water resources of these rivers. But the ruling Awami League Government did not raise the issue at the United Nations for fear of losing India's support toward economic development.

Also, local movement in the Rangpur Division gained momentum, which appealed to the authorities to take solid steps to reinvigorate the "Buri Teesta" which is one of the main distributaries of Teesta River. The protesters urged the government to hasten the Treaty to get the river water for the rejuvenation of the dwindling economy of the northwestern part of the nation.

The Center and West Bengal conflict in India regarding the Teesta River agreement persists, while the Central Government is in favor of the agreement, but the state government of West Bengal is opposing the finalization of the agreement. This conflict between Indian government and the West Bengal government is putting stress on India-Bangladesh bilateral ties.

What is needed is that there should be a cooperative understanding between India, West Bengal, and Bangladesh over the sharing of the Teesta River water. Let there be a joint operation by the involved nations with the consensus of West Bengal and to initiate new approach to governance and management of the Teesta River water resources.

15.12 Recommendations for Equitable Governance

Transboundary water governance and management is a very complex issue because of its international characterization that each nation riparian with its jurisdiction within the basin struggles and strives for more control over the water resources of the basin, and more advantages at the cost of the other riparian nation(s). This individual struggle on the part of riparian nations makes the nature of governance and management more conflictual and complex. The governance and management of water resources of transboundary rivers however incorporates individualistic regulation, management, conservation, and supply of water, operating over several political, social, economic, and administrative systems, scales, and realities at the same time. Nonetheless, the challenges of governance and management of the Teesta water resources as mentioned earlier in this chapter simultaneously also represent various spectrums of opportunities to improve governance of this transboundary river. Therefore, meeting these challenges is necessary for the good governance and management of Teesta water resource, but since, each stakeholder has its own mechanisms of governance and management of the water resources, e.g., Sikkim's approach of governance and management hinges on harnessing the resources for generating hydropower, whereas West Bengal's approach of governance and mechanism pivots around irrigation, and more water, while the approach of Bangladesh, with regard to governance and management of the Teesta water resources hinge on flood control, and drought control, besides irrigation. Even within India, Sikkim and West Bengal have different perceptual dimensions on the use of the Teesta waters, whereas West Bengal and Bangladesh have same objectives in their perceptual dimension, about governance and management of the Teesta waters. Nevertheless, West Bengal is the only riparian state on the Indian side which has been opposing any concrete deal with Bangladesh since the beginning. Under the existing scenario, it seems extremely impossible to draw up an all-embracing strategy for governance and management, which would be acceptable to all three stakeholders.

However, some recommendations for equitable management of Teesta water resources include, firstly, to increase the transparency among the stakeholders of the Teesta Governance communication, information, and participation are the key aspects. Free flow of communication, easy and quick access to information, and participation of the local shareholders at the grass-root level decision-making will make the governance and management of the Teesta waters fairer and more equitable. Secondly, rather than a top-down approach that is driven by the intentions to gain specific political ambitions, a bottom-up approach to the governance and management of the Teesta waters will be more inclusive, and, thus, more successful. Thirdly, Environmental issues and concerns must be at the center of the governance and management process of the Teesta waters. For holistic and sustainable management of Teesta River and its ecosystem, all shareholders (state governments, central governments, and local people) must be directly engaged in restoring and preserving the ecological and environmental balance in the basin area. Fourthly, the government on both sides of the borders must recognize poverty in the Teesta basin area as a separate socioeconomic issue, rather than just a side effect of the Teesta conflict. Simply solving the water equation will not make this epidemic of modern society disappear. Solid steps like creating more economic opportunities for laborers, providing funds, loans, and/or trainings to small farmers and entrepreneurs, aquaculture, alternative farming techniques, use of new agricultural technologies and cropping patterns, cultivating drought-resistant crops, etc., coupled with rainwater harvesting, flood water channeling, groundwater level recharging, restoration of aquifer, etc., are essential for uplifting the economy of the area. Fifthly, the people and the stakeholders at all scales must be conscious that many unfavorable socioeconomic and political realities of the region are projected on the Teesta water-sharing issue, that makes the issue unnecessarily complex, our main concern should be reviving of the Teesta River. And for that, a joint committee should be created with the experts from both the country for evaluating the health of the river and river basin from the perspective of a whole hydrological unit. Sixthly, the formation of the Teesta River Basin Organization (TRBO) with the purpose of initiating processes for Integrated Teesta River Water Management (ITRWM) mechanisms for the whole basin of the river. The ITRWM will be a comprehensive participatory and implementation tool for managing, governing, and developing water resources of the Teesta River in a balanced way so that equity in water resource distribution in the basin could be maintained. The entire Teesta basin is required for the ITRWM mechanism, and for that India and Bangladesh shall have to abandon part of their sovereign right over the basin that will ensure joint administration of the basin. The functioning of the ITRWM would be based on the principles of *social equity*, *economic efficiency*, and *environmental sustainability*. Social equity will ensure equal access to water resources for all users (particularly marginal, and poor regardless of their national identity. Economic efficiency is the greatest benefit to the greatest number of users possible with the availability of joint Indo-Bangladesh finance and water resource, while environmental sustainability is the maintenance of the Teesta aquatic ecosystem as acknowledged as users and adequate allocation will be made to sustain their natural functioning.

The aforesaid recommendations may be considered in unison with one another to find out a comprehensive but all-embracing way toward an amicable solution to the vexed problem of the equitable distribution of the Teesta water resources.

15.13 Conclusion

The Teesta River is one of those transboundary rivers out of such 54 rivers, between India and Bangladesh which play a very decisive role in the *genre de vie* of 30 million people, living in the river basin. It is a historical river, because the people living in the basin area over generations through different political and social systems however have acquired an environmental-based lifestyle, with historical habitation and experience. *Geo*political division of the Teesta River basin disrupted the homogenizing effect of the basin ecology, with impacts upon the people of the basin. It becomes a political river, with transboundary character, flowing across the two sovereign and independent nations: India and Bangladesh. Earlier, it used to flow through Sikkim (a quasi-sovereign Kingdom), India, and Bangladesh (former East Pakistan), and it was its *geo*political characterization so long Sikkim was *independent*. In 1975, Sikkim became a part of the Indian Union. With Sikkim's accession to India, the Teesta's multinational character changed into a bilateral character, i.e., between India and Bangladesh.

Nevertheless, there are three stakeholders, which put claims over the water resources of the Teesta River. The *river* is rich in water resources as it originates on the higher reaches of the Himalayas, in a glacial source, with the perennial supply of water, the river flows with a very high speed across Sikkim. Since, the Teesta flows across a very rugged mountainous topography, and it cuts into deep gorges and flows through narrow valleys, creating excellent conditions for the generation of hydropower projects. Sikkim has large number of dams and hydropower projects over the Teesta River; thus, Sikkim uses the bulk of the Teesta's water resources. The flow of the river becomes little bit sluggish once it enters West Bengal, and starts flowing on the plains, and becomes more sluggish, when it enters Bangladesh, where its braided channel becomes wider. By the time the Teesta enters Bangladesh the flow of the river becomes relatively slower with considerable seasonal variations.

The bulk of water resources of the Teesta is consumed in upper reaches of the *river*, Sikkim, and West Bengal, but *little* is left for Bangladesh. This is the crux of the interstate problem between India and Bangladesh. All the stakeholders have their own objectives and mechanisms of governance and management of the Teesta water, which are structurally different from each other. Sikkim's objective is fuller utilization of the potentials of the water resources for hydropower generation, and this is manifested by many dams and hydropower projects. Sikkim went ahead of its hydropower projects, despite protests and demonstrations by the ecologists and social activists. West Bengal's approach toward governance and management of the Teesta water resources appears to be sustainable to the extent as to manage the bulk of water resources of the *river* from being wasted. The Teesta flows across the north Bengal plain that depends upon agriculture, but the natural water supply is not assured rather it depends on the monsoon. The water flowing through the Teesta during the lean period declines substantially therefore irrigation is a necessity for the region. The Teesta water resources are used for irrigational purposes. The Gajoldobha Barrage on the Teesta in West Bengal is a multipurpose river valley project with multifaceted objectives. This barrage has improved the quality of life of the people of north Bengal plains of West Bengal. Since, much of the water resources of the Teesta is used and consumed upstream, Bangladesh faces the problem of river water scarcity, particularly during the lean period, lasting for seven or more months, depending upon the discharge and flows. Bangladesh also faces the wrath of floods during the rainy season, when excess water is released from the Gajoldobha Barrage in West Bengal. Bangladesh also constructed a huge Teesta Barrage, with large capacity at Dalia, which is 16 km downstream from the Indian border at Duani. It is also a multipurpose project. There is no doubt that the construction of the Gajoldobha Barrage in West Bengal has necessarily reduced average annual flow of the Teesta River in Bangladesh largely affecting the crop productivity and the traditional *genre de vie* of people of the area. The Teesta River seems to be a *pride* for Sikkim, and envy for Bangladesh, because whatever it gets, it is too small for its requirement. West Bengal however lies in between pride and envy.

Lack of agreement between the stakeholders largely impedes the prospect of any solution to the water-sharing problem between the *involved nations*, India and Bangladesh, which results in *geopoliticized* conflictual between them. Therefore, the Teesta *geo*politics pivots around the question of equity in water-sharing between India and Bangladesh, and for little more than three decades, no permanent solution to the Teesta water disputes have been found out. There is an urgent need to convert the *Interim Deal* to a permanent Teesta River Water Treaty, as a possible resolution to the conflict. Too much of geopolitics on water disputes is bad for regional stability and security. But, unless a concrete solution is found out the problem of conflictual, about governance and management of the Teesta water resources will continue between India and Bangladesh, otherwise the regional geopolitics will remain instable, and disturbed as well.

References

Afroz R, Rahman Md A (2013) Transboundary River water for Ganges and Teesta Rivers in Bangladesh: an assessment. Global Sci Technol J 1(1):100–111

Ahmed I (2015) Geopolitics of trans-boundary water management: case of Nile Basin riparian. Modern Res Stud Int J Humanit Soc Sci 2(2):355–374

Banik S (2016) Socio-economic impact of Gazaldoba Teesta Multiple River valley project. J Soc Environ 5(2):26–35

Bauer A, Quiroz JC (2013) Resource governance. In: Goldthau A (ed) The handbook of global energy policy. Wiley-Blackwell, Hoboken

Chan NW, Roy R, Chaffin BC (2016) Water governance in Bangladesh: an evaluation of institutional and political context. MDPI, Basel

Islam Md F, Higano Y (1999) International Environmental Issue Between India and Bangladesh: Environmental and Socio-Economic Effects on The Teesta River Area. Paper presented at the 39th Congress of the European Regional Science Association: "Regional Cohesion and Competitiveness in 21st Century Europe", European Regional Science Association (ERSA), Dublin, Ireland, Louvain-la-Neuve, August 23–27, 1999

Islam MF (2016) Water use and poverty reduction. New Front Reg Sci Asian Perspect 8. https://doi.org/10.1007/978-4-431-55172-0_2

Khan Md SS, Islam AR, Md T (2015) Anthropogenic impact on morphology of Teesta River in northern Bangladesh: an exploratory study. J Geosci Geomatics 3(3):50–55

Lockwood M et al (2010) Governance principles for the natural resources management. Soc Nat Resour 20(10):986–1001

Mirchandani M (2016) The Teesta water dispute: geopolitics, myth and economics. Observer Research Foundation, New Delhi

Mondal Md H, Islam Md S (2017) Chronological trends in maximum and minimum water flows of the Teesta River, Bangladesh, and its implications. Jàmbá J Disaster Risk Stud 9(1):373

Mukerjee B, Saha UD (2016) Teesta barrage project – a brief review of unattained goals and associated changes. Int J Sci Res 5(5)

Noolker-oak G (2017) Geopolitics of Water Conflicts in The Teesta River Basin. Available at https://gaurinoolkar.files.wordpress.com/2017/12/teesta-conflicts_gno.pdf. Accessed 5th October 2018

Prasai S, Surie MD (2013) Political economy analysis of the Teesta River basin. The Asia Foundation, New Delhi

Rudra K (2003) Taming the Teesta. Ecol Asia 11(1)

SFG Report (2013) Rivers of peace: restructuring India Bangladesh relations. Strategic Foresight Group, Mumbai

Singh R (2010) Trans-boundary water politics and conflicts in South Asia: towards 'Water for peace'. Centre For Democracy and Social Action (CDSA), New Delhi. Available at http://in.boell.org/sites/default/files/downloads/water._Final.pdf

Syed A, Afroza H, Arfan U et al (2017) The Teesta Basin: enough water for power and agriculture for all? Himalayan Adaptation, Water and Resilience Research (HI-AWARE), Kathmandu

16 Governance Issues for Sustainable Water Management in Rapti River Basin, Uttar Pradesh

Narendra Kumar Rana and Neha Singh

Abstract

During the last two decades, various key concepts emerged in the field of water management both in developed as well as developing countries. In the developing world compulsion for the integration of ecological, social, and economic aspects of sustainable development in water management led to several debates on the notions of integration, good water governance, and participatory water management. The present empirical study on a small river basin shows that even if integration, good governance, and participation have many potential benefits, these are difficult to achieve in practice. In this context, a critical analysis of water resource management is pertinent. The Rapti is a hill-fed river basin shared by two riparian nations, i.e., India and Nepal. Due to inadequate management, the resource potential of the river basin is not fully utilized, rather the river became a symbol of underdevelopment in the region. With the increasing concentration of anthropogenic activities both at upstream as well as downstream part, river ecology has been continuously degrading and numbers of environmental and social conflicts are emerging. With the help of primary and secondary data, the study highlights how integration is difficult in the case of a river shared by two riparian nations and identified the complexity caused by multiple stakeholders at the basin level. The study also identifies a number of governance issues like, management of floodplains and their resources, compliance to flood forecasting and warning, public utility management within the active channel zones, annual maintenance of river banks, illegal sand mining, integration of development schemes within the context of floodplain environment, livelihood issues and incorporation of community expectations that need to be prioritized for sustainable water management at basin scale at micro-level. Besides, the study highlighted several potential issues for future research.

Keywords

Floodplains · Governance · Rapti River · Sustainable Water Management

N. K. Rana (✉)
Department of Geography, Institute of Science, Banaras Hindu University, Varanasi, India

N. Singh
Department of Geography, H. R. P. G. College, Khalilabad, India

B. Thakur et al. (eds.), *Resource Management, Sustainable Development and Governance*, Sustainable Development Goals Series, https://doi.org/10.1007/978-3-030-85839-1_16

16.1 Introduction

With the beginning of the twenty-first century, it was becoming apparent, and a growing consensus had emerged among the global leaders and water managers that the world was in a state of critical water crisis. Early analysis on the causes of the crisis pointed towards resource scarcity out of demand and supply imbalances caused by rapid population growth and associated anthropogenic activities. As remedial solutions, different approaches to manage scarce but finite resource at different scales were proposed in the late 1970s. At the conceptual level ecological, economic, behavioral, participatory, and, lately, integrated approaches were gained prominence, whereas remote sensing and GIS approach dominated the methodological ones (Thakur 2003). The pressure on water resources highlighted the hydrological, social, economic, cultural, and ecological interdependencies in aquatic as well as terrestrial ecosystems. These interdependencies demanded more integrated approaches to developing and managing water and land resources. Integrated River Basin Planning and Management is being considered widely as a process, dynamic in nature, for managing this dynamic resource along with other static one using basin or subbasin as a manageable unit. The objective is to mitigate the natural hazards and use resources for productive and social purposes with due care to the environment to ensure the sustainability of the system (Parmanik 2014). Two decades later however another analysis on the cause of the crisis had taken center stage (Walters 2013). While it was still largely agreed that some areas of the globe were certainly water-stressed, it had come to be believed that the water crisis was a consequence of politics, poverty, and the way water is managed and allocated (UNDP 2006). In short, it had been determined and decided that the crisis of water was an outcome of a crisis of governance (UN WWAP 2003). Since then, critical evaluations of policies, institutions, and management principles within the lens of governance and sustainable water management have emerged. Notable studies in the Indian context are on institutional framework and governance (South Asian Network on Dams, Rivers and People 1999; Khemka 2016), river basin management and governance (Pant 2002; Blackmore 2010; Kumar and Prakash 2018), and institutional restructuring and water governance (Kumar and Bharat 2014; Joy 2016; Kumar et al. 2017; Shah 2016, 2018a, 2018b; FAO 2018). All these studies are at macroscale either dealing with institutional and policy changes at national and basin level and most cases lacked an ecosystem perspective to sustainable water management.

In this context, the present paper examines the emergence of the concept of governance in river basin planning in India. To that extent the concept has been integrated into river basin planning at microlevel and local issues involved in sustainable water management are analyzed, considering the Rapti River basin as the case study. The most fundamental elements in successful water governance are (1) sustainability, (2) inclusiveness, and (3) institutional culture. The scope of the present paper is limited to sustainability only which has three dimensions (1) natural, (2) financial, and (3) human resources (Nayar 2013).

Box 16.1 Sustainable Development Goals
Transboundary water governance for Rapti River basin in Nepal and India is a significant factor for sustainable development and is widely acknowledged. Rapti river's water resources are catalytic agents through transboundary cooperation for improved international trade, economic development, navigation, energy generation, wildlife conservation, and broader basin-based regional integration. The authors of this chapter agree and observe that appropriate water governance through agreements and institutional arrangements can offer an important means by which transboundary waters can be managed in an equitable and sustainable way, furthering support to biodiversity and the ecosystem, and maintaining peace and security. These steps towards international water governance could also offer opportunities to realize wider benefits

such as strengthening climate change and resolve conflicts over water resources. The UN sustainable development goals (# 6) recognizes the significance of transboundary cooperation as a prerequisite for realizing water-related SDG targets and indicators. Thus, its significance goes beyond water sharing itself. The authors recognize and appreciate the integration of riparian river basin approach into transboundary water cooperation to meet the goals of SDGs. This approach represents enormous benefits for upstream activities on land and water influencing nutrient loads from unmanaged agricultural run-off and inadequate wastewater treatment which continue to cause eutrophication. Not to mention that unsustainable practices upstream lead to a water-food-energy-ecosystem nexus crisis. To tackle these challenges the authors suggest an integrated perspective that takes a holistic approach considering the linkages across the full continuum.

16.2 Objectives

This chapter has the following two objectives (1) to examine the emergence of the concept of governance in sustainable water resource management and (2) to investigate the governance issues for sustainable water management in Rapti River basin.

16.3 Study Area: The Rapti River Basin

The Rapti River basin is an integral part of the middle Ganga plain, well known for its recurring flood events. Located on the eastern part of Uttar Pradesh, the basin is formed by the Rapti River and its tributaries, namely, Burhi Rapti, Ami, Rohin, and Gorra. The River Rapti flows in the subhumid to humid monsoon region of the middle Ganga plain. It is the largest tributary of the river Ghaghara which, in turn, is a major constituent of the Ganga. The Rapti River basin extends from 26°18′ 00″ N to 28°33′06″ N and 81°33′00 E to 83°45′06″ E and covers an area of 25,793 km^2 out of which 44% 11,380 km^2 lies in Nepal and 58% (14,413 km^2) in the eastern part of Uttar Pradesh. The Rapti rises at an elevation of 3048 m in Dregaunra range of Shiwalik in Nepal and covers a total distance of 782 Km, (of which 331Km lies in Nepal) before joining Ghaghara at Barhaj in the Deoria district of Eastern Uttar Pradesh. The River Rapti is fed by numerous tributaries and effluents. Those of the northern or left bank originated from Shiwalik and *Bhabar* region. Those on the south represent merely old beds of the river. Important left bank tributaries on the Rapti are Burhi Rapti, Ghonghi, Kain, and Rohin. The Bhakla, Ami, and Taraina are the noted right bank tributaries (Fig. 16.1). The basin consists geologically of two distinct portions: structurally, it is a segment of the great Indo-Gangetic trough, and it has also some marginal portion of the Himalaya's foothill region of the Shiwalik (Mohindra et al. 1992).

The lower Rapti basin, administratively, consists of Gorakhpur, Deoria, Sant Kabir Nagar, and Maharajganj districts of the eastern part of Uttar Pradesh. A large number of perennial lakes like Ramgarh tal, Narhai tal, Karmaini tal, Amiar tal, and Chillua tal are found in the abandoned channels of rivers, which have become blocked by the accumulation of silt, or by the accumulation of water in deep natural depression. The lower Rapti River basin records annual mean rainfall of 1238.9 mm and annual mean temperature of the district is 25.30°C. The southwest monsoon usually arrives over the basin by the middle of June and withdraws by the end of September. The basin is characterized by flat floodplains with net sown area as the dominant land use. Extensive deforestation has occurred along the upper catchment of the basin (Lal 1995). The basin has often suffered from fluvial hazards like floods, landslides along embankments, channel migration accompanied by croplands, and erosion of fertile soils (Rana and Verma 2009; Rana and Tyagi 2008). Sometimes,

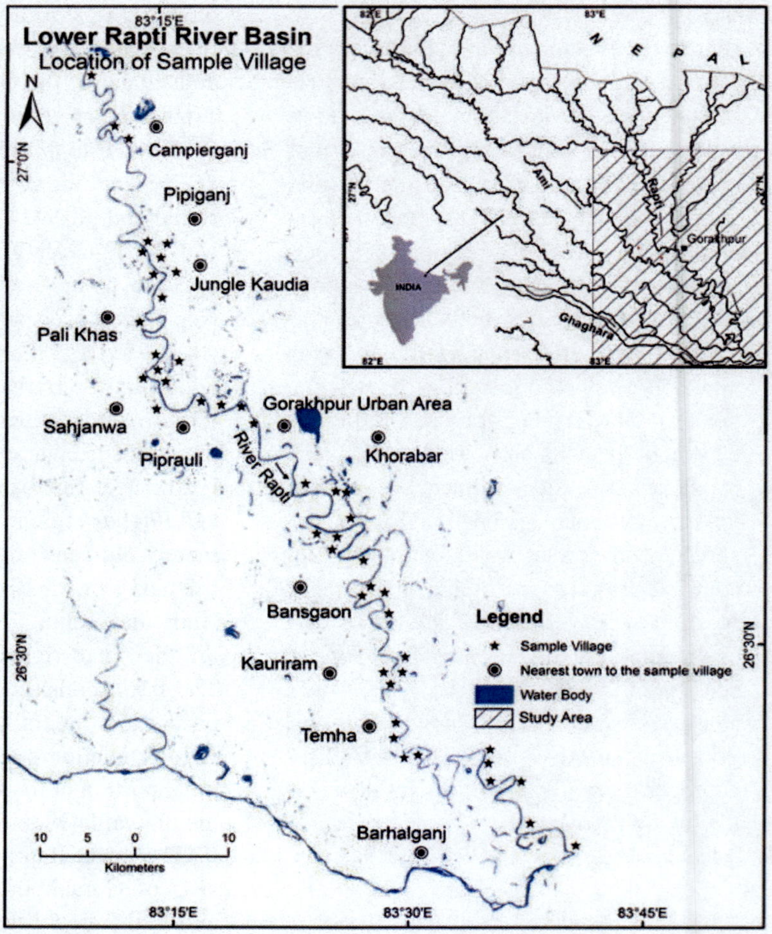

Fig. 16.1 Location of sample villages

due to waterlogging crops are damaged and crop cycle and production get severely affected. Further, people have become victims of water-borne diseases like diarrhea, cholera, dengue, and encephalitis (Shukla and Rahman 2012).

16.4 Data and Methods

This study is based on both primary and secondary data sources. Primary data is collected through household questionnaire survey of program beneficiaries, flood-affected community, and interview with NGO officials working with local community, *Gram Panchayat Adhikari*, *Gram Vikas Adhikari*, and *Gram Pradhan*. A stratified sampling method was used for the selection of 54 representative villages. Starting from the confluence of river Rapti with river Ghaghara, nine representative regimes were selected every 10 km apart towards upstream of the river. Two villages each from places experiencing high, moderate, and low flood frequency (Fig. 16.2). Around 260 households were selected based on purposive sampling. Representative households from various program beneficiaries, households with BPL cards, small and marginal farmers, women self-help groups, and socially marginal groups are also included. Secondary data is obtained from disaster management authority and from the office of the District Magistrate; Gauge level data pertaining to flood forecasting and warning from the office of the Central Water Commission, New Delhi. Demographic and land use data is collected from District Statistical handbooks.

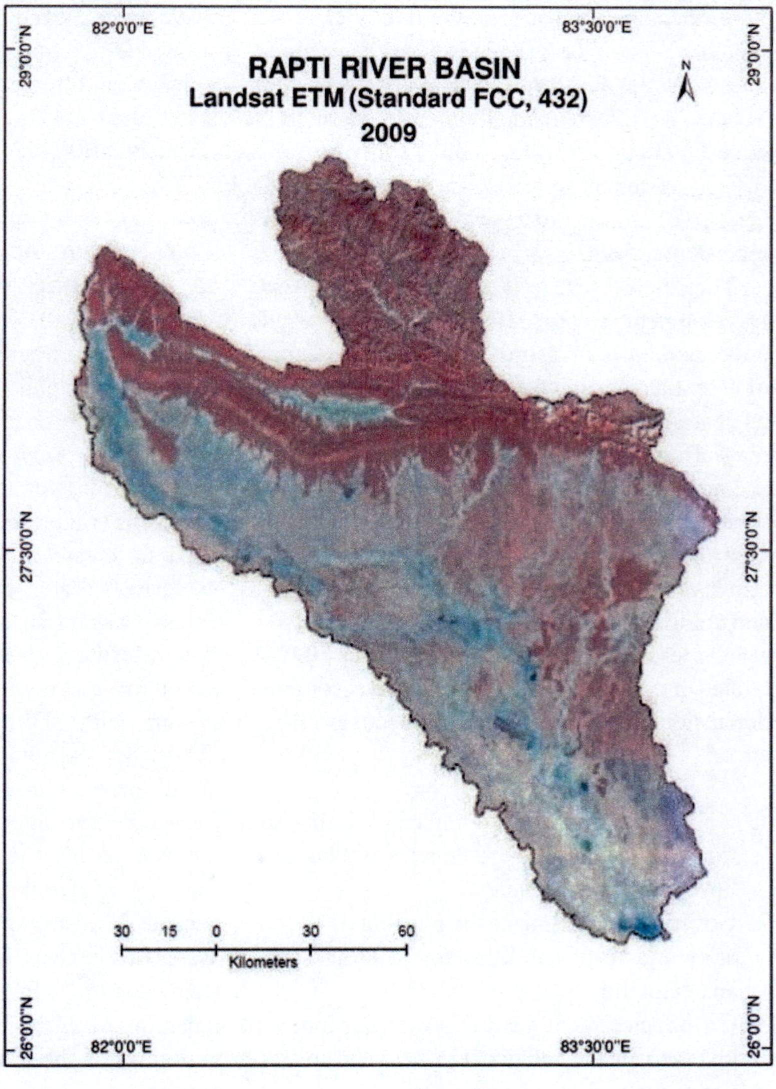

Fig. 16.2 Rapti River Basin

16.5 Emergence of the Concept of Governance

The emergence of the concept of governance in resource management is associated with the introduction of liberal economic policies in 1991. With the induction of new economic policies, governance has emerged as a key element in policy outcomes. Led by international and bilateral development agencies, in the early 1990s, it began to be asserted that the failure and crisis of development throughout the second half of last century was a consequence of bad policies and governance practices within the developing countries (World Bank 1989; World Bank 1991a; World Bank 1991b). Earlier, state was considered as an engine of growth and so was the ownership and custodian of basic natural resources. With the failure of development, performance accompanied by ineffective policies, poor implementation, monitoring, and massive degradation of the natural resource base, triggered the search for an alternate policy framework, and new institutions for delivering developmental outcomes and public services. In this process, the role of state was redefined. The World Bank using the term governance for the first time in its Report and Sub-Saharan Africa redefined the role of state. It advocated that the state should shift the focus from government to governance (Mathur 2016). The role of the state was not reduced but reconceptualized.

The role of state was redefined in the World Development Report 1992. The state would undertake basic investment in, and management of essential social and physical infrastructure but its central role was to encourage the free and fair play of market forces in an impartial, open, and accountable manner. Good governance came to be associated with the capacity of the state to provide these conditions and create a climate that attracts international capital through cooperation and aid through international agencies.

A brief review of literature by Stoker 1998 as quoted by Mathur (2016) reveals five common dimensions of governance perspectives. They are:

1. Governance refers to a complex set of institutions and actors that are drawn from but also beyond the government.
2. Governance recognizes the blurring of boundaries and responsibilities for tackling social and economic issues.
3. Governance identifies the power dependence involved in the relationship between institutions involved in collective action.
4. Governance is about autonomous self-governing networks of actors.
5. Governance recognizes the capacity to get things done which does not rest on the power of the government to command or use its authority. It sees government as able to use new tools and techniques to steer and guide.

In what can be termed the "good governance" discourse, it is now insisted that if international development ambitions and aspirations of the early part of the twenty-first century are to be achieved, such as the Millennium Development Goals (MDG) of reducing by half the number of people without access to water by 2015, or the Sustainable Development Goals (SDG) of clean water and sanitation, then, the institutions of governance, and in particular the institutions of the state need to be transformed and reformed (Walters 2013).

16.6 Governance and River Basin Management in India

India's water wealth is inequitably distributed, both spatially and seasonally. Besides, its geographical location along with physical, socioeconomic, and environmental conditions create major hurdles for water management at basins, subbasins (watersheds) as well as local scale. The need to create a separate central ministry to exclusively cater the water resources of this country was felt in the year 1985. The Ministry of Water Resources was formed following the bifurcation of the then Ministry of Irrigation and Power. With shifting priorities and emerging challenges, the ministry was renamed as Ministry of Water Resources, River Development, and Ganga Rejuvenation in 2014. Subsequently, in the year 2019, the ministry was merged with Ministry of Drinking Water and Sanitation to became Ministry of Jal Shakti, Department of Water Resources, River Development, and Ganga Rejuvenation. Frequent reformulation and renaming of this ministry reflect India's seriousness towards the mounting water challenges the country has been facing over the past few decades.

Presently, Ministry of Jal Shakti (Waterpower) is the nodal agency for the optimal sustainable development, maintenance of quality, and efficient use of water resources to match with the growing demands of this precious natural resource of the country (Ministry of Jal Shakti 2020). Under the Ministry, there are a number of departments, boards, and technical agencies to frame rules and regulations, and to monitor water resource development in India. Most prominent among them is Central Water Commission (CWC). With the establishment of CWC in 1945, India's engagement with water resource management at the basin level has a chequered history.

The CWC envisioned to promote integrated and sustainable development and management of India's water resources by using state-of-the-art technology and competency and by coordinating all stockholders (Central Water Commission 2020b). The Commission has entrusted with17 core activities including techno-economic appraisal of irrigation, flood control and multipurpose projects, collection, maintenance, and dissemination of hydrological data, flood forecasting, basin-wise development of water resources, morphological study of river behavior, dam safety, etc. The CWC has divided the country into 20 river basins comprising of 12 major and 8 composite one for the convenience of river basin planning. The Central Ground Water Board (CGWB) is another nodal agency for assessment and augmentation of groundwater, regulation and protection of groundwater, and maintenance of water quality and standards. The National Water Development Agency (NWDA) was set up in 1982 to primarily carry out water resource development of peninsular rivers; later the Himalayan River components were added. Presently, the agency is responsible for the implementation of ILR projects and water resource projects under Pradhan Mantri Krishi Sinchai Yojana (PMKSY). Central Soil and Materials Research Station, Central Water and Power Research Station, National Institute of Hydrology are some of the important research institutes, whereas Brahmaputra Board, Upper Yamuna River Board, Ganga Flood Control Commission are important boards and commissions assigned with area-specific tasks.

Those organizational setup and functioning of different institutions came across harsh criticism on the ground of complex institutional setup and heavy dependency on bureaucracy. Most of the institutional structures are based on large dam-centered water resource development that ignores to address the socioeconomic inequalities. Besides, techno-centric approach misunderstood the comprehensive approach and has several operational and institutional drawbacks.

Many of these so-called RBOs were either subject-oriented or project-oriented organization. They were guided with an engineering perspective of river management, and they completely ignored community involvement. This perspective lacked in comprehensive understanding of the river system for evolving various options to manage the river. Nor were they able to learn lessons from the cumulative impacts of the project. None of the organization has done any pre- or post-facto analysis of the project they implemented to assess the real costs, benefits, and impacts, especially in the downstream (South Asian Network on Dams, Rivers and People 1999).

16.7 Governance Issues in Rapti River Basin

A brief survey of literature and secondary database on physical and socioeconomic characteristics of Rapti River basin along with governance and river basin management issues revealed the following governance and management challenges. The findings were verified and further strengthened by limited field visit and questionnaire survey.

16.7.1 Management of Floods and Floodplain Resources

State governments are responsible for floods and floodplain management in India. Central funding only supports state programs. In the Post-Independence period, different committees were set up and partial studies were undertaken to minimize the flood risk. Even if those committees have suggested an elaborate plan of structural and nonstructural flood mitigation measures from time to time, a generalized framework of disaster management process was visualized only after the enactment of Disaster Management Act 2005. With this act a major paradigm shift in public policy (from relief and rehabilitation to prevention, preparedness, response, and recovery) on disaster management took place. Accordingly, the states develop their own plans at state and local level for different types of disasters including flood. In the Rapti River basin, the state agencies adopted different types of measures to manage the flood risk. They can be grouped and discussed into two principal categories: (1) Structural flood mitigation measures and (2) Nonstructural or flood emergency measures. Presently, the overall responsibilities rest with district disaster management authority headed by District Magistrate and four other departments, namely, Police, Health, Public Construction, and Irrigation. The basin area is partially shared by nine districts; thus, nine district disaster management authorities are working independently in the basin. A closer scrutiny of the district disaster management and reduction plan prepared by the respective district disaster management authority revealed that they have a vibrant emergency support function policy and team that can be activated during the time of emergency. A clear mandate and work division among the sister departments and their integration during other phases of disaster management cycle, i.e., preparedness, redevelopment, and mitigation phase is lacking. Representations from crucial departments like agriculture, rural development, primary and basic education, transport, electricity, and local community are missing. Thus, execution of programs related to rural development, sanitation and drinking water, irrigation, basic and primary educations are executed by their respective department without incorporating changes to comply with the ongoing flood management measures within the floodplain.

16.7.1.1 Compliance to Flood Forecasting and Warning

Flood forecasting is a scientific evaluation of an event in real-time leading to the issue of a general alert about hazardous conditions. The practical aim of flood forecasting is to reduce the loss of life and the economic damage caused by floods. Thus, the accurate forecasting of flood conditions is an essential prerequisite for the provision of reliable flood warning schemes. In recent years, flood forecasting accuracy has increased greatly with improvements in telecommunications and computerized data handling and processing. The need for reliability in flood forecasting has been stressed frequently. Errors in the forecast of flood stage or of the time of arrival of flood conditions may lead to under preparation and loss of avoidable damage or to over preparation, unnecessary expense and anxiety, and a subsequent loss of credibility. Besides, social acceptability of the flood forecasting and warning is becoming a major issue. The flood forecasting network of the Rapti River basin comes under the supervision of Middle Ganga Division No.1, Lucknow. The entire network of the basin is administered through two subdivisional headquarters. The upper regime of the Rapti comes under the upper Rapti-Ghaghara subdivision with headquarters at Gonda. The Subdivision has four gauging sites on the River Rapti, namely, Kakardhari Bhinga, Balrampur and Bansi, and one gauging site at Kakarauli on the river Burhi Rapti. Similarly, the lower Rapti regime is administered by Lower Rapti-Ghaghra Subdivision with headquarters at Gorakhpur. It has two sites (Regauli and Birdghat) on the River Rapti and only one site on the River Rohin (Trimohanighat). All these sites are operated and maintained by Central Water Commission. Even if, the basin has a good network of gauging stations and flood forecasting and warning performance, their practical utility remains a cause of concern during emergency management. This is because of low level of social acceptability, lack of faith, accompanied rumor, unsatisfactory evacuation, rescue and relief camps, and sense of insecurity to property and belongings among the floodplain dwellers (Rana 2005) (Table 16.1).

16.7.1.2 Integration of Sectorial Plans

Although the basin has an established floodplain management programs consisting of structural measures and flood emergency plans (Rana 2017), it has several drawbacks. Other structural and nonstructural measures like land use planning controls, development and building controls were not included in the ongoing floodplain management measures. Due to the encroachment of wetlands, water-retaining capacities of those natural detention basins are decreasing. It needs proper implementation of land use policy. Similarly, there is a need for standardization of

Table 16.1 Rapti River Basin; details of sites for flood forecasting and warning

Name of the river	Gauging site	Type of site[a]	Danger level in m.	Highest flood level recorded	Recorded year
Rapti	Kakardhari	G	131.00	131.35	1979
	Bhinga	GDQ	119.50	120.10	1997
	Balrampur	GDSQ	104.62	105.25	2000
	Bansi	G	84.90	85.82	1998
	Regauli	GDSQ	80.30	82.12	2000
	Gorakhpur (Birdghat)	GDSQ	74.98	77.54	1998
Burhi Rapti	Kakrahi	G	85.65	88.97	1998
Kunhra	Uska bazar	G	83.52	85.62	1998
Rohin	Trimohini Ghat	G	82.44	85.43	2001

Source: Central Water Commission, 2002
[a]G = Gauge, D=Discharge, S=Silt, and Q = Water Quality

embankment constructions and maintenances to minimize corruption. Public institutions like schools, primary health centers, Gram Panchayat buildings, fertilizer distribution centers, etc., in the floodplains are in low-lying areas thus facing the twin problems of flooding and waterlogging even in flood-free years. Availability of minimum facilities, resources, and trained manpower are lacking at the grassroots level emergency centers thereby making their presence meaningless. Lack of integration of roles and responsibilities was the main hindrance for relief and rescue operations during the time of emergency and rescue operations. Besides, there is an urgent need for a contingency plan for emergency operation in case of failure of embankments.

16.7.2 Public Utility Management in the Channel Migration Zones

Most of the rivers in the Gangetic plains experience severe channel erosion; the situation is alarming in the middle and lower Ganga Plains. Extremely low gradient with alluvial bedding and fluctuation in river discharge are the primary cause of severe channel movements and erosion in these areas. Recent studies (Rana et al. 2009; Singh and Awasti 2011; Singh and Rana 2014; Rana 2015a; b) confirmed that flooding during the high discharge period and lateral erosion during the low discharge period are the most dominant fluvial hazards of this region. The study further observed that lateral erosion is an independent fluvial hazard that operates during a low discharge period. Low degree of compaction due to the presence of sandy and silty facies in the river valley deposits, mass movement, paleocurrent pattern, and fracture initiates and enhances the lateral erosion. Extensive studies, policy planning, and project implementation have been carried out on flooding, but not much attention has been paid to the phenomenon of lateral erosion and their possible impacts. Construction activities within the floodplains particularly in the active channel migration zones are a common practice observed during the field visit. Road and bridge constructions by the public works department, public buildings like schools, panchayat Bhawan's by the local administration, lying down of railway lines and power grids by railways and electricity departments are implementing their projects work without acknowledging the river behavior. In most of the cases active channel migrations making their project redundant even before their completion (Rana 2005).

Impact of channel migration on settlements is another dominant problem. The Rapti River basin is a most densely populated area of eastern Uttar Pradesh because of the availability of fertile land and irrigation facilities. Frequent occurrences of floods increase the natural fertility of the soil. The local community supports the view that flood provides bonus to the farmers living in the region. Thus, there is a tendency to reclaim and occupy the floodplains within the active channel migration zones for settlements and agriculture (Singh 1966; Singh 1975). The most serious problem regarding the safety of settlements is uncontrolled and unplanned, ever expansion of village and urban settlements, which do not have any master plan for development. Absence of any regulating authority further complicated the problems.

16.7.3 Maintenance of Embankments

Most of the notable rivers of the lower Rapti basin are fully embanked (Table 16.2). There are 64 embankments in the basin having a total length of 446 km. Major chunk of resource has been allocated every year in the name of annual maintenance of embankments and it is a major source of corruption. The mud embankments were endangered by rat holes and the damage caused by porcupines. As embankments are close to human habitation, rats and porcupines live within bounds just above the water level. As the water rises the rats and porcupines also change their shelters with the rising water level. The change is not necessarily in a straight line but invariably in a zigzag manner. This ultimately affects the strength of the embankments throughout the basin. The embankment also provides a false sense of flood protection, because of which new settlements are coming up close to the embankments and sometimes between the river channel and the embankments (Rana 2014).

16.7.4 Prevention of Sand Mining

Sand deposited along the riverbed is an important resource of construction materials. Moreover, it creates more revenue for the state government. Although there is permission for sand mining in limited areas through legal permit, illegal sand mining is going on in most of the cases. It not only affects the general morphology of the river thereby intensifying the flood furry but also poses a serious threat to the local ecosystem. An intensive survey of sand mining areas revealed that illegal mining is going on in almost every village located near river embankment. Villagers blame the nexus between contractors and local administration for the unlawful activities and suggested an urgent need to prepare a legitimate policy, implementation, and monitoring with community participation to check the illegal sand mining. It is significant to note here, this illegal practice is going on along the embankments which initiate further lateral erosion and ultimately bank failure.

16.7.5 Livelihood Issues

The basin is inhabited mostly by small and marginal farmers. Agriculture is the major livelihood. Absence of major industries and lack of ancillary economic activities limited the scope for livelihood diversification. During the time of floods, lack of works due to waterlogging and hardship in mobility makes the daily life miserable almost every year. This serious issue has been neglected in policy formulation and integration. In recent times, different nongovernment organizations (NGOs) are working in the basin to support livelihood diversification (Rana and Singh 2011). During field visit and interaction with the NGO officials six major thrust areas are identified, on which different NGOs are working. They are diversification of livelihoods, public health, capacity building and community awareness, sanitation and drinking water, women empowerment through national livelihood mission, and formation of village disaster committee. Different local

Table 16.2 Length of Embankments in different rivers, lower Rapti River Basin

Name of the river	Rapti	Ami	Ghaghra	Rohin	Kuwano	Gurra
Flow length in the basin (in km)	134	77	77	30	23	17
Extent of embankment	Partially	Fully	Partially	Fully	Fully	Fully

Source: District Disaster Management Authority, Gorakhpur (2015)

Table 16.3 Major development activities undertaken by NGOs in lower Rapti River Basin

S. no	Name of the NGO	Selected community	Selected area (Community Development Blocks)	Ongoing activities
1.	Gorakhpur environmental action group (GEAG)	Small and marginal farmers	Jungle Kaudia, Camperganj, Kaudiram, Mehdawal	Strengthening of livelihoods through • Promotion of flood-resistant paddy. • Distribution of agricultural implements and diffusion of innovative techniques. • Promotion of high yielding variety of vegetables. • Women empowerment through self-help group. • Formation of village disaster committee and awareness campaign. • Linkages of government programs with target groups.
2.	Poorvanchal Gramin Vikas Sansthan	Dalit, minority, small, and marginal farmers	Jungle Kaudia, Khorabar	• Promotion of kitchen garden. • Promotion of flood-resistant agriculture. • Training of tailoring and stitching for women. • Construction of elevated hand pumps with local body (Panchyat).
3.	Sohratgarh environmental society	Small and marginal farmers	Campareganj	• Livelihood diversification. • Regional demonstration of flood-resistant crops. • Awareness for disaster risk reduction.
4.	Poorvanchal Granin Sewa Samiti	Women and children	Brahmapur	• Strengthening of livelihoods. • Health care and sanitation.

Source: Based on personal interview with NGO officials

NGOs working in the basin with different target groups with a mandate to ensure inclusive development is given in Table 16.3. One of the reasons for the successful development and working of NGO institutions is that they have emerged with a comprehensive understanding of the watershed region. They engaged with constant and "interactive approach" with the local community in understanding their problems and solution.

16.7.6 Synchronization of Forest, Land, and Water Policy

The Rapti River basin is shared by Nepal and India. The upper catchment area of the basin is in the Shiwalik Himalaya of Nepal, whereas the lower catchment is a part of the middle Ganga plain. Sustainable management of the basin needs proper synchronizations of policies between these riparian countries. Land use change in the upstream may jeopardize the irrigation and water resource development projects downstream; thus, there is an urgent need for equitable and reasonable agreement between India and Nepal to synchronize their land, water, and forest policies with respect to Rapti River basin.

16.7.7 Incorporation of Community Expectations

Ideally, the community can expect that floodplains will be developed and used in an ecologically, economically, and socially sustainable fashion and in accordance with the broader principle of sustainable natural resource and environment management and of integrated or total catchment management. Thus, in every stage of planning, the community's aspirations and local needs should be incorporated through their direct involvement and participation in an inclusive manner. Floodplain management needs to ensure that the following expectations of the community are met. People wish to live and work on floodplains at no untoward risk to life and health or unacceptable risk of damage to goods, possession, and infrastructure because of flooding. They need site-specific integrated management measures for existing, future, and residual flood problems. People can be secure in knowing that in the event of inevitable future floods, effective arrangements will be made to alleviate the economic and social costs of flooding, on both an individual and community basis, and recovery of the flooded area and its residents and occupants will be fostered. The community is to be actively involved in the floodplain management process, both in developing management plans and in meeting their obligations under those plans (Rana 2015a).

16.8 Concluding Remarks

This study has demonstrated that there are several governance and management issues at the microlevel of basin management that must be addressed at the apex level during the formulation and implementation of plans. Management of floodplains and its resources, compliance to flood forecasting and warning, integration of sectoral plans, public utility management within the active channel zones, annual maintenance of floodplains, illegal sand mining, integration of development schemes within the context of floodplain environment, livelihood issues, and incorporation of community expectations are some of the governance issues that need to be prioritized for sustainable water management at basin scale.

The research findings contribute to the future advocacy of governance in sustainable water management particularly on issues like livelihood diversification should be promoted through self-help groups with active community participation and civil society, integration and synchronization of plans and policies among rural development, public constructions, basic educations, and district disaster management authority.

The research further outlines some of the crucial issues on which research should be carried out in the future. Ways to synchronize land, water, and forest policies among the upstream and downstream riparian state of Nepal and India with respect to small river basins like Rapti need to be addressed. Livelihoods in the floodplains are characterized by uncertainty and risk thus diversification of livelihoods through community awareness, skill training, and capacity buildings should be explored. Secondary hazards coming out of floods and flood inundation like sand casting, creation of wetlands and waterlogging, erosion of croplands should be investigated to formulate risk reduction measures.

References

Blackmore D (2010) River basin management: opportunities and risks. eWater CRC and Asian Development Bank

Central Water Commission (2020a) Mission Statement, available at http://cwc.gov.in/mission-statement. Accessed on 25 Oct 2020

Central Water Commission (2020b) River Basin Planning available at http://cwc.gov.in/river-basin-planning. Accessed on 30 Mar 2020

District Disaster Management Authority (2015) District flood management and reduction Plan-2014-15. Gorakhpur, Office of the District Magistrate

FAO (2018) Water Accounting for Water Governance and Sustainable Development, available at http://www.fao.org/3/I8868EN/i8868en.pdf. Accessed on 02 Nov 2020

Joy KJ (2016) An important step in reforming water governance. Econ Polit Wkly 51(52):30–33

Khemka R (2016) From policy to practice principles of water governance. Econ Polit Wkly LI(52):27–30

Kumar SV and Bharat KG (2014) Perspectives on a water resource policy for India, The Energy and Resource Institute, TERI, available at https://www.teriin.org/

sites/default/files/2017-12/persp.pdf. Accessed on 02 Nov 2020
Kumar TMS, Prakash KL (2018) River basin management in India: contemporary issues and future governance. Int Res J 7(3):49–54
Kumar MD, Prathapar SA, Saleth RM, Reddy VR, Narayanamoorthy A, Bassi N (2017) New 'water management paradigm' outdated concepts? Econ Polit Wkly LII(49):89–94
Lal H (1995) Impact of Deforestation on Ecosystem of Indo-Nepal Tarai: A Case Study of Gorakhpur Region, Unpublished PhD thesis submitted to the Deptt. of Geography, D.D.U Gorakhpur University
Mathur K (2016) From government to governance, A brief Survey of the Indian Experience, NBT, India, New Delhi
Ministry of Jal Shakti, Department of Water Resources, River Development and Ganga Rejuvenation (2020) Mission/Vision/Functions, available at http://mowr.gov.in/about-us/functions. Accessed on 25 Oct 2020
Mohindra R, Prakash B, Prasad J (1992) Historical geomorphology and Pedology of the Gandak Megafan, Middle Gangetic Plains, India. Earth Surf Process Landforms 17:643–662
Nayar V (2013) The water crisis-rethinking water governance. J Land Rural Stud:75–94. https://doi.org/10.1177/2321024913487271
Pant N (2002) Groundwater Issues in Eastern and Western Alluvium of Ganga Basin, IWMI-TATA Water Policy Research Program available at https://publications.iwmi.org/pdf/H029657.pdf
Parmanik K (2014) Integrated River Basin Planning and Management: Concept and Issues, available at https://nwa.mah.nic.in/coursematerial/IRBPM2014/COURSEWARE/IRBPMpercent20Conceptspercent20&percent20issues1.pdf. Accessed on 25 Oct 2020
Rana NK (2005) Role of stream dynamics and hydrological modeling in flood mitigation: case study from Rapti River basin. Unpublished Ph. D thesis submitted to DDU Gorakhpur University, Gorakhpur
Rana NK (2014) Flood hazards, risk and floodplain Management in Rapti River Basin India, The Geographer 61(1):38–51
Rana NK (2015a) Floodplain management in rapti river basin. In: Dutt AK, Noble AG, Costa FJ, Thakur SK, Thakur RR, Sharma HS (eds) Spatial diversity and dynamics in resources and urban development (regional resources), vol I. Springer, Netherlands, pp 379–398
Rana NK (2015b) Delimitation of channel migration zones in the Rapti River basin, middle Ganga Plains. Natl Geogr J India 61(1):99–112
Rana NK (2017) Identification & Mapping of floodplains for sustainable development in the ganga Sub-Basin. In: Thakur B, Mishra HS, Sharma S, Chattopadhyay S, Singh S (eds) Regional development: Theory & Practice, vol 2. Concept Publishing Co. Pvt. Ltd., New Delhi
Rana NK, Singh D (2011) Role of NGOs in disaster management. Yojana 55(11):39–40
Rana NK, Tyagi N (2008) Assessment of flood risk zone using R. S. & G.I. S.: a study of Gorakhpur, U.P. Geogr Rev India, Kolkata 70(4):305–314
Rana NK, Kumar R, Kumar D (2009) Nature of channel shifting of a foothills-Fed River in the alluvial setting: a case study of river Rapti Basin. Indian J Geomorphol 13&14(October):83–98
Shah M (2016) A 21st century institutional architecture for water reforms in India, final report submitted to the Ministry of Water Resources. Government of India, New Delhi
Shah M (2018a) Resistance to reforms in water governance. Econ Polit Wkly LIII(6):60–63
Shah M (2018b) Reforming India's water governance to meet 21st century challenges, discussion paper, IWMI-TATA water policy program, available at http://www.iwmi.cgiar.org/iwmi-tata/PDFs/iwmi-tata_water_policy_discussion_paper_issue_01_2018.pdf. Accessed on 25 Oct 2020
Shukla AK, Rahman A (2012) An assessment of impact of flood on infrastructure facilities in lower Rapti River Basin. The Deccan Geographer 2(50):77–89
Singh SC (1966) Changes in the courses of Rivers and their effects on urban settlements in the middle ganga plain, unpublished Ph. D thesis submitted in the Department of Geography, Banaras Hindu University, Varanasi
Singh U (1975) Changes in the courses of the Rapti and their effects on human settlements, minor research project report submitted to the UGC, Gorakhpur
Singh DS, Awasti A (2011) Natural hazards in the Ghaghara area, ganga plain, India. Nat Hazards 57:213–225
Singh KN, Rana NK (2014) Delimitation of channel migration zones in ganga Sub-Basin using remote sensing & GIS. ASR Publication, Ghaziabad
South Asian Network on Dams, Rivers and People (1999) River Basins Organizations in India – Institutional Frameworks and Management Options. A Case for Fundamental Review, available at https://sandrp.files.wordpress.com/2018/03/rvrbsin_org_wcd.pdf. Accessed on 25 Oct 2020
Thakur B (2003) Perspectives in resource management in developing countries, vol 1. Concept Publishing Company, New Delhi
UNDP (2006) Beyond scarcity: power, poverty and the global water crisis. Palgrave Macmillan, New York
UNWWAP (United Nations World Water Assessment Programme) (2003) The 1st UN world water development report: water for people, water for life. UNESCO, Paris
Walters V (2013) Water, democracy and neoliberalism in India, the power to reform. Routledge Contemporary South Asia Series, New York
World Bank (1989) Sub-Saharan Africa: from crisis to sustainable development. World Bank, Washington, DC
World Bank (1991a) Managing development: the governance dimension. A discussion paper. World Bank, Washington, DC
World Bank (1991b) World development report 1991: the challenge of development. Oxford University Press, New York

Part V

Exploring Human Dimensions

17 Social Transformation, Ecosystem Services, and Resource Sustainability in Nepal Hills

Krishna Prasad Poudel

Abstract

This chapter is based on social transformation and its impacts on ecosystem services with a view to explain its implication on sustainability of the resources supply, the case from the hill region of Nepal. The hill region of Nepal has been widely experiencing blended economic activities, resources extraction, and utilization practices. In the last couple of decades, Nepal has been experiencing social transformation and that has direct impacts on ecosystem services. This chapter attempts to find out such impacts in the local contexts of the middle part of the hill region in the case of three Village Development Committees (VDCs) located at the Kaski, Syangja, and Tanahun Districts of Nepal. The primary sources of data were collected from the field survey from July to November 2015. Secondary data were collected from different published sources. From the study it was revealed that the social transformation results in the changes of traditional resources extraction patterns where primary production at rural areas has been declining. Large chunk of local inhabitants have experienced food insufficiencies from their farm products. Despite that, they have a good amount of bank balance. That is basically supported by the remittances from those who go out of the country as a foreign labor. Therefore, remittance is turning to a good source of resources for livelihood in the area. Local people are less worried about the management of their local resources.

Keywords

Ecosystem Services · Labor Migration · Remittance · Resource Extraction · Social Transformation

K. P. Poudel (✉)
Department of Geography, Central Department of Education, Tribhuvan University, Kathmandu, Nepal

17.1 Introduction

The pace of social transformation was quite slow over decades under the direct rule of the King during the period of Partyless Panchayat governance system. The year 1990 is marked for the restoration of democratic revolution after the "People's Movement I" against the Partyless Panchayat System. It changed the governing system from unitary to multiparty. That has resulted in several social and customary rules prevailed in the past. Immediately after the multiparty ruling system was established in 1991, the so-called Maoist's insurgency headed by the

B. Thakur et al. (eds.), *Resource Management, Sustainable Development and Governance*, Sustainable Development Goals Series, https://doi.org/10.1007/978-3-030-85839-1_17

Maoist's Revolutionary Forces (MLF) started civilian war against the then ruling governance system in 1996. That situation had remained for over a decade from 1996 to 2006 (Schultzjan 2017).

Nepal is predominantly an agrarian country. People used to practice various types of agricultural activities in different terrain. Conventionally, they used to follow farm-forest-livestock as a dominant linkage for livelihood. However, in the recent years, the social transformation has caused changes in the traditional practices. To get a better opportunity, migration (i.e., rural to urban and to international labor force) is increasingly becoming an important livelihood strategy for farm households in rural Nepal. Maharjan et al. (2013) concluded their empirical study that in the context of the rural area the impact of migration on subsistence farming is quite high. But, whenever remittances are high enough to substitute income from subsistence farming, the farm households are more likely to neglect farming than being engaged in commercial farming. Despite the rise in this phenomenon, little is understood about the impact of migration on farm production. Along with changes in agriculture labor force and livelihood supports coming from remittances, impacts have been visualized in the ecosystem services. There are not exactly how the ecosystem services are changing and how much the conventional livelihood support mechanism is under the impact of such change, these are still the research questions.

In the present context of global change, the ecosystem seems to be affected by several human activities. For instance, the reduction of biodiversity, exploitation of natural resources, pollution, land use change, and climate change are among the major causes of ecosystem degradation (Walker et al. 2004). The factors of social transformation have direct linkages to such changes. To attain the goal of sustainable development, there is a need for greater understanding of the ecosystem with a higher level of awareness (Brand 2009).

Box 17.1 Sustainable Development Goals

Despite evidence that ecosystem services are relevant to every SDG goal they remain chronically undervalued and are largely missing in the targets of the SDGs. In the context of the hill regions of Nepal, through his research Poudel recognizes the interdependencies of ecosystem health and social well-being with the hope that the same would build resilience of the poor and vulnerable people to economic, social, and environmental shocks, disasters, and climate-related extreme events. We need to better integrate ecosystem services into the SDG targets in Nepal. As we connect the dots of social transformation, ecosystem services, and resource sustainability with SDG goals, targets, and indicators in the context of hilly regions of Nepal it is evident that there is a need to protect natural resources and eliminate poverty. The achievement of these goals cannot happen merely with the effort of the government of Nepal but will need full cooperation and support of both the civil society and private sector as well. The issues and challenges in implementing the SDGs in Nepal include mainstreaming the SDGs into periodic plans and annual budgets, localization of SDGs at subnational levels, up-scaling implementation, mobilizing financial resources, capacity development at the national and subnational levels, post-disaster recovery and reconstruction challenges, and strengthening governance and service delivery.

17.2 Objectives and Justification

This chapter is based on the study of local resource extraction practices of the respective communities in a traditional way from where they usually get supports for their livelihood. While the transformation takes place in the social systems, various driving forces in the society exert impacts on the local ecosystem services. In such contextual ground, this study is confined to fulfill the intended goals of the identification of the intensity of impacts over the ecosystem services and its implications on resource supply systems within the sustainability perspective.

The impact of social transformation on resource extraction and practices has produced a series of studies in developing countries. Such practices have direct linkages on the ecosystem services of the area. The massive changes have been pointed to the hill districts where the proportion of active labor force migration from the origin in terms of the total population is comparatively very high. As of 2013/14 data in the four hilly districts, i.e., Syangja, Makwanpur, Dhading, and Tanahun were among the top ten districts sending the labor to the foreign countries (MoL&E 2016). Within such typical contextual ground, the empirical research towards the impact of transformation on ecosystem services and its implications on the sustainability of resources supply has been carried out in the selected locations of three districts, i.e., Kaski, Syangja, and Tanahun, the hills of the middle part of the country.

17.3 Methodology

Distribution of survey locations are extending in three different cross-sections of the Seti Gandaki River basin, i.e., upper, middle, and lower part, representing different geographical environments, such as physical, ecological, sociocultural, and economic. Machhapuchchhre VDC of Kaski district in the upper part, Takasar VDC of Syangja district in the middle part, and Bandipur VDC of Tanahun district in the lower part were purposively entertained for the empirical study (Fig. 17.1).

Field data were collected at two levels, i.e., local level and district level. The data from local level were collected based on the administration of household questionnaire (in total of 35 in each VDC), participatory rural appraisal (PRA), Key Informants Interview (KII), ecosystem plot survey, expert observation, and local level consultation meetings with local stakeholders. District level information was collected from the consultation meetings with district level representations of different sectors, line agencies, organizations, and stakeholders. Data acquired from different methods and sources were tabulated and computed in the form of tables, figures, and maps and, finally, interpretation was carried on.

17.4 Social Transformation

After the restoration of multiparty democratic system in 1991, widely noted as "People's Movement I" Nepal has been experiencing several ups and downs in various aspects of the society. The decade-long Maoist insurgency spawned a large literature, mostly of a political nature but Yadav (2016) has analyzed the impacts and implications of the Maoist insurgency on contemporary socioeconomic change and transformation (Sharma 2016). Nonlinear transformation of the Nepalese society had been experienced (Dahal 2010). The discourse among the academic arena had been put forth during that period. Dahal (2010) further pointed out, the explanation was towards the collective consciousness of Nepalese citizens, political, economic, social, and ecological trends and causes and correlation of forces providing a sound basis for the appraisal of the goal of transformation and the legitimacy of its means (Dahal 2010, p. 1). However, it is noticed that because of the so-called war against feudalistic society along with the governance system resulted in the fragmentation of civil society, polarization of corporate climate and poor governance have eroded the political capacity of leadership to sustain nonlinear transformation (Dahal 2010, p. 3). As Dahal (2010) termed the "creative destruction" of Nepal's public sector, industries, education, health, and communication has been experienced. It has been directly internalized by the market forces for the transformation of the welfare state into the financial capitalism since the late 1990s. Nepali citizens entered the international labor market apart from Indian market especially in Malaysia and Gulf countries (MoL&E 2016). The labor migration systemi-

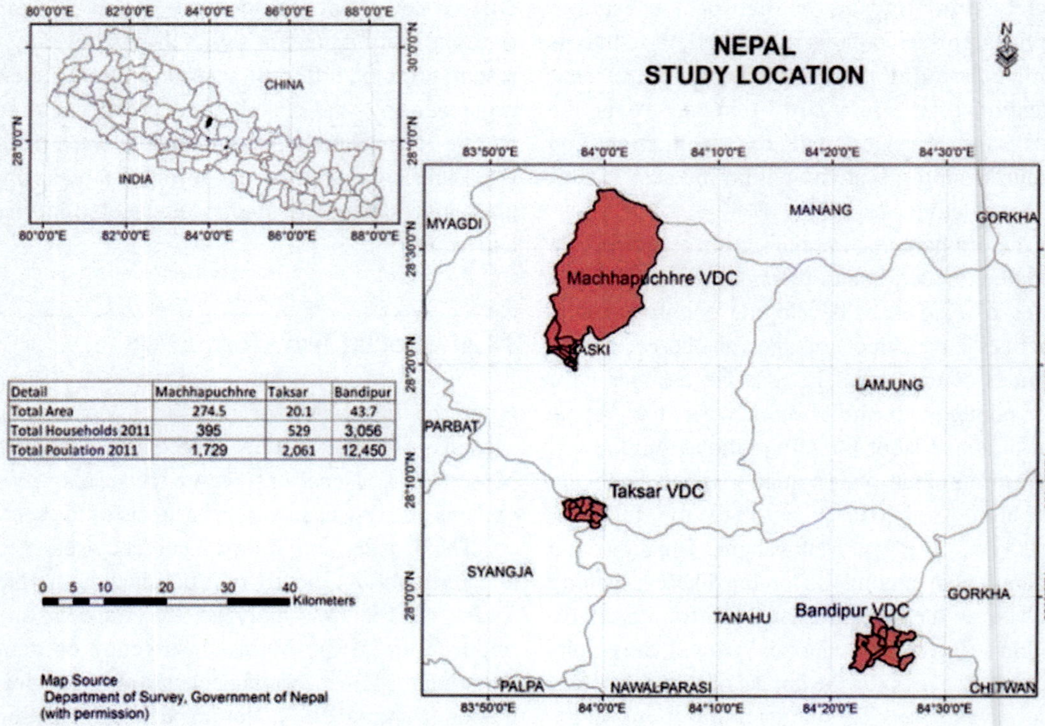

Detail	Machhapuchhre	Taksar	Bandipur
Total Area	274.5	20.1	43.7
Total Households 2011	395	529	3,056
Total Poulation 2011	1,729	2,061	12,450

Fig. 17.1 Location of Study Area

cally broke the state's linkages with society, economy, and ecology (Dahal 2010, p. 3). Because such changes appeared in the country, the direct impacts have been experienced in the people's mobility, consumer market, and population dynamics. Unexpectedly, foreign labor migration has been swelled up in such a way that it has shifted the agricultural-based Nepalese economy towards remittance-based economy. The country receives remittances of over 25% relative to the GDP per year (Maskay et al. 2015). The Government statistics show that more than 3.8 million permits to work abroad (excluding India) were issued by the Government within the period of 1993/94–2014/15 fiscal years, which represents almost 14% of the current population. According to the census data (2011), nearly 71% of the total absent population (1,921,494), or people living out of the country (including living in India) cited private and institutional jobs abroad as the reasons for leaving. Similarly, the report mentions that there has been a huge increase in the inflow of remittances, contributed 10.9% share of the gross domestic product (GDP) in 2003/04 and 27.7% in 2014/15. The remittance flow therefore is a major contributor to development financing in Nepal (MoL&E 2016, p. 1).

The study sites are in the rural geographical domain. The census records between 2001 and 2011 show the total population decrease in Machhapuchchhre and Taksar and slightly increase in Bandipur (Fig. 17.2).

The inhabitants in those areas have followed basically the agrarian practices for their livelihood. But in the meantime, the youngsters have a trend to go out from their origin for different purposes. From the field observation and interaction with the local resource persons, it was also clearly visible that the new generations in the villages were quite less in numbers. The major reasons for that were widely reported because of foreign labor migration, education in the cities, and salaried job out of the origin (Plate 17.1). In between two census periods (2001 and 2011), the decrease in both sex and total number is high in Machhapuchchhre and Taksar, whereas in

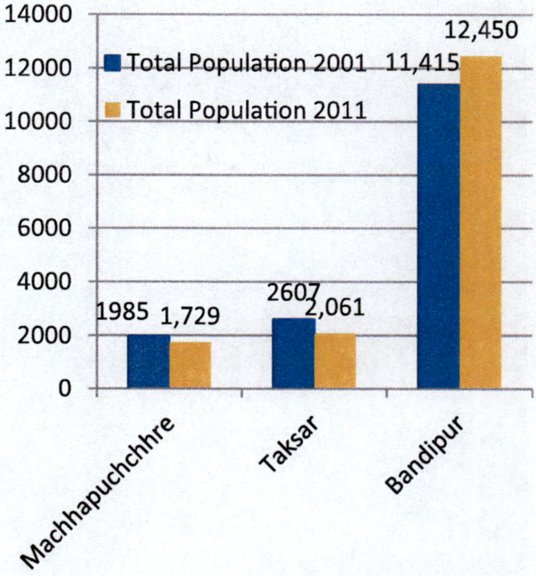

Fig. 17.2 Population change between 2001 and 2011 (Source, CBS Government of Nepal, 2001 and 2012)

Bandipur total population has a slight increment. The case of Bandipur is an emerging town and the other two are entirely rural in nature. Therefore, the population change has a different scenario (Fig. 17.3).

From the two period census data it is evident that population growth within a decade is either negative or minimal. These show the rural areas have such factors which have no attraction for the population concentration. From the field survey, it was revealed that basically active age group (20–40 years) left the villages for foreign labor. The key informants of all villages repeatedly talked about the scarcity of human resource for the conventional farming practices. So, the previously cultivated lands were left fallow. Such fallow lands are being covered by shrubs and bushes. Many locations of such previously cultivated lands are being covered by new plants and weeds and most of those weeds are invasive in nature.

In terms of food sufficiency, a large (58.2%) proportion of respondents reported food insufficiency from their own production. Among the respondents, only 41.8% reported enough food for 12 months, 4.0% reported food sufficiency for 10–11 months, 14.6% reported food sufficiency for 7–9 months, and 31.2% reported food just enough for 4–6 months, while 8.4% reported food sufficiency only for 3 or less months. This shows that livelihood is in a vulnerable situation.

Despite the poor level of food sufficiency, respondents reported satisfactory savings. Among the total, 65.7% of respondents reported that they saved in a bank account. Among them, 63.1% of households were found their saving less than 1000 rupees (approx. US$10) per month and 1.7% of households saved 20,000–50,000 rupees (US$200–500) per month. This scenario has been proved by the statement of the local key informants who mentioned that the family members were out of the country and the money was sent by them and kept that money in the bank. Although the local product is not enough for their annual food requirements, they have bank balance that money usually remits their family members from out of the country.

In the rural areas, there are new dynamics of social transformation by which different social organizations were formed and active and performing local level activities on their own. For organizing conventional rituals and even the small-scale development activities, mothers (females) of the communities were usually active. In every community, they have formed Mother's Group (MGs). It was found that Mother's Groups were very active in every sector and they were eight to ten groups in each VDC having 10–20

Plate 17.1 House left by the owner in Taksar VDC (photo: Author April 2016)

members in one group. This shows that Mothers' Groups are one of the strong and reliable social channels delivering development aids to the grass-root community. There were also few Father's Groups and active in social activities. Similarly, children were also organized into groups through the Children's Club (CC). They also focus mainly on children's welfare. But Mothers' and Fathers' groups usually being involved directly related to local resources extraction and management, development activities, social awareness, and infrastructure building in their communities. Volunteer fundraising; construction of community buildings, drinking water supply system; construction and maintenance of community schools, foot-trail, and motor road; running informal education classes for elders, women, and poor; introducing income-generating activities; running cooperatives, and kind and cash support to the needy local household/persons were some of the noticeable activities in the community. Several MGs were also running vol-

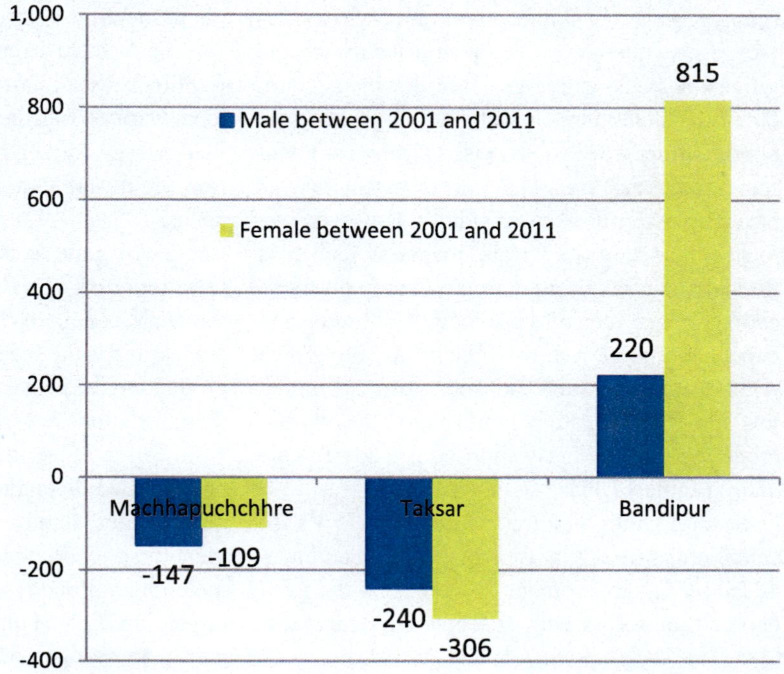

Fig. 17.3 Population change by sex between 2001 and 2011 (Source, CBS, Government of Nepal 2001 and 2012)

unteer small (micro) saving and credit activities within their group to solve cash needs. Tole Improvement Committees (Tole Sudhar Samiti) were also formed in some clustered settlements. Their activities were also focused on their local area development. In addition to these volunteer social community-based organizations (CBOs), VDCs have some sector-based organizations like Community Forest User's Group (CFUGs), Leasehold Forest Users Groups (LFUGs), and Drinking Water User's Groups (DWUGs), Farmers Group, Road Construction and Maintenance Groups, and so on. It is observed that all these CBOs are directly related to maintain the health of the ecosystem and social well-being in the area were coming into existence mainly after the 1990s politico-social development.

17.4.1 Ecosystem Services

The Millennium Ecosystem Assessment Report (2005) recognizes the interdependencies of ecosystem health and social well-being. Based on this approach four distinct ecosystem services like provisioning services, regulating services, supporting services, and cultural services have been categorized (Reid et al. 2005). Using the same approach, ecosystem analysis has been carried on.

17.4.2 Provisioning Services

Provisioning services include agriculture production, such as the production of cereal crops, legumes, fruits, and vegetables. These provisioning services are necessary day-to-day food items for the people. The well-being of the system is determined by the amount of production. However, the impact of social transformation is one of the driving factors of production with the diversity of such services. In the selected VDCs, cereals are recognized as the major staple food. Among them, paddy (rice), maize, and millet are common along with wheat, barley, and buckwheat as other subsidiary crops. In all VDCs selected for the study have maize, millet, and paddy where maize and millet are grown in the dry area (bari) and paddy is grown in the wetland (khet). Some small patches have the upland rice (*ghaiya*) grown in sloping areas and dry river terraces *(tar)*. In some parts, upland rice used to be grown in an area by the slash-down and burn (*khoriya*) method in the past, but this practice has almost stopped due to increased awareness about forest conservation. The cereal production is basically determined by irrigation. However, due to steep and elevated terrain, irrigation facility is limited and is resulting in minimal crop diversification. Since cereals production is monsoon dependent, only a few cereals can be grown within a short summer monsoon period. Limited lowland with irrigation facility produces winter crops, wheat, and early monsoon paddy (*chaite dhan*) (Table 17.1).

Several high-yielding varieties (HYV) of cereal crops (paddy and maize) have been introduced by the government as well as by the local people themselves with an objective of increasing production. Consequently, several indigenous varieties have reportedly disappeared. In Bandipur VDC, farmers have sown HYV and local variety together (Plate 17.2). During the field survey farmer reported local variety has a low yield and that is grown only for festival needs.

Table 17.1 Provisioning Services (Cereals)[a]

VDCs	Cereals	
Machhapuchchhre	Maize, Rice, millet, barley, wheat	
Taksar	Buckwheat, maize, Rice, millet, wheat	
Bandipur	Ghaiya, maize, Rice, millet, wheat	
Local and High Yielding Varieties (HYV)		
Crop	Indigenous variety	HYV
Paddy	Dudhe marsy, thulo marsy, sano marsy, jarneli, jethobudho, pahele, gauri, kalo marsy, battisara, manamuri madese, mansara, anadi, juwari, raisalu, patle, rabijar	mansuli, japanese, jurmakhule, khumal-4, sabitri, radha-4, chhomrong-7
Maize	Seto makai, sunpahelo, pahelo, sathiya	Khumaltar, Makawanpur, Americane, Kakani pahelo
Millet	Kalo, jhapre, dalle, kuwakote	

[a]Local names are given because no scientific names are available for each species type

Source: Field survey, November. 2015

HYV is good for yield hence several problems were reported with it. Local people reported that the HYV cereals have a low resistance to pests, are less tasty, the seeds cannot be used for further production and there are problems in storing them for a longer period.

In terms of vegetable production, the VDCs have limited diversification mainly due to inadequate water supply. During the rainy season, there is a good diversity of green vegetables including various types of gourds, brinjal, ladies finger, spinach, pumpkins, lettuce, cucumber, garlic, and ginger. Most of these vegetables are common in all selected VDCs. During dry winter, tomato, potato, cauliflower, cabbage, radish, carrot, black-eye beans, etc., are produced only where water supply is possible. The upland dry area has less vegetable diversity (spinach, radish, tomato, potato, and beans). In the lowland, the major vegetables are spinach, radish, tomato, brinjal, ladies finger, pumpkins, lettuce, peas and beans, carrot, potato, onion, etc. (Table 17.2).

The production of oilseeds and legumes are also considered to be provisioning services providing protein-rich nutrition. They are grown either as mono-crop, multi-crop, or intercrop. In the Machhapuchchhre and Taksar VDCs, soybean and black-eye pulse are grown with paddy as intercrop. They are particularly grown in terrace bund. Specifically, rice bean *(Masyang)* is grown as an intercrop with maize and millet. In Bandipur black-eye pulse (*kalo mas*), horse gram lentil (*gahat)*, and *masyang* are common. This stretch is famous for the black-eye pulse, mainly grown after upland rice (*ghaiya)* in *tar and bari*.

Among the fruits, orange, pear, and plum are common in elevated areas, while guava, papaya, and banana are grown at lower elevations. Orange is produced at higher elevations and north-facing slope of the hill areas. In the lowland mango,

Plate 17.2 The black is local and the yellow is HYV in the picture, both local and HYV paddy grown in the single plot at Bandipur VDC (Photo by Author, November 2015)

Table 17.2 Provisioning Services (Vegetables)

VDC	Vegetables
Machhapuchre	Garlic, onion, potato, Colo-cassia, radish, lettuce, cauliflower, green mustard, sponge guard
Taksar	Potato, cabbage, pumpkin, Colo-cassia, garlic, tomato
Bandipur	Black-eye bean, brinjal, chayote, mustard green, sponge guard, potato, cucumber, radish, cauliflower, lettuce

Source: Field survey November 2015

papaya, guava, litchi, and banana are common (Table 17.3).

17.4.3 Regulating Services

Regulating services comprise regulation of natural processes and events like flood, drought, land degradation, and diseases. Such services have direct linkages with ecosystem health and human well-being. Events like drought, landslide, and floods can be recorded physically and people from outside the area can observe and monitor those events. However, gradual effect of changes caused by the climate can be well perceived by the people living in the area for a long time and their observation and feelings signify substantially. In the study area, the local community expressed different experiences about regulating services. For instance, mosquito appearance in higher altitude, new invasive species of plants, change in rainfall pattern, dying out of water

Table 17.3 Provisioning Services (Fruits)

VDCs	Fruit species
Machhapuchre	Banana, Guava, Orange, Pears
Taksar	Banana, Orange, Mango, Pear, Plum, Peach
Bandipur	Mango, Orange, Guava, Bayberry, Pear, Pineapple

Source: Field survey, November 2015

Table 17.4 Regulating Services

VDCs	Services
Machhapuchchhre	Appear mosquito, changes in seasonal pattern, decrease water resources, increase of temperature, new weed and pest
Taksar	Effect in agricultural system, increase of hotter days in summer and cold days in winter, shift of season
Bandipur	Appearance of mosquito, decrease of water sources, increase hotter days in summer, increases no of invasive species

Source: Field survey November 2015

springs, cloud burst, shifting in season, increasing warmness in summer months, flash flood, introduction of new pests in crop and plant are some of their critical observations (Table 17.4). These events have important linkages with the climate change.

17.4.4 Supporting Services

In connection to the supporting services, plants, wild animals, and birds are the major categories. These services show a wide diversity in the selected VDCs. Dominant plant species like reeds (*A. falcata)*, bamboo (Dendracalamus *strichts*), rhododendron (*Rhododendron arboretum*), birch (*Bitula utilis*), utis (*Alnus nepalansis*), various herbs, and flowering plants are common in the elevated area of the Machhapuchchhre VDC and extend up to the High Himalayas. At lower elevations, katus (*Castanopsis indica),* chilaune (*Schima wallicht*), maleto (*Macaranga indica*), rakta chandan (*Daphnephylum sp.*), chutro (*Berberis aristata*), guenlo (*Elaeagruts parvifloria*), aiselu (*Rubus elliptica*), *khirro*, chanp (*M. champaca*), and mauwa (*Engelhardtia spicata*) are common. In the Taksar VDC, sal (*Shorea rubusta*), simal (*Bombm ceiba)* are common at lower elevations and katus (*Castanopsis indica),* chilaune (*Schima wallicht*), *phaledo*, aiselu (*Rubus elliptica*), *archal*, and *kagiyo*, are common in elevated areas. Likewise, in the Bandipur VDC sal *(S. robusta)* is the dominant species and katus (*Castanopsis indica),* chilaune (*Schima wallicht*) are common at the elevated part (Table 17.5).

Some plant species like *mainkanda*, *bhalayo, tuni*, dalchini (sinkauli) *(C. zeylanicum)*, *sipligan (Crateva unilocularis)*, *rukh bayar* are less frequently observed. With respect to wild animals, there are wide ranges of wildlife from the low elevation to the upland. Similarly, different bird species are reported from the area. Local people also reported the existence of some supporting services (plant, animal, and birds) in the past, which are rare and have now disappeared (Table 17.6).

17.4.5 Cultural Services

The selected study area is important for local cultural and esthetic resources. Many of them are closely related to religious beliefs and worships of the people. Similarly, archeological sites and esthetic sceneries are famous. The hot water spring of Machhapuchchhre, close view of Snow-clad Annapurna and Machhapuchchhre peaks, Siddha Cave, and historical/archeological sites of Taksar are famous sites. Similarly, historical sites, old hamlet and market center, old Newari culture, and viewpoint of Bandipur are very

Table 17.5 Reported supporting services (plant)

VDCs	Services
Machhapuchchhre	Chilaune (*Schima wallicht*), Maleto (*Macaranga indica*), Uttis ((*Alnus nepalansis*)
Taksar	Sal (*Shorea rubusta*), simal (*Bombm ceiba*), Katus (*Castanopsis indica)*, Chilaune (*Schima wallicht*), Jamun, Kyamun, Dhairo, Chanp, Tuni
Bandipur	Sal (*Shorea rubusta*), Chilaune (*Schima wallicht*), Katus, Phaledo (*Erythrina strica/varigata)* (decreasing) Chanp (*M. champaca*), Kafal, Archal, Eiselu, Kaiyo, Thakal, Liligurans, Sungava, Khirro, Mainkanda (not commonly found) Valayo, Bayer, Sindure

Source: Field survey, November 2015

Table 17.6 Reported Supporting Services

VDCs	*Species Type (Animal)*
Machhapuchre	Leopard, Monkey, Porcupine, Rabit, Ghoral, Black Beer
Taksar	Monkey, Leopard, Porcupine, Ghoral, Water Oot (otter), Malsapro, Jackal, Rabbit, Deer, Gohoro, Bharse, Bear
Bandipur	Wild bore (Badel), Bear, Rabbit, Squirrel, Leopard, Monkey, Porcupine, Tiger, Bear, Wild cat, Snake
VDCs	Species Types (Birds)
Machhapuchre	Bankhukhura, Kalij, khome Karjung (Paraglyding), Pyura, Danphe
Taksar	Luiche, Kalij, Dove, Kakakul, Jureli
Bandipur	Bankhukhura, Kalij, Kakakul, Jureli, Dhukur, Vangera, Kalij Gidhha
VDCs	*Species Types (not frequently appear)*
	Rare occurrences
Machhapuchchhree	
Wild animal	Jackal, *Gohoro*, Water Oot, Bear
Wild bird	Owl, Vulture, *Kalij (Pheasant)*, *Hoochil*
Wild plant	Mainkanda
Taksar and Bandipur	
Wild animal	Ghoral, Red dear (mirga), Bear, *Kalij(Pheasent)*, Jackal
Wild bird	Vulture
Wild plant	*Mainkanda*, *rukh bayar*, dar (cedar)

Source: Field survey, November 2015

famous esthetic resources of the selected area. In addition, the various ethnic groups residing in the area have their unique sociocultural practices (Table 17.7).

In the study area, the ecosystem services have direct linkages within the communities. Ecosystem services are the basic sources of livelihood of local people. The endowment, availability, diversification, and stability of the different types of ecosystem services provide the economic benefits of local inhabitants as well as the health of the local environment. There is a close interaction of people and environment in the communities. However, proper utilization and management of those resources are mandatory.

17.4.6 Sustainability of Resource Supply

From the evidence, some of the notable facts are that population growth in the area is not excessive, young generations are going out for the earning, and remittances have been supporting the household economy as well as the national economy. Despite that, a substantial percentage of the local people have reported that local production was not enough to meet their daily needs. During the field survey, it was also observed that several ecosystem services are degrading due to labor shortage in the traditionally practiced resource extraction activities. From the land use/cover map analysis of two successive years, i.e., 1995 and 2010, the major regulating services, i.e., cultivated land and the forest land within the VDCs under study were found that cultivated land in Taksar and Bandipur VDCs with large shrinkage. It was 42.1% in 1995 which reduced to 26.7% by 2010 in Taksar VDC. It was 48% in 1995 which reduced to 41.2% by 2010 in Bandipur VDC (Table 17.8). Both VDCs have

Table 17.7 Reported cultural services

VDC	Services
Machhapuchchhre	Bhumedanda, Thulodunga, hot water spring, and Annapurna gorge viewpoints
Taksar	Barahi temple, Narayan temple, Saraswoti temple, Shiva, ram and Saraswoti temple, Siwalaya temple, Taksar Thum (old palace), siddha cave and archeological coin location
Bandipur	Bindawasini temple, Khani mai, Khadgamai temple, Mukundeshwori, Newari dance, siddha gupha, Tindhara, mountain viewpoints

Source: Field survey, November 2015

increments in forest land. In case of Machhapuchchhre VDC, the case did not find the same trend that might be because of its location. Both Taksar and Bandipur are located at the middle part of the hill, whereas Machhapuchchhre VDC is located close to the high-altitude Himalayas, where glacial coverage was also confined. Because of that, mapping methodologies might vary in two different data.

Having all these ecosystem services, several new developments and changes in extraction of resource activities have appeared. During the field survey, key informants reported different situations of resource extraction activities. In the Machhapuchchhre VDC people were reporting local high-value supporting services have massive loss in recent years. Although the area is under the Annapurna Conservation Area (ACA), a number of animal species like musk deer, leopard, and wild black beer were not seen frequently which were used to be in the past. In the lower part of Taksar and Bandipur, wild red monkey population was increasing. This increase has been causing the loss of local farmers' product. Monkeys frequently destroy the crops because of that local people were compelled to leave their cultivated area to fallow. However, it was evident that the wild fruits and roots are gradually disappearing in the natural forest. Because of that not only monkey, porcupine, leopard, red deer, but rabbit-like wild animals also gradually used to appear in the farm areas to search for their food. Biodiversity of natural forest areas are being degraded and there the wild predators and prey behavior are changing. Interestingly, it was reported in all VDCs that jackal population disappeared in the last few years. The reason behind such report was the introduction of modern animal veterinary medicine which used to eat animal carcasses by jackal, may be that was not suited to jackal. Because of that the jackal population is either dead or not reproducing the new generation. Scientific verification is needed for this logic.

Since the 1990s, forest conservation has been given to the community management system by the Government of Nepal. The communities formed the User's Group and management responsibilities have been taken by them. Basically, those User Groups used to extract firewood, timber, and grass from the community managed forest. Local users' group usually operates forest weeding every year. During the time of weeding, people lopped branches and cut many unrecognized species in the name of weeding. In such practices, several species got lost from the forest areas and thinning of biodiversity (Plate 17.3).

In many cases, the new plant species also disappeared from the natural forest. The reason behind such disappearance was forest fire. Forest fires in those areas were common due to the domination of deciduous forest species and seasonal dry period. The natural process of deciduous forest leaf-out during winter and after winter there is a dry month from February to May. From the month of February to May most of those forests used to come under the forest fire. Due to frequent fire several species used to disappear.

Table 17.8 Distribution of area within different land use/cover types in 1995 and 2010 (area in km^2)

Land use/cover types	VDCs under study											
	Machhapuchhre				Taksar				Bandipur			
	1995	%	2010	%	1995	%	2010	%	1995	%	2010	%
Cultivation	3.1	1.1	3.2	1.2	8.5	42.1	5.3	26.2	21.0	48.0	17.9	41.2
Built up	0.0	0.0	0.0	0.0	0.0	0.0	0.0	0.0	0.0	0.0	0.0	0.0
Forest	91.6	33.4	74.6	27.2	11.3	56.2	13.6	67.4	21.5	49.3	24.2	55.7
Bush	5.4	2.0	18.7	6.8	0.0	0.1	1.1	5.6	0.8	1.8	0.0	0.0
Grass	72.4	26.4	24.2	8.8	0.0	0.0	0.1	0.6	0.0	0.0	0.4	1.0
Barren land	72.6	26.4	47.9	17.5	0.0	0.0	0.0	0.0	0.0	0.0	0.7	1.6
Cutting	4.4	1.6	0.0	0.0	0.0	0.0	0.0	0.0	0.0	0.0	0.0	0.0
River/Stre-	0.0	0.0	0.7	0.2	0.1	0.3	0.0	0.0	0.2	0.4	0.2	0.6
Pond or lake	0.0	0.0	0.3	0.1	0.0	0.0	0.0	0.0	0.0	0.0	0.0	0.0
Sand	1.7	0.6	0.0	0.0	0.3	1.5	0.0	0.0	0.2	0.5	0.0	0.0
Glacier	23.4	8.5	104.6	38.1	0.0	0.0	0.0	0.0	0.0	0.0	0.0	0.0
Total	274.5	100.0	274.1	100.0	20.1	100.0	20.1	100.0	43.7	100.0	43.5	100.0

Source: Computed from the map (Source land use/cover 1995 is based on the Topographical map 1;25,000 scale published by Department of Survey, Government of Nepal and Land use/cover 2010 is published by ICIMOD 2010)

Plate 17.3 Community forest weeding and extracting firewood from the forest during weeding (Photo by Author, February 2016)

17.5 Conclusion

Since the early 1990s, Nepal is experiencing its fast move towards social transformation. The adaptation of liberal open economy in 1991 by the multiparty democratic system-led government opened the international labor market for the Nepalese youths. In 1996, the Maoist-led war against the government spread the unrest and unsecured situation even in the local village. Due to such reasons, Nepalese youth left the country en-masse for foreign wage labor. The Government statistics show a total of 2,723,587 labor permits were issued by the Department of Labor and Employment from 2008/09 to 2014/15 (MoL&E 2016). The ILO website shows that the number of migrants leaving Nepal for work is increasing every year. During the last fiscal year 2014, more than 520,000 labor permits were issued to Nepalese planning to work abroad (Ilo n.d.). Because of such trend of labor migration after the social transformation, the sustainability of resource supply in the rural mid-hill of the country has two different observations. The first observation is that the population growth is either stagnant or declining due to outmigration. That has caused the depopulation in many settlements. Because of this, the so-called high population pressure on resources is now turned off. The second most obvious observation is that human resource to maintain the resource is decreasing. That results in management issues in the rural areas. Because of that local production is not enough for the daily requirements of inhabitants even though population growth is not high. Local people are compelled to maintain their livelihood from the remittances. There is a horn dilemma for the young generations. Do they have compulsion to go out of the country for wage labor due to unavailability of their local resources or are they following that trend because of easy earnings? Further studies on these issues with wide geographical coverage are recommended.

Acknowledgments This chapter has partially used the fieldwork data collected from the funding support of "Window of Opportunity" Hariyo Ban Program, WWF Nepal to the Center for Disaster Management Studies (CDMS), June 2015 to May 2016. The author sincerely acknowledges the field study team, CDMS and Hariyo Ban Program, WWF Nepal.

References

Brand F (2009) Critical natural capital revisited: ecological resilience and sustainable development. Ecol Econ 68(3):605–612

CBS (2001 and 2012) National Population Census 2001. Government of Nepal, Central Bureau of Statistics, Kathmandu

Dahal DR (2010) Social transformation in Nepal: a personal reflection. Paper presented at a high-level dialogue on 'transformation for peace' organized by General Federation of Nepalese Trade Unions, (GEFONT) in cooperation with Friedrich-Ebert-Stiftung (FES) on July 8–9, 2010, Nagarkot

Ilo (n.d.). (http://www.ilo.org/kathmandu/areasofwork/labour-migration/lang%2D%2Den/index.htm

Maharjan A, Bauer S, Knerr B (2013) Migration for labor and its impact on farm production in Nepal. Centre for the Study of labor and mobility. Social Science Baha, Kathmandu

Maskay NM, Steinkamp S, Westermann F (2015) The impact of remittances on central Bank balance sheets and inflation in Nepal. Nepal Rastra Bank Economic Review 27(2):1–18

MoL&E (2016) Labor migration for employment: a status report for Nepal, 2014/2015, Ministry of Labor and Employment, Government of Nepal, Kathmandu

Reid WV et al (2005) Millennium ecosystem assessment, 2005. Ecosystems and human Well-being synthesis. Island Press, Washington, DC, World Resources Institute

Schultzjan K (2017) A decade after Nepal's Maoist rebellion, little justice for victims. The New York Times, https://www.nytimes.com/2017/01/29/world/asia/a--decade-after-nepals-maoist-rebellion-little-justice-for-victims.html

Sharma P (2016) Book review: resources Himalaya foundation, in Poonam Yadav, social transformation in post conflict Nepal: a gender perspective. Routledge, London

Walker B, Holling CS, Carpenter SR, Kinzig A (2004) Resilience, adaptability, and transformability in social-ecological systems. Ecol Soc 9(2):5

Yadav P (2016) Social transformation in post conflict Nepal: a gender perspective. Routledge, London

Check for updates

18 Determinants of Land Use Dynamics and Its Ecological Implications in India: A State-Level Analysis

Shahab Fazal, Nasrin Banu, and S. K. Azharuddin

Abstract

The population in India has increased significantly which has put pressure on its economy as well as on land and other natural resources. The land-man ratio is decreasing and will continue to intensify further in future. The demand for land is enormous and it results in large-scale land transformation. The changes in land use have a significant impact on livelihood of population as well as on ecological conditions. This chapter examines the land use dynamics during the period of 1990–1991 to 2010–2011 and its ecological implications by budgeting different category of land use in India as well as among its states. The present study finds that India is passing through a critical phase of land transformation. The net sown area is decreasing along with land under pastures and miscellaneous trees. Similarly, the result shows that the urban population growth had also significant negative association with agricultural land. On the contrary, the urban population growth rate had strong positive association with nonagricultural land use change. This means that increase in urban population in the country results in increase in the nonagricultural land use mainly through urban expansion. Land under nonagricultural sector has increased substantially at the cost of agricultural and ecological sector. This makes the country ecologically fragile.

Keywords

Ecological Implication · Land Use Budgeting · Land Use Dynamics · Nonagricultural Land Use

18.1 Introduction

Global change refers to changes that alter the land, water, and atmosphere globally. It also refers to local change that has global importance (Rockström et al. 2009). Herein we focus on land use which is an important component of global change. Some estimates suggest that humans have modified more than 50% of the earth's land surface (Hooke and MartínDuque 2012).

Humans are land animals, as their survival depends largely on resources which come primarily from land resources upon which we survive. Marsh (1864) has pointed in the middle of the nineteenth century that the earth's ability to provide many ecosystem services upon which humans depend is rapidly exhausting. This change is unabated and in recent times the pace of land transformation has accelerated. The changes in land use are greatly attributed to

S. Fazal (✉) · N. Banu · S. K. Azharuddin
Department of Geography, Aligarh Muslim University, Aligarh, India

B. Thakur et al. (eds.), *Resource Management, Sustainable Development and Governance*, Sustainable Development Goals Series, https://doi.org/10.1007/978-3-030-85839-1_18

increases in population which create demand for resources.

Unfortunately, potential ecological consequences are not always considered in making a decision regarding land use. In recent times, numerous impacts of changes in land use have been identified (Lambin and Geist 2006). Earth is moved and the landscape modified, commonly degraded by many of our activities. Degrading the land resource in many ways degrades our life support system as land is finite as well as essential resource for our future generations.

Evidently, there have been two important trends, first, the total land area dedicated to human uses (e.g., settlement, agriculture, forestry, etc.) has grown significantly, second, and increasing production of goods and services has intensified both use and control of land. Much of the earth's habitable surface is dedicated to human use, mostly for production of food and fiber (Richards 1990). It is estimated that for the eighteenth century the global area of forests and woodlands have declined by about 30% while world croplands have increased by 450% (Richards 1990).

Significantly, these changes have been more pronounced in the last 50 years. Interestingly during this recent period of rapid transformation, tropical regions especially Asia are experiencing the highest rates of change (Houghton et al. 1991; FAO 1993; 1996). The pace, magnitude, and spatial research of human alterations of earth's land surface are unprecedented. Land use/land cover changes are so pervasive that the vulnerability of places and people to climatic, economic, or sociopolitical perturbations (Kasperson et al. 1995).

Land use/land cover change has always been at the center of human civilization, but in recent times, a lot of emphases was given to know the pace, direction, and driving force (Liu et al. 2005, Hurtt et al. 2006, Liu and Hanqin 2010). LU/LC changes were found to be a major driving force for climate change, food production both at regional and global scales (Feddema et al. 2005; Jain and Yang 2005; Tian et al. 2012; Tao et al. 2013). In the Indian context, LU/LC change analysis also has great importance because of the pressure on land due to the growing population.

India occupies 2.4% of the world's land area but carries more than 1200 million people over its land (18% of the world population). India with great physical diversity, its location, and vastness along with large population have significant bearing on LU/LC of the country. The increasing population is putting pressure on forest and agricultural land. Significantly, 85% of India's culturable land is in agriculture sector but with transformations in economic sector there is an increasing trend of its conversion to nonagricultural sectors.

India being an agrarian country, studies have shown that there is tendency for land shifts to the agricultural sector but in recent times increases are also found to be in fallow lands which further transform into cultivable wastes (Pandey and Tewari 1987). State-level studies have also reported the same pattern of transformations like Ramaswamy et al. 2005 found in Tamil Nadu and Pandey and Tewari 1987 for Uttarakhand, while Ratnareddy (1991) found that LU/LC changes were governed by availability of resources.

Land being one of the most basic natural resources in India had always been the subject of debate regarding its effective use. Several studies have highlighted the phenomena of land transformation. These studies have pointed out that the rapid pace of economic development along with population growth, urbanization, and industrialization exert tremendous pressure on the limited natural resource base of the country (Bardhan and Tewari 2010). The changes in land use affect the ecosystem of an area in terms of vegetation, local weather effects, land quality itself, and the quality of life that can be sustained (Pandey and Tewari 1987). The changes in land use over time have important implications, the preeminent being the effect on ecology which ultimately impacts the quality of people's lives.

This chapter attempts to examine the land use dynamics and its ecological implications by budgeting different categories of land use in India as well as among its States and Union Territories. The study also analyzes the trends and patterns of land use change during the period of 1990–1991 to 2010–2011. India with a geographical area of 328.7 million hectares, its tropical location results in great diversity and dynamism in land use/land cover. Land use, a dynamic phenomenon has witnessed accelerated changes in recent times. These changes in land use were driven mainly by biophysical factors and human needs. This poses a serious challenge to researchers and policymakers to strike a balance in the use of natural resources, keeping this in mind the analysis of LU/LC data. They are analyzed for their conservation as well as sustainable development and livelihood security.

Box 18.1 Sustainable Development Goals
Sustainable land use management is crucial in India to maintain and restore land resources, tackle climate change, securing biodiversity, and maintaining crucial ecosystem services, while ensuring shared prosperity and well-being. Thus, land can play an important part in accelerating the achievement of many Sustainable Development Goals (SDGs). Recognizing the above challenges, the authors in this chapter observe that a growing urban population in India has led to a change in land use resulting in a significant impact on livelihoods of the population, as well as the ecological conditions rather dramatically. Understanding and articulating these land use transformation challenges through policy and capacity building is crucial to eradicate poverty, reduce inequality, and advance inclusive growth, all of which are responsible for maintaining the ecosystem and the goods and services they provide. A major concern presented by the authors is the transformation in land use away from agriculture and ecology, towards nonagricultural sector. This negative trade-off in the Indian context will have to be addressed with the help of the UN SDGs and India's planning wing *Niti Aayog's* effort if the country is to address food security in the future, agricultural systems will have to respond to change drivers such as population growth, changing dietary habits, and climate change. Achieving sustainable development in land use will require strengthening capacity at all levels, leveraging the catalytic potential of SDG 15 with multiplier effects on other goals, and promoting an economic development model that discourages land degradation.

18.2 Data and Methods

For the present study, data were compiled from land use archives at state level for three time periods, i.e., 1990–1991, 2000–2001, and 2010–2011 in India. The study uses secondary data sets which have been collected from publications of the official agricultural and economic statistics; land use statistics, published by Directorate of Economics and Statistics, Ministry of Agriculture, Government of India.

The land use statistics in India is generally reported (R = *Total reported area*) under the following broad classes:

$$
\begin{aligned}
&F_r = \textit{Area under forests},\\
&P = \textit{Area under permanent pastures},\\
&M = \textit{Area under miscellaneous tree crops and grooves},\\
&B = \textit{Barren lands},\\
&W = \textit{Culturable wastelands},\\
&F_c = \textit{Current fallows},\\
&F_o = \textit{Fallow lands other than current fallows},\\
&C = \textit{Net sown area, and}\\
&N = \textit{Area under nonagricultural uses}.
\end{aligned}
$$

The first accounting identity linearly summed up the area under all land use classes which equal to the total reporting area, consists of forest, permanent pasture, land under miscellaneous trees and groves, culturable wasteland, current fallows, other fallow lands, net sown area, and land under nonagricultural uses.

$$R = F_r + P + M + B + W + F_c + F_o + C + N$$

The dynamics of land use shifts were examined with the help of a simple identity of linearly additive land use changes. The net change in reporting is counted by calculating the changes in different land use classes.

$$\Delta R = \Delta F_r + \Delta P + \Delta M + \Delta B + \Delta W + \Delta F_c + \Delta F_o + \Delta C + \Delta N$$

The present study categorizes the total land endowment (reporting area) into three broad sectors, i.e., ecological sector (*E*), agricultural sector (*A*), and nonagricultural sector (*N*).

$$\Delta R = \Delta E + \Delta A + \Delta N$$

The ecological sector was further divided into two subsectors, viz., the enviable ecology (EE) and unenviable ecology (UE) which comprises several minor sectors (Table 18.1).

$$\Delta E = \Delta EE + \Delta UE$$

The change of enviable ecological sector consists of changes in forest, permanent pasture, and land under miscellaneous trees and groves, while unenviable ecology represents changes in land under barren and uncultivable uses.

The study assumes that land use shifts have ecological implications. Thus, if the land transformation takes place from unenviable to enviable ecology, then, it would have favorable ecological effects, but, if the land of enviable ecology shifts to unenviable ecology, then it may result in an adverse impact on ecology. Even land use dynamics within enviable ecology may also have ecological implications. If land shifts from permanent pasture and miscellaneous trees and groves to forest, this change is considered to have positive ecological implications. On the other side, negative ecological implications would be inflicted if the forestland turned into permanent pastures.

$$\Delta EE = \Delta F_r + \Delta P + \Delta M$$

$$\Delta UE = \Delta B$$

The agricultural sector also consists of four sub-land use classes, i.e., culturable wasteland, other fallows, current fallows, and land under net sown area. The net change in agricultural sector, if positive (+**Δ*A***), it will be at the cost of ecological sector which means that it draws land from forest, permanent pastures, and land under miscellaneous trees. When the net change in agricultural sector is negative ($-\Delta A$), the land use shift may occur to either the ecological sector (enviable/unenviable ecological sector) or nonagricultural sector or both. The shift of land from agricultural sectors to unenviable ecological sector would have an adverse impact on the agricultural sector. Land transformations within the agricultural sector have important implications depending upon the dynamics of change among the classes. If there is a positive change to the agricultural sector (+**Δ*A***), and an increase in the net sown area (+**Δ*C***), this situation would be in favor of agricultural sector and considered as desirable change. Even there is negative net change (−**Δ*A***), but positive change in net sown area (+**Δ*C***), the dynamics are considered favor-

Table 18.1 Dynamics of land use changes in different sectors

Changes in reported area (ΔR)	Changes in ecological sector (ΔE)	Changes in enviable ecological sector (ΔEE)	Changes in forest (ΔF_r)
			Changes in permanent pasture land (ΔP)
			Changes in miscellaneous tree crops and groves (ΔM)
		Changes in unenviable ecological sector (ΔUE)	Changes in barren lands (ΔB)
	Changes in agricultural sector (ΔA)	Changes in culturable wastelands (ΔW)	
		Changes in current fallows (ΔF_c)	
		Changes in fallow lands other than current fallows (ΔF_o)	
		Changes in net sown area (ΔC)	
	Changes in nonagricultural sector(ΔN)	Changes in nonagricultural sector (ΔN)	

able for agricultural sector and assumed as desirable land use dynamics as the cropped area is increasing. But decreases in net sown area (−ΔC) by shift of cultivated land towards fallow land and culturable wasteland have negative impact on agricultural sector. Further, this situation would require larger investments and efforts to reclaim such wasteland.

$$\Delta A = \Delta W + \Delta F_c + \Delta F_o + \Delta C$$

The nonagricultural sector consists of land under nonagricultural land uses. This sector is another critical sector of land use change. The increases in land under nonagricultural sector can be in three ways, it may occur at the cost of ecological sector or agricultural sector and even drawing land from both. If the land under enviable ecological sector transformed into nonagricultural sector, this will have adverse effects for the ecology. Thus, better utilization is assumed to be when nonagricultural sector draws land from unenviable ecological sector. Similarly, increases in nonagricultural sector gaining land from agricultural sector may also be detrimental as this may result in food shortage.

$$\Delta N = \Delta N$$

This study further evolved the index of desirable and undesirable land use dynamics separately along with their composite index based on their positive or negative changes.

18.2.1 Index of Desirable Land Use Dynamics

Desirable land use category has four land use classes, i.e., forest, permanent pasture, and land under miscellaneous trees and grooves (subsector of enviable ecology) and net sown area (subsector of agricultural sector). If the increases are observed in these four (forest, pasture, miscellaneous trees and groves, and net sown area) subsector of land use class then each subsector is assigned the score of +1. While the decreases among these subsectors of land use class are assigned with the score of −1. The cumulative score of +3 or more was classified as most favorable land transformation, the score of +1 to +2 was coded as favorable and score of 0 was classified as no change (neither favorable nor unfavorable), similarly the score of −2 to −1 coded as unfavorable and the score of −3 or more as most unfavorable land transformation.

$$\text{Desirable Index} = \Delta F_r + \Delta P + \Delta M\left(\text{Enviable Ecology}\right) + \Delta C\left(\text{Subgroup of Agricultural Sector}\right)$$

18.2.2 Index of Undesirable Land Use Dynamics

The index of undesirable consists of barren land (land under unenviable ecology), land under nonagricultural uses (nonagricultural sector), culturable wasteland, and fallow land[1] (subsectors of agricultural sector). If there are increases in these four subsectors of land use classes, then, each subsector has been assigned the score −1. Here, increases among these land use classes were assigned negative scores because of their negative impact on ecology, while decreases have been assigned positive scores. Thus, undesirability score was classified with a score of +3 or more as most favorable, score of +1 to +2 was coded as favorable and score of 0 was classified as no change (neither favorable nor unfavorable), score of −2 to −1 was coded as unfavorable and the score of −3 or more as most unfavorable.

[1]Fallow land includes both the current fallows and fallow other than current fallow land ($F = F$).

$$\text{Undesirable Index} = \Delta B + (\text{Unenviable Ecology}) + \Delta W + \Delta F(\text{Sub} - \text{group of Agricultural Sector} + \Delta N)(\text{Non} - \text{agricultural sector})$$

18.2.3 Composite Index of Desirable and Undesirable Land Use Dynamics

The land use dynamics index (both desirable and undesirable) ranges from +4 (most favorable) to −4 (most unfavorable). Thus, the composite land use index range varies from +8 to −8. The scores were further used to categorize states into five classes of most favorable (composite score of +5 or more), favorable (composite score+1 to +4), no change (composite score of 0), unfavorable (composite score 0f −4 to −1), and most unfavorable (composite score of −5 or more).

$$\text{Composite Index} = \text{Desirable Index} + \text{Undesirable Index}$$

18.3 Discussion and Results

Land use pattern in India relates to the physical characteristics of land, the institutional and other resources framework like livelihood, labor, and available capital. All these aspects are associated with the economic as well as social and cultural development. India with a geographical area of 328.7 million hectares, of which land utilization statistics are available for almost 93% of the entire area is around 306 million hectares.

18.3.1 Land Use Pattern

18.3.1.1 Total Reported Land

The land for which the data on classification of land use is available is known as Reporting Land. This reported land, where the land use pattern figures are reported on land records are based on village records or papers. These records are preserved by village revenue agency and here the data is completely based on details of entire areas. In case, where the records are not preserved, the estimates are mostly based on sample survey. Thus, the statistics of land use pattern are based on the abovementioned two methods. The total reported area in India is about 305 mha during the study period of 1990–1991, 2000–2001, and 2010–2011 (Table 18.2).

18.3.1.2 Forests

This includes all lands classed as forest under any legal enactment dealing with forests or administered as forests, whether state-owned or private, and whether wooded or maintained as potential forest land. The area of minor crops raised in the forest and grazing lands, or areas open for grazing within the forests are included under the forest area. The total forested area in India is recorded as little over 67.6 mha in 1990–1991; however, it has increased to 69.8 mha in 2000–2001 and 70 mha in 2010–2011 (Table 18.2). However, literature indicates that this is not suggestive of a real increase of area under forest but is due to incremental increase of reporting area under forest. The states of Manipur and Sikkim reported highest increase in forest cover while Jammu and Kashmir and Assam reported maximum decreases.

18.3.1.3 Permanent Pastures and other Grazing Lands

This category includes all grazing lands whether they are permanent pastures or meadows. Village common grazing land is also included under this category. The land use statistics have shown a decreasing trend in India. The area under permanent pastures and other grazing lands was 11.3 mha in 1990–1991 which decreased to 10.6 mha in 2000–2001 and further declined to 10.3 mha in 2010–2011 (Table 18.2). The land under permanent pastures is despondently low and suggests a remarkable population pressure on the land. It constitutes an important component of rural livelihood as these are primary sources of fodder and animal grazing as well as

Table 18.2 Trends in different land use classes in India

Land use class	Area ('000 ha.) 1990–1991	2000–2001	2010–2011	Compound growth rate (%) 1990–2000	2000–2010	1990–2010	Change (%) 1990–2010
Total reported area	*304,933*	*305,196*	*305,918*	*0.009*	*0.024*	*0.016*	*0.323*
Area under forests	67,658 (22.28)	69,843 (22.88)	70,028 (22.89)	0.318	0.026	0.172	3.503
Permanent pastures and other grazing lands	11,336 (3.71)	10,662 (3.49)	10,305 (3.36)	−0.611	−0.340	−0.476	−9.095
Land under misc. tree crops	3792 (1.24)	3445 (1.12)	3204 (1.04)	−0.955	−0.723	−0.839	−15.506
Total land under enviable ecological sector	*82,786 (27.14)*	*83,950 (27.50)*	*83,537 (27.30)*	*0.140*	*−0.049*	*0.045*	*0.907*
Barren and unculturable land	19,492 (6.39)	17,483 (5.72)	17,175 (5.61)	−1.082	−0.178	−0.631	−11.887
Total land under unenviable ecological sector	*19,492 (6.39)*	*17,483 (5.72)*	*17,175 (5.61)*	*−1.082*	*−0.178*	*−0.631*	*−11.887*
Total land under ecological sector	*102,278 (33.54)*	*101,433 (33.23)*	*100,712 (32.92)*	*−0.083*	*−0.071*	*−0.077*	*−1.531*
Culturable waste land	15,032 (4.90)	13,631 (4.46)	12,646 (4.13)	−0.974	−0.747	−0.860	−15.873
Fallow lands other than current fallows	9959 (3.26)	10,267 (3.36)	10,323 (3.37)	0.305	0.054	0.180	3.655
Current fallows	14,069 (4.61)	14,777 (4.84)	14,275 (4.66)	0.492	−0.345	0.073	1.464
Net area sown	142,280 (46.65)	141,336 (46.31)	141,563 (46.27)	−0.067	0.016	−0.025	−0.504
Total land under agricultural sector	*181,340 (59.46)*	*180,011 (58.98)*	*178,807 (58.44)*	*−0.074*	*−0.067*	*−0.070*	*−1.397*
Area nonagricultural use	21,314 (6.98)	23,752 (7.78)	26,399 (8.63)	1.089	1.062	1.076	23.858
Total land under nonagricultural sector	*21,314 (6.98)*	*23,752 (7.78)*	*26,399 (8.63)*	*1.089*	*1.062*	*1.076*	*23.858*

Source: Directorate of Economics and Statistics, Ministry of Agriculture, Government of India

source of fuelwood. Any decrease in pastures and grazing land can be seen as a threat to rural livelihood. The states of Andhra Pradesh, Madhya Pradesh, Odisha, and Rajasthan have recorded maximum decrease, significantly, these states also have larger nomad and pastoral population.

18.3.1.4 Land under Miscellaneous Tree Crops

This includes all cultivable land which is not included in "Net area sown" but is put to some agricultural uses. Lands under Casuarina trees, thatching grasses, bamboo bushes and other groves for fuel, etc., which are not included under "Orchards" are also classed under this category. This category has also shown a declining trend in India. The area under miscellaneous tree was reported to be 3.7 mha in 1990–1991 which decreased to 3.46 mha in 2000–2001 and further declined to 3.2 mha in 2010–2011. It reveals that much of the tree crops and pastures representing common property resources have reduced in significance over time. Common property resources are important in terms of providing fuel supplies, grazing area, employment, and income generation options for rural poor as they depend on it for their sustenance. This is also against the interest of reducing pollution and improving the environmental status of the country. The states of Odisha, Uttar Pradesh, Assam, and Maharashtra have shown a maximum decline.

18.3.1.5 Barren and Unculturable Land

This category includes all barren and unculturable land like mountains, deserts, etc. Land which cannot be brought under cultivation except at an exorbitant cost were classed as unculturable and barren, whether such land is in isolated blocks or within cultivated holdings. Again, this category of land has also decreased during the study period in India. The area under barren and unculturable land was reported to be 19.4 mha in 1990–1991 which decreased to 17.4 mha in 2000–2001 and further declined to 17.1 mha in 2010–2011 (Table 18.2). The decline in barren and unculturable land can essentially be attributed to the increase in area under nonagricultural use. Uttar Pradesh has recorded maximum increase in this category, while Rajasthan, Andhra Pradesh, and Madhya Pradesh have shown decreases under barren and unculturable land. In some way it indicates that our better lands are transforming poor and efforts are that poor land to be brought under cultivation. But these lands are not so productive and tend to fluctuate between culturable waste, fallow, and cultivated lands.

18.3.1.6 Area under Nonagricultural Use

This includes all lands occupied by buildings, roads, and railways or underwater, e.g., rivers and canals and other lands put to uses other than agriculture. The statistics suggest significant increase in land under nonagricultural use in India during the study period. The total land under this category was little over 21.3 mha in 1990–1991, which steadily increased to 23.7 mha in 2000–2001 and 26.4 mha in 2010–2011 (Table 18.2).

This does not augur well in our economy which is predominantly agricultural. The increase may be attributed to rise in human population as well as launching of development programs/projects for boosting the economy of the country and urbanization as well. But at the same time, the country is losing finite land resource impacting the livelihood of people and threat to food security. The states of Andhra Pradesh, Odisha, Maharashtra, and Rajasthan have reported a maximum increase, interestingly these states have also larger investments for secondary and tertiary sectors.

18.3.1.7 Culturable Wasteland

This includes land available for cultivation, whether not taken up for cultivation or taken up for cultivation once but not cultivated during the current year and the last 5 years or more in succession for one reason or the other. Such lands may be either fallow or covered with shrubs and jungles, which are not put to any use. They may be assessed or unassessed and may lie in isolated blocks or within cultivated holdings. Again, this category of land has also decreased during the study period in India. The area under culturable wasteland was reported to be about 15 mha in 1990–1991 which decreased to 13.6 mha in 2000–2001 and further declined to 12.6 mha in 2010–2011 (Table 18.2). The decrease in this category also attributed to pressure on land due to economic and social transformation. The land is being abandoned and later used for urban, industrial, and other infrastructural developments. Uttar Pradesh has recorded a maximum increase in this category, while Rajasthan, Andhra Pradesh, and Madhya Pradesh have shown decreases under culturable wasteland.

18.3.1.8 Fallow Lands Other than Current Fallows

This includes all lands, which were taken up for cultivation but are temporarily out of cultivation for a period of not less than 1 year and not more than 5 years. Fallow lands witnessed an increase of 3.7% during the study period, which is a matter of concern. The total fallow land was over 9.9 mha in 1990–1991, which increased to 10.2 mha in 2000–2001 and 10.3 mha in 2010–2011. The increase in this category is also attributed to demand for land for urban, industrial, and other infrastructural developments. So, gradually, the marginal agricultural land was converted to fallow and later used for nonagricultural uses. The states of Tamil Nadu, Odisha, Maharashtra, Bihar, and Uttar Pradesh have reported increase in this category while the state of Rajasthan has shown decline in fallow lands (other than current fallows). This indicates the increase is largely attributed to vagaries of monsoon. Evidence is also that agricultural lands are degraded due to excessive and unsustainable use of inputs, and, thus, these lands are left idle.

18.3.1.9 Current Fallows

This represents cropped area, which is kept fallow during the current year. For example, if any seeding area is not cropped against the same year, it may be treated as current fallow. The Fallow lands have also showed an increasing trend during the study period The total current fallow land was over 14 mha in 1990–1991, which increased to 14.7 mha in 2000–2001 but declined marginally to 14.2 mha in 2010–2011 (Table 18.2). There are several interrelated reasons like poverty of farmers, inadequate supply of water, irregular monsoon, silting of canals and rivers and unremunerative nature of farming, etc., for putting land as fallow. The marginal decrease between 2000 and 2010 period may be a sign of better utilization of land in the form of bringing it back into cultivation.

18.3.1.10 Net Area Sown

This represents the total area sown with crops and orchards. India has the second largest agricultural land in the world. It has a net sown area of 142.2 mha in 1990–1991 which is reduced to 141.3 mha in 2000–2001 and remained almost stagnant in 2010–2011, occupying 141.5 mha (Table 18.2). The agricultural lands are the backbone of the country as it plays an important role in inclusive economic development by providing rural livelihood and ensuring food self-sufficiency. India is observing decrease in land-man ratio, so we need to maintain our agricultural land and preserve it for future generations.

18.3.2 Land Use Changes

To highlight the salient features of land use changes at the state level, we evaluated land use trends during the 1990–1991 to 2010–2011 and the results are discussed in the following subsection.

18.3.2.1 Substantial Increase in the Share of Nonagricultural Lands

The loss of farmlands to other uses is an unavoidable phenomenon during economic development, population growth, and urbanization periods. India is passing through this phase and economic growth stimulates the demand for land for nonagricultural uses. This simulation is observed in two aspects. Firstly, higher economic growth leads to an increase in housing requirements, transportation infrastructure, and industrial development. Secondly, with the high economic growth, share of urban population is also rising during the last decade.

At the all India level, the area under nonagricultural use has increased by an average 2.36 lakh hectares per year. The land use dynamics during the study period especially in the case of area under nonagricultural use suggests that total area under nonagricultural use expanded by more than 23%. However, the rate of area expansion was not uniform during two subperiods (i.e., 1990–1991 to 2000–2001 and 2000–2001 to 2010–2011) and among different states. Area under nonagricultural use has increased in most of the states during both periods. However, the increase was faster during the 2000s compared with the 1990s. The states of Andhra Pradesh and Assam had an addition of more than 1 lakh ha to nonagricultural use during the 1990s and the number of states having more than 1 lakh ha addition to nonagricultural use increased to seven (Uttar Pradesh, Odisha, Madhya Pradesh, West Bengal, Tamil Nadu, Andhra Pradesh, and Rajasthan) during 2000–2001 and 2010–2011 (Fig. 18.1a, b). The rate of nonagricultural land expansion also varied among different states. For example, Himachal Pradesh recorded the highest increase (138.3%) in area under nonagricultural use between 1990–91 and 2010–2011. Other important states, which witnessed a significant increase in area under nonagricultural use, included Odisha, Haryana, Kerala, Punjab, Uttar

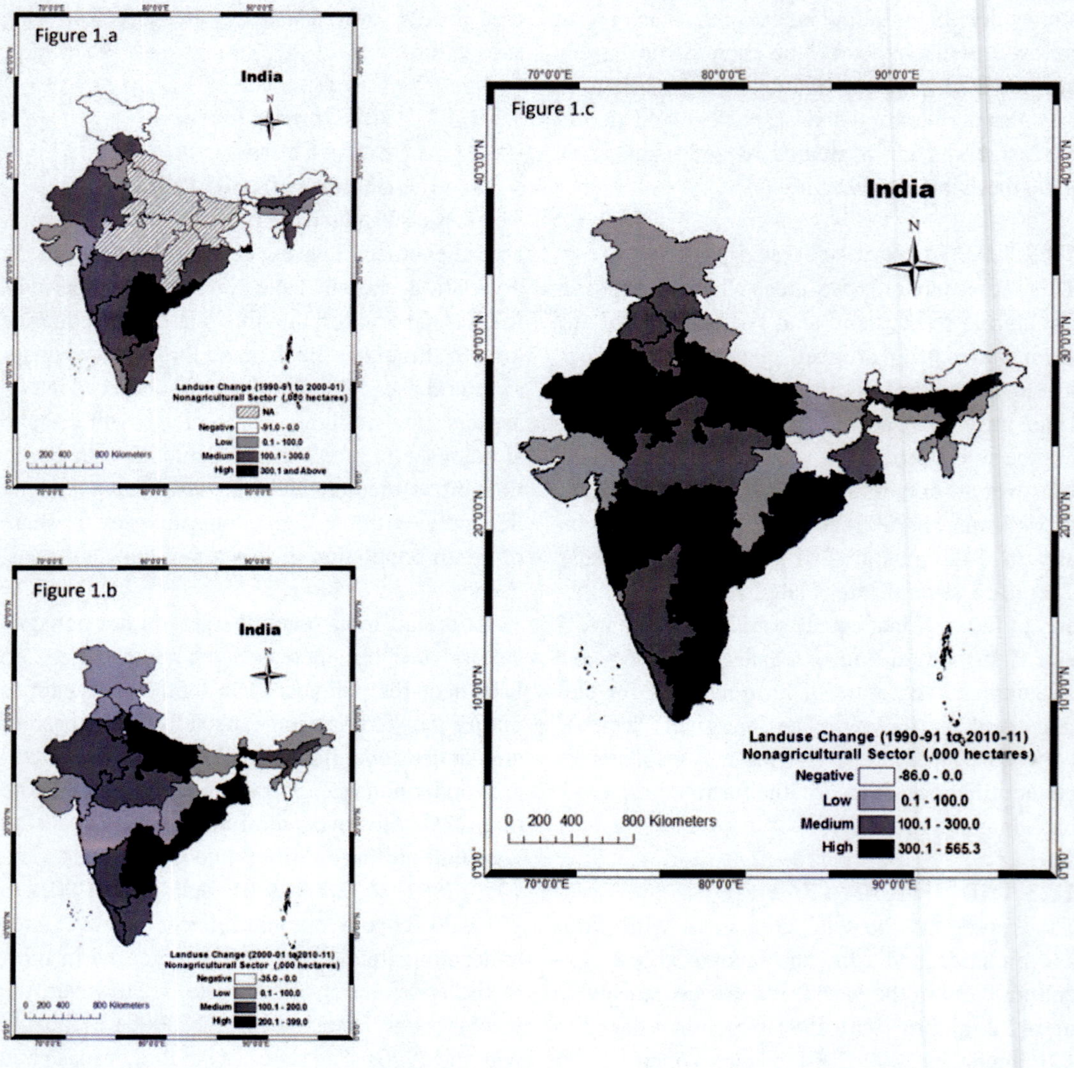

Fig. 18.1 Land use dynamics in nonagricultural sector

Pradesh, Maharashtra, Rajasthan, Tamil Nadu, Andhra Pradesh, and Assam (Fig. 18.1c). The number of states having higher increase in area under nonagricultural use than the all India average increased from six (Andhra Pradesh, Haryana, Himachal Pradesh, Kerala, Maharashtra, and Odisha) in 1990s with inclusion of states of Bihar, Madhya Pradesh, Punjab, Rajasthan, and West Bengal in 2000s. During the last decade, Haryana, Punjab, Uttar Pradesh, and Bihar, country's major agrarian states, experienced more than 20% increase in area under nonagricultural use, which is a cause of concern.

18.3.2.2 Significant Decline in Agricultural Lands

Agriculture plays an important role in providing rural livelihood, ensuring food self-sufficiency and inclusive economic development. India is one of the most land scarce country in the world, with average per capita availability of arable land being 0.15 ha in 2011, down from 0.22 ha in 1991. More than 45% of the total land area (305.8 million ha) is used for crop production. Despite a growing population and increased demand for agricultural products, the net sown area in the country has declined during the last two decades.

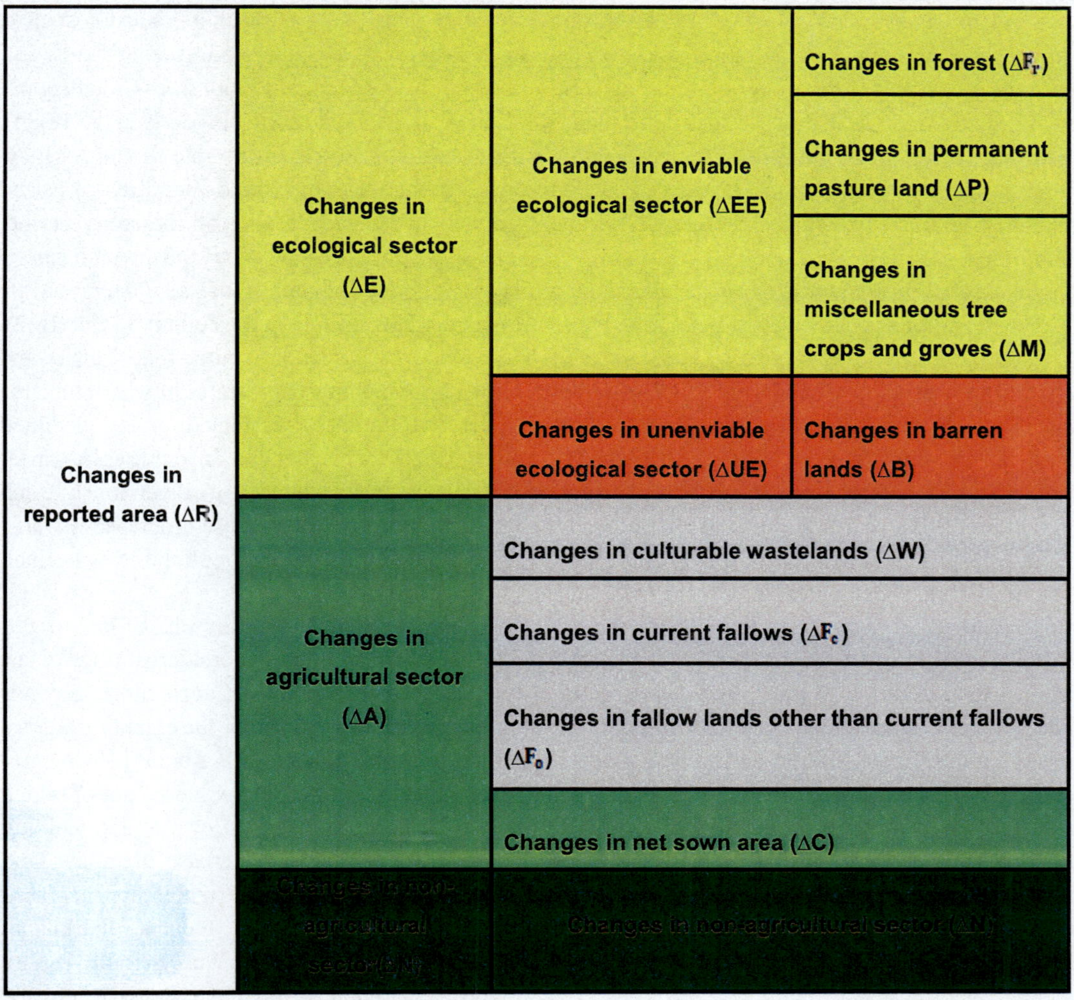

Fig. 18.2 Dynamics of land use changes in different sectors

In fact, over 3 million ha of productive arable land has been lost to other sectors in the country. There was about 1.4% decline in the agricultural land over the last 20 years (Table 18.2). While new technologies, including irrigation, high yielding varieties, better management practices, etc., have led to an increase of crop production in the country. However, in a country like India, which is already experiencing high pressure on its productive agricultural land resources, a decline in the availability of agricultural land could have serious implications for maintaining food self-sufficiency, as well as in ensuring household food security and rural livelihoods.

Over the last 20 years, the availability of agricultural land has been declining. In case of total agricultural land, the land area declined from about 181.3 million ha in 1990–1991 to about 178.8 million ha in 2010–2011. Conversion of prime agricultural land to nonagricultural uses has become a matter of serious concern in the country during the last two decades, and about 3 million ha of agricultural land has been lost during the study period in the country. However, it is interesting to note that some states have been able to bring additional land under agriculture. For example, in Gujarat, total agricultural land increased by about 3 lakh hectares (Fig. 18.2f). Other states which recorded an increase in agricultural land included Madhya Pradesh, and Jammu and Kashmir. On the other hand, Odisha lost about 1.2 million ha of agricultural land and Maharashtra and Uttar Pradesh lost more than 5 lakh ha. (Fig. 18.2f). Other states, which registered a decline in agricultural land (>1 lakh ha) included Tamil Nadu, Bihar, West Bengal, Kerala, and Punjab.

In case of net sown area, Rajasthan added nearly 2 million ha to net sown area, while Gujarat increased net sown area by about 9.5 lakh ha during the study period (Fig. 18.2a). Other states which witnessed an increase in net sown area during the last two decades are Madhya Pradesh (4.15 lakh ha) and North-Eastern states (excluding Sikkim and Tripura). All other states have lost prime agricultural land during post-reforms period. The highest loss in net sown area was recorded in Odisha and Maharashtra (more than 1 million ha), followed by Bihar, Maharashtra, Tamil Nadu, Jharkhand, Uttar Pradesh, and West Bengal. The trends in the loss of agricultural land differed during two subperiods of 1990s and 2000s. In total, an estimated 1.71 million ha of prime farmland was converted to nonagricultural uses during the 20-year period 1990–1991 to 2010–2011. This translates into an average annual loss of over 85,000 ha of productive land (Table 18.2).

18.3.2.3 Substantial change in Other Land Use/Land Cover Categories

The agricultural sector consists of net sown area, land under miscellaneous tree crops and groves (not included in net area sown), culturable wasteland, and total fallow land (current and other fallows). Most of the land use studies generally focus on net sown area, but it is equally important to examine trends in other categories of arable land to explore possibility of increasing area under cultivation. Permanent pastures and other grazing lands constitute an important component of rural livelihoods especially in agriculturally poor states. Moreover, pasture lands are under constant threat of conversion to other uses and encroachment. India has lost about 1.03 million ha of pastures and grazing lands during the last decade (Table 18.2). Himachal Pradesh lost close to 3 lakh ha of pastures and grazing lands, followed by Madhya Pradesh, Chhattisgarh, Maharashtra, and Andhra Pradesh (Fig. 18.3b). However, it is important to note that at all India level, about 5 lakh ha of land under tree crops and groves have decreased during the study period. The share of cultivable wastelands, which can be brought under cultivation, has also decreased by about 2.4 mha hectare in the country (Table 18.2). Interestingly, Rajasthan (more than 1 mha) and Gujarat are the important states in which the area under culturable wasteland has declined (Fig. 18.2b). This decline in culturable wastelands could be partly due to diversion of such lands to nonagricultural use, increase in area under trees, and bringing culturable wasteland under cultivation.

In India, about 24.6 million ha of land is fallow and it has marginally increased during the last two decades. Since Indian agriculture depends on the vagaries of monsoon, the extent of fallows largely depends on rainfall. About 14% of arable area is under total fallow lands, out of which about 8.2% are current fallows and remaining other fallow lands. The increase in fallow land in the country can be attributed to increased fallow land in states like Odisha, Jharkhand, Bihar, Tamil Nadu, Kerala, West Bengal, Andhra Pradesh, Karnataka, and Uttar Pradesh. On the other hand, the area under fallow land declined in Rajasthan (12.51 lakh ha), Gujarat, Madhya Pradesh, and Assam (Fig. 18.2e).

18.3.3 The Sectoral Land Use Dynamics

The analysis of sectoral share among different land use classes in India reveals that there is marginal decrease of −1.5% within the ecological sector (consisting of forests, pastures, and miscellaneous trees in enviable ecological sector and barren land in unenviable ecological sector). The enviable ecological sector possessing little over 27% of total reported area with major share from forest of 22.89% area (an increase of 3.5%), while during the study period permanent pastures

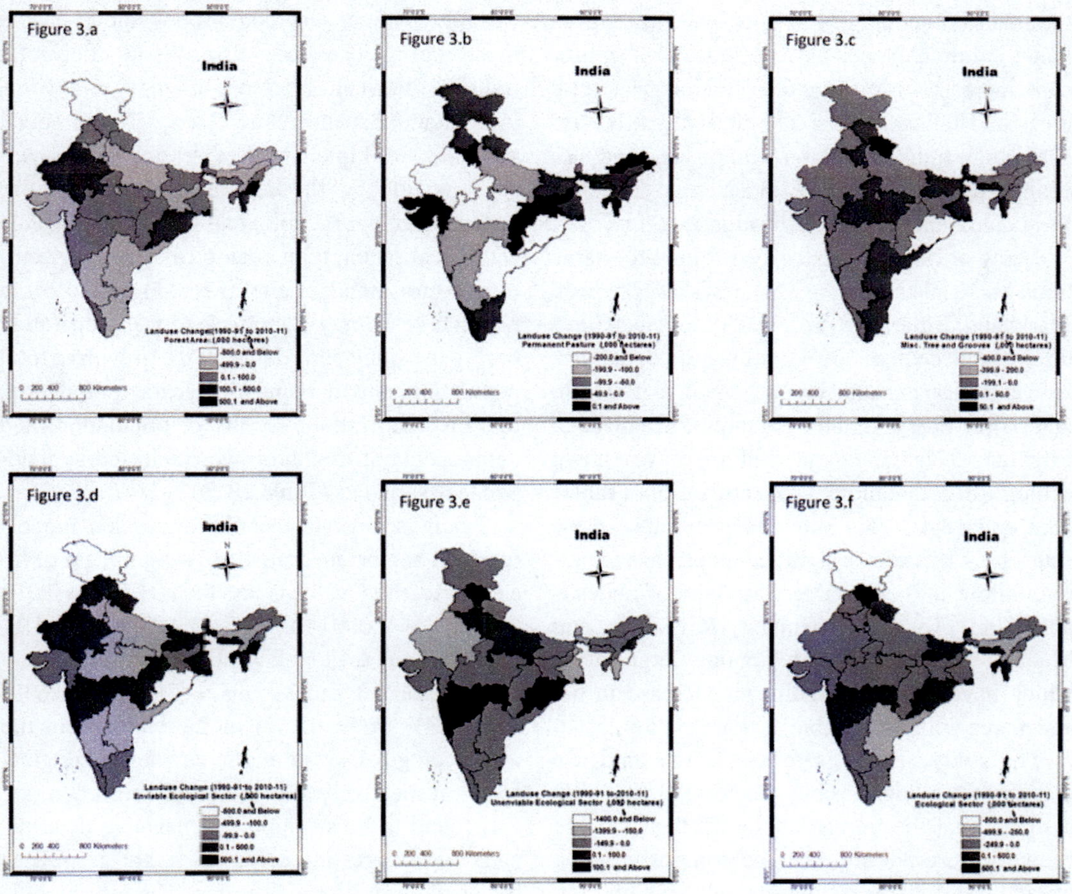

Fig. 18.3 Land use dynamics in ecological sector

and land under miscellaneous trees have decreased. The enviable ecological sector is increasing by 0.9% while unenviable ecological sector decreased by almost 12%. This decrease is caused by utilizing these lands for other uses mainly by improving its status. Most striking trend observed within ecological sector was in reduction in share of permanent pasture and other grazing lands (−9.09%). This is an important component of rural livelihoods as these provide fodder to grazing animals and fuelwood to rural poor. But these lands are under threat of conversion and encroachments.

The state-wise analysis shows increase within enviable ecological sector in hill states of Manipur, Himachal Pradesh, Mizoram, and Sikkim and noticeable decrease in the states of Jammu and Kashmir, Odisha, Uttar Pradesh, Andhra Pradesh, and Karnataka (Fig. 18.3d). The unenviable ecological sector increases in states of Maharashtra, U.P, Odisha, Himachal Pradesh, and Haryana were most striking as these are agricultural dominated states reporting increase in barren and unculturable land (Fig. 18.3e). Strikingly, most of the larger mainland states have reported decrease in the composite ecological sector (Fig. 18.3f).

The overall land use budgeting indicates that we are rapidly transforming land which is ecologically undesirable. At all India level, the index of desirable land use category is unfavorable because of massive decrease in net sown area, pastures, and miscellaneous trees. On the other hand, the index of undesirable land use category the status is encouraging as poor lands like barren lands, culturable lands have decreased but significant increase was observed in nonagricultural areas and fallows lands. The present study suggests that India based on composite land use change in (desirable and undesirable category) lies in unfavorable land utilization category.

Many of the large states also follow the same trend as Maharashtra, Odisha, Uttar Pradesh, Jharkhand, Bihar, Haryana, and Tamil Nadu have unfavorable ecological land utilization dynamics. However, Rajasthan, Gujarat, Madhya Pradesh, and North-Eastern States (Manipur, Meghalaya, Mizoram, Nagaland, and Sikkim) have shown ecologically favorable land utilization (Tables 18.3 and 18.4). This situation is because these states have recorded a significant decrease in barren, fallow, and culturable wasteland of undesirable land use sector. Gujarat, Rajasthan, and Madhya Pradesh are only among large states which have shown a significant increase in net sown area within desirable sector.

The study also tries to assess the land use dynamics among various sectors. The study assumes that the trends among all the land use categories are not similar. The composite index of land use dynamics suggests that Manipur and Meghalaya have better scores mainly because of favorable transformation in undesirable land use categories. Among the mainland states, Chhattisgarh and Gujarat also showed same trend but their scores were lower than the North-Eastern States due to unfavorable land transformation in desirable category of land utilization (Tables 18.3 and 18.4). Most of the states have recorded unfavorable land use transformations. Uttar Pradesh, Delhi, Tamil Nadu, and Odisha have very high negative composite scores recording unfavorable land transformation in both desirable and undesirable land use classes.

18.3.4 Determinants of Land Use Change

The twentieth century has been a century of unprecedented population growth and its resultant impact on Earth. Land is one of the important natural resources where anthropogenic impact is most glaring. The study also tries to stitch relationship between land use change and some important independent demographic and social variables. Multiple linear regression model was executed to find the determinants of land use change in enviable ecological sectors, unenviable ecological sector, total ecological sector, agricultural sector, nonagricultural sector, and also the net change in inter-sectoral land uses separately during the study period, where the growth of total population, urban population, rural population, rate of urbanization, density of population, and components of migration were selected as independent variables (Table 18.5).

Findings indicate that the regression models for each sector are statistically significant as in each case, the F value is less than 0.05 ($F = 0.014$ for ΔE, $F = 0.0001$ for ΔEE, $F = 0.006$ for ΔUE, $F = 0.001$ for ΔA and F = 0.0001 for ΔN).

The model 1 and 2 explain 63% ($R^2 = 0.630$) and 96.3% ($R^2 = 0.963$) of the variance in the total ecological sector and in enviable ecological sector, respectively. The increased migration rate was found to be an important negative determinant for changes in ecological sector as well as for enviable ecological sector. The beta coefficient of −7.014 makes the total population growth as strongest determinant for enviable land use change. Population in India is increasing and the demand for land is never-ending. Thus, land under forest, pastures, and miscellaneous trees and grooves are under threat. This is evident from regression analysis as the population growth was found to be responsible for decrease in the land use classes categorized within enviable ecological sector. Similarly, the urban population growth had also significant negative association with agricultural land in model 4 where 77.4% of variance was explained. It suggests that increase in urban population was strong predictor for conversion of agricultural land to urban land use. On the contrary, the coefficient (14.59) for urban population growth rate had strong positive association with nonagricultural land use change in model 5 ($R^2 = 0.982$). This means that increase in urban population in the country results in increase in nonagricultural land use mainly through urban

Table 18.3 Desirable and undesirable land use dynamics

	Index of desirable landuse	Status	Index of undesirable landuse	Status	Composite index of desirable and undesirable landuse	Status
A&N Island	−2	Unfavorable	2	Favorable	0	No change
Andhra Pradesh	0	No change	2	Favorable	2	Favorable
Arunachal Pradesh	0	No change	0	No change	0	No change
Assam	−2	Unfavorable	2	Favorable	0	No change
Bihar	0	No change	0	No change	0	No change
Chandigarh	−4	Most unfavorable	2	Favorable	−2	Unfavorable
Chhattisgarh	2	Favorable	2	Favorable	4	Favorable
D & N Haveli	−2	Unfavorable	0	No change	−2	Unfavorable
Daman and Diu	−2	Unfavorable	2	Favorable	0	No change
Delhi	−4	Most unfavorable	−2	Unfavorable	−6	Most unfavorable
Goa	2	Favorable	0	No change	2	Favorable
Gujarat	2	Favorable	2	Favorable	4	Favorable
Haryana	0	No change	−2	Unfavorable	−2	Unfavorable
Himachal Pradesh	2	Favorable	−2	Unfavorable	0	No change
Jammu and Kashmir	−2	Unfavorable	0	No change	−2	Unfavorable
Jharkhand	−2	Unfavorable	2	Favorable	0	No change
Karnataka	−2	Unfavorable	2	Favorable	0	No change
Kerala	−2	Unfavorable	0	No change	−2	Unfavorable
Lakshadweep	−2	Unfavorable	0	No change	−2	Unfavorable
Madhya Pradesh	0	No change	2	Favorable	2	Favorable
Maharashtra	0	No change	−2	Unfavorable	−2	Unfavorable
Manipur	2	Favorable	4	Most favorable	6	Most favorable
Meghalaya	2	Favorable	2	Favorable	4	Favorable
Mizoram	4	Most favorable	2	Favorable	6	Most favorable
Nagaland	0	No change	2	Favorable	2	Favorable
Odisha	−2	Unfavorable	−2	Unfavorable	−4	Unfavorable
Puducherry	0	No change	−2	Unfavorable	−2	Unfavorable
Punjab	−2	Unfavorable	2	Favorable	0	No change
Rajasthan	0	No change	2	Favorable	2	Favorable
Sikkim	0	No change	2	Favorable	2	Favorable
Tamil Nadu	−2	Unfavorable	−2	Unfavorable	−4	Unfavorable
Tripura	−2	Unfavorable	0	No change	−2	Unfavorable
Uttar Pradesh	−4	Most unfavorable	−2	Unfavorable	−6	Most Unfavorable
Uttarakhand	0	No change	0	No change	0	No change
West Bengal	0	No change	0	No change	0	No change
All India	−2	Unfavorable	0	No change	−2	Unfavorable

Table 18.4 State-wise landuse dynamics

	Index of desirable land use	Index of undesirable land use	Composite index of land use change
Most favorable	Mizoram	Manipur,	Manipur, Mizoram
Favorable	Gujarat, Meghalaya, Manipur, Chhattisgarh, Goa, Himachal Pradesh	Andhra Pradesh, Gujarat, Meghalaya, Nagaland, Rajasthan, Sikkim, Mizoram, A&N Island, Assam, Daman and Diu, Karnataka, Punjab, Chandigarh, Chhattisgarh, Jharkhand, Madhya Pradesh	Andhra Pradesh, Chhattisgarh, Goa, Gujarat, Meghalaya, Nagaland, Rajasthan, Sikkim, Madhya Pradesh
No change	Andhra Pradesh, Nagaland, Rajasthan Sikkim, Arunachal Pradesh, Uttarakhand, West Bengal, Haryana, Maharashtra, Puducherry, Bihar, Madhya Pradesh	Goa, Arunachal Pradesh, Uttarakhand, West Bengal, D & N Haveli, Jammu and Kashmir, Kerala, Lakshadweep, Tripura, Bihar, All India	A&N Island, Arunachal Pradesh, Assam, Bihar, Daman and Diu, Himachal Pradesh, Karnataka, Punjab, Uttarakhand, West Bengal, Jharkhand
Unfavorable	A&N Island, Assam, Daman and Diu, Karnataka, Punjab, D & N Haveli, Jammu and Kashmir, Jharkhand, Kerala, Lakshadweep, Tripura, Odisha, Tamil Nadu, All India	Delhi, Himachal Pradesh, Haryana, Maharashtra, Odisha, Puducherry, Tamil Nadu, Uttar Pradesh	Chandigarh, D & N Haveli, Haryana, Jammu and Kashmir, Kerala, Lakshadweep, Maharashtra, Odisha, Puducherry, Tamil Nadu, Tripura, All India
Most unfavorable	Chandigarh, Uttar Pradesh, Delhi		Delhi, Uttar Pradesh

expansion. It corroborates the result that nonagricultural land is increasing at the cost of agricultural and ecological sectors. Further, roads were assumed as proxy variable for infrastructure development, a prerequisite for sustained economic development of the country through strong linkages between rural and urban resource areas. The regression analysis finds that the increase in the total road length contributes to the increase of nonagricultural land. The coefficient of roads was 0.5067 and statistically significant at 1% level of significance. The result also means that the increase in road network indirectly results in loss of agricultural land as road projects require large tracts of land mostly through conversion of agricultural land and other lands from desirable sector.

18.4 Conclusion

Land is a finite natural resource. It becomes critical in India because of rapidly shrinking land-man ratio. Thus, the use of land is paramount importance in India. The trends of land utilization are critical because of changes in economic sector in the country. The pressure exerted by India's growing economy on land and other natural resources has intensified in the post-liberalization phase and will further intensify in the future in the face of burgeoning population and the demand for conversion of agricultural land to nonagricultural use. The changes in land use over time have important implications; the preeminent being the effect on ecology which ultimately impacts the quality of peoples' lives and livelihood.

The present study finds that India is passing through a critical phase of land transformation. The net sown area is decreasing along with land under pastures and miscellaneous trees, etc. This makes the country ecologically fragile. Significantly, most of the states have recorded same trend except a few hilly North-Eastern states or agriculturally poor states like Rajasthan, Madhya Pradesh, and Gujarat.

More significant land use change dynamics were observed in undesirable land use category.

Table 18.5 Multiple Regression Model

	Dependent variable	R^2	ANOVA (Sig[a])	Independent variable	Standardized coefficients (beta)	Sig[b]
Model 1	Ecological sector ΔE	0.630	0.014	CAGR of Total Population-1991–2011	−8.351	0.074
				CAGR of Rural Population-1991–2011	0.452	0.091
				CAGR of Urban Population-1991–2011	14.139	0.106
				CAGR of Rate of Urbanization-1991–2011	−6.389	0.142
				CAGR of Density-1991–2011	0.364	0.508
				CGR of Distribution of ST Population-1991–2011	0.450	0.297
				Inmigrants-1991–2001	0.807	0.523
				Outmigrants-1991–2001	−1.452	0.285
				Migrants from Othercountries-1991–2001		
				Net Migrants-1991–2001	−1.003	0.490
				Migration Rate-1991–2001	−0.495	0.013**
Model 2	Enviable ecology ΔEE	0.963	0.0001	CAGR of Total Population-1991–2011	−7.014	0.043*
				CAGR of Rural Population-1991–2011	−0.117	0.539
				CAGR of Urban Population-1991–2011	11.150	0.083
				CAGR of Rate of Urbanization-1991–2011	−4.904	0.125
				CAGR of Density-1991–2011	0.092	0.818
				CGR of Distribution of ST Population-1991–2011	0.299	0.342
				Inmigrants-1991–2001	0.507	0.584
				Outmigrants-1991–2001	−0.540	0.584
				Migrants from Othercountries-1991–2001		
				Net Migrants-1991–2001	−0.410	0.699
				Migration Rate-1991–2001	−0.896	0.000***
Model 3	Unenviable ecology ΔUE	0.699	0.006	CAGR of Total Population-1991–2011	−1.993	0.694
				CAGR of Rural Population-1991–2011	−0.206	0.481
				CAGR of Urban Population-1991–2011	6.703	0.484
				CAGR of Rate of Urbanization-1991–2011	−3.932	0.413
				CAGR of Density-1991–2011	−1.232	0.055
				CGR of Distribution of ST Population-1991–2011	0.289	0.547
				Inmigrants-1991–2001	−1.261	0.377

(continued)

Table 18.5 (continued)

	Dependent variable	R^2	ANOVA (Sig[a])	Independent variable	Standardized coefficients (beta)	Sig[b]
				Outmigrants-1991–2001	1.328	0.382
				Migrants from Othercountries-1991–2001		
				Net Migrants-1991–2001	1.351	0.409
				Migration Rate-1991–2001	−0.352	0.101
Model 4	Agriculture sector ΔA	0.774	0.001	CAGR of Total Population-1991–2011	0.926	0.829
				CAGR of Rural Population-1991–2011	0.194	0.433
				CAGR of Urban Population-1991–2011	−2.934	0.017**
				CAGR of Rate of Urbanization-1991–2011	1.305	0.747
				CAGR of Density-1991–2011	0.800	0.135
				CGR of Distribution of ST Population-1991–2011	−0.003	0.993
				Inmigrants-1991–2001	−0.980	0.417
				Outmigrants-1991–2001	0.810	0.528
				Migrants from Othercountries-1991–2001		
				Net Migrants-1991–2001	1.098	0.428
				Migration Rate-1991–2001	−0.106	0.548
Model 5	Nonagricultural sector ΔN	0.982	0.0001	CAGR of Total Population-1991–2011	−8.712	0.019**
				CAGR of Rural Population-1991–2011	−0.199	0.321
				CAGR of Urban Population-1991–2011	14.591	0.034**
				CAGR of Rate of Urbanization-1991–2011	−6.771	0.048
				CAGR of Density-1991–2011	0.025	0.952
				CGR of Distribution of ST Population-1991–2011	0.362	0.275
				Inmigrants-1991–2001	0.292	0.763
				Outmigrants-1991–2001	−0.369	0.720
				Migrants from Othercountries-1991–2001		
				Net Migrants-1991–2001	−0.186	0.867
				Migration Rate-1991–2001	−0.900	0.000***
				CGR Length of Total Road (1991–2011)	0.507	0.000***

**Significant at the 0.05 level
***Significant at the 0.01 level

Land under nonagricultural sector has increased, substantially. This trend is obvious in changing economic scenario of the country with dominating urbanization process. But equally important land transformations are among other sectors like barren land, culturable waste, and fallow land. The study finds that all India level, there is considerable decrease in barren and culturable wasteland but increase in fallow land. This means that efforts by different agencies to improve this poor land are not effective because the heavy investments have also not transformed poor land to productive agricultural land and that is why share of fallow land is increasing. This is evident from the land use trend among agriculturally poor states of Rajasthan, Gujarat, North-Eastern Hill States, where the share of barren land and culturable waste have decreased, significantly. The culturable wastelands that are brought under cultivation are not very productive, have lower crop yield, and may not be cultivated every year due to lack of appropriate technologies, crops, and irrigation facilities. Therefore, there is a need to bring such lands under cultivation on a sustainable basis.

Overall, the study stresses for formulating long-term sustainable land use planning. The land is a crucial resource, and it should be judiciously utilized for the economic development and prosperity of general mass of the country. The demand for land by the urban sector is enormous, but it should not be met by productive agricultural lands. The study suggests that there should be formulation of SAZ (Special Agricultural Zone) to protect the productive agricultural lands. The other side of this land is that SEZ's (Special Economic Zone) should be diverted to less productive land preferably to wasteland.

References

Bardhan D, Tewari SK (2010) An investigation into land use dynamics in india and land under-utilization. Indian J Agric Econ 65(4):658–676

FAO (1993) Tropical Resources Assessment 1990. FAO Forestry Paper 112. Food and Agricultural Organization of the United Nations, Rome

FAO (1996) Forest Resources Assessment 1990: Survey of Tropical Forest Cover and Study of Change Processes. *FAO Forestry Paper* 130, Food and Agricultural Organization of the United Nations, Rome

Feddema JJ et al (2005) The importance of land-cover change in simulating future climates. Science 310(5754):1674–1678

Hooke RL, MartínDuque JF (2012) Land Transformation by Humans: A Review. Geol Soc Am Today 22(12):4–10

Houghton RA, Skole DL, Lefkowitz DS (1991) Changes in the landscape of Latin America between 1850 And 1985, II: a net release of CO_2 into the atmosphere. Forest Ecol Manag 38:173–199

Hurtt GC et al (2006) The underpinnings of land-use history: three centuries of global gridded land-use transitions, wood harvest activity, and resulting secondary lands. Glob Chang Biol 12:1208–1229

Jain AK, Yang X (2005) Modeling the effects of two different land cover change data sets on the carbon stocks of plants and soils in concert with CO_2 and climate change. Global Biochem Cycles 19(2):1–18

Kasperson JX, Kasperson RE, Turner BL II (eds) (1995) Regions at risk: comparisons of threatened environments. United Nations University Press, Tokyo

Lambin EF, Geist HJ (eds) (2006) Land-use and land cover change: local processes and global impacts. Springer, Berlin

Liu M, Hanqin T (2010) China's land cover and land use change from 1700 to 2005: estimations from high-resolution satellite data and historical archives. Global Biochem Cycles 24(3):1–18

Liu J et al (2005) Spatial and temporal patterns of China's cropland during 1990–2000: an analysis based on landsat TM Data. Remote Sens Environ 98(4):442–456

Marsh GP (1864) Man and nature or physical geography as modified by human action. Belknap Press, Harvard

Pandey VK, Tewari SK (1987) some sociological implications land use dynamics in Uttar Pradesh. Indian J Agric Econ 43(2):388–394

Ramaswamy C, Balasubramanian R, Sivakumar SD (2005) Dynamics of land use pattern with reference to fallow lands: an empirical investigation in Tamil Nadu. Indian J Agric Econ 60(4):629–643

Ratnareddy V (1991) Under-utilization of land in Andhra Pradesh: extent and determinants. Indian J Agric Econ 46(4):555–567

Richards JF (1990) Land Transformation. In: Turner BL II et al (eds) The earth as transformed by human action: global and regional changes in the biosphere over the past 300 years. Cambridge University Press, Cambridge, pp 163–178

Rockström J et al (2009) Planetary boundaries: exploring the safe operating space for humanity. Ecol Soc 14(2):32

Tao B et al (2013) Terrestrial carbon balance in tropical Asia: contribution from cropland expansion and land management. Glob Planet Chang 100:85–98

Tian HQ et al (2012) century-scale responses of ecosystem carbon storage and flux to multiple environmental changes in the Southern United States. Ecosystems 15(4):674–694

19 Land Use Change and Its Impact on Ecosystem Services: Food, Livelihood, and Health Security in Kumaon Himalayas

Prakash C. Tiwari and Bhagwati Joshi

Abstract

This chapter attempts to assess the impacts of land use dynamics on ecosystem services, food, livelihood, and health security in the Himalaya with a case illustration of The Upper Kosi catchment in the Kumaon Himalaya, India. The study used remote sensing and field-based techniques along with qualitative and quantitative empirical methods. The results indicate that population growth and resultant increased demand for natural resources have brought about rapid land use changes, decreasing forests (4.36%), extending cultivation (14.33%), and increasing waste, and degraded land (2.18%). These land use dynamics have adversely affected primary ecosystem services, such as water, biodiversity, and biomass productivity. Nearly 33% of natural springs have dried and 11% have become seasonal, and 736 km stream-length has dried. Consequently, supply of biomass manure to agriculture has declined (41%), irrigation potential has reduced (18%), and food productivity has decreased (25%) increasing deficit levels in food, fodder, and fuelwood by 32%, 20%, and 27%, respectively. The region registered 38%, 24%, 15%, and 28% decline of rural livelihood opportunities, respectively in forest, agriculture, livestock, and traditional handicraft sectors due to loss of forests and biodiversity and decline in agricultural and livestock production. Nearly 47% males and 65% females have been affected by a variety of water-borne diseases in the region.

Keywords

Groundwater Recharge · Population Growth · Resource Deficit · Subsistence Agriculture · Water-Borne Diseases

19.1 Introduction

During recent years, land use changes have emerged as one of the powerful drivers of global change affecting mountain landscapes and ecosystems (Zierl and Bugmann 2005; Löffler et al. 2011). Economic globalization and population growth are likely to have far reaching impacts on the natural ecosystems as well as on human sustainability in mountains (Tasser et al. 2005). The mountain regions of the world, particularly in developing countries have experienced drastic changes in land use patterns during the last some decades (FAO 2008). These changes in ecosystem structure and functions are causing great loss

P. C. Tiwari (✉)
Department of Geography, Kumaun University, Nainital, India

B. Joshi
Department of Geography, Government Post-Graduate College, Rudrapur, India

B. Thakur et al. (eds.), *Resource Management, Sustainable Development and Governance*, Sustainable Development Goals Series, https://doi.org/10.1007/978-3-030-85839-1_19

of biodiversity and disrupting biogeochemical cycles and hydrological processes in high mountains (Haigh 2002; Borsdorf et al. 2010; Huddleston et al. 2003).

The Himalaya being ecologically fragile, economically underdeveloped, and the most densely populated mountain ecosystem on the planet is highly vulnerable to the impacts of environmental changes (ICIMOD 2010; 2017). The nature of terrain imposes severe limitations on the scale of productive activities as well as on the efficiency of infrastructural facilities in the region (Maithani 1986). As a result, biomass-based subsistence agriculture constitutes the main source of rural livelihood and food security (Tiwari and Joshi 2017). More than 75% population of the region depends on traditional subsistence agriculture even though the availability of arable land is severely limited and crop productivity is considerably low (Tiwari and Joshi 2017). This traditional agriculture is interlinked with forests and pastures and flow of biomass energy from forests to agroecosystem is mediated through livestock (Moench 1989; Singh et al. 1984). During the recent past, a variety of changes have emerged in traditional resource use structure mainly in response to population growth and resultant increased demand for natural resources (Haigh 2002; Ives 1985). Besides, during recent years, the Himalaya has experienced rapid urban growth and emergence of many rural service centers which are intensifying the land use and disrupting hydrological regimes of the Himalayan watersheds (Tiwari and Joshi 2016). A large proportion of arable land is being encroached upon by urban growth and expansion of infrastructure, services, and economic activities in the region (Tiwari and Joshi 2016). Moreover, the improved access to market and growth of tourism has also transformed traditional rural resource utilization patterns, and as a result, there is a regional shift from traditional crop farming and animal husbandry system to village-based production of fruits, vegetables, flowers, and milk for sale and this has a large impact on the traditional resource development process and land use pattern (Singh et al. 1984). Consequently, the critical natural resources, such as land, water, forests, and biodiversity have depleted steadily and significantly leading to their conversion into degraded and wastelands in the region (Singh 2007; Tiwari 2000). These land use changes have unprecedented adverse impact on basic ecosystem services, particularly freshwater, biomass, soil nutrients leading to decline in productivity of rural ecosystem and undermining food and livelihood insecurities in the Himalaya (Ives 1989; Valdiya and Bartarya 1991; ICIMOD 2017).

The main objective of this chapter is to interpret the trends of land use dynamics in the Himalaya in its ecological and socioeconomic backdrop and to assess their impacts on ecosystem services, community food, livelihood, and health systems securities with a case illustration of the Upper Kosi catchment in the Himalaya, India.

Box 19.1 Sustainable Development Goals

In the Upper Kosi catchment of the Kumaon Himalayas, one of the critical challenges is the dramatic and dynamic change in land use resulting in a change of its ecosystem services, food availability, livelihood, and the health of the local population. The authors of this chapter observe that for the sustainable development of the local population, it is important to conserve the ecosystems and alleviate poverty in the low agricultural productivity or ecologically vulnerable areas. This can only happen if stakeholders carefully consider the relationship between land use changes and the economic development in the vulnerable areas of the Kumaon Himalayas. The authors recognize that this region is significant for its enriched biodiversity and immense ecosystem services. However, it emerges from their research that the same is under pressure from rapid population growth, developmental activities, unplanned urbanization, climate change, and the associated changes in land use and land cover. These challenges can be addressed with the help of targets and indicators enshrined in SDG-15. India's planning organization *Niti Aayog,* mirrors the *SDG* as it moves forward to build capacity to provide integrated sustainable development in Kumaon Himalayas by understanding and articulating the current food system along with climate change as together they will determine food security, livelihood, and health security in the region. This chapter's key findings include the need to preserve Kumaon Himalayas ecosystem, conservation and restoration of degraded habitat, and an understanding of the local and regional drivers of change and its impact on biodiversity and ecosystem loss.

19.2 The Study Area

The Upper Kosi Catchment (upstream Someshwar) which encompasses a geographical area of 107.94 km^2 between 1500 and 2650 m above mean sea level in the Kumaon Lesser Himalaya in Uttarakhand was taken as the case study site (Fig. 19.1). The Kosi is one of the major rain-fed rivers of Kumaon which ultimately drains into the Ganges system. The catchment is one of the densely populated and agriculturally colonized tracts of the Kumaon Himalaya. There are total 65 villages in the catchment and the density of population has been calculated to be 149 persons km^2, whereas the availability of per capita cultivated land is merely 0.17 ha, and more than 90% of land holdings are of less than 1 ha. As in other parts of the Kumaon Himalaya, the traditional process of natural resource development has been changing rapidly mainly in response to the growth of population and resultant increased demand of natural resources for the past few decades. The study area has been divided into four micro-watersheds for a detailed study of various research parameters (Fig. 19.1).

19.3 Data Collection and Methodology

The information and data required for the study have been generated and collected from various primary and secondary sources. Necessary data and information required for the assessment of the impacts of land use dynamics on ecosystem services, drinking water supply, irrigation, agricultural productivity, food supplies, employment, and health have been generated through primary sources employing field observations and monitoring methods, and through conducting comprehensive socioeconomic surveys using exclusively designed schedules and questionnaires, conducted during 2011–2015. The relevant secondary information was derived from high-resolution satellite images (1985 and 2015), Survey of India (SOI) Topographical Maps, forest maps, cadastral maps, government land records, local Drinking Water, and Irrigation Departments during 2011 and 2015 and used in the present work. The land use dynamics of the study area have been monitored between 1985 and 2015 using high-resolution satellite data. Digital interpretation techniques supported by on-screen visual recording and rectification have been used for this purpose. The first step involved in preparation of visual interpretation is key based on preliminary interpretation of satellite data and extensive ground truth collection. This was followed by the digital classification of land cover/ land use through screen visual recording and rectification. To enhance the interpretability of the remote sensing data for digital analysis, several image enhancement techniques, such as Principal Component Analysis (PCA) and Normalized Difference Vegetation Index (NDVI), were employed. The land use changes that took place in the region between 1985 and 2015 were detected using change detection techniques in GIS environment (Fig. 19.2).

Water flow in springs and streams in the region is monitored by the government departments of drinking water supply and irrigation,

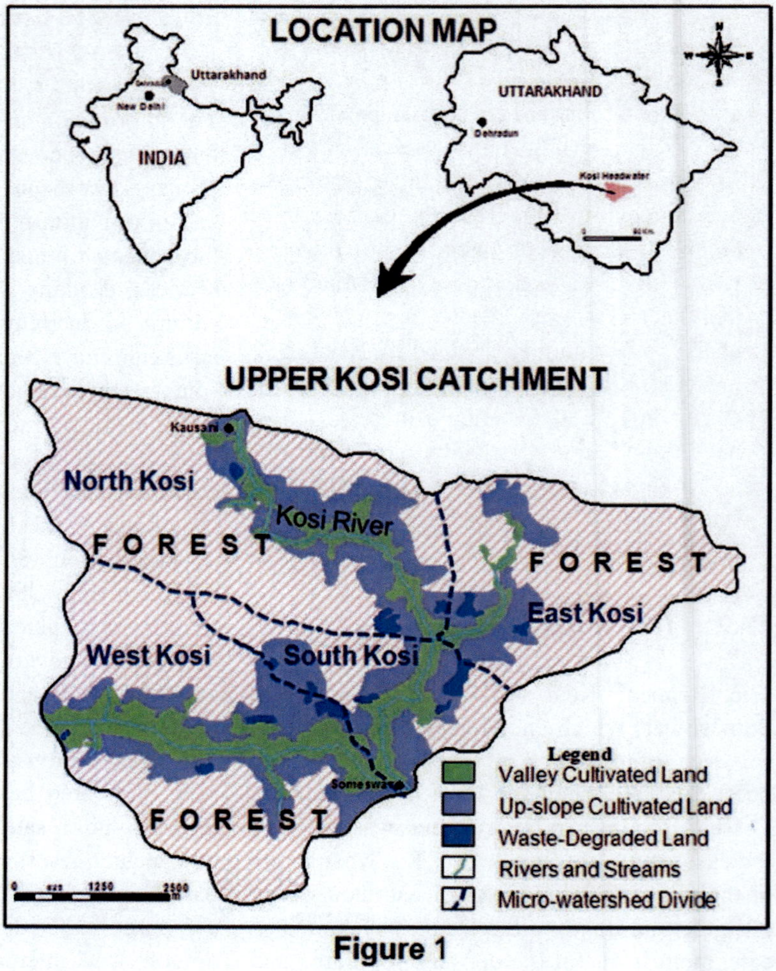

Fig. 19.1 Location Map of the Study Area

and the relevant information on these hydrological parameters have been obtained from these agencies for the period during 1985 and 2015 and used in the present study. Besides, the information with respect to environmental status of water resources (e.g., streams and springs) has been generated through intensive field surveys, mapping, and interviewing elderly people in each of the villages of the study region using exclusively framed schedules and questionnaires. In order to make the outcomes of the study more applicable and community-oriented, detailed appraisal and mapping of the land, water, and forest resources in all 65 villages of the Kosi headwater have been carried out with the involvement of local people.

19.4 Current Land Use

The current land use pattern of the catchment has been broadly classified into: (1) reserved forests, (2) community forests, (3) cultivated land, (4) degraded and wasteland, and (5) water bodies which respectively constitute 68.21%, 3.12%, 25.73%, 2.18%, and 0.76% of the total area (107.94 km^2) of the region (Table 19.1 and Fig. 19.3). The reserved forests are State Property Resource which are situated outside village boundaries and supposed to be completely free from all kinds of resource use pressures and encroachments. However, traditionally, the rural communities living interspersed the reserved forests have enjoyed limited rights and concessions, but now these facilities have been withdrawn or

Fig. 19.2 Methodology of land use change detection

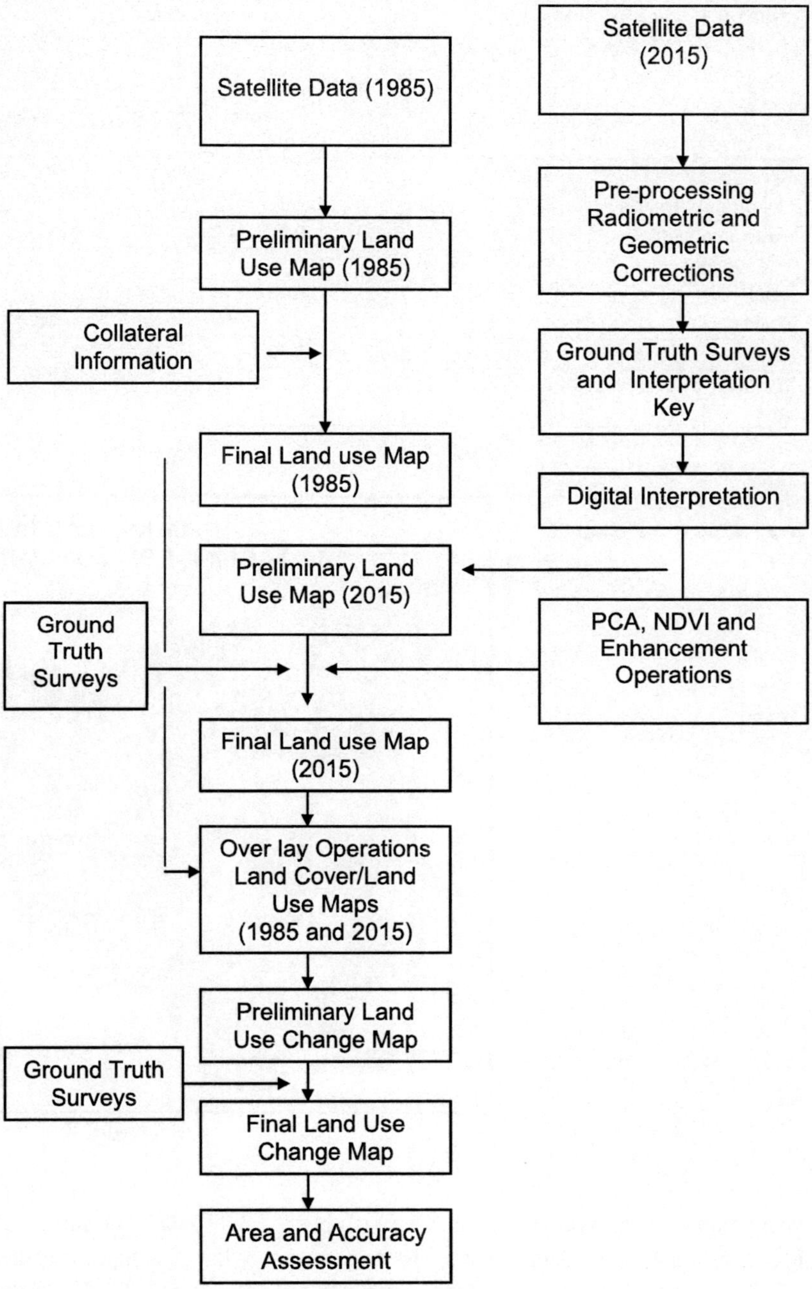

limited in most of the reserved forests of Uttarakhand, particularly after the creation of network of protected areas and prohibition of green felling above an altitude of 1000 m in the Himalaya. Community forests which broadly include all those forests and forest land which fall inside the village boundary, except the private forests, are under Common Pool Resources (CPR). To involve the local people in the protection and conservation of these forests, the control of some of the community forests has now been transferred to the respective villages after reviving the system of Forest *Panchayats* (Forest *Panchayat* is a constitutional village level institution created for participatory management of village forests in India during the British regime). During recent

Table 19.1 Land use changes in upper Kosi catchment between1985 and 2015 (in km^2)

Land use classes in 2015	Land use classes in 1985					
	Forests area	Cultivated land	Degraded and wasteland land	Water bodies	Total (2015) in km^2	Total (2015) % of total area
Reserved forests	73.63	–	–	–	73.63	68.21
Community forests	2.07	–	1.30	–	3.37	3.12
Cultivated land	3.34	24.23	0.20	–	27.77	25.73
Waste and degraded Land	1.47	0.06	0.82	–	2.35	2.18
Water bodies	–	–	–	00.82	0.82	0.76
Total (1985) in km^2	80.51	24.29	2.32	00.82	107.94	100.00
Total (2015) % of total area	74.58	22.50	02.14	0.76	100.00	

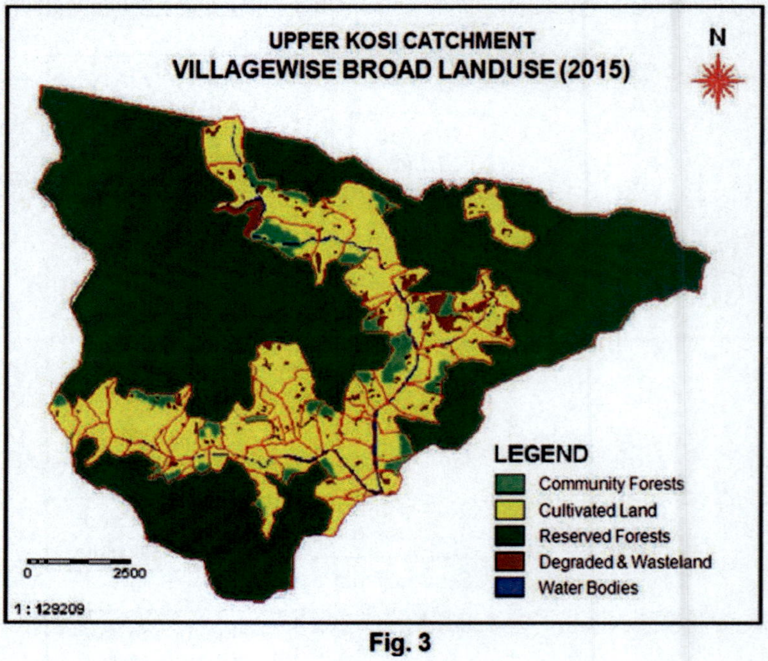

Fig. 19.3 Broad Land Use Pattern of the Study Area

years, some of the village forests have also been brought under the Joint Forest Management (JFM) in the region. This is a recent experiment in the process of participatory forest management in the region which is currently being implemented under project mode with financial support of the World Bank. The JFM projects are being executed by Nongovernmental Organizations (NGOs) through community participation. Interestingly, only 3.37 km^2 or 3.12% of the total area of the catchment is under community forests (Table 19.1). But the availability of merely 3.37 km^2 of community forests for as many as 65 villages is highly inadequate in the region, where forest-based subsistence farming constitutes the main source of rural food and livelihood. Practically, this is the only forest available to the local population for the fulfillment of their all forest-based resource needs. Nevertheless, as many as 36 villages (out of total 65) in the region do not have any community forest. As a result, more than 55% of villages of the region are practically dependent on reserved forests for the fulfillment of their various resource needs.

As mentioned in the preceding sections, the Catchment represents one of the densely populated and intensively cultivated regions of the Kumaon Himalaya. An area of 27.77 km^2 or 25.73% is under cultivation of which only 15% is irrigated. The remaining cultivated land, mainly lying upslope and ridges are never irrigated because of the nonavailability of water and its inappropriate management. Although the availability of arable land is severely limited, yet, in the absence of other viable means of livelihood, dependence on agriculture is considerably high. As a result, the intensity of cropping is very high (150%). The higher cropping intensity in low agricultural potential areas symbolizes distress cultivation of land (Maithani 1986). Out of total 65 villages, 47% are intensively cultivated with more than 75% of their total area under cultivation, 15 have cultivated land ranging from 45% to 75%, and only 3 villages of the watershed have less than 45% of their area under cultivation. Out of the total area of the watershed (107.94 km^2) 2.35 km^2 or 2.18% was identified as wastelands, and 0.82 km^2 or 0.76% is under waterbodies that mainly include streambeds and tiny mountain canals.

19.5 Resource Use Dynamics

Traditional resource utilization pattern in the region has been changing fast mainly in response to population growth and increasing economic and social marginalization (Maithani 1986). The impacts of changes in community resource utilization structure are clearly discernible in terms of rapid land use changes (Tiwari 2000). Agriculture is being extended to forests, and marginal and submarginal lands, and pastures are turning into waste and degraded land due to overexploitation and resultant decline in productivity. With the rapid growth of population, the pressure on cultivated land has increased, and more than 90% of land holdings are of less than 1 ha. Consequently, availability of cultivated land is merely 0.14 ha/person against a minimum of 0.2 ha/person as required for practicing agriculture on a sustainable basis in high Himalayan Mountain ecosystem (Ashish 1983; Singh et al. 1984).

Out of total 65 villages of the Upper Kosi Catchment, as many as 36 have no forests within their boundaries, and per capita availability of forests in the remaining 29 villages is below 0.15 ha. In the Himalaya, 5–10 ha well-stocked forest is required to support biomass requirement of 1 ha of agricultural land on a sustainable basis (Singh et al. 1984). Whereas, in the Upper Kosi Catchment, the availability of forest per ha of cultivated land ranges between 0.10–2.07 ha. The grazing pressure is also very acute as only 0.02–1.67 ha/cattle grazing area is available in the catchment against the ecologically recommended norm of minimum 3.5 ha/cattle (Singh et al. 1984).

19.6 Land Use Changes

Results of land use change detection exercise revealed that out of the total area (107.94 km^2) of the region 8.44 km^2 or 7.81% has changed from one land use to other during 1985 and 2015. Table 19.1 shows that in contrast to the general conception, the agricultural land in the region has not increased much during the last 30 years. The total cultivated land has increased 24.29 km^2 in 1985 to 27.77 km^2 in 2015, and, thus, registered an overall increase of 14.33%. This increase in cultivated land has been brought through the extension of cultivation in forests (3.34 km^2) and wastelands (0.20 km^2). The area under forests in the catchment has declined from 80.51 km^2 in 1985 to 77.00 km^2 (73.63 km^2 reserved forest and 3.37 km^2 community forests) in 2015 mainly due to diversion of 3.34 km^2 forestland to agriculture and turning of 1.47 km^2 community forests into degraded and wastelands thus registered a total decrease of 4.36%. Wasteland has increased from 2.32 km^2 or 2.14% in 1978 to 2.35 km^2 or 2.18% in 2008 as 1.47 km^2 forests and 0.06 km^2 cultivated land was converted into wasteland in the catchment. But, at the same time 0.20 km^2 and 1.30 km^2 wasteland have been brought, respectively, under cultivation and community forests in the region during the last 30 years (Table 19.1).

19.6.1 Impact of Land Use Changes on Ecosystem Services

These land use changes are of great significance in ecologically fragile, tectonically live, and economically underdeveloped The Himalaya, as the ecosystem services, that it provides support one of the most densely populated regions of the world, dependent on subsistence agriculture. Excavation of fragile slopes for road and house construction, removal of vegetal cover, extension of agriculture to marginal and submarginal areas and forests, and intensification of land use under the impact of changing resource use practices are leading to rapid environmental and climate changes and resultant loss of ecosystem services in the Himalaya (ICIMOD 2017; Joshi et al. 2003).

The decrease in forests has disrupted the hydrological regime of the Himalayan watershed. Studies indicate that amount of surface run-off from cultivated (80%) and barren land (85%) is much higher compared to that from forests (25%) (Tiwari 2000). These hydrological disruptions are now clearly discernible in: (1) long-term decreasing trend of stream discharge, (2) diminishing discharge and drying of springs, and (3) biotic impact on surface run-off flow system and channel network capacity (Ives 1985; Tiwari and Joshi 2017). Table 19.2 shows that more than 33% of natural springs have completely dried, nearly 11% of springs have become seasonal, and a stream-length of 736 km has dried during the last 30 years mainly due to deforestation and resultant decreased recharge of groundwater in the region. Table 19.3 shows that as many as 61% of villages have been facing great scarcity of water for all purposes, where the situation turns into a severe water crisis during dry months (Tiwari and Joshi 2017).

Table 19.2 Changes in status of water resources in upper Kosi catchment (1985–2015)

Micro-watersheds	Total area (km^2)	Springs dried (in %)	Springs become seasonal (in %)	Stream-length dried (in m)
North Kosi	44.23	41	17	311
East Kosi	29.18	36	11	227
West Kosi	23.37	47	21	114
South Kosi	11.16	11	05	84
Total	107.94	33.75	10.80	736

The study revealed that depletion of forests and the resultant hydrological disruptions have caused 29% (East Kosi) to 58% (South Kosi) decline in supply of biomass to agroecosystem and loss of 14% (East Kosi) to 21% (South Kosi) irrigation potential (in terms of irrigated area) in different micro-watersheds of the Upper Kosi Catchment during the last 30 years (Table 19.3). The loss of primary ecosystem services, particularly water and biomass, have direct adverse impact on the productivity of the subsistence agricultural system. The different micro-watersheds of the Upper Kosi catchment have lost their agricultural productivity ranging from 19% in West Kosi to 25% in North Kosi with an overall decline of 25% (Table 19.3).

The studies carried out in different parts of the middle Himalayan mountains investigated that the changes in land use pattern are one of the important factors responsible for hydrological disruptions and depletion of water resources in the Himalaya. Pathak et al. (1983) while studying the partitioning of rainfall by certain forest stands in the Kumaon Himalaya observed that well stock oak forest infiltrates nearly 70% of the total rainfall, and, thus, significantly increase the groundwater recharge. Rai and Sharma (1995) and Sharma et al. (2007) explored that degradation of forests and their conservation into degraded land contributed to reducing groundwater recharge and resultant decreased water generating capacity of soil in several parts of the Sikkim Himalaya. The works of Haigh and Rawat (1990), Tiwari (2008, 2010), and Verma and Kothyari (2005) proved that the amount of overland flow is considerably high in agricultural, barren, and degraded land compared to the areas under forests in the catchments of Kosi and other rivers in the Uttarakhand Himalaya. Investigations conducted in several other mountain ecosystems of the world also substantiated these findings (FAO 2005; Wasson et al. 2008).

Table 19.3 Changes in water availability, biomass supply, and irrigation potential in upper Kosi catchment (1985–2015)

Micro-watersheds	% Villages facing water scarcity	% Decrease in biomass supply to agriculture	% Irrigation potential reduced	% Agricultural productivity declined
North Kosi	67	35	14	25
East Kosi	51	29	17	33
West Kosi	69	41	21	19
South Kosi	57	58	19	24
Total	61	41	18	25

Table 19.4 Changes in food, Fodder, and fuelwood deficit Situations in upper Kosi catchment (1985–2015)

Micro-watersheds	% Food deficit	% Fodder deficit	% Fuelwood deficit
North Kosi	38	19	37
East Kosi	27	13	15
West Kosi	39	25	31
South Kosi	23	24	26
Total	32	20	27

19.6.2 Impact on Food and Livelihood Security

The region has been facing deficit situations in food, fodder, and fuelwood for long period mainly due to constraints of subsistence economy and population growth. However, the loss of ecosystem services has further enhanced the resource deficit levels. The region recorded, respectively, 23–39%, 13–25%, and 15–37% deficit of food, fodder, and fuelwood in different micro-watersheds of the region between 1985 and 2015 (Table 19.4). A huge proportion of the rural population, particularly, landless, marginalized, and poor, largely depend on agricultural labor, village-based processing of agricultural and livestock products, making agricultural tools and traditional handicraft items and collection of herbs, fruits, and medicinal plants from forests for livelihood. But, due to the depletion of forests and biodiversity, livelihood in forest-based and medicinal plant collection activities has, respectively, increased from 37% to 40% and from 20% to 29% in different micro-watersheds. Similarly, livelihood opportunities in agriculture, livestock, and handicraft sectors have also increased, respectively, between 19% and 29%, 9% and 22%, and 22% and 40% in different micro-watersheds of the Upper Kosi Catchment during the last 30 years owing to reduced agricultural and livestock productivity (Table 19.5).

19.6.3 Impact on Community Health

Owing to reduced availability of water for various uses people are not able to take proper care of their sanitation and personal hygiene affecting the health conditions of population in rural areas. The study revealed that a large proportion of population of all age groups, particularly the rural women are affected by several kinds of water-borne diseases (Tiwari and Joshi 2013). It was observed that 47% male and 65% female population in all age-categories are under severe threat of a variety of health risks generated mainly due to unhygienic conditions and lack of sanitation in homes and environs (Tiwari and Joshi 2007). Children below the age of 15 years and aged people above 55 years are most affected by water-generated health risks. Further, decline in food and livestock productivity has reduced nutrient supplies to the rural population, particularly the children which are already malnourished and deficient in nutrients, and, thus, affecting the overall health of the rural population in the region.

19.7 Conclusion

During the last three decades, there has been a significant conversion of forests into cultivated and degraded land in the Himalaya. As a result, the proportion of both agricultural and wastelands has increased, while the area under forests

Table 19.5 Impact of land use dynamics on rural livelihood in upper Kosi catchment (1985–2015)

Micro-watersheds	% Decline in forest-based activities	% Decline in agro-based activities	% Decline in medicinal plant collection activities	% Decline in livestock production activities	% Decline in traditional handicraft and agricultural tool making activities
North Kosi	40	24	29	14	26
East Kosi	34	29	22	15	24
West Kosi	39	19	23	22	22
South Kosi	37	24	20	09	40
Total	38	24	24	15	28

has declined in the region. The main driving forces of these land use dynamics are population growth and the resultant changes in community resource use structure. These land use changes have shown unprecedented adverse impact on water generating capacity of the land to springs and streams, biomass supplies to agroecosystem, and productivity of natural resources in the region. As a result, considerably large proportion of natural springs and heads of several perennial streams have dried affecting rural water supplies, leading to loss of irrigation potential, and rendering rural areas highly deficit in food, fodder, and fuelwood. These ecological impacts of ongoing land use changes have not only undermined community health, threatened the livelihood and food securities of rural poor, but have also increased the trends of outmigration of entrepreneuring rural male youths, and, thus, have contributed to further worsening the quality of rural life in the region. A comprehensive land use policy considering both natural and socioeconomic parameters is therefore imperative for the restoration of ecological services and attaining community sustainability in the entire Himalayan region.

Acknowledgment The authors are grateful to the Department of Science and Technology (DST), Government of India for providing generous financial support for carrying the study.

References

Ashish M (1983) Agricultural economy of Kumaon Hills: a threat to ecological disaster. In: Singh OP (ed) The Himalaya: nature, man and culture. Rajesh Publications, New Delhi, pp 233–245

Borsdorf A, Tappeiner U, Tasser E (2010) Mapping the Alps. In: Borsdorf A, Grabherr G, Heinrich K, Scott B, Stötter J (eds) Challenges for mountain regions: tackling complexity. Böhlau, Vienna, pp 186–191

FAO (2005) Forest and Water. FAO Forestry Paper No.155

Food and Agricultural Organization (2008) Food Security in Mountains: High Time for Action. Brochure of the International Mountain Day. Weblink: http://www.mountaineering.ie/documentbank/uploads/IMD08percent20brochure.pdf

Haigh M (2002) Headwater control: integrating land and livelihoods. Paper presented at the international conference on sustainable development of headwater resources. United Nation's International University, Nairobi, Kenya, September

Haigh MJ, Rawat JS (1990) Hydrological impact of deforestation in the central the Himalaya. Hydrology of mountainous. Proceedings of the Strbské Pleso workshop, Czechoslovakia, June 1988. IAHS Publ. No. 190, pp. 419–433

Huddleston B, Ataman E, d'Ostiani L (2003) Towards a GIS-based analysis of mountain environments and populations. Environment and natural resources working paper no. 10. FAO, United Nations, Rome

ICIMOD (2010) Mountains of the world: ecosystem Services in a Time of global and climate change: seizing opportunities: meeting challenges. Framework paper prepared for the Mountain Initiative of the Government of Nepal by ICIMOD and the Government of Nepal, Ministry of Environment

ICIMOD (2017). Framework For Integrated Ecosystem Management In The Hindu Kush The Himalaya, ICIMOD Working Paper 2017/10

Ives JD (1985) The Himalayan environmental regeneration: an overview what are the problems and how can they be tackled. In: Singh JS (ed) Environmental regeneration in the Himalaya. Gyanodaya Prakashan, Nainital, pp 1–11

Ives JD (1989) Deforestation in the Himalaya: the cause of increased flooding in Bangladesh and northern India. Land Use Policy 6:187–193

Joshi PK, Yang X, Agarwal SP, Das K, Roy PS (2003) Impact of resource utilization in the Himalayan watershed a landscape ecological approach for watershed development and planning. Asian J Geoinfo 3(4):1–9

Löffler J, Anschlag K, Baker B, Finch O-D, Diekkrüger B, Wundram D, Schröder B, Ape P, Lundberg A (2011)

Mountain ecosystem response to global change. Erdkunde 65(2):189–19.3

Maithani BP (1986) Towards Sustainable Hill area development. Himalaya Man Nat Cult 16(2):4–7

Moench M (1989) Forest degradation and the structure of biomass utilization in a the Himalayan Foothill Village. Environ Conserv 16(2):137–146

Pathak PC, Pandey AK, Singh JS (1983) Partitioning of rainfall by certain Forest stands in Kumaun the Himalaya. Trop Plant Sci Res 1:123–126

Rai SC, Sharma E (1995) Land-use change and resource degradation in Sikkim the Himalaya: a case study from Mamlay watershed. In: Singh RB, Haigh MJ (eds) Sustainable reconstruction of highland and headwater regions. Oxford & IBH Publishing Co., New Delhi, pp 265–278

Sharma E, Bhuchar S, Xing M, Kothyari BP (2007) Land use change and its impact on hydro-ecological linkages in the Himalayan watersheds. Trop Ecol 48(2):151–161

Singh SP (2007) The Himalayan Forest ecosystem services: incorporating in National Accounting. Central the Himalayan environment association. Nainital, Uttarakhand

Singh JS, Pandey U, Tiwari AK (1984) Man and forests: a central the Himalayan case study. Ambio 13:80–87

Tasser E, Tappeiner U, Cernusca A (2005) Ecological effects of land-use changes in the European Alps. In: Huber UM, Bugmann HKM, Reasoner MA (eds) Global change and mountain regions: an overview of current knowledge, advances in global change research, vol 23. Springer, Dordrecht, pp 409–420

Tiwari PC (2000) Land use changes in the Himalaya and their impact on the plains ecosystem: need for sustainable land use. Land Use Policy 17:101–111

Tiwari PC (2008) Management of Common Pool Wasteland for Rural Livelihood Improvement in The Himalayas, *Sustainable Solutions: A Spotlight on South Asian Research*, Proceedings of 10th International Sustainable Development Conference, December 2007, Sustainable Development Policy Institute, Islamabad, Pakistan, pp. 75–101, 2008

Tiwari PC (2010) Land use changes and conservation of water resources in the Himalayan headwaters. Proceedings of the 2nd German Indian conference on research for sustainability: Energy & Land use, pp. 170–74

Tiwari PC, Joshi B (2007) Rehabilitation and Management of Wasteland in the Himalayan headwaters: an experimental study of Kosi headwater in Kumaon lesser Himalayas in India. J World Assoc Soil Water Conserv 2:39–62

Tiwari PC, Joshi B (2013) Rainfall variability and its impact on water resources and rural health in Kumaon Himalaya, India. In: Grover V (ed) Impact of climate change on water and health. CRC Press, Taylor and Francis Group, pp 286–307

Tiwari PC, Joshi B (2016), Rapid urban growth in mountainous regions: the case of Nainital, India, urbanization and global environment change (UGEC) viewpoints, march 2016, Global Institute of Sustainability, Arizona State University, Arizona, USA

Tiwari PC, Joshi B (2017) Local institutions and Forest governance in the Himalaya. In: Datta K (ed) Man Forest in the Himalayas and Northeast. B. R. Publishing Corporation, Delhi, pp 33–60

Valdiya KS, Bartarya SK (1991) Hydrological studies of springs in the catchment of Gaula River, Kumaon lesser Himalaya, India. Muntain Res Dev 11:17–25

Verma PK, Kothyari BP (2005) Assessment of rainfall, runoff and sediment losses of the Bheta gad watershed, India. In: White R, Bhuchar SK (eds) Resource constraints and management options in mountain watersheds of the Himalayas. ICIMOD, Kathmandu, pp 155–166

Wasson RJ, Juyal N, Jaiswal M, McCulloch M, Sarin MM, Jain P, Srivastava P, Singhvi AK (2008) The mountain-lowland debate: deforestation and sediment transport in the upper ganga catchment. J Environ Manag 88(1):53–61

Zierl B, Bugmann H (2005) Global change impacts on hydrological processes in alpine catchments. Water Resou Res 41:W02028. https://doi.org/10.1029/2004WR003447

20 Impact of Land Use Change on Livelihood Options: A Case Study of Upper Pasolgad Watershed, Uttarakhand

Bindhy Wasini Pandey, Abhay Shankar Prasad, and Jitendra Kumar Mahto

Abstract

Land is basic and significant among natural resources. The fundamental aim of land use study is to understand the distribution of land under various uses in different socioeconomic and environmental conditions. In India, livelihood culture is directly or indirectly related to, geographical terrain, local climatic conditions, and social structure. In mountain or hilly terrain where livelihood security is directly dependent on changing climatic conditions, it has become a very important issue for investigation. This study is an attempt to understand the impact of climate change, physiographic, slope, drainage pattern, land use practices, vegetation, altitude, and anthropogenic interference (viz., road construction and settlement, etc.) on sustainable livelihood and common property resources. There is a close linkage between livelihoods, anthropogenic activities, and climate change. This chapter's focus is on managing the forest plantation and watershed conservation using remote sensing and GIS in Pasolgad Watershed in Pauri Garhwal, Uttarakhand. Findings indicate that soil erosion is more prone to the area where the plantation is very sparse with high runoff. Research posits that local communities have undertaken plantation of forest with the local leaders. The Van Panchayat ensures the conservation of local forest through effective implementation of programs and policies of the govt. Besides, it demonstrates that the community towards sustainable use of forest through awareness and campaign. It also illustrates that the community forestry helps in a greater extent to conserve and maintain the forest cover at local level.

Keywords

Community-Based Management · Land Use Change · Livelihood Security · Resource Sustainability

20.1 Introduction

Land is vital for human survival and is significant among natural resources. Its study is to discern the distribution of land under various uses in different socioeconomic and environmental conditions (Singh and Singh 2013). With the increase of population, the importance of studies and research on land utilization has become an urgent need to reduce the regional imbalance in development and to attain sustainable development

B. W. Pandey (✉) · J. K. Mahto
Department of Geography, Delhi School of Economics, University of Delhi, Delhi, India

A. S. Prasad
Department of Geography, Dyal Singh College, University of Delhi, New Delhi, India

B. Thakur et al. (eds.), *Resource Management, Sustainable Development and Governance*, Sustainable Development Goals Series, https://doi.org/10.1007/978-3-030-85839-1_20

through the enhancement in global food security and human habitations (Pandey 2014). Its analysis is necessary as it ensures the best possible advantage of land in an enduring manner to its people. Therefore, it is believed that land resource shows a reciprocal relationship between the prevailing ecological conditions and people in a region. In Indian sustainable livelihood context, it has become more relevant as more than 60% of people depend upon agriculture. In India, livelihood culture is directly or indirectly related to the social structure, geographical terrain, local climatic conditions, and social structure. In mountain or hilly terrain where livelihood security is directly dependent on changing climatic conditions, it has become an important issue for investigation. This study addresses the need to understand the impact of, physiographic, slope, drainage pattern, climate change, land use practices, vegetation, altitude, and anthropogenic interference (viz., road construction and settlement, etc.) on sustainable livelihood and common property resources. This appears to be the result of a close linkage between livelihoods, anthropogenic activities, and climate change (DFID 2001).

This study provides insights on how climate change can influence the socioeconomic setting in the Upper Pasolgad watershed in a few ways. It can influence the economy (e.g., agriculture, livestock, forestry, tourism, fishery, etc.). Potential of tourisms (natural, adventure, wildlife, cultural, and pilgrimage) are enormous in Pasolgad. Tourism has become a significant part of occupation and a major source of income. This region has diversity in nature of tourism. Plenty of natural places, places for adventure, and pilgrimages present a unique combination and provide opportunity for extensive tourism activity. It has opened new avenues for the local people to rise above poverty. Natural resources management and outmigration are main problems in this region. There are many push factors which are responsible for the outmigration, i.e., climate change, land degradation, water shortage but also inundations, poverty, famine, and population pressure. Moreover, insecure livelihoods drive millions of vulnerable people to move from their original habitat to another place where the pull factor is for better economic opportunities. Therefore, a strong focus on this is needed to address the problems at their roots (Kumar and Negi 2017). It is also evident that most of those on the move from rural areas also head towards urban agglomerations and where they often concentrate in slum areas. Migration, if managed properly (legal, well prepared, etc.) could be beneficial to both source and target places (remittances, brain circulation, skills, and labor mobility). Viewing migration as an adaptation rather than a curse needs considerable conceptual rethinking (Walsh-Meyer and Crews 2002).

Massive decline in the coverage of common property resources and disruption of traditional management system are the results of improper policies and side effects of development strategies. Land use/land cover map and hazard's risk map are great forward steps in proper planning of economic activities in the areas and prevention of hazards-related disaster resulting from climatic extremes (Singh 2002). Increasing anthropogenic activities along the slopes have altered the existing land use pattern posing risk from different types of hazards. Environmental hazards and subsequent loss of life, property, and disruption in biophysical systems have become a common feature in mountain environments at national and global scale (Turner et al. 1995). Land use/land cover (LULC) changes are affected by human intervention and natural phenomena, such as agricultural demand and trade, population growth and consumption patterns, urbanization and economic development, science and technology, and other factors (Thakur 2011). Consequently, information about LULC is essential for any natural resource management and action planning (Singh et al. 2001). Timely and precise information about LULC change detection of the earth's surface is extremely important for understanding relationships and interactions between human and natural phenomena for better management of decision-making (Anderson et al. 1976). Understanding of the causes of land use change has moved from simplistic representations of two or three driving forces to a much more profound understanding that involves situation-specific interaction among many factors at different spatial and temporal scales.

Box 20.1 Sustainable Development Goals

Hill regions in Uttarakhand Himalayas face numerous and severe environmental, social, and economic challenges. Sustainable land use is central to many of the environmental and socioeconomic issues facing the communities today. The authors of this chapter point out that agricultural and forest produce, which is fundamental to human well-being in the hills, has profound consequences for biodiversity and climate change. These challenges are acknowledged by the UN sustainable development goals (SDG-15) through their framework which focuses on the impact of natural resources on social and economic development, and therefore on the importance of conservation and sustainable use of land and forest resources leading to the protection of biodiversity, ecosystem, and wildlife. The authors in studying the mountainous Pasolgad watershed in Pauri Garhwal, Uttarakhand recognize the benefits derived by people from the surrounding landscape and the crucial role played by the mountain ecosystem in providing water resources, which further contributes to agriculture in hilly landscapes, reduces land abandonment thereby reducing land degradation, and mitigates climate change minimizing hydrogeological risks. Along with the above, the governance process from the local to the national also acknowledges the twin challenges of reversing biodiversity declines and mitigating climate change, while producing sufficient food to ensure zero hunger, which must be tackled together. Making land use systems sustainable is central to achieving these—and other—Sustainable development goals. According to the authors, these challenges can be addressed by better understanding and managing the synergies and trade-offs inherent in the hilly landscapes of Uttarakhand, such as eradication of poverty, providing food security, nutrition and sustainable agriculture, access to freshwater, sustainable energy, which will collectively help improve livelihoods of mountain people and to conserve mountain ecosystems for the sustained prosperity of the present and future generations.

20.2 Study Area

Upper Pasolgad watershed is situated in Pauri Garhwal district, Uttarakhand. Its headquarters is at Pauri. The Pauri Garhwal district is ringed by Haridwar, Dehradun, Tehri Garhwal, Rudraprayag, Chamoli, Almora, and Nainital districts from three sides. Within Garhwal Himalaya, the Pasolgad watershed comes under the Pauri Garhwal district between 29°52′23″ N to 29°54′28″ N latitude and 79°1′48″ to 79°4′43″ E longitude (Fig. 20.1).

It comes under Bironkhal block and Thailisain tehsil with the elevation from mean sea level ranging between 1100 and 2300 m. The total geographical area of the district is 5329 km^2 (Bandooni 2004; Bandooni et al. 2015). A major part of this Himalayan state comes under rainforests and alpine forests that are home to some of the highly endangered wildlife species. The forest area of the district comprises whole of the Garhwal and Lansdowne Forest Divisions and greater part of the Kalagarh Forest Division. Nearly 60% area of the district is covered under forest. Forest cover is dwindling rapidly due to indiscriminate felling, which is causing denudation of hillsides and consequent landslips. The extension of cultivation also led to the loss of forests. The forest in the district extends from about 250 m to over 3000 m above sea level and is divided into three main altitudinal floristic divisions, outer foothill forests, warm temperate forests, and cool temperate forests. Agriculture is the main occupation of the people of the district. From a study of occupational structure of the economy of the district, it is obvious that about 66% of the working force was engaged in agriculture and allied activities at the 2011 census.

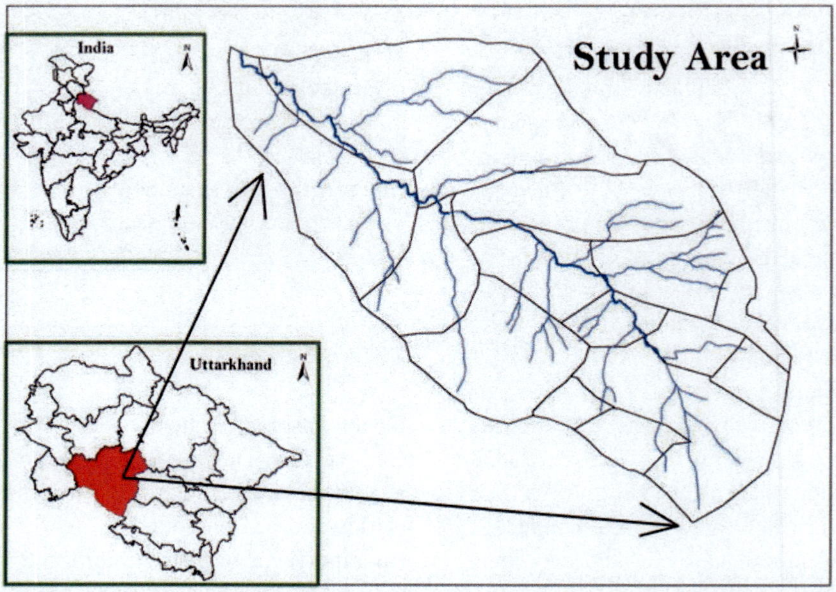

Fig. 20.1 The Study Area, Upper Pasolgad, Uttarakhand

Level land being scarce, cultivation is generally done on terraced fields. The land of Pauri Garhwal is blessed with a splendid view of snow-bound peaks of the Himalayas, scenic valleys and surroundings, meandering rivers, dense forests, and hospitable people with a rich culture. Diverse in topography, the district of Pauri Garhwal varies from the foothills of the "Bhabar" areas of Kotdwar to the soul-lifting meadows of Dudhatoli, sprawling at a high altitude which remains snow-bound during the winter months. Filled with places of tourist interest, most locations in Pauri Garhwal offer a breath-taking view of the snow-laden Himalayan splendor.

20.3 Database and Research Methodology

Both qualitative and quantitative methods have been used in the study. The study is based on both primary and secondary data. The data has been obtained from satellite imageries for mapping forest cover, water bodies, and land use/land cover. Satellite imageries have been obtained from Global Land Cover Facilities and Earth Science Data and Google Earth 2013. Information related to forest cover and water resources/water bodies have been extracted from the Survey of India Toposheet. The land use/land cover data has been obtained from satellite images. Data on agriculture and land use pattern has been obtained from District Agriculture Statistics Handbook.

20.3.1 Interview and Questionnaire

First, 100 household surveys were conducted to obtain information on socioeconomic conditions. A structured questionnaire was prepared to record relevant information. The households were selected based on Stratified Random Sampling (SRS). Convenient random sampling method was adopted for carrying out the survey; by picking random samples from different villages. The opinion of the respondents was measured on their statements and the analysis was done based on final data trends that are generated through individual responses of the Respondents. The methodology followed for carrying out this research was a questionnaire-based survey among various households. The questionnaire consists of questions focusing on the livelihood security through forest. The questions included in the questionnaire focused on the perception of people, level of awareness about the forest resources, utilization, and government policies and schemes regarding forest conservation and management.

20.3.2 Collection of Primary Data through Conduct of PRA Exercises

The PRA approach was adopted wherein the local people enthusiastically involved themselves in planning for their respective Panchayats. PRA exercises were conducted for ensuring the participation of the various stakeholders. These exercises were conducted during the months of October and November in which most of the local community attended and participated in it.

20.4 Results and Discussions

20.4.1 Land Use and Land Cover Pattern and Its Impact on Livelihood Sustainability

Land use change is always caused by multiple interacting factors originating from different levels of organization of the coupled human-environment systems. The mix of driving forces of land use change vary in time and space, according to specific human-environment conditions (Pandey 2005). Sustainable development and management of land resources have become a vital issue in view of meeting the needs of food, fodder, and fiber to the growing population. Land use is the utility of land for various purposes by people like agriculture, industries, residential, recreational, and other purposes (Pandey 2013; Thakur et al. 2014). The more recent focus on issues related to ecosystem goods and services, sustainability, and vulnerability has led to a greater emphasis on the dynamics coupling between human societies and their ecosystems at a local scale (Pandey 2011). Changes in land use pattern, application of scientific methods of cultivation, commercial exploitation of the forest cover, illicit felling of trees, and uncontrolled growth of tourism are important factors affecting geo-ecology of the valley. The continuing disruption associated with the extreme events of nature, careless application of technology, and reckless utilization of natural resources have induced environmental threats. Livelihood culture is directly or indirectly related to the geographical terrain, local climatic conditions, land use pattern, social structure, and its community participation. In mountain or hilly terrain livelihood security is directly dependent on changing climatic condition. Impact of physiography, slope, altitude, lithology, soil type, rainfall, drainage pattern, land use practices, vegetation, and anthropogenic interference (viz., road construction and settlement) is important on sustainable livelihood and common property resources. Land is considered one of the most valuable and important natural resources and is the basis of survival for all human beings. It is also an important determinant for the development of a particular region as it influences the socioeconomic conditions of the local people (Pandey and Prasad 2016).

20.4.2 Land Use and Land Cover Pattern in Pasolgad Watershed

Land use/land cover pattern of the Pasolgad watershed indicates that maximum area is under dense forest (42.59%) and agriculture (43.82%) (Fig. 20.2). Agriculture is the main occupation of the people, engaging around 90% of the total workers. Agriculture occupies 43.82% of the watershed which is an important activity and main occupation of the village (Table 20.1). It shows that agriculture is still in primitive stage and there is enough scope for modernization.

The forest area in Pasolgad is of following types: (1) Subtropical Forest Zone: Found at the elevation of about 1100 m. It is mainly dominated by deciduous, sub-deciduous species, such as *Kachnar, Amla, Semal, and Haldu* (2) Warm-Temperate Zone: Found at the elevations between 900 and 1600 m. Pine Forest dominate this *zone* with *chir* pine as the dominant tree. (3) Cool-

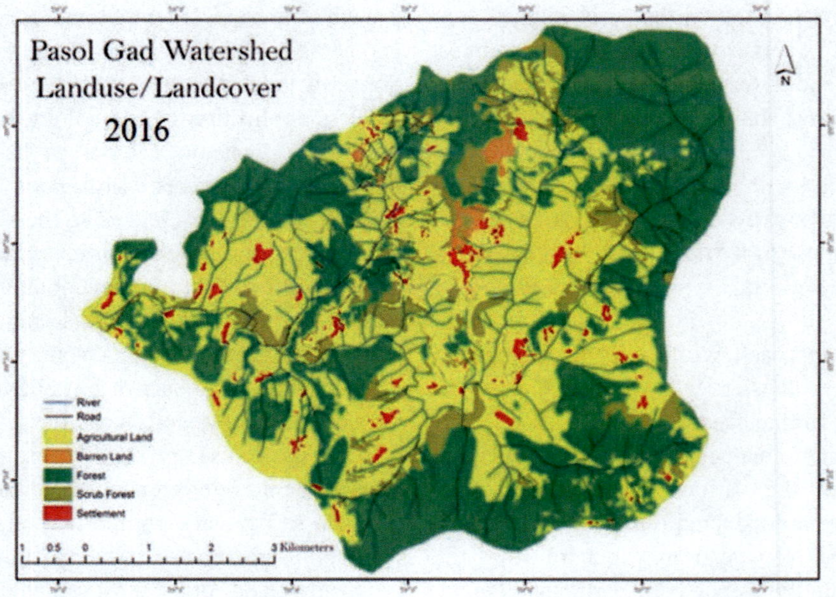

Fig. 20.2 Land Use and Land cover, Pasolgad Watershed Region, 2016. Source: Landsat TM, 2016

Table. 20.1 Land use/Land cover—2016, Pasolgad Watersheds

S. no.	Land use/Land cover	Area in hectare	Area in percent
1	Settlement	74	1.53
2	Agricultural land	2132	43.82
3	Dense forest	2072	42.59
4	Scrub Forest	376	7.73
5	Barren land	47	0.64
6	Road	31	0.64
7	Water bodies	134	2.76
	Total	4866	100

Source: Compiled by the authors, Landsat TM Satellite Image, 2016

Temperate Zone: It extends between 1600 and 2100 m and is dominated by *oak, Rhododendron, melu, anyar, Utis, fir, cedar, spruce,* etc. (4) Cold-Temperate Zone: It is found above 2100 m. Main trees are Cedar and spruce (Bandooni et al. 2015).

20.4.3 Land Use/Land Cover Change Pattern in Upper Pasolgad Watershed

Forest area increased from approx. 648.62 ha. (2016) as compared to 636.48 ha. (2002), Figs. 20.3 and 20.4). There was an increase of approximately 12 ha forest cover in Upper Pasolgad and this increase is not due to human effort or any government initiative but due to natural factors (Figs. 20.3 and 20.4).

Forest in Upper Pasolgad mainly consists of dense forest and scrub forest. Dense forest at present is well grown in upper part of the mountain. These forests are rich in flora and fauna species (Table 20.2 and Plate 20.1). There was a decrease in the overall state forest cover of Uttarakhand, as reported in 2016 by Forest Survey of India (FSI). The forest cover of Uttarakhand has found to be decreased 268 km^2 in the last 2 years having a total area of 24,240 km^2.

The main reason for the decrease in forest cover has been described as rotational felling and diversion of forest land for development activities in the report. Also, the cases of forest fire naturally and man-made result in significant loss of flora and fauna in the forest. Due to increase in developmental activities like building road, hotel, govt. infrastructure led to the removal of green forest especially in the lower valley of the mountain. That is why majority of degradation takes place at the bottom of the mountain. Lower valley

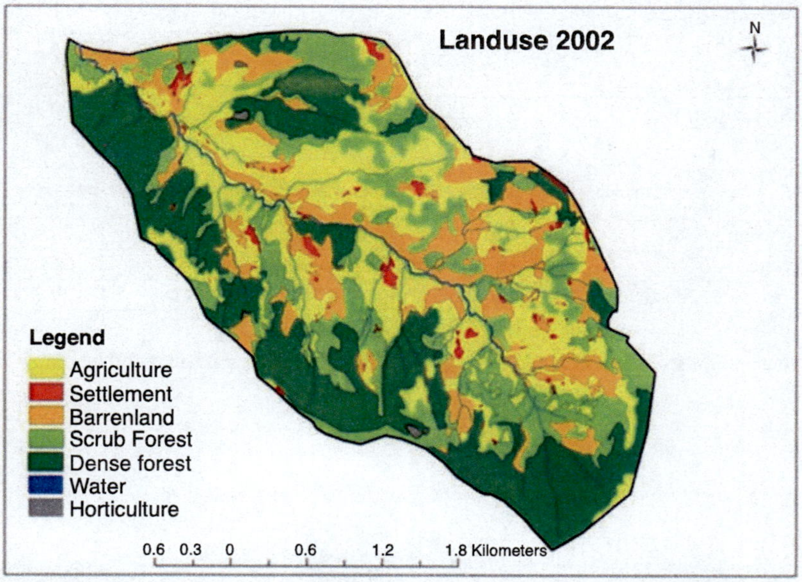

Fig. 20.3 Land Use and Land Cover pattern, Upper Pasolgad, Pauri Garhwal. Source: Landsat TM, 2016

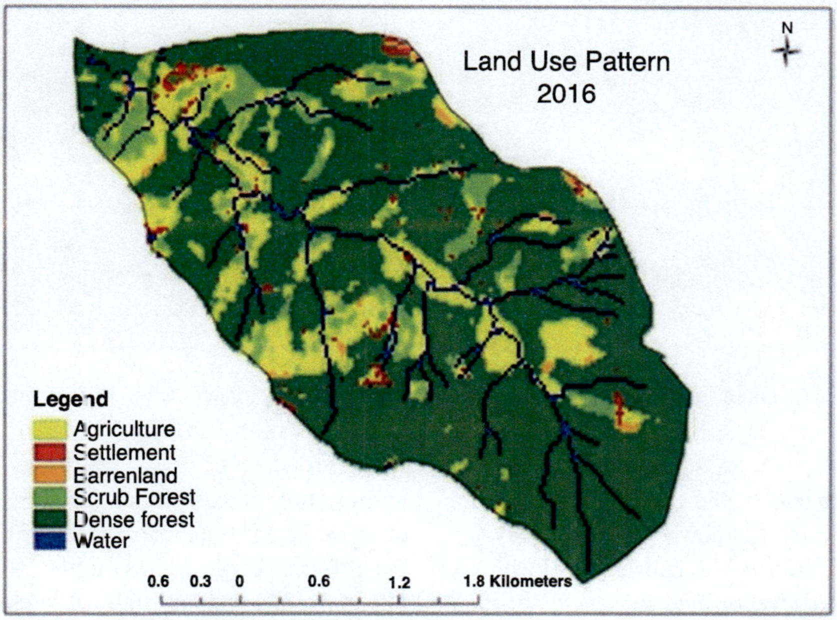

Fig. 20.4 Land Use and Land Cover pattern, Upper Pasolgad, Pauri Garhwal Source: Landsat TM, 2016

is highly encroached by the human due to easy accessibility through well-connected roads.

However, the scene is completely different in case of upper mountain region like Pasolgad watershed area where there is an increase in forest areas (Fig. 20.5). Due to less interference by humans, forests are gradually increasing in the upper areas of the mountains. Taking the case of Upper Pasolgad watershed there is a natural growth of plant species in the adjoining areas of the mountain and most of the species are wild in nature.

Another reason for increase in forest cover is the community participation of the people in judicious use forest. Government has designated this forest as community-owned forest and form

Table 20.2 Land use and Land cover/change in Upper Pasolgad Watershed

	2002		2016	
Classes	Area (ha)	Area (%)	Area (ha)	Area (%)
Agriculture	152.28	14.53	120.60	11.50
Settlement	47.32	4.51	48.71	4.64
Scrub forest	91.42	8.72	95.41	9.10
Dense Forest	636.48	60.74	648.62	61.90
Barren, fallow	71.31	6.80	79.90	7.62
Water	15.03	1.43	14.63	1.39
Fruits trees/horticulture	33.97	3.24	39.95	3.81
Total	1047.85	100	1047.85	100

Source: Forest Survey, PWD, Department of Horticulture (Uttarakhand), 2016

Plate 20.1 Dense Forest in Pasolgad, and Scrub Forest in Upper Pasolgad, Pauri Garhwal, 2016. Source:-Primary Survey, 2016

a group for managing the affairs of forest and forest resources locally known as Van Panchayat. In each village, there is Van Panchayat which looks upon the conservation and management of the designated forest. The Upper Pasolgad watershed has well-developed community forestry where they educate and aware the people of the village about the importance of forest. Community forest makes better and effective utilization of the forest products and produce. They know the requirement of the local population and can be easily fulfilled without much obstruction and problem faced. This is the reason why regions where forest is owned by the community cause less damage to the forest. They also conserve and initiate afforestation process which increases the extent of forest cover. In all tribal areas of the country, the government gave power to the local community in maintaining the affairs of forest. So, forest is owned by the community of that region which helps in the protection of the forest. The village community and Van Panchayat played an important role in conservation and management of the forest. They provide a framework and makes rules and regulation keeping in mind the need of local requirement without harming and degrading the forest much.

The above Figs. 20.3 and 20.4 highlighted an increase in forest cover in the Upper Pasolgad watershed. Forest in the Upper Pasolgad is

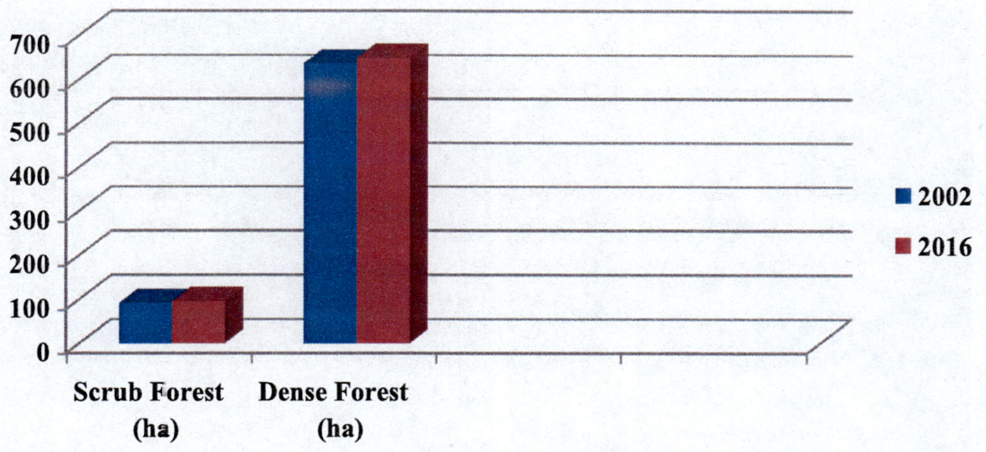

Fig. 20.5 Forest areas from 2002 to 2016 in Upper Pasolgad, Pauri Garhwal, UK. Source: Primary Source, 2016

mainly divided into two areas locally known by two different names, i.e., *Ringnal jungle and Maosikhali jungle*. Both these forests cover an area of 6.47 km^2. in Upper Pasolgad watershed. Both these forests are dense having different varieties of flora and fauna species. The most important trees are *Chir, Baanjh, Ayar, Rhodendron, Kafal, Bheu,* etc. *Rhodendron tree* is locally named as Buras. Among them, *Chir, Baanjh, Buras* dominated the forest and cover major areas of forest. These forests gradually increase due to their natural growth and the efforts made by local and Van Panchayat in conservation of forest highlights the level of awareness created among the local villagers. In the coming decades, there might be a further rise in area under forest cover.

20.4.4 Livelihood Options, Forest Resource, and its Sustainability through Community-Based Forest Management

Livelihood in the Upper Pasolgad watershed is totally dependent on primary activities. They practice subsistence agriculture for their own consumption however in horticulture due to surplus production of *Malta* and *Almond* they sell in the market. These domestic animals play a major role in their livelihood opportunities, such as goat, sheep, cattle, buffalo, hen for milk, wool, egg, and meat (Plate 20.2). All these domestic animals helped people in getting their nutritional diet from egg, meat, fruits, etc. Due to this reason children from the upper mountain areas do not have cases of malnutrition, and they are less affected by diseases of lowland like lungs disease, water-borne disease, viral infections due to better, fresh, and polluted free environment. Also due to polluted free environment, the life expectancy of mountain people is greater than lowland. The food items they consume are of good nutritional value leading to good health status. This highlights that people of mountain areas are totally dependent on nature and especially forest for getting their livelihood for all requirements.

Livelihoods in this watershed are directly connected to nature and natural resources. Almost all families do grow at least part of their food and often have cattle heads. The survival of forests, ecosystems, and rivers are inherent sustainability of livelihoods. People cultivate twice a year, i.e., during *Kharif* season and *Rabi* season and in between them *zaid crops* are grown. As agriculture in upper areas totally depends on rainfall and no alternative system of irrigation is available to the farmer, this increases the risk of agricultural productivity of farmer. For practicing agriculture, people also need the help of forest wood for fencing the agricultural field. Woods protect agricultural field from eating away by animals. Large and thick trunk are used as a natural barrier for storing water and carrying irrigation on the field.

Plate 20.2 Livelihood Options in Pasolgad, Uttarakhand. Source: Primary Survey, 2016

This highlights the judicious and effective use of local resources available to them. Thus, forest plays a major role in facilitating livelihood to the local people and support agriculture sector. It should be noted that the entire population depends on forest and forest produce to carry out their day-to-day business.

Overall, in the results presented here, it is evident that timber products of softwood are of great demand in the international market which made them very costlier. The Himalayan state is rich in softwood trees like *pine, spruce, deodar, poplar, cedar, fir, maple, walnut,* etc. In the *Ringnal* and *Maosikhali* jungle trees like *Chir, Baanjh, Ayar, Rhodendron, Kafal,* and *Bheuetc* have dominated. A well-grown *Chir* has enough wood for making a house in watershed. Usually, one must take permission from the Van Panchayat for cutting *chir* tree and the Van Panchayat allows dry and old *chir* tree for construction of houses and animal shelter. Due to high economic value, better strength, and durability of *chir's wood*, cases of illegal selling of timber woods had been seen in the past. With the help of forest officers, the Van Panchayat largely control these problems. Now, the incidence of illegal marketing of timber fell tremendously. Use of GPS and satellite for monitoring the forest cover by govt. and proper vision by the Van Panchayat helped in the growth of forest. Villagers collect the minor forest products for their self-utilization and consumption like fuelwoods. Almost from every household of the village, one male or female member goes to the forest for the collection of fuel woods (Plate 20.3). Their collection of fuel depends upon the season of the year. In the winter season they fre-

Plate 20.3 Collection of Fuel Wood for Consumption. Source: Primary Survey, 2016

quently go to the forest on daily basis as the consumption of fuel is more during this season. On the other hand, before the rainy season, the villagers collect as many as non-timber fuelwoods and store them in the house. This was done because the forest products get wet due to rainwater and the wet fuelwoods do not burn easily. These fuelwoods are used in cooking food. They are allowed to collect only dry woods of the forest which are naturally dried. They are not allowed to cut fresh woods of forest. Besides, all the woods of the forest are used in making household materials like table, chair, furniture, etc. During our survey, we also found woods often used as a tub for drinking water by the animals. The local people feed water in this tub for their cows, buffaloes, sheep, goat, etc. This technique highlights the judicious use of resource and the level of dependency by the local people on the forest. It is impossible to even imagine agriculture in the hills without forests. Thus, it should be noted that forest ensures food security and livelihood opportunity to the local population.

20.4.5 Medicinal Plant as Forest Resource: Livelihood Potential and Sustainability

Several medicinal and disease curing plants are grown in the forest that are wild in nature and the local population depends on them for getting cure for disease. *Buras Juice* which is prepared from Buras flower of Buras tree is good for heart patients. For a TB patient, *chir* forest can be used for a morning walk which, in turn, helps in getting better recovery from TB. *Chir* forest acts as a boon for TB patients. People also grow *Ramdana* which is locally known as Cholai which is good for health. Pulses like *Gehat, Kala bhat, Rajma* helped in curing stomach-related problems and increasing the digestion capacity of an individual. *Lakhora* is one variety of *Chilli* which is antibiotic in nature help in the treatment of cancer. All the above species of trees act as a boon directly and indirectly for the livelihood security of the local people. All the essential things which support lives in upper mountain regions are almost fulfilled by these species. Apart from these, they are of high ecological and environmental importance. Ayurvedic medicine, local rural medicine along with other Complementary and Alternative Medicine (CAM) are popular but restricted within few regions. Though, they are also used as home remedies by the local people. People want to continue with this tradition but cannot manage due to unavailability of the local healers. Policymakers have come up with the biodiversity register which is under preparation containing special section on the traditional healers considering them the custodians of the traditional knowledge. Holistic health care and preventive care are now an important aspect of being healthy. The focus is shifting from the *illness* to the *wellness*. Thus, to elevate health services, cross-sectoral and multilevel intervention should promote untapped rural medicine and Medical Tourism because it has lot to offer in aging, chronic degeneration, dermatological, and respiratory problems.

20.4.6 Resource and Skill Management for Livelihood Securities Promotions and Sustainability

There are numerous possibilities of enhancing livelihoods in the area by introducing new technologies and skills. To popularize new livelihoods, related to technologies and skills, it is necessary to demonstrate and set up a skill development center in the village. The center can have demonstrations on energy, components of farm management and small production units, small fabrication unit, and skill development units for training, and technical support for setting up enterprise and further follow-ups and "hand holding." During the survey it was found that there are numerous opportunities for creating new employment and enhancing income of people by improving farm management, cultivating organic crops and spices, aromatic and medicinal plants, agro-processing, wool processing, woolen felt making, harnessing water power, and using low-cost power for energizing machinery, local milling of grains, oilseeds and spices, incense making, using bamboo for basketry and related articles, such as plates for temple "Puja," floriculture to meet requirements of temples and making powder/pigment from dried flower petals, beekeeping, setting up local bakeries, and training people for providing improved services to tourism infrastructure . The above initiatives are the indicators of women empowerment in livelihood securities. To introduce these skills and technologies for enhancing livelihoods there is a need to set up a self-contained Vocational Skill Development Centre (VSDC) as a first step that would be essential for demonstration of technologies and provide training to local people and promoting technical innovations for livelihoods. Once trained, people can be guided for further improvement of skills and absorption of technology, establishing enterprises, and developing marketing linkages (Prasad et al. 2016). To introduce these skills and technologies for enhancing livelihoods there is a need to set up a self-contained Vocational Skill Development Centre (VSDC) as a first step that would be essential for a demonstration of technologies and providing training to local people and promoting technical innovations for livelihoods in the Upper Pasolgad. Once trained, people can be guided for further improvement of skills and absorption of technology, establishing enterprises, and developing marketing linkages.

20.4.7 Beekeeping

Beekeeping is a tradition in various households of the area. However, improved methods need to be introduced, such as usage of improved boxes for beehives, improved methods of honey collection and purification, and care of honeybees in upper areas of the Pauri Garhwal district. These areas have a good potential for Beekeeping. A training program on beekeeping would help many families in enhancing income. Honey is readily sold. It is purchased by various Ayurvedic medicine manufacturers.

Based on the survey, assessment of local resources and discussions with people the following categories of livelihoods were identified: (1) Production-oriented livelihoods based on local resources; products suited for local consumption, sale in the local religious tourism market, and having outside marketing potential and (2) Service-skill based livelihoods for strengthening maintenance, reconstruction, and improving communication processes to improve the quality of life and to support local market and tourism infrastructure. Many of these livelihoods and new sources of income will be useful with rising tourism in the future.

Participatory forest management approach should be implemented throughout the country.

Marketing cooperatives of fruits, off-season vegetables, and flower growing farmers need to be encouraged. Sustainable tourism and medical tourism need to be promoted (Pandey and Prasad 2016). The Government plays an important role in providing a social protection to its citizens to increase income and assets, enhance capabilities and ensure access to entitlements and claims by programs like National Rural Livelihoods Mission (NRLM) and MGNREGA. Livelihood security means adequate and sustainable access to capabilities, assets, and activities for a means of living.

20.4.8 Cultivations for Sustainable Livelihood Diversification

The sustainable livelihood security of such region may best be achieved by enhanced land uses aided by technologies of bioengineering and small-scale engineering involving the renaturalization of degraded mountain ecosystem (Carney 1998). Taking the ecosystem approach and evaluating the interactions between atmospheric, biological, physical, and anthropogenic components the mitigation measures may be investigated. Local knowledge systems have a vital role to play in the implementation of land resource management policies. Use of Remote Sensing data, supplemented by Geographical Information System (GIS) will enhance the techniques of hazard mitigation particularly monitoring and forecasting of hazards and land use pattern. Land use planning through the inculcation of hazard zone mapping may prove an effective tool for enhancement of land uses and prevention from hazards by reducing the probability of risk (Pandey 2002). Soil erosion, deforestation, landslides, loss of biodiversity, and degradation of land and water resources are common problems throughout the watershed. Landslides can be caused by poor ground conditions, geomorphic phenomena, and natural physical forces and quite often due to heavy spells of rainfall coupled with impeded drainage (Prasad et al. 2016).

Agricultural farming is carried out mainly on the narrow patches of terraced field. Subsistence cereal farming is dominating in the entire farming system, which is quite insufficient for the livelihood of the people inhabited in this watershed. As a result of this, poverty and malnutrition are very common phenomenon among the inhabitants and the struggle for livelihood is higher than the other watersheds of the Himalaya (Prasad and Pandey 2017). The high rate of outmigration to the foothills and plains is the consequence of these prevailing factors. The terrain is also not fit for intensive cultivation of agriculture crops due to the high rate of soil erosion, instability, and fragility of the landscape (Walingo et al. 2009). As it is already mentioned earlier that the basin possesses suitable geo-environmental conditions to produce off-season vegetables therefore high variety of vegetables, such as onion, ginger, garlic, capsicum, cauliflower, ladies finger, cucumber, pumpkin, tomato, and potato can be produced here. It is noticed that the land under vegetable crops is proportionally very low. With the efforts done by the governmental agencies and innovation in the field of cultivation, the land under cereal crops should be transformed either into off-season vegetables or fruit trees (Prasad and Pandey 2017).

Pauri Garhwal is the largest hill district of the state. Small villages are scattered at different heights. Due to heavy rainfall, there is a lot of soil erosion. This district faces serious problems in terms of seed quality and lack of credit, drinking water, power supply, and milch animals; there are few milk routes and there has been reckless deforestation. The irrigated area is only 8.5% of the net cropped area. Farming is the main activity and organic cosmetic ingredients are produced. One hundred and five villages have been declared organic villages (Sati 2004). Thus, there is a need for biofertilizers and biopesticides. With increasing climate variability, there is a need to understand livelihood dynamics as part of efforts to deal with the aftermaths of extreme natural events. The following points must be considered in initiating the forestry and common property resource development plans in the Pasolgad watershed, such as Social Forestry Project should aim at motivating people to raise different broad species, Mixed groups of field staff should be sent for training for knowing the problems/limitations of each functionary collectively, The University of Horticulture and Forestry should provide more scientific information pertaining to social forestry program, The concept of Integrated Rural Nursery Program (IRNP) should be put into practice. Some community-based Kishan Nurseries should be encouraged with an understanding that such beneficiaries would use to raise plants in that area and private entrepreneurs who wish to raise plantation for industrial/packing uses should be encouraged.

20.5 Conclusion

The local community of Pasolgad watershed understands the necessity to manage water, forest, and other resources of the area for sustainable development. They are doing work on forest management since 1980; environmental awareness camps since 1985, pits for water recharge since 1990, and to develop villages since 2015. Out of 25 villages under these moments around 8 villages are from Pasolgad watershed. After the active participation, the forest and water resources of these villages are better as compared to surrounding villages. All these works are done by the locals under the guidance of notable environmentalist Mr. Sacchidanand Bharti. The need for understanding forest and water relationship is necessary especially in the areas where the runoff is very high, and source of income is the only agriculture. The focus of the present study is on managing the forest plantation and watershed conservation using remote sensing and GIS in Pasolgad Watershed in Pauri Garhwal. The results suggest that soil erosion is more prone to the area where the plantation is less with possibilities of very high runoff. Local communities have undertaken plantation of forest with the local leaders. The Van Panchayat ensures the conservation of local forest through effective implementation of programs and policies of the government. Besides, it also mobilizes the community towards sustainable use of forest through awareness and campaign. Community forestry helps in a greater extent to conserve and maintain the forest cover at local level. Although mountainous region is vulnerable to various hazards, like earthquakes, landslides, snow avalanche, etc., these can be minimized if the human encroachment on natural environment is also minimized. This can be decreased only when humans are satisfied that their own action can someday lead to the destruction of humankind. Although many govt. initiatives are started for the conservation of forest but most of them are not well targeted due to lack of awareness among the locals about the programs and policy.

References

Anderson JR, Hardy EE, Roach JT, Witmer RE (1976) A land use and land cover classification system for use with remote sensor data. Geological Survey Professional Paper No. 964, U.S. Government Printing Office, Washington, D.C. pp 28

Bandooni SK (2004) Land resource management and development in hill areas. Research India Press, Delhi

Bandooni SK, Devi LM, Prasad AS (2015) The use of remote sensing and GIS for managing Forest plantation and watershed conservation, in Pasolgad watershed in Pauri Garhwal, Uttarakhand. In: Sati VP et al (eds) Climate change and socio-ecological transformation. Today and Tomorrow's Printers and Publishers, New Delhi, pp 381–400

Carney D (1998) Sustainable rural livelihoods: what contribution can we make? Department for International Development (DFID), Papers presented at the Department for International Development's natural resources advisers' conference, July 1998, London

DFID [Department for International Development] (2001) Sustainable livelihoods guidance sheets. Nos. 1–8, London

Kumar A, Negi MS (2017) Assessment of land capability and suitability classification for agriculture using geospatial techniques in Rudraprayag District (Garhwal Himalaya). In: Sati VP, Lalmalsawmzauva KC (eds) Natural resources management for sustainable development and rural livelihoods, vol 2. Today and Tomorrow's Printers and Publishers, New Delhi, pp 443–463

Pandey BW (2002) Geo-environmental hazards in Himalaya: assessment and mapping. Mittal Publication, New Delhi

Pandey BW (2005) Natural resource management: approaches and techniques. Mittal Publications, New Delhi

Pandey BW (ed) (2011) In search of sustainability: global perspective on marginality, technology, politics, and planning of land use. Research India Press, Delhi

Pandey BW (2013) Land use change and marginality of livelihood in Himachal Himalaya. J Land Use Water Manag 13(1):15–20

Pandey BW (2014) Land use change, sustainability, planning and technology for land resource management. In: Pandey BW (ed) Dynamics of land use changes, sustainability, and management. Research India Press, Delhi

Pandey BW, Prasad AS (2016) Land use pattern and hazards risk assessment in Kullu Valley along the Beas River in Western Himalaya. J Land Use Water Manag 16(1):1–18

Prasad AS, Pandey BW (2017) Climate change, land use and land cover change detection and its impact on hydrological hazards and sustainable developments: a case study of Alaknanda River basin, Uttarakhand, India. Riscuri si Catastrofe J 20:55–68

Prasad AS, Pandey BW, Leimgruber W, Kunwar RM (2016) Mountain Hazard susceptibility and livelihood security in the upper catchment of the river Beas, Kullu Valley, Himachal Pradesh, India. Geoenviron Disasters 3:3

Sati VP (2004) Systems of agricultural farming in the Uttaranchal Himalaya. J Mountain Sci 2(1):76–85

Singh RB (2002) Himalayan environment, land use/land cover changes and Highland lowland interactive system with Indus-Ganga-Brahmaputra Plains. In: Abrol YP et al (eds) Land use-historical perspectives: focus on Indo-Gangetic plains. Allied Publication, New Delhi

Singh M, Singh RB (2013) Change in land use and land cover through watershed development program: a case of Nakti Nadi sub-watershed in Haryana Shiwaliks. Horizon J Soc Sci 4(1):11–19

Singh RB, Fox J, Himiyama Y (eds) (2001) Land use and land cover change. Science Publication, Inc., Enfield (NH), USA and Oxford & IBH Publication, New Delhi

Thakur B (2011) Depletion and degradation of land resources in India. Series Editor: L.R. Singh R. N. Dubey Foundation Monograph, Allahabad

Thakur B, Sharma HS, Mishra S, Chattopadhyay SK, Singh S (eds) (2014) Regional development theory and practices: emerging perspectives. Concept Publications, New Delhi

Turner BL, Skole D, Sanderson S, Fischer G, Fresco L, Leemans R (1995) Land use and land cover sciences/ research plan. IGBP glob, change Rep. 35/HDP Rep. 7, Int. Geosph-Biosph program., hum., dimension, global environment. Change program. Stockholm, Geneva

Walingo MK, Liwenga E, Kangalawe R, Kabumbuli R (2009) Perceived impact of land-use changes and livelihood diversification strategies of communities in the Lake Victoria Basin of Kenya. J Agric Biotechnol Sustain Dev 1(3):69–78

Walsh-Meyer S, Crews K (eds) (2002) Linking people, place, and policy: a GI-science approach. Kluwer, Dordrecht

Rural Livelihood and Women: Glimpses from an Indian Tribal Village

21

Purva Yadav, Shreya Akarshna, and Anuradha Shankar

Abstract

In the twenty-first century, livelihoods will be needed by perhaps two or three times the present human population, and it is therefore necessary to understand the changing dynamics of it. A livelihood comprises people, their capabilities, and their means of living. Livelihood is environmentally sustainable when it maintains or develops the local and global assets on which means of livelihood depend. The livelihood framework is also applicable in the context of rural areas. This covers capital both natural and social. Indian villages (both tribal and non-tribal) are unique in their own way, and very intriguing if you try to understand it from the emerging interrelationship between resource and livelihood from the women perspective. This chapter draws on insights from a small tribal village called Audali located in Sitarganj Tehsil of Udham Singh Nagar district in Uttarakhand. Tharu tribe, who has migrated from the state of Rajasthan centuries back to this area, dominates Audali village. It is interesting to see how development process has transformed their means of livelihood and how role of women became paramount at the household and community level with the changing socio-economic profile of the household and the village. The study throws light on the experiences of the group of motivated Tharu women in a more informal and fluid personal-cum-work space, such as home-based small-scale activities like handicraft, and capture the shifts and continuum in social relations, particularly in the context of means of livelihood and gendered division of work.

Keywords

Livelihood · Resource · Rural Spaces · Sustainability · Tribal Women · Uttarakhand

21.1 Introduction

In the twenty-first century, livelihoods will be needed by perhaps two or three times the present human population, and it is necessary to understand the dynamics of it. A livelihood comprises of people, their capabilities, and their means of living, including food, income, and assets. Tangible assets are resources and stores, and intangible assets are claims and access. A liveli-

P. Yadav (✉) · S. Akarshna
Centre for the Study of Regional Development, School of Social Sciences, Jawaharlal Nehru University, New Delhi, India
e-mail: purvayadav@mail.jnu.ac.in

A. Shankar
Department of Geography, Shyama Prasad Mukherji College, University of Delhi, New Delhi, India

B. Thakur et al. (eds.), *Resource Management, Sustainable Development and Governance*, Sustainable Development Goals Series, https://doi.org/10.1007/978-3-030-85839-1_21

hood is environmentally sustainable when it maintains or enhances the local and global assets on which livelihoods depend and have net beneficial effects on other livelihoods (Chambers and Gordon 1991). Therefore, for changing structures and processes for sustainable outcomes, effective local institutions are required, such as government agencies, civil organizations as well as the private sector. Community capacity-building and strengthening of these institutions are key to any livelihood strategy.[1]

Across countries, particularly in the developing world participation of women in community initiatives can have long-lasting benefits for women and children, as well as for the entire community. UNICEF[2] has aptly noted, "women who are empowered to take action, whether through programs led by governments, non-governmental organizations or those driven by the community, often have a positive influence on the lives of other women." On similar lines, the UN on marking the International Day of Rural Women stressed that women play a key role in the progress of rural households, local as well as national economies.[3] UN Secretary-General, Ban Ki-Moon has stressed that women are very crucial to the successful attainment of almost all the 17 Sustainable Development Goals (SDGs). It is in this backdrop, the present chapter is an attempt more in the line of sharing the experiences of rural women from a small tribal village of India, and their crucial role in sustainable livelihood, along with their strong community-driven efforts towards making their village a "smart village."

Box 21.1 Sustainable Development Goals

Natural resource endowments and livelihoods are connected very intimately. This relationship impacts rural women in Indian villages gravely as their dependence in rural economy on natural endowments emerges in different roles, i.e., as farmers, wage earners, and entrepreneurs taking responsibilities for the well-being of the members of their families, including food provision and care for children and the elderly. This context is well captured by the authors in their research in the southwestern region of Uttarakhand state of India where tribal women in a small rural village *Audali* face the challenge of unpaid work, particularly in poor households, often collecting wood and water. These women from indigenous and grassroots communities are often also the custodians of traditional knowledge, which becomes key for their communities' livelihoods, resilience, and survival. While the UN sustainable development goals recognizes that women are very crucial to the successful attainment of almost all the 17 SDGs, yet the current reality for some women in *Audali* village is one where they face constraints in engaging with economic activities because of gender-based discrimination and social norms, disproportionate involvement in unpaid work, unequal access to education, healthcare, property, and financial and other services. The authors are concerned that women in this village are also disproportionately vulnerable to the impacts of environmental disasters and climate change. These vulnerabilities can be addressed by ensuring gender equality, and empowering rural tribal women through decent work and productive employment. Thus, not only contributing to inclusive and sustainable economic growth but also enhancing the effectiveness of poverty reduction and food security initiatives and climate change mitigation and adaptation efforts.

[1] http://www.fao.org/3/X9371e/x9371e22.htm

[2] https://www.unicef.org/sowc07/docs/sowc07_panel_5_5.pdf

[3] https://www.un.org/sustainabledevelopment/blog/2015/10/rural-women-are-the-backbone-of-sustainable-livelihoods-ban-declares-on-international-day/

21.2 Literature Review

We begin with a review of relevant literatures by summarizing the findings. According to Murray (2001), livelihood research can also be undertaken at the micro-level, such as households and communities. It constitutes an empirical investigation of the mixed livelihood modes, and more so of the relationships existing between them and changes have taken over the temporal scale. In this scenario, studying livelihood becomes important. De Haan and Zoomers (2005) pointed out the importance of livelihood studies in understanding the lives of poor people over the years. They dealt primarily with the challenges faced in livelihood studies, accessibility, and the relationship between accessibility and the decision-making process. Furthermore, the complexity of strategic, unintentional behavior, and structural factors poses grave difficulties.

Several explanatory variables govern access to livelihood opportunities, such as the role of institutions, various organizations, and social relations, but under-estimation of these factors within livelihood studies downplay the structural features giving an ambiguous picture of livelihood trajectory.

The livelihoods framework also seems to be applicable in the case of rural areas, which include human capital, physical capital, social capital, financial capital, and natural capital. Ellis (1999) points out that natural capital relates to the natural resource base. It is also noted that livelihood strategies composed of a range of activities, such as the access to both the assets and its utility interceded by social factors (like social relations, institutions, and organizations), exogenous trends (for example, economic trends), and calamities (drought, floods, disease). Similarly, Shangpliang (2012) has emphasized that livelihood primarily concerns human beings and their resources—natural, economic, and social. These resources, however, cannot be dissociated from the concerns about accessibility—physical and social, particularly in the context of the changing political, economic, and socio-cultural circumstances. Therefore, she further added about forest economy, that an "understanding of livelihood is necessary as it clarifies some important issues about the role and importance of the forest as a Common Property Resource (CPR) in the economic life of rural women."

Along with the multifarious changes experienced by the economies in the region, gender relations have also been altered in a diverse and complex manner. This is mainly because of the reason that gender relation is an inherent feature of any economy and more so the power dynamics existing over the control and accessibility of resources make it more interesting and useful to understand. United Nations Development Program (UNDP 2011) has mentioned in its report that the most disadvantaged people and communities "carry a double burden of deprivation and most vulnerable to the wider effects of environment degradation…" If this deprivation is checked and ended, it could enhance capabilities, expand people's choice, and boost human development. It is further noted how some countries and communities have improved their environmental sustainability and equity, which, in turn, has positively affected their human development dynamics. This is primarily made possible by transforming the gender roles and empowerment issues in their society. In this context, Mishra (2007a; b) noted that despite its relevance, gender relation dynamics are "often neglected in the mainstream discussions on economic development and change." Furthermore, Aggarwal (2018) concerning Sustainable Development Goal (SDG) 5 on gender equality has indicated "SDG 5 has both potential and limitations. The potential lies in its focus on women's access to land and property, and natural resources. Secure land rights for women can improve both their productivity as farmers and family nutritional allocations."

Further, Mishra (2018) highlighted sustainable ways to secure forest diversity in her study on the experiences of Kondh women in the Rayadaga district of Odisha. The realization of Kondh women for the land rights which can provide them with a decent livelihood and dignified life has led them demanding for the same from the state. Female participation can be seen in agricultural activities, but in domestic work, women are contributing their substantial labor. Though women can share their opinion on village affairs of their community, the head of the village council *"kutumb"* would always be a male. Female inclusion in political space and providing them equal rights on the economic resource is gradually being raised. Similarly, Sengupta (2018) has emphasized the existing gender-based inequality in the Reang tribe of Tripura. Lack of rights over land and natural resources has aggravated the existing inequality. Further, state-led developments, such as rubber plantation schemes, bamboo missions, tea-coffee plantation, etc. are discussed considering their visible impact on the livelihood of women mainly because they are discouraged from either practicing subsistence agriculture or the decline of the *"jhum cultivation."* The need for providing political space to women to strengthen their bargaining power in every field considering livelihood options is also stressed. Regarding the ownership, women can obtain land via, the family (especially inheritance), the market, and the state. Also, SDG 5 talks about financial services. Affordable credit can help women farmers invest in necessary input. SDG 5 also emphasizes natural resources. Moreover, the target stresses women's participation in public life. Although it focuses on legislation and Village Councils, this can be extended to community institutions managing forests and water. Sengupta (2018) further adds "the failure of SDG 5 to explicitly recognize access to forests and fisheries, or the challenges of climate change, restricts its potential." Hence, it is quite explicit that the issue of gender equality has caught a lot of attention at the global and national level, in recent years, yet much needs to be done in this direction.

Recent studies have highlighted that not only the participation of women is significant, but it is equally essential "how they participate" and "how much." It is widely accepted that the livelihood needs of men and women are quite distinct because of their differential utilization of the existing resources. Gender differentiation in using, accessing forest products, and some common current assumptions regarding these are tested. Sunderland et al. (2014) compared to what extent findings on countries or regional levels are consistent with the global dataset. They have considered three regions, namely, Africa, Asia, and Latin America, and compared them with the global level trend. They argued that in collecting unprocessed forest products, men and women contribute roughly equal, which is a deviation from the generalization that unprocessed forest products are collected mainly by women, but except for Africa. They questioned the notion of women collecting more diversified products than men and claimed that in aggregated categories, men seem to collect more diversified products. Their major concern revolved around whether a gendered equation in using, accessing, managing, etc. forest products is evolving or is it just a part of a continuum?

The link between land resources and rural livelihoods with the increasingly multiplying role of women as providers in the slow-moving rural economies makes it essential to redefine the gendered division of control, accessibility, and utilization of the resources particularly in the context of climate change. Rural economy, particularly tribal, inextricably depends on the livelihoods based on forest resources and minor produce. The collection of firewood, timber, edible plants and herbs, broom shrubs is used to cater to their household need of fuel, food, and fodder. Village commons and forests are essential and supplementary sources of livelihood of the tribal communities, ensuring female participation in it requires strengthening of their bargaining power. A study completed by Agarwal (2001) in India and Nepal minutely assesses the exclusion of women from participation in development agenda. Fieldwork done on community forestry groups of these countries shows though there is regional variation in women's status in terms of representation in decision-making, a lack of inclusivity can be seen everywhere. She further examines the participation scenario, in defining the rules for forest use, its protection, and distribution of benefits among people. It was found that the nominal participation of women is even less than 10%, passive participation is almost negligible, and their opinion rarely sought for forest development plans.

On similar lines, Mitra and Mishra (2011) studied the impact of household-level heterogeneity on the degree and nature of forest dependency under conditions of declining forest cover and institutional diversity. They covered randomly selected 12 villages of five districts in Arunachal Pradesh based on forest coverage and form of property rights structures over the forest, i.e., community ownership and private ownership. It is noted that the total consumption expenditure, household size, and land operated expected to be positively related to the amount of forest consumption. The study further shows that as the income of the households increases the percentage consumption of timber increases, along with a decline in minor forest products like bamboo and leafy vegetables. Common property resources (CPRs) play a key role in the rural economy of Arunachal Pradesh, although the income from CPRs is higher for the rich in absolute terms and access to the forests is crucial for the survival of the poor. Forests have always been a key source of survival in the north-east region of India. Various studies on the livelihoods and employment diversification processes in the region suggest that major factors affecting livelihood diversification are the natural resources and the institutional support that define and differentiate the level of the accessibility and utilization of such resources, level of infrastructural development-physical and social, along with the state intervention in the local economy (Mishra 2007a; b). Studies also highlight that the conflict-prone/vulnerable regions experience a different kind and process of livelihood diversification (Fernandes and Melville 2005; Barbora 2002).

Women, particularly in poor rural households, are burdened with household chores and family responsibilities. Shimray (2004) looking into women's status in the Naga society of Tangkhul Nagas of Manipur state, tried to understand it through the lens of household responsibility, workforce participation, and division of labor. This study is based on primarily on the field survey and reveals that even though women's status is very high and assumes to be having a happy and free life considering cultural level, they are still tied with household chores, even non-household activities, e.g., construction work and repairing are often taken by women. The work burden on women is comparatively more than men. Gender disparity is significant in household chores where female participation is more. Agarwal (1989) has highlighted that gender disparity that exists access to resources takes different forms, such as intra-family differences in the distribution of necessities, women's disadvantaged position in the labor market, their limited access to the major mode of production, and related technology. Increasing degradation and privatization of the country's common property resources adversely affect the poor in general, and women in particular, because of their dependence on it for sustenance. Further, she also highlighted that there have been instances when women are not "passive" but have also raised against their marginalization, and in today's time emerged as key actors at the grass-root level initiatives. In this connection, she says "In particular, in response to a growing crisis of survival, poor peasant and tribal women have emerged in the forefront of many ecology initiatives."

Rural women spend more time and labor in the forest for subsistence and for additional income to support the household. It will not be wrong to point out that the alarming rate of climate change and environmental degradation will affect women, directly and indirectly, more than men. Dasgupta (2001) affirmed that "rural poverty, resource degradation, and institutional failure pull in different directions and are not unrelated to an intellectual tension between the concerns people share about. Global warming and acid rains sweep across regions, nations, continents and about those matters (such as the decline in firewood or water sources) that are specific and the needs and concerns of the poor in as small a group as a village community." It will not be wrong to say that the problem related to the environment may present differently to different people and places over time.

According to UN data, between 2010 and 2015, "the world has lost 3.3 million hectares of forest areas. The poor rural women depend largely on common-pool resources and are especially affected by their depletion." One in six persons globally depends on forests for supplementary food, green manure, fodder, firewood, etc. Women and girls are the main gatherers of forest products, especially food and firewood; the latter continues to be the primary cooking fuel in most of the rural India. Poffenberger's (1990) study pointed out that in the state of West Bengal in India; tribal women collect *sal* leaves for about 6 months in a year and earn a minimal amount of Rs. 72/month. Ghatak (2013) assessed the livelihood pattern of the Savara and Lodha tribal community from seven villages of two sub-divisions of two districts of South Bengal, North 24 Parganas and Paschim Medinipur. Women of the Lodha tribe are mainly indulged in the collection of fuelwoods, while the Savara tribe devotes their more time in collecting medicinal herbs. Thus, women of these tribes are, still, highly involved in subsistence and traditional activities of tribal society. Forest products are, still, the most important source of income and non-forest produces are the main employment providers for women. In recent times, they are also working as maidservants in wealthy households of the village or the neighboring areas. Mishra and Upadhyay (2007) have reflected on the emerging gender relation and structurally transforming the economy through diversification of the workforce and rise of non-farm sectors in Arunachal Pradesh which is predominantly inhabited by tribal people. They have investigated the changing status of women in this tribal state through some selective indicators of well-being. More specifically, sex ratio explains through gender gap in education, work participation rate, freedom in mobility and decision-making, participation in political affairs, etc. They found a change in the trend with the improvement in female education, workforce participation, and education levels. There is still a lack of female representation in political space, role in decision-making, etc. making it difficult to summarize in favor of the state's position tilting towards the well-being of women. There have been several attempts to lend support to higher the dependency of tribal households on the forest economy, the higher is the poverty incidence.

It is important to note that women and livelihood related issues are multifarious and highly complex intricately linked with the socio-economic fabric that varies across space and time. For example, covering all studies on this issue and studying each of these aspects in detail are beyond the gamut of our study, yet an attempt is made to capture the nuances of this very crucial topic especially in the era of SDGs.

21.3 Purpose of the Study

The present study attempts to highlight the sustainable relationship between (wo) man, nature, and technology/institutions. The study draws majorly on the effective role of women in the small tribal village in India. The utility of natural resources is not uniform and equally effective across time and space. The characteristics of the natural, political, economic, and socio-cultural milieu determine the scale of effective utilization of the resources at different scales, namely, at the individual, household, and community level. The increasing requirements for fuel and raw material-based resources can be satisfied only by the full use of the existing resources, and that, in turn, require adequate equipment and technologies be in place. It is an established belief that resources have distinct use and pose different challenges to both men and women. The gendered roles and responsibilities at the household and village community have a differential impact on the rural economy. As discussed earlier in this study, the primary providers of daily necessities, such as water, food, and fuel at the household and community levels, women in rural areas are mainly dependent on natural resources for their livelihoods, and, hence, are particularly highly resource-dependent for their livelihoods, and are, therefore, more vulnerable to changes in the availability and accessibility of these resources over the years.

Hence, the present chapter throws light on the changing means of livelihood from the women's perspective. The examples and challenges discussed in the chapter are based on the experiences of a group of tribal women from a small

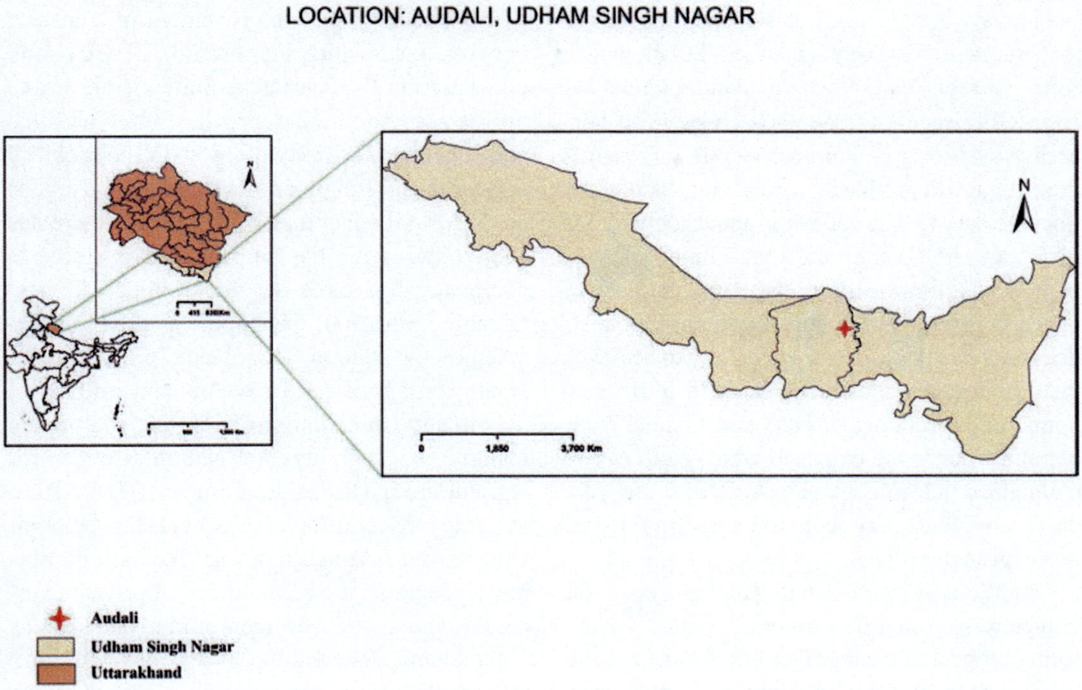

Fig. 21.1 Location of Study Area (Audali village)

village called Audali in the Sitarganj Tehsil of Udham Singh Nagar, Uttarakhand[4] (Fig. 21.1). To understand the village dynamics, it is essential to glance through the ongoing process of development experienced by the region at the macro-level and the same is dealt with subsequently in the chapter.

21.4 Study Area

Figure 21.1 shows location of the study area. According to Census of India 2011, the Udham Singh Nagar district ranks third in terms of population in the state and is one of the highly urbanized districts in the state with more than 36% of the urban population. The Sitarganj Tehsil has the highest number of inhabited villages, i.e., 127 and with approximately 21% has the second-highest share of Scheduled Tribe rural population in the district (Census of India 2011). Audali is a small tribal village located in Sitarganj Tehsil and comes under Bharauni Gram Panchayat. Based on the village survey, and transect walk, it was noted that this village has a total of 118 households, and about 30% of the total residing population has no land. This tribal village is dominated by Tharus, and the average size of landholding is approximately 3.8 acres. Though farming remains an important activity due to seasonality, changing agro-climatic and socio-economic dynamics, most of the rural households are forced to diversify their means of livelihood. Also, a gendered division of work was visible with women overwhelmingly engaged in agriculture-related activities while their male counterparts work in the non-farm occupations.

Audali is an interesting village to consider as the women here along with their conventional role, attempt to address their need to have secure

[4]Uttarakhand became the 27th state of the Republic of India on 9th November 2000. To secure greater economic and political autonomy, the state was created from the state of Uttar Pradesh (U.P.) and is in the northwestern part of U.P. Udham Singh Nagar did not exist as a separate administrative entity until the year 1995. The seven tehsils or Development Blocks of Udham Singh Nagar, namely, Khatima, Bajpur, Kashipur, Gadarpur, Rudrapur, Jaspur, and Sitarganj had been part of Nainital district for more than 100 years (Robinson 2001).

livelihood options and sustenance strategies. According to Hogan & Stewart (2013), it is a form of localism, "where communities are required to secure their socio-economic sustainability by taking an entrepreneurial approach to developing their local assets and resources. Societies are facing questions about the viability of both interventionist and market-based approaches for ensuring the continued sustenance of given communities. But, since localism does not have the capacity effectively to deal with the question of the carrying capacity of the economic or environmental base at a societal level, one must query the extent to which faith can be maintained in localism as an effective strategy for the future." This may seem to be true in the context of Audali village.

Traditionally, rural small landholders in the village were primarily farmers, essentially cultivating crops and raising livestock for their livelihoods (Annexure 1). In Udham Singh Nagar, over the years with the rapid pace of urbanization and coming up of State Infrastructure and Industrial Development Corporation of Uttarakhand, i.e., SIDCUL Ltd. industrial area in 2002, there has been a shift in the livelihood options of the local people residing in the area and the surrounding village. The state-sponsored and supported industrial development drive is now in its second phase of expansion. Under the new industrial policy that aims to provide an encompassing framework to enable sustainable industrial development in the state of Uttarakhand[5] will boost an investor-friendly milieu to encourage a rapid pace of industrial development in the region and generate employment opportunity for the local populace and widen resource base of the state making it more resilient.

Development and improvement of transport network connecting villages to SIDCUL and other parts of the district are quite visible as one moves across the area enabling better physical and social accessibility along with the mobility of people and freight across the region. There is no denying the fact that with SIDCUL there has been a shift from the farm-based workforce to companies and mills developed under the sustainable industrial development drive undertaken by the state of Uttarakhand. Furthermore, in the context of rural development initiatives, the effective interventions by public and private institutions, such as the National Bank for Agriculture and Rural Development (NABARD), Institute of Social Development (ISD), GB Pant University of Agriculture and Technology outreach programs, CSR initiatives of private companies, and also government initiatives, such as Uttarakhand State Rural Livelihoods Mission,[6] etc. are making a meaningful difference in the process of rural development and promotion of sustainable rural livelihoods. Similarly, the role of institutions—formal and informal could make a meaningful difference. Arrow (1974) has noted that there are institutions that serve all economic purposes that are neither the market system nor the state. Nisha and Asokhan (2015) studied the Nilgiri District of Tamil Nadu, in India, about the livelihood development and upliftment of tribal women vis-à-vis the communities. Purposive selection of Nilgiri District was done because it has a high percentage share of tribal population. From the selected two blocks of this district, they picked four revenue villages of each block. The study noted that developmental programs are helping these women immensely. The impact of these livelihood developmental programs is visible in the rising educational, economic, and innovativeness in adopting the technology. A positive sign can be seen in joint decision-making which has improved the self-confidence and leadership-

[5] Sector-wise contribution to GSDP of Uttarakhand reveals the increasing contribution of the secondary and tertiary sectors in recent years. The contribution of the primary sector, 14% during 2004–05, has come down to 13% in 2015–2016. Whereas the contribution of the secondary sector recorded at 52% in 2011–2012, has come down to 46% in 2015–2016. On the other hand, contribution of the tertiary sector which is reported 34% in 2011–2012 has gone up to 40% in 2015–2016 (des.uk.gov.in)

[6] Uttarakhand State Rural Livelihoods Mission, Government of Uttarakhand (n.d.) Accessed on http://usrlm.uk.gov.in/pages/display/57-livelihoods

skill of these tribal women. The current chapter draws on the experiences of Tharu women of Audali village and their means of livelihood. In this backdrop, it is important to throw light on how they perceive the changing dynamics of interrelationships existing between the natural and social capital, and of course their own identity. The positive impact of institutional support on the lives of tribal women of Audali is worth sharing.

21.5 Data and Methods of Analysis

This work is based on both primary and secondary data. Initially, a transect walk and village survey were done to understand the village morphology and profile. It is worth noting that given the purpose of this chapter the focus is only on a specific aspect of rural life and, hence, presents a partial view of the tribal community. In the current context, we have drawn largely from our observations and information collected via semi-structured interviews and focus group discussions. Snowball sampling technique was used to select the sample group.

For socio-economic parameters, we have used data from the Census of India, Primary Census Abstract (2011). Besides, information related to self-help groups and livelihood initiatives undertaken in Audali is collated from the ISD website, and other related information is extracted from the state government sources.

21.6 Results and Discussion

The change in the specificity of the region is triggered by several geographical and non-geographical factors, and therefore, it is quite interesting to locate women in the larger context of sustainable rural livelihood. Audali is unique in the sense that it presents a good example where motivated women can make a meaningful difference at the household and community level and can be part of the decision-making process at the grass-root level. The chapter will sketch out a narrative and experiences of the group of women who with the support of their family and institutional support took a bold initiative and paved a way for others to join. However, the challenges they face are real and need to be handled effectively at different levels. The initial interaction in the village started with our main contact Sunita. She seems to be an ordinary woman with an extraordinary will to grow and is determined to create her own identity. The pride with which Sunita and her family shared their ancestral lineage seem to fade in a hesitant smile while sharing how the family and others in the village lost their land due to flooding of the river, along with increased control of the non-tribal elites on their agricultural lands, particularly by Punjabis. In Udham Singh Nagar, the Sikh community has tremendous economic dominance as owners of large farms and industries, and their multiplying number is quite pronounced in the district. Subhash (1998) reported that Udham Singh Nagar is often referred to as "little Punjab" or the "Canada of U.P." In this context, Robinson (2001) has written "Over the decades since partition, many other Sikhs and Hindu Punjabis (from within India) have also resettled in the Tarai. Some of these Punjabi settlers, especially those who are influential or well connected, have also acquired vast tracts of farmland (some have thousands of acres). In 2000, many of the Tarai's original inhabitants, the Tharu and Bhuksa tribals have been reduced to the status of small-time farmers, or landless laborers who now work for others the land that they once owned." However, as the Punjabi settler community has a strong political patronage network, and with the lack of an effective political organization for the tribal community to voice their interests, the chance for the tribals to secure adequate redress was meager. Furthermore, the low level of literacy among the tribals in the Tarai has made it even more challenging for them to successfully protect their claim to land that was once theirs.

The power dynamics have changed in the village and so is the socio-economic mosaic of the village spatial economy. Therefore, once owners of the land irrespective of the size, Tharus are now left with either small landholdings or no landholdings and are forced to work as paid manual labor for rural elites during the sowing and harvesting season. Alternatively, working in the mills/companies located in the SIDCUL industrial area. The shift in the means of livelihood with the changing socio-economic structure of the village and rapid urbanization process has also led to a more pronounced gendered division of labor. Men of the household are moving away towards wage paid manual labor—farm or nonfarm. Quite many Tharu men across age groups started working in SIDCUL, Sitarganj. It was interesting to see that with the change in the economic base of the Audali village, the community has distinctly resecured their livelihoods amidst changes to the economic profile of their household and emerging rural-urban interlinkages with the rapid pace of the urbanization process. It is seen in the case of Audali village dominated by Tharu tribe that women have successfully expanded their spaces of activity and continue to get trained in traditional craft skills. However, they are still confined to what Raju (2011) calls "feminine tasks. As women enter paid employment and significantly expand their activity spaces, they continue to get trained in traditional crafts, remain bunched in 'feminine tasks."

The village has considerably benefited from the institutional support in different forms and scales over the years. For example, NABARD has always been in favor of helping people especially rural women and youth in finding alternatives of livelihood opportunities in the non-farm sectors, such as skill development, rural entrepreneurship in various fields of activities, handicrafts, handlooms, Institute of Social Development (ISD), an Indian Social Organization, in collaboration with NABARD and TATA Co. encouraged women of the villages to create a Self-Help Group (SHG) in Udham Singh Nagar district. In this direction, Venkatesh (2009) emphasizes the significance of microfinance and SHGs in a diverse country like India with glaring socio-economic hierarchies. SHGs and micro-finance have become one of the factors to bridge the existing gap and help the rural poor, weaker sections of the society. It was mentioned that the concerted effort of the society and government is reflected in the multiplying number of SHGs, and steep rise in loan disbursal by the government under the development scheme. According to the annual report of NABARD, there is an increase in the number of SHGs from only 255 in 1992–1993 to 620,109 in 2005–2006. These initiatives have worked well and, still, contribute to empowering the rural population across the country.

In Audali, GB Pant Institute's intervention program has also been instrumental in imparting knowledge and awareness among tribal women to further enhance healthy, medicinal, and high yielding agri-products through its training sessions and encouraged SHGs (Box 21.2) in the village to adopt traditional trades for diversification of livelihood means and income generation. Broadly, it can be inferred that all these institutions together have helped and encouraged women of Audali village to understand their entrepreneurial capabilities, optimize the use of materials collected from nature, and benefit from working as a collective that will boost the preexisting community feeling or participatory approach at the individual and community level.

Box 21.2 Self-Help Groups(SHGs) in Audali

There are about nine self-help groups in Audali and all of them have female members only and the total number varies from 10 to 14 per group. Among the existing SHG, Jamuna stands out with the persistent effort to keep the SHG active and functional. Jamuna SFG is an ideal and extraordinary example of a group of visionary and brave women in a typical Indian rural setting. Sunita is the catalyst of the SFG Jamuna, and it is important to note that she has been the village head in the past. The group has 10 woman members, and the reason that brought them together is social homogeneity and proximity. The major source of livelihood is predominantly wage manual labor.

Farming on its own, however, still of significance, as Sunita said "Farming is increasingly becoming insufficient means of sustenance for us. Hence, the idea of diversifying means of livelihood got the center stage of our interactions over common social space. Therefore, we started looking for possible diversification of livelihoods, to encourage independent income-generating capabilities." She further adds, "the handicraft skill is acquired by our succeeding generations and have practiced it over generations for our own use. However, with the changing occupational structure and socio-economic profile at the household and community level, we are considering taking it up as an income generating livelihood option." In this connect with sparkling eyes Sunita and other Tharu women said, "The arrival of the ISD in Audali brought a new ray of hope and things started to happen." They have been forthcoming in communicating their need-based training programs, for example, they requested ISD to organize skill development program particularly for handicraft products, as they are already doing it with some traditional skills and with some more formal training, they wanted to use their inherent skills as means of livelihood and income generation. ISD officials also encouraged these women how they can use their free time after doing household chores to make handicraft products and start selling them in local *haats* (markets) to begin with.

Given the enthusiasm and determination of these women, ISD organized skill development programs in Audali on January 2018, and about 50 women participated in the training sessions. The opportunity has differential impact across women group. Primarily because of the lack of the stable remunerative prices for the handicraft products, women members of the eight SHGs (excluding Jamuna) did not pursue it further as a yearlong alternative livelihood and continue to work as causal wage labors on farms. Other possible factors, such as the differences in their relative role in household, stage-of-life identities, class position of their respective households, abilities to mobilize social capital, and access to earnings might have also affected the decision-making capacity for diversification of livelihood in the context of these women. A female member of the SHG, who wish not to be named, said "I have a family of five without any landholding. My husband and myself are daily wage laborers. He earns 250–350 INR and I get 200–250 INR per day. With this amount we could barely sustain ourselves. The vocational training was organized, and it was useful, but it would be extremely difficult for me to manage both activities simultaneously, i.e., farm-based work that is the major means of livelihood for my family and handicraft work, besides the household chores. It is risky and time consuming. Although I have my husband's support, but we lack capital. This activity requires investment initially and the return will come only once we sell the entire produce. The market for the handicraft product is unreliable and it is difficult for me to take the risk at this stage. There is no surety and fixed amount that we will earn by selling these products as it is purely demand driven and market size does make a lot of difference in the profit that we will earn. On the other hand, the daily wage work is getting us some money to support our household on day-to-day basis, and it is crucial at this point of time as my children are growing". Discussions with other womenfolk in the village brought forth similar concerns. It was predominantly highlighted in their narrative that the absence of perennial source of employment has resulted in confining them to the less productive agriculture and other non-farm activities. Therefore, the skewed participation in the alternative livelihood options under SHG mainly reflects the complexity inherent in the livelihood and sustenance issues in the household and village economy, then just being a matter of choice (Table 21.1).

In this backdrop, it is important to note that Jamuna SHG is the only group that has pursued their determination to diversify their livelihood options. The group could actively function primarily due to the juxtaposition of multiple factors, such as social homogeneity, proximity, a perennial source of household income from the diversified means of livelihood, access to financial capital, strong social network, entrepreneur skills, political participation at the grass-root level, etc. The members used training skills in

Table 21.1 Details of Self-Help Groups (SHGs), Audali

S.no.	Name of SHGs	Number of members			Name of Bank	Year of opening Bank account
		Male	Female	Total		
1	Jamuna	0	10	10	UGB, Sitarganj	20.05.2014
2	Maheep	0	10	10	UGB, Sitarganj	22.05.2014
3	Savita	0	14	14	UGB, Sitarganj	22.05.2014
4	Gayatri	0	11	11	UGB, Sitarganj	22.05.2014
5	Shakti	0	11	11	UGB, Sitarganj	23.05.2014
6	Jagrity	0	10	10	UGB, Sitarganj	11.07.2014
7	Raaj	0	12	12	UGB, Sitarganj	11.07.2014
8	Kalpana	0	10	10	USDC, Sitrganj	24.09.2014
9	Pavani	0	10	10	USDC, Bidaura	18.11.2015

Source: http://www.isduk.in/shg.php

making handicraft products for commercial purposes. They did concede that with the training they could make more quantities of handicraft products in relatively less time. It is worth noting by March 2018 which is within a short period since their initial skill development training program (i.e. in January, 2018) they could manage to sell their products in the localized haats held in the proximate areas, such as Pantnagar, Ram Nagar, and TATA company. Besides, organic produce like garlic, lentils, and turmeric they also sold handicraft products, such as baskets, mirrors, dustbin, mat, etc. They also concede that earlier they relied heavily on the spatially localized resources for the small-scale production, but with the increasing scale of production, they are procuring raw materials from markets in Bareilly district of Uttar Pradesh. The handicraft work is primarily confined for the period from the month of February to October as they have no other farm-related activities. They can now make a variety of handicraft products, such as mat, vase, basket (different shapes), mirror, dustbin, vegetable stand and stool, etc. The members are shareholders in the group which have given them a sense of empowerment and achievement. Further, the profit earned is shared amongst the producers and the balance goes to the share reserves of the SHG. It is viewed as mutually beneficial engagement (Figs. 21.2 and 21.3).

Although the present scale of production is small, Sunita and other women members feel a sense of pride and are determined to further increase the scale of production and extend the market area that goes beyond local and regional areas. When asked about how much this activity contributes to their household income, Sunita replied promptly, "My family owns the land and, of course, land is in the name of male member of the family, like any other household of the village. My son earns about 75,000 INR per year, and his income from farming is above 1 lakh per annum approximately. The handicraft work is very recent, and till now we have sold our products in three "Kisan Mela" organized in Pantnagar. The profit from the sale is distributed equally among members and 10% of it goes to "*Krishi Utpadak Sangh*." We will use the saved amount to buy raw materials in the future. I got around 9000/– INR from all three sales. However, farming is the major source of income, but it is seasonal, so my son's salary is utilized for daily expenditures and acts as a perennial source of income for my family. Income from SHG Jamuna is an additional source of income, which is obviously helpful."

Sunita has high expectations from this group and if it works, it will become a better livelihood option for women of this village. She further adds, "though not every member of the SHG is active, they are now willing to participate with rigor. The hesitation is mainly due to the lack of money in the account of the group. I have organized a meeting in January 2019 and encouraged members and other womenfolk in the village to participate in making handicraft products. Given the lack of capital, we all decided to contribute about 100 INR, and to which all members and non-members agreed. We have ensured transparency in the functioning of the group and all

Fig. 21.2 Handicraft products made by Tharu women
Source: By First Author

members will equally share the profit. We are now targeting city markets to sell our products at suitable prices."

Jamuna SHG presents an encouraging example of how with meaningful institutional intervention, women can play an active role in the livelihood diversification that is sustainable in nature. It is important to highlight that these women can manage to work effectively and get tremendous support from their family members—particularly their male counterparts. Be it procuring raw materials, processing, making, or selling the product in the market. As noted earlier in the chapter, the scale of activity is very small, and at the initial stage, therefore, other villagers and local leaders seem to not care much about it. Further, it will not be wrong to say that this activity is perceived as a non-threatening exercise to others, so no opposition was seen from non-participating rural folks. Yet there is much that needs to be done and understand in this direction.

The experiences drawn and narrated in the paper pertain to a small group of tribal women (Jamuna members) who were led by an informed, empowered, and literate women leader. This is an example of how women with decision-making roles can positively influence the socio-economic profile of the household and village for the betterment of the community, and, of course, sustainable livelihood. Although there are also challenges faced by these women, and if appropriately managed might optimize the institutional support and women's role in the diversification of rural livelihood. Following are a few of the challenges faced and reported by them: (1) Lack of a stable market and remunerative prices, (2) Absence of better marketing facilities, (3) Need to sensitize self-help groups to market dynamics, (4) Require systematic alignment with the value

Fig. 21.3 Tools used for making the handicraft products by Tharu women
Source: By First Author

chain, and to improve forward and backward linkages, (5) To enable rural women to sell their products throughout the year, (6) Need to improve and expand existing livelihood means and identifying, and (7) Identifying New opportunities within the key livelihood options existing in the village.

Based on the micro-level case study, broadly, it can be inferred that livelihood is indeed a gendered process and is based on the gendered ownership of livelihood assets, including disparities in access to institutions, information, and technology (Krishna 2004). Hence, the likelihood of success is more, when (1) decision-making and initiatives are locally controlled in a bottom-up framework, (2) the activities are strongly linked to the existing knowledge and skill base of the women as well as the locally available resources, and (3) external interventions are often crucial in bridging the skill-gaps and expanding the scope for livelihood diversification through enabling market access.

21.7 Findings and Conclusion

In the backdrop of the discussions, it is essential to locate women in the rural livelihood context, and more concrete effort is required to bring women at the center from the periphery of the development process. There is an urgent need for improvement in the social norms and environment that often act as barriers to women's motivation to come out of the stereotypical restrictive norms and mindset. It is established in the literature how active and productive participation of women in the diversification of livelihood has infused positive change in the socio-economic development at the household as well as the community level. Further, it leads to an improvement in the resilience of household livelihoods to ensure sustainable availability of food and necessities. The effective intervention of formal and informal institutions helps reduce the structural vulnerability of the means of livelihood by focusing on rural women. The micro-level experiences of Tharu women presented in the chapter have reiterated how crucial is their role in the ongoing development drive at different spatial scales.

It is important to note, that the scope of the study is limited, and it requires further detailed analysis to understand the complexity of the issue, and to capture the underlying socio-economic dynamics existing in Indian villages. A comprehensive research is needed on how to optimize the incentives of benign initiatives, such as SHGs, and to look for other alternatives to improve the condition of tribal women in rural spaces. In the case of Audali, apart from handicrafts, though at a small scale, they have started commercializing produce from their kitchen garden, such as selling turmeric, lentils, garlic, etc. Women of this tribal village are making an honest effort to find a ground that would provide them better options to get a balance between household responsibility and encourage sustainable livelihood options. The examples given in the chapter are not intended to act as the basis for any generalization. It is a preliminary attempt to throw light on how informed and empowered women can contribute positively to shaping the sustainable livelihood in villages of India and will hopefully encourage further detailed studies in this direction (Fig. 21.4).

Figure 21.5a–d provides further insight into the occupational structure of Udham Singh Nagar and Sitarganj from the gender perspective. Figure 21.5a shows the distribution of total workers (main and marginal) in four major economic activities. It is seen that at the district level, about half of the male workers are engaged in other

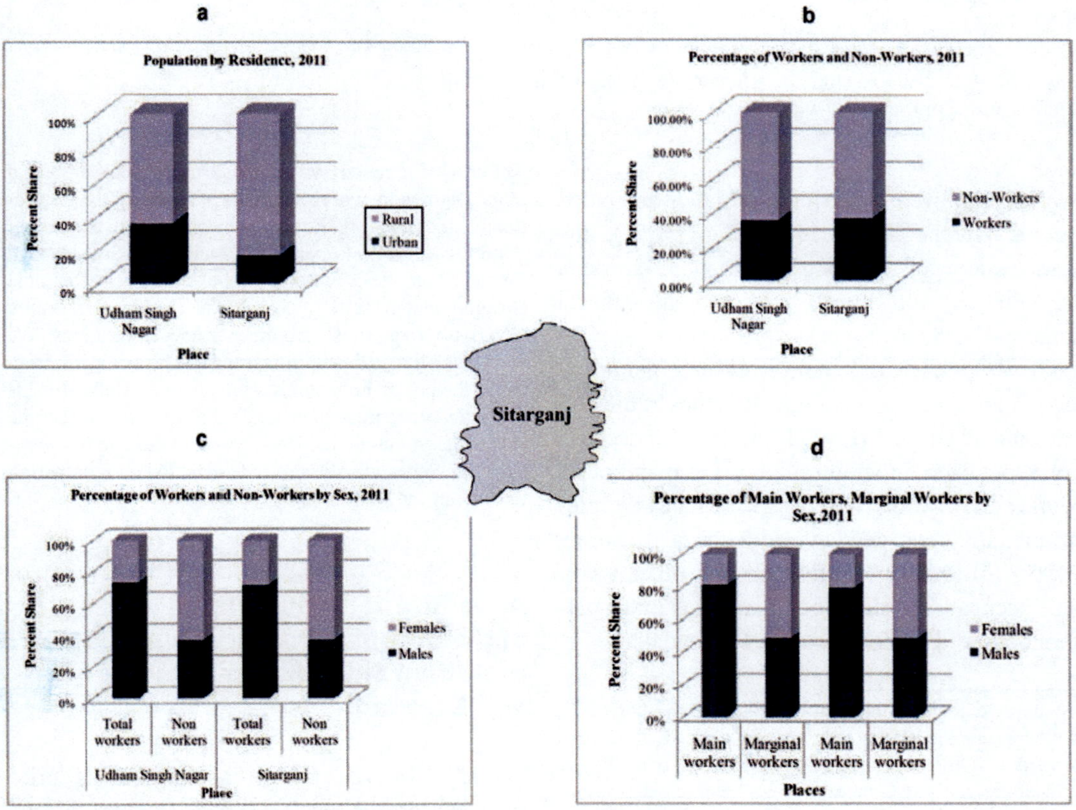

Fig. 21.4 Workforce participation (Udham Singh Nagar and Sitarganj)
Source: Based on Census of India, 2011

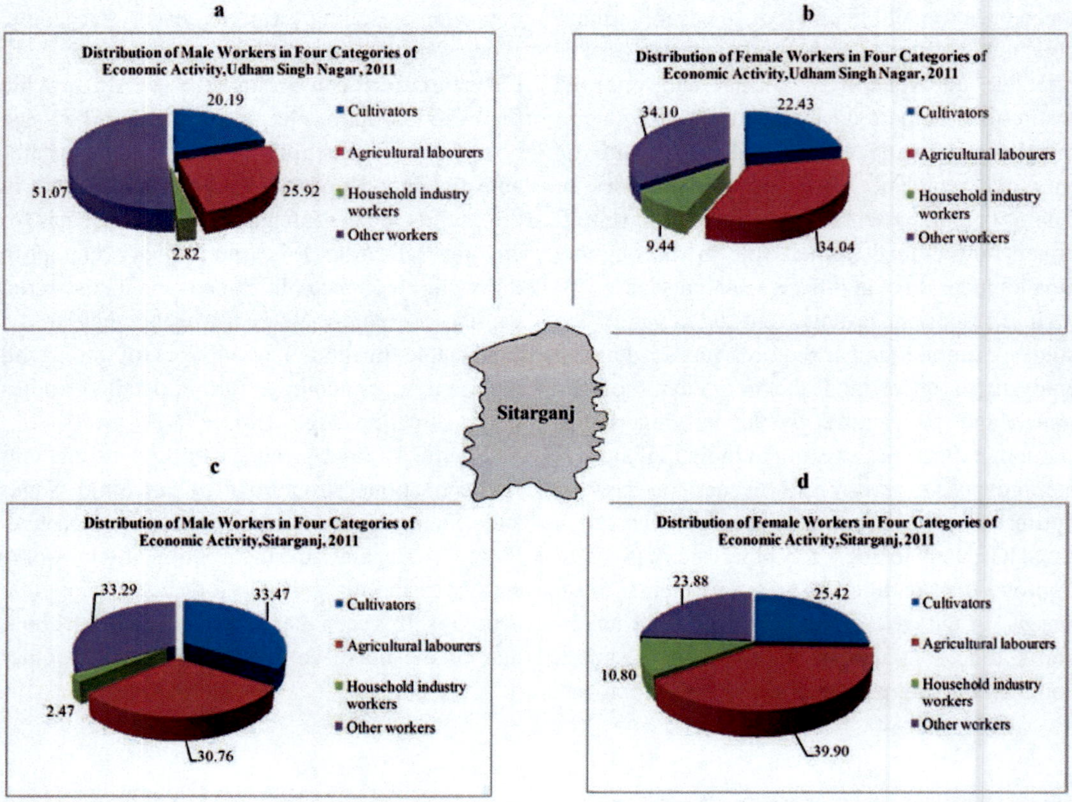

Fig. 21.5 Occupational structure (Udham Singh Nagar and Sitarganj)
Source: Based on Census of India, 2011

works,[7] followed by agricultural labor and cultivators. At the Tehsil level (Fig. 21.5c), the females are not only engaged as other workers but also as cultivators (33%) and agricultural laborers (31%).

If we look at the female workforce participation in these major economic activities at the district and tehsil level (Fig. 21.5b), females account for about 34% agricultural labor force and other workers. However, in Sitarganj, females are predominantly engaged as agricultural laborers (40%) followed by cultivation and other work. The occupational structure broadly fits in the pattern observed at the micro-level in Audali.

[7]Workers other than cultivators, agricultural laborers, or workers in Household Industry, as defined above are termed as "Other Workers" (OW). Examples of such type of workers are government servants, municipal employees, teachers, factory workers, plantation workers, those engaged in trade, commerce, business, transport, banking, mining, construction, political or social work, priests, entertainment artists, etc. (Census of India, 2011)

Acknowledgements We are thankful to Mrs. Sunita and other Tharuwomen of Audali village for being the inspiration behind this work. Further we would like to extend our gratitude to Prof. Deepak Mishra and Prof. Sachidanand Sinha from the Centre for the Study of Regional Development (CSRD), Jawaharlal Nehru University (JNU) for their useful comments and suggestions. We would also like to thank Professor Baleshwar Thakur from Delhi School of Economics, University of Delhi, for his consistent encouragement and helpful comments on the chapter. Last but not the least, we would like to acknowledge Ms. Sudeshna Mitra from CSRD, JNU, for preparing the map of the study area.

Annexure 1: An overview of the occupational structure of Udham Singh Nagar and Sitarganj

Figures 21.4a–d and 21.5a–d present a brief snapshot of the occupational structure of the Udham Singh Nagar and Sitarganj. Figure 4a reveals that the share of rural population to total population is higher than the urban for Udham Singh Nagar and Sitarganj. The figure presents a comparative picture of the workforce profile. Figure 4b shows that the percent share of non-workers is greater as compared to the total workforce in both district and tehsil. The gender composition indicates that at both the spatial units, females predominantly constitute the non-worker category with about 80%. Similarly, Figure 4d reveals the gendered division of main and marginal workers. It is quite clear that the females participate mainly as marginal workers in the labor market than as main workers. This is in line with the stereotypical pattern of the gender composition of the workforce at various spatial scales.

References

Agarwal B (1989) Rural women, poverty and natural resources: sustenance, sustainability and struggle for change. Econ Polit Wkly 24(43):WS46–WS65

Agarwal B (2001) Participatory exclusions, community forestry, and gender: an analysis for South Asia and a conceptual framework. World Dev 29(10):1623–1648

Aggarwal B (2018) She is the answer. The Indian Express, December 10. Weblink: www.indianforest.com

Arrow K (1974) The limits of organization. W.W. Norton and Company, New York

Barbora S (2002) Ethnic politics and land use: genesis of conflicts in India's north-east. Econ Polit Wkly 37(13):1285–1292

Chambers R, Gordon RC (1991) Sustainable rural livelihoods: practical concepts for the 21st century, IDS discussion paper 296, December

Dasgupta P (2001) Human well-being and the natural environment. Oxford University Press, New Delhi

De Haan L, Zoomers A (2005) Exploring the frontier of livelihoods research. Dev Chang 36(1):27–47

Ellis F (1999) Rural livelihood diversity in developing countries: evidence and policy implications. Natural resource perspectives, number 40, Department of International Development

Fernandes W, Melville P (2005) Land relations and ethnic conflicts: the case of Northeast India. North-Eastern Social Research Centre, Guwahati

Ghatak P (2013) Changing livelihood pattern of Savara and Lodha of West Bengal. J Indian Anthropol Soc 48:177–189

Hogan A, Stewart L (2013) The coupling of rural communities with their Economic Base: agriculture, localism and the discourse of self-sufficiency. Policy Stud 34(4):441–454

Krishna S (ed) (2004) Livelihood and gender: equity in community resource management. Sage, New Delhi

Mishra DK (2007a) Gender, Forest, and livelihoods: a note on the political economy of transition in North-East India. Soc Change 37(4):65–90

Mishra DK (2007b) Rural non-farm employment in Arunachal Pradesh: growth, composition and determinants, NLI research studies series, no. 07.5/2007, VV Giri National Labor Institute, NOIDA, Uttar Pradesh, India

Mishra N (2018) Women and Forest: changing gender relations among the Kondhs in Odisha. Soc Sci Spectrum 4(2):86–93

Mishra DK, Upadhyay V (2007) In the name of the community: gender, development, and governance in Arunachal Pradesh. In: Krishna S (ed) Women's livelihood rights: recasting citizenship for development. SAGE Publications, New Delhi

Mitra A, Mishra DK (2011) Environmental resource consumption pattern in rural Arunachal Pradesh. Forest Policy Econ 13(3):166–170

Murray C (2001) Livelihoods research: some conceptual and methodological issues. Chronic poverty research Centre, working paper no. 5, September

Nisha R, Asokhan M (2015) Empowerment of tribal women through livelihood development. IOSR J Humanit Soc Sci 20(8):6–10

Poffenberger M (1990) Joint Management for Public Forests: experiences from South Asia, in research policy for community forestry: Asia-Pacific region (158–82). Regional Community Forestry Training Center, Bangkok

Raju S (ed) (2011) Gendered geographies: space and place in South Asia, 1st edn. Oxford University Press, New Delhi

Robinson J (2001) Regionalizing India: Uttarakhand and the politics of creating states. South Asia J South Asian Stud 24(2):189–212

Sengupta M (2018) Gender, tribe and development: a case study of the Reang tribal Community in Tripura (Northeast India). Asian Ethnicity 19(1):106–126

Shangpliang RM (2012) Forest legislations and livelihood strategies: Khasi women in rural Meghalaya. Sociol Bull 61(3):479–493

Shimray UA (2004) Women's work in Naga society, household work, workforce participation and division of labor. Econ Polit Wkly 39:1698–1711

Subhash T (1998) Plainsmen fear minority status in hill state. Times of India, 13 July

Sunderland T, Achdiawan R, Angelsen A, Babigunira R, Ickowitz A, Paumgarten F et al (2014) Challenging perceptions about men, women and forest product use: a global comparative study. World Dev 64(SI):S56–S66

UNDP (2011) Human Development Report, sustainability and equity: a better future for all. Palgrave Macmillan, New York

Venkatesh R (2009) Micro-finance and self-help groups. Catalyst for rural transformation. Indian J Polit Sci 70(1):131–138

Land Degradation and Agricultural Sustainability in Kendujhar District, Odisha

22

Rajesh K. Abhay and Punyatoya Patra

Abstract

Land degradation is one of the important challenges in achieving the agricultural sustainability in a region. Land degradation affects soil fertility and reduces agricultural production. In this way, productivity and profitability from the agriculture and food security of the region become a challenging task. The economy and livelihood of people of Kendujhar District, Odisha depend much on agriculture and associated activities. The first two sustainable development goals of United Nations on 'end poverty' and 'zero hunger' are closely linked with agricultural sustainability of the world. Therefore, to achieve agricultural sustainability in a region, it is needed that the land should be sustainably used so that it remains productive for a long period. The present paper investigates these issues in the north-eastern parts of Odisha, where land degradation is the function of socio-economic activities. Some of the factors, which are indicators of agricultural development, have turned up into responsible factors for land degradation. The study analyses these factors and suggests ways to achieve agricultural sustainability in the region. Increasing resilience of land has been seen as a future approach to solve this problem. The study concludes that resilience approach can improve the current situation and can contribute to achieve the sustainable development goals.

Keywords

Cropping Intensity · Irrigation · Resilience · Soil Erosion · Sustainability

22.1 Introduction

Modern agriculture faces many issues for its sustainability. Degradation of land resources is one of the prime environmental issues while its sustainability for agriculture is another major concern in India. Agricultural use of land is more than one-third, which is largest on the terrestrial surface of the Earth (about 38%). In the present situation, agriculture is a major driving force behind environmental problems like land degradation, climate change, deforestation, biodiversity loss, groundwater depletion, and so on (Foley et al. 2011). Agricultural expansion is taking place to feed the increasing population of the World and, consequently, due to which various other problems are faced by human beings. In

R. K. Abhay (✉)
Department of Geography, Dyal Singh College, University of Delhi, New Delhi, India
e-mail: rkabhay@dsc.du.ac.in

P. Patra
Department of Geography, Aditi Mahavidyalaya, University of Delhi, Delhi, India

B. Thakur et al. (eds.), *Resource Management, Sustainable Development and Governance*, Sustainable Development Goals Series, https://doi.org/10.1007/978-3-030-85839-1_22

addition, the major problems of agriculture are land degradation and soil erosion, which are now considered to be serious problems worldwide (Hitzhusen 1993). The exponential increase of population with limited land and water resources has created great concern in the World today. An increase in degraded lands threatens the food security of regions also (Conacher 2001). However, the sustainability of land depends on its optimum utilization.

Land degradation is the process initiated by human interference, in which the quality of land reduces to its lower grade. The quality reduction of land may be one degree, two degrees, or many more depending upon the time and intensity of the degradation process. The reduction of grades can be measured through production or usefulness. When the quality reduces, the production or usefulness also decreases. The land is fixed; therefore, there is no alternative option except to take remedial measures to normalize the process of land degradation. It helps in enhancing the food production and restoration of fragile ecosystems as well. Land management is important for several reasons. It is helpful in meeting the needs of the growing human population for food, fodder, biomass energy, and timber. It helps in biodiversity preservation and improves economic development and so on (Daily 1995). The adoption of various management practices enhances the resilience capacity or the ability of the land system to return to its pre-altered state following the change. If the management techniques are locally suitable, applicable, and cost-effective, they are easily adopted by the local land users and, therefore, will contribute to the agricultural sustainability in the region.

Resilience is the characteristics of the property of a system to absorb change and utilize change, including resistance to a shock. According to Stocking and Murnaghan (2001), it refers to the ability of a system to return to its pre-altered state following the change. Further, it has also been defined as the 'ways in which people adapt to changing circumstances to get by and "make do" through the exercising of the autonomous initiative' (Katz 2004). Therefore, it can be emphasized that the resilience of land means that

Box 22.1 Sustainable Development Goals

Like many other places in the world, agriculture in rural Odisha of India provides livelihood to a large population. Unfortunately, agriculture in Kendujhar district of Odisha is facing challenges because of land degradation. This challenge is crucial as harnessing agriculture is significant for achieving sustainable development. Land degradation adversely affects crop production and reduces the production potential of existing fertile lands. The increasing degradation is intensifying desertification in Kendujhar district of Odisha leading to a decline in ecosystem function and productive capacity, as well as deterioration of soil health by affecting the associated biodiversity, natural ecological processes, and ecosystem resilience. The authors in this chapter recognize that agriculture and its allied sectors are crucial to the achievement of the 17 SDGs adopted by the United Nations referred to as Agenda 2030. More significantly, the SDGs are interconnected in many contexts and a link with agriculture is clear for many of them. As the poor and vulnerable population of this region are dependent on agriculture, ending poverty in all its forms is linked to increasing returns from agriculture. To that end it is important that ownership and control over land and natural resources are significant indicators which are also essential endowments for practicing sustainable agriculture. Thus, while SDG 1, 2, 3, 5, 6, 7, 8, 10, 12, 13, 14, 15 are all collectively linked with agriculture, the authors recognize that SDG 1, 2, 13, and 15 better inform sustainable agriculture and land management by articulating the status of poverty, hunger, climate change, and land degradation. Urgent and concerted action will be required to avoid worsening land degradation. Managing land resources is critical for ensuring our vision for sustainable food, livelihood, and agriculture and this can be best addressed by stakeholders at every scale putting their commitment to SDG through best practices.

how easy it is to restore the land. It means if a land system undergoes the degradation process, the response of land to human efforts to regain its fertility will be determined by the resilience capacity of the land. High resilience reduces the risk of serious degradation and vice versa. Therefore, it depends on the natural quality of land and resilience which refers to the human initiation to upgrade land and to which the inherent characteristics of land also respond. The resilience of land helps the farmers to restore the land capacity for beneficial production from agriculture. Therefore, resilience and agricultural sustainability are positively related.

Land resource has always been the backbone of agrarian economies because of its socio-economic benefits (Singh and Singh 2019). The local economy of the people of Kendujhar District (Odisha State) also depends on land for agriculture and minerals exploration. The present study seeks to evaluate the aforesaid discussed issues in the Kendujhar District.

22.2 Aims and Objectives

Based on the preceding discussion, the present chapter examines the existing condition of land degradation, its responsible factor in the study area, issues in achieving agricultural sustainability and evaluates the role of resilience approach in it. Therefore, an attempt is made to highlight the magnitude of the problem and suggestions are made to improve the condition for the sustainable livelihood of the people in the study area.

22.3 Geographical Profile of the Study Area

The Kendujhar District (earlier known as Keonjhar) is in the north-eastern part of Odisha State. It covers an area of 8303.43 km^2 and stretches between 21°1′ and 22°10′ North latitude and 85°11′ and 86°22′ East longitude. Administratively, there are 13 Community Development (CD) Blocks which are Anandpur, Banspal, Champua, Ghasipura, Ghatgaon, Harichandanpur, Hatadihi, Jhumpura, Joda, Kendujhargarh, Patana, Saharapada, Telkoi in the District (Fig. 22.1). Kendujhargarh is the district headquarters. The district is bounded by Mayurbhanj and Bhadrak District on the east, Jajpur District on the south, Dhenkanal and Sundargarh District on the west. The Kendujhar District makes a state boundary with West Singhbhum District of Jharkhand State on the north (Census of India 2011).

The Baitarini is the major river of the district that drains more than 75% of the area of the district. The major right bank tributaries of Baitarini river joining within the study area are Orarai *nadi*, Mermeda *nadi,* Kanjhari *nadi*, Kukurkata *nadi*, Musal *nadi* and Kusai *nadi*. The major left bank tributaries are Mohalda *nadi*, Bhirol *nadi* and Tel *nadi*. The soils of the region are mainly divided into four broad categories, namely, black soil, red soil, mixed black and red soil, and alluvial soil. The climate of the district is tropical, characterized by high temperatures and medium to high rainfall in most of the year. The district is the most thickly forested part of the state.

The study area has an agrarian economy where most people are engaged in agriculture and are mainly dependent on land-centred activities for their livelihoods like agriculture, forestry, mining, and forest products, etc. The agro-climatic condition, altitude, and soil type of Kendujhar District provide amiable conditions for growing a wide range of horticultural crops. Paddy is the staple crop which is grown in both the agricultural seasons (rabi and kharif). The production of paddy decides the food security of the farmers for the whole year and the economic condition in the region. Besides, Maize, Til, Niger, Arhar. etc. are also grown in the district. The mineral resources are rich and extensive in the study area, the important being iron, manganese, bauxite, china clay, etc. (Kendujhar 2020). The district is the home of about 18 lakh people with a population density of 217 persons per km^2. The literacy level is 69% and sex ratio are 987 females per 1000 males. The district has large proportion of Scheduled Tribe population (45%) (Census of India 2011). Transport network is moderately

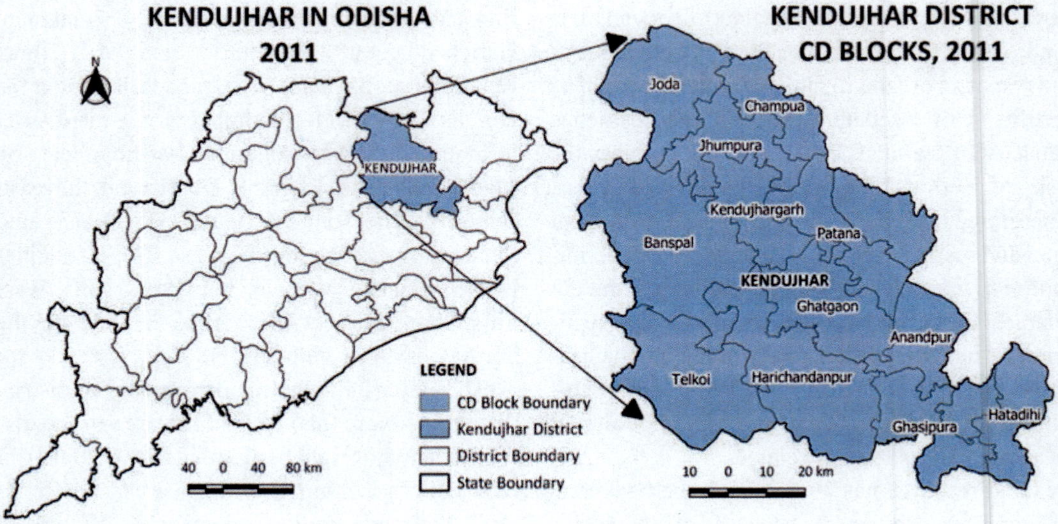

Fig. 22.1 The Study Area. Source: Prepared by Authors

developed in the district. The National Highway No. 6 passes across the region from west to east and the State Highway No. 10 is the main road for communication in the northern part. The communication network in southern part is totally backward. The social development is not much because people belong to a traditional-agricultural society.

22.4 Data Source and Methodology

The study is based on secondary source of data. The data has been collected through various reports, journals, Census Handbook, etc. The demographic data has been collected from the Census of India, District Census Handbook reports of 2011 for the Kendujhar District. The data on land degradation, soil erosion, cropping intensity, forest cover, irrigation, and agricultural land use have been collected from District Irrigation Plan (DIP) of Kendujhar, which was prepared by District Level Implementation Committee, Kendujhar in 2016. To represent the data in a better way, the maps for land degradation have been prepared based on data collected from DIP (2016). The block is the spatial unit which are 13 in number. All the blocks have been digitized using QGIS 3.8 software. The choropleth technique has been used to represent the data into various spatial categories. The collected data have been processed and classified into various categories. Further, the collected data are arranged, tabulated, processed, and analysed with the help of software like MS-Excel and QGIS software. Further, in the last section, suggestions are made to increase resilience which can improve their capacity to mitigate the impact of land degradation and help in achieving sustainability of land degradation.

22.5 Results and Discussion

22.5.1 Land Degradation Profile of the District

Land degradation is a serious issue for agricultural sustainability in the Kendujhar District. Odisha Remote Sensing Application Centre, Bhubaneswar (DIP 2016) estimated that 59.22% geographical area of the Kendujhar District is degraded. Out of the total degraded area, 98.65% region is under the slightly degraded, while 1.35% is severely, and the rest is moderately degraded (Table 22.1 and Fig. 22.2). After combining all the three categories of land degrada-

Table 22.1 Block-wise Status of Land Degradation in Kendujhar District

S. no.	CD block	Total geographical area	Area in hectare			Total (% of Block)
			Slight	Moderate	Severe	
1	Anandpur	36,824	30,133.65	–	38.12	30,171.8 (81.94)
2	Banspal	163,784	62,202.08	–	2112.54	64,314.6 (39.27)
3	Champua	37,382	27,180.03	–	218.71	27,398.7 (73.29)
4	Ghasipura	40,938	28,512.92	–	158.11	28,671.0 (70.04)
5	Ghatgaon	74,098	42,821.65	–	499.05	43,320.7 (58.46)
6	Harichandanpur	89,771	66,125.09	–	1798.07	67,923.2 (75.66)
7	Hatadihi	44,488	10,864.98	–	–	10,865.0 (24.42)
8	Jhumpura	56,657	34,254.21	–	862.87	35,117.1 (61.98)
9	Joda	51,900	32,042.44	–	59.2	32,101.6 (61.85)
10	Kendujhargarh	56,073	39,371.53	–	395.28	39,766.8 (70.92)
11	Patana	46,208	33,877.97	–	–	33,878.0 (73.32)
12	Saharapada	41,371	30,355.16	2.67	–	30,357.8 (73.38)
13	Telkoi	91,506	47,354.95	–	498.51	47,853.5 (52.30
	Kendujhar District	830,343	485,096.66	2.67	6640.46	491739.8 (59.22)

Source: District Irrigation Plan of Kendujhar, 2016, p. 32

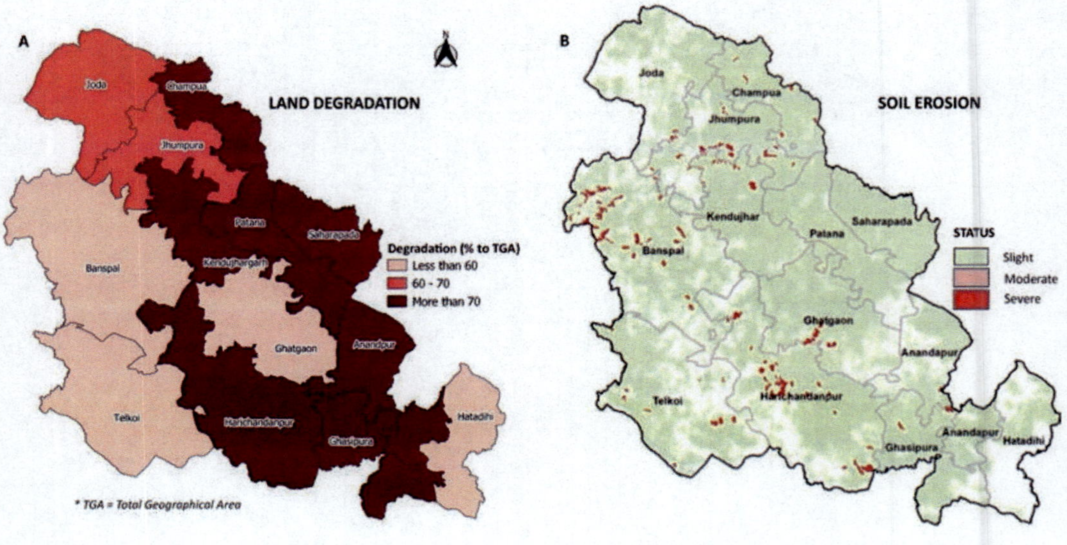

Fig. 22.2 Land Degradation in Kendujhar District. Source: Prepared by Authors and ORSAC based on District Irrigation Plan of Kendujhar, 2016

tion, it is found that the highest percentage of geographical area of Anandpur block is degraded (82%), while the least is in Hatadihi block (24%). Out of 13 blocks, 7 blocks experience more than 70% area under different forms of land degradation (Fig. 22.2b). These areas are mostly found on the eastern side of the district which is dominantly an agricultural area. The blocks in the western part (Banspal) and southwestern (Telkoi) have less degradation. The underlying causes of these spatial patterns are discussed in the following sections.

22.5.2 Factors Contributing to Land Degradation

There are many factors that affect land resources by aggravating the process of land degradation and create hindrance in achieving sustainable agriculture (Sharma et al. 2015). These factors reduce the capacity of land to produce effectively. The quality depends upon natural as well as anthropogenic factors. Natural factors like climate, slope, soil, drainage density, geological factors, forest cover, etc. and, on the other hand, population pressure, overuse of agricultural land, poor management, deforestation, government policies like anthropogenic factors contribute to increasing and decreasing the soil quality which ultimately lead to land degradation. With reference to the present study, four indicators, discussed below, are taken to analyse their spatial correlation with land degradation (Fig. 22.3).

22.5.2.1 Irrigated Area

Irrigated area is the area over which various artificial means of irrigating the crops are applied to grow crops. In the study area, major sources of irrigation are canals, ponds, tanks, reservoirs, tube-wells, and bore wells. Timely irrigation ensures higher production and productivity of lands. Assured irrigation facilitates farmers to take more than one crop from the same patch of land in different agricultural seasons. Due to the continuous use of land over time, it becomes less productive or degraded, if efforts are not taken to keep the land productive. It has been found in the study area that the net irrigated area (120,803 ha) constitutes one-third of the total cropped area of Kendujhar District (370,598 ha). It ranges from 10.5% (Banspal block) to 54.08% (Hatadihi block). It is found that the areas with a higher percentage of net irrigated areas like Patana, Kendujhargarh, Ghasipura, also show a higher degree of land degradation (Fig. 22.3a). On the

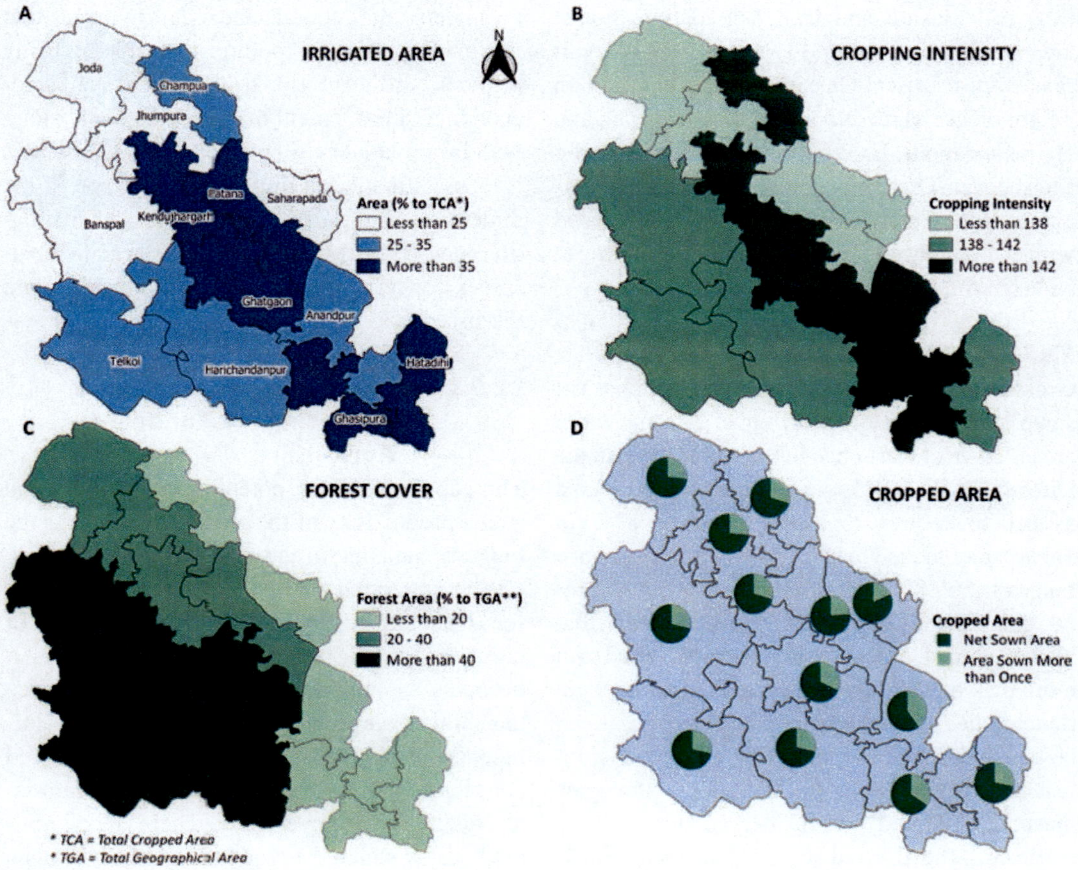

Fig. 22.3 Factors Contributing to Land Degradation in Kendujhar District, 2016. Source: Prepared by Authors

other hand, western parts of the district—Banspal, Joda and Telkoi blocks—have low irrigation as well as low percentage of land degradation. It concludes that irrigation is high, but the farmers are not making sufficient efforts to maintain their land, and thus, experiencing the problem of land degradation.

22.5.2.2 Cropping Intensity

Cropping intensity is the ratio of gross cropped area to net sown area. High cropping intensity ensures that the net area is being cropped more than once during a particular agricultural year. Increasing cropping intensity is important as the expansion of agriculture in different areas is not viable because it will be at the cost of other land use categories. Simultaneously, the population is also rapidly increasing, therefore, the intensity of cropping needs to be increased with production as well as productivity. The cropping intensity of the district is 140.49. It ranges from 122.6 in Saharpada block to 166.3 in Anandpur block. It is very high (more than 142) in the blocks of Anandpur, Ghasipura, Ghatgaon, Champua, and Kendujhargarh. Among them, Champua, Kendujhargarh, Ghasipura blocks show high cropping intensity as well as high land degradation (Fig. 22.3b). Therefore, the cropping intensity emerges as one of the important factors in increasing the intensity of land degradation in the study area. Thus, there is a need to pay proper attention to soil fertility with higher cropping intensity and maintaining yield gaps.

22.5.2.3 Forest Cover

Forests are very important to check soil erosion on steep slopes. Due to forest cover, soil remains intact and does not carry away and, hence, controls soil erosion and land degradation. Forest area of the district is 311,717 ha (37.54%). It ranges from 14% in Ghasipura block to 64.6% in Telkoi block. Three blocks (Telkoi, Banspal, and Harichandanpur) have more than 50% area under forest cover. These blocks also show less percentage of land degradation which can be correlated with the quantity of forest cover. Ghasipura can be taken for example.

22.5.2.4 Cropped Area

Gross cropped area includes the area which is net sown and sown more than once. The net sown area is sown at least once in the whole year. In the District, all the blocks have more than 60% area as net sown area to a maximum of 81% in Saharpada block. The area sown more than once ranges from 18% to approximately 40%. A careful examination of pie diagrams explains that blocks with a higher percentage of area sown more than once are spatially found in areas experiencing high land degradation. Therefore, it can be concluded that cropping the same area is not feasible in the district. Thus, suitable strategies must be developed to check degradation.

Besides the above discussed factors responsible for land degradation in the study area, mining activities and shifting cultivation also contribute. Mining activities in the Joda block and shifting cultivation in the Banspal block aggravate the problem of land degradation and puts sustainability on the threat. In other areas, rural population density and urbanization were found to be positive and significant determinants of land degradation (Baba et al. 2019).

22.5.3 Agricultural Sustainability: Challenges and Ways to Achieve

From the preceding discussion, many sustainability challenges have emerged related to land degradation and agricultural sustainability. The discussion has raised multi-dimensional challenges related to degradation. These challenges need to be considered by policymakers while planning for the agricultural sustainability of the area. Here, agricultural sustainability means that it provides sufficient food to its people by using resources, efficiently; and it can improve socio-economic development of the local people along with improving environmental quality. However, it looks difficult, but quite possible if one of the dimensions is ensured properly, then other situations can be easily achieved. Some specific challenges concerning agricultural sustainability are discussed below.

22.5.3.1 Improving Yield than Increasing Area under Agriculture

The gap between the potential productivity and actual productivity of the area is identified as the biggest challenge in the study area. Foley et al. (2011) suggested that to grow more food, the focus should be on increasing the number of crops harvested per hectare. In this way, it becomes an important part of agricultural sustainability as it promotes farmers to produce sufficiently for their livelihood security and contributes to improving the socio-economic status of the farmer over time. There are various methods by which the productivity of land can be increased. In doing this, the role of government is very important for being a facilitator in providing proper knowledge about the usage of ideal quantity of fertilizers, methods of nutrient fixation in the soil, keep the patch of land as fallow sometimes, regularly practicing crop rotation methods for increasing the agricultural productivity in the country or state in general and Kendujhar District. Asgher and Sharma (2018), through fertilizer yield analysis, revealed that with the increase in fertilizer consumption, the yield of crop per hectare increased in the Doda District of Jammu and Kashmir. But at the same time, it is recommended that the rational use of fertilizer is quite important. Otherwise, the overuse of chemical fertilizers can have negative impacts on the land and soil of the area. Therefore, the suitability of each method should be carefully examined and executed.

22.5.3.2 Irrigation Efficiency

In the previous discussion, it has been found that some areas with irrigation facilities are also experiencing higher land degradation. Therefore, in the Kendujhar District, irrigation facilities are not giving expected returns to farmers. Various studies have also affirmed that the benefits and impacts of irrigation are not evenly distributed and water needed for crop production varies greatly across the world (Foley 2011). Therefore, for agricultural sustainability, no doubt irrigation is required but efficiency of irrigation needs to be verified. Excess amount of water given to crops is not going to help in producing more, but the conjunctive use of water can help a lot in this regard. Further, the use of water-saving devices like drip and sprinkler irrigation is the need of the hour.

22.5.3.3 Soil Conservation Measures

Soil conservation measures are used to control soil erosion and prevent it by using curing methods where it has not reached its threshold limits. These include vegetative cover, engineering and agronomic practices, contour cultivation, bunding, graded bunding, contour strip cropping, etc. In the Kendujhar District, farmers are less aware of these conservation methods. Therefore, irrigation facilities are also not able to provide sufficient benefits. In this way, the adoption of soil conservation measures is another important challenge for agricultural sustainability in the region. There are many methods which can help in this regard. Forest cover is one of the important ways of this. The plant roots bind the topsoil and prevent it from being drifted so easily with water or air. The vegetative cover also slows up runoff water and causes much of it to percolate into the ground that promotes more vegetative growth. Further, engineering, and agronomic practices can be applied to counteract the erosive force of the water on the soil. These include contour cultivation, bunding, graded bunding, contour strip cropping, etc. used alone or in combination with terracing like bench terracing. When terraces are properly constructed and adequately supported by approved cropping and tillage practices, they prove to be one of the most effective erosion-control measures applicable to cultivated sloping lands. Terraces are helpful in the management of agricultural lands which are in the hilly area with moderate to steep slopes. The study area has mostly plateau topography, terraces are important at the margins. Terraces are also used to reduce the runoff and soil erosion. Terrace farming is currently becoming very popular in the hilly parts of the study region. Crop rotation is another one that is discussed in detail in the next section. Overall, the adoption of various soil conservation measures in the Kendujhar District is low. The farmers are using their indigenous knowledge to conserve soil, but this solution remains temporary. Therefore, agricultural sustainability in the district is very much dependent on soil sustainability which mostly depends on soil conservation measures.

22.5.3.4 Crop Rotation

In the Kendujhar District, paddy is cultivated in both the agricultural seasons of *kharif* and *rabi* on the irrigated farms. Cultivating the same patch of land, again and again, reduces soil nutrition and production also decreases. It has been mentioned that soil sustainability can be maintained by implementing the crop rotation methods (Ouda et al. 2018). In the study area, few villages practice it, but not in a scientific manner, therefore, the adoption of crop rotation methods is a significant challenge. Most people do not use these methods as well as their benefits. Rotating crops is one of the key principles of conservation agriculture. Crop rotation means changing the type of crops grown in the field each season or each year. It keeps the soil fertile with rich nutrients and plays an important role in increasing crop production as well as increases land productivity (Zohry and Ouda 2018). It helps in controlling weeds, pests, and diseases also. It has been found that the yields of crops were lower in the case of monoculture than in crop rotation (Zoltán et al. 2000). Therefore, a well-planned crop rotation makes farm management more flexible and provides enough scope in increasing or decreasing the area under different crops according to fluctuations in the physical, cultural, and ecological conditions, and socio-economic constraints. Thus, crop sustainability and profitability in long run can be ensured by crop rotation to a maximum extent.

22.5.3.5 Alternative Cash Crops to Increase Farmers' Income

There are various factors in the study area due to which farmers are getting low benefits from their agricultural fields. Like dependence on mines and minor forest products for extra income and illegal selling of valuable woods from the forest, villagers neglect their agricultural field. Further, sludge from mining areas also reduces the productivity of agricultural fields and the humus content of soil reduces due to deforestation (Abhay 2015). Therefore, poor income from agriculture does not motivate farmers to improve the quality of land. Thus, alternative agriculture during the off-season can contribute to increasing the per capita income of farmers. For example, after the rabi season crop (winter season from month of November to March), mostly agricultural land remains fallow or vacant due to low rainfall. This time can be utilized for those crops, which use minimum water and give maximum cash in hand. In the Kendujhar District, the cultivation of turmeric and ginger crops is suitable and should be promoted sustainably because there is very small area under these crops at present. These two crops require less effort and give better economic returns. The increased income will contribute to improving the farming practices, the social life of the farmers and their living standard. The earned money will also promote the farmer to improve their agricultural land for maximum economic returns and will support the economic dimension of agricultural sustainability.

Agricultural sustainability has space and time dimensions. In the long term, equal emphasis will be put on economic, environmental, and socio-institutional developments at national, regional, and local levels (Zhen and Routray 2003). The aforesaid discussed challenges and ways have a big role to play in achieving agricultural sustainability. Today, concerns about sustainability in agricultural systems centre on the need to develop technologies and practices that do not have adverse effects on environmental resources and, simultaneously, on economic returns along with improvement in food productivity.

22.6 Findings and Conclusions

In this paper, we have analyzed the severe impacts of land degradation in the Kendujhar district, Odisha. Land degradation is spatially widespread throughout the district, though it is most severe in the eastern part. Various factors like irrigated area, cropping intensity, and gross cropped areas are showing negative impacts on land resources and pose a challenge to the sustainability of soil and, ultimately, to the agricultural sustainability of the region. These factors should have contributed in increasing the production and productivity of land but they have become responsible for degrading land because of careless and inappropriate application. Therefore, various challenges have emerged in ensuring the agricultural sustainability of the region. The research revealed that challenges can be easily taken care, if suitable methods are adopted and implemented with the help of government policies and local farmers.

Further with reference to Kendujhar, resilience approach is recommended to land management, based on the farmers' viewpoint on which future studies shall be organised in different physio-socio-economic conditions. Actually, resilience is the characteristics or a property of a system to bounce back. According to Stocking and Murnaghan (2001), it refers to the ability of a system to return to its pre-altered state following the change. Further, It has also been defined as the "ways in which people adapt to changing circumstances to get by and 'make do' through the exercising of the autonomous initiative" (Katz 2004). It means if a land system undergoes the degradation process, the response of land to human efforts to regain its fertility will be determined by the resilience capacity of the land. High resilience reduces the risk of serious degradation and vice-versa (Abhay and Patra, 2018). Also, it is reiterated that farmer's evaluation of local conditions and decision-making must be taken into consideration while designing policies for agricultural development for the study area which helps in improving the quality of soil, but, in long run, it will contribute in achieving agricultural sustainability in the study area. Therefore, it depends on the natural quality of land and resilience which refers to the human initiation to upgrade land and to which the inherent characteristics of land also respond. The resilience of land helps the farmers to restore the land capacity for beneficial production from agriculture. Therefore, resilience and agricultural sustainability are positively related.

References

Abdel Kawy WAM, Ali RR (2012) Assessment of soil degradation and resilience at Northeast Nile Delta, Egypt: the impact on soil productivity. Egypt J Remote Sens Space Sci 15:19–30

Abhay RK (2015) Land degradation and its impact on agricultural production in the Kendujhar plateau. Annal Natl Assoc Geogr India 35(1):71–88

Abhay RK, Patra P (2018) Measuring resilience of land against degradation from a Farmer's perspective approach in Kendujhar plateau, Odisha (India). Eur J Geogr 9(1):23–37

Asgher S, Sharma S (2018) Pattern and growth of agriculture in the hilly region of middle Himalayas: a case study of Doda District, J&K. Geographer 65(1):91–99

Baba SH, Wani MH, Zargar BA, Bhat IF (2019) Determinants of land degradation in Jammu and Kashmir: implications for land governance. Agric Econ Res Rev 32(2):271–281

Census of India (2011) Final population totals. Odisha, Data Sheet, Directorate of Census Operations, Odisha

Conacher AJ (ed) (2001) Land degradation. Kluwer Academic Publishers, Dordrecht

Daily GC (1995) Restoring value to the World's degraded lands. Science (New Series) 269(5222):350–354

DIP [District Irrigation Plan] (2016) District irrigation plan of Kendujhar (Odisha), Pradhan Mantri Krishi Sinchayee Yojana (PMKSY). District Level Implementation Committee, Kendujhar

Foley JA (2011) Can we feed the world sustain the planet? Sci Am 305(5):60–65

Foley JA et al (2011) Solutions for a cultivated planet. Nature 478:337–342

Hitzhusen FJ (1993) Land degradation and sustainability of agricultural growth: some economic concepts and evidence from selected developing countries. Agric Ecosyst Environ 46(1–4):69–79

Katz C (2004) Growing up global: economic restructuring and Children's everyday lives. University of Minnesota Press, Minneapolis, MN

Kendujhar (2020) About the District. https://kendujhar.nic.in/about-District/. Accessed on February 28, 2020

Lal R (1997) Degradation and resilience of soils. Philos Trans R Soc Lond B 352:997–1010

Ouda S, Zohry A, Noreldin T (2018) Crop rotation maintains soil sustainability. In: Ouda S, Zohry A, Noreldin T (eds) Crop rotation: an approach to secure future food. Springer, Cham, pp 55–76

Seybold CA, Herrick JE, Brejda JJ (1999) Soil resilience: a fundamental component of soil quality. Soil Sci 164(4):224–234

Sharma S, Burark SS, Meena GL (2015) Land degradation and sustainable agriculture in Rajasthan, India. J Ind Pollut Control 31(1):7–11

Singh J, Singh B (2019) The extent of agricultural land conversion in Punjab: impact on production conditions of agriculture. Indian J Econ Dev 15(3):360–368

Stocking M, Murnaghan N (2001) Handbook for the field assessment of land degradation. Earthscan, London

Zhen L, Routray JK (2003) Operational indicators for measuring agricultural sustainability in developing countries. Environ Manag 32(1):34–46

Zohry A, Ouda S (2018) Crop rotation increases lands productivity. In: Ouda S, Zohry A, Noreldin T (eds) Crop rotation: an approach to secure future food. Springer, Cham, pp 39–54

Zoltán B, Béla G, DangQuoc L (2000) Effect of crop rotation and fertilization on maize and wheat yields and yield stability in a long-term experiment. Eur J Agron 13(2–3):225–244

23 Diversification of Agriculture Through Medicinal Plants in Kangra District, Himachal Pradesh

Nitu and R. B. Singh

Abstract

This study offers a new approach to diversification of agriculture as an alternative livelihood. It investigates the livelihood situation in changing socio-economic environments of Kangra district of Himachal Pradesh. Kangra is a repository of medicinal and aromatic plants but many of the species are at the verge of extinction due to over-exploitation. The article draws on data from primary survey to analyze herbal medicine system in the Dharamshala region of Kangra district. Kangra is important for the growth of medicinal plants with high potency. The place is frequently visited by foreign tourists for *Ayurvedic* treatments. The results show that herbs, health, livelihood, and rich medical tourism are interlinked in the region. The findings have important implications for theoretical developments and policy formulation. It is concluded that the region has great potential in different enterprises, in the area from traditional crop may be diversified towards cash crops and medicinal plant production. It is suggested that the efficient management, timely harvesting, and processing of medicinal plants can conserve biodiversity, sustain human health, and raise economic opportunities to local people. It is also advocated for mapping of fruits and medicinal belts growing area.

Keywords

Alternative Livelihood · Diversification · Medical tourism · Medicinal plants · Nutrition · Web Portal

23.1 Introduction

Rural livelihood security is enhanced due to diverse portfolio of activities. Rural Non-Farm Employment (RNFE) contributed significantly to the rural employment and income. Diversity promotes greater flexibility as it allows more possibilities for substitution between opportunities that are declined and those that are expanding (Venkateswarlu 2005; Frank 2000). Off season vegetable cultivation in Chota Bhangal region provides ample economic opportunities to the farmers and help in raising their livelihood (District Human Development Report 2009). Government schemes are focusing on the horticulture production for livelihood, but the current trend does not show any direct benefit from this. The primary survey reveals the subsistence nature

Nitu (✉)
Indian Institute of Public Administration, Delhi, India

R. B. Singh
Department of Geography, Delhi School of Economics, University of Delhi, Delhi, India

B. Thakur et al. (eds.), *Resource Management, Sustainable Development and Governance*, Sustainable Development Goals Series, https://doi.org/10.1007/978-3-030-85839-1_23

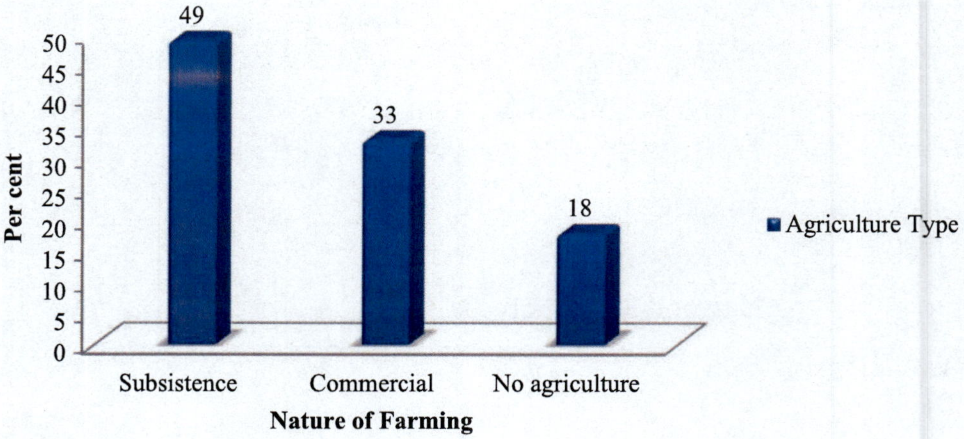

Fig. 23.1 Nature of Agriculture in Surveyed Villages. Source: Primary Survey

of agriculture. The main staple food comprises wheat, maize, and rice. There has been considerable increase in the cropping pattern intensity revealed from the given statistics and it is achieved because of efforts made by the agriculture department in the state. The primary survey reveals that about 50% of the people are engaged in subsistence type of agriculture and about 33% are engaged in commercial farming (Fig. 23.1). More than 60% of agriculture land is utilized for growing fodder and dairy farming which has emerged as a major source of non-farm activity in the study region. The community placed the problem of monkey menace contribution to about 75% of the total loss of agriculture. More than 40% of the community has almost left agriculture in all the three study villages except Zamanabad which is popular as good quality vegetable growing village. There is also an additional assistance from the agriculture department to promote this village. As per community response, the success of agriculture in this village persists because the village is free from monkey menace. In all the other surveyed villages practice of agriculture is limited owing to production of few crops. There has been shift from maize production to few grains like wheat and rice only for self-consumption and most of the time it is not sufficient for the whole year.

Box 23.1 Sustainable Development Goals

Preservation of agricultural diversity along with ecosystem and biodiversity is crucial for the health security and livelihoods of local population. In the context of Kangra, Himachal Pradesh of India, the benefits and services that biodiversity and ecosystems can provide to the health of people is largely unrecognized and unappreciated; however, they have tremendous impacts on our wellbeing and daily lives. Even SDG-3 of Agenda 2030 aims at reducing mortality and improving global health, particularly for pregnant woman, newborns, and children. The indirect benefits of promoting local agricultural diversity and preserving ecosystem and biodiversity is that it promotes the eradication of epidemics and communicable disease, provides better prevention and treatment for narcotic drugs, alcohol, and tobacco consumption. Besides, it also promotes the implementation of universal healthcare and gives access to affordable and essential medicines for all. Given the ecologically fragile nature of Kangra, Himachal Pradesh, conserving medicinal plants for sustenance and livelihoods, not only promotes ecosystem and biodiversity but are essential for sustainable development overall. This will need partnerships between stakeholders at national, regional, and local scale and will serve as a great example of how to localize SDGs, besides mainstreaming conservation, and sustainable use of medicinal plants.

23.2 Diversification of Agriculture: An Alternative Livelihood

The share of area and production of fruits has increased from 1990 to 2015. The major increase has been noticed during 2005 with highest production of 42% fruits while the area during this year was recorded lowest. The production of fruit in 2015 has recorded to be low compared to 2010 (Figs. 23.2 and 23.3). The present situation calls for a new model of agricultural development in hill region where medicinal plants are seen to be a viable livelihood opportunity in changing socio-economic environments. Several medicinal plant species along with medicinal valued fruits like *Kiwi* are two ways beneficial as, on the one hand, these are best crops which cannot be destroyed by stray animals while, on the other hand, good financial returns are expected.

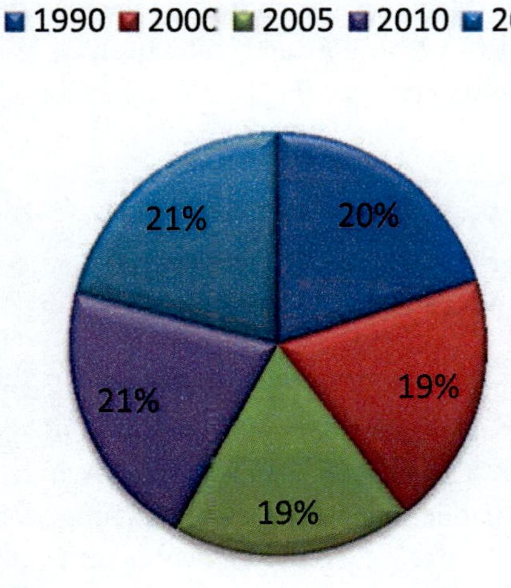

Fig. 23.2 Year wise Total Area under Fruits in Kangra. Source: Statistical Abstract (2014–2015)

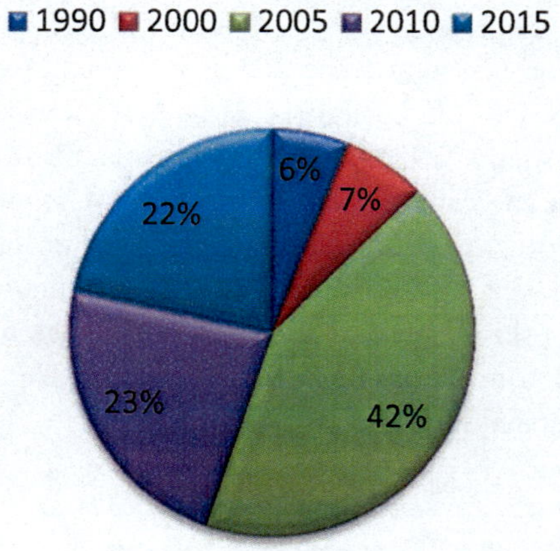

Fig. 23.3 Year wise Fruit Production in Kangra. Source: Statistical Abstract (2014–2015)

23.3 Indigenous Wisdom and Globalization Impact on Economic Opportunity

Traditional healing practitioners are diffused in different parts of India. Herbal medicines are traditionally associated with nature's worship, in which preservation of local cultures and ecologies is intertwined. With an impact of globalization there has been a tremendous increase in the material culture. Local knowledge is being abandoned for the sake of modern technology in the rural society. The impact of globalization on indigenous wisdom of health care is reflected in various forms. Herbal commodities, products ranging from beauty to daily using stuffs have become a part and parcel of the people. The tourist destination like Dharamshala in the Kangra district shows variety of herbal products and services including spas, yoga, and message centers. India has only 0.5% shares in the global export market of medicinal plants and as projected by the Planning Commission of India. There is an urgent need to enhance Indian herbal export by Rs. 50 billion annually to generate employment and income to approximately 20 million families. Recent research and development have shown the increasing importance of Ayurveda as a base for nanotechnology based NDDS (Novel Drug Delivery System). The domestic demand for the medicinal plants during 2014–2015 has been estimated at 195,000 MT and the total utilization of the medicinal plants (raw form) has been estimated as 512,000. The total trade value for the same year has been estimated at 55 billion. About nine-fold increase in trade value has been registered from 2005–2006 to 2014–2015 (National Medicinal Plant Board 2017).

23.4 An Overview of Indian Traditional System of Medicine in Kangra District of Himachal Pradesh

Traditional system of medicine along with *Yoga* and acupuncture has become global products of tourism as evident from a study by Harvard and Johns Hopkins universities which indicate a three-fold surge in journal articles on yoga therapy in peer reviewed journals worldwide. The study on yoga in therapeutic intervention during 1967–2013 addressed top three diseases including cardiovascular disease, mental health, and respiratory problems (Pamela et al. 2015). In India, about 7000 plants prepare the base of traditional system of medicine that account for the livelihood of the community and create a parallel mode of healing for the common masses. For them, the synthetic modern treatment and drugs are unaffordable and beyond reach.

23.5 Medicinal Plant Richness and Untapped Potential for Livelihood Opportunity

Medicinal plant richness and healing practices of Himalaya find special mention in the ancient scriptures. Medicinal plants, an important subset of minor forest produce play significant role in health and livelihood security of people. The Himachal Pradesh Forestry policy highlights the importance of minor forest produce requirements of rural and tribal population. Himachal Pradesh lies in North-West Himalaya having 1.7% of the country's geographical area. It showcases medicinal plant richness and diversity over different agro-climatic and altitudinal zones. It harbors 3500 flowering plants in which about 800 medicinal plants are used for various health purposes.

Health and herbs are associated phenomena in India. Medicinal plants are used for human and veterinary health care purposes and the people of Himachal Pradesh know the use of more than 800 plant species. The rapid depletion of these potential bioresources has started affecting the use pattern of traditional know how on these plants which negatively impact the health and livelihood security of rural masses.

During primary survey, about 95% of Tibetans and 70% of Ayurvedic healers reported its popularity mainly for lifestyle diseases, for instance blood pressure, diabetes, arthritis, obesity, and liver problems. Two major trends on herbal medicine system have emerged in the Dharamshala region of the Kangra district. It can be categorized into two streams: local *Himachalis'* initiatives and the Tibetans' initiatives in promotion of herbal medicine system. For local *Himachalis*, the government subsidy has helped few local progressive farmers to establish small-scale herbal unit. But the benefits are restricted, and reach is limited. Tibetan community attracts tourists in the region and provides livelihood to the local people. Market-oriented livelihood options help in the emergence of *Ayurvedic* spas and health clinics along with the influence of Tibetan medicine. But the Indian System of Medicine is not attracting the local people due to its high cost and, subsequently, lesser awareness and practice (Figs. 23.4 and 23.5).

23.6 Preference of Local Population in Surveyed Villages

The primary survey depicts the unclear occupation of medicinal plants in the surveyed villages. About 78% of respondents prefer both Ayurvedic and Allopathic mode of treatment among which they first prefer Ayurvedic mode due to its quality. About 86% respondents have admitted that there is an erosion of traditional knowledge since

Fig. 23.4 Medicinal Plant Seller at Mc. leodganj

Fig. 23.5 *Lungru* (herbal Plant) Seller at Dharamshala

the past two decades. In terms of collection of medicinal plants, hardly 1% of the respondents were engaged in the collection of medicinal plants from high hills. But they hesitate to admit that they temporarily work for some traders as petty business due to fear of forest officials. Kangra, being an agrarian district, also possesses wasteland and when asked about its utilization for growing medicinal plants; less than half percent of marginal farmers agreed if they receive profit, whereas 53% had no wasteland available (Figs. 23.6 and 23.7).

About 88% respondents have reported that some local and Tibetans are engaged in medicinal plant sector on commercial basis and earn good money. But the local people's involvement with these private herbal owners is negligible. Those working in their herbal factories do not belong to village and the district itself. Therefore, it is worth to mention here that livelihood opportunities from medicinal plant and healing sector are negligible in the surveyed villages, whereas there are few unprofessional local healers diffused in the district who do not charge money for

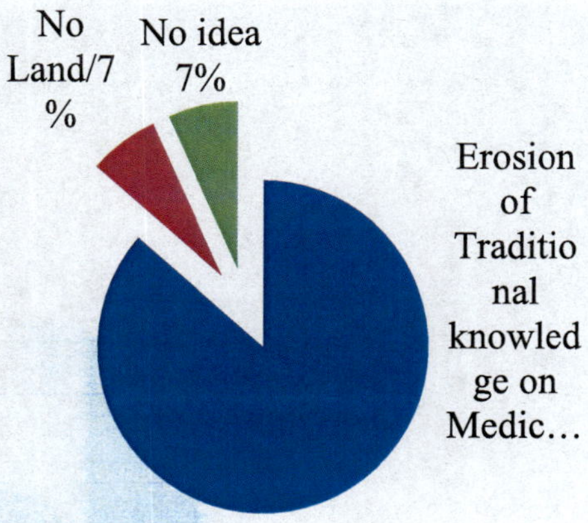

Fig. 23.6 Problems in Picking up Medicinal Plant

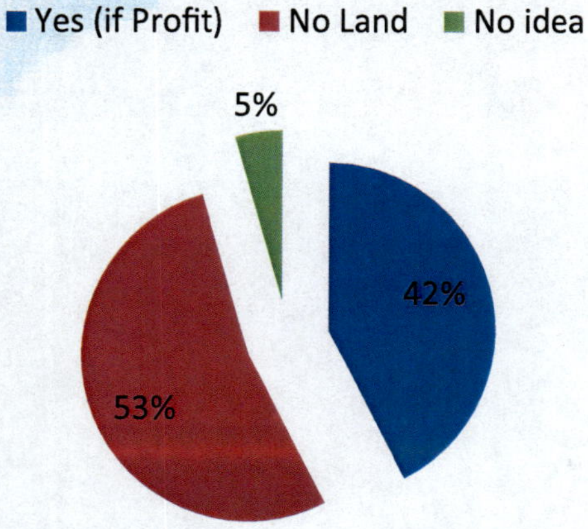

Fig. 23.7 Interest in Growing Medicinal Plant on Wasteland

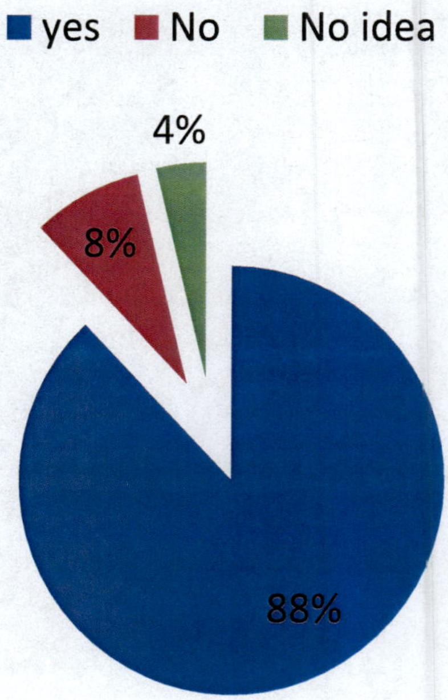

Fig. 23.8 People's Awareness on anyone Earning Livelihood from Medicinal Plants

their treatment and are hardly acknowledged in the local community (Fig. 23.8).

23.7 Geography, Medicinal Plants, and Medical Tourism

The concept of medical geography and place (*Desha)* is mentioned in Charak Samhita, thousands of years back before Hippocrates. Drugs indicate its origin, drugs in different regions have different names, and places in relation to physicians are some important facts. The Himalaya is the most important place in the book and the first symposium of ancient Indian physicians/sages on disease and Ayurveda took place near the Himalaya. The medicinal plants found in the Himalaya are unique in its properties as grapes and pomegranate are found to be sweet, whereas these fruits are found to be sour in other places of India. It is important to note that Kangra has all types of hill zones as well as agro-climatic conditions favorable for the variety of medicinal plant growth (Fig. 23.9). Basically, low hill constitutes elevation up to 1000 m., mid hill constitutes elevation between 1000 and 2000 m. and high hill is above 2000 m. The chapter of *udara chikitsa* was expounded on mount Kailas and the reason being the water in the rivers originating from the Himalaya is like nectar useful for the health and well-being. The concept of *Desha* is important in drug identification and discovery. The indigenous medicinal plants are more potent and help in potent drug formation. Research has shown that the potency and genotypes of Withania Somnifera of Kashmir and some other regions differ significantly (Bhavana and Shreevathsa 2014). Kangra harbors typical geographical situation with valleys, low hills, middle hills, and extreme high hills along with snow-laden mountains, on the one side, while sunshine and extreme temperature and rainfall, on the other. The geographical loca-

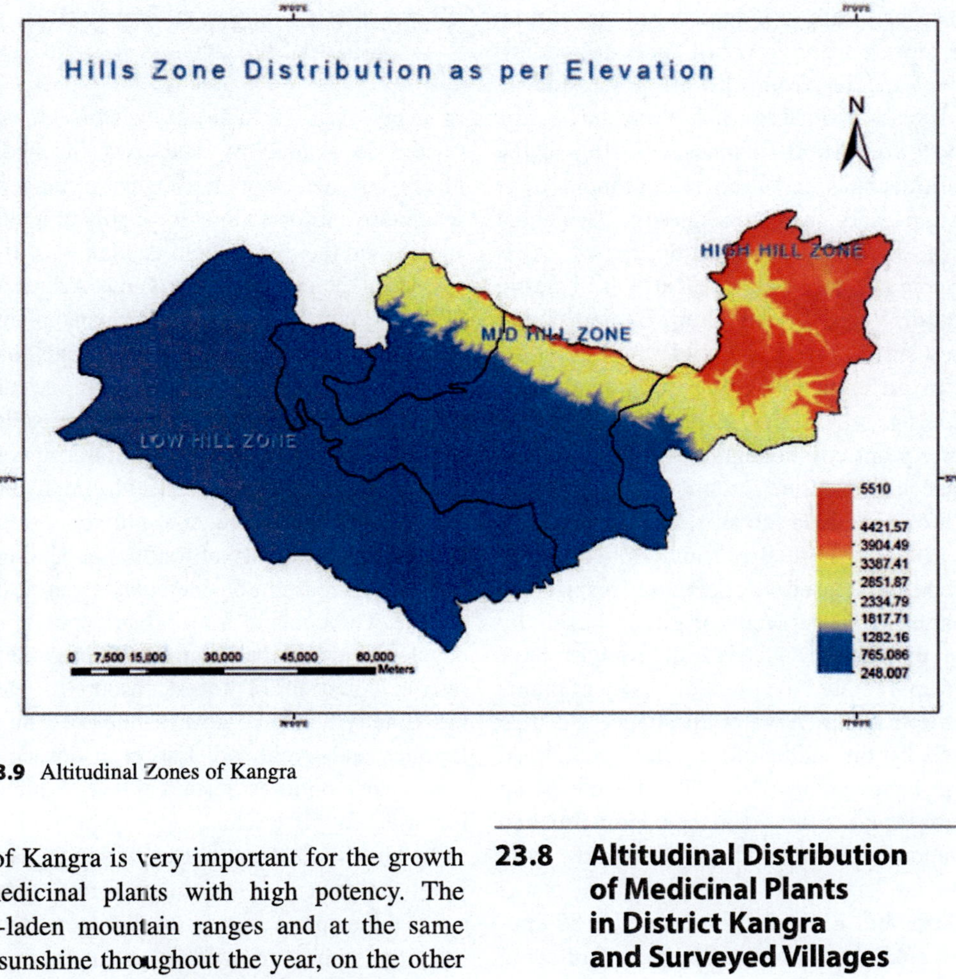

Fig. 23.9 Altitudinal Zones of Kangra

tion of Kangra is very important for the growth of medicinal plants with high potency. The snow-laden mountain ranges and at the same time sunshine throughout the year, on the other side of the mountain help the growth of variety of vegetation. These are potential herbs for the treatment of chronic disease. Majority of farmers use manure instead of fertilizers and live simple life which make it pollution free space. The clean and chemical free environment makes it the potential destination for Ayurveda. Due to the purity and sanctity of the environment, majority of foreign tourists visit the place for Ayurvedic treatment more specifically Ayurvedic purification and rejuvenation therapy. Ayush Kama and Dr. Sibbi clinics at Mcleodganj and Namlang Himal Ayurvedic resort at Bir village of Kangra are some Ayurvedic centers frequently visited by foreign tourists. As per the primary survey, there has been an increase in the number of foreigners after getting Ayurvedic treatment once. More than 50% of them revisit Kangra at least once in a year to purify their body.

23.8 Altitudinal Distribution of Medicinal Plants in District Kangra and Surveyed Villages

Several studies have reported the ethnobotanical uses and conservation need of various medicinal plants in lower hill zone of the Himalaya (Parkash and Aggarwal 2010). Himachal Pradesh harbors 3500 species of medicinal plants and provides livelihood among which 800 species are used for medicinal purposes and provide health benefits as well as livelihood to the rural masses. Asthma and bronchitis are commonly found among rural masses of high hill region. Several medicinal plants with best of their qualities including Rhododendron (endemic to the Himalayan belt) are found at high hills. Researchers have also highlighted the medicinal plant diversity being highest in the low hill zone of sub-tropical areas, while it decreases with increasing altitude (Choudhary

2011). There are 36 species of medicinal plants reported from low hill zone, 56 medicinal plant species from mid-hill zone, and 47 plant species from high hill zone during the survey. About 1000 species occurring in the Shiwalik ranges have been documented as medicinal and aromatic plants. Himachal Pradesh is the largest supplier of Atis, Salampanja, Dhoop, Chora, Kutki, Talispatra, Daruhaldi, Vach, Revand chini, Bankakri, and Somlata entirely from the wild which is diversified in different hill zones (Chauhan 2003; Fig. 23.10).

Many plant components have been used in pharmaceutical preparations, such components are periwinkle (an anticancer drug), digitalis (a heart regulator), and ephedrine (a bronchodilator used to decrease respiratory congestion). These were originally used by people of Kangra. Natives of Kangra have always appreciated the great diversity of plants available to them (Arya et al. 2012). It is also reported by the stakeholders that specialized healers have great potential of converting medicinal plants into finished/semi-finished value-added products before export from Himalayan states to boost medical tourism (Arya and Kaur, 2012). In total, 139 species of medicinal plants have been collected from primary and secondary sources in three altitudinal zones. The medicinal plant use pattern also varies and about 81 medicinal plant species are used (Fig. 23.11).

23.9 Policy Issues of Medicinal Plant Sector

In many parts of India, indigenous knowledge related to availability and uses of medicinal plants has not been thoroughly cataloged and quantitative information on the role of medicinal species in the rural socio-economy is limited (Dobriyal et al. 1997; Farooque Ashraf et al. 2012). Today, the demand of medicinal plants has increased to such an extent that it leads to overexploitation of natural resources. Medicinal plants sector is highly unorganized leading to degradation of the Himalayan bioresources and demand supply gap. Herbs, health, and livelihood are always interlinked and provide promising future. The popularity of traditional healing system has increased considerably in the past one decade. The demand from urban population and developed world has alarmingly increased the over-exploitation of forest resources. Herbal-based activities have always impacted on local families and livelihood. Earlier, a decade back dependence on forest resources was ample in the forest-based villages and, consequently, villagers earned livelihood from it. But, within the last few years the extreme pressure on forest resources stopped the open access to local people. Thus, the entire pressure is on wild population and only a few crops like Kuth, Kala jeera, Kesar, and Hops are cultivated in Lahaul-Spiti and Kinnaur. In this context, commercial cultivation along with value addition for economic prosperity of this hill state

Fig. 23.10 Medicinal Plants of High Hill Zone in Chota Bhangal Region, Kangra

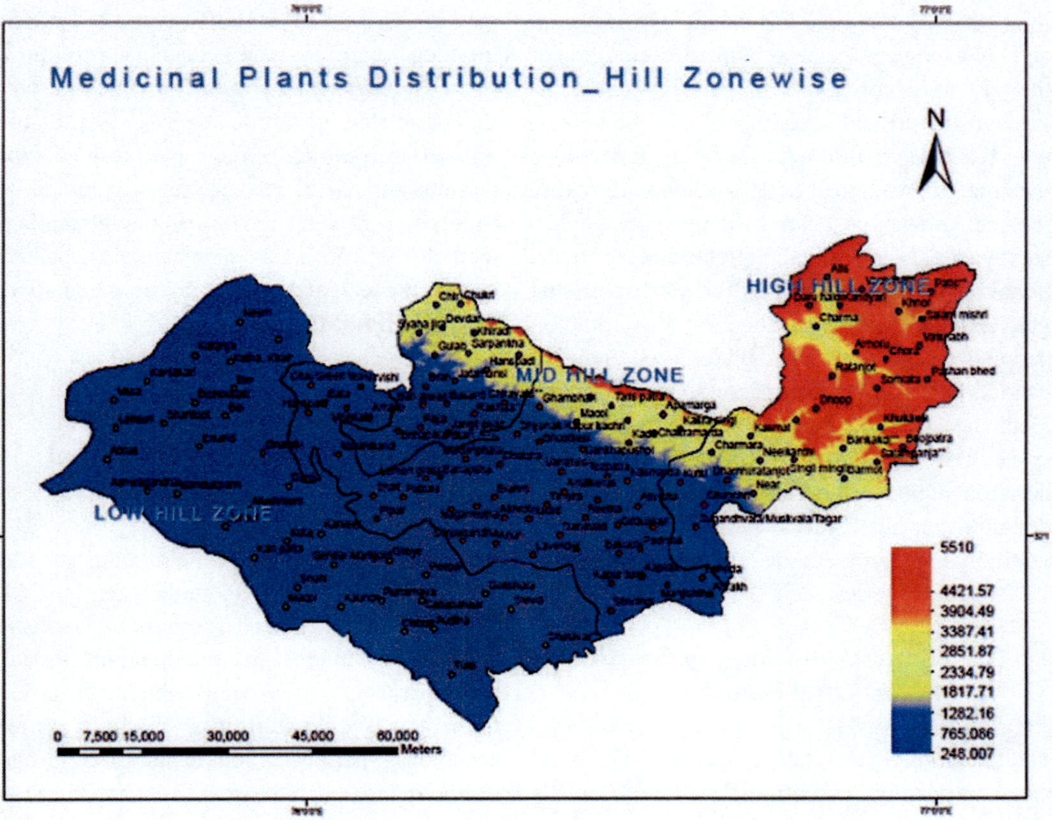

Fig. 23.11 Hill Zone wise Medicinal Plants Distribution

is highlighted (Chauhan 2003). The prioritization of species for cultivation in each altitudinal zone of Himachal Pradesh was determined according to medicinal uses, its trade values, pharma preparations, and demand. However, appropriate agro-techniques and post-harvest methods are unavailable (Prakash 2001; Sultan and Singh 2006). The situation is grim today because the pressure on medicinal plants is a problem since decades. It is referred in several research (Table 23.1).

India is endowed with rich tourist destinations. The globe sees India as the center for spiritualism, *Yoga,* and alternative medicine, and yet it figures at the bottom of the tourism with merely 0.5% share out of the global tourism. Pulse diagnosis skill of Ayurvedic doctors of India is considered unique by many foreign health practitioners. There is possible threat that ancient medical traditions in India may fade into obscurity if we ignore to preserve and fail to deliver

Table 23.1 Demand and Supply of Medicinal Plants in North-West Himalaya

Botanical name	Demand (tonnes)	Supply (tonnes)
Orchis latifolia	More than 5000	Less than 100
Rauwolfia serpentina	More than 5000	Less than 1000
Gentiana kurroo	More than 5000	Less than 100
Aconitum heterophyllum	More than 1000	Less than 100
P. zeylanica	More than 1000	Less than 100
Onosma bracteatum	More than 5000	Less than 100
Picrorhiza kurroa	More than 5000	Less than 100
Dioscorea deltoidea	More than 5000	Less than 100

Source: Ministry of Health 1987

their services in a judicious manner. The wisdom and knowledge of the ancient medical tradition of the past must continue to deliver its health services to the present society so that it would survive as a living tradition for the future generations to come. There is great need to listen to the experienced knowledge bearers of the society. The surveyed data reveals the preference of traditional health care in India. Efficacy of traditional medicine is well established by the common people's use pattern. Lack of research and data validation is the prime concern about traditional medicine system indigenous to geographical space. Therefore, bringing opportunity in traditional medicine and building capability of traditional healers shall lead towards sustainability of traditional medicine sector.

23.10 Demand and Supply Position of Medicinal Plants

The global market of India in terms of medicinal plants is negligible with 0.3% share. India's export share (8%) is low as compared to China (24%) and USA (11%). Total commercial demand for the herbal-based medicines within India as well as rural household during 2014–2015 is estimated to be 512,000 and 167,500 metric tons, respectively, whereas the estimated export was 134,500 metric tons (estimated value in INR was 3211 crore or 32.11 billion). The domestic consumption of herbal medicine was 195,000 metric tons (value in INR. is 1950 crore or 19.5 billion) during 2014–15. The trade value of herbal raw drugs was of value 5500 crore INR or 55 billion. E-charak portal has been launched as virtual marketplace for medicinal plants to create trade linkage between buyers and sellers. The share of cultivation of medicinal plants is negligible which is evident from the data (HPFSMPP 2006) (Table 23.2).

The collection of herbal plants from major herbal mandis located in cities like Delhi, Amritsar, Bengaluru, Kolkata, etc. has been initiated by NMPB for the benefit of stakeholders like NMPB. In context of Himachal Pradesh, about 2500 tons of medicinal plants are officially exported every year from forest region, whereas the illegal parallel trade is also suspected in the state. The legal annual trade is worth 10 crore in INR. The state government earns 40 lakhs (or 4 million) per annum through the issue of export permits. Annual trade, price range, demand, and supply vary from species to species and market to market. The irony is that the chain of herbal medicine is unclear, and price is highly speculative in Kangra (HPFSMPP 2006).

23.11 Reasons for Lack of Cultivation of Medicinal Plants

Besides research, development, domestication, and propagation of MAPs undertaken by CSIR, IHBT, Palampur the general mass of the district is not taking interest in its cultivation is due to several reasons. The reasons which restrict local farmers to take up cultivation of MAPs are long gestation period, non-standardized agro-technologies (when to grow MAP crops). Stevia has great potential to be cultivated in low, mid, and temperate zones, but due to lack of large-scale processing units and marketing facilities it is not picking up. Sixty dispensaries and eight pharmacies of Government Ayurveda college, Paprola medical college provide livelihood to the people, but not to satisfactory level because there are no permanent recruitments in the government sectors since decade and the salary of Ayurveda personnel is low compared to Allopathic streams in the district. This issue is also raised by the Ayurvedic associations of doctors as well as personnels. The salaries of temporarily appointed Ayurvedic doctors' range between INR 8000 and 15,000 (US$107.54 to 201.64 approximately), whereas it is more than INR 40,000 (US$537.71) for allopathic temporary appointed doctors in the government hospitals/dispensaries.

Herbal cultivation and procurement of herbs are not picking up in the state due to lack of minimum support price of herbs. The collection of herbs is unscientific and there is exploitation of collectors because they are not professionals and carry on this collection as petty side business to

Table 23.2 List of Some High Traded Medicinal Plants with Supply Source, 2008 (More than 100 MT/Year)

Name of the species	Trade name	Major supply source	Estimated annual trade (MT)	Price range (Rs. per kg)
Abies spectabilis	Talispatra	HF	500–1000	30–50
A. precatorius	Gunja	W	200–500	10–15
A. catechu	Katha/Khair[a]	TF	200–500	10–15
A. ferox.	Vachnag	HF	100–200	150–250
Aconitum heterophyllum	Atis	HF	200–500	2000–4000
A. calamus	Vach	C	500–1000	30–35
Adhatoda zeylanica	Adusa	C	2000–5000	10–15
B. monnieri	Brahmi	W	2000–5000	30–35
Berberis aristata	Daruhaldi	HF	500–1000	15–35
Bergenia ciliata	Pashanabheda	HF	200–500	15–20
B. diffusa	Punarnava	W	2000–50,000	30–40
C. roseus	Sadabahar	C, locally available	200–500	40–55
C. deodara	Devdar	HF	500–1000	25–35
Celastrus paniculatus	Malkangani[a]/ Jotishmati	TF	200–500	48–55
C. asiatica	Brahmi booti/ mandukparni	W	500–1000	30–35
C. tamala	Tejpatta	HF	500–1000	15–35
C. colocynthis	Indrayan	W	2000–5000	15–20
Commiphora wightii	Guggul[a]	HF	500–1000	140–160
Convolvulus microphyllus	Shankhapushpi	W	1000–2000	15–20
Emblica officinalis	Amla[a]	TF	16,000	30–35
Ephedra gerardiana	Somlata	HF	200–500	25–35
Jurinea macrocephala	Dhoop	HF	1000–2000	60–150
Nardostachys grandiflora	Jatamansi	HF	200–500	110–150
Nilgirianthus ciliatus	Kurinji	TF	200–500	15–20
O. americanum	Ban tulsi	W	500–1000	20–25
O. basilicum	Kali tulsi	C	1000–2000	20–25
O. tenuiflorum	Tulsi	C	2000–5000	20–25
Onosma hispidum	Ratanjot	HF	500–1000	50–60
Oroxylum indicum	Tetu chhal	TF	1000–2000	20–30
Parmelia perlata	Chadila	HF	1000–2000	80–90
P. harmala	Harmal	W	200–500	40–45
P. amarus	Bhumi amla/Tamalaki	W	2000–5000	20–25
Picrorhiza kurroa	Kutaki	HF	200–500	220–230
Pistacia integerrima	Kakar singi	HF	150–200	90–110
P. zeylanica	Chitrak[a]	W	2000–5000	20–25
P. pinnata	Karanji	C and HF	2000–5000	20–25
P. corylifolia	Bawachi/Bakuchi	W	200–500	10–20
Rheum australe	Revan chini	HF	500–1000	25–30
Rhododendron anthopogon	Talispatra	HF	100–200	15–30
S. album	Chandan[a]	TF	200–500	700–850
S. mukorossi	Reetha[a]	TF	200–500	30–40

(continued)

Table 23.2 (continued)

Name of the species	Trade name	Major supply source	Estimated annual trade (MT)	Price range (Rs. per kg)
Sphaeranthus indicus	Gorakh mundi/ Munditika	W	200–500	15–20
S. urens	Karaya	TF	500–1000	80–100
Taxus wallichiana	Talispatra	HF	100–200	75–90
Tinospora cordifolia	Giloy[a]	W	2000–5000	10–15
T. terrestris	Gokhru	W	2000–5000	10–20
T. cucumerina	Patol panchang	W	500–1000	15–20
Valeriana jatamansi	Musakbala/Tagara	HF	100–200	95–100
Viola pilosa	Banapsha	HF	200–500	300–350
V. negundo	Neergundi	HF	200–500	10–15
W. somnifera	Ashwagandha	HF	2000–5000	60–70
Woodfordia fruticosa	Dhai phool/Dhataki	W	2000–5000	10–15

Note: [a] Also Found in Low and Mid hills; *TF* Tropical Forest; *HF* Himalayan Forest; *W* Wastelands; *C* Largely Cultivated
Source: Compiled and Modified from Foundation of Revitalization of Local Health Traditions (FRLHT), 2016

earn some extra income. Few local collectors of medicinal plants from Deol and Bhaded village are engaged in medicinal plant collection. The commercial firms from Amritsar, Delhi buy from collectors and sell at throw away prices. There is no fixed price for several medicinal plants, such as *Shatavari, Brahmi, Ashwagandha, Harad, Baheda,* etc. In Kangra, no commercial herbal cultivation is carried except for Amla which is commercially grown in nearby places, such as Purva. The stakeholders also reported that *Harad* is destroyed by the forest personnel to clean the area. Hardly any pharmacy promotes commercial cultivation of medicinal plants, and they also have rate fluctuation. Some major cause of lack of commercial cultivation of medicinal plants is the small size of land holdings, involvement of local people in alternative income generating activities like MGNREGA.

Medical tourism with respect to Panchakarma is not picking up in the district due to lack of government subsidy and trained professionals. There is lack of infrastructure in attaining quality services for medical tourists which also restricts foreign tourists in terms of absence of advanced hospitals. Geographically, Kangra is isolated, weather conditions are extreme which makes Panchkarma difficult to carry and, therefore, tourists do not stay more than 2 days. On the other hand, Tibetan medicine is preferred by foreign tourists and for that they stay in Mcleod Ganj.

23.12 Erosion of Medicinal Plant Wealth Due to Climate Change

Studies by Indian Agricultural Research Institute, India, and other agencies indicate greater expected loss in rabi crop. A rise in 1° of temperature will bring reduction in wheat production by 4–5 million tons, lower yields from dairy cattle, and reduction in fish breeding. Lower yield of cash crop, fruits, cereals, aromatic, and medicinal plants is also attributed to climate change. Hill and Mountain Farmers' confusion over the climate change is now recognized where temperate fruit belt has moved upwards by about 50–100 km with increased intensity of disease (Pratap 2003). Broom cultivation has shifted 12 km northwards altitudinally in Multhan, Kangra and this all happened because forest resources were removed in large scale (Singh 2011). Basically, 1 hectare of farming land requires 7 hectare of forest area which means 1:7 but the Beas region has 1:1 ratio. Earlier research reveals that major deforestation in Himachal has started since 1970s and 80% was destroyed for the commercial purpose (SPRA 1984). In the past few years, the occurrence of landslide has increased due to road construction and major infrastructural development works. The Mcleodganj region experienced severe land sliding in the past years which affected the overall growth in the tourist region. When there is landslide, it not only blocks the road, but several important medicinal herbs, shrubs, and trees are destroyed (Nitu, Singh and Singh 2016).

23.13 Opportunities and Prospects

Holistic health is achieved when human being realizes full play of mind, intellect, and the elusive element of soul. Many Himalayans medicinal plant species have become extinct, and others are on the way of extinction and, thus, herbs should be grown by popularizing agronomy of herbs under Bioindustrial Watershed mechanism (Bali 2005). Indigenous medicine has all potential to serve as a promising soft industry because efficacy of traditional medicine is well established by the common people's use pattern. For achieving this, there is great need to listen to the experienced knowledge bearers of the society. The surveyed data reveals the preference of both the traditional and modern health care. There is a growing demand for Indigenous Health Care which provides an employment opportunity under Medical Tourism in the Himalayan destination too.

23.14 Measurement of Crop Diversification

Herfindahl Hirschman Index was named after Orris, C. Herfindahl and Alber, O. Hirschman. Herfindahl Index (HI) is used to measure crop diversification. The value of HI varies between 0 and 1. It takes the value of 1 when the field is under monoculture but when its value comes zero, it indicates high agricultural diversification. Therefore, as HI increases, the diversification in a particular region decreases and as HI decreases, the diversification increases (Sindhu and Govindaru 2014).

$$HI = \sum_{n}^{ix1} Pi^2$$, here Pi is share of each crop

Here, $Pi = Ai/\sum n$, i = 1 Ai
where Ai = Area Under ith Crop
$\sum Ai$ = Total Cropped Area

In terms of vegetables, the Herfindahl Index is 0.60 which means diversification in vegetables is relatively low compared to crops (HI = 0.33) and fruits. It is therefore suggested that there is a need of more emphasis on vegetable production in the district (Table 23.3).

HI of fruit is 0.39 means high fruit diversification but it is lesser in terms of crop diversification and higher in terms of vegetable diversification (Table 23.4). There should be more emphasis on fruit diversification in the potential areas. The agricultural extension services and other government and non-government agencies should promote the fruit production.

Table 23.3 Area of Vegetables including Spices in Kangra, 2008–2009

	2008–2009	Pi	Pi^2
Potato	2874	0.28	0.0784
Other veg./spices	7222	0.72	0.5184
Total area	10, 096		0.60

Source: Compiled from Census of India (2011)

Table 23.4 Diversification of Area under Fruits with Medicinal Value

	2008–2009	Pi	Pi^2
Apple	454	0.01	0.0001
Plum	411	0.01	0.0001
Peach	214	0.01	0.0001
Apricot	41	0.00	0
Pear	420	0.01	0.0001
Almond	402	0.01	0.0001
Walnut	203	0.01	0.0001
Mango	21,147	0.59	0.3481
Litchi	2757	0.08	0.0064
Guava	667	0.02	0.0004
Orange/Kinnow	5561	0.16	0.0256
K. Lime	2778	0.08	0.0064
Galgal	490	0.01	0.0001
Pomegranate	70	0.00	0
Total area	35,615		0.39

Source: Compiled from Census of India (2011) Himachal Pradesh

23.15 Livelihood Concerns in Kangra District through Spices/Medicinal Plants

The implementation of Horticulture Technology Mission (HTM) is playing an important role in the growth of area and production of fruit in the district. Garlic is the second most widely consumed spice. In India, the production of garlic is quite low compared to other countries which pose a challenge in Indian economy. The reason for the low yield is the lack of awareness among farmers on new technologies like HYV seeds, plant protection measures, and overuse of chemical fertilizers. But at the same time, Himachal Pradesh poses suitable climatic condition to produce quality garlic as well as providing good economic opportunity to the farming community if biofertilizers and organic inputs are used (Sharma and Chauhan 2013) (Fig. 23.12). The medicinal value of the spices grown in hill region provides potential benefit to raise the income of farmers if it is taken as mission mode.

There is a variation in share of miscellaneous crops. Fruits and pastures are widely distributed in the southern tehsils. The potato and mustard growing region is restricted to mid and high hills due to suitability of climate. The low regions of Nurpur and Indora show variety of production due to suitability of climate for the citrus fruits. The Demand of oranges of Nurpur was high but in the recent years there is tough competition from other state. The local demand is high in Kangra which makes it a viable source of income for the farming community (Fig. 23.13).

23.16 Protected Agriculture and Medicinal Plant Cultivation

Climate change mitigation is the need of the hour. Some of the mitigation strategies include development of drought and pest resistant crop varieties, soil and water conservation, revival of traditional source of water, demonstration of new agro-techniques, and financial support to enable them to adopt new technologies, farmers to invest in and adopt relevant technologies to overcome climate related stresses (Rana and Randhawa 2016). Also mixed cropping in higher reaches of Chota and Bara Bhangal will enable farmers to take numerous products in one season, help retaining soil fertility as well as preventing soil erosion. Fishponds can provide excellent retention of storm runoff and will not only help combatting climate change but also promote rural tourism simultaneously (ADB Report 2013). The Simpson Diversity Index of 0.71 shows the preference of diversification but the process of crop diversification in the district is, however, facing important challenge and to overcome this there is a need of building knowledge and skill among farmers (Nitu and Singh 2016).

In protected agriculture, vulnerability factors are controlled and altered to the advantage of plants to combat climate change (NAAS 2010). The Indian Himalaya have the second or third

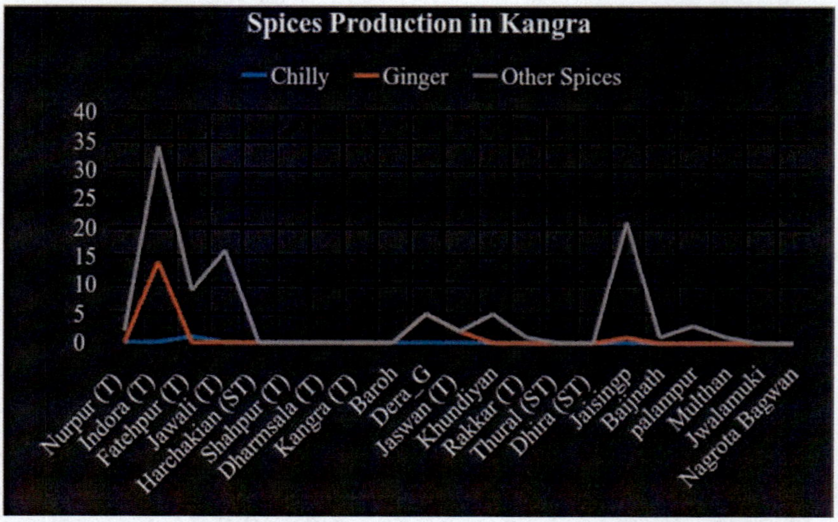

Fig. 23.12 Spice Production in Kangra. Source Compiled from District Revenue Office, 2011–2012

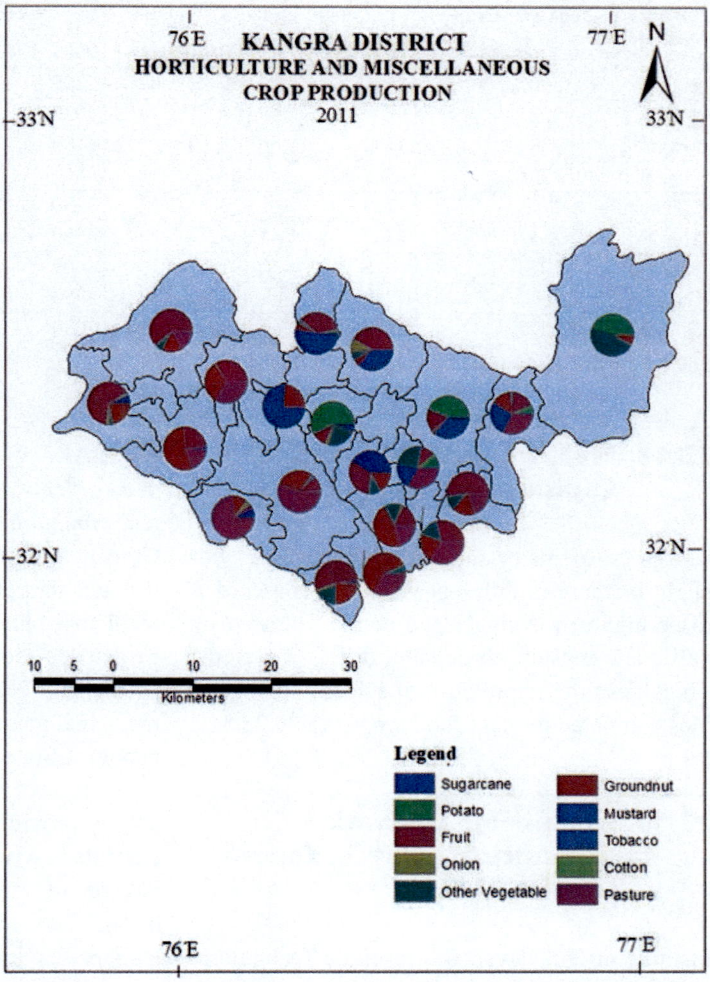

Fig. 23.13 Horticulture and Miscellaneous Crop Production, Kangra. Source Compiled from District Revenue Office, Kangra 2011–2012

richest resources of sea buckthorn (30,000–40,000 ha) and other medicinal herbs in the world. The fruits and leaves of sea buckthorn are very rich sources of Vitamin C and other bioactive substances (Choudhary and Jaggi 2002). Tea cultivation is another opportunity for a region like Kangra which is highly suitable for its production with new value addition for the European market (Sankhian 2007).

23.17 Prospects of Growing Vegetables and Medicinal Plants

A tool called Niche Extrapolator using sample polygon has been developed to locate the potential good area for growing crop and raise the income level of farmers by using technology. This tool first extracts the biophysical conditions to extrapolate the small area into the larger extent. For instance, wheat, paddy, vegetable (including medicinal value spices, e.g., ginger, onion, garlic, etc.) area can be extended to other potential regions to secure livelihood (CSK Palmpur 2008) (Box 23.2).

Box 23.2 Potential Enterprise for Gainful Employment

Farm: Diversified cultivation, organic farming, sericulture, agro-tourism, medicinal and aromatic plants, mushroom, nursery raising, flower nursery.

Non-Farm: Vermi-compost, mushroom rearing, rural handicraft, fruits and vegetables processing, beekeeping, dairy farming, tourist service, petty business

Table 23.5 Year wise Crop Price in Kangra (in INR./Quintal)

Crops	2006–2007	2007–2008	2008–0209	2009–2010	2010–2011	2011–2012
Rice	800	860	960	1200	1290	1357
Wheat	902	950	950	1020	1169	1245
Jowar	978	1100	145	1100	1381	1432
Maize	830	858	858	986	1000	1035
Chickpea	2791	3200	3200	3250	3488	3556
Mustard	2837	3362	3362	3550	5000	5128
Potato	1020	1016	1016	1400	860	11,120
Garlic	4600	4750	4750	4750	5800	6854
Ginger	2500	2600	2600	3000	3343	3522

Source: District Revenue Officer, Kangra (2011)

23.18 Measurement of Financial Capital of Farmers

The price of major food crops is varying with slight increase in different years and is recorded to be highest for ginger and potato followed by garlic and mustard throughout different periods. This draws the importance of spices and vegetables compared to other food crops (Table 23.5).

23.19 Sustainable Cultivation of Medicinal Plants in Kangra District

Institute of Himalayan Bio-resource Technology (IHBT) situated at Palampur, Himachal Pradesh is devoted to ex-situ cultivation of Himalayan medicinal plants. Picrorhiza kurroa is used for preparation of several drugs. IHBT has collected several accessions from different agro-climatic zone of HP and maintaining them in the farm under controlled condition. The plant has been successfully domesticated under Palampur condition. Research and development are also being carried out to understand the regulation of picroside biosynthesis. There are several medicinal plants which are at brink of extinction due to over-exploitation. *Aconitum heterophyllum* is one among them which has been domesticated under ex-situ conservation technique and efforts are under way for cryopreservation of the seeds. *Ginkgo biloba, S. sclarea*, C. sativus, and *Arnebia euchroma* are under domestication at IHBT. The Institute has released its own variety called HIMBALA by years of study on Valeriana jatamansi, *Himkachri* from Hedychium spicatum and *Himhaldi* from Curcuma aromatic. Herbal Garden Joginder Nagar situated at Mandi is devoted particularly to ex-situ conservation of medicinal plants. The prime objective of herbal garden is to develop Agro-Techniques of Medicinal Plants for making cultivation of medicinal plants to supplement the income of the people of State, generating awareness on the various aspects of medicinal plants among the people and by providing practical demonstration to the students of Ayurvedic College, Paprola for identification of medicinal plants. Varieties of tree, herbs, and climber species are cultivated in the garden. For instance, about 28 tree species are maintained in the garden among which C. tamala has 312 maintained plants and Emblica officinalis Gaertn. has 144 plants for conservation. Herb species contain 24 types among which O. kilimandscharicum Guerke have about 1200 plants maintained and 230 plants maintained for P. zeylanica L. species.

23.20 Indigenous Medicine: A Promising Soft Industry

Today, all around the globe, we can witness immensely increasing diseases affecting human body in all manners. The age-old saying "in healthy body rests healthy mind," clearly indicates that health is a major issue, and this necessitates physical exercise. Sidhbari is a suburb (small town) of Dharamshala town in the

Kangra district which is a center of yogic healing and tantric meditation practices. People from across the world and India itself visit the place for health benefit. The indigenous medicines or CAM like Acupuncture, Yoga, Ayurveda, Korean Oriental Medicine, spiritualism, meditation, and various other forms have proved to be highly effective in changing socio-economic environments. Tourism industry is the largest employment generator of the world. Yoga tours, CAM therapy, Ayurveda resorts, Yoga-themed retreats have become one of the most popular types of recreation generating livelihoods for millions. The indigenous Patanjali Ayurveda sales turnover in the year 2006–2007 was merely INR 0.5 million which jumped and reached up to 200 million in the year 2013–2014. Those who practice Yoga prefer Ayurveda products not only as medicine but also as health and food supplements. There are other big brands like Dabur and Himalaya which reach global population. Still, there is need to boost more indigenous companies based on traditional system under the *Make in India* initiative. The prices of these products are higher as per the purchasing capacity of the common man. This needs to be seen in a bigger frame to bring holistic health to all. The policies need to be planned and implemented in a phase-wise manner with reference to lifestyle diseases. Better geriatric care by opening *Ayu* geriatric regional care centers, training of medical and para medical professionals in *rasayana tantra*, better involvement at PHCs by trained people, domiciliary visits even at village level because 8% of above 60 years and 27% of above 80 years are confined to beds while 70% of elderly live in villages. Referral system, using IEC-media, folk-media for propagating and popularizing *Rasayana trantra* is important to achieve holistic health. Public–private partnership, tele-conferencing enabling medical insurance companies to recognize Ayurveda for re-imbursement in a bigger way, encouragement to families which support elderly by subsidized/ free medicines, nutrients, food, aids required for day-to-day activities and day care facility with Ayurvedic diet. Developing medicinal gardens which serve the dual role of primary health care medicine and giving health friendly, eco-friendly vegetation should be encouraged. Monitoring the schemes at district and state health societies goes a long way in helping elderly, diabetic, etc.

23.21 Promotion of Medical Tourism to Boost Livelihood

Tourism is the major source of economic growth. Many countries have transformed their economies through tourism development. Medical tourism is quite popular in various parts of the world. Tourism is found to be playing significant role in livelihood security, health and services, revitalization of local cultures, preservation of traditional skills, environmental management, and pro-poor growth. The demand for medicinal plants in global market provides economic opportunities to the farmers. Dharamshala in Himachal Pradesh has become a popular destination for the foreign tourists. The ecology and Tibetan culture of the place along with diversified forms of health care system (CAM) attract both domestic and foreign tourists. The convergence of spiritualism and yoga is a move forward towards unity and peace.

23.22 Conclusion

The existing agriculture scenario of the district is subsistence in nature but has great potential in different enterprises if the area from traditional crop could be diversified towards cash crops and medicinal plant production. Kangra is a repository of medicinal and aromatic plants but many of the species are at the verge of extinction due to their unscientific over-exploitation and climate change. Scientific cultivation, efficient management, and timely harvesting and processing of medicinal plants can conserve biodiversity and sustain human health and raise economic opportunities to local marginalized farmers. Today, we need healing physicians instead of medico technocrats. The value-added medicinal plants-based materials have great demand in the domestic as well as international market.

The development of herbal medicines from the rich traditional source requires an integrated approach. These institutions should be motivated at large scale and the awareness should reach not only to the progressive but also the general farmers, too. The advantages of herbal gardens and herbal home garden would be a practical approach in finding solution to conservation of medicinal plants. Every geographical region has its unique identification, thus, promoting geo-garden will certainly uplift the existing situation of herbal stream. The need for coordinated conservation action based on both in situ and ex-situ strategies is the answer to current depleting situation. In case of Himachal Pradesh, mere subsidizing to few progressive farmers and leaving behind huge marginal farmers is not going to work for long. Rather it will lead to loss of government money without any positive output. These are certain loopholes acting as challenge in the growth of herbal sector. In a nutshell, infrastructural development and herbal products cultivation, and manufacturing are ways forward to promote rural economy vis a vis livelihood security.

Mountainous areas of the Himalaya with rapid expansion of tourism and other economic activities have led to a spate of unplanned change in land use. Approximately, nearly 86% believed that there has been considerable loss of traditional knowledge system in Kangra. The use of Remote Sensing and Geographic Information System (GIS) along with communication tool, i.e., herbal portal is very essential for boosting marketing. KSIS software has been developed by IHBT Palampur as a database portal for the medicinal plants of Kangra.

Communication of traditional wisdom of farming practices, health management through medicinal plants should be disseminated by developing district level portal for medicinal plants. The portal for medicinal plants including producer and buyer and stakeholders, etc. can be connected through mobile SMS services to provide market information.

References

Arya V, Kaur R (2012) Kangrian Medicinal Flora. Pranav Prakashan, Kangra

Arya V, Thakur R, Kashyap CP (2012) Ethnomedicinal Flora of Kangra Valley of Himachal Pradesh, India. Journal of Tropical Medicinal Plants 13(2):177–187

Asian Development Bank (ADB) Report (2013) Tourism in Himachal Pradesh and The Way Ahead, KPMG, Source: www.scribd.com. Accessed on 9 Sept 2016

Bali JS (2005) Bioindustrial watershed management concepts and strategies, soil conservation Society of India, NAAS complex. I.A.R.I, New Delhi

Bhavana KR, Shreevathsa (2014) Medical geography. Charaka Samhita 35(4):371–377

Census of India (2011) District census handbook, part–XII a, village and town directory, Kangra, Series-03, Himachal Pradesh

Chauhan NS (2003) Important medicinal and aromatic plants of Himachal Pradesh. Indian Forester 129(8):979–998

Choudhary AK (2011) Impact of excessive export and exploitation on herbal biodiversity of Himachal-Himalayas. Department of Agroforestry and Environment, CSK HPKV, Palampur (HP)

Choudhary AK, Jaggi RC (2002) Role of Seabuckthorn in the conservation of fragile ecosystem of temperate Himalayan region. Tigerpaper 29(1):3–4

CSK Himachal Pradesh Krishi Vishwavidyalaya (2008) Mountain Agriculture System Information Files for Planning Niche-Based Agriculture Development in Kangra and Mandi

District Human Development Report, Kangra (2009) Planning Department, Himachal Pradesh

District Revenue Office, Kangra (2011) Government of Himachal Pradesh, Shimla

Dobriyal RM, Singh GS, Rao KS, Saxena KG (1997) Medicinal plant resources in Chhakinal watershed in the North-Western Himalaya. J Herbs Spices Med Plants 5:15–27

Farooque Ashraf MD, Mazumder AS, Mazumder R (2012) Review of Plumeria Acuminata. Int J Res Pharm Chem. Source: www.ijrpc.com. Accessed on 13 Sept 2015

Foundation of Revitalization of Local Health Tradition (FRLHT) (2016) Report Submitted to National Medicinal Plant Board, Government of India, Source: www.medicinalplants.in. Accessed on 27 Aug 2016

Frank E (2000) Rural livelihoods and diversity in developing countries. University Press, Oxford

HPFSMPP (2006) A Policy for Conserving and Strengthening Medicinal Plant Resource Base in Forest Areas as well as out of Forests for Use Towards Enhancing Health and Livelihood Security of The People of The State on Sustainable Basis, Forest Department, Dharamshala, Himachal Pradesh Forestry Sector Medicinal Plants Policy, Annexure "A" Notified Vide No. FFE-B-C (16)-8/2005 dated 27 Nov.

Ministry of Health (1987) Report

NAAS (2010) Protected cultivation in north-west Himalaya, policy paper 47. National Academy of Agricultural Sciences, New Delhi, p 16

National Medicinal Plant Board (2017) Source: www.nmpb.com. Accessed on 9 Oct 2017

Nitu, Singh RB (2016) Herbal-Faith Healing Tradition: A Spatial Study of Gaddi Healers in Kangra District of Himachal Pradesh. In: Tripathy B, Prasad DV (eds) Ethnoscience and traditional technology in India, vol 2. Aayu Publications, New Delhi

Nitu, Singh RB, Singh A (2016) Impact of climate change on tourism and medicinal plant wealth of Kangra, Himachal Pradesh. Nam Today LV(02):22–25

Pamela EJ, Slutsky Jeremiah, Singh Nilkanth, Khalsa SBS (2015) Yoga as a therapeutic intervention: a bibliometric analysis of published research studies from 1967–2013. J Complement Altern Med. Source: updates@liebertpubmail.com. Accessed on 4 June 2015

Parkash V, Aggarwal A (2010) Traditional uses of Ethnomedicinal plants of lower foothills of Himachal Pradesh-I. Indian J Tradit Knowl 9(3):519–521

Prakash V (2001) Indian medicinal plants: current status. In: Samant SS, Dhar U, Palni LMS (eds) Himalayan medicinal plants: potential and prospects. Nainital, Gyanodaya Prakashan, pp 45–65

Pratap T (2003) Himachal Farmers' livelihood concerns and opportunities in agriculture. CSK Himachal Pradesh Hill University, Palampur

Rana Isha, Randhawa SS (2016) Impact of Rainfall on Agriculture in Himachal Pradesh, Source: www.hpccc.gov.in. Accessed on 11 Apr 2016

Sankhian A (2007) Tea Cultivation and Sustainable Development: A Case Study of Kangra Valley, Ph.D. Thesis, CSK, Palampur Agriculture University, Source: www.Shodhganga.Inflibnit.Com. Accessed on 9 Apr 2016

Sharma HS, Chauhan SK (2013) Agricultural transformation in trans Himalayan region of Himachal Pradesh: cropping pattern, technology adoption and emerging challenges. Agric Econ Res Rev 26:2013

Sindhu PS, Govindaru V (2014) Spatio -temporal change of crop diversification in Kerala: an economic review, landscape ecology and water management. Springer, New York

Singh A (2011) Diversification in agriculture. Source: http//www.eoearth.org/view article/15175. Accessed on 15 July 2017

SPRA (1984) Deforestation in Himachal Pradesh, Society for Participatory Research in Asia, New Delhi

Statistical Abstract, Kangra (2014-2015) Directorate of economics and statistics, Himachal Pradesh

Sultan J, Singh CJ (2006) Herbal market Spectrum, Sirmor, Himachal Pradesh: Grameen Vikas Evam Paryavaran Jagran Samiti. Paonta Sahib 1(2):1–7

Venkateswarlu A (2005) Rural non-farm employment in Andhra Pradesh: 1961-2001. J Rural Dev 24(4):459–480

Part VI

Response to National, Regional, and Global Change

Sanitation Challenges and Policy Options in Developing Countries: A Critical Review

24

Nuthan Maharaj and Brij Maharaj

Abstract

Sanitation is a looming crisis with many policy challenges in the global South. Many developing countries struggle to cope with issues of sanitation which is exacerbated by water shortages, behavioural issues, and rapid urbanisation with limited resources. Sanitation challenges include poverty, lack of political will, limited or no community participation, inadequate gender inclusion, unreliable data and, finally, a lack of an integrated approach between the various stakeholders - government, private sector and civil society. A critical review of these challenges demonstrates that moving from the guidelines of Millennium Development Goals (MDG) to the Sustainable Development Goals (SDGs) has brought marginal improvements to solve sanitation related issues. In this context, this chapter presents a critical review of the endemic sanitation challenges across the developing nations and assesses various policy options to address sanitation challenges. This chapter also recognises that sanitation is a human right incorporated in the SDGs with the aim to ameliorate the conditions of those without access to basic sanitation and associated challenges. In order to reduce sanitation challenges, this chapter proposes the adoption of a multi-stakeholder, inclusive approach, comprising local government, communities, and the small enterprises sector with a view to achieving community empowerment to promote equitable access to hygiene needs, and advocating for political commitment, promoting gender equity, and enhancing youth involvement.

Keywords

Developing countries · Ecological sustainability · Millennium Development Goals (MDG) · Multi-stakeholder collaboration · Sanitation challenges · SDG · Swachh Bharat Abhiyan (Clean India Mission)

24.1 Introduction

The sanitation crisis is a major challenge in most developing countries. It is exacerbated by increasing urbanisation, poverty, lack of political will, poor institutional response, and limited financial resources to address the cumulative demands. Despite greater commitment to provide adequate sanitation and water following the adoption of the Millennium Development Goals (MDGs),

N. Maharaj
Economic Development Unit, Durban Municipality, Durban, South Africa

B. Maharaj (✉)
University of KwaZulu-Natal, Durban, South Africa
e-mail: maharajB@ukzn.ac.za

B. Thakur et al. (eds.), *Resource Management, Sustainable Development and Governance*, Sustainable Development Goals Series, https://doi.org/10.1007/978-3-030-85839-1_24

and subsequently, the Sustainable Development Goals (SDGs), marginal improvement is noted. Vulnerable groups, especially women and children, bear the consequences. Open defecation remains a scourge in the twenty-first century. Poor sanitation is a risk to human health, the economy, and the environment.

The provision of sanitation has been plagued with two critical challenges in the developing countries for decades, and this continues in the twenty-first century. Firstly, the urgency to satisfy a human need through adequate sanitation access for the millions who are forced to resort to primitive and unhygienic methods in the absence of improved ablution facilities. Secondly, the governance and institutional reforms in the sector have not resolved the sanitation crisis, due to fiscal constraints, insufficient capacity, and the unresponsiveness to local context and societal-specific sanitation problems.

Box 24.1 Sustainable Development Goals

Sustainable Development Goal 6 aims to ensure the availability of sustainable management of water and sanitation worldwide. Sanitation is essential for the survival and development of children and adults, more so in the developing world. In the twenty-first century both Millennium Development Goals (MDGs) and Sustainable Development Goals (SDGs) have aimed to reduce the demographics drastically and dramatically with no sustainable access to water and sanitation. So far, and as we enter the last decade of Agenda 2030 the effectiveness of the targets and indicators of SDG 6 has raised concerns among both academics and practitioners. While the urban population in developing countries has more than doubled between 1950 and 2000, its rural population has no access to improved sanitation. This has caused operational challenges of mammoth proportion and made developing regions vulnerable due to population growth, urbanization and increased industrialisation, and water competition which has been threatening agricultural production and food security affecting water quality. In this chapter, the authors not only review this emerging context vis-à-vis sanitation challenges but also recognise that sanitation issues are further compounded by the impacts of climate change, which hamper the achievement of SDG 6 if the broader issues are left unaddressed. This chapter as it assesses different policy measures also recognises that water scarcity, poor water quality, and inadequate sanitation all affect the health of ecosystems, societies, and economies and in the end will negatively impact the achievements of the other SDGs as well.

This chapter presents a critical review of the endemic sanitation challenges experienced in the developing countries in the twenty-first century. It also assesses various policy options to address the sanitation challenge. The chapter is divided into five sections and begins with an outline of sanitation challenges with specific reference to vulnerable groups, followed by a review of saniation and the Millennial Development Goals. The right to sanitation is the theme of the third section, the sustainable development goals are discussed in the fourth section. The final section assesses different policy measures which include multi-stakeholder collaboration and partnerships, recycling human waste and ecological sustainability; supply-driven sanitation solutions; innovation and enterprise, and alternative pro-poor sanitation options.

24.2 Sanitation Challenges and Vulnerability

In 2000, one-sixth (1.1 billion people) of the global population did not have access to a safe water supply. About 2.4 billion (two-fifths) lacked access to improved sanitation. The majority who lacked access to these basic services were in Asia and Africa, where there was also a sharp rural-urban divide. Eighty percent of those in rural areas (2 billion) lacked satisfactory sanitation (WHO/UNICEF, 2000, p. 1). The UN (2019, p. 1) has emphasised that: "A toilet is not just a toilet. It is a life-saver, dignity-protector and opportunity-maker".

Women who are the primary care givers are burdened with the responsibility of managing household sanitary needs. Furthermore, primitive methods of defecating place women and children at risk of disease and even death. According to Ramachandraiah (2001, p. 620), of the 37 most fatal ailments in developing countries, 21 are caused by water and sanitation related diseases, with 1.5 million children under the age of 5 years dying annually. Similarly, there is a statistically significant relationship between maternal, infant, and child mortality due to the lack of access to water and sanitation (Cheng et al. 2012).

Poor sanitation contributes to blindness causing Trachoma. Helminth infections transmitted mainly through exposure to faeces which are exacerbated by open defecation. Schistosomiasis resulting in debilitated growth and impairment is contracted through exposure to contaminated faeces and urine. Although these diseases occur in adults as well, children are most susceptible to these fatal illnesses. While medical treatment through antibiotics and other medicines provides mitigation, improved sanitation shows greater promise of prevention (Mara et al. 2010, p. 1). The turn of the century witnessed greater focus on goals and targets to improve sanitation for the poor.

24.3 Sanitation and the Millennium Development Goals

The Millennium Development Goals (MDGs) were the universal framework advanced by the global community to improve the quality of life of people around the world. The MDGs gained impetus through pledges made at the 2000 Millennium Summit at the General Assembly of the United Nations (UN). The UN sealed the pledge of all countries to meet specific targets aimed at addressing critical human development

Table 24.1 Water and Sanitation—Advancing the MDGs

MDGS	Contribution of improved drinking water and sanitation
Goal 1: Eradicate Extreme Poverty and Hunger	• The security of household livelihoods rests on the health of its members; adults who are ill themselves or must care for sick children are less productive. • Illnesses caused by unsafe drinking water and inadequate sanitation generate high health costs relative to income for the poor. • Healthy people are better able to absorb nutrients in food than those suffering from water-related diseases, particularly helminths, which rob their hosts of calories. • The time lost because of long-distance water collection and poor health contributes to poverty and reduced food security.
Goal 2: Achieve universal Primary education	• Improved health and reduced water-carrying burdens improve school attendance, especially among girls. • Having separate sanitation facilities for girls and boys in school increases girls' attendance, especially after they enter adolescence.
Goal 3: Promote gender Equality and empower women	• Reduced time, health and care-giving burdens from improved water services give women more time for productive endeavours, adult education and leisure. • Water sources and sanitation facilities closer to home put women and girls at less risk of assault while collecting water or searching for privacy.
Goal 4: Reduce child mortality	• Improved sanitation and drinking water sources reduce infant and child morbidity and mortality
Goal 5: Improve maternal health	• Accessible sources of water reduce labour burdens and health problems resulting from water portage, reducing maternal mortality risks. • Safe drinking water and basic sanitation are needed in health care facilities to ensure basic hygiene practices following delivery.

(continued)

Table 24.1 (continued)

MDGS	Contribution of improved drinking water and sanitation
Goal 6: Combat HIV/AIDS, Malaria and other diseases	• Safe drinking water and basic sanitation help prevent water-related diseases, including diarrhoeal diseases, schistosomiasis, filariasis, trachoma and helminths. • The reliability of drinking water supplies and improved water management in human settlement areas reduce transmission risks of malaria and dengue fever.
Goal 7: Ensure environmental Sustainability	• Adequate treatment and disposal of wastewater contributes to better ecosystem conservation and less pressure on scarce freshwater resources. • Careful use of water resources prevents contamination of groundwater and helps minimise the cost of water treatment.
Goal 8: Develop a global Partnership for Development	• Development agendas and partnerships should recognise the fundamental role that safe drinking water and basic sanitation play in economic and social development.

Source: WHO/UNICEF (2004, p. 9)

problems and eradicating extreme poverty by 2015.

Water has intrinsic value in improving sanitation, health, and poverty reduction and was formally recognised in MDG seven. The target was to reduce the population with inadequate and unsustainable access to safe drinking water and basic sanitation by half by 2015 (UNDP 2003). However, improving access to water and sanitation was intrinsic to the realisation of all the MDGs as illustrated in Table 24.1. For example, unsafe water and poor sanitation contribute to poor hygiene, illness, infant and child mortality, poor school attendance especially for teenage girls. Improved water services and access to sanitation result in more productive time for women, reduced child mortality and better maternal health (Table 24.1).

Some progress was made in terms of access to safe drinking water. In 1990, (the MDGs baseline year), 76% of the global population had access to safe drinking water, but this had increased to 90% in 2012. However, there were regional variations and unevenness, especially between urban and rural, and the affluent and poor (WHO 2018). However, progress in terms of provision of basic sanitation was disappointing:

> In 2012, 2.5 billion people did not have access to improved sanitation facilities, with 1 billion these people still practicing open defecation. The number of people living in urban areas without access to improved sanitation is increasing because of rapid growth in the size of urban populations (WHO 2018, p. 1).

According to World Health Organisation (2012), a good indicator of improved sanitation in urban and rural areas to meet MDG targets must translate into a step-up in sanitation facilities. There must be a shift from a range of primitive mechanisms used for defecation. These less hygienic methods include bucket toilets, flush or pour-flush that deposit sludge into rivers or drains, open pit latrine, hanging toilets, i.e. using packets to relieve oneself, or the use of open fields where no other option is available. The step-up will include facilities which allow least exposure and handling of faecal matter. This may include a flush system that is transported by piped sewer, the use of a septic tank, a ventilated pit latrine with a slab to cover faecal matter, as well as composting toilets where the contents are allowed to dry over a period and then, used for agricultural purposes (WHO 2012).

As illustrated in Fig. 24.1, there was significant reduction in the global rate of open defecation. In the year 2000, 1.3 billion people (or 21% of the global population) were forced to practice open defecation. In 2017, this was reduced to 673 million (or 9%). Nevertheless, it is evident from Fig. 24.1 that many African and Asian countries were still struggling to reduce the number of people with inadequate access to sanitation. Notwithstanding, the flagship Swachh Bharat Abhiyan (Clean India Mission) sanitation project (Jain et al. 2020), India still has the largest number, with 344 million people practicing open defecation, followed by Nigeria and Indonesia (Kashiwase 2019).

An important development in the first decade of the twenty-first century was recognition of access to sanitation as a human right.

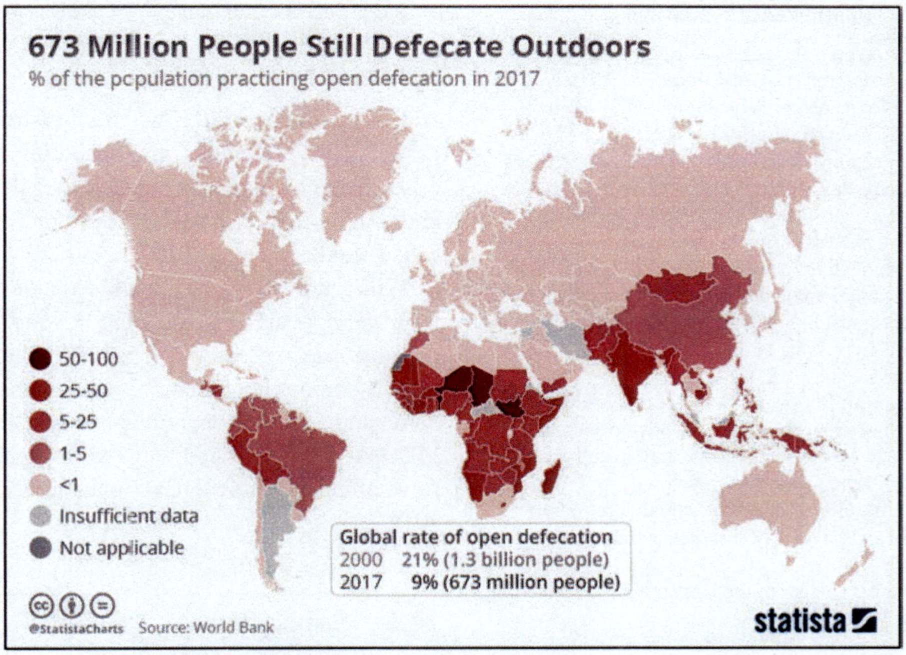

Fig. 24.1 Percentage of Population Practicing Open Defecation - 2017. Source:https://www.statista.com/chart/18419/progress-against-open-defecation/(accessed 10/02/2020)

24.4 Right to Water and Sanitation

Access to water and sanitation was recognised as a fundamental human right by the United Nations General Assembly on 28 July 2010 (United Nations 2010). The principles underlying rights to sanitation include equality and non-discrimination, right to information, participation, and accountability (Baer 2017). The realisation of the right to water and sanitation depended on availability, quality, accessibility, and affordability (Table 24.2). While the right is afforded to all citizens, the most vulnerable are women, children, people with disabilities and the aged, who require special infrastructure provision, especially regarding basic services, such as sanitation. Non-provision of sanitation facilities for vulnerable groups is a contravention of human rights (Mwebaza 2010; Mara et al. 2010; Mehta and Movik 2010; Reddy and Batchelor 2012, Bhanushali 2019).

Sanitation legislation and policy has failed to meet the practical sanitation requirements of the disabled. The inability to integrate the needs of disabled is not only discriminating in terms of the human rights of the individual, but also encroaches on family members or caregivers. Family members are constrained by the lack of adequately designed facilities at household level, restricting their human and economic engagements (Matsebe 2006).

The failure to provide adequate sanitation contributes to the triple discrimination and exploitation of women (Mehta and Movik 2010; Adams et al. 2019; Koonan 2019; McFarlane 2019). Women are more susceptible to infection in the absence of proper sanitation (Mara et al. 2010). In most of the poor households, women are burdened with the maintenance of sanitation facilities and provision of water consuming many hours of their day (Azeez et al. 2019).

Due to increased responsibility of family and household sanitation demands, women are restricted from engaging in productive income-generating activities, thereby perpetuating poverty and hardship (De Albuquerque and Winkler 2010). There are wider social repercussions, including "reduced school attendance, inconvenience, wasted time, and lack of privacy and

Table 24.2 Human Rights to Sanitation

Principles underlying the human right to sanitation:
1. **Non-discrimination and equality:** All people must be able to access adequate sanitation services, without discrimination, prioritising the most vulnerable and disadvantaged individuals and groups.
2. **Participation:** Everyone must be able to participate in decisions relating to their access to sanitation without discrimination.
3. **The right to information:** Information relating to access to sanitation, including planned programmes and projects must be freely available to those who will be affected, in relevant languages and through appropriate media.
4. **Accountability (monitoring and access to justice):** States must be able to be held to account for any failure to ensure access to sanitation, and access (and lack of access) must be monitored.
5. **Sustainability:** Access to sanitation must be financially and physically sustainable, including in the long term.
The normative content of the human right to sanitation is defined by:
1. **Availability:** A sufficient number of sanitation facilities must be available for all individuals.
2. **Accessibility:** Sanitation services must be accessible to everyone within, or in the immediate vicinity, of household, health and educational institution, public institutions and places and workplace. Physical security must not be threatened when accessing facilities.
3. **Quality:** Sanitation facilities must be hygienically and technically safe to use. To ensure good hygiene, access to water for cleansing and handwashing at critical times is essential.
4. **Affordability:** The price of sanitation and services must be affordable for all without compromising the ability to pay for other essential necessities guaranteed by human rights such as water, food, housing and health care.
5. **Acceptability:** Services, in particular sanitation facilities, have to be culturally acceptable. This will often require gender-specific facilities, constructed to ensure privacy and dignity.

Source: WHO (2018, p. 3)

security for women" (Asian Development Bank 2009a, b, p. 11). When safe, usable ablution facilities are not at hand,

> women and girls face three types of toilet insecurity: (1) the material reality for many women and girls that they do not have access to a toilet; (2) the risk of venturing out for open defecation if there is no toilet; and (3) having access to a public toilet, but one that is unusable (e.g., filthy) or unsafe (e.g., insufficient lighting), so that women and girls accept the risk of going for open defecation (O'Reilly 2016, p. 19).

Hence, in addition to the lack of infrastructure or availability of facilities, the sanitation crisis is exacerbated by outdated, superstitious traditions, and discrimination based on religion, caste, or tribe (Mukherjee et al. 2020).

While not all targets were met, the MDGs were very focused on reducing poverty and the progress was measurable. There is agreement in the UN community of nations that a global development agenda must continue beyond 2015. The MDGs was replaced by the Sustainable Development Goals (SDGs) for the next 15 years, 2016–2030.

24.5 Sanitation and the Sustainable Development Goals

Sustainable Development Goal 6 focused specifically on water and sanitation. The emphasis was on universal access to basic services, especially safe and affordable drinking water, as well as the elimination of open defecation (Table 24.2). The objective for SDG 6 was to: "ensure availability and sustainable management of water and sanitation for all". More specifically, the following targets were set for 2030:

- 6.1 achieve universal and equitable access to safe and affordable drinking water for all.
- 6.2 achieve access to adequate and equitable sanitation and hygiene for all, and end open defecation, paying special attention to the needs of women and girls and those in vulnerable situations.
- 6.3 improve water quality by reducing pollution, eliminating dumping, and minimising release of hazardous chemicals and materials, halving the proportion of untreated wastewater, and substantially increasing recycling and safe reuse globally.
- 6.4 substantially increase water-use efficiency across all sectors and ensure sustainable withdrawals and supply of freshwater to address

Table 24.3 Water, Sanitation and Hygiene Challenges in 2015

Drinking water	Sanitation	Hygiene
• 71% of the global population (5.2 billion people) used a **safely managed** drinking water service; that is, one located on premises, available when Needed and free from contamination. • Eight out of ten people (5.8 billion) used improved sources with **water available when needed**. • Three quarters of the global population (5.4 billion) used improved sources **located on premises**. • Three out of four people (5.4 billion) used improved sources **free from contamination**. • 844 million people still lacked even a **basic** drinking water service. • 263 million people spent over 30 min per round trip to collect water from an improved source (a **limited** drinking water service). • 159 million people still collected drinking water directly from **surface water** sources, 58% lived in sub-Saharan Africa.	• 39% of the global population (2.9 billion people) used a **safely managed** sanitation service; that is, excreta safely disposed of in situ or treated off-site. • 27% of the global population (1.9 billion people) used private sanitation facilities connected to sewers from which **wastewater was treated**. • 13% of the global population (0.9 billion people) used toilets or latrines where excreta were **disposed of in situ**. • Available data were insufficient to make a global estimate of the proportion of population using septic tanks and latrines From which excreta are **emptied and treated off-site**. • 2.3 billion people still lacked even a **basic** sanitation service. • 600 million people used a **limited** sanitation service. • 892 million people worldwide still practised **open defecation**.	• 70 countries had comparable data available on handwashing with soap and water, representing 30% of the global population. • Coverage of **basic handwashing** facilities with soap and water varied from 15% in sub-Saharan Africa to 76% in Western Asia and northern Africa, but data are currently insufficient to produce a global estimate, or estimates for other SDG regions. • Iin least developed countries, 27 per cent of the population had basic handwashing facilities with soap and water, while 26% had handwashing facilities lacking soap or water. The remaining 47% had no facility. • In sub-Saharan Africa, three out of five people with basic handwashing facilities (89 million people) lived in urban areas. • Many high-income countries lacked sufficient data to estimate the population with basic handwashing facilities

Source: WHO/UNICEF (2017, p. 66)

water scarcity, and substantially reduce the number of people suffering from water scarcity.

- 6.5 implement integrated water resource management at all levels, including through transboundary cooperation as appropriate.
- 6.6 protect and restore water-related ecosystems, including mountains, forests, wetlands, rivers, aquifers, and lakes (United Nations 2016).

Water–Sanitation–Hygiene (WASH) was critical for the realisation of the 2030 SDGs (UN-Water 2016). The SDGs were interlinked and mutually reinforcing, and access to water and sanitation was integral to the realisation of several other goals:

> Examples of synergies include increasing access to water supply, sanitation and hygiene (WASH) [6.1, 6.2] in homes, healthcare facilities, schools, and workplaces, complemented by wastewater treatment [6.3], as a way to reduce risk of water-borne disease [3.1–3.3, 3.9] and malnutrition [2.2]; support education [4.1–4.5] and a productive workforce [8.5, 8.8]; and address poverty [1.1, 1.2, 1.4], gender inequality [5.1, 5.2, 5.4, 5.5] and other inequality [10.1–10.3] (UN-Water 2016, p. 6).

Several systemic challenges impeded developing countries in their progress towards achieving their MDG targets and are likely to remain an obstacle in realising the SDGs as well. A key concern was the insufficient investment in water and sanitation programmes (Herrera 2019).

Water cannot be substituted, it is at the forefront of sustainable development and a key factor for socio-economic development and food production. The unavailability of water impacts negatively on personal and sanitation hygiene practices. The absence of water for hand washing promotes ill health. In 2016, a survey of 36

Table 24.4 SDG 6 Global goals, targets and indicators for drinking water, sanitation and hygiene

Wash sector goal	SDG global target	SDG global indicator
Ending open defecation	6.2 By 2030, achieve access to adequate and equitable sanitation and hygiene for all and end open defecation, paying special attention to the needs of women and girls and those in vulnerable situations	6.2.1 population practising open defecation
Achieving universal access to basic services	6.4 By 2030, ensure all men and women, in particular the poor and vulnerable, have equal rights to economic resources, as well as access to basic services	6.4.1 population living in households with access to basic services (including basic drinking water, sanitation and hygiene)
Progress towards safely managed services	6.1 By 2030, achieve universal and equitable access to safe and affordable drinking water for all 6.2 By 2030, achieve access to adequate and equitable sanitation and hygiene for all and end open defecation, paying special attention to the needs of women and girls and those in vulnerable situations	6.1.1 Population using safely managed drinking water services 6.2.1 Population using safely managed sanitation services 6.2.1 Population with a basic handwashing facility with soap and water available on premises

Source: WHO/UNICEF (2017, p. 2)

African countries revealed that 45% of households did not have sufficient clean water, and 51% had to leave their accommodation to obtain water. One-third did not have access to piped water, and two-thirds did not have access to sewer systems (Walker 2016). According to report prepared by the United Cities and Local Governments (UCLG 2018, p. 76), "844 million people around the world still lack basic access to water services and 2.3 billion people lack access to sanitation…".

The absence of adequate sanitation contributes to contamination and the rise in water-borne diseases, which inevitably impact negatively on the economy. The importance of clean water and adequate sanitation on productivity cannot be overemphasised for every dollar spent on the provision of adequate water and sanitation, nine dollars' worth of productive activity is yielded (Tissington 2011). The impact of inadequate water and sanitation services burdens the economy through low productivity exacerbated by absenteeism and a sickly workforce that are living under unhygienic and diseased conditions (Ramachandraiah 2001). The final section of this chapter assesses strategies to improve sanitation (Table 24.4).

24.6 Strategies to Improve Sanitation

Sanitation is about people. The need for dignity is inherent in all human beings. According to Mwebaza (2010: 10) there are four important factors to consider when providing basic sanitation:

> …accessibility on a sustainable basis; the ability to meet the basic human needs of safety, hygiene, and convenience; a service provision for both excreta and sullage disposal; and culmination in a clean and healthy living environment.

Furthermore, in 2018, the WHO proposed new guidelines on sanitation and health which can be summarised as follows: firstly, sanitation interventions should ensure entire communities which have access to toilets that safely contain excreta; secondly, the full sanitation system should undergo local health risk assessments to protect individuals and communities from exposure to excreta–whether this be from unsafe toilets, leaking storage or inadequate treatment, thirdly, sanitation should be integrated into regular local government-led planning and service provision to avert the higher costs associated with retrofitting sanitation and to ensure sustainability, fourthly, the health sector should invest more and play a coordinating role in sanitation planning to protect public health (WHO 2018, p. 1).

Improved sanitation facilities should prevent human contact with waste, make sure that methods of disposal are controlled and environmentally friendly, thus ensuring maximum protection of human health and well-being.

Successful sanitation interventions also hinge on the user's acceptability of inventions to better manage human waste. Any innovation which compromises human dignity is bound to fail. A substantial increase in national budget allocations to sanitation and enhanced political will amongst most local government institutions, together with the need for an overhaul in governance mechanisms, are critical for improving global access to sanitation by 2015 (Mwebaza 2010; McFarlane and Silver 2017). Improved governance through better coordination between national, regional, and local government and community levels will enhance the possibility of improving sanitation for the poor (Mwebaza 2010; Ako et al. 2010; Kennedy-Walker et al. 2015; McFarlane and Silver 2017). The following policy options will be assessed in this section: "supply driven" sanitation solutions; multi-stakeholder collaboration and partnerships for improved sanitation; recycling human waste and ecological sustainability; innovation and enterprise in sanitation provision; and alternative pro-poor sanitation solutions.

24.6.1 Government-Led Sanitation Solutions

Government-led or "supply-driven" sanitation projects have arguably had limited success amidst scarce resources, in meeting the varied and enormous demands for sanitation worldwide (De Albuquerque and Winkler 2010; Hueso and Bell 2013). During the International Sanitation Decade 1980–1990, India launched the subsidised Central Rural Sanitation Programme (CRSP) aimed at improving the lives of people and saving the dignity of women. However, almost two decades into implementation, neither funding aid nor good policy has enabled expeditious delivery or the expected success rate (Ganguly 2008).

Ganguly (2008) adds that despite technical assistance and advice from WHO, UNICEF, and the UNDP, the 6-year review of the CRSP revealed that cultural practices and perceptions have impacted on people's use of the facilities. It was apparent that user rejection was due to the lack of information and education about the use of the facility. Community participation was minimum or non-existent. The review confirmed that the subsidised supply-driven, top-down model managed and guided by government did not work (Ganguly 2008). Hence, improving access to sanitation is not merely a matter of improving the physical facilities, but also requires intensive community education and sensitisation (Reddy and Batchelor 2012). New approaches to sanitation provision generally have low or no subsidies, for several reasons, including: firstly, improvements in sanitation coverage typically stop once subsidy budgets run out, secondly, subsidies lead to inappropriate facility designs that are often too expensive, thirdly, subsidies are often not captured by the poor, who need sanitation most, fourthly, subsidies can potentially destroy a developing sanitation market by creating perverse incentives; and finally, households often do not use and maintain latrines that are heavily subsidised (Graham 2011, p. 23).

24.6.2 Multi-Stakeholder Collaboration and Partnerships for Improved Sanitation

Participatory approaches have emerged in response to the challenges associated with supply-driven strategies. The governance approach in the sanitation sector has been increasingly evolving to accommodate an array of stakeholders, including the private sector and community organisations, and is a shift the purely government-led intervention (Graham 2011; Van Vliet et al. 2011; Adams and Boateng 2018). The goal is to ensure that all stakeholders are consulted and participate in all phases of sanitation projects to ensure that the needs and choices are taken seriously, and solutions are suitable to local environmental conditions (Graham 2011).

Partnerships between the private sector, non-governmental sector, communities, and the state are recommended for resource mobilisation and sustainable sanitation provision (Tukahirwa et al. 2010; Powell and Yurchenko 2020). Partnership networks are a conduit for scaling up of pro-poor sanitation as well as exploring effective options for sustainable systems (Asian Development Bank 2009a, b; Van Vliet et al. 2011). Tukahirwa et al. (2010, p. 12) observe the emergence of a "modernized mixture model", where various sectors work in tandem to meet pro-poor sanitation needs, but also note the limited success of a private sector market-led approach due to profit orientation.

A study by Tukahirwa et al. (2010) observed greater success when civic organisations drive sanitation programs. Emphasis should be placed on the need for districts and communities to participate in decision-making to resolve problems, and to reduce the cancer of corruption in the delivery of sanitation (Mwebaza 2010).

24.6.3 Recycling Human Waste and Ecological Sustainability

Access to improved sanitation has a positive impact on the environment, health, social and economic status of people in developing countries (Mara et al. 2010; Kumar et al. 2011; Saleem et al. 2019). In India, for example, poor sanitation systems, shoddy sludge management, and unhygienic sanitation practices have grave environmental impacts. Sewage effluence deposited in rivers and streams is the main source of water contamination (Ramachandraiah 2001). Only 30% of the wastewater is being treated, with the balance deposited into rivers, streams, and open fields, exacerbating the challenge of clean water provision, and the risk of disease from faecal contaminated water. Innovative human waste management could avert environmental and health impact on poor communities (Asian Development Bank 2009a, b).

In Bangladesh, the impact of climate change, with seasonal flooding in slum settlements, exposed communities to unhygienic swampy living conditions which was contaminated by untreated stagnant sewerage (Rahman and Rahman 2015. In these desperate conditions, communities resorted to "hanging toilets" which emptied into the drains and rivers which are main source of water for washing and drinking, thereby exacerbating the crisis of human health and environmental integrity (Münch et al. 2009).

The most common means of human waste disposal practised in Kiberia were the defecation in polythene bags which were subsequently flung into the open fields and, hence, dubbed "flying toilets" (Corburn and Karanja 2014). This was a primary method of excreta disposal, and more than 60% of people in Kiberia engaged in this practice, which posed immense environmental and human health risks, as plastic bags blocked drains promoting flooding and exposure to the contents caused disease and illness (Münch et al. 2009,p. 3).

There is significant potential for sustainable ecological practices for energy and nutrient production through the recycling of human waste. Biogas and nutrients for agricultural use could be derived from processing human waste (Asian Development Bank 2009a, b). However, common human habits are difficult to break. Introducing innovation, therefore, meant that users needed to embrace new technology and use them correctly to improve environmental integrity and their personal health. In the slums of Kenya and Bangladesh, the use of a biodegradable sanitation "peepoo" bag was piloted. The technology is simply a packet which allowed the user privacy, minimal contact with the faeces and safe disposal. This sanitation technology is a scientifically developed ammonia-based bag which reacts to urea, and, in turn, acts as a catalyst for destroying dangerous pathogens and decomposing the content for use as fertiliser (Münch et al. 2009).

According to Factura et al. (2010), scientific methods of converting faecal matter into biowaste for agricultural use could also solve societal food security and faecal management challenges. Their studies have shown that faecal waste may be converted to highly fertile material hygienically and sustainably.

The application of anaerobic vermi-composting and lacto-fermentation through the "tera petra sanitation" solution yields an odourless product suitable for urban agriculture. This application was tested in Brazil and shown to be ideal in areas where upgrades of pit latrines, urine diversion, and even bucket toilets are utilised. Factura et al. (2010) stressed that the success of the on-site application, however, depends on effective participatory planning, well-guided fermentation of the product, and effectively organised professional operations and maintenance for optimal, hygienic, and pollution free recycling of faecal matter.

24.6.4 Innovation and Enterprise in Sanitation Provision

In some developing countries, the market-driven model proved to be "demand responsive", yielding greater success and customer satisfaction. In certain countries, even the poor preferred a market-driven approach which gives them options with the choice of facilities they could access (De Albuquerque and Winkler 2010).

A study conducted in ten African countries by the UNDP-World Bank Water and Sanitation Programme between 1998 and 1999 recorded that peri-urban sanitation systems in African cities did not have bulk water-borne infrastructure. Sanitation services were unregulated and informal, with reliance on public toilets as the only facility in certain areas. Being outside the mandate of government, the cleaning of latrine systems was largely done by small-scale entrepreneurs who also worked in an unregulated and untaxed informal sector, which employed up to 90% of the urban workers. These entrepreneurs worked in a highly competitive market as their services were unsubsidised and customer satisfaction was the only criterion to keep them in business. They were independent and were therefore able to innovate around the type of service and facility they supported and maintained (Baskovich 2008, p. 2).

A study on sanitation entrepreneurship in rural Indonesia concluded that "insufficient customer demand, inadequate capacity building opportunities, lack of financing options for entrepreneurs and their customers, and limited government support" undermined the success of sanitation enterprises (Murta et al. 2018, p. 343).

Solo (1999) also noted success with the small-scale entrepreneurship and NGO driven services segment, which he coined the "other" sector, in sanitation provision. The "other" sector initiatives introduced a paradigm shift in countries like India, China, Tanzania, and Brazil, in providing services for the poor, including sanitation. Its proven success lies in its ability to "produce appropriate models and fill every circumstance and need" (Solo 1999, p. 121). Such models evolved to suit user needs. They have become a preferred choice of service providers due to their good customer relations and service quality, their ability to respond and grow with the demands, their capacity to reach the poor with flexibility in choice of technology and pricing of services. Scholars have iterated that the flexible and affordable sanitation solutions yield greatest satisfaction through improved services (Solo 1999; Reddy and Batchelor 2012). These strategies were subsequently incorporated in alternate pro-poor policies.

24.6.5 Alternative Pro-Poor Sanitation Solution (APSS)

The Alternative Pro-Poor Sanitation Solutions (APPS) approach views the poor as "customers" rather than "beneficiaries" waiting for government to deliver. This was a pilot project in Peru which offered a market-related solution for poor communities, with opportunities for the poor to enter the informal sector market through private sector driven sanitation solutions. The pilot study was mindful of the objectives of social inclusion, equality, and solidarity which have a bearing on societal behaviour and practices (Baskovich 2008).

The APSS integrated market-related, partnership-driven model introduced behaviour

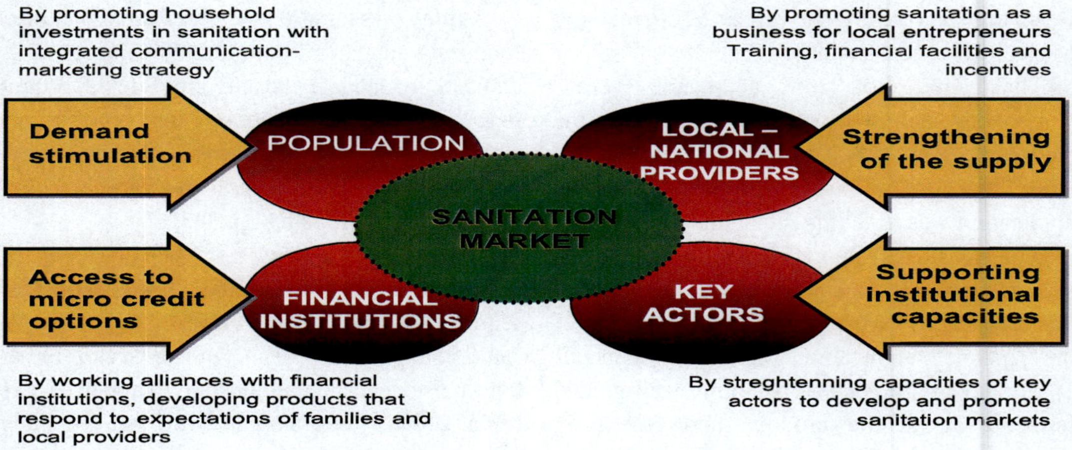

Fig. 24.2 APPS Integrated Market-Related Partnership-Driven Model. Source: Adapted from Baskovich (2008, p. 4)

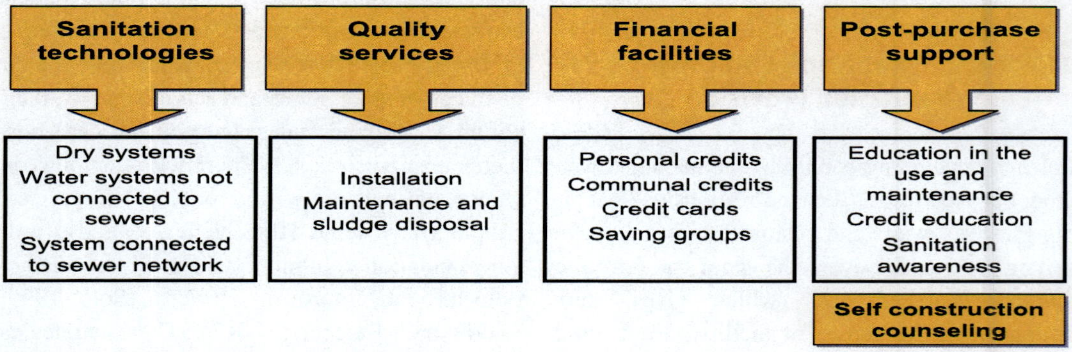

Fig. 24.3 Integrated Sanitation Package. Source: Adapted from Baskovich (2008, p. 8)

change in local communities seeking a local response to a local problem. Figure 24.2 illustrates the processes in introducing and marketing the APSS market approach. Communication, social marketing, promoting behavioural change, and the offer of financing options encouraged poor communities to see business initiatives in sanitation provision. It was viewed as an opportunity to improve their living standards, well-being, and environmental conditions, and restoring a sense of dignity. Sanitation options gave users a choice of a desired affordable system through an integrated sanitation package illustrated in Fig. 24.3.

Several lessons emerged from the adoption of the APSS model, and these included: firstly, commitment to activities of lower income groups or smaller enterprises increased through engagement in the larger economy; secondly, larger private sector companies increased their interests in social corporate responsibility; thirdly, the initiative shed new perspectives on restoring macroeconomic stability, peace, and democracy in Peru (Baskovich 2008).

The APSS model offers a new approach for market-related provision and increased choice for "customers" providing an opportunity for growing a business-like mindset for the sanitation market. However, these come with a series of challenges, namely:

1. Meeting people's demands require on-going innovation at low cost.
2. Endorsing behavioural change as a medium to long term task requiring financial support.

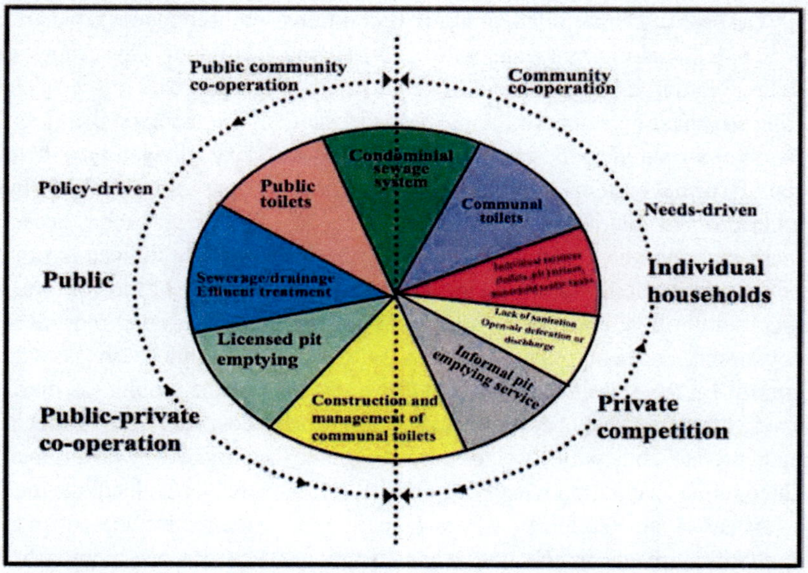

Fig. 24.4 The "Sanitation Wheel". Source: Adapted from Allen et al. (2006: 14)

3. Sustained private sector involvement required optimal public sector support regarding regulation and promotion of market-related services.
4. Impact of international financial sector on the micro-financiers.
5. A recommended government subsidised model does not augur well for the sanitation market and could disintegrate the APSS purpose of market-related sanitation provision.
6. The market-related APSS approach calls for a change in paradigm, roles, and functions of the different actors in sanitation governance (Baskovich 2008, p. 8).

The APSS market approach focused on quality, sustainable sanitation services. It responded to people's expectations, creating a sense of social inclusion and satisfaction, and promising improved basic services for the poor. The private sector engagement also provided an opportunity for skills transfers and knowledge building in local communities and emerging entrepreneurs, with emphasis on customer satisfaction. Improved product quality, branding and marketing, including research on innovation and environmental sustainability, were brought to the fore when local communities engaged as partners.

According to Michelutti (2008) community-driven projects are aimed at empowering local communities, while delivering water and sanitation projects. In Tanzania, for example, the success of such projects was dependent on the communities' ability to develop efficient projects together with an effective governance plan. Most often, community freedom in prioritising project intervention focused largely on water and neglected the need for proper sanitation. The institutional systems in sanitation (and water) governance in Tanzania operate within a formal, informal, and intermediate mechanism, as follows: (1) The Formal Sector comprises the policymakers, regulator, and private companies hired by the services authority to provide the services to all areas, including the informal settlements. Co-operative organisations formed partnerships with the formal sector and provided support with local intervention in terms of finance and consultation of local actors; (2) The Informal System served as a means for service acquisition by low-income settlements that are not reached by formal means of distribution; (3) The Intermediate System refers to the negotiators or facilitators between the formal and informal systems. They may be legal or illegal actors. They may include the NGO sector (Michelutti 2008, pp. 1–3).

The case of Dar es Salam presented conditions which by analogy, resonate with Sub-Saharan cities. Numerous systems and different blurred roles assumed by actors in the provision of sanitation contributed to the institutional fragmentation. Informal systems provided services in areas where formal distribution was not available. An increasing number of diverse actors from the non-governmental sector begin to work with local authorities as partners, advancing a more networked and complicated system with less control by the state (Michelutti 2008). In many ways, the fuzzy roles between formal and informal, private, and public sectors are accommodated in the "sanitation wheel" approach.

Allen et al. (2006, p. 3) developed "The Sanitation Wheel" which is a schematic representation of a strategy to incorporate public, private, and informal strategies to improve sanitation options for the poor (Fig. 24.4).

There are two sides of the wheel: "formal" on the left side, which represent the policy driven mechanisms and, the right side, represent the "informal", more localised strategies adopted by the poor for the provision of sanitation services. With both sides of the wheel working in tandem, an active spectrum of stakeholders from government, NGOs, private sector, and communities themselves can jointly develop strategies and implement them as a multi-sectoral co-operative solution to urban and peri-urban contexts. This multi-agent co-production proved to be effective in changing community perceptions and response to sanitation solutions in cities like Caracas, Mumbai, and Tiruchirappalli (Allen et al. 2006). There are flexible delivery systems with appropriate standards.

24.7 Conclusion

There is a global challenge to meet the basic needs of an increasing population due to rapid urbanisation, insufficient infrastructure, and inability of the local government structures to upscale and sustain innovative community-driven sanitation solutions. Inadequate sanitation facilities impact most on vulnerable groups, especially women and children. This chapter reviewed the endemic sanitation challenges in the developing countries. It also assessed the global benchmark towards poverty alleviation and improved sanitation conditions as was set out in the MDGs and SDGs. In both approaches access to sanitation was one of the key indicators of an improved and dignified quality of life, and was inextricably linked for the realisation of most of the goals and targets.

While there is evidence of some progress, a major problem is the lack of access to sanitation coupled with ineffective physical infrastructure provided by governments. Numerous strategies to deliver sanitation to the poorest communities prove ineffective without an integrated multi-stakeholder governance approach to sanitation. Innovation regarding sanitation technology bears no fruit if too much emphasis is placed on infrastructural issues, neglecting the softer issues of education, social acceptability, and behaviour change.

There are also problems in sanitation governance. Supply side challenges include institutional incapacity, shortage of resources, lack of political will, and tokenistic participatory governance in the sanitation sector. The chapter also underscored the economic potential of sanitation for poor communities through entrepreneurial initiatives regarding human waste management. It also illustrated that poor waste management could be mitigated through innovative recycling of human waste.

Any attempt to resolve the sanitation challenges must adopt a multi-stakeholder, inclusive approach, comprising local government, communities, and small enterprises sector. Amidst abject poverty, communities are more concerned about survival than practicing hygienic living. There is an opportunity for NGOs, CBOs, and training institutions to jointly engage in educating peri-urban and rural communities about dignified sanitation practices and health care. The collaboration of all stakeholders is pivotal in addressing the sanitation challenge worldwide. Working collectively, local government, communities, and the private sector are key to providing sustainable sanitation solutions.

References

Adams EA, Juran L, Ajibade I (2019) Spaces of exclusion in community water governance: a feminist political ecology of gender and participation in Malawi's urban water user associations. Geoforum 95:33–142

Adams EA, Boateng GO (2018) Are urban informal communities capable of co-production? The influence of community–public partnerships on water access in Lilongwe, Malawi. Environ Urban 30:461–480

Adams EA, Sambu D, Smiley SL (2019) Urban water supply in sub-Saharan Africa: historical and emerging policies and institutional arrangements. Int J Water Resour Dev 35:240–263

Ako AA, Shimada J, Eyong GET, Fantong WY (2010) Access to potable water and sanitation in Cameroon within the context of millennium development goals (MDGS). Journal for Water and Science Technology 5:1317–1339

Allen A, Davila JD, Hofmann P (2006) Governance of water and sanitation Services for the Peri-Urban Poor: A framework for understanding and action in metropolitan regions. Development Planning Unit, London

Asian Development Bank (2009a) India's sanitation for all: how to make it happen? Philippines, Asian Development Bank

Asian Development Bank (2009b) Investing in sustainable infrastructure to improving lives in Asia and the Pacific. Philippines, Asian Development Bank

Baer M (2017) The human right to water and sanitation: champions and challengers in the fight for new rights acceptance in A Brysk and M Stohl (eds), expanding human rights twenty-first century norms and governance. Edward Elgar Publishing, London, pp 94–114

Baskovich MR (2008) Building inclusive sanitation Markets for the Poor. International Resource Centre, Netherlands

Bhanushali K (2019) Assessable sanitation and toilets for persons with disabilities: a human rights approach. SSRN: https://ssrn.com/abstract=3361615 or https://doi.org/10.2139/ssrn.3361615

Cheng JJ, Wallace CJS, Watt S, Newbold KB, Mente A (2012) An ecological quantification of the relationships between water, sanitation and infant, child, and maternal mortality. J Environ Health 11:1–8

Corburn J, Karanja I (2014) Informal settlements and A relational view of health in Nairobi, Kenya: sanitation, gender and dignity. Health Promot Int 31:258–269

De Albuquerque C, Winkler IT (2010) Neither friend nor foe: why the commercialization of water and sanitation services is not the Main issue in the realization of human rights. Brown J World Affairs 17:167–179

Azeez EPA, Negi DP, Mishra A (2019) Women's experiences of defecating in the open: a qualitative study. Indian J Gend Stud 26:160–170

Factura HT, Bettendorf C, Buzie H, Pieplow J, Reckin J, Otterpohl R (2010) Terra Preta sanitation: re-discovered from an ancient Amazonian civilisation-integrating sanitation, bio-waste management and agriculture. J Water Sci Technol 31:2673–2679

Ganguly SC (2008) India's National Sanitation and Hygiene Program: From Experience to Policy West Bengal and Maharashtra Models Provide Keys to Success. www.wateraid.org (Accessed on 12 Oct 2019)

Graham J (2011) Sanitation and hygiene: taking stock after three decades in JMH Selendy (ed) water and sanitation-related diseases and the environment: challenges, interventions, and preventive measures. Wiley-Blackwell, London, pp 17–27

Herrera V (2019) Reconciling global aspirations and local realities: challenges facing the sustainable development goals for water and sanitation. World Dev 118:106–117

Hueso A, Bell B (2013) An untold story of policy failure: the Total sanitation campaign in India. Water Policy 15:1001–1017

Jain A, Wagner A, Snell-Rood C, Ray I (2020) Understanding open defecation in the age of swachh bharat abhiyan: agency, accountability, and anger in rural Bihar. Int J Environ Res Public Health 17(4):1384

Kashiwase H (2019) Open Defecation Nearly Halved Since 2000 But is Still Practiced by 670 million. https://blogs.worldbank.org/opendata/open-defecation-nearly-halved-2000-still-practiced-670-million. Accessed 10 Feb 2020

Kennedy-Walker R, Amezaga JM, Paterson CA (2015) The role of power, politics and history in achieving sanitation service provision in informal urban environments: A case study of Lusaka, Zambia. Environ Urban 27:489–504

Koonan S (2019) Sanitation interventions in India: gender myopia and implications for gender equality. Indian J Gend Stud 26:40–58

Kumar GS, Kar SS, Jain A (2011) Health and environmental sanitation in India: issues for prioritizing control strategies. Indian J Occup Environ Med 15:93–96

Mara D, Lane J, Scott B, Trouba D (2010) Sanitation and health: policy forum. J PLoS Med 7:1–7

Matsebe G (2006) Sanitation policy in South Africa: does it address people with disabilities? Paper presented at the 32nd WEDC international conference on sustainable development of water resources, water supply and environmental sanitation, Colombo, Sri Lanka

McFarlane C (2019) The urbanization of the sanitation crisis: placing waste in the City. Dev Chang 50:1239–1262

McFarlane C, Silver J (2017) The political city: "seeing sanitation" and making the urban political in Cape Town. Antipode 49:125–148

Mehta L, Movik S (2010) Shit matters: the potential of community-led Total sanitation. Practical Action Team, Warwickshire

Michelutti L (2008) The Venularization of democracy: comparisons across India and Venezuela. Economic and Social Research Council, Swindon

Mukherjee S, Sundberg T, Schütt B (2020) Assessment of water security in socially excluded areas in Kolkata,

India: an approach focusing on water, sanitation, and hygiene. Water 12(3):746. https://doi.org/10.3390/w12030746

Münch EV, Wirseen C, Patel D, Wheaton A, Jachnow W (2009) Urban Slum Dwellers in Kenya and Bangladesh Benefit from Using Peepoo Bags which are Self-Sanitising and Biodegradable. https://www.susana.org/_resources/documents/default/2-900-en-benefit-from-using-peepoo-bags-kenya-bangladesh-2009.pdf

Murta JCD, Willetts JRM, Triwahyudi W (2018) Sanitation entrepreneurship in rural Indonesia: a closer look. Environ Dev Sustain 20:343–359

Mwebaza R (2010) Sustaining good governance in water and sanitation in Uganda. Monograph. Institute for Security Studies: Uganda

O'Reilly K (2016) From toilet insecurity to toilet security: creating safe sanitation for women and girls. Wiley Interdiscip Rev Water 3:19–24

Powell J, Yurchenko Y (2020) The evolution of private provision in urban drinking water: New geographies, institutional ambiguity and the need for political economy. New Polit Econ, 25(1):91–106

Rahman S, Rahman MA (2015) Climate extremes and challenges to infrastructure development in coastal cities of Bangladesh. Weather Clim Extrem 7:96–108

Ramachandraiah C (2001) Drinking water as a fundamental right. Econ Polit Wkly J 36:619–621

Reddy VR, Batchelor C (2012) Cost of providing sustainable water, sanitation and hygiene (WASH) services: an initial assessment of a life-cycle cost approach (LCCA) in Rural Andhra Pradesh, India. J Water Policy. 14: 409–429

Saleem M, Burdett T, Heaslip V (2019) Health and social impacts of open defecation on women: a systematic review. BMC Public Health 19(1):158

Solo TM (1999) Small-scale entrepreneurs in the urban water and sanitation market. Environ Urban 11:117–131

Tissington K (2011) Basic sanitation in South Africa: A guide to legislation, policy and practice. Socio-Economic Rights Institute of South Africa (SERI). www.itldesign.co.za. Accessed in 8 July 2012

Tukahirwa JT, Mol APJ, Oosterveer P (2010) Civil society participation in urban sanitation and solid waste Management in Uganda. J Local Environ 15:1–14

UCLG (2018) Towards the localization of the SDGS. United Cities and Local Governments, Barcelona

UNDP (2003) Human development report 2003 - millennium development goals: A compact among nations to end human poverty. OUP, New York

United Nations (2010) The human right to water and sanitation. UN Doc. A/RES/64/292. https://undocs.org/A/RES/64/292

United Nations (2016) Sustainable development goals. https://www.un.org/sustainabledevelopment/water-and-sanitation/. Accessed 10 Jan 2020

UN-Water (2016) World Water Development Report. https://www.unwater.org/publications/world-water-development-report-2016/

UN (2019) Eight Things You Need to Know about The Sanitation Crisis. https://www.un.org/sustainabledevelopment/blog/2019/10/eight-things-you-need-to-know-about-the-sanitation-crisis/. Accessed 10 Jan 2020

Van Vliet BJMV, Spaargaren G, Oosterveer P (2011) Sanitation under challenge: contributions from the social sciences. J Water Policy 13:797–809

Walker C (2016) Lack of safe water, sanitation spurs growing dissatisfaction with government performance. Afrobarometer dispatch no. 76. http://afrobarometer.org/sites/default/files/publications/Dispatches/ab_r6_dispatchno76_ water_and_sanitation_in_africa1.pdf. Accessed 12 Jan 2020

WHO/UNICEF (2000) Global water supply and sanitation assessment 2000 report. World Health Organization, Geneva

WHO/UNICEF (2004) Meeting the MDG drinking water sanitation target: A mid-term assessment of Progress. World Health Organization and United Nations Children's Fund, Geneva

WHO/UNICEF (2017) Progress on drinking water, sanitation and hygiene. World Health Organization and United Nations Children's Fund, Geneva

World Health Organization (2012) Water Sanitation Hygiene – Key Terms. https://www.who.int/water_sanitation_health/monitoring/jmp2012/key_terms/en/. Accessed 10 Jan 2020

World Health Organization (2018) Guidelines on sanitation and health. World Health Organization, Geneva

Solid Waste Management for Environmental Sustainability in India

25

Surya Tewari

Abstract

The problem of solid waste, though not unique in India, is so huge that it hurdles the achievement of environmental sustainability. Solid waste is most visible of all the wastes generated in the country. The problem is so acute that even after one and a half decades of implementation of Municipal Solid Waste Rules (2000), the country has not got rid of the heaps of waste generated. Unattended solid waste on the roadside and open lands is a normal scene in the country. Though the issue has now come in the public discourse with the implementation of Swachh Bharat Abhiyan, achieving the mammoth task of cleaning up India is a concern. In this context, the chapter analyzes the waste generated and handled by computing the same at the level of States/ Union Territories and Cities/Towns in the country. The best practices at the inter-country and intra-country level have been discussed and policies have been suggested. It is found that the problem is not with respect to quantity per se, but, with poor mishandling which starts at the doorstep of waste generation. The success in handling solid waste depends on mix of legal practices, processing technologies, recycling, public awareness, and involvement

S. Tewari (✉)
Institute for Studies in Industrial Development, New Delhi, India

Keywords

Public Private Partnership · Solid Waste Management · Swachh Bharat Abhiyan · Waste-to-Energy

25.1 Introduction

Solid waste is most visible of all the wastes generated. Concentration of people, level of economic activity, incomes, climate, and culture influence the quantity and variety of solid waste generated. However, the most important factor is economic which culminates in high mass production, consumption, and generation of waste. This explains high per capita solid waste generation in the countries of OECD (Organization for Economic Cooperation and Development) compared to Africa and South Asia (World Bank 2012). The difference, however, exists within a country between urban and rural, on the one side and among urban, on the other. The urban areas excel rural areas with respect to quantity and variety of waste. Similarly, urban areas higher up in the hierarchy of towns generate more and varied waste in comparison to lower order towns.

As per Waste Atlas (2016), the average amount of municipal solid waste (in urban areas under municipal government) generated per capita per year is 777 kg in Canada, 733.7 kg in the USA, 747 kg in Denmark, 617 kg in Germany, 640 kg

B. Thakur et al. (eds.), *Resource Management, Sustainable Development and Governance*, Sustainable Development Goals Series, https://doi.org/10.1007/978-3-030-85839-1_25

in Australia, 340 kg in Russia, 620 kg in UAE, 229.4 kg in China, 182.5 kg in India, 255.5 kg in South Africa, and 153.9 kg in Nigeria to name a few countries across different regions in the world.[1] Importantly, the problem is not the generation of waste per se, but, the capacity of treating it without compromising the health and environmental aspect. Due to technological and financial weaknesses the comparatively small quantity generated became a big environmental and health issue in India. The cities in India are faced with acute problem with respect to solid waste. Heaps of solid waste plying on roads, streets, open areas, water bodies are common sites in our cities. In fact, it is important to note that even when the most dominant activity of municipalities in the country is disposal of solid waste, we have not been able to get rid of solid waste and the situation looks precarious and bleak.

Solid Waste Management (SWM) is one of the neglected areas in India. Under Jawaharlal Nehru National Urban Renewal Mission (JNNURM), the importance was accorded, however, in that also the success was little. By March 31, 2014, in chosen 63 cities/towns under JNNURM, only 12 solid waste management projects out of 42 sanctioned had been completed with approved cost of Rs. 582.89 crore (5828.9 million) (GOI 2014a). The total cost approved for all infrastructure projects for selected cities/towns was Rs. 14,281 crores (142.81 billion). In other cities and towns under Urban Infrastructure Development of Small and Medium Towns (UIDSMT) component of JNNURM, 19 projects had been completed with approved cost of Rs. 125.86 crores (1258.6 million) by the same time (GOI 2014b). The total approved cost for all the projects under the program was Rs. 5870 crores (58.7 billion).

Now, in place of JNNURM, Smart City, AMRUT (Atal Mission for Rejuvenation and Urban Transformation) and HRIDAY (Heritage City Development and Augmentation Yojana) programs have been running which have solid waste management as one of the focus areas. Specifically, Swachh Bharat Mission (SBM) which focuses on sanitation alone has SWM as one of the main components. SBM asks for detailed project report on solid waste management. Under this mission central government will provide grant/viability gap funding to the maximum of 20% to each project for processing of solid waste (MoUD 2014).

[1] The atlas was accessed in February 2016.

Box 25.1 Sustainable Development Goals

In India, solid waste management is crucial to achieving environmental sustainability. With the rise in population, urbanization, and standards of living, the generation of solid waste has increased significantly. As the third largest generator of solid waste in the world, management of solid waste has emerged as a major challenge for sustainable environment. According to the author, a significant aspect of this challenge emerges from collection, storage, and disposal of solid waste in inefficient ways. The UN Agenda 2030 recognizes solid waste management as vital for sustainable development. Sustainable Development Goals (SDGs) broadly and SDG # 12 cannot be met without managing solid waste management as a priority. To that end, SDG 1, 2, 4, 5, 6, 7, 8, 9, 10, 16, and 17 are interrelated and interdependent for the deliverance of SDGs overall and for solid waste management. SDG # 12 cannot be achieved without responsible consumption and production, including in the Indian context decentralized solid waste management solutions into sub-national and national policies and practice, improving waste segregation, collection, recovery, and final disposal, ensuring inclusive planning and implementation, and finally establishing programs, guidelines for monitoring, reporting, verification, and documentation of good practices. Sustainable waste management within the framework of SDGs in India provides opportunities for collaboration and partnerships between the various stakeholders—the wealthy and poor, nonprofit, government, formal and informal sector, business and communities at different scales.

Coming to the rules governing specifically the solid waste generated in municipal areas, the municipal solid waste in the country was, until recently, governed by Municipal Solid Waste Management (MSW) Rules 2000 for residential, commercial, and treated bio-medical waste (exclude industrial waste) (MoEFCC 2000). Now, the MSW Rules 2000 have been replaced by SWM Rules, 2016 (MoEFCC 2016).

Since SWM Rules 2016 have recently been applied, the whole task of SWM, until recently, was handled by MSW Rules, 2000. As per SWM Rules 2016, local government (earlier only urban local government) are responsible for implementation of these Rules within the area under their jurisdiction. The whole task from the points of generation to points of disposal is a responsibility of Local Governments. Central Pollution Control Board (under Ministry of Environment, Forests, and Climate Change, MoEFCC) in coordination with State Pollution Control Board or Committees is responsible for monitoring and reviewing of implementation of the rules.

Since it has been more than one and a half decades since the MSW Rules came into being, it is important to take stock of the situation in waste generation and handling in the country together with various issues in the discourse of dealing with waste. This paper is divided into 6 sections, the following section analyzes the state of waste generation together with waste collection and treatment efficiency at the level of the States/UTs and Cities/Towns. The Sect. 25.3 describes the handling of the waste in the country. The Sect. 25.4 presents the intra-country and inter-country best practices. The Sect. 25.5 puts forth suggestions, and Sect. 25.6 concludes.

25.2 Waste Generation, Collection, and Treatment Efficiency

Five groups of States/UTs are observed considering the total municipal solid waste generated in the states in 2013–2014 (Table 25.1). Group I in the Table comprises states with highest generation of municipal solid waste. There are four states in this category, viz. Maharashtra (26,820 Tons Per Day, TPD), Uttar Pradesh (19,180 TPD), Tamil Nadu (14,532 TPD), and Andhra Pradesh (11,500 TPD). The least generators in Group V are Lakshadweep, Sikkim, Andaman and Nicobar Islands, Dadar Nagar Haveli, and Daman and Diu. There is a strong correlation between generated municipal solid waste and population of statutory towns (excluding outgrowths).[2] The correlation coefficient comes out to be 0.971 at 1% significance level. Overall, there are few large generators. The number of states and UTs combined increase with the falling group number. Eleven States and UTs (Group IV) have waste generation in the range of 100–1000 TPD.

Table 25.1 Volume of Generated Municipal Solid Waste (2013–2014)

Group	No. of States/No. of UTs	Volume (TPD)
I	4	>10,000
II	5/1	5000–10,000
III	9	1000–5000
IV	9/2	100–1000
V	1/3	<100

Source: Computed from the data obtained from CPCB (2015)

Note: Data of two UTs, viz. Daman and Diu and Dadar Nagar Haveli is given together. Both have therefore been considered as one. TPD is Tonnes Per Day

If we see the collection efficiency, we realize that the states with most municipal solid waste generation are not the ones collecting all that waste generated. All the states and UTs generating lowest (Group V in Table 25.1) municipal solid waste have 100% collection efficiency.[3] In highest generators (Group I in Table 25.1) it is only Uttar Pradesh which collects most of the generated municipal solid waste. Four other states, viz. Gujarat, Jharkhand, Uttarakhand, and Tripura and one UT, i.e., Puducherry also has its best waste collection. The waste collection is less than 50% in Rajasthan and Kerala. Overall, most

[2]There is no statutory town in Lakshadweep. There are 6 census towns as per Census 2011 handled by Village Panchayats which are responsible for municipal solid waste management. The population of census towns of Lakshadweep is considered for correlation.

[3]Leave aside Lakshadweep whose data on waste collection is not available.

Table 25.2 Proportion of generated municipal solid waste collected (2013–2014)

Percent waste collected from total waste generated	No. of states/No. of UTs
<50	2
50–75	8
75–100	11/2
100	6/3
NA	1/1

Source: Computed from the data obtained from CPCB (2015)

Note: Data on waste collection is not available on Bihar and Lakshadweep

Table 25.3 Proportion of waste treated from the collected waste (2013–2014)

Percent waste treated to collected	No. of states/No. of UTs
Nil	3/2
1–10	6/1
11–29	9
32–63	5/1
76–100	3/1
NA	2/1

Source: Computed from the data obtained from CPCB (2015)

Note: Data on waste treatment is not available on Bihar, Manipur and Lakshadweep

of the States/UTs (22 in number) have waste collection of more than 75% (Table 25.2).

Coming to treatment part, only Goa records 100% treatment of collected waste (Table 25.3). Only 1% collected waste is treated in Punjab, Odisha, and Sikkim, while no treatment takes place in Uttarakhand, Tripura, Mizoram, Puducherry, Dadar Nagar Haveli, and Daman and Diu. The configuration of different states and UTs is as below (Table 25.3). Most of the states and UTs have poor treatment of collected waste.

Coming to waste generation in Cities/Towns, the CPCB survey of 2010–2011 of 59 Cities/Towns (CPCB 2013) shows that the range of waste generation varies from 2 TPD in Kavaratti (Lakshadweep) to 6800 TPD in Delhi (CPCB 2013). Other than Delhi, the largest generator is Mumbai with tonnage of 6500 per day. Table 25.4 classifies cities/towns based on volume of waste generated. In the first category of 6000–7000 TPD, it is Delhi and Mumbai, followed by

Table 25.4 Volume of waste generated in cities/towns (2010–2011)

TPD	No. of cities/towns
6000–7000	2
4000–5000	2
3000–4000	2
1000–3000	5
500–1000	10
300–500	10
100–300	16
1–100	12

Source: Computation based on CPCB (2013)

Chennai and Hyderabad in the next category of 4000–5000 TPD. In the third of 3000 to 4000 TPD it is Bengaluru and Kolkata. The 5 cities with waste generation between 1000 and 3000 TPD are Ahmedabad, Kanpur, Lucknow, Pune, and Surat. All these cities are among the largest cities in terms of population in the country.

From the Table 25.4 it is clear that around 64% of the Cities/Towns generate less than 500 TPD. It is important to note that from these 59 cities, 40 are million cities and 19 others are not million but capital cities.[4] Further, these cities except for 9 were under JNNURM as well. Having information of all the important and capital cities, it may be assumed that rest of the cities and towns in India generate lesser tonnage per day.

The survey of the same cities was conducted in 2004–2005 as well (CPCB 2013). The computation of growth in waste generation in these cities from 2004–2005 to 2010–2011 reveals that most of the Cities/Towns have low to negative growth (Tables 25.5 and 25.6). Nine cities have reported highest growth in waste generation. Individually, Itanagar (Arunachal Pradesh) reported highest growth by whopping 750%. The increase is due to small base as tonnage increased from 12 to 112 TPD. Four other cities (Kohima, Imphal, Shillong, and Gangtok) from northeast have shown more than 100% growth. Silvassa (Dadar Nagar Haveli) has also summarized the same level of increase. The major cities in this

[4]As per 2011 Census there are 53 million cities in the country.

Table 25.5 Growth in waste generation in cities/towns (2004–05 to 2010–11)

Per cent growth	Cities
100 and above	9
75–100	7
50–75	7
25–50	12
Less than 25	10
Negative	14

Source: Computation based on CPCB (2013)

category are Lucknow, Bengaluru, and Gandhinagar. The cities with largest generation of solid waste, viz. Delhi and Mumbai have witnessed increase of 15 and 22%, respectively.

25.3 Waste Handling in India

Drawing on the previous perspective, it is reported that waste generation as such is not very high in the country. Then the question is why we see huge waste around us? Where does the problem lie? Is it in the collection and treatment of waste? It is therefore important to explore how waste is handled in the country.

The whole chain of events to handle waste involves segregated storage at source, primary door to door collection, primary transport, secondary storage, secondary transport to processing facilities and final disposal. However, this ideal chain from beginning to end is not followed satisfactorily. In India, the primary task of segregation is not followed or if followed, then, scantily and door to door collection is practiced partly. People are indeed free to dump their waste the way they choose. They may put it in the vacant areas or in community bins, if accessible. It is observed that the community bins are located at distance, are mostly overflown with stray animals poking into it and repugnant odor, flies and rodents proliferating. The problems are very well documented by Kumar et al. (2009).

Processing of waste involves material and energy recovery. This is done either through separating recyclable waste (paper, plastic, glass, metal, fabric) or composting (aerobic of windrow type that involves arranging waste in rows or windrows) or vermicomposting (earthworm-based) or bio methanation (biogas/anaerobic digestion) or Refuse Derived Fuel (RDF)Plants (waste plants make fluffs or pallets for use as secondary fuels in high energy consumption industries like cement, steel, aluminum, etc. and in Waste-to-Energy boilers to generate power) or Waste-to-Energy (WtE) plants (thermal decomposition/combustion of waste using technology of incineration, gasification, or pyrolysis) and bioremediation (interventional process involving humans), and landfill mining (after natural decomposition) at the site of landfills. From CPCB Reports and other studies, it is worth noting that waste processing is only partially done in India. The recycling is basically done by informal waste pickers, itinerant waste buyers, and informal waste dealers. The analysis revealed that these informal workers recycle 56% of recyclable material generated in large cities (quoted in Annepu 2012).

With respect to construction and demolition (C&D) waste that can be recycled effectively, MSW Rules (2000) ask for segregated collection and proper disposal. This waste can be utilized for making tiles, bricks, blocks, etc. and can divert a lot of waste reaching the landfills, the aspect noted in SWM Rules 2016. However, the opportunity has not been used, there is only one treatment plant operating for C&D waste in country that is at Burari in Delhi (Nath 2014). As per Centre for Science and Environment (2014) the estimated C&D waste generated in India in 2013 was 530 million tons.

The compost and vermicompost plants—processes most convenient and suitable to handle large organic waste generated in the country—are also not sufficient. For example, in Rajasthan, only one plant is in operation at Jaipur. The composting plants at Pali, Bhilwara, and Bharatpur are not in operation.

In Uttar Pradesh, the state with largest population (total 200 million with urban 44.4 million second only to Maharashtra with 50.8 million), 16 waste processing facilities exist (CPCB 2015). In Maharashtra, 65 composting, 38 vermicomposting, 24 biomethanation with 16 Nisargruna biogas projects set up by Bhabha Atomic

Table 25.6 Municipal solid waste generated in cities (in TPD)

Cities	2004–2005	2010–2011	Percent growth (2004–2005 to 2010–2011)
Itanagar	12	102	750
Kohima	13	45	246
Imphal	43	120	179
Lucknow	475	1200	153
Bengaluru	1669	3700	122
Gandhinagar	44	97	120
Silvassa	16	35	119
Shillong	45	97	116
Gangtok	13	26	100
Dhanbad	77	150	95
Pondicherry	130	250	92
Hyderabad	2187	4200	92
Aizawl	57	107	88
Jabalpur	216	400	85
Ahmedabad	1302	2300	77
Nashik	200	350	75
Bhubaneshwar	234	400	71
Vadodara	357	600	68
Dehradun	131	220	68
Daman	15	25	67
Madurai	275	450	64
Vijayawada	374	600	60
Faridabad	448	700	56
Chennai	3036	4500	48
Thiruvananthapuram	171	250	46
Kanpur	1100	1600	45
Jammu	215	300	40
Kolkata	2653	3670	38
Agartala	77	102	32
Coimbatore	530	700	32
Indore	557	720	29
Nagpur	504	650	29
Srinagar	428	550	29
Shimla	39	50	28
Amritsar	438	550	26
Guwahati	166	204	23
Mumbai	5320	6500	22
Raipur	184	224	22
Surat	1000	1200	20
Ludhiana	735	850	16
Delhi	5922	6800	15
Rajkot	207	230	11
Pune	1175	1300	11
Varanasi	425	450	6
Asansol	207	210	1
Chandigarh	326	264	−19
Agra	654	520	−20
Panjim	32	25	−22

(continued)

Table 25.6 (continued)

Cities	2004–2005	2010–2011	Percent growth (2004–2005 to 2010–2011)
Allahabad	509	350	−31
Ranchi	208	140	−33
Kavaratti	3	2	−33
Bhopal	574	350	−39
Port Blair	76	45	−41
Vishakhapatnam	584	334	−43
Patna	511	220	−57
Kochi	400	150	−63
Jaipur	904	310	−66
Meerut	490	52	−89
Jamshedpur	338	28	−92

Source: CPCB (2013)

Research Centre (BARC)[5] exist. However, it is reported that almost all of 110 ULBs using these plants processing the waste partially (CPCB 2015). In Gujarat, the number of such facilities also looks attractive. The state has 40 composting and 38 vermicomposting facilities.

The states in the southern part of the country have also quite several basic waste processing facilities. In Kerala, for example, there are 30 compost, 9 vermicompost, and 16 biogas plants. Tamil Nadu, on the other hand, has 157 composting and 24 vermicomposting plants (CPCB 2015). In all other states, either there is no such facility or very few. Again, even if there are waste processing facilities, they are partially working, intermittently working or non-working. This is a situation across the country. Mix waste together with presence of hazardous items, due to lack or poor segregation makes the processed item unusable and non-marketable and thereby affects the operation of the plant (Basu 2013).

Turning on to Waste-to-Energy plants, they are basically RDF plants. As per CPCB Report, there are 22 RDF plants with 8 in the State of Maharashtra (CPCB 2015). Only Delhi has proper WtE plant—incineration-based mass combustion plant—at Okhla. However, it is about to be closed due to complains of emissions of toxins by the residents. The plant utilizes 1950 Metric Ton[6] waste to generate 16 MW energy (Bhatnagar 2015). Most of the states and UTs do not have the facility to turn waste into energy. The small-scale biogas plants are also energy plants but are not the one making commercial production and handling large amount of waste. The status of bioremediation or landfill mining is poor in the country. The scope of processing landfill gases (methane and nitrous oxide) is high as 90% of the waste in the country is simply landfilled (Kumar 2005).

There are very few sanitary landfills in the country. The procedures for final disposal of the waste are very well documented but are not followed. Rather than landfilling dumping is done as it is easy to dump than to treat or fill scientifically. However, starting a landfill is a very problematic task. The problem begins first in identifying, then acquiring and, finally, using the

[5]Nisargruna biogas plants are different from other biogas plants in the sense that they have two operational phases, one decomposition of waste in the presence of air in the first stage and in second processing in the absence of air. It thus produces 70–80% methane and 30 to 20% CO_2 in comparison to 55–65% methane and 45 to 35% CO_2 in conventional biogas plants. The high methane content is useful in direct heating or in generating electricity (Prasad 2012). At present, 160 plants are in operation in the country. One TPD capacity plant involves cost of Rs. 16 lakhs for set up. See BARC (2015).

[6]Metric Ton and Tone are same with only difference of first one adopted as a metric unit of mass in USA and the later one in Britain. Metric Ton or Tone is equivalent to 1000 kg.

land for the said purpose. The news reports show that 15% of the Municipalities in Uttarakhand do not have required land to meet the solid waste disposal commitments (Azad 2015). Again, there are issues of use of prime agricultural and forest land for landfilling. Very recently, Delhi Development Authority (DDA) has cleared the conversion of two green spaces for landfill in Dwarka which has been vehemently opposed by the residents (Chitlangia 2015). Due to unavailability of land many of the existing landfills are continued to be used even when they have exhausted their capacity. All the three landfills of Delhi at Ghazipur, Okhla, and Bhalaswa are overflowing but are continuously used. The ban imposed by Delhi Pollution Control Committee in 2009 has not thwarted the municipalities in using these sites (Nair 2014). There is acute scarcity of land to find alternative landfill sites in Delhi.

Acquisition of land for landfilling is very stiff across the country. The opposition becomes so strong that sometimes the plan must be shelved. The opposition is since intentions as mentioned in MSW Rules (2000) have not been implemented in practice. Absence of scientific management of landfills creates tremendous environmental and health impact and, thus, is strongly objected by people. The primary survey of urban areas in Uttarakhand, for example, shows that municipalities have tough time in acquiring and using land (Ramachandran et al. 2016). In the primary survey of Chamba, a municipality in Uttarakhand, it has been found that the municipality has purchased a land which it is unable to use. Similarly, there is news that villagers in Taloja, Navi Mumbai the proposed site for shifting of Deonar landfill site which has exhausted its capacity and is facing eruption of fire are opposing any such move (India Today 2016). Such examples can multiply.

25.4 Inter- and Intra-Country Best Practices

In the previous section we have seen that there is mishandling of waste in the country and if this situation is not mended it will be monstrous in future. But question is how? Are there examples which can be emulated? Coming to inter-country level, it has been found that there have been countries which have large-scale waste generation, facing acute shortage of land, but have been successfully able to deal with waste.

Japan passed through the same phase as India is currently undergoing but it thwarted the problem through concerted efforts. The period after Second World War was characterized by reconstruction, mass production, consumption, and consequent high generation of waste as a byproduct in Japan. The waste continued to rise till the bubble economy phase in late 1980s and early 1990s. Though efforts were started to be made in 1950s but real beginning was made in late 1980s and early 1990s. The incineration plants were made technologically efficient, and fears of dioxins and other hazardous gas emission were removed from public mind. Collection of segregated waste started. The incineration plants operate in the middle of the cities and towns in residential and commercial areas in the country. In fact, with respect to incineration plants the doubts regarding hazardous pollution have been effectively dealt by Japan through technology advancements. Japan uses incineration technology at large scale but has effectively curbed the level of hazardous emission from such plants.

Together with incineration technology, Japan is very much focusing on 3R principles of Reduce, Reuse and Recycle. The twenty-first Century is declared as century of creating Sound Material Cycle Society in Japan. 3Rs are promoted in a big way in the country. 3R promotion months, awards, environmentally friendly shopping campaigns, recycling centers and plazas, etc. are popularly demonstrated. Eco-towns have also been developed. In view of the type of waste generated in Japan, the government has formulated regulations dealing with all such wastes separately. There are separate rules/laws for containers and packaging material, home appliances, food, construction, automobile, and small home appliances. The consumers while discarding the items like home appliances, automobiles are required to deliver them to retailers who deliver them to manufacture who is responsible for recycling. The consumer bears all the cost related to collection, transportation, and recycling (Japan-Ministry of the Environment 2014).

Like Japan, Singapore also focuses on 3Rs. Way back in 2000 with the exhaustion of onshore landfills (wetlands), the country turned to 3R. Currently, 60% of the solid waste is recycled, 38% incinerated, and 2% landfilled in the country. The landfill has been developed by reclaiming land from sea. Around 3% electricity demand of Singapore is met by 4 WtE plants (Yep 2015).

Similarly, The Netherlands has focused on recycling. Recycling reached 50% in 2009 in the country earlier by 11 years from the target year. Out of 9.8 million tons waste generated in 2010, five million tons were recycled, 3.2 million tons incinerated, and 0.03 million tons landfilled (Milios 2013). Since 1995 ban is operative in the country on landfilling of 35 items pertaining to combustible and biodegradable items. Landfill tax was also imposed. The tax has been raised many a times until, finally, repealed in 2012. The government strictly enforces waste segregation, charges in proportion to quantity and extended producer responsibility. The country operates 12 incineration plants and treats imported waste as the plants have overcapacity. Incineration is practiced instead of gasification and pyrolysis due to lack of homogenous waste. In 2012, 4014 GWh of electricity and 14.1 petajoule of heat were produced by waste incineration plants in the Netherlands (Wong 2014).

The point is not utilizing any of the technique but the one which is most appropriate. To illustrate, The Netherlands presents an example where despite no financial dearth, the gasification and pyrolysis techniques are not practiced due to non-homogenous nature of waste. With technology advancement, the above-mentioned countries have achieved the control of toxic gases and ash outpour. The dedicated waste management wing in the primer engineering institutes can be established in India as done in Japan. In Japan, the students earn engineering in waste management and pollution control. This can be started in India as we face a huge problem of waste and dedicated personnel would help a lot in dealing with waste problem.

It is also to be seen that the USA, the economic powerhouse of the world also faces a crisis with respect to solid waste. In 2012, the USA recycled 26% of its municipal solid waste, compost 8%, incinerate 12%, and landfill the rest, 46% (US-EPA 2014). So, despite becoming negative by seeing the mammoth task at hand, the thinking and work should be started in a proper direction. The USA, for example, is continuously raising the amount of solid waste diverted away from landfills. In 1980, the quantity of solid waste landfilled was 90% which has now reached to half. Industrial waste is strictly regulated in the USA unlike India, where it also ends up in landfills. In fact, in the USA there is a prevalence of "Not in My Backyard" attitude of people which makes landfilling difficult.[7]

Ocean landfilling has also been practiced in countries like Japan and Singapore due to technological advancement. Japan carries out ocean landfilling in the Osaka region. A landfill island—Osaka Bay Phoenix Center—has been developed by reclaiming land from Ocean. The same is being done by Singapore. The waste is scientifically managed in these landfills unlike majority of countries who are dumping dredged material, industrial waste, sewage sludge, and radioactive wastes (solid nuclear waste is dumped in concrete drums) indiscriminately into the oceans. The dumping is so high in the North Pacific Sub-Tropical Gyre that it is being regarded as largest landfill of the world and is known as *Great Pacific Garbage Patch*. The gyre is full of plastic waste. Such garbage patches can be seen in Atlantic and Indian Ocean as well as in the zone of horse latitudes (Parker 2014). As per one estimate, 8 million tons of plastic waste were dumped in the ocean in 2010 (Parker 2015). Since 3/4 part of the world is ocean and lands have competitive uses, the oceans may be systematically chosen for filling of waste. One needs to explore ways of dumping waste in oceans other than through reclamations. Would it be possible to dump inert waste scientifically in deepest trenches? Here, it should be noted that India allowed filling up of

[7] See HWS Colleges Edu (n.d.).

inert waste in specific locations in hills as per MSW Rules (2000). The question arises can this be done in oceans. The United States Committee on Pollution (1966) hinted on filling the canyons in the ocean way back in 1966.

In India, change is noticeable. The change has come through government itself. The SWM Rules 2016 are step ahead from MSW Rules 2000. Their spatial coverage is very large. The new Rules bring both the municipal towns and census towns (under village panchayats), outgrowths, notified areas, notified industrial townships, state and central organizations, airports, airbases, ports, harbors, defense establishments, special economic zones, places of religious, pilgrimage and historical importance area in the ambit of Rules and, hence, dropped the word Municipal.

With overall monitoring of Ministry of Environment, Forest, and Climate Change, the new Rules involve seven other Central Ministries, viz. Ministry of Urban Development, Rural Development, Chemicals and Fertilizers, Agriculture, Power, New and Renewable Energy, and States and Local Governments. Industrial bodies, viz. FICCI (Federation of Indian Chambers of Commerce and Industry) and CII (Confederation of Indian Industries) are also involved. The involvement of Ministry of Chemicals and Fertilizers and of agriculture is very welcoming for promotion and sale of city compost together with fertilizers in a ratio of 3 or 4 bags: 6 or 7 bags, respectively. Further, the ministry of power is supposed to ensure compulsorily purchase of energy generated from waste by distribution company. If this is strictly done, then, half of the problem of the solid waste will be resolved. The composition of Indian waste shows that it is 30–45% organic, 6–10% recyclable and rest inert (Kumar et al. 2009). The new SWM Rules 2016 have, in fact, mandated segregation of waste at source, charging user fees from waste generators, fines, integration of Kabadiwals in the waste handling process. The intention to effectively deal with waste is reflected from countrywide taking up of Swachh Bharat Abhiyan and interest by the highest functionary of the government. However, it would be seen whether these rules will be flouted or followed in coming years when most of the requirements to deal with waste would be fully adopted.

The change though at some places is encouraging. In big cities people have started engaging waste collectors by paying nominal charges. The formation of committees or Resident Welfare Associations (RWAs) has helped in this direction. In some areas, NGOs working in the field of waste management have involved waste pickers, destitute and such other people have led to the segregated collection at source, processing and dumping in the secondary storage site. Municipalities which face perpetual financial crunch and are short of staff found it easy to support public and civil society initiatives through Public Private Participation (PPP). The involvement of waste pickers in some wards in Pune, Mumbai, and Patna has been very successful. In Vrindavan, widows, destitute, and disabled have been involved in waste management (NIUA 2015). The family size biogas plants have been adopted in states like Kerala. Biotech is a renewable energy company based in Kerala (formed in 1994)which has installed 22,000 such plants for household organic waste management. The company has also set up 280 institutional plants and 52 wastes to electricity biogas projects in collaboration with local-self government and private institutions (information received in an email response). The company had received Ashden, the global green energy award in 2007.

Further, interest in the field of waste management as profitable business has been generated in the country. The Okhla compost plant in Delhi has gained carbon credits under United Nations Framework Convention on Climate Change (UNFCC) and Clean Development Mechanism (The Hindu March 6 2013). Various startups have emerged in waste management such as Ecowise, Synergy Waste Management Pvt. Ltd., Green Power System, Lets Recycle, Pom Pom, Encashea, etc. to name a few of such companies (Jha 2014; Sriram 2015). It has been reported that the startups are raising funds and expanding their operations. The apps have been developed to make waste management more effective and efficient. Though such active efforts are like a drop of waste removed from the sea of waste, but they indeed show that the resolution of the issue can be achieved if decentralized way is followed.

25.5 Suggested Measures for Sustainable Waste Management

Perhaps certain aspects in waste management just need the inculcation of spirit in the individuals. For example, how difficult it is to store waste in segregated form? Further, how hard it is to inform, educate, and communicate to people? Likewise, what costs in taking support of waste pickers, itinerant waste buyers, and other informal workers in not just collection but also in material recovery via, recycling? Similarly, what is the problem in charging user fees from households for the services? These four things are highly achievable. Such initiatives in various parts of the country have yielded positive outcomes (NIUA 2015). SWM Rules 2016 have, in fact, involved all these processes. Now what is required is strict implementation of these practices across the country.

The 3-bin storage focusing on biodegradable, non-biodegradable, and domestic hazardous waste as mentioned in SWM Rules 2016 should involve a provision for providing such bins by local body itself by charging the waste generator. Providing access to waste pickers and such other informal waste workers for recycling is a welcome step. But this pool of human resource can also be engaged in collection of all kinds of waste plus local processing as evident in cities like Pune and Mumbai (NIUA 2015). However, it is very important to ensure medical facilities with regular body checkup to waste management personnel. These personnel should also be provided with advance personal equipment. This will induce confidence in the workforce and give them a sense of work for a national cause.

In treatment aspect, the SWM Rules 2016 ask the RWAs, market associations, gated communities, institutions, and other non-residential complex with over 200 dwellings or land size of more than 5000 m^2 to set up processing facilities within their premises "as far as possible." It is important to note that the limit of over 5000 m^2 can be flouted easily by keeping the dwellings below 200 and land size below 5000 m^2. Further, for new residential, commercial, institutional, or industrial unit of whatever size the treatment of biodegradable waste in the healthier way should be compulsorily enforced by taking feasible group of households as a unit. This will serve two purposes, first, waste processing unit sharing the same space will make the community to think for reduction, and secondly, it will promote research for better technologies.

For older housing complexes and institutions, a mechanism should be devised to treat the waste in the complex itself. Wherever, it is impossible due to unavailability of land, the community biodegradable treatment plants as also suggested in SWM Rules 2016 should be constructed. In fact, building up of such community plants should be made an important work to be undertaken by local government as building up of community centers. Private and PPP based companies and working platforms should be encouraged and strengthened. The help should be provided to these companies in terms of technology, equipment, finances, and manpower.

Since waste management is such a work where money is difficult to come, the financial resources need to be explored both at the national and international level. The Research and Development in the field of waste management and making viable projects will attract angel and venture investors in this area. The funds can also be procured from *Pradhan Mantri* MUDRA Yojana (Micro Units Development and Refinance Agency Limited). International funding has been made available for solid waste projects from GEF (Global Environment Facility), UNDP-SGP (United Nations Development Programme-Small Grants Programme) and GIZ ASER (German International Cooperation- Advisory Service for Environmental Management, Program of Government of India, and Germany), JICA ODA (Japan International Cooperation Agency-Official Development Assistance). Such funding can be enhanced via, active G2G (Government to Government) interaction in this direction. The guidelines should, therefore, be clearly formulated.

The dedicated workforce should be created in the country to deal with waste management. Engineering colleges and Industrial Training Institutes should introduce full-fledged course on waste management. The collaboration with other universities, foreign institutions/agencies excelling in this field should be introduced to develop techniques specifically customized to handle Indian situation.

It is further needed to carry out study on this issue in each local body area by involving the students from engineering, geology, geography, environmental science background. They should study existing generation, collection, transport, disposal points of solid waste issue and give comprehensive report to local body which, then, work upon it and submit action taken at district level.

Auditing of the local government with respect to their targets and achievements in SWM should be undertaken and provision should be made to award best performing local body and to fine laggard local government. Award for best performing local body in SWM should be given at district, state, and national level to give fillip to local bodies to act.

Information, Education and Communication (IEC) though find mention in SWM Rules 2016, but it should be made one of the main planks of work in this direction. The realization by everyone of his/her action with respect to their sustainable living is very crucial. This would not only encourage treatment in decentralized way but will also lead for the demand from government to work for the proper treatment and disposal. People support and active participation is a key for efficient waste handling. The momentum and psychological internalization of the importance of the issue of sanitation and hygiene created by Swachh Bharat Abhiyan must be harnessed to make everyone more sensitive towards issue of solid waste management. The research also reveals that the success in waste management would be achieved when mix of legal practices, financial resources, processing technologies, recycling, public awareness, and involvement is ensured.

25.6 Conclusion

The study has examined the level of waste generation and handling in India. From the analysis, it has been found that across all States/UTs, four states are generating over 10,000 TPD of waste and 11 states are generating in the range of 100 to 1000 TPD. With respect to collection, 22 States/UTs have waste collection of over 75%. The treatment of collected waste is, however, poor. From the analysis of Cities/Towns it is noted that most of them generate less than 500 TPD. The problem in India appears to be not the quantity of waste per se, but that of waste handling. Both the collection and treatment part need to be raised. The recovery of material and energy is in a very poor state in the country. There are very few WtE plants in the country, and they are basically RDF-based. Many of them are working below their potential. The land filling is not done scientifically, thereby raising vehement opposition wherever planned.

The paper brings forth the issue that despite worrying about size of population and quantity of waste we need to concentrate on the collection and processing part of the same. India lacks in processing facilities and advanced technologies, but even then, focus has consistently remained on that. The first step of segregated storage of waste is very crucial. It is quite probable that if this step is effectively implemented, then the task of processing will be easier. This requires wholehearted participation by each waste generator in the country. The capabilities of private companies, NGOs, community groups, waste pickers should be effectively utilized.

As solid waste management has become one of the focused programs of the government under Swachh Bharat Abhiyan, further research in this area need to focus on the constraints and success in the SBM through primary data. Furthermore, through field-based research there is a need to explore the feasibility of small-scale waste treatment facility in the vicinity of waste generation area. The objective assessment of the capacity of local government should also be focused to improve living condition.

References

Annepu RK (2012) Sustainable Solid Waste Management in India. http://www.seas.columbia.edu/earth/wtert/sofos/Sustainable%20Solid%20Waste%20Management%20in%20India_Final.pdf

Azad S (2015) Land Scarcity Hits Ambitious Urban Waste Disposal Plan, July 17. http://timesofindia.indiatimes.com/city/dehradun/Land-scarcity-hits-ambitious-urban-waste-disposal-plan/articleshow/48117503.cms

Basu S (2013) Wasted Wealth. Down to Earth, August 31. http://www.downtoearth.org.in/coverage/wasted-wealth-41917

BARC (2015) Nisargrunabiogas Plant: Based on Biodegradable Waste Resource. Bhabha Atomic Research Centre. http://www.barc.gov.in/technologies/kitchen/index html. Accessed on 18 July 2015

Bhatnagar GV (2015) Controversy Dogs Waste-To-Energy Plant. The Wire. http://thewire.in/2015/05/19/waste-to-energy-plant-in-delhi-under-a-cloud-2054/

CPCB (2015) Consolidated Annual Review Report on Implementation of Municipal Solid Waste (Handling & Management) Rules 2000. Central Pollution Control Board. http://www.appcb.ap.nic.in/main/Status%20Report%20on%20MSW.pdf

Centre for Science and Environment (2014) Construction and Demolition Waste. http://www.cseindia.org/userfiles/Construction-and%20-demolition-waste.pdf

Chitlangia R (2015) DDA Clears Landfill Plan for Dwarka Green Site. ET Realty, November 14. http://realty.economictimes.indiatimes.com/news/industry/dda-clears-landfill-plan-for-dwarka-green site/49776239

CPCB (2013) Status Report on Municipal Solid Waste Management. Central Pollution Control Board. http://www.cpcb.nic.in/divisionsofheadoffice/pcp/MSW_Report.pdf

GOI (2014a) Completed Projects Under JNNURM. http://jnnurm.nic.in/wp-content/uploads/2014/04/Completed-Projects-29-04-20141.pdf

GOI (2014b) Completed Projects Under UIDSSMT.http://jnnurm.nic.in/wp-content/uploads/2014/05/completed-projects-under-UIDSSMt.pdf

HWS Colleges Edu (n.d.) Landfills: Where Does Our Trash Go? http://people.hws.edu/halfman/Data/PublicInterestArticles/Landfills.pdf

India Today (2016) Villagers in Taloja Oppose Shifting of Deonar dumping Site. India Today, February 3. http://indiatoday.intoday.in/story/villagers-in-taloja-oppose-shifting-of-deonar-dumping-site/1/586447.html

Japan-Ministry of the Environment (2014) History and Current State of Waste Management in Japan. https://www.env.go.jp/en/recycle/smcs/attach/hcswm.pdf

Jha K (2014) 12 Notable Waste Management Startups in India. April 9. http://thetechpanda.com/2014/04/09/12-waste-management-startups-in-india/

Kumar S (2005) Municipal Solid Waste Management in India: Present Practices and Future Challenge. http://saahas.org/wp-content-saahas/uploads/2015/02/SWMindia10.pdf

Kumar S, Bhattacharya JK, Vaidya AN, Chakrabarti T, Devotta S, Akolkar AB (2009) Assessment of The Status of Municipal Solid Waste Management in Metro Cities, State Capitals, Class I Cities, And Class II Towns in India: An Insight. Waste Manag 29:883–895

Milios L (2013) Municipal Waste Management in The Netherlands. http://www.eea.europa.eu/...waste/netherlands-municipal-waste-management.pdf

MoUD (2014) Guidelines for Swachh Bharat Mission Urban. Ministry of Urban Development, Government of India. http://moud.gov.in/sites/upload_files/moud/files/SBM_Guideline.pdf

MoEFCC (2016) Solid Waste Management Rules. http://www.moef.gov.in/sites/default/files/SWM%202016.pdf

MoEFCC (2000) Municipal Solid Waste Rules 2000. Ministry of Environment, Forest, and Climate Change, Government of India. http://www.moef.nic.in/legis/hsm/mswmhr.html

Nair HV (2014) Dumping Ground Too Small for Delhi's Garbage. India Today, March 2. http://indiatoday.intoday.in/story/dumping-garbage-overflow-delhi-dda-mcd-south-north-corporation/1/346378.html

Nath D (2014) India's First Plant That Recycles Construction Waste. http://www.thehindu.com/news/cities/Delhi/indias-first-plant-that-recycles-construction-waste/article6362727.ece

NIUA (2015) Urban Solid Waste Management in Indian Cities. National Institue of Urban Affairs, New Delhi. http://pearl.niua.org/sites/default/files/books/GP-IN3_SWM.pdf

Parker L (2014) The Best Way to Deal with Ocean Trash. National Geographic. April 16. http://news.nationalgeographic.com/news/2014/04/140414-ocean-garbage-patch-plastic-pacific-debris/

Parker L (2015) Eight Million Tons of Plastic Dumped in Ocean Every Year. National Geographic. February 13. http://news.nationalgeographic.com/news/2015/02/150212-ocean-debris-plastic-garbage-patches-science/

Prasad R (2012) Efficient Way to Turn Waste into Resource. The Hindu. October 18. http://www.thehindu.com/sci-tech/science/efficient-way-to-turn-waste-intoresource/article4006137.ece

Ramachandran H, Tewari S, Begum A (2016) Study of Urbanization of Uttarakhand. Report submitted to the Fourth Uttarakhand State Finance Commission. Institute for Studies in Industrial Development, New Delhi

Sriram S (2015) These 2 Indian Startups Will Pay Cash for Your Trash. December 22. http://gadgets.ndtv.com/internet/features/these-2-indian-startups-will-pay-cash-for-your-trash-781033

The Hindu (2013) Okhla Composting Plant Earns Rs. 25 Lakh as Carbon Credits. The Hindu, March 6. http://www.thehindu.com/news/cities/Delhi/okhla-composting-plant-earns-rs25-lakh-as-carbon-credits/article4481168.ece

United States Committee on Pollution (1966) Waste management and control: a report to the federal council for science and technology. National Academy of Sciences-National Research Council, US, Washington, DC

US-EPA (2014) Municipal Solid Waste Generation, Recycling and Disposal in The United States: Facts and Figures For 2012. United States-Environmental Protection Agency. https://www3.epa.gov/epawaste/nonhaz/municipal/pubs/2012_msw_fs.pdf

Waste Atlas (2016) Waste Generation Per Capita (Kg/yr). Accessed in February 2016. http://www.atlas.d-waste.com/

Wong D. (2014) Waste Management Policy in The Netherlands. Research Office Legislative Council Secretariat. http://www.legco.gov.hk/yr13-14/english/sec/library/1314in10-e.pdf

World Bank (2012) What A Waste: A Global Review of Solid Waste Management. http://siteresources.worldbank.org/INTURBANDEVELOPMENT/Resources/336387-1334852610766/What_a_Waste2012_Final.pdf

Yep E (2015) Singapore's innovative waste disposal system. Wall Street J, September 13

Social Impact Assessment of Indian Water and Allied Policies and Programs

26

Dipankar Roy, Shobha Kumari,
Akhilesh Kumar Mishra, and Suresh Chand Rai

Abstract

Social impact assessment (SIA) is having a noteworthy degree of independent applicability apart from being a sub-process of an environmental impact assessment (EIA) scheme. The Indian water resource planning sector is a specific example of the defected practice of SIA and consequent visible problems of inefficiency and underperformance. This link between poor SIA planning and the said underperformance is a prime focus of this article. It has discussed the existing scope of SIA in Indian water resource management during both pre-feasibility and post-development stages.

Keywords

Audit · Community · Environment · Policy · Social Impact Assessment (SIA) · Water

D. Roy · S. Kumari
Department of Geography, Institute of Science, Bananas Hindu University, Varanasi, India

A. K. Mishra · S. C. Rai (✉)
Department of Geography, Delhi School of Economics, University of Delhi, Delhi, India

26.1 Introduction

Social impacts are the impacts on societies on account of government policies on marketed and non-marketed goods and services. These impacts are essential indicators for assessing the costs and benefits of public policies (Harper and Price 2011).Social impact assessment (SIA) involves characterization of the social aspects of the environment, forecasting the nature of impacts on the population due to the implementation of any given action, and developing mitigation strategies for the likely adverse consequences (Goldman and Scott 2000). In India, the National Resettlement and Rehabilitation Act (2007) have recognized SIA as a central part of resettlement planning and implementation process. Thus, measuring displacements and managing rehabilitations are considered as two key aspects of SIA in India. However, SIA schemes are currently treated merely as sub-processes of environmental impact assessment (EIA) practices, and, thus, not given its due importance. In case of water resources development and management sector, scope of SIA is not just limited to the question of displacement or rehabilitation. The linkages between management efficiencies and livelihood security are visibly robust in case of water resources (Cummins et al. 2007). Here, development interventions invariably include mechanisms of water allocation and redistribution, and thus, bring unwarranted social impact for downstream

B. Thakur et al. (eds.), *Resource Management, Sustainable Development and Governance*, Sustainable Development Goals Series, https://doi.org/10.1007/978-3-030-85839-1_26

appropriators and groundwater users (Cummins et al. 2007). So, the scope of SIA application in this sector is considerably broad.

Impact assessment schemes should be based on neutral information and sound science that can aid objectivity in decision-making process (Keskinen 2008). However, it rarely happens in practice, and both planning process and impact assessments are bound with the values and interests of different social groups and associated socio-political contexts (Keskinen 2008). These contexts have a robust presence in water resource management paradigms across the globe. Nature of a water management system is primarily a product of the interactions between these social and political undercurrents. In India, the National Environmental Policy (NEP) 2006 has comprehensively dictated the needs and procedures of awarding environmental clearance (EC) for various development activities including the water resource management sector. However, the policy certainly lacks an inclusive character with respect to the aspirations of this resource management sector. A very narrow context of economic development is dominating the policy-making think-tanks, and consequently, unproductive investments are receiving environmental clearances (Thakkar 2012). Structural planning and management philosophy has central importance, and non-structural solutions like "no development" option are summarily rejected from the scope. In many developed nations, this "no development" option is equally considered in case of water resources planning, and social impacts of the development activities are used as justified parameters for measuring performances of the operational infrastructure (Reuss 1992). In contrast, in India, most of the environmental impact assessment studies do not measure the vulnerability of different social and economic groups towards any prospective impacts of the development projects (Pandey et al. 2013). Consequently, the performance level of most of the water management infrastructures is below the expected standard, and their social impacts are negative (Thakkar 2012).

The present paper is established on two-fold objectives: (i) Highlight the importance of social impact assessment in the context of the water and allied policies in India, and (ii) Assess the major social impacts of the said policies in terms of their relative strengths and weaknesses.

Box 26.1 Sustainable Development Goals

Ensuring the availability of clean water and sanitation for all in India is crucial for achieving overall sustainable development within the framework of UN-SDGs. The linkages between water and sustainable development are numerous, complex, and often subtle. In so much, water is not only integral to maintaining an ecosystem and biodiversity, but it is also a key driver of economic and social development, including health, gender equality, resilience, inclusive cities, and societies. More recently, the Government of India has begun to recognize the need to align water-related SDGs with other major water policies and programs such as *Swachh Bharat Mission* with its ambitious goal of providing universal sanitation while trying to address the gap in the water and sanitation sector. The authors in this chapter recognize the significance of social impact assessment (SIA) vis-à-vis water policies and programs and their connections with the other SDGs so that stakeholders may comprehend shortcomings, contradictions, and lack of alignment between SDG #6 and the other SDGs and can make conscious choices, prioritizations, and optimizations in implementing the programs and policies. SIA of water programs and policies in the Indian context is also significant as India has several comprehensive national level laws, policies, and programs whose formulation and implementation are an ongoing process. Moreover, since water is a state subject and India's diversity and the UN sustainability framework demands water policies and programs to be addressed holistically so as to achieve universal and equitable access to safe and affordable drinking water, achieve access to adequate and equitable sanitation and hygiene for all, improve water quality by reducing pollution, eliminate dumping and minimizing release of hazardous chemicals and materials, substantially increase water-use efficiency across all sectors, implement integrated water resources management at all levels, including through transboundary co-operation as appropriate, protect and restore water-related ecosystems, and support and strengthen the participation of local communities in improving water and sanitation management.

26.2 Methodology

The study has adopted the Global Water Partnership (GWP) Toolbox for the assessment of visible social impacts of Indian national water and allied policies. The said toolbox has laid out a comprehensive list of 55 tools organized in three-tier hierarchical structure for framing the IWRM perspective at large. Here, the Assessment Instruments tool (Serial No. C2) including its six sub-order tools has been selected for a comprehensive assessment of the social impact of Indian water and allied sectors. These sub-tools are risk assessment (C2.01), vulnerability assessment (C2.02), social assessment (C2.03), ecosystem assessment (C2.04), environmental impact assessment (C2.05), and economic assessment (C2.06).

Here, the sub-tools, viz. risk and vulnerability assessment have been considered as synonymous which is basically the level of threats to the social and economic stability of the population. Social assessment helps to measure the degree of community participation and empowerment following the implementations of the government and non-government funded policies and programs. Ecosystem and environment impact assessment both bring in focus the effectiveness of the said policies and programs in addressing the environmental challenges in face of rapid population growth and agro-ecological deteriorations. Economic assessment measures merely the productivity and economic gains of the investments over the water infrastructures and programs.

With all these tools of reference, the existing ground realities of the Indian water policies and programs and allied sectors have been divided into the following four sections and subsequent 12 sub-sections in consideration of the implications of these selected six sub-tools. Thus, a contextual profile of SIA in the water sector has been prepared. A discussion based method has been adopted to qualitatively assess the relevance of the current policies and programs based on the selected GWP tools. Finally, the significant observations concerning each of the tool elements have been tabulated.

26.3 Risk and Vulnerability Assessment

26.3.1 Issues of Livelihood Risk

The issue of livelihood security is not considered a stated parameter in conventional impact assessment process in India (Pandey et al. 2013). Thakkar (2012) has discussed many case studies to prove the linkages between poverty and vulnerability towards changing water availability patterns consequent upon the variable climatic conditions. He has mentioned explicitly about the fragile and delicate balance between natural resource endowment levels and associated livelihood securities of the indigenous populations. An impact assessment should ideally be conducted with a bottom-up approach that may identify and empower such vulnerable groups. The evaluation of vulnerability in many public financed schemes in some states gives an outstanding example (Table 26.1). In each of the cases, the livelihood security of the population was under threat owing to increasing climatic variability and consequent water shortages. The governments and other concerned agencies conducted a meso-scale assessment of the problems based on sound scientific tools. This information was shared with the

Table 26.1 Some Community Run Water Resource Management Ventures in India

Project village	State	Year	Coordinating agency/organization
Laporiya	Rajasthan	1973	Gram Vikas Navyuvak Mandal Laporiya (GVNML)
Kodikitunda	Odisha	1993	Agragamee (an NGO)
Hiware Bazar	Maharashtra	1995	Respective village panchayat
Palangarai	Tamil Nadu	2004	Tamil Nadu Water
Kui	Uttarakhand	2004	Government of Uttarakhand
Lakhwar and Chhotau	Uttarakhand	2006	The Energy and Resources Institute (New Delhi) and the Society for Motivational Training and Action (SMTA)

respective communities, and a range of mitigation options was finally shortlisted. The local self-governance (Village panchayats) was incentivized to maneuver the projects. The communities themselves took on the task of action plan development and implementations. The success of these projects was remarkable on two accounts, viz. timing and sustainability. Unlike the slow and incremental improvements in the government-run projects, here, the changes could be measured within a very short interval of time from 1 to 2 years. At the same time, the beneficiaries were encouraged to mobilize their individual sources of resources for better operation and management (OandM) of the built-up infrastructure. It is an alternative example of the stereotype social impact assessment where the scope of people's participation is limited only to the public hearings or written consent papers (MoEF 2006).

26.3.2 Water Disputes

The Constitution of India does not explicitly state anything regarding the issues of water right apart from declaring the inter-state rivers as a union subject and other aspects as a state subject (Union Planning Commission 2007). Thus, it has accepted and somehow encouraged the growth of political control and decision making over issues of water allocation and use. Such legislative authority has superseded some other most important facets of the resource management agenda, viz. riparian, extractive, and distributive water rights. This had brought the legislations and those informal right holding communities into a state of conflict. Till date, only the surface water sector has received the heat of these conflicts (Table 26.2). Krishna-Godavari water dispute (1950–1969), Cauvery water dispute (1968–1990), and Ravi-Beas dispute (1976–1986) are worthwhile for this discussion. One problematic aspect of such conflicts has been the apparent impairments of the "*ad hoc* water tribunals" (commissioned after the Inter-State Water Dispute Act 1956). In some instances, many states have declined to accept the decisions of the tribunal, and thus, exposed their lack of binding power (Richards and Singh 2002). In most of the

Table 26.2 Cases of Inter-State Water Disputes in India in the Post-Independent Period

River/basin	Disputant states	Time length of dispute	
		From	To
Ravi and Beas	Punjab and Haryana	1955	Ongoing
Godavari	Maharashtra, Madhya Pradesh (including Chhattisgarh), Orissa, Karnataka, and Andhra Pradesh	1969	1980
Krishna River	Maharashtra, Karnataka and Andhra Pradesh	1969	1976
Cauvery	Kerala, Karnataka, Tamil Nadu, and Union Territory of Pondicherry	1990	2007
Periyar	Kerala and Tamil Nadu	2000	2014
Mahadayi/ Mandovi River	Goa, Karnataka, and Maharashtra	2002	Ongoing
Vansadhara River	Odisha and Andhra Pradesh	2006	2013

cases, the point sources of such disputes have been the river valley development projects (irrigation or multipurpose). These projects facilitate the large extractive use of the river water and give an advantage to the upper riparian states. Understandably, the lower riparian states seek political redemption via, filing lawsuits, and, thus, the conflict is created. For example, the Krishna Tribunal (1973) permitted the large-scale diversion of river water outside the natural boundary of the basin if the respective diversion channel is located within the political boundary of the upper riparian states (Richards and Singh 2002). It may be considered as a violation of rights of the riparian communities, and negligence of the geography and ecology of the entire basin. Here, society or community should have received the priority and not the political boundary. Ironically, any such "development intervention" should have been grounded on clearly perceived social impact assessments based on the riparian and distributive principles, and not just for extractive use.

26.3.3 Chronic Poverty and Poor Accessibility

About 21.9% of the total population in India is still below poverty level that respectively includes 25.7% of the rural and 13.7% of the urban population (Planning Commission 2014). Their existing access to safe drinking water and sanitation has a dim status (Jha 2010). Now, this poor population has been used a rallying cry for market-led reforms of water supply and sanitation sector which has been financially supported by the international financial institutions (IFIs) like the Asian Development Bank (ADB) or World Bank (WB). The very first step of such reform practices starts with the de-politicization of that sector when political elements are outrightly denied any entry to policy-making circle (Wit 2002). This task is generally carried out in the name of curbing corruptions. These political elements are actually the voice for the poor population who hardly has any other medium of expression. This strong political compulsion has so far forced the public agencies to ensure a minimum standard of water supply and sanitation services to the slum dwellers in India (Wit 2002). In countryside, this compulsion is visible in the form of irrigation-electricity subsidy and loan waivers announcements for the drought or flood-affected farmers (Shah et al. 2006). The IFIs are seeking a complete exclusion of these political elements and, thus, indirectly sidelining the interests of the poverty-ridden mass. Paradoxically, the IFIs itself is citing its own policy as pro-poor like that of the ADBs stated policy vision (Withanage et al. 2007). At the same time, terms like *full cost recovery* and *tradable water rights* are quite enthusiastically highlighted in their market propaganda. ADB is yet to give any clarification on how these apparently divergent aims of poverty reduction and full cost recovery and tradable water entitlements would work together in Indian context? In cities like Mumbai (2007), Bengaluru (2007), or Delhi (2011), such inequities have resulted in incidents of public confrontations against such business ventures (Budhya 2007; Jafri 2007; Koonan and Sampat 2012). Neighboring China may also give some explanation of the possible social impacts of such policies. In China, the private water utilities are primarily in business since enactment of the Water Law 1988. In between 1989 and 2003, there were 9 reported cases of water rates hikes as much as up to 23 times (Yu 2007). Till 2000, at some instances, even the poorest households had to account for 4.2% of their monthly expenditure for domestic water bills (Yu 2007). Since 2005, the water market in China is generating an annual revenue of nearly 50 billion Yuan, and 53% of the water enterprises are categorized as profitable (Yu 2007).

26.4 Social Assessment

26.4.1 Inclusions of Social Impacts in Current Policies

Around 55% of the net irrigated area in India is dependent on groundwater sources where the primary challenge is over consumption. Micro-

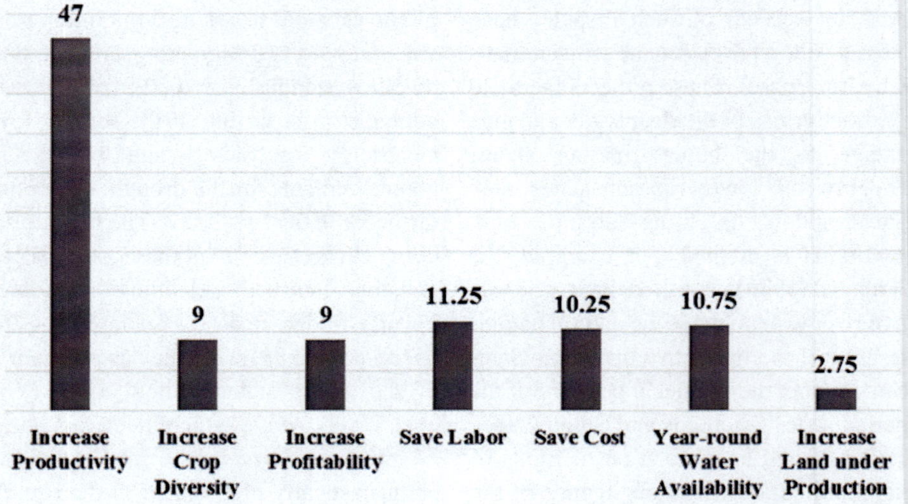

Fig. 26.1 Farmers' perception (percent) towards adaptability of micro-irrigation tools. (Source: India Micro-irrigation Program Survey, 2011)

irrigation systems like drip or sprinkler are a viable solution to this problem of overuse. These systems have a potential to save water by 10–69% on various crops and increase the yield of those crop by 7–54%. However, one observed limitation of these micro-irrigation tools is the high operation and management costs on account of high energy requirements. At the same time, farmer's adoption of micro-irrigation (MI) technology is directly related to an increase in production, and the water using efficiency of these MI tools is considered only as subsidiary factors. The Indian micro-irrigation program survey (2012) funded by the Gates Foundation across the four states, viz. Bihar, Maharashtra, Odisha, and Tamil Nadu has found that an average 47% of the surveyed farmers have purchased micro-irrigation equipment for crop productivity enhancement; rather than for water savings (Fig. 26.1). These aspects of farmers' behavior need particular policy focus. Still, micro-irrigation measures in India appear more as a part of sideline irrigation policies and less a regular part of the agricultural policies. Currently, the National Water Policy (NWP) 2012 considers the component of micro-irrigation (clause 6.5) only as mean of irrigation water savings. Even the National Mission on Micro-irrigation (2010) program is originally a hybrid approach to integrate the successive government schemes like the National Food Security Mission (NFSM), Integrated Scheme of Oilseeds, Pulses, Oil palm and Maize (ISOPOM), Technology Mission on Cotton (TMC), etc. which had have a common component of micro-irrigation. Clearly, here the micro-irrigation measures are used as a function of crop diversification, and the least attention is paid towards linking these measures to the targeted agricultural growth rate of 5% as cited in the national agricultural policy 2000. Unless such measures are promoted as a central thrust of the agricultural growth strategy, positive changes would be hard to achieve concerning the groundwater depletion scenario.

26.4.2 Need of Better Water Audit-Plans

Good water management systems generally have provisions for water auditing and service level benchmarking (SLBM). Both these processes are interdependent and interconnected in nature. In India, some state water policies like that of Maharashtra (2003–2004) are having provisions for water audits or benchmarks, but, national policy level applications of benchmarking or auditing have been a slightly misguided process. For example, the Central Water Commission (CWC) is claimed to have been conducting performance evaluation studies of irrigation projects since the eighth FYP (1992–1997), but it took the commission till 2002 to include the component of SLBM for irrigation projects (General Guidelines for Water Audit and Water Conservation, 2005). Further, it could have pipelined the objective of water auditing for irrigation sector only in 2005 (on voluntary basis by the state irrigation departments). So, in either situation, they were evaluating performance without any referred benchmarks, or putting benchmarks without audit assessments. It is an act without any clear policy vision. Similarly, the Ministry of Urban Development nodded for SLBM for urban water and sanitation sector of late in 2008. So, the policy-making circle seems to be disaggregated entirely on the issue of water auditing and benchmarking. Post-audits are considered as a "leading practice" and a key part of monitoring stage of an EMP. In India, the monitoring phase of any project heavily depends on 6 months' progress reports submitted by the proposing authority of any project which is to be assessed by the six selected centers of the MoEF. It is a clear indication of a one-size-fits-all approach. Currently, the MoEF does not invite any external environmental audits process by an independent agency for the submitted progress reports. So, the entire process is highly centralized in nature that ultimately puts a question mark on quality control mechanism. The nation has a large scope of policy learning from countries like New Zealand where an "open door policy" followed by industries like mining has made provisions for a fortnight publication of newsletters in local newspapers regarding environmental quality of surrounding areas (Laurencont et al. 2009).

26.4.3 Social Monitoring Framework

Mitigation and adaptation are complementary strategies in water resource management. In the adaptive water management practices, a central role is given to interactive social learning processes to impart more flexibility in institutional structures (Tracey 2009). Engaging the community in the mitigation plan through a socio-economic monitoring framework (SMF) has some advantages in comparison to the conventional expert-driven practices (Tracey 2009). It is one of the end products of community partnership programs (CPP). Practically, SMF minimizes the risk of oversampling during the process of indicator short listings for monitoring tasks (Tracey 2009). Instead of wasting resources on developing the best combination of indicators (for impact assessments) by the experts, SMF makes use of the locally available information that has already been collected by any local agency or organization (for example, employment or livelihood related data collected by any local NGO or CBO). Hence, SMF is comparatively a cost-effective process. Without this, the policy intervention options may acquire an uneconomically large shape (Rajabi et al. 2001). This happens especially under the circumstances where many of the popular options are actually interdependent in nature (Rajabi et al. 2001). Here, the Madurai-based civil society organization called Development of Humane Action (DHAN) foundation can be considered as one of the pioneer agencies in this direction. The said agency has a long experience of working across 12 states and 10, 47, 924 households in micro-level livelihood improvement through better resource management (Table 26.3). Community-based micro-financing and insurance cover, indigenous knowledge utilization, and information sharing mechanisms are some of the critical strategies for their success. Their programs have eventually received recognition and support from several national schemes and mainstream public

Table 26.3 Network Structure of DHAN

Outreach network	2010	2011	2012
Households	810,185	923,865	1,047,924
Primary groups	33,039	37,071	45,525
Clusters	1648	1432	1728
Federations	260	283	303
Villages	9757	10,755	12,406
Districts	51	54	66
States	12	12	12

(Source: DHAN 2012)

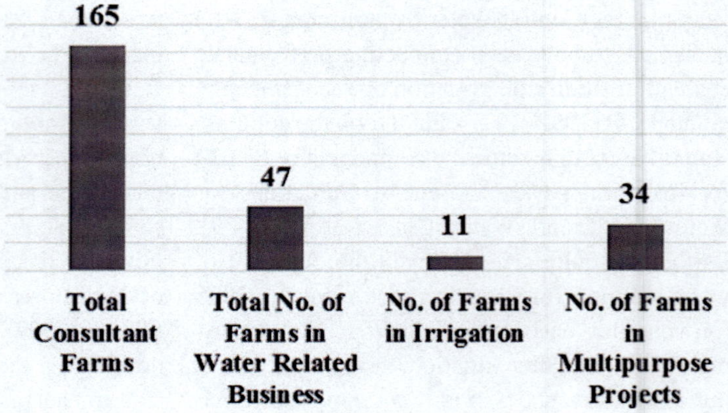

Fig. 26.2 Business profile of the ESIA consultancy farms in India (2015). (Source: https://moef.gov.in)

agencies like National Watershed Development Program for Rainfed Areas, Drought Prone Area Programme, as well as schemes of the National Bank for Agriculture and Rural Development like Indo-German Watershed Development Programme (IGWDP) or Watershed Development Fund (WDF). It is an emergent form of social monitoring framework in India where the administration side is deliberately seeking partnerships with these civil society organizations in disaster management tasks. Still, the scope of such partnerships has not yet been fully explored by the current environmental policy at the national level, and the impact of assessment task has remained an expert managed job.

26.5 Ecology and Environmental Impact Assessment

26.5.1 Current Institutional Capability

Indian agro-economy is heavily dependent on the groundwater resources, and more than 60% of the net irrigation is drawn from this source (Thakkar 2012). Presently, this fresh water source is at a stage of peril. Till 2011, nearly 60% of all the country's districts are suffering from either quantity or quality problems of groundwater reserves. However, the ESIA (EIA and SIA) business sector can address only the surface water resources. The Ministry of Environment and Forests (MoEF 1994) has earmarked the river valley projects for compulsory impact assessments (MOEF 2006). Groundwater guzzling industries and enterprises like soft-drink or bottled-water procurements, water parks, golf courses, etc. are not explicitly accounted at all for their visible negative environmental and social impacts (Shiva 2005; Haider 2010). Currently, a total 47 accredited EIA consultant farms have expertise in selected three areas, viz. irrigation projects, multipurpose projects, and river interlinking (Fig. 26.2).Then, out of these 47 farms, a majority, i.e., 35 have specialization in multipurpose projects. Existing legislation also does not account these agencies for any compulsory monitoring or environmental or social auditing at the post-development stage of a project. Unlike in countries like New Zealand, these agencies generally do not have any active public outreach cells and sharing of information between the concerned agency and civil society is minimum (Laurencont et al. 2009). Interestingly, none of the public owned agencies is directly involved in EIA consultancy business, and participation by the government is virtually nil in defining the "market-standards" of ESIA processes and outputs. It may mostly reduce the technical competency of the public agencies to undertake any such detail and complicated ESIA studies.

26.5.2 Challenges of Groundwater Management

The union and state governments provide high rates of subsidy during the *rabi* season (winter cropping) on electricity and diesel to support the extensive groundwater irrigation across the country which virtually encourages the farmers for deep groundwater boring which is often ecologically disastrous. This problem is most dangerous in the "groundwater hotspot" zones which comprise the southern peninsula and western parts of the country (Thakkar 2012). Available data from these eight states shows that annually, an average 25.56% of the total procured electricity is used by the agriculture sector in exchange for an average flat tariff rate of just Rs. 0.41/KWh/year (Table 26.4). These areas have been at the upfront during the agro-production boom following the Green Revolution in 1967–1968. Indiscriminate groundwater misuse and overuse have been fueling this production boom at that period which soon turned their regional groundwater ecology vulnerable. Additionally, this unsustainable groundwater-energy nexus has made inroads for a comprehensive change in cropping practices across the country which is not always in conformity with the regional groundwater ecology, but still, such crops are receiving high government subsidies in terms of minimum support price and farm inputs like seeds, fertilizers, pesticides, etc. (Planning Commission 2007). In all the said eight states, the average subsidy on agriculture is up to the scale of US$ 42 million/year (Table 26.4). In Punjab, for example, minimum support price (market price subsidies) is provided for paddy or chick-pea cultivation which is a prime cause of groundwater depletion (Planning Commission 2007). Degrading quantity and quality status of the groundwater reserve is also affecting the drinking water security of the country as an 85% of the rural water supply is primarily met from groundwater sources (Planning Commission 2007). Impacts of the degraded groundwater quality on human health are a matter of great concern as a recent report entitled "Water Pollution in India" (2011) by the Comptroller and Auditor General of India (CAG) across 25 states has found that at least 19 states do not have the necessary infrastructure to assess the risk to human health on account of drinking contaminated water.

26.5.3 Progressive Assessment of the Impacts

The NEP (2006) gives considerable thrust on environment management plan preparations which entails a list of measures to ensure the proper mitigation of the negative impacts of the concerned project on the physical and social environment. In case of water resource management planning, application of this process may be problematic. This particular resource management sector has scope for both physical and social planning (Reuss 1992). The new generation of water planner is credited to be better equipped to predict the associated risks of the resource development activities than the social planners (Crase 2010). Still, their expertise is concentrated to handle a range of robust stochas-

Table 26.4 Status of Agricultural Power Consumption and Subsidies in Selected States

State	Percent consumption of electricity by agriculture in 2003–2004	Average tariff for agricultural electricity in 2001–2002 (Rs./KWh/year)	Subsidy for agriculture in 2001–2002 (US$ million)
Haryana	28.20	0.48	30
Punjab	19.9	No tariff	35
Uttar Pradesh	10.9	1.19	20
Gujarat	28.5	0.62	68
Andhra Pradesh	27.4	0.14	62
Karnataka	28.7	0.39	39
Tamil Nadu	18.8	0.01	47
Rajasthan	14.6	0.46	35
Average	***25.56***	***0.41***	***42***

tic models (Crase 2010). These empirical models can barely combine environmental and social data sets, and there is always a chance of misgiving results. In case of water resources, where it is somewhat feasible to devise adaptation plans for short-term impacts, the long-term impacts are challenging to conceptualize (Keskinen 2008). So, there remains a greater scope for assessment of the impacts throughout a more extended monitoring period. At each stage, there is a need for rigorous measurements and analysis of monitored data that can be used for future impact assessments (Crase 2010). Without such a historical database, an assessment plan is likely to give flawed results on possible environmental (and social) impacts (Crase 2010). If the available data source is discontinuous and non-existent, then, field-based research projects should be undertaken to plug in this gap (Perkins et al. 2009). Presently, in India, if an EIA Report is assessed to be of the inferior standard, the project proponents are just required to submit some more additional project details, but it happens during the post-project stage when the damage has already been set to gear (Pandey et al. 2013).

26.6 Economic Efficiency

26.6.1 Irrigation Performance of the River Valley Projects

There are more than 4300 large dams in India, and primary function of 96% of them is irrigation. With this numerical strength, India ranks as the third top nation regarding the number of dams (Dandekar 2010; GIZ and GoI 2011). Till 2007, these dams used to irrigate about 33.74 million hectares (around 30%) of the total gross irrigated area of the country (Planning Commission 2009). Further till 2011, approximately 91% of these functional dams were concentrated within five states only (MoWR 2011). The ultimate irrigation potential created by these dams has remained entirely unchanged in between 1972 and 2007; despite a massive investment of INR. 18254 billion which was around 69% of total plan expenditure on irrigation (Planning Commission 2009). So, the noteworthy and medium irrigation policies have been plagued by interregional and intraregional inequality and inequity, and a very negligible success rate. Although primarily designed for irrigation, of late these dams have increasingly been used for non-irrigation commercial activities. The non-irrigation revenue (including selling water to domestic and industrial users) of the irrigation authorities has almost tripled in between 1978 and 1988 from INR. 30 million to INR 99 million (SANDARP 1999). The states of Maharashtra and Madhya Pradesh together account for 70% of this non-irrigation revenue. The Maharashtra State Ordinance No. 11, 17 September 2010, for example, was particularly issued for increasing the non-irrigation revenue of the irrigation authority (Dandekar 2011). Noticeably, dams are being used by a powerful lobby within the government in favor of the bulk water using entities like industries, amusement parks, golf courses, urban centers, etc.

26.6.2 Flood Mitigation Efficiency of Existing River Valley Projects

One major objective of RVD project is flood control management (Clause 5.5, NWP 2012). Controversially, RVD project management authorities themselves have augmented flood hazard intensities in India (Table 26.5). The prime reason has been unscientific and below standard operations by the dam authorities (Thakkar 2012). The National Commission on Flood (NCF), constituted in 1980, observed that the Damodar River Valley (DVC) project installed in 1948 has become a prime factor behind intensification of flood damage within the basin in 1978 (Report of the National Commission on Floods, 1980). The report concluded that between 1958 and 1978, no scientific studies were undertaken to assess the impact of the observed low level discharge on the downstream drainage caused by the dams which is also the cause of drainage congestion and consequent flooding. Even the Ministry of Environment and Forests

Table 26.5 Recent RVD Projects Induced Flood Events. (Source: Based on Bhattacharya and Kumar, 2010, Thakkar, 2010)

River	Year of flood	Major cause(s)
Damodar	1978	Steady deterioration of the downstream drainage system by weak flush and consequent siltation due to the obstruction created by the dams of the Damodar Valley Corporation
Satluj	1986	Operational negligence in Bhakra dam
Kosi	1993 2008	Operational negligence to the maintenance of Kosi embankments
Tapi	2006	Operational negligence in Ukai dam
Ghaggar Basin	2010	Ill-maintenance of Sutlej-Yamuna Link canal
Ganga	2010	Operational negligence in Tehri dam
Mahanadi	2008 2011	Operational negligence in Hirakud dam

has also acknowledged the need of more transparent and participatory management of the existing dams and reservoirs; with an optimal real-time, control on technical issues like live storage conditions and discharge timings (Thakkar 2012). Contrariwise, free-flowing undammed rivers can mitigate the intensity of flood hazard in a better way and also maintain a more productive and sustainable aquatic ecosystem health (Jhunjhunwala 2010). Even the NCF (1980) has noted that numerous saucer-shaped depressions occur in the chronically flood-prone plains in eastern Uttar Pradesh, North Bihar, West Bengal, and Assam. They are locally known as *Jheels* in Uttar Pradesh, *Chaurs* in Bihar, *Beels* in West Bengal and Assam. These depressions can significantly reduce the flood peaks and delay the flow of water into main rivers downstream (a case of free-flowing river). However, most of these systems are in poor shape due to neglect of maintenance, silting up of the links with the rivers, proliferation of weeds and also some attempts of draining these depressions for farming. In Bihar, construction of embankments has cut-off many depressions from the rivers. The NCF observed that these bodies can be rich source of fisheries and should be preserved as such they also have flood moderating potentials. In a number of nations like the USA, Canada, Australia, and Sweden have enacted concerned legislation to promote this value of free-flowing rivers (Dandekar 2010). But this undammed river was not considered at all as a policy alternative in India (Dandekar 2010).

26.6.3 Community Partnership Programs

In water management planning, the associated risk factor is predominantly borne by the communities developed around the use of surface or groundwater (Crase 2010). It involves a substantial adjustment on their parts regarding their daily and long-term water requirements (Crase 2010). Still, in comparison to the environmental impacts, monitoring mechanisms for socio-economic impacts may have instead been in an infant state. In India, each year a substantially large volume of data is procured on physical, institutional, and socio-economic dimensions, but the data is not always easily accessible, and their quality is also variable (Rao et al. 2003). A few community baseline studies have been conducted as part of various government-run livelihood development programs, but, they could not efficiently be compiled into a consolidated data base. As a viable alternative to this information deficiency, a thrust is currently given on "community partnership programs" (Pattenden et al. 2009). This kind of programs is helpful in building a structured dialog process between the community and concerned proponent of a project. Due to the close proximity of the communities to the resource, they have a better information base in comparison to the public agencies, and hence, they can manage the administration costs of program implementation with greater efficiency. The ambitious "Democratization of Water Management'" project launched in Tamil Nadu in 2004 in 153 village panchayats (local village administration) of 29 districts is such an example (Suresh 2007). It was a collaboration between the village communities and Tamil Nadu Water

Table 26.6 Ground realities of SIA and scope of improvement

Toolbox element	Tool specification	Strengths	Weaknesses	Suggestions for improvement
C2. Assessment Instruments	C2.01. Risk assessment C2.02. Vulnerability assessment	1. Community engagement efforts are already underway as an alternative to the conventional end-pipe-solution approach by the public agencies.	1. Ground feedback mechanism is very weak to adopt the good practices and convert them into national level policy guidelines. 2. Immature institutional structure which cannot safeguard the interests of marginalized population in face of increasing marketization of water governance. 3. Legal prioritization of extractive water rights over riparian rights which is a prime cause of water disputes.	1. Strengthening the existing information sharing network across the various tiers of water governance which can promote meaningful community partnership. 2. Effective decentralization of resource and decision-making power to the lowest tier of governance.
	C2.03. Social Assessment	2. Enactment of the water audit and bench marking guidelines at the national level.	4. Lack of scoping mechanism for audit provisions and community empowerment at the grass-root level.	3. Ensuring legal accountability of the project promoters throughout the project life cycle and post-closure period for regular auditing and information sharing.
	C2.04. Ecosystem assessment C2.05. Environmental impact assessment	3. Institutionalization of the EIA provision at the national level.	5. Incentivization (unintended and indirect) of bad practices in case of water over consumption. 6. Over dominance of structural solution measures. 7. Weak public sector capacity and market based approach by the private sector with a clear commitment for business	4. Policy inclusion of the "no development" option with a positive thrust on ecological management of flood and drought. 5. Strengthening the main stream public agencies to carry out research and development tasks.
	C2.06. Economic assessment	4. High degree of development of surface and groundwater resource across the state.	8. Equating development only in terms of indiscriminate extraction and a minimum care for conjunctive wise use. 8. Asymmetrical distribution of developed potentials.	6. Alignment of the mainstream agro-policies with the irrigation policies with effective incentives and disincentives to stop wastages and over usage.

Supply and Drainage Board (TWAD)-based on full information sharing on water availability, appropriate technology and sustainable water supply systems (WSS). The results were awe-inspiring with registered savings up to US$ 5 million in scheme implementation for TWAD. By community choices, 41% of villages opted for new WSS, and another 47% decided to continue with existing WSS. So, the end users of the resources have the better capacity to assess the social impacts of policies and programs, and thus can make a better choice regarding the acceptance or rejection of such programs or policies.

In India, the water and allied policy drafts include detailed sets of preventive and suggestive measures. The problem lies in the administrative perception that they take these measures as an end. This trend has somehow reduced these policy provisions into meager categorical compliance jobs. Table 26.6 gives a comparative assessment of the relative strengths and weakness of these said policies in terms of their social impacts. There lies a more significant role for their grass-root level scoping. Innovative and strategic approaches can help in bringing these wonderfully articulated policy guidelines to a state of the optimal application.

26.7 Conclusion

Social impact assessment schemes have an independent range of applicability during both the pre-feasibility study and post-development phase of water resource development projects. But, under current management practices, the schemes are treated merely as subsidiary parts of entire EIA programs. Its scope has been left limited to legislative compliance level of the EIA notification 2006. The components of community participation in final EIA reports serve the terms of reference (ToR) requirements of the environmental clearance. Hardly, any incentive is given towards an equity-based engagement between project proponents and the community. It has been a particular reason behind the weak nature of most of the environmental monitoring and mitigation plans. Such plans are not adequately supported by mandatory ecological audit (social audit) provisions, as, existing legislation do not account the project promoters to undergo any such audit. Thus, the questions of transparency and accountability are severely compromised. Below standard performances by most of the existing RVD projects are evidence of this condition. Further, the feedback loops of the entire monitoring framework become obsolete, as, if any severe breach of environmental quality standards occurs, the Impact Assessment Authority would not be having any clue towards this. Owing to this, the national water resource management mechanisms have failed to improve upon their observed shortcomings, and consequently, contributed to the growing interregional imbalances and conflicts around the resources usages. Comparatively, community-led management ventures in this sector have yielded more sustainable and efficient results because of a strong commitment to social impacts assessments. However, legislation should not be perceived as an end in itself against this observed limitation. As there will always be a need for the leading practices which may necessarily go beyond meager legislative compliances. Aspects of policy learning need to be encouraged not only at governance level but also at community initiatives levels.

References

Bhattacharya S, Kumar V (2010) Unprecedented floods in Ghaggar basin. Dams Rivers People 8(5–6):2–6

Budhya G (2007) Struggles against failing privatization, for people-centred model – case of Bangalore, Karnataka. In: Manahan MA, Yamamoto N, Hoedeman O (eds) Water democracy: reclaiming public water in Asia, pp 21–24. Retrieved from http://www.tni.org/books/publicwater.htm

Crase L (2010) A cautionary note on the use of socio-economic analysis in water planning. Econ Pap 29(1):41–47

Cummins T, Watson A, Barclay E, Reeve I (2007) The Economic and Social Impacts of Water Trading: Case Studies in The Victorian Murray Valley. Rural Industries Research and Development Corporation Publication No. 07/121, Government of Australia

Dandekar P (2010) Free-Flowing Rivers Around the World. Dams Rivers People 8(5–6):7–11

Dandekar P (2011) Maharashtra Gifts Farmers' Water to Industries. Dams Rivers People 9(2):1

DHAN (2012) Livelihood Matters: Annual Report 2012. DHAN Foundation, Madurai

General Guidelines for Water Audit and Water Conservation (2005) Central Water Commission. Government of India

GIZ and GOI (2011) Adaptation to Climate Change with A Focus on Rural Areas and India. German International Cooperation and Government of India

Goldman L, Scott B (2000) Introduction. In: Goldman LR (ed) Social impact analysis: an applied anthropology manual. Berg, Oxford, pp 1–31

Haider F. Teri Green Mission Has a Golf Course, Gurgaon, February 20, (2010). http://indiatoday.intoday.in/story/TERI+green+mission+has+a+golf+course/1/84879.html

Harper G, Price R (2011) A Framework for Understanding the Social Impacts of Policy and their Effects on Wellbeing. Government of United Kingdom

Jafri A (2007) Countering water privatisation in Mumbai: evolving a public-public model. In: Manahan MA, Yamamoto N, Hoedeman O (eds) Water democracy: reclaiming public water in Asia, pp 27–30

Jha N (2010) Access of The Poor to Water Supply and Sanitation in India: Salient Concept, Issues, and Cases. Working Paper number 62 (International Policy Centre for Inclusive Growth)

Jhunjhunwala (2010) Economics of River Flows: Lessons from Dam Removals in America. Kalpaz Books, New Delhi

Kelkar U, Narula KK, Sharma VP, Chandna U (2008) Vulnerability and Adaptation to Climate Variability and Water Stress in Uttarakhand State, India. Glob Environ Change 18(2008):564–574

Keskinen M (2008) Water resources development and impact assessment in the Mekong basin: which way to go? Ambio 37(3):193–198. https://doi.org/10.1007/s13280-012-0261-x

Koonan S, Sampat P (2012) Delhi Water Supply Reforms: Public-Private Partnerships or Privatisation? Econ Polit Wkly XLVII(17):32–39

Laurencont T, Thompson J, Slight M, Eglinton K, Mason K (2009) Community engagement which extends beyond compliance. In: Evaluating performance monitoring and auditing: leading practice sustainable development program for the mining industry. Department of Resources, Energy and Tourism. Australian Government, Canberra, pp 59–61

Mishra DK (2011) Bihar Floods: Breaching Embankments and Suppressing Information. Dams Rivers People 9(2):8–9

MOEF (2006) Notice S.O. 1533, Gazette of India, Ministry of Environment and Forest, Government of India, New Delhi. September 14th

MoWR (2011) Report of the Working Group on Major and Medium Irrigation and Command Area Development for the XII Five Year Plan (2012–2017). Ministry of Water Resources. Government of India, New Delhi

Pandey KM, Debbarma A, Das H, Roy A, Nath W (2013) Environmental impact assessment and management: protecting ecology in North-East India. J Environ Res Dev 7(4):1459–1466

Pattenden C, Barclay MA, Martin F (2009) Evaluating the Effectiveness of Community Partnership Programs. In: Evaluating performance monitoring and auditing: leading practice sustainable development program for the mining industry. Department of Resources, Energy and Tourism, Australian Government, Canberra, p 29

Planning Commission (2007) Report of the expert group on groundwater management and ownership. Government of India, New Delhi

Planning Commission (2009) Report of the task force on irrigation. Government of India, New Delhi

Planning Commission (2014) Report of the expert group to review the methodology for measurement of poverty. Government of India, New Delhi

Rajabi S, Hipel HW, Kilgour DM (2001) Multiple criteria screening of a large water policy subset selection problem. J Am Water Resour Assoc 37(3):533–546

Rao MRM, Batchelor CH, James AJ, Nagaraja R, Seeley J, Butterworth JA (2003) The Andhra Pradesh Rural Livelihoods Programme Water Audit. Government of Andhra Pradesh

Report of the National Commission on Floods (1980) Vol. I Government of India, New Delhi

Reuss M (1992) Coping with uncertainty: social scientists, engineers, and federal water resources planning. Nat Resour J. 32(1):101–135

Richards A, Singh N (2002) Inter-state water disputes in India: institutions and policies. Int J Water Resour Dev 18(4):611–625. https://doi.org/10.1080/0790062022000017437

SANDARP (1999) Assessment of Water Supply Options in Urban India: Large Dams Have No Case, Paper submitted to the World Commission on Dams: Thematic Reviews: Options Assessment, South Asia Network on Dams, Rivers and People, New Delhi, India. November 1999

Shiva V (2005) India: soft drinks, hard cases. Le Monde diplomatique. Retrieved from https://mondediplo.com/2005/03/14india

Suresh V (2007) Solution for the water crisis – democratisation, not privatisation! Promising stories from Tamil Nadu, India. In: Manahan MA, Yamamoto N, Hoedeman O (eds) Water democracy: reclaiming public water in Asia. Focus on the Glob South, Bangkok, pp 31–33. Retrieved from http://www.tni.org/books/publicwater.htm

Thakkar H (2010) Bhakra reservoir is being operated in casual, adhoc manner? Need for clearly defined norms of accountability in reservoir operations. Dams Rivers People. 8(7-8):5–6

Thakkar H (2012) Water Sector Options for India in a Changing Climate. South Asia Network on Dams, Rivers and People

Thakkar H, Dharmadhiary S (2010) CWC Is abdicating its responsibility on Kosi disaster: Kosi enquiry

commission must persist with CWC to find the truth. Dams Rivers People. 8(7-8):5–6
Tracey M (2009) Indigenous Community Involvement in Monitoring. In: Evaluating Performance Monitoring and Auditing: Leading Practice Sustainable Development Program for the Mining Industry. Department of Resources, Energy and Tourism. Australian Government, Canberra, pp 57–58
Withanage H, Manahan MA, Hoedeman O (2007) The Asian Development Bank Water Policy: Privileging Private Sector Investment Over Pro-Poor Access. In: Manahan MA, Yamamoto N, Hoedeman O (eds) Water democracy: reclaiming public water in Asia. Focus on the Global South, Bangkok, pp 3–8. Retrieved from http://www.tni.org/books/publicwater.htm
Yu AL (2007) China's road to the commodification of water. In: Manahan MA, Yamamoto N, Hoedeman O (eds) Water democracy: reclaiming public water in Asia. Focus on the Global South, Bangkok, pp 16–18. Retrieved from http://www.tni.org/books/publicwater.htm

Web sources

Wit, JW. De. (2002) Urban Poverty Alleviation in Bangalore: Institutional and Community - Level Dilemmas, Economic and Political Weekly, https://www.epw.in/journal/2002/38/special-articles/urban-poverty-alleviation-bangalore.html
India Micro-Irrigation Program Survey (2012) Summery of Findings. https://files.givewell.org/files/DWDA%202009/IDEI/IMRB-%20Report%202012%20IDE%20India%20Survey%20Findings.pdf
National Water Policy (2012) https://jalshakti-dowr.gov.in/sites/default/files/NWP2012Eng6495132651_1.pdf

27 Challenges and Opportunities Towards Management of Solid Wastes in Indian Cities: Beyond the Rhetoric of Convenience

Ruchira Ghosh and M. Satish Kumar

Abstract

Sustainable solid waste management is a global challenge, but with 2030 agenda for sustainable development and livelihood security, it creates opportunities for human settlement with gender equality, and sanitation, considering resource resilience and sustainability. India is a country experiencing rapid urbanisation rate (from 27.82 in 2000 to 31.1 in 2011, Census of India, 2011), and human inflow to cities has made landfill sites super-saturated. Moreover, the concept of reduction of waste does not appear in the Indian context of waste management (WM) scheme, pushing backward the resilience and livelihood security. Only 2R (recycling and reuse) occurs effectively from 3Rs strategy in the waste management hierarchy. Besides, recycling is done with economic benefit and reuse in selective personal use; however, reduction has lost its vitality in Indian waste management sector. Thus, the role of informal sector is crucial to recycling industries. But, within the informal sector, women are critical to WM and are exposed to health hazards from improper waste handling, posing greater life threats to not only them but to the coming generations as well. Furthermore, with more smart city development schemes, there comes economic opportunities for cities resulting in increased migration, land use change, and enhanced waste generation but lack of scientific disposal sites. Thus, recycling units have huge scope in near future, making roles of these informal sector and specially women decisive towards sustainable resource management and livelihood security. This sector has been neglected for long and public health will be more than compromised with further increase in the waste generation and management issues. However, with massive issues of human settlement, centralised WM is no longer a viable option in the long run. This calls for a decentralised management approach towards solid waste management (SWM), and informal sector playing vital role. Thus, the chapter aims to voice challenges and opportunities for women in the informal settlements within decentralised management scale addressing health and hygiene and livelihood security.

R. Ghosh
Centre for Sustainable Technologies, Belfast School of Architecture and the Built Environment, Ulster University, Jordanstown, UK

M. S. Kumar (✉)
School of Natural and Built Environment, Queen's University Belfast, Belfast, UK
e-mail: s.kumar@qub.ac.uk

B. Thakur et al. (eds.), *Resource Management, Sustainable Development and Governance*, Sustainable Development Goals Series, https://doi.org/10.1007/978-3-030-85839-1_27

Keywords

Livelihood · Security · Recycle · Reuse · Reduce · Solid Waste Management · Sustainable Development Goals (SDGs) · Unorganised-Informal sector · Women

27.1 Introduction

Formal urban structure in modern independent India was a reflection from the colonial times, where the developmental dynamics was largely a portrayal of an imperialist system. The colonial economy initiated a significant flow of people and commodities across its strategic port cities and their hinterland.

This had inevitably weakened the regional centripetal forces associated with dynamic economic activities established during the medieval period (Kundu 2011). The pre-existing urban-rural interactive events were gradually replaced by colonially minded export–import oriented commodity flows. Thus, urban–rural interaction was not an outcome of significant economic dynamism, rather, urban development was more aligned to socio-political organisations of the imperial space. At the same time the very population shifts necessary to sustain the emergent new urban centres led to aggravating the core–periphery dissensions, thereby reinforcing the centrifugal forces in the country. This brought about greater **centralisation** over the delivery of urban service unlike **decentralised** delivery of utility services in the pre-colonial era. Over time, the dysfunctionality of the cities in the context of the regional economy increased producing serious socio-economic distortions and affecting the basic functionality of urban centres in general and imperial cities in particular. These cities were planned with an agenda of providing high-quality utility services and civic amenities to the elites linked to the ruling class or those who could afford high prices (Willcocks 1984).

Moreover, India was getting gradually drawn into the orbit of the capitalistic society during two centuries of colonial rule, its political economy became the major obstruction to technological advancement in agriculture and industry. This led to more population flow into secondary and tertiary services from primary sector.

With the Indian "Raj" being fully integrated into the imperial economic system, with constant famines and natural disasters, there was an increased demand to keep cities safe and clean. To be free of diseases, such as plague, malaria, cholera, smallpox, leprosy, syphilis, and influenza became a key imperative of the colonial regime. Loss of productive human power meant less profit to be siphoned off to the metropole. At the same time "the British had a narrow view of their own responsibility towards health and welfare of Indians" (Arnold 1993; Pati and Harrison 2001).

Over time, changes to characteristics of Waste as a category from homogenous to heterogenous entity resulted in the distortion in the **3R** system-***Reuse, Recycle, Reduce*** towards waste management. Lack of investments in the Presidency cities of Kolkata, Chennai, and Mumbai meant that despite increased cordoning off the elite parts of the city, the boundary between *White and Black Towns* was breached (Kumar 2002, 2006a, b; 1994) and miasma persisted thereby resulting in major epidemics affecting health and hygiene around urban centres.

The Independence in 1947 brought political and economic restructuring with focused public sector investments resulting in the eventual modification of the urban hierarchy (Chandramouli 2011). However, regional disparities persisted, despite the public sector playing a major role in investments in 'backward' areas (Kundu 2011). The reason for this was that 'urban' India was still seen as degenerate and lack of continued investments meant a not so viable urban settlement pattern to emerge. Thus, intransigent rural-urban disparities continued to persist with ***urban bias*** and a despairingly low *'Hindu rate of GDP growth'* of less than 2% becoming the norm. There was an appalling lack of urban amenities and services.

However, the differences in utility service delivery between urban core and peripheries continued to create huge garbage dumpsites in the outskirts of towns and cities (Kumar et al. 2009). The idea of 'hyper-urbanisation' (Bhagat 2011), and the supposition that this trend would continue for the next few decades in independent India, was based on the idea of absolute population growth (Kundu 2011). The urban growth scenarios in India showed the rates were reasonably high in the fifties, which fell sharply during the sixties (largely due to definitional anomalies) and again reached its peak in the 1970s, thereby, portraying a 'hyper-urbanism' trend, which did not turn out to be so in the next decade.

There was the possibility for the urban structure to instigate utility service provisioning to become more coherent and unified rather than succumbing to the limitations imposed by differences across urban cores and their peripheries. Today, the management of solid wastes is largely dependent on the availability of unencroached landfill sites (Sharholy et al. 2008). This has resulted in increased underground water contamination due to uncoordinated and unregulated disposal of solid wastes (Mor et al. 2006). It has been plagued by endemic air pollution from displaced construction and demolition rubble from continued and sporadic infrastructure developments, the plying of beyond-the-shelf-life automobiles on Indian roads, the running of criminally tampered auto rickshaws flouting all norms of anti-pollution regulatory frameworks. All of these have major implications towards health, hygiene, and greenhouse gas emission considerations.

The chapter is divided into the following. The first section investigates the urban transition and evolution of the concept of smart city, the second section consists of paradigm shift in the conceptualisation of SWM from the pre-colonial, colonial to the postcolonial period and drivers for this. The third section highlights the importance of the role of informal sector in the waste management industry and interventions required to align with UN mandated Sustainable Development Goals (SDGs). The final section presents the conclusion with strategies and recommendations for a sustainable waste management and livelihood strategy.

Box 27.1 Sustainable Development Goals

Indian cities have not learnt to manage their solid wastes. Open dumping is a common practice in most Indian cities, apart from the metro-cities. Indeed, this practice poses significant environmental and health risks due to toxic and greenhouse gases emission through direct combustion and/or decay of wastes. India's solid waste management is a challenge associated with burgeoning urbanisation which in turn impacts on various dimensions of sustainable development goals such as livelihood security, gender equality, sanitation, among others. The authors recognise that increasing volume and complexity of waste associated with growing Indian economy pose a serious risk to ecosystems and human health. They further argue that poor waste management—ranging from non-existing collection systems to ineffective disposal—causes air pollution, water, and soil contamination. Open and unsanitary landfills contribute to contamination of drinking water and can cause infection and transmit diseases. The dispersal of debris pollutes ecosystems and dangerous substances from electronic waste or industrial garbage places a strain on the health of urban dwellers and the environment. Sustainable solid waste management is not only concerned directly with **SDG 12,** i.e. 'responsible consumption and production' which includes targets focused on environmentally sound management of all waste through prevention, reduction, recycling, and reuse (targets **12.4 and 12.5**) and reduction of food waste (target **12.3**), but it also touches several other goals, targets, and indicators of Agenda 2030. Thus, the chapter captures challenges and opportunities for sustainable solid waste management in India.

27.2 Urban Transition and Evolution of the Concept of 'Smart City'

27.2.1 Urbanisation

Historically, urbanisation in India can be characterised by a lopsided urban growth with a predominance of primate cities (Kundu 2011; Bhagat 2011). The intensity of internal migration in India is expected to increase with future economic, political and environmental instabilities (Deshingkar and Sandi 2012) as manifested by the recent pandemic Rapid regional urban transformation has resulted in a proliferation of cities and towns in India. India has the unique distinction after China to account for 7935 cities and towns according to the 2011 Census. At the same time, 70% of the urban population lives in 468 Class I Urban Agglomerations (UAs, with population of 100 thousand and above). The number of Class I UAs increased from 384 in 2001 to 468 in 2011. Furthermore, there are 53 million-plus UAs, which comprise 43% of India's urban population.

The number of million-plus UAs in India increased from 35 in 2001 to 53 in 2011—an addition of 18 UAs during the period 2001–2011. This demonstrates the fact that the nature and pattern of urban population is heavily concentrated in large cities. There were eight megacities with populations of more than 5 million in 2011. Of these megacities, three with population of more than 10 million were Greater Mumbai UA (18.4 million), Delhi UA (16.3 million), and Kolkata UA (14.1 million). The growth of **second tier** cities and towns have increased the level of mobility among the Indian population. It has also exacerbated the key challenges faced by the cities in terms of housing, poverty and access to critical services, such as health care, education, safe drinking water and sanitation, etc. As a result, there is widening of inequality and deepening social exclusions across the rural and urban landscape of India. This resulted in the creation of a disjointed urban form and services. Globally, 1 million people are moving into cities every week, but in India alone, every minute, 30 country dwellers are permanently moving to a city. When more than half of the world (54%) population are urban dwellers, India currently account for 31% urban. This is expected to rise quite rapidly by the next census (UK Trade and Investment 2015). At the same time, continued flow of migrants to cities and the concomitant urban transformations from the application of emergent information and communications technologies ('ICT') are redrawing the boundaries of urban governance. With economic growth, the positioning of India as an emerging economy presents new dynamism and challenges due to changes in lifestyles and increasing risk of a sustainable habitat.

Such changes compound existing and endemic urban problems and traditional port cities, such as Mumbai, Kolkata, and Chennai face continued challenges to human health and safety. This is more so in terms of contaminated water, air pollution, and outbreak of communicable diseases, such as dengue, malaria, flu, zika, COVID-19 among others. Moreover, with poor sanitation there is an increase in pulmonary diseases, such as asthma, flu, TB etc., among urban working-class populations, resulting in loss of precious working days and poses a threat to urban and national productivity. Therefore, the key challenge in urban India, in both the formal and unorganised settlements is the inadequate management of solid waste, decrease in per capita availability of water, unreliable water quality, insufficient sewage coverage and degrading ambient air quality (Ahluwalia et al. 2014). These are problems not specific to India alone, but the drivers and pressures behind these are distinct. The provisioning of urban utility services in India is largely outsourced from the formal to informal establishments within the municipalities and their extensions.

Resource inefficiency, resistance to adopting upcoming novel technologies, lack of regulatory framework, weak institutional capacity, non-compliance, ad hoc legal regulatory framework and low financialization of services are some of the key challenges for the sector (Ghosh and Kansal 2014). This calls for a need to rethink urban design and city developmental plans to enable the attainment of SDGs. Increased community participation in urban governance and the development of urban spaces calls for social and financial investments (Kumar 2003).

27.2.2 Evolution of Urban Spaces and Smart City Concept

The purpose of new city design aims to deal with increased traffic congestion, ineffective solid waste management and treatment, endemic air pollution, and lack of sustainable livelihood thereby impacting on quality of life and well-being. Although, there is no universally accepted definition of Smart City (Kumar and McKenna, 2018), such a definition changes with the level of development, behavioural change, and attitude of the city residents. In pursuance of sustainable smart cities, the objective is to promote cities that can provide efficient utility provisioning and thereby ensure a decent quality of life to its citizens.

Smart City is about improving efficiency and reducing the quantum of waste whereby sustainable efficiency becomes the norm. At the same time, 'smartness' also implies liveability and well-being. Cities therefore use the 'smart' concept to promote global standard efficiency in terms of energy use, resource use, reduction of waste generation and thereby ensuring future urban sustainability.

The concept of smart city in India was first conceptualised during BJP election manifesto in 2014 (Datta 2015). The Smart cities draft was circulated for the first time focusing on retrofitting projects in existing cities in 2015 and not on green field or satellite township schemes (Awasthy et al. 2018). '*Smart City is defined as a city where structure and function of various urban systems are clearly defined, simple, highly responsive and malleable via, contemporary technology and design*' (Smart Cities 2010; MoUD 2014). The Government of India under the leadership of Honourable Prime Minister Mr. Narendra Modi initiated the 'Smart Cities Mission'—a scheme for urban renewal and retrofitting program with a mission to develop 100 smart and sustainable cities by the end of the financial year 2019–2020. The Mission summarises three basic models of development—*Area-based Development* (ABD), *Pan-City,* and *Greenfield Developments*. ***Area-based*** development aimed at urban development by encouraging cities to concentrate their finances towards re-energising their cities as a form of urban renewal. The ***Pan-City*** developments focused on smart solutions for the entire city, whereas ***Greenfield*** developments referred to extensions to an existing urban built environment, requiring huge capital investment (Hoelscher 2016). The core infrastructure elements in a Smart City included: (1) adequate water supply, (2) assured electricity supply, (3) sanitation, including solid waste management, (4) efficient urban mobility and public transport, (5) affordable housing, especially for the poor, (6) robust IT connectivity and digitalization, (7) good governance, especially e-Governance and citizen participation, (8) sustainable environment, (9) safety and security of citizens, particularly women, children, and the elderly, and (10) health and education (MoUD 2015). All these core elements are exactly in line with UN Sustainable Development Goals (UN 2015). Thus, we can say smart cities objectives are encouraging the country towards achieving these sustainable goals.

The smart city project in India is part of a $1.5tn global market opportunity (UKTI 2015). The government of India planned to invest INR 1.91 trillion (US$ 30 billion) for the 89 selected smart cities under the mission in 2015, and as per the proposed investment, 80% of the funds would be spent on *Area-based development* (ABD) and remaining 20% of the funds on ICT Solutions (UKTI 2015). The challenge for attaining the smart city targets is highly dependent on adequate and generous financial investments, with no silver bullet for technology-intensive approaches.

Indeed, most municipalities are cash strapped and lack institutional structure and capacity to generate revenue from waste due to low tax base and poor tariff system, as well as being compounded by chronic corruption (Narayana,

2009). This further creates a dissonance towards attaining smart city goals. Today, municipalities have increased the adoption of their technological fixes, such as incineration-based waste-to-energy, which are made attractive by liberal concessions provided through Smart Cities and at the same time, unlocking the inherent potentialities of recycling offers comparatively wide range of value-adding economic opportunities. This is particularly true for all stakeholders in the city, the local communities, businesses including start-ups. The embedding of the circular economy framework creates new opportunities towards reduction in greenhouse gas emissions in lieu of incineration or landfill sites as an option (Kumar 2019a, b, c; Oates et al. 2018). Recycling, therefore, can save energy, create jobs, and provide a sustainable basis for the planet threatened by overconsumption and wastage. But recycling as a concept will be successful only if we are engaged with **'reduction'** and **'reusing'** our scarce resources, which is significantly missing from the Indian context (Kumar 2019a, b, c). Thus, reducing the ecological burden calls for strategic and creative use of technology, coupled with incentivised financial instruments. Smart City goals, therefore, call for a decentralised approach towards recycling and skill upgradation of informal workers (recyclers).

27.3 Paradigm Shift in Solid Waste Management with Urbanisation Transition

Solid Waste Management (SWM) in India has seen huge change in terms of its scale and characteristic. This can be attributed to consumer behavioural change towards management of domestic and industrial/ commercial segregation and treatment of wastes. The emergence of new innovative business models in the waste management sector presents an optimistic transition from traditional to advanced practices.

In ancient India, solid waste management was mostly related to disposal of municipal solid waste near a water body, i.e., river. Here, waste generated were mostly agriculture-based organic forms of food, and materials discarded from construction activities. In the medieval era, with the advancement of technology, rudimentary forms of town planning emerged which again reiterated the use of traditional waterways to remove solid waste from the urban environment. The annual cycle of monsoonal activities aided in this endeavour.

Characteristics of the waste included mostly cloth-based bags, dresses, and other home utility items (plant-based fabric like cotton and silk to last long), fresh or dry leaf-based plates and containers, rugs, furniture from wood and utensils from mud, and toiletries were natural extracts from fruits, flowers, soil, and leaves. Medicines were also plant-based so there was no scope for chemical discharge into the environment. Some items were made from wax, glass, and metals (jewellery and house decorative items). Majority of the households had utensils made up of clay and metal (iron, aluminium, gold, silver, and bronze). However, all these items were largely reused and recycled. The food was usually freshly grown, cooked, and wastage was minimal. At the same time, there were no fast-food outlets, thereby decreasing waste output. Transportation was mostly based on animal power and again waste from fuel was non-existent. The traditional houses were based on basic construction materials, such as bamboo, clay and mud, stones, namely granite, marbles, including wooden furniture carvings and some binding materials, such as traditional chalk, etc. to hold the bricks together. The paint was also organic in nature, which was of natural colours and mixed with mud, cow-dung based slurry and stone (both precious and non-precious) chips to cover the walls. The fuels used were mostly wood and the ashes were also used as building material to cover the bricks on walls. Hence, the construction and demolition waste were either degradable or reusable in nature. Looking into the system of management in pre-colonial era, segregation of waste was conducted at the household or waste origin level, greatly aiding reuse, and recycling opportunities. Minimal wastes were burnt openly but at a distance from human habitation to avoid outbreak of disease. Indeed, segregation was a household-centered activity in the context of ancient India, where the ownership of the waste was vested at the household level. An incentive to that could be the product and by-product generated from recycling of the household waste was utilised by them in some form or the other, such as creation of natural biodegradable fertiliser for their gardens, etc., thereby enhancing savings in monetary terms. As a rural village-based economy, the pre-capitalist system was in operation and, therefore, management of key utilities were individually and at times communally managed at a decentralised level (Kumar 2019a, b, c; Bhave and Sadhwani 2016; Deshkar 2010; Muniapan and Shaikh 2007). With clear principles of *Arthashastra Politics of Statecraft* in place, the management of disputes were dispensed easily. The above system of management worked efficiently for the existing societal structure of the time.

The advent of colonial period in India witnessed introduction of various new concepts, namely, centralised waste management, packaging materials, metal popularity, synthetic and chemical introduction into cosmetics, medicines and decorative items, mining of petroleum, coal, iron, and change in cooking and transportation fuel due to the use of large-scale fossil fuel into the societal system. This was a huge transition from biomass dependency and was advanced by obsession with efficiency, demand-led production and consumption, increased concentration of people in relatively small spaces in the cities as labour for the imperial economy.

Indeed, the characteristic of the waste changed from the being merely **organic** to being supplanted increasingly by **non-organic** wastes. This ran in conjunction with radical improvements in living standards and amenities with the advent and the establishment of colonial rule in India.

Changes in the societal framework made vital differences in lifestyle from concrete houses with wooden and glass decorations, increased use of electronics like (watches and bulbs) to adaptation of new generation of cosmetics, fertilisers, ceramic utensils, leather shoes, paints for walls and chemical colours. Slowly, the convenience and luxury took over the necessity, so the society started adopting westernised modes of culture and existence. In time, the fruits of western industrial revolution introduced new forms of labour into the subcontinent. The labour time was expanded to include lighted factories beyond the diurnal agricultural rhythms. Except for agro-based products, majority of the industrial products were imported from Great Britain thereby shifting the terms of trade. This resulted in the articulation of Indian economy into the world economic system with silver replacing gold as the major legal tender. The characteristics of waste include ceramic, glass, paper, wood, cloth, and diverse forms of metals. Thus, the incentive to reuse and recycle started to decline, and this was the first time when people had to consider about heterogenous solid waste. With the advent of planned intervention in the cities of India, a centralised management of waste emerged via, the Public Works Department (PWD). Sanitary provisions meant centralised preventative control for seasonal and unseasonal outbreak of epidemics. However, the network of management was coordinated by bringing together influential stakeholders from the elites in the society. The colonial period also saw the introduction of a centralised tax system for the first time for managing the city amenities and utility services (Sengupta 1985).

This arrangement worked well for those who were rate payers in the city, however, the lack of affordability for the majority in the city's unorganised quarters and slums meant that ***cleanliness*** had to be imposed based on charitable interventions by enlightened Indians, such as *Behram Malabari* (Behramji Merwanji Malabari JP (1853–1912) who was an Indian poet, publicist, author, and social reformer best known for his ardent advocacy for the protection of the rights of women and for his activities against child marriage (Burton 1998; Ramanna 2002). Waste became contentious issue when epidemic, such as cholera, plague, and smallpox ravaged the cities of colonial India. At this time, the idea of segregation of waste which was a norm in traditional societies on subsistence living income became irrelevant. The draconian colonial laws meant that stringent regulation of abattoirs in *'Black Towns'* and wastes emanating from the slums resulted in discarding of wastes into water bodies, and in areas distant from habitation. There was no proper planning in the disposal of solid wastes in the cities. Only the *'White Town'* had any semblance of waste collection to be either burnt or dumped into water bodies or discarded in wastelands (Kumar 2006a, b).

However, with the imperatives of Independence, waste management issues remained peripheral. With increased migration of people into the cities there was a quantum leap in the waste being generated. The postcolonial period saw increased technological interventions and with economic growth, increased urbanisation new materials, such as plastics, electronics, packaging materials, low-cost building materials, chemical-laden fertilizers, paints, etc. came into the ambit of solid wastes (Fortuna, 2018; UN Environment, 2019; Ray, 2018). With the expansion of food industry and its commercialisation meant an increase in the volume of food wastes. The expansion of the manufacturing sector, saw toxic elements, such as chemicals and non-biodegradable wastes, concrete, bricks, grit, and metal wastes from the construction industry dominate the SWM landscape. India generates 62 million tonnes of waste (mixed waste containing both recyclable and non-recyclable waste) every year, with an average annual growth rate of 4% (Swaminathan, 2018; PIB 2016). Of the total waste collected only 20 percent is processed and the remaining 80 percent is dumped in landfill sites (Shrivastava, 2019). In other words, 30 million MT of 'highly polluting' unprocessed solid waste is dumped in landfill sites in India.

Such a huge increase in heterogeneous waste characteristics made waste management even more complicated. The local municipalities pursued a centralised management approach towards waste disposal and at the same time priorities on top of the agenda for independent India was provision of food security, employment for all and the defence of national sovereignty. Utility management remained at a standstill with huge landfill dump yards across every Indian city (Narayana 2009). What we can observe was the increasing heterogeneity of waste, increasing population, lack of education (Bolaane 2006) and a ***trust-deficit*** in terms of urban governance. The reluctance towards Reuse, Recycling and Reduction meant there was no ownership towards waste created either at the household or non-domestic level (Kumar 2019a, b, c). The passing of the buck and blame-game between the civil society and the local municipalities continues to this day. Lack of awareness and education regarding wastes management, importance of segregation, significance of recycling and reusing meant that pollution of the cities continue unabated. At the same time, with the proliferation of slums and informal settlements in cities, waste management has now secured a new lease of life. The unorganised sector in the slums of India took over the charge of segregation and management of recycled materials.

This provided alternative livelihood options for the dwellers. The downside was that these activities were undertaken in extremely hazardous environment with no protection and no cover for health and safety and with minimal remuneration below subsistence level. These emerged as the least dignified job- 'rag pickers' (O'Connell 2011) in the society. However, with economic development, rapid urbanisation, and expanding population, the problem of solid waste management was further aggravated despite the support from informal sector workers. The management could not attain much success as they lacked proper formal institutions towards waste management (Zhu et al. 2008).

For long, solid waste management has always been considered in conjugation with wastewater management and it is only recently that the characteristics, need, and management technologies for dealing with solid wastes have started to develop in isolation to wastewater management. There has been continued expansion of cities to include rural-urban fringes or extended metropolitan regions [EMRs] (McGee and Robinson 1995). This has invariably increased risks to public health emanating from the peripheries of the metropolitan regions. Increasing lack of additional space to provide landfill options has become acute in Indian cities thereby exacerbating the volume of heterogeneous waste in circulation (Narayana 2009). This propelled the need for a scientific approach towards sustainable solid waste management in India. Water provisioning and dealing with wastewater were considered as a priority so that laws and regulations were framed to monitor and manage water resources. The increasing and unabated pollution of river Ganges, Brahmaputra, and Yamuna meant that the ecosystem of these rivers was heading towards extinction. The rules for waste management first came out in 2000 (CPCB 2000) and the galvanised response came in 2014, with the NDA coming to power. The first amendment was initiated in 2016, but it lacked the critical technological intervention to enable the establishment of a rewarding circular economy in India.

With the announcement of the UN SDG targets, action to save the environment and thereby the planet meant that India lacked the long-term perspective in dealing with waste management. This had more to do with the cultural inhibitions which promoted attitudes and practices harmful to the pristine environment (Kumar 2019a, b, c). India was still struggling with lifting endless population out of the quagmire of poverty below the breadline. There were few municipal councils and city developers who presented a well thought out strategies towards the management of waste. Waste needed to be a ***resource*** to become successfully integrated into the circular economy. At present, waste mismanagement is still associated with air pollution along with water and land pollution (ESCAP 2017). There is a need to go beyond the generic perceptions of 'waste'. Emerging economies like India face the challenge of syncing with sustainability goals (Mahadevia 2001).

The lack of source segregation in Indian cities and mixed waste was separated from recyclables by middlemen namely, rag-pickers at the dump yards. This has radically reduced the quality and quantity of recyclables that could be retrieved from the waste (Swaminathan 2018). For segregation to take a central place in India's waste management scenario, it is important to eliminate the mixing of waste during door-to-door waste collection. The planning for waste management channelization moves in vicious circle, where the waste collectors claim that since there is no separate channel for recyclables, i.e. organic and inert wastes, so there is no point to collect them separately. On the other hand, waste managers claim that since waste collectors' mixes wastes therefore the efficiency of the waste management decreases.

Now, India is in an alarming state with no land left to dispose waste. The distressing situation of lack of land was indicated two decades earlier (Venkateswaran 1994) but there has been no significant improvement since then. Presently, the collection efficiency is less than 60% from households and only 15% of the collected waste is processed. With approximately 50% of the total waste being organic, the volumes of recyclables tend to grow each year with India's high urbanisation rate, introducing threat to the land required for disposal and introducing further health risks (Swaminathan 2018). Hence, the situation is even more disastrous and there is an emergent need to associate waste management with sustainability goals for future viability.

27.4 Role of Informal Sector in the Waste Management Industry

The SWM framework relates to the collection, segregation, transportation, treatment, and disposal of waste (MoEF, 2000; Bhushan et al, 2018). The SWM formal management structure as practiced in India is vastly different from the established framework. But the irony is that very few cities follow this structure and most cities and towns in India lack resources and capacity to deal with the challenge. The disconnected jurisdictional priorities meant that discordant and multiple local bodies remain at logger head and unable to work towards a common purpose. Increased politicisation of these bodies meant that there is unnecessary repetition of activities resulting in reduced efficiency and outcomes. This is visible across all major metropolitan centres in all regions. Another major constraint is weak urban institutional structure in almost all developing countries (Marshall and Farahbakhsh 2013).

In practice, there is in operation SWM in terms of collection and transportation, occasional and intermittent treatment of biodegradable wastes to form garden composts or to produce methane gas to add to the national power grid or simply to undertake unscientific mass dumping. It is suggested that the management of solid waste in Indian smart cities is somewhat restricted to about 70% of its collection and transportation, of which only 33% is processed and the rest disposed off into landfills (Kansal 2001). Landfill sites have become a major source of contamination of the natural resources, thereby affecting the generic atmosphere in Indian cities. The air and water pollution are exacerbated due to increased suspension of particulates emanating from these sites, especially during temperature inversions after the monsoons. Thus, instigating urgent control and management of landfills, and biodegradable waste is critical for maintaining a Healthy Cities Liveable agenda (EC 2015). Given that this is the case with Smart Cities under direct Central and State supervision, there is little hope for the Small and Medium Towns which remain under the mercy of State government jurisdiction with limited to non-existent funds. These towns are under grave danger of adding to the urban way of living, becoming unhygienic and unsustainable (Mittal and Sethi 2018).

There are several reasons for the mismanagement of MSW (Ghosh and Kansal 2017) largely attributed to regional diversity associated with the social-economic-geographical and financial attributes of a city. However, the primary reason being lack of a sense of ownership of waste generated and ensuring due diligence in terms of segregating wastes at the household at the level of the neighbourhood and indeed at the commercial and industrial level. It is important to note that in the Indian context of waste management, of the 3R strategy (Reduce, Recycle, Reuse), only 2R (Recycling and Reuse) exist (Kumar 2019a, b, c). Recycling is usually attempted to gain monetary benefit. Reuse, on the other hand, is largely selective and individualised as part of personal choice. Indeed, ***Reduction*** has lost its social, cultural, and political imperative in the Indian waste management sector. In the case of recycling, the informal sector plays a critical role. Hence, reuse and recycling towards circular economy is the need of the hour. In this sense informal sector contributes to the city's recycling rates as well as reduces the costs of managing solid waste substantially (UN-HABITAT 2010). The unorganised urban rag pickers are the most vital part of the recycling industry responsible for not only segregating, but also for influencing greenhouse gas reduction in terms of cost and scale of waste treatment (Pappu et al. 2007).

Rag pickers are the most important part of the informal waste management strategy, yet the critical role of rag pickers has neither been recognised nor acknowledged. They are basically the marginalised sections of the Indian society. India is also grappling the problem of shelter in urban areas in addition to other socio-economic problems as in other Asian countries (Schübeler et al. 1996). They are a vital resource segregator in the waste recycling industry, moving from community bins to treatment units, to landfill sites to collect/segregate recyclables (paper, plastic, glass, metal, rare minerals, etc.). They risk their lives to sell them to scrap merchants to earn their barely minimum livelihood. Usually, the middlemen (scrap purchaser) earn the maximum profit and these poor hard-working rag pickers are left as destitute and devoid of all basic amenities of life (Chintan 2012). Rag picking as a profession comes with its own set of utterly inhuman challenges. They struggle for basic sanitation, water supply and power supply as the neighbourhood is an unauthorised settlement disconnected from the city's basic services (Troschinetz and Mihelcic 2009).

The community of the waste pickers (people whose main job is to collect and segregate wastes) usually resides near the waste dumping as low-cost access to land and devoid of all urban services and amenities. The entire family, including children takes part in the collection, segregation and selling of scrap materials, a profession that has been endured by many generations in the family, where the collection and segregation of waste is carried out by the community. Therefore health risks and livelihood challenges have been inherited by generations. Even though ragpickers save almost 14% of the municipal budget annually, their role is largely unrecognised, and are generally deprived of the right to work (UN-HABITAT 2010). According to an estimate, the ragpickers reduce up to 20% load on transportation and on landfill (Pappu et al. 2007).

These workers are also affected by serious problems, such as alcoholism, drug abuse, domestic violence, and illiteracy. Most wastes are not segregated and often leaking batteries containing mercury, broken glasses, used medicine bottles or syringes and vegetable peels are often thrown in together in the same bins. It then becomes the responsibility of the waste collector to pick up trash and separate them manually without gloves, masks, and devoid of proper cleaning facilities. Social isolation and unsanitary working conditions result in serious illness, injuries, and critical infections. They work a minimum of 10–12 h every day and are paid meagrely below the subsistence level. Women as main breadwinner in these slums face major health and social challenges and abuses.

The *Swachh Bharat* guidelines (2017) highlight the informal sector as a special focus group helping to contribute to the streamlining and formalisation of solid waste management systems and practices. The municipalities, therefore, need to ensure that "the informal sector workers (rag pickers) in the waste management sector are given priority to upgrade their working conditions 'with full safety protocols in place' and are enumerated and integrated into the formal system of solid waste management in cities" (SBM 2017; EPA, 1988).

27.4.1 Women as Key Provider

Women living in informal settlements form the bulk of the self-help groups who segregate the waste for waste management industries from risky areas like landfill sites. However, they are minimally aware of the health risks resulting from improper handling of waste. Moreover, the informal settlements they reside in lack effective waste management services, usually not within the jurisdiction of the urban local bodies. This compromises their health due to lack of proper sanitation in the informal settlements in which they live, and due to the nature of their job as waste segregators (UN-HABITAT 2010). Although, city planners and urban local bodies claim they do provide health services and education to rag pickers, however, the general scenario for the informal sector remains unchanged. Lack of health and sanitation facility makes these slums a vector and harbinger of diseases, such as dengue, cholera, tuberculosis, polio, typhoid, and malaria, thereby meeting the SDGs becomes a critical challenge (UN-HABITAT 2010, 2017). The 2030 agenda for sustainable development (UN 2015) towards safety and dignity to women states that all forms of discrimination against all women and girls everywhere should come to an end. Moreover, elimination of all forms of violence against women in the public, space, in public service, or in any private sphere including trafficking and other forms of exploitation need to be promoted. Also, SDG 11 (UN 2015) mentions safety and affordable housing with the provision of basic amenities and the upgradation of slums as being an imperative goal.

Furthermore, the agenda also emphasises sustainable urbanisation and increased participation, including integrated and sustainable human settlements for all. At times, in the fashioning of installations about waste-to-energy plants and sanitary landfilling, etc. has deprived the poor rag pickers of the much-needed alternative earnings. An example from Ahmedabad, is of self-employed women's association as a part of the trade union that who have negotiated improvements for its members in terms of wages and working conditions. This can be a way to include informal sector activities into formal waste management structures and the circular economy (Oates et al. 2018). Indeed, management of solid wastes cannot be achieved in the absence of sustainable livelihood security. This calls for the upgradation of city's capacity towards resource efficiency and dealing with the challenge of climate change, especially for vulnerable children, women, and aged population. Therefore, if India is targeting to attain SDG 2030 targets, security and respect towards women and children as a part of respecting wellbeing and livelihood options need prioritisation.

27.4.2 Women Engaged in Waste Recycling

Women's share in informal waste recycling is generally higher as compared to men (UN Environment 2019), and most of them are sole bread winners. Mostly, the female participation is limited to unregulated employment, bringing more social challenges for women to earn a living. Even though both women and men face similar challenges of health issues and the social stigma, women are further burdened by domestic responsibilities and concerns around their physical safety. India has the lowest score on the World Economic Forum's 2018 Gender Gap Index, ranked 108 of the 149 countries evaluated (World Economic Forum 2018), with low political empowerment and economic participation opportunity for women. Also, waste work ranks lowest in the hierarchy of urban informal occupations. Moreover, in India there is the added aspect of the caste system, deepening the waste management related social stigma.

27.4.3 Initiatives to Empower Lives of Women Waste Pickers in India

27.4.3.1 Self-Help Women's Association (SEWA) in Ahmedabad

Ahmedabad has one of the largest and oldest self-help group in India (World Bank 2004) established in1972. SEWA supports more than 25,000 women waste pickers to voice the empowerment of women recyclers for their basic living standards and working conditions.

27.4.3.2 Body Shop

The Body Shop is a big name in cosmetics and beauty industry. The Body Shop together with tech business partner Plastics for Change and, a local NGO and a social enterprise, purchased 250 tonnes of plastic collected by pickers in 2019. The recycled plastic was used to package hair care ranges having high selling rates. This will not only empower women waste pickers in segregating wastes and getting a higher pay but will also help these companies to source plastics when the demand is high (Thompson 2019).

27.4.3.3 Women in Informal Employment Globalising and Organizing (WIEGO)

WIEGO is a global network with the vision for securing livelihoods for the working poor, especially women, in the informal sector. This network besides ensuring financial empowerment to women waste pickers also provides self-health care and childcare facilities in slums of ragpickers. They are more focused to voice the challenges faced by women waste pickers in addition to other responsibilities in raising visibility for women workers (WIEGO 2014).

27.5 Conclusion

Solid waste management indeed is a global issue and is responsible for up to 5% of global greenhouse gas (GHG) emissions along with continuous waste generation, and its mismanagement in cities. This problem is further aggravated now post the pandemic (Oates et al. 2018). The 2030 agenda for sustainable development (UN 2015) gives significant role to cities, human settlements aligning to questions of livelihood security, gender equality, provision of safe drinking water and sanitation. This reiterates the urgency for a populous country like India with a billion plus population. With increasing rural to urban and urban to urban migration, land use patterns have changed, and their little free space remaining continue to be as designated as landfill sites. At the same time, existing sites have become super-saturated thereby adding to air and water pollution at the expense of healthy and liveable cities. Therefore, with large booming million cities of India, a centralised waste management policy is no longer a viable option in the long run.

Decentralised waste management is the only alternative. Recycling units have a huge scope for the immediate future, and in this context the role of the informal sector becomes decisive in meeting the demands of a circular economy. A decentralised waste management strategy based on effective governance and through inclusive community participation, with women taking the lead and benefitting from the set-up, will go a long way in finally achieving the smart city goals of India. There is a need to change urban management and governance to a decentralised scale and incentivise the urban population, especially the economically weaker sections with bonus of regular income and added health benefits. **Community participation** then becomes the fulcrum for a successful decentralised waste management model in India. The decision about ensuring a sustainable habitat should be taken by and with the community. It is generally difficult for planners and Urban Local Bodies (ULBs) to track and manage waste generated from across the city, from formal and informal settlements. As a result, the rate of mismanagement is high with increasing inefficiency, low productivity, and low outcomes.

Schemes, which are State-led, or centrally sponsored schemes supporting municipal authorities to integrate and include informal waste pickers into the formal system, can deliver competitive recycling rates at lower costs. This will help meet **income, health, and job security** targets thereby reducing urban poverty and enhancing scope towards meeting climate resilience targets (SDG 11). The Ministry of Housing and Urban Affairs recently released another set of guidelines that links *Swachh Bharat Abhiyan* (or Clean India Campaign) to the *National Urban Livelihoods Mission*, a **poverty alleviation scheme** that aims to improve access to employment for the urban poor. Individual motivation, community participation, and incentives from central, state, and local authorities, including urban waste managers to decentralise SWM, will provide opportunities for the future (Bolaane 2006; Mrayyan and Hamdi 2006; Milea 2009; O'Connell 2011; McAllister, 2015).

With the number of centrally sponsored missions for urban and rural regeneration, namely, *Atal Mission for Rejuvenation and Urban Transformation (AMRUT), Pradhan Mantri Awas Yojana, Prasad Mission and the Shyama Prasad Mukherji National Rurban Mission (NRuM)*, it becomes imperative that institutional capacity for solid waste management within a circular economy framework need to be established to ensure productive contribution by all sections of the Indian population. We all must individually and collectively contribute to the **Swachh Bharat campaign**. Effective financial devolution will also help to build capacity of ULBs towards effective service delivery. Indeed decentralised, devolve and capacity development will ensure urban sustainability of central-led programmes. Technological interventions will only succeed if urban institutions are strengthened to deliver effective and productive services (Ahluwalia 2019) and are able to negotiate the 'waste-value dialectic' within and beyond the peripheries of rampant urban transformation, which some refer to as 'lumpen urbanisation' (Gidwani and Maringanti 2016).

References

Ahluwalia IJ (2019) Urban governance in India. J Urban Aff 41(1):83–102

Ahluwalia IJ, Kanbur R, Mohanty PK (2014) Urbanization in India: challenges, opportunities and the way forward. Sage India. New Delhi. http://icrier.org/pdf/urbanisation_in_india.pdf. Accessed on 13 Oct

Arnold D (1993) Colonising the body: state medicine and epidemic disease in nineteenth century India. University of California Press, Berkley

Bhagat RB (2011) Emerging pattern of urbanization in India. Econ Polit Wkly 46(34):10–12

Bhave P, Sadhwani K (2016) Solid waste management legislation: a review. Environ Policy Law 46(2):168–175

Bhushan C, Sambyal SS, Walani N (2018) Model framework for segregation: guidelines for managing municipal solid waste through segregation, reuse and recycling in Zanzibar. Centre for Science and Environment, New Delhi. Weblink: https://www.cseindia.org/model-framework-for-segregation-guidelines-for-managing-municipal-solid-waste-through-segregation-reuse-and-recycling-in-zanzibar-8848. Accessed on 13 Sept 2019

Bolaane B (2006) Constraints to promoting people centered approaches in recycling. Habitat Int 30(4):731–740

Burton A (1998) At the heart of the empire. University of California Press, Berkley

Chandramouli, C (2011) Provisional population totals paper 1 of 2011 India series 1. India: Registrar General and Census Commissioner of India

Chintan (2012) Give Back our waste: what the Okhla waste-to-energy plant has done to local Wastepickers. Chintan Environmental Research and Action Group, New Delhi. http://www.indiaenvironmentportal.org.in/files/file/chintan-report-give-back-our-waste.pdf. Accessed on 13 Sept 2019

CPCB (2000) The municipal solid wastes (management and handling) rules. Central Pollution Control Board, New Delhi. http://www.indiaenvironmentportal.org.in/content/257913/the-municipal-solid-wastes-management-and-handling-rules-2000/. Accessed on 13 Sept 2019

Datta A (2015) New utopias of postcolonial India: entrepreneurial urbanization in Dholera Smart City, Gujarat. Dialog Human Geogr 5(1):3–22

Deshingkar P, Sandi M (2012) Migration and human development in India: new challenges and opportunities. National Workshop on internal migration and human development in India. 6–7 December 2011, Workshop Compendium, vol. 2, Workshop papers, UNESCO, and UNICEF. https://ideas.repec.org/p/hdr/papers/hdrp-2009-13.html. Accessed on 28 Aug 2019

Deshkar SM (2010) Kautilya Arthashastra and its relevance to urban planning studies. Inst Town Plan India 7(1):87–95

EPA (1988) Waste Minimization Opportunity Assessment Manual. https://cfpub.epa.gov/si/si_public_record_report.cfm?Lab=NRMRL&dirEntryId=456 67. Accessed on 28 Aug 2019

ESCAP (2017) Sustainable solid waste Management in Secondary Cities and Small Towns in Asia Pacific. United Nations, Thailand

European Commission (2015) Directive of The European Parliament and of The Council Amending Directive 1999/31/EC on the landfill of waste. https://eur-lex.europa.eu/legal-content/EN/TXT/PDF/?uri=CELEX:52015PC0594&from=EN. Accessed on 28 Aug 2019

Fortuna C (2018) Plastics ban protects World's Oldest & Largest Inhabited River Island. *Cleantechnica*. https://cleantechnica.com/2018/07/10/plastics-ban-protects-worlds-oldest-largest-inhabited-river-island-cleantechnicaexclusive. Accessed on 28 Aug 2019

Ghosh R, Kansal A (2014) Urban Challenges in India and Mission for Sustainable Habitat. Interdisciplina, 2(2):281–304

Ghosh R, Kansal A (2017) Implications of municipal solid waste Management in Developing Countries on greenhouse gas emissions. In: Singh S, Goyal R, Jain A (eds) The urban environment crisis in India: new initiatives in safe water and waste management. Cambridge Scholar Publications, New Delhi, pp 190–202

Gidwani V, Maringanti A (2016) The waste-dialectic: lumpen urbanisation in contemporary India. Comp Stud South Asia Afr Middle East 36(1):112–133

https://www.deccanherald.com/national/indo-uk-team-preserve-cultural-684869.html. Accessed on 08 Aug 2019

Hoelscher K (2016) The evolution of the smart cities agenda in India. International Area Studies Review, 19(1):28–44

Kansal A (2001) Critical appraisal of solid waste disposal technologies. Indian J Environ Protect 19(3):83–96

Kumar MS (1994) Plague: a catastrophe of ignorance. Ann Natl Assoc Geogrs India XIV(2):94–99

Kumar MS (2002) The evolution of spatial ordering in colonial Madras. In: Blunt A, McEwan C (eds) Postcolonial geographies. Continuum Press, London, pp 85–98

Kumar MS (2003) Urbanisation. Encyclopaedia of community: from the village to the virtual world, vol 4. Sage, Thousand Oaks, pp 1433–1437

Kumar MS (2006a) Idioms, symbolism and divisions: beyond the black and white towns in Madras, 1652-1790. In: Raju S, Kumar MS, Corbridge S (eds) Colonial and postcolonial geographies of India. Sage, Delhi, pp 23–48

Kumar MS (2006b) Stakeholders' visions towards community-based sustainable development. Northern Ireland environment link environmental fact sheet 5I1): p. 11

Kumar MS (2019a) Cleaning India, catch the waste at the site of its creation. India CSR Network. Noida. https://indiacsr.in/cleaning-india-catch-the-waste-at-the-site-of-its-creation/

Kumar MS (2019b) Swachh Bharat-Mission possible? Cleaning India, managing waste. The CSR Journal, New Delhi. https://thecsrjournal.in/cleaning-india-managing-waste/. Accessed on 18 Sept 2019

Kumar MS (2019c) Chapter 10: What is right for the City or right to the City: criticality of citizenship and governance? In: Kumar A, Meshram DS (eds) City prosperity and sustainability. Routledge, Milton Park. (In Press)

Kumar MS, McKenna N (2018) Smart city and its relevance to the global south. Mimeograph

Kumar S, Bhattacharyya JK, Vaidya AN, Chakrabarti T, Devotta S, Akolkar AB (2009) Assessment of the status of municipal solid waste management in metro cities, state capitals, class I cities, and class II towns in India: an insight. Waste Manag 29:883–895

Kundu A (2011) Trends and Processes of Urbanisation in India. International Institute for Environment and Development (IIED) and UNFPA

Marshall RE, Farahbakhsh K (2013) Systems approaches to integrated solid waste management in developing countries. Waste Management, 33(4):988–1003

Mahadevia D (2001) Sustainable urban development in India: an inclusive perspective. Dev Pract 11(2–3):242–259

McAllister J (2015) Factors influencing solid-waste Management in the Developing World. All graduate plan B and other reports. Paper 528

McGee TG, Robinson IM (1995) The mega-urban regions of Southeast Asia. UBC Press, Vancouver

Milea A (2009) Waste as a social dilemma: issues of social and environmental justice and the role of residents in municipal solid waste management, Delhi, India. Master's Thesis, Lund University, Lund, Sweden

Ministry of Urban Development (MoUD)-(2015) Smart Cities: Mission Statement & Guidelines. http://smartcityrajkot.in/Docs/SmartCityGuidelines.pdf

Mittal S, Sethi M (2018) Smart and livable cities: opportunities to enhance quality of life and realize multiple co-benefits. In: Sethi M, Puppim de Oliveira J (eds) Mainstreaming climate co-benefits in Indian cities. Exploring urban change in South Asia. Springer, Singapore

Mor S, Ravindra K, Dahiya RP, Chandra A (2006) Leachate characterization and assessment of groundwater pollution near municipal solid waste landfill site. Environ Monit Assess 118:435–456

Mrayyan B, Hamdi MR (2006) Management approaches to integrated solid waste in industrialized zones in Jordan: a case of Zarqa City. Waste Manag 26(2):195–205

MoEF-Ministry of Environment and Forests (2000) The Municipal Solid Wastes Management and Handling Rules 2000; The Gazette of India, New Delhi

MoUD (2014) Smart Cities as envisioned by MoUD. ArcIndia News, 9:20–22

Muniapan B, Shaikh JM (2007) Lessons in corporate governance from Kautilya's Arthashastra in ancient India. World Rev Entrepreneurship Manag Sustain Dev 3(1):50–61

Narayana T (2009) Municipal solid waste Management in India: from waste disposal to recovery of resources? Waste Management, 29(3):1163–1166

Oates L, Sudmant A, Gouldson A, Gillard R (2018) Reduced waste and improved livelihoods for all: lessons on waste management from Ahmedabad, India. Coalition for Urban Transitions. London and Washington, DC. http://newclimateeconomy.net/content/cities-working-papers. Accessed on 18 Aug 2019

O'Connell EJ (2011) Increasing public participation in municipal solid waste reduction. Geogr Bull 52(2):105–118

Pappu A, Saxena M, Asolekar SR (2007) Solid wastes generation in India and their recycling potential in building materials. Build Environ 42:2311–2320

Pati B, Harrison M (2001) Health, medicine and empire: perspectives on colonial India. Hyderabad, Orient Longman

Ramanna M (2002) Western medicine and public health in colonial Bombay, 1845–1895. Orient Longman, New Delhi

Ray K (2018) Indo-UK team led by Dr. M. Satish Kumar to Preserve Cultural Heritage of Majuli. Deccan Herald:

Registrar General and Census Commissioner (2011) Census of India. Ministry of Home Affairs. Government of India, New Delhi

SBM (2017) Guidelines for Swachh Bharat Mission Gramin. MoUD, Government of India, New Delhi. Accessed on 18 Aug 2019

Schübeler P, Wehrle K, Christen J (1996) Conceptual framework for municipal solid waste Management in low-Income Countries. Working paper no. 9. World Bank

Sengupta N (1985) Irrigation: traditional vs modern. Econ Polit Wkly 20(45–47): 1919–1938

Sharholy M, Ahmad K, Mahmood G, Trivedi RC (2008) Municipal solid waste management in Indian Cities: a review. Waste Manag 28:459–467

Shrivastava R (2019) India's trash bomb: 80% of 1.5 lakh metric Tonne daily garbage remains exposed, untreated. https://www.indiatoday.in/india/story/india-s-trash-bomb-80-of-1-5-lakh-metric-tonne-daily-garbage-remains-exposed-untreated-1571769-2019-07-21. Accessed on 18 Aug 2019

Swaminathan M (2018) How can India's waste problem see a systemic change? Econ Polit Wkly 53(16)

Thompson A (2019) The body shop initiative That's changing the lives of waste pickers in Bangalore. Marie Claire News Report. https://www.marieclaire.co.uk/reports/waste-pickers-652770. Accessed on 18 Aug 2019

Troschinetz AM, Mihelcic JR (2009) Sustainable recycling of municipal solid waste in developing countries. Waste Manag 29(2):915–923

UN (2015) Transforming our World: The 2030 Agenda for Sustainable Development. A/RES/70/1″. United Nations. https://sustainabledevelopment.un.org/content/documents/21252030%20Agenda%20f or%20 Sustainable%20Development%20web.pdf. Accessed on 18 Aug 2019

UN Environment (2019) The Role of Gender in Waste Management: Gender Perspectives on Waste in India, Indonesia, The Philippines, and Vietnam. UN Environment in partnership with Ocean Conservancy, the New Plastics Economy Initiative of the Ellen MacArthur Foundation, and GRID Arendal (GEF Project ID No. 9681). Published by GA Circular. 54D

South Bridge Road, Singapore, 058685. Accessed on 19 Sept 2019

UN-Habitat (2010) Solid waste Management in the World's cities: water and sanitation in the World's cities. London, Washington. Accessed on 18 Aug 2019

UKTI-UK-Trade and Investment (2015) India's Smart Cities Programme: The UK offer to build together. https://assets.publishing.service.gov.uk/government/uploads/system/uploads/attachment_data/file/460151/UKTI_-_The_UK_offer_to_build_together__1_.pdf

Venkateswaran S (1994) Managing waste: ecological, economic and social dimensions. Econ Polit Wkly 29(45/46):2907–2911

WIEGO (2014) The Urban Informal Workforce: Waste Pickers/Recyclers (IEMS Sector Summary). http://www.wiego.org/informal_economy_law/waste-pickers-india. Accessed on 29 Sept 2019

Willcocks W (1984) Ancient system of irrigation in Bengal. BR publishing corporation, Delhi. https://archive.org/details/in.ernet.dli.2015.62230/page/n3. Accessed on 19 Sept 2019

World Bank (2004) India's self-employed Women's association (SEWA) empowerment through mobilization of poor women on a large scale: a case study from reducing poverty, sustaining growth what works, what does not, and why a global exchange for scaling up success and scaling up poverty reduction: a global learning process and conference. Shanghai, May 25–27, 2004, http://web.worldbank.org/archive/website00819C/WEB/PDF/INDIA_SE.PDF. Accessed on 29 Sept 2019

World Economic Forum (2018) World Economic Forum Gender Gap 2018. http://reports.weforum.org/global-gender-gap-report-2018/data-explorer/#economy=PHL19. Accessed on 19 Sept 2019

Zhu DA, Asnani PU, Zurbrügg C, Anapolsky S, Mani S (2008) Improving municipal solid waste management in India: A sourcebook for policy makers and practitioners. Washington, DC: World Bank

28 Natural Resource Evaluation for Ecotourism and Geotourism Destination in Hong Kong

Lawal M. Marafa

Abstract

Ecotourism and geotourism have triggered a need to explore natural places that are devoid of human impact and have pristine environments that will provide lasting experiences while attracting substantial numbers of visitors. For a destination to provide such experiences, it needs to be identified, evaluated for such potential, and managed for sustainable use. This study attempts to evaluate an outlying Island devoid of human habitation with potential to attract ecotourists, geotourists, and nature lovers. Initially, the basic theory and definitions of ecotourism and geotourism were reviewed putting significance in the local context and the need for designation and evaluation of potential sites was underscored. Although different methods of evaluation exist in the literature, no single methodology has gained international currency for evaluation of natural areas specifically for geotourism, ecotourism, or nature-based tourism. A modified version of the simple additive weighting method (SAWM) was selected for this study. Four checkpoints (P1, P2, P3, and P4) were identified for evaluation and assessment along the major loop layout trail on the Tung Ping Chau Island near Hong Kong. Weights indicating importance and availability were assigned on indicators grouped as natural, cultural, infrastructure, and level of human impacts. Overall attractions were then calculated to identify which of the checkpoints provide the most attractions or special interests. Although the method is simple, it can be adopted and improved to be used elsewhere, as its simplicity implied that (1) all criteria were subjectively selected and measured on an interval; (2) all indicators and values are expressed in comparable units; and (3) weights are assigned to each criterion. This initial simplicity will make it possible for adaptation as it can easily be interpreted by non-academic personnel responsible for decision-making processes and, thus, serve as basis for planning and management of resources at destinations.

Keywords

Ecotourism · Evaluation · Geotourism · Hong Kong · Resource Management · Resource Evaluation

L. M. Marafa (✉)
Centre for Environmental Policy and Resource Management, Department of Geography and Resource Management, The Chinese University of Hong Kong, Sha Tin, N.T., Hong Kong
e-mail: lmmarafa@cuhk.edu.hk

B. Thakur et al. (eds.), *Resource Management, Sustainable Development and Governance*, Sustainable Development Goals Series, https://doi.org/10.1007/978-3-030-85839-1_28

28.1 Introduction

Although tourism has grown to be one of the world's leading economic sectors, concern is being raised as to the negative impacts that it brings to host destinations. When the tourism industry received a boost in the 1950s (due to rising standards of living, increases in leisure time and development in infrastructure), it quickly assumed the context of mass tourism that was characterized as fulfilling the "three S" type of experience of sun, sand, and sea (Boyd 2001; Rotariu 2006; Weaver and Lawton 2006; Cameron and Gatewood 2008).

In the past few decades and especially from the 1980s, mass tourism has often been viewed as inherently negative and unsustainable (Valentine 1993; Wunder 2000; Lawton and Weaver 2001; Weaver and Lawton 2006). Scholars, practitioners, and various interest groups have sought to explore and develop alternative types of tourism (Ceballos-Lascurain 1988; Blamey 2001; Page and Dowling 2002; Kiss 2004), targeting specific niche markets (Fennell 2002). Consequently, a niche is being carved out as ecotourism and most recently, geotourism (Farsani et al. 2011). Indeed, Fennell (1999) had postulated ecotourism as being distinct from mass tourism as it develops based on stipulated ethical and responsible principles. Geotourism was seen as a distinct sector of natural area tourism (Dowling and Newsome 2010) and one which concentrates on connection to abiotic landscape features. Consequently, it can be argued that ecotourism and geotourism, variously defined as responsible travel to natural areas to view and study, help conserve such environments and in some cases help improve the welfare of the local people, is now arguably the fastest growing sector in the overall tourism industry. As a result, the challenge is now put on proponents, researchers, and stakeholders to identify sites that can serve both as ecotourism and geotourism destinations.

Although Hong Kong is a cosmopolitan city with most of the population completely urbanized, there is a growing trend of nature appreciation that is transforming into visitation to natural areas or ecological site visits (eco-site visits). There is also an increasing interest in such visits to sites of geological interests, such as geoparks. While Hong Kong is famous for mass tourism (but not necessarily in the context of the "three S"), often referred to as a heaven for shopping, dining, and sightseeing, the emergence of nature-based tourism, such as ecotourism and now geotourism, is making the countryside more attractive. Annually, over 10 million people visit the country parks (Wong 1997; Marafa and Fung 2004) and proportionally, the number increases in times when people prefer to visit the countryside for leisure and recreation. This is a testimony to a growing interest. While there is an increasing awareness of the development of ecotourism and geotourism worldwide, in Hong Kong, it is similarly becoming important (Ng and Li 2000; Marafa and Ng 2008). Because of the increasing number of visitors to ecologically and geologically sensitive areas and the growing interest in nature-based tourism, there is the need to further understand the rise in this trend. It is, therefore, worthwhile to understand ecotourism and geotourism in Hong Kong context as it will affect Hong Kong natural areas.

As ecotourism, geotourism, and nature-based recreation become more important, there is the need for tools, frameworks, and strategies that will make such activities sustainable. Indeed, the ecotourism and geotourism trend and the booming interest in eco-site visits will likely pose a great deal of concern and create challenges for the identification, selection, and management of destination sites. Because of mass tourism and growing environmental concern, scholars have argued for the integration of tourism and environmental development programs like interpretation and other value-added activities (Fung and Jim 2015; Wang et al. 2015). This can be done with specific case studies, such as this one. Identifying, selecting, and planning sites and destinations for ecotourism, geotourism, and nature-based recreation will help to lead to minimum negative impact and help develop this growing interest. While there has been a proliferation of alternative tourism trends, the natural environment (especially at the periphery of urban areas), when well presented as an attraction, can have a profound effect on psychological and physical relationship that the visitors can experience.

Many scholars have postulated that for ecotourism, geotourism, and nature-based tourism to be viable, there must be an abundance and access to natural environments (Koh et al. 2014). If this is the case, then these aspects of tourism can be affected by the factors of supply and demand. Although Mitchell (1989) has indicated that (natural) resources are an expression of appraisal and represent a subjective concept, this study examined the extent to which natural resources are available and accessible to facilitate a growing special interest in tourism. The availability of the natural resource base, therefore, represents the supply component.

As natural, geological, and protected areas are becoming attractive to recreationists and ecotourists, there is the need to evaluate and assess whether certain areas can provide the experiences sought by these visitors. In resource management, ideas of evaluating the significance of natural areas for conservation or other purposes are evolving and are finding wide spectra of applications (Smith and Theberge 1986; Priskin 2001, 2003). While applications of such evaluations are widespread in EIA, land use planning and planning for protected areas, the use in nature-based recreation and in particular ecotourism and geotourism have not gained universal acceptance. Hence, no standardization of any methodology has been reported.

Having an inventory of nature-based resources is important to planners as well as to the decision-makers of the tourism industry (Priskin 2001). This inventory, evaluation and assessment will identify landscapes suitable for nature-based recreation, ecotourism, and geotourism activities. As these special interest types of tourism depend on the environment (both natural and human-made) the evaluation of such landscapes and environments poses a challenge to resource managers. This type of evaluation will help identify what should be preserved and why (Lowenthal 1981). The outcome will also identify policies regarding the way in which preserved landscapes and pristine environment should be used.

To this end, the main thrust of this study is to evaluate and assess the natural resource base and a potential site in Hong Kong for ecotourism and geotourism. The study reports a simple and effective method in identifying and assessing resources for ecotourism, geotourism, and nature-based tourism on Tung Ping Chau (an outlying Island, shown in Fig. 28.1) that can be replicated elsewhere. Such a methodology can identify attractions and inventory relevant resources for sustainable use. Specifically, the objectives of this study are (1) to review the current development of ecotourism and geotourism in Hong Kong; (2) to use a simple and effective way to identify and assess such relevant resources on Tung Ping Chau Island, and (3) to formulate recommendations on planning and management for sustainable development of ecotourism and geotourism on Tung Ping Chau (TPC). When this is done and is successful, it will help to plan geotourism within the sustainable tourism domain like ecotourism. While ecotourism focuses on biotic resources, geotourism tends to focus on abiotic features.

Box 28.1 Sustainable Development Goals

The conceptual notion of ecotourism and geotourism has become popular in the last few decades and can reciprocate the pursuit of the SDGs. In the context of Hong Kong, the local government has become conscious of the need to invest in sustainable tourism and related policies that support the SDGs. According to Marafa if tourism is managed sustainably, it can provide a powerful incentive to support conservation of nature as well as provide both political and financial support to outlying islands of Hong Kong devoid of human habitation potentially attracting ecotourists, geotourists, and nature lovers. Both ecotourism and geotourism are influenced by numerous goals, targets, and indicators of Agenda 2030 and can serve as part of drivers for realizing the SDGs. Although mention was specifically made of the SDGs 8, 12, and 14 (sustainable economic growth, sustainable consumption and production, and the sustainable use of oceans and marine resources), ecotourism and geotourism can also contribute directly to SDG 15 (protect, restore, and promote sustainable use of terrestrial ecosystems). As we evaluate the future of both ecotourism and geotourism in terms of sustainability it is evident that both are influenced by population growth and increasing consumption, urbanization, and other imponderables such as recreation preferences, pandemics, an

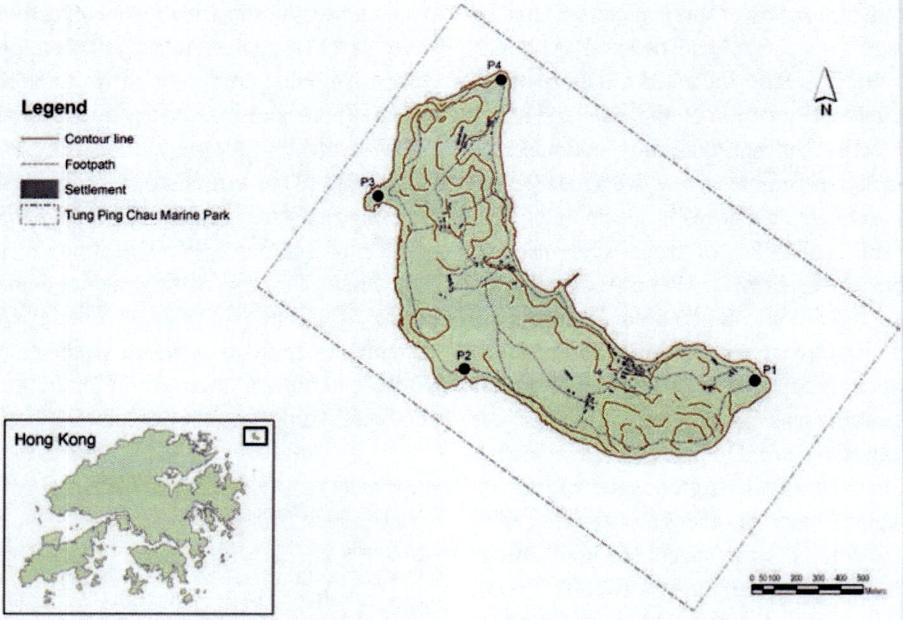

Fig. 28.1 Tung Ping Chau showing checkpoints on the majority country tail

aging population, immigration and access to information and technology among others. Thus, ecotourism, geotourism, and indeed sustainable tourism are necessary and call for identification, evaluation, and management to achieve conservation goals and support to the local economy.

28.2 Adopting Definition and Resource Inventory Methodology for Ecotourism and Geotourism

As an environmental phenomenon, tourism seeks to sell or promote the environment as its product (Taylor and Stanley 1992). With increasing environmental awareness, the countryside and natural areas have become attractive destinations providing numerous opportunities including recreational fishing (Zwirn et al. 2005), wildlife tourism (Curtin 2005), ecotourism, (Buckley 2003), geotourism (Dowling and Newsome 2010), and often amenities that can add to experience and make the visit memorable. As a result, potential environmental and social impacts of tourism are being highlighted, making it pertinent to find a way of allowing tourists enjoy the environment that attracted them in the first place, but with little or no negative impact. This can be achieved only when specific destinations are identified and promoted as nature-based tourism, ecotourism, and geotourism destinations.

Indeed, while ecotourism has received much attention in the past decades, there is still considerable debate over what the term really means in various communities (Valentine 1993; Sirakaya et al. 2002; Donohoe and Needham 2006). Ecotourism and geotourism have been identified as concepts that forge a relationship with conservation, sustainability, and biological diversity. It can similarly be agreed that ecotourism recognizes intrinsic values of the natural environment, help develop ecologically and culturally sustainable tourism, and educate and inspire visitors and local communities to participate and appreciate the importance of natural and cultural assets of a given site. On the other hand, geotourism is a relatively new concept that has emerged as a rapidly growing industry that can help sustain or enhance the geographical character of a place, its environment, culture, aesthetics, heritage, and the well-being of its residents (Heggie 2009).

From the numerous descriptions of geotourism, it is clear that this kind of tourism is complementary to the current trend of environmental awareness in Hong Kong and elsewhere. But does Hong Kong have the potential to develop both ecotourism and geotourism? Are there any places suitable for this development? We need to explore, use, and possibly adopt resource inventory or resource evaluation methodologies in order to promote and develop both ecotourism and geotourism in Hong Kong. In a place like Hong Kong where about 70% of the territory is countryside and about 41% of the land area falls under the protected area system, it is necessary to embark on a comparative evaluation of landscapes that have potential to serve as destinations for ecotourism and geotourism.

Landscape evaluation research and natural resource management have numerous implications for nature-based recreation and ecotourism and geotourism. Such research can lead to inventories of scenic attributes, which can then be considered along with other information normally available to decision-makers. Landscape assessments may also serve as a significant input to environmental impact assessments since an overriding concern in such work is to determine the impact of human activities on natural systems when they are exposed to various types of tourism. Such impacts caused by humans on a site that have amenities and attractions can constitute a severance to pleasant activities and the site can become less attractive to visitors. Therefore, landscape evaluation can decide whether a specific site, corridor, or area deserves to be preserved, protected, maintained, or improved (Mitchell 1989) as it is set to accommodate ecotourists and geotourists.

Evaluating natural areas involves making measurements based on a set of criteria and deciding which areas are most significant (Smith and Theberge 1987; Deng et al. 2002; Priskin 2001). In most methodologies for evaluation, different criteria are used as a basis for assessment. Smith and Theberge further stated that overall assessment is often done by summing subjective scores assigned to different indicators and criteria. The evaluation of landscapes indeed involves the exercise of value judgment, because regarding such evaluations there will always be subjective judgment involved (Phillips 1997; Burger 2000; Gulinck et al. 2001). In most evaluations, different criteria are used as a basis for assessment.

As overall assessment is often done by summing subjective ranks, and scores are assigned to different criteria, it has generated widespread criticism (Nelson et al. 1997; Obua 1996). However, it is clear in the literature that there is no standard approach for evaluating the potential of a natural area or resource for nature-based recreation, ecotourism, or even geotourism (Obua and Harding 1996). Given this reality, the feasibility of an area to provide amenities for ecotourism can fundamentally be based on its natural environmental potential (Parker and Khare 2005). As the ecotourists and geotourists seek specific attributes as an outcome of experience, resource inventory can inform the proponents, planners, or stakeholders about what the site can offer. Together with specific inventory studies, there is, thus, a need to incorporate appropriate multicriteria evaluation (Deng et al. 2002) in ecotourism and geotourism whether explicitly or implicitly.

Although Priskin (2001) has pointed out that natural resource assessment may be criticized for subjectivity in assigning values to indicators and criteria, such an approach leaves open the potential for the use of matrices to complete or execute such resource assessment tasks. Consequently, as we designate places for nature-based recreation, ecological site visits, ecotourism, and geotourism, we need to inventory, assess, and even classify sites that will have appropriate indicators to warrant such visits. Patterson (2002) had earlier indicated that an inventory for a potential site can provide initial information related to water, climate, topography and biodiversity, geodiversity which will provide a basis for the development of such tourism products.

For a site like Tung Ping Chau to be developed as a destination for nature-based recreation and provide ecotourism and geotourism product, an assessment of its potential to offer such opportunities is necessary. As this site is considered by many locals as a viable destination, it is pertinent to note that most visitors to Tung Ping Chau are local and day visitors. Although there is a dearth of information on methodology, this study has identified similar studies and methodologies that have been used and adopted in assessing, evaluating, and classifying natural resource areas for recreation purposes (Bastedo et al. 1984; Cocklin et al. 1990; Priskin 2001; USFS 2002), geological sites (Wojtowicz et al. 2011; Serrano and Gonzalez Trueba 2011).

To bring to light the perceived potentials of an area, a qualitative method can be used to derive a potential index, which when adequately interpreted, will be useful for further research, awareness, and decision-makers. Furthermore, to indicate the potential and demand for nature-based recreation, Priskin (2001) used checklists and matrices to simultaneously inventory and assess the resources that will complement experiences at natural area destinations.

28.3 The Adopted Methodology

In this study, the inventory, assessment, and evaluation of Tung Ping Chau involved the checklist approach modified from Priskin (2001), Cocklin et al. (1990), and Smith and Theberge (1987). In addition, the additive weighting method (SAWM) developed by Smith and Theberge (1987) was preferred as it is simple and can be applied when (1) all criteria are measured on an interval or ratio scale; (2) all values are expressed in comparable units; and (3) weights can be assigned to each criterion.

The SAWM considers a set of criteria to which weights are assigned. The weights reflect the importance and whether the item of the criterion exists and when summed up, an overall value for a particular use is obtained. The checklist and scoring system are modified to standardize and suit the resource assessment in the local case study. This eventually allows for data collection systematically. Four criteria are used in this study to assess the sites, and these are (1) natural attractions (NA); (2) cultural attractions (CA); (3) tourism infrastructure (TI); and (4) human impacts (HI). The criteria NA, CA, and TI are considered amenities that are positive and attract special interest visitors while HI is considered as a negative factor capable of presenting visitors with a negative experience. Each criterion consists of a set of indicators in a matrix form. Each matrix includes a different set of indicators, relevant to the criterion being assessed. Since the study site is an island, four checkpoints were selected along the loop layout trail. Each checkpoint received a weighted score to reflect the importance of the indicators. Classification of each criterion is then introduced to reveal the importance of the criterion at each checkpoint. Lastly, (5) overall attractiveness (OA) is calculated using the weights assigned to the indicators to find out the most attractive checkpoint on TPC.

28.4 Study Area

Tung Ping Chau (TPC) is an island with an area of about 1.1 km^2. The island is located at the northeast part of Hong Kong (Fig. 28.1). Part of TPC was included in the Plover Cove country park (Extension) in June 1979, thereby bringing it under the government's protected area system. In November 2001, about 2.7 km^2 of marine area around TPC was designated as a marine park (Chow and Ngar 2002a), bringing the total number of marine parks to four in the territory. This development has made the island one of the important ecological and geological sites that continues to attract visitors.

Although TPC is relatively flat, it rises to a height of 48 m at the southern tip. In addition to the marine environment for which the island primarily got its protected area status, the physical environment, divided into geological and biological features, can similarly be distinguished. Geological landforms include wave-cut platforms that extend to more than 33 m inwards from the sea. Sea caves, sea cliffs, and stacks have formed because of coastal erosion. In addition, honeycomb weathering, chert veins, and elongated sandy beach form attractive natural landforms typical of geotourists attractions.

The biological or ecological environment is represented by woodland, shrubland, and grassland. The woodland is primarily located on gentle slopes and valleys; shrublands are located at higher regions, while the grasslands are in the central region where agricultural fields were abandoned (Chow and Ngar 2002b). The fauna is characteristically a concert of butterflies, dragonflies, and several species of avifauna. Furthermore, TPC is known for its marine life that includes numerous species of stony corals (Chow and Ngar 2002b).

As TPC is an island that attracts visitors, there is the need to investigate, evaluate, and analyze the resources that create such an attraction and provide valuable experience from a supply component. To this end, this research focused on the supply component that dealt with resources suitable for nature-based recreation, ecotourism, and geotourism. Although these evaluation and assessment borrow heavily from methodologies used in landscape resource evaluation and analysis, Pigram (1983); Mitchell (1989); Cocklin et al. (1990); Obua (1996) and Priskin (2001) had all criticized such methodologies as being too subjective. Nonetheless, Mitchell (1989) indicated that resources are an expression of appraisal and indeed can only be a subjective concept. However, this kind of evaluation and appraisal can be approached systematically by using a systematic assessment methodology (Smith and Theberge 1986), as in this study.

28.5 Resource Evaluation and Assessment

In order to achieve the above objectives, checkpoints along the loop trail (Fig. 28.1) were identified where clusters of visitors consistently were observed. After a series of visits and observations, the checkpoints were selected for in-depth investigation. At the identified checkpoints, inventories of potential resource attractions were completed to establish the potential of such points to provide special interest touristic experience. The checkpoints are identified in Fig. 28.1 and are referred to as P1, P2, P3, and P4.

Natural Attractions (NA): Investigations were based on 20 indicators, grouped into three categories: floral diversity (grassland, scrubland, woodland, feng shui wood, mangrove), fauna diversity (amphibians, reptiles, fish, butterflies, dragonflies, coastal life, birds), and special geologic or geomorphologic features (shoreline, beaches, wave-cut platforms, coastal cliff, headland, bay, special shaped stone, caves). Natural attractions are usually the prime attractions for ecotourism and geotourism.

Cultural Attractions (CA): These represent one of the important components in destinations providing both ecotourism and geotourism experiences. In order to assess the cultural attractions, three indicators were considered for assessment (archeological sites, traditional lifestyles, villages).

Tourism Infrastructure (TI): This is vital in a destination as it enhances the visitor's enjoyment (Priskin 2001). Ten indicators were taken into consideration relative to this criterion (picnic sites, BBQ sites, campsites, rubbish bins, shelters, public phones, information boards, toilet facilities, seats, park signs).

Level of Human Impact (HI): Impacts caused by ecotourists, nature-based recreationists, and geotourists often result in environmental degradation at a destination. Only visually outstanding elements of degradation are investigated (litter, noise pollution, fire, graffiti, erosion of landforms).

In the evaluation matrix for natural and cultural attractions and tourism infrastructure, each indicator is assigned a number of 0 or 1 to show the importance of such an indicator. While 1 represents the "presence" of the indicator, 0 represents the "absence" of that indicator. The maximum score that each checkpoint can get for natural attractions is 20, while a maximum score of 3 and 10 could be achieved for cultural attractions and tourism infrastructure, respectively. In addition, a five-class rating is created to determine the degree of natural attractiveness.

Because impacts have a negative connotation, in the criteria 5 indicators are studied and similarly, a value of 0 and − 1 are assigned to each indicator. 0 indicates absence while −1 is assigned where impacts on such indicators are visible. The score of low (0 to −1), medium (−2 to −3) ,and high (−4 to −5) denotes the extent to which human impact is discernible, representing a 3-class rating category.

Overall Attractiveness (OA): The overall attractiveness of a particular checkpoint is calculated by simple addition of the scores (a modification of the SAWM, proposed by Smith and Theberge in 1987) from the four categories of indicators. The sum of the total scores of tourism infrastructure, natural attraction and cultural attraction, minus the scores of human impacts, constitutes the overall degree of attraction as indicated in the following model:

$$\text{Overall Attractiveness} = (\text{Tourism Infrastructure} + \text{Natural Attractions} + \text{Cultural Attractions} - \text{Level of Human Impact})$$

For simple empirical representation, the model can be expressed as follows:

$$OA = TI_{w1} + NA_{w2} + CA_{w3} + HI_{w4} \quad (28.1)$$

where

A = Attractiveness

TI = Tourism Infrastructure

NA = Natural Attractions

CA = Cultural Attractions

HI = Human Impact

w1, w2, etc. = Weights (scores) assigned to each index

In this study, the fieldwork constituted a major component and involved visitation, observation, and inventory of resources. The first visit was a reconnaissance survey where the sites and the activities occurring therein were observed and recorded. Potential sites for further inventory studies were selected. Four visits were then arranged over a period of 1 year with each visit dedicated to P1, P2, P3, and P4, respectively. Although most visitors to the island allude to ecotourism and geotourism given the popular trend of such activities and the attraction generated after the creation of the Hong Kong Global Geopark, this study investigated the sites based on their land use capability as almost 90% of the visitors are day trippers. The implication, therefore, is that if prudent land use resource planning is understood and implemented, the use of the sites could be sustainable.

28.6 Results of Resource Evaluation

28.6.1 Natural Areas

Natural resources in TPC are evaluated based on the floral diversity, faunal diversity, and special geological features. Generally, the four checkpoints receive a higher-than-average level on natural attractions relative to the first point of entrance which is the Ferry Pier. P2 and P3 have a high level of natural attractions, while P1 and P4 received a medium score in the floral diversity category. For faunal diversity, all four sites received relatively high-ranking scores.

From observations and field checks, it was found that for floral diversity, except for P1 where only scrubland was identified, P2 and P3 contain both scrubland and woodland, while grassland and scrubland predominated in P3. All sites have a high level of attraction in special geologic or geomorphologic features. Shoreline, beaches, special shaped stones, coastal cliffs, evidence of honeycomb weathering, and caves can be found at all sites (Table 28.1).

28.6.2 Cultural Attractions

Generally, cultural attractions reflect past civilizations. Checkpoints investigated show little or no cultural attractions in all four sites. Although archeological sites, traditional lifestyles and major villages are absent, there are some abandoned villages in the inner area of TPC. However, while they reflect evidence of past civilization, none is found at any of the checkpoints.

Table 28.1 The checkpoint indicator result for natural attractions

	P1	P2	P3	P4
Fauna				
Amphibians	0	0	0	0
Reptiles	0	0	0	0
Fish	1	1	1	1
Butterflies	1	1	1	1
Dragonflies	0	1	0	0
Coastal life	1	1	1	1
Birds	1	1	1	1
Sub-Total	***4***	***5***	***4***	***4***
Flora				
Grassland	0	0	1	0
Scrubland	1	1	1	1
Woodland	0	1	0	1
Fung Shui wood	0	0	0	0
Mangrove	0	0	0	0
Sub-total	***1***	***2***	***2***	***2***
Special geologic/geomorphologic features				
Shoreline	1	1	1	1
Beaches	0	0	0	0
Wave-cut platforms	1	1	1	1
Coastal cliff	1	1	1	1
Headland	1	1	1	0
Bay	0	0	1	0
Special shaped stone	1	1	1	1
Caves	1	1	1	1
Sub-total	***6***	***6***	***7***	***5***
Total	11	13	13	11

Table 28.2 The checkpoint indicator result for adequacy of tourism facilities

	P1	P2	P3	P4
Picnic sites	1	0	0	0
BBQ sites	1	0	0	0
Campsites	1	0	0	0
Rubbish bins	1	1	1	1
Shelters	1	0	0	0
Public phones	0	0	0	0
Information boards	1	1	1	1
Toilet facilities	1	0	0	0
Seats	1	1	1	1
Park signs	1	0	1	1
Total	9	3	4	4

28.6.3 Tourism Infrastructure

In general, infrastructure and facilities make destinations more attractive to visitors. While some visitors (eco-specialists) can be comfortable with minimum infrastructure and facilities, most visitors (eco-generalists) expect a modicum of comfort to make their experience memorable (Porter 2003). To this end, the supporting infrastructure and facilities complementing nature-based recreation, ecotourism. and geotourism in TPC are now acceptable if compared to the existing structure in most of the country parks in Hong Kong. Basic facilities, such as rubbish bins, information boards, benches can all be found at the four checkpoints. However, other facilities, such as shelters, toilets, and signposts are sparsely found.

From the fieldwork, checkpoint P1 received a relatively high score (9), denoting high adequacy of convenience facilities, while the other sites (P2, P3 and P4) received low scores, as seen in Table 28.2. Indeed, P1 is located closer to the ferry pier and, as a result, it has good accessibility and serves as a point for embarkation and disembarkation. In this sense, P1 faces a higher demand for facilities from the visitors.

28.6.4 Human Impacts

The score ratings of four checkpoints are ranked from medium to low. In general, the human impact on Tung Ping Chau is not serious as indicated in Table 28.3. The high visitor numbers visiting sites, such as P1 and P3 cause a relatively higher level of human impacts among the four sites. This probably indicates that the human impacts are related to the number of visitors. Visually, graffiti and litter are the common problems associated with human impact on checkpoints studied in TPC.

Results of the empirical representation of four checkpoints P1, P2, P3, and P4 derived from the model {1} show that P1 has more potential as a destination for visitation as follows: {P1 = 11 + 9 + 0–3 = 17}; {P2 = 13 + 3 + 0–1 = 15}; {P3 = 13 + 4 + 0–2 = 15}; {P4 = 11 + 4 + 0–0 = 15}. Results in Table 28.4 provide a simple reliability analysis of the indicators. Although this appears to be the case, it might be that the experience of visitors reflects the cumulative effects of all four sites. The sites (P1, P2, P3, and P4) chosen for this study were investigated for similarity or dif-

Table 28.3 The checkpoint indicator result for the level of human impacts

	P1	P2	P3	P4
Litter	−1	−1	−1	0
Noise pollution	−1	0	0	0
Fire	0	0	0	0
Graffiti	−1	0	−1	0
Erosion of landforms	0	0	0	0
Total	−3	−1	−2	0

Table 28.4 Reliability analysis and Alpha Scale of Scores

Covariance Matrix:	
P1	0.263
P2	0.239
P3	0.239
P4	0.260
Sum of square	0.587
Degree of freedom	79
Mean square	0.245
F value	0.422
Relativity coefficient	0.899

ferences according to the indices tabulated in Tables 28.1, 28.2, and 28.3, and the attractiveness index derived. Although the reliability coefficient measured is consistent as Alpha value is 0.899 making it an acceptable significance level, a larger sample size of checkpoints and results would be more useful in authenticating indices of attractiveness.

28.7 Discussion and Conclusions

Although attractiveness (and indeed potential) of a place to serve as ecotourism and geotourism destination is a characteristic primarily determined by the physical environment, also referred to as the amenity, such attractiveness depends largely on the human response to that environment. As a result, therefore, results and outcomes of investigations reflect some subjectivity in interpretation. Investigations and evaluation of attractions, particularly for ecotourism and geotourism, are not a straightforward process as they involve some subjectivity and for ecotourism and geotourism to succeed and flourish, a site will have to provide a multidimensional array of experiences. Evaluation and assessment of natural areas for nature-based recreation, ecotourism, and geotourism tend to involve subjective as well as objective factors. Although the two may not be easily separated (Marion and Reid, 2007), Roome (1984) has underlined the importance of both factors in decision-making. If we need to designate and develop destinations to provide amenities for ecotourism, geotourism, and nature-based tourism, it is necessary to begin the task with evaluation and assessment of such sites for the potential to offer such opportunities. This is the task undertaken in this study.

While there is no standardized universal methodology for evaluating such potentials, subjective methodologies mentioned in the literature (Smith and Theberge 1987; Obua and Harding 1996; Priskin 2001) and partly applied in this work can be a starting point. It can then be further improved and modified to suit individual destinations. Simplicity is required to make the methods accessible to non-specialists in government, industry, and the public. The need for simplicity no doubt results in the exclusion of complex management science methods that have the potential of unraveling unique problems and their possible solutions. Where these are feasible and adaptable, they can be incorporated. These will be the next stage of the research undertaking. When well-conceived, defined, and implemented, the principles underlying such methods are not difficult and can easily be communicated. Specific reference to the model being used and justification of the underlying assumptions seems a reasonable demand (Janssen and Nijkamp 1985; Bosak et al., 2010).

As this study has attempted to assess natural areas for ecotourism and geotourism purposes, the four criteria used were clearly defined. The method used here is better than the other methods for reasons enumerated earlier although this is an assumption, and it might be different if other more complex analyses are involved. Evaluations should give more weight to ecological and ecosystem integrity where ecotourism is discussed, particularly where the site is a protected area or pristine natural environment. In addition, cultural, infrastructure, and the degree of human impact are similarly important where there is such evidence and deserve attention when ecotourism and geotourism as themes are involved. Given the acceptance of the adoptable definition that both ecotourism and geotourism represent sustainable forms of tourism, it might be better to have a final score for a site (for example: ecology and economic indicators) versus several other indicators which might include cultural and human impacts, etc. While some studies identified the measurement of criteria as consisting of simply measuring environmental variables, such as counting the number of rare species present (Smith and Theberge 1987) this study involved assigning scores to measure variables based on locational availability and evidence of severance or destruction of such amenities by human impacts.

As further evaluation and assessment are needed, in the wake of the rising demand for ecotourism, geotourism, and nature-based recreation sites especially in Hong Kong, simplicity is required to make the method discernible to decision-makers, business people, and the general public, a fact similarly observed by Smith and Theberge (1987); Priskin (2001) and indeed the USFS (2002) as highlighted earlier.

Although the methodology adopted in this study cannot be devoid of criticism for its simplicity, there is enormous potential for its improvement and possible universal applicability. The main features of the methodology used are numerous and can be identified in the following six points: (1) the methodology is based on subjective assessment, but can be enhanced by use of questionnaires to relate to visitor needs; (2) the method can be used specifically for evaluation of scenic, conservation, and landscape values of a given countryside destination; (3) it provides a framework for identification of criteria for further assessment; (4) the criteria identified can form a basis for planning and management of the available resources; (5) this method can further be developed by the use of geographical tools like remote sensing and Geographical Information System; and (6) finally, it is simple and easily interpreted by non-academic personnel saddled with decision-making responsibility.

However, for such methodologies to be effective and provide a lasting significant impact in decision-making, separate evaluations that will concentrate on ecological, geological, cultural, and human impacts will be most useful. This can be improved when such studies based on this, and an improved methodology is tested and applied elsewhere.

Lastly, the relative importance and potentials of a site are not easy to determine as this can vary (sometimes) significantly from person to person. A further study on this theme and at this site will be to survey the visitors by using questionnaires. This approach will further elucidate and improve the efficiency of the analysis as large number of variables will be considered.

References

Bastedo JD, Nelson JG, Theberge JB (1984) Ecological approach to resource survey and planning for environmentally significant areas: the ABC method. Environ Manag 8:125–134

Blamey RK (2001) Principles of ecotourism. In: Weaver DB (ed) The encyclopedia of ecotourism. CABI Pub, Wallingford

Bosak K, Boley B, Zaret K (2010) Deconstructing the 'crown of the continent': power, politics, and the process of creating National Geographic's Geotourism map guides. Tour Geogr 12(3):460–480

Boyd S (2001) Cultural and heritage tourism in Canada: opportunities, principles, and challenges. Tour Hosp Res 3(3):211–233

Buckley R (2003). Environmental Inputs and Outputs in Ecotourism: Geotourism with a Positive Triple Bottom Line? Journal of Ecotourism 2:76–82.

Burger J (2000) Landscapes, tourism, and conservation. Sci Total Environ 249:39–49

Cameron CM, Gatewood JB (2008) Beyond sun, sand and sea: the emergent tourism program in the Turks and Caicos Islands. J Herit Tour 3(1):55–73

Ceballos-Lascurain H (1988) The future of ecotourism. Mexico Journal 13–14

Chow WK, Ngar YN (2002a) Exploring Tung ping Chau. Cosmos Books Ltd., Friends of the Country Parks, Hong Kong

Chow WK, Ngar YN (2002b) Tung ping Chau eco-exploration. Cosmos Books Ltd., Friends of the Country Parks, Hong Kong

Cocklin C, Harte M, Hay J (1990) Resource assessment for recreation and tourism: a New Zealand example. Landsc Urban Plan 19:291–303

Curtin S (2005) Nature, wild animals and tourism: an experiential view. J Ecotour 4(1):1–15

Deng J, King B, Bauer T (2002) Evaluating natural attractions for tourism. Ann Tour Res 29(2):422–438

Donohoe HM, Needham RD (2006) Ecotourism: the evolving contemporary definition. J Ecotour 5(3):1–19
Dowling RK, Newsome D (2010) Geotourism. Routledge
Farsani NT, Coelho C, Cosat C (2011) Geotourism and Geoparks as novel strategies for socio-economic development in rural areas. Int J Tour Res 13:68–81
Fennell D (1999) Ecotourism: an introduction. Routledge, London
Fennell D (2002) Ecotourism program planning. CABI Pub, New York
Fung CKW, Jim CY (2015) Unraveling Hong Kong Geopark experience with visitor-employed photography method. Appl Geogr 62:301–313
Gulinck H, Vyverman N, Bouchout KV, Gobin A (2001) Landscape as framework for integrating local subsistence and ecotourism: a case study in Zimbabwe. Landsc Urban Plan 53:173–182
Janssen R, Nijkamp P (1985) A multi-criteria evaluation typology of environmental management problems. In: Haines Y, Chankong V (eds) Decision making with multiple objectives. Springer-Verlag, Berlin
Kiss A (2004) Is community-based ecotourism a good use of biodiversity conservation funds? Trends Ecol Evol 19(5):232–236
Koh YK, Oh KH, Youn ST, Kim HG (2014) Geodiversity and Geotourism utilization of islands: Gwanmae Island of South Korea. J Marine Island Cult 3:106–112
Lawton LJ, Weaver DB (2001) Modified spaces. In: Weaver DB (ed) The encyclopedia of ecotourism. CABI Publishing, Wallingford
Lowenthal D (1981) Conclusion: dilemmas of preservation. In: Lowental D, Binney M (eds) Our past before us: why do we save it? Temple Smith, London
Marafa LM, Fung T (2004) Changes in participation in leisure and outdoor recreation activities among people living in Hong Kong during SARS outbreak. World Leisure 46(2):38–47
Marafa LM, Ng SL (2008) Geology and Geo-Conservation as the Causal Aspect of Development: A Framework for Understanding. Paper presented in the Asia-Pacific Seminar on Geological Conservation and Sustainable Development 2008, organized by Geotechnical Division & Young Members Committee, The Hong Kong Institution of Engineers, Department of Geography & Resource Management, The Chinese University of Hong Kong, Association for Geo-conservation, Hong Kong
Mitchell B (1989) Geography and resources analysis. Longman, New York
Nelson G, Serafin R, Skibicki A (1997) Land use and decision-making for National Parks and protected areas. In: Nelson G, Serafin R (eds) National parks and protected areas: keystones to conservation and sustainable development. Springer, Berlin
Ng CN, Li Y (2000) Ecotourism in Hong Kong: Its Potentials and Limitations. Paper presented in Ecotourism Conference in The Dominican Republic 20–22 July 2000. Internet source : http://www.kiskeya-alternative.org/publica/diversos/hong-kong.htm
Obua J (1996) The potential, development and ecological impact of ecotourism in Kibale National Park, Uganda. J Environ Manag 50:27–38
Obua J, Harding DM (1996) Visitor characteristics and attitudes towards Kibale National Park, Uganda. Tour Manag 17(7):495–505
Page SJ, Dowling RK (2002) Ecotourism. Prentice Hall, Harlow
Parker S, Khare A (2005) Understanding success factors for ensuring sustainability in ecotourism development in southern Africa. J Ecotour 4(1):32–46
Patterson C (2002) The business of ecotourism, 2nd edn. Explorer's Guide Publishing, Rhinelander, WI
Phillips A (1997) Landscape approaches to National Parks and protected areas. In: Nelson G, Serafin R (eds) National Parks and protected areas: keystones to conservation and sustainable development. Springer, Berlin
Pigram JJ (1983) Outdoor recreation and resource management. St. Martin's Press, New York
Porter JF (2003) World heritage sites – what are they? Environmentalist 23:7–8
Priskin J (2001) Assessment of natural resources for nature-based tourism: the case of the central coast region of Western Australia. J Tour Manag 22:637–648
Priskin J (2003) Physical impacts of four-wheel drive related tourism and recreation in a semi-arid, natural coastal environment. Ocean Coast Manag 46(1–2):127–155
Roome NJ (1984) Evaluation in nature conservation decision-making. Environ Conserv 11:247–252
Rotariu I (2006) The new tourism communication: from 3S to 3E with case study on Romania. Unpublished Thesis
Serrano E, Gonzalez Trueba JJ (2011) Environmental education and landscape leisure. Geotourist map and Geomorphosites in the Picos De Europa National Park. GeoJ Tour Geosites 8(2):295–308
Sirakaya E, Teye, V, Sonmez S (2002) Understanding Residents' Support for Tourism Development in the Central Region of Ghana. J Travel Res 41:57–76
Smith PGR, Theberge JB (1986) A review of criteria for evaluating natural areas. Environ Manag 10(6):715–734
Smith PGR, Theberge JB (1987) Evaluating natural area using multiple criteria: theory and practice. Environ Manag 11:447–460

Taylor GD, Stanley D (1992) Tourism, sustainable development and the environment: an agenda for research. J Travel Res 31(1):66–67
USFS (2002) The US Forest Service visual management system (VMS) http://www.idwr.state.id.us/planpol/watplan/planning/litttle_salmon_draft/final_plan/chapter5.pdf
Valentine PS (1993) Ecotourism and nature conservation: a definition with some recent developments in Micronesia. Tour Manag 14(2):107–115
Wang L, Tian M, Wang L (2015) Geodiversity, geoconservation and Geotourism in Hong Kong Global Geopark of China. Proc Geol Assoc 126(3):426–437
Weaver D, Lawton L (2006) Tourism management, 3rd edn. John Wiley & Sons, Singapore
Wojtowicz B, Strachowka R, Stryz M (2011) The perspectives of the development of tourism in the areas of Geoparks in Poland. Procedia Soc Behav Sci 19:150–157
Wong F-Y (1997) Recreational use of Hong Kong Country Parks: An Analysis of Patterns and Demand. Unpublished Ph.D.Thesis, The University Hong Kong
Zwirn M, Pinsky M, Rahr G (2005) Angling ecotourism: issues, guidelines and experience from Kamchatka. J Ecotour 4(1):16–31

29 Climate Change Knowledge: Comparison of People's and Scientists' Perceptions in Western Himalayas

Pallavi V. Das

Abstract

Climate change in the Himalayan region is an important component of global climate change. Several studies have provided evidence for climate change in this region as well as insights into its dynamics. However, most of these studies have focused on the physical aspects of climate change paying relatively less attention to its social and economic dimensions. This chapter takes a historical approach to the examination of the socio-economic impacts of climate change. The geographical context of my study is the Shimla district of Himachal Pradesh in India. I focus on how apple farmers in this region have been experiencing climate change and responding to it over time. I also compare the apple farmers' knowledge of climate change to that of scientists in Himachal Pradesh. In making such a comparison, I shed light on the fact that lay people (e.g., farmers) and scientists who are the experts have actually similar views about climate change and adaptive strategies. This goes against the widely held view that they are different.

Keywords

Adaptive strategies · Apple farming communities · Climate change · Himalayas · Shimla

29.1 Introduction

It is an established fact that although climate change is a global phenomenon not all parts of the world and communities are equally impacted by it. Ecologically sensitive places are especially vulnerable to the impact of climate change (Baer 2012, White 2012, Marino and Ribbot 2012, Diemberger et al. 2012). Within these places, it is indeed men and women, who are poor in terms of income and/or assets as well as social-political connections[1] that are more vulnerable to the impact of climate change in the form of floods, droughts, and so forth (IPCC 2007). Therefore, in order to understand the impacts of climate change on vulnerable sections of society living in places undergoing drastic climate change and to mitigate these impacts, it is necessary to examine how these people experience, and respond to climate change. I call this knowledge of ordinary

[1] Social capital in academic terminology.

P. V. Das (✉)
Department of History, Lakehead University, Orillia, ON, Canada
e-mail: pvdas@lakeheadu.ca

B. Thakur et al. (eds.), *Resource Management, Sustainable Development and Governance*, Sustainable Development Goals Series, https://doi.org/10.1007/978-3-030-85839-1_29

people's perception which is shaped by their experience of climate change and which, in turn, shapes their response to climate change, as people's climate knowledge. This knowledge is built on their past experiences and adaptations to the changing climate. It is a part of the local knowledge or traditional knowledge system. My view of people's climate knowledge or local knowledge of climate change builds on the concept of local knowledge developed by Nakashima et al. (2012). It is defined as the understandings, skills, and philosophies developed by societies with long histories of interaction with their natural surroundings.

Existing literature on understanding climate change and developing adaptation strategies clearly highlights the fact that there is a need for adequate climate change knowledge to address the climate change crisis and that local or traditional knowledge is an important source of climate knowledge and adaptation (Usher 2000; Berkes and Berkes 2005; Salick and Ross 2009; Berkes and Davidson-Hunt 2010; Berrang-Ford et al. 2011; Bawa and Ingty 2012). These studies also underscore the perception that scientific knowledge by itself is inadequate and that it should be complemented by local knowledge so that adequate adaptation strategies can be devised (Huntington 2000; Bridges and McClatchey 2009; Byg and Salick 2009; Chaudhary and Bawa 2011). The underlying assumption is that both the systems of knowledge are different and that therefore their adaptation strategies are different. But are they really different? The aim of this chapter is to examine the above assumption by comparing the local climate change knowledge of apple farmers in the Western Himalayan state of Himachal Pradesh, India and that of the scientists.

Among the different geo-climatic environments in which the dynamics of climate change has been studied, the mountains are the most sensitive. Mountains, as a result of climate change, are hotspots of threatened loss of biodiversity (Kollmair et al. 2005). They cover 27% of the earth's land area and account for 17% of the world population. Moreover, 90% of the global mountain population lives in transition and developing countries (Swiss Agency for Development and Cooperation 2012). These mountain communities have long faced challenges from a range of social-ecological processes such as physical fragility, geographical accessibility, and relative political marginality. And the threat from these factors has only intensified due to current climate change (Jodha 1992). Also, people of mountainous regions in least developed countries who are poor are recognized to be among the most vulnerable to climate change globally (McDowell et al. 2013). One such region is the Himalayas. The Himalayas are referred to as the Water Tower of Asia as their glaciers and snow feed some of the largest river systems.[2] The Himalayas are experiencing elevation-dependent warming which means higher elevation areas are getting warmer than the plains (Chakravorty 2017). Rising temperatures and climate change are thus affecting the livelihood and survival of both the upstream and downstream communities in the region. Most studies on this region have focused on the physical dimensions of climate change such as glacier retreat, precipitation and temperature changes and their impact on the physical landscape[3] and biodiversity (Xu et al. 2009; Chaujar 2009; Chawla et al. 2012). However, there have been relatively fewer studies on the socio-economic dimensions of climate change in the Himalayan context.

In this chapter, I apply the term people's climate knowledge to the apple farmers' perceptions and understanding of climate change which, in turn, shape their response to the impacts of climate change, in the Shimla district of Himachal Pradesh in India. Further, this chapter presents apple farmers' climate change knowledge and compares it to that of the experts' or scientists' knowledge.

---

[2] For example, the Indus, Ganga, and Brahmaputra river systems.

[3] Such as floods.

Box 29.1 Sustainable Development Goals

As part of the Himalayan Mountain ecosystem, Himachal Pradesh is home to a wide range of natural resources. As the threat of climate change is now an established fact, it is not surprising that the apple orchards in Shimla of Himachal Pradesh are vanishing. This region of Himachal Pradesh has emerged as a climate hotspot and is the most affected both physically and from a socio-economic perspective. The author's research makes it evident that the Shimla region of Himachal is not only vulnerable to climate change as is expressed by changing weather patterns and rising temperatures but also as the same leaves the apple orchard farmers vulnerable to extremes of consequences. Global warming in Himachal has led to erratic rainfall, shifts in snowline and extreme weather events (including frequent flooding), in some places according to the state's action plan on climate change, while other regions could see an increase in the intensity of rainfall, coupled with storms. Thus, the impact of climate on agriculture and food security confirms the assumptions and fears of Agenda 2030 and its goals, targets, and indicators as has been displayed by SDG-13. As a result, the Shimla region of Himachal and its farmers are left with no choice but to consider adaptive strategies for sustainable and efficient management of water resources in the context of a growing economy, increasing population size, fast urbanization, and a relatively backward agriculture.

29.2 Study Area and Methodology

Himachal Pradesh is largely a mountainous state in the Western Himalayas of India with a population of 6.86 million. Compared to India's overall rural population percentage of 70%, in Himachal Pradesh 90% of the people live in rural areas and are engaged in agriculture, forest, and horticulture-based livelihoods. It is known as the "Horticultural State of India." Apple accounts for 50% of the area under horticultural production, 76% of the total fruit production and is the most important cash crop of the province (State Department of Horticulture Himachal Pradesh 2016a, b). Himachal Pradesh is divided into four agro-climatic zones.

- Zone 1 is low hills with altitude from 365 m to 914 m.[4]
- Zone 2 is mid hills with altitude from 915 m to 1523 m.[5]
- Zone 3 is high hills with altitude from 1524 m to 2742 m.[6]
- Zone 4 is high hills of cold and dry zone with altitude above 3000 m.[7]
- (State Department of Horticulture Himachal Pradesh 2016a, b).

According to the National Horticultural Board of India, apple can grow at an altitude between 1500 m and 2700 m above sea level. In recent years, the production of apple has been negatively affected by poor snowfall, unseasonal precipitation, and drought (IANS 2017).

In 1904, Samuel Evans Stokes, an American missionary introduced the commercial variety of apples largely grown in Himachal Pradesh today. Prior to cultivation of apples, subsistence crops, such as wheat, corn, millets, and barley were grown (Punjab Government 2012). It was in the 1950s that apple cultivation became widespread in the Shimla district as it was more remunerative than food grains for the farmers. Thus, there was a shift from subsistence farming to commercial farming in Himachal Pradesh in general. About 80–90% of apple cultivation is done by small and marginal farmers (landholding <1 hectare) who are almost completely dependent on apple cultivation for their livelihood. So, this makes them

[4] 1200 ft. to 2999 ft.
[5] 3000 ft. to 4999 ft.
[6] 5000 ft. to 8966 ft.
[7] Above 9000 ft.

not only vulnerable to the vagaries of the market, which work like forces of nature, but also to changing climatic conditions.

The area under apple cultivation in Himachal Pradesh grew from 400 hectares in 1950–51 to 88,560 hectares in 2005–2006 and to 110,679 hectares in 2015–2016. However, its productivity declined at the rate of 0.016 tons per hectare annually between 1985 and 2009 (Rana et al. 2012). More recently, between 2005 and 2014 apple productivity dropped by 0.18 tons per hectare every year. The overall drop in productivity was 9.4 tons per hectare in the past two decades (Sharma 2017).

Shimla district accounted for about 60% of total apple production of Himachal Pradesh in 2015–2016 (Government of India 2017). In this district, apples accounted for 98% of the total fruit production in 2005–2006. Thus, Shimla district is an ideal place to study climate change from the perspective of apple farmers whose apple cultivation is experiencing production-related difficulties due to adverse impacts of climate change.

I use a combination of primary and secondary data in this study. In the summers of 2015, 2016, and 2017, 70 semi-structured interviews were conducted with apple farmers in Rohru and Jubbal districts to examine how they perceive climate change. During the same period, scientists or experts were interviewed on their views on climate change. I also consulted scientific reports, meteorological data, newspaper accounts, and government publications to corroborate farmers' accounts of climate change.

29.3 People's Climate Change Knowledge

All apple farmers interviewed believed that climate change was taking place. In addition, farmers observed climate change specifically in terms of categories such as snowfall, rainfall, temperature, and hail.

29.3.1 Snowfall

Almost all the apple farmers interviewed said that there is a decrease in snowfall over time. One farmer noted that.

In the past 3–4 years we experienced good snowfall which is between 2 ft. and 3 ft. but not that much in quantity as we experienced 15–20 years ago which was 7–8 ft.

Another farmer who was in his 80s noted that.

> I experienced here 8 ft. to 9 ft. snowfall (30 years ago) and more importantly it was on time and our prediction never proved wrong. There are some festivals that we correlate with weather like on *Shivratri* (in February) we expect snowfall.

The farmers consider snow to be "white manure" for the fruit orchards: snow helps in meeting the minimum chilling requirement, also sustains the required level of moisture in the soil during summer when the fruit is maturing. The apple tree requires 400–1200 h of chilling with the average temperature at seven degrees Celsius or less, during dormancy and before flowering begins by March-end, so the quantity and timing of snowfall are important. According to the farmers, in the past (i.e., 15–20 years ago), the normal winter season, i.e., from December 15 to February 15, used to have heavy snowfall. Heavy snow in the early part of winter would last longer by not melting fast and would contribute to soil moisture and also act as fertilizer for the apple tree. However, farmers observed that now most of the heavy snowfall takes place in late February or March when it is warmer compared to December so the snow does not stay on the ground but melts away. Farmers felt that they would be happy with just average amount of snowfall provided it happened on time, i.e., in the early part of winter rather than later.

29.3.2 Rainfall

Farmers mainly noted that the timing of the rainfall has changed. In the past during the period of March to May, the pre-monsoon showers were scarce and light while there was good rainfall during the monsoon period from June to August. But nowadays it rains heavily in the March to May period, and there is less rain in the monsoon period. Because heavy rainfall is in the pre-monsoon period, farmers feel that monsoon is coming earlier every year. Thus, the rainfall pattern *has* altered. Rainfall has not only become irregular but unpredictable (Gautam et al. 2014). In the past both rain and snow had a fixed time but now their timings have changed.

29.3.3 Temperature

Most farmers noticed that there has been an overall rise in temperature compared to 30 years ago (the 1980s). In the past, even during the summer season, temperatures did not go beyond 20 °C, but nowadays the maximum temperature has reached 30 degrees Celsius. As one female farmer noted

> Yes, temperature is increasing as compared to 15–20 years ago. It is because of cutting of forest. I used to work in field for whole day but didn't feel as hot as it is today, it is impossible to sit in sunlight in summer season. We don't expect this kind of temperature in [the] Shimla region.

There is also the perception among farmers that the temperature distribution has undergone a significant shift in addition to an overall increase in temperatures. The hottest period of the year is no longer from 15 May to 15 June. It is now July. Because of the late snowfall in March and heavy pre-monsoon showers from March to May, those months have lower temperatures than usual.

29.3.4 Hail

A change in the occurrence of Hail[8] was another major indicator of climate change according to all the apple farmers interviewed in the region. They noted that:

> Yes there is definitely a change in hail storms. They have become more frequent as compared to the past 15–20 years. Now, hail fall occurs every week [during summer] that has adverse effect on our crops.

Another farmer observed that hail occurred once in 4–6 years but now they occur every year. Farmers also noted that hailstorms used to happen during a particular time in a year, for example, in November, and, that too, at higher elevations but now hail storms occur in March and at lower elevations as well. Now it occurs mostly during flowering (budding period), affecting their apple crop (both qualitatively and quantitatively). One farmer reported "Our crop has been affected by the hailstorms for the last 5 years consecutively." In response to the frequent hail fall, the Himachal Pradesh government installed anti-hail guns. But the farmers felt that these guns were not useful as their impact was limited to a small area. During my fieldwork trips, I noticed that some of the bigger apple orchard owners had covered their trees with hail resistant nets. However, the marginal farmers did not use anti-hail nets as they were expensive. Local newspapers, such as *Giriraj Weekly* during the mid-1980s discuss the increased frequency of hailstorms in the region in April and the government's promotion of hail resistant nets during that time. There were reports of hailstorms damaging apple production in 1998–1999, 1999–2000, 2004–2005, and more recently in 2015 (Sharma 2017).

29.4 Impact of Climate Change

Apple farmers keenly observe climatic changes especially those that affect apple production. Overall, farmers noticed changing climatic conditions discussed above including untimely rainfall, increase in temperature that adversely affected budding, flowering, fruit setting, and apple production. The fruit growth and development stages of apple-growing season are divided into four stages: dormancy or pre-flowering stage (January–March), flowering, fruit-set and fruit developmental stage (April–June), fruit develop-

[8] Locally known as Olaa.

mental stage (July–September), and post-harvest stage (October–December) (Bhattacharya et al. 2018). As one farmer remarked:

> Now the stages of apple formation, such as budding, flowering and fruit setting have changed completely because of climate change. Sometimes, it is early and sometimes, it is late. Earlier apple flowering time was in the month of late March and early April but now it has changed to late April and early May. This is because snowfall nowadays is taking place even as late as end of March and early April, whereas in the past it used to fall on 15 December to 15 February period. The flowering is uneven; sometimes it is timely but mostly later or earlier and causing less production.

Farmers noted that shorter winters or delay in snowfall in December and January lead to fewer chilling hours and affect apple production.

Another important observation by the farmers of the impact of climate change on apple production is that in the past the apple tree used to bear fruits within 5–6 years of its planting, but now it takes 12–15 years to become fruit-bearing. Farmers attributed it to the lack of soil moisture caused by uneven/untimely rainfall, higher temperatures, and drought like conditions. According to one farmer:

> In the past even if it didn't rain for 2 months the soil had enough moisture to sustain plant growth but nowadays drought like conditions start to appear within a week of no rainfall.

Apart from climate change directly affecting apple productivity, farmers reported an increase in existing diseases and the occurrence of new diseases, in apple trees due to climate change. This, in turn, impacts apple quality and production. Some diseases, such as Canker[9] have increased in the month of March and April due to heavy snowfall. Insects, such as borers, wooly apple aphid as well as red mites have increased due to heavy rainfall. These affect the size and color of fruits. Higher temperatures have resulted in an increase in borer insect which affects the roots of the apple tree. The increased incidence of the above noted diseases has led to a higher frequency of pesticide spraying by farmers when compared to in the past. This has affected the quality of apples and its storage time. As one farmer noted "Previously apple could be stored up to 8–10 months at room temperature after harvesting. But, nowadays, it can be stored for 2 months only." Farmers also recalled that scab[10] disease was quite prevalent about 30 years ago. This corresponds to the reports in *GiriRaj Weekly* of increased incidence of Scab disease in apples in the 1980s that led to the Horticulture Department of the government promoting the use of pesticides. Similarly, in 1995 Marssonina leaf blotch disease[11] affected the apple trees due to heavy rainfall and high temperatures.

29.5 Farmers Understanding and Response to Climate Change

Most farmers attributed the change in climatic conditions to human activities like deforestation taking place locally. They also attributed it to local phenomena, such as urbanization, industrialization, soil, and water pollution (due to increased use of pesticides and fertilizers) and increase in population.[12] The apple farmers are not aware of climate change as a global environmental phenomenon.

Many farmers believe that they cannot do anything to mitigate climate change as it is in the hands of God. They have accepted climate change as fait accompli: that it is inevitable and that they cannot change it, just as they are helpless about changes in apple prices. Yet, those who have the resources are coping with climate

[9] Canker is a fungal disease which attacks the bark of apple trees causing a sunken area of dead bark and eventually the death of the branch.

[10] Scab is a fungal disease in apple trees that causes dark patches on the fruits and leaves making the fruit non-marketable.

[11] Marssonina leaf blotch is a fungal disease where the leaves show blotches and over time cause defoliation of the apple trees.

[12] It will be interesting to see if Himachal Pradesh experiences higher rate of population increase, urbanization, industrialisation, etc. than India since the 1980s.

change in two ways: (1) Technological fix and (2) Geographical fix

1. Some apple farmers believe in technological fixes as responses to the impact of climate change. These include the use of fertilizers to improve productivity, the use of pesticides to get rid of diseases, and the use of hail nets to stop hail damage to the apple crop. However, a field agronomist of a fertilizer company pointed out, for the fertilizer to work a lot of water is required, otherwise it just ends up in the soil. Rainfall is unpredictable, droughts are recurrent in the area, so the use of fertilizers cannot really help improve apple productivity. Moreover, although fertilizers have improved productivity in the short term for some farmers, they have damaged the fertility and soil quality in the long term (Partap and Partap 2009). Increased use of pesticides in the long term has led to pests developing resistance to the pesticide. This also resulted in the decline of bees and other beneficial insects affecting the pollination of apple flowers and therefore apple production. Anti-hail nets are another technological fix to deal with the frequent hailstorms that the region is experiencing due to climate change. However, anti-hail nets lead to increase in temperature and, therefore, have increased the incidence of mites in apple trees for which the farmers have to use mite sprays. Poor farmers cannot afford anti-hail nets. Yet, since the apple farmers are dependent on a single crop for their livelihood, they are forced to buy the anti-hail nets. The demand for anti-hail nets has led to 100% increase in their sales between 1990 and 2017, and it has emerged as a profitable commodity for anti-hail net manufacturers and sellers. In fact, in 2017, there was 50–60% shortage in supply of anti-hail nets in the Shimla district. Many of the anti-hail net stores said that farmers were placing orders for the nets much in advance so that these would be available and be installed at the right time. Similarly, there have been increased sales of fertilizers, pesticides, and insecticides produced by local as well as multinational corporations. Thus, the adverse impact of climate change on apple productivity has created opportunities for corporations to profit from the climate crisis (Fletcher 2012). In the future, as climate change intensifies its impact on apple productivity, other technological fixes, such as disease-resistant varieties of apple, high density plantations, early yielding varieties of apple, may become sources of profit for biotech corporations. In a market society, every problem is a business opportunity. So, the technological fix for the people is also a commercial fix for business.
2. Farmers are resorting to a geographical fix in moving apple cultivation from lower hills to higher hills, as more moisture is retained there due to more snowfall and lower temperatures. In fact, the president of the Himachal Seb Utpadak Sangh,[13] Mr. Rakesh Singha confirmed that the geographical shift in the apple belt to upper elevation had indeed taken place. During an interview he gave me, he said:

   In the 1950s, elevation of 5000 ft. to 6000 ft. above sea level was best for apple cultivation but by 2000s the suitable elevation for apple orchards is 7000 ft. to 8000 ft.

As temperature rises, and with low, uncertain and untimely snowfall as well as rainfall, apple farming has indeed shifted to higher altitude areas. Those who can afford to buy additional land in higher altitudes are able to shift some of their cultivation there. Most of the farmers cannot do this as they have already invested a lot in their existing apple orchards. The farmers in the previously apple-growing low altitude areas, such as Kullu have adopted crop diversification as a response to climate change. In other words, they have started growing other fruit crops, such as pomegranate, kiwi, and vegetables as apple farming is not viable anymore. However, given climate change and uncertain climatic conditions, we do not know how long the crop diversification in the lower altitudes is going to last in the long term because climate change may affect the other crops as well (Partap and Partap 2009). Also, with climate change the question arises as to

[13]The Himachal Apple Growers Association.

what height will the apple farming belt finally shift to, beyond which, it will not be cultivable any more. High altitude (around 10,000 ft) apple farming in the Kinnaur district, about 235 km from Shimla, is also experiencing the impact of climate change that is affecting apple production there (Newsgram 2017). Given that in the mountainous terrain there is relatively less land available for cultivation,[14] many of the small and marginal farmers have expanded their apple orchards into state-owned forest lands which are considered illegal. This has brought them into conflict with forest officials who have started evicting the farmers from the encroached lands and have cut the apple trees growing there (Himachal Watcher 2016). There is a state-led attempt at what Harvey (2007) would call accumulation by dispossession, i.e., removing farmers from state-owned land.

29.6 Scientists' Climate Change Knowledge

Scientists were interviewed at Himachal Pradesh Horticultural University at Nauni, Department of Horticulture, Regional Research Centre at Mashobra and State Center of Climate Change for their views on climate change and adaptation to the impacts of climate change. The scientists had some of the same indicators as the apple farmers for climate change, such as variability of rainfall pattern or distribution, alteration in snowfall pattern, decrease in snowfall amount, etc. However, unlike the farmers, scientists pointed to extra-local phenomena, such as glacial retreat, population decline in some species of animals, birds, and trees, rise in sea levels, and increase in extreme events, such as droughts and floods. They attributed these extra-local changes to global warming caused by increased greenhouse gases in the atmosphere, deforestation, and urbanization around the world.

Scientists recognized that apple farmers were mainly affected by the presence of extreme weather events like heavy rainfall events, increased hail fall, reduced snowfall, change in snowfall pattern, and the reduction in the chilling requirements due to rise in temperature and reduced snowfall. Dr. S. S. Randhawa, Senior Scientific Officer at the State Council for Science, Technology and Environment agrees that heavy snowfall in December and January has declined while a lot of snowfall takes place in February and March (Newsgram 2017). Horticultural scientists have also noted that from 1973 to 1990, the average snowfall was 190 cm but in the last decade it was 95 cm (Dogra 2009). Scientists have also observed that the winter temperatures have increased by 2–3° centigrade from 2000 to 2007 when compared to winter temperatures between 1973 and 1990 (Dogra 2009). A scientific report by the State Centre on Climate Change also confirmed an increasing trend in pre-monsoon rain in the Shimla district from 1901 to 2012 (Rana and Randhawa 2013). The Meteorological Department noted that although rainfall is spread out over a longer period in a year, the total rainfall amount has decreased (Sharma 2017). Decadal analyses of snowfall pattern by the Regional Meteorological Department in Shimla for the period from 1977 to 2013 show that both total amount of snowfall and total number of days of snowfall have decreased in the last decade as compared to the previous two decades (Singh et al. 2014). Local newspaper *Amar Ujaala* reported heavy snowfall in March 2011, 2012 in Shimla. According to the Meteorological Department, this happened after 5 years confirming that snowfall in March is indeed not the norm but was happening frequently in the past decade.

Apart from the above observations of climate change in the Shimla district, scientists also noticed the impact of these changes on apple productivity. In their studies they observed that chilling hours of less than 1000 in apple trees lead to erratic flowering and poor fruit formation (Jindal and Mankotia 2004). Horticultural scientists have also confirmed that overall apple cultivation in Himachal Pradesh has shifted from lower altitude regions experiencing warmer temperatures, such as Kullu valley to higher altitude regions, such as Lahaul and Spiti which used to be cold deserts in the past (Rana et al. 2012). Scientists have noted that for every one degree

[14] Only 10% of the total land in Himachal Pradesh is available for cultivation.

rise of temperature, apple cultivation shifts upward by about 984 ft (Dogra 2009). The incidence of diseases in apples has increased and become widespread as in 1983, the scientists recalled, when Apple Scab was an epidemic disease due to drought. In addition, scientists also noted that some new diseases have emerged such as Apple chlorotic leaf virus, Apple mosaic, Apple canker due to climatic changes of temperature and moisture.

About the mitigation of the impact of climate change, scientists argued that at the global level it can be reduced by adopting several measures like controlling the emissions of GHGs, improving micro-climatic environs by adopting afforestation measures, increasing forest cover which will act as carbon sinks, conserving water resources, creating awareness about climate changes, adopting various renewal energy sources, etc. At the local level, scientists stated that farmers should adopt disease-resistant and climate change-resistant varieties of apple trees. New varieties have actually been developed in India through the use of biotechnology. These varieties of apples require less chilling hours and can withstand higher temperatures and have been already introduced to farmers. Thus, according to scientists, technology will help the apple farmers to adapt to climate change. Also, in lower altitude areas, other crops, such as pomegranates, kiwi, etc. are being promoted among farmers so that they can adapt to the impacts of climate change.

29.7 Conclusion

All the changes in climate observed by the farmers are closely tied to the cycle of the apple tree. This is not surprising given that apple cultivation is their main livelihood. Since the natural cycle of precipitation and temperature has been affected by climate change, it has become increasingly out of sync with the natural cycle of the apple tree. Also, apple cultivation which replaced food grain cultivation has made the farmers dependent on the market for their subsistence: they have to sell apples to buy their farm inputs (apple seedlings, fertilizers, etc.) and necessities, including food. So any fluctuation in apple production apart from increase in input prices and output prices can impact their ability to sell apples for a profit and earn their living. Sudden changes in weather can contribute to fluctuations in apple sale, and thus in farmers' income. The farmers' reference point for discerning climate change are the climatic conditions that existed in the region two to three decades ago. Earlier conditions were ideal for apple cultivation and for the production of good quality apples. Overall, the apple farmers' accounts of climate change, objectively rooted in their struggle to earn a living, match with the trends of average temperature, precipitation based on meteorological records, scientific reports, and newspaper accounts. In fact, farmers' perception of climate change complements meteorological data by providing details that are relevant to their livelihood, such as timing of rainfall and snowfall seasons that influence the health of apples and productivity of apple farms (Marin and Berkes 2013). Apple farmers' observations of climate change in the Shimla district correspond with those observed by the horticultural scientists and other experts.

While much of the existing literature on traditional knowledge and scientific knowledge points to fundamental contradiction between them, my research unpacks fundamental similarities between these forms of knowledge. Apple farmers' view of climate change is similar to the experts' view. This is because their views are rooted in practice. Here, practice refers to the empirical examination of climate data by scientists and the real-world experience of climate change by farmers. Now, my thesis is that when our views of nature or of society's interaction with nature are rooted in practice, and not in irrational thinking, scientists and lay people are bound to have overlapping views especially at the local level. However, farmers do not seem to be aware of the extra-local instances of climate change nor are they aware of the deeper reasons, or underlying mechanisms for climate change itself.

As a result, farmers' micro-scale adaptation to climate change seeks to resort to two kinds of fixes that the experts and government people *also* recommend: a geographical and a technological fix. The geographical fix involves moving apple cultivation from lower hills to higher ones, while the technological fix involves using technologically manipulated new variety of apple trees. Thus, common people's and scientists' views on what can be done to cope with climate change are similarly limited. Existing literature on local or traditional climate change knowledge emphasizes that traditional adaptive strategies are useful and should be incorporated into present day climate change adaptive strategies. While common people's climate knowledge certainly provides a nuanced understanding of climate change at the local level and complements scientific knowledge, however it has limitations in terms of adaptive strategies to address current climate change. This is because contemporary climate change is not only unprecedented, unpredictable but is also happening fast. There are also limits to what scientists are suggesting in terms of adaptive strategies, such as technological fix and geographical fix.[15] As I have said not everyone can buy expensive technology or a lot of land in the upper altitudes to cope with the crisis of apple farming in the lower altitudes caused by climate change. Mountain environment provides ample and unique opportunities for studying climate change with its climatically sensitive physical conditions in relation to socio-economic processes (IPCC 2007).

Acknowledgement This research was funded by an Insight Development Grant of the Social Sciences and Humanities Research Council (SSHRC), Canada. A version of the chapter has been published in *GeoJournal* https://doi.org/10.1007/s10708-021-10371-z

References

Baer H (2012) Global capitalism and climate change: the need for an alternative world-system. Altamira Press, Lanham

Bawa KS, Ingty T (2012) Climate change in Sikkim: a synthesis. In: Arrawatia ML, Tambe S (eds) Climate change in Sikkim: patterns. Impacts Initiatives, Information and Public Relations Department, Government of Sikkim, Gangtok, pp 413–424

Berkes F, Berkes MK (2005) Ecological complexity, fuzzy logic, and holism in indigenous knowledge. Futures 41:6–12

Berkes F, Davidson-Hunt IJ (2010) Innovating through commons use: community-based enterprises. Int J Commons 4:1–7

Berrang-Ford L, Ford JD, Paterson J (2011) Are we adapting to climate change? Glob Environ Change 21:25–33

Bhattacharya B, Gupta P, Sharma VK (2018) Impact of varying climatic conditions on apple, a case study of Shimla District, Himachal Pradesh. Int J Appl Pure Sci Agric 4(2):16–26

Bridges KW, McClatchey WC (2009) Living on the margin: ethnoecological insights from Marshall Islanders at Rongelap Atoll. Glob Environ Chang 19:140–146

Byg A, Salick J (2009) Local perspectives on a global phenomenon — climate change in eastern Tibetan villages. Glob Environ Chang 19:156–166

Chakravorty A (2017) Climate change: water crisis engulfs World's 'highest' village https://indianexpress.com/article/india/climate-change-global-warming-water-crisis-himachal-pradesh-spiti-mountain-valley-virbhadra-singh-himachal-pradesh-assembly-elections-himachal-farmers-4939235/. Accessed 20 Nov 2018

Chaudhary P, Bawa KS (2011) Local perceptions of climate change validated by scientific evidence in the Himalayas. Biol Lett 7:767–770

Chaujar RK (2009) Climate change and its impact on the Himalayan glaciers - a case study on the Chorabari glacier, Garhwal Himalaya, India. Curr Sci (00113891) 96(5):703–708. Academic Search Premier, EBSCOhost Accessed 16 Jan 2014

Chawla A et al (2012) Long-term ecological and biodiversity monitoring in the Western Himalaya using satellite remote sensing. Current Science (00113891) 102(8):1142–1156. Academic Search Premier, EBSCOhost Accessed 16 Jan 2014

Diemberger H et al (2012) Communicating climate knowledge: proxies, processes, politics. Curr Anthropol 53(2):226–244

Dogra CS (2009) The worm has turned: climate change has driven apple cultivation from Mid-Heights to the higher reaches https://www.outlookindia.com/magazine/story/the-worm-has-turned/240619. Accessed 14 June 2018

Fletcher R (2012) Capitalizing on chaos: climate change and disaster capitalism ephemera. Ephemera Theory Polit Org 12(1/2):97–112

Gautam HR, Sharma IM, Kumar R (2014) Climate change is affecting apple cultivation in Himachal Pradesh. Curr Sci 106(4):498–499

Government of India (2017) Horticultural statistics at a glance 2017. Controller of Publication, New Delhi

Harvey D (2007) Neoliberalism as creative destruction. Ann Am Acad Polit Soc Sci 610:22–44

Himachal Watcher (2016) Himachal small farmers reach secretariat to stage 24 hour protest to stop eviction, Claim Community Forest Rights. https://himachalwatchercom/2016/05/04/himachals-small-

[15] Which some farmers are already adopting.

farmers-reach-secretariat-to-stage-24-hour-protest-to-stop-eviction-claim-community-forest-rights. Accessed 10 Jan 2017

Huntington HP (2000) Using traditional ecological knowledge in science: methods and applications. Ecol Appl 10:1270–1274

IANS (2017) How climate change shifts apple cultivation in Himachal Pradesh. http://zeenews.india.com/environment/how-climate-change-shifts-apple-cultivation-in-himachal-pradesh-2019048.html. Accessed 10 Aug 2017

IPCC (2007) Climate change 2007: impacts, adaptation and vulnerability. In: Parry ML, Canziani OF, Palutikof JP, van der Linden PJ, Hanson CE (eds) Contribution of working group ii to the fourth assessment report of the intergovernmental panel on climate change, Cambridge

Jindal KK, Mankotia MS (2004) Impact of changing climatic conditions on chilling units, physiological attributes and productivity of apple in Western Himalayas. Acta Horticul 662:111–117

Jodha NS (1992) Mountain perspective and sustainability: a framework for development strategy. In: Jodha NS, Banskota M, Partap T (eds) Sustainable Mountain agriculture: perspectives and issues volume 1. ICIMOD, New Delhi, pp 41–82

Kollmair M et al (2005) Mountains: special places to be protected? An analysis of worldwide nature conservation efforts in mountains. Int J Biodivers Sci Manag 1:1–9

Marin A, Berkes F (2013) Local People's accounts of climate change: to what extent are they influenced by the media? WIREs Clim Chang 4:1–8

Marino E, Ribbot J (2012) Editorial. Glob Environ Chang 22:323–328

McDowell G et al (2013) Climate-related hydrological change and human vulnerability in Remote Mountain regions: a case study from Khumbu, Nepal. Reg Environ Change 13:299–310

Nakashima DJ et al. (2012) Weathering uncertainty: traditional knowledge for climate change assessment and adaptation. UNESCO and United Nations University, Paris and Darwin, Australia

Newsgram (2017) Climate Change in Himachal Pradesh: How Are Kinnaur's Apples Being Affected by It? https://www.newsgram.com/climate-change-in-himachal-pradesh-how-are-kinnaurs-apples-being-affected-by-it. Accessed 14 July 2017

Partap B, Partap T (2009) Climate change impact on hill agriculture and farmers' adaptive strategies: a case study of Kullu Valley in Himachal Pradesh. http://wwwindiaenvironmentportalorgin/files/Climate%20Change%20Impactpdf. Accessed 15 July 2017

Punjab Government (2012) Gazetteer of the Shimla District 1904. Reprint B.R. Publishing Company, Delhi, p 58

Rana I, Randhawa SS (2013) Changing trends of monsoon in Himachal Pradesh. State Centre on Climate Change, Shimla

Rana RS et al. (2012) The impact of climate change on a shift of the Apple Belt in Himachal Pradesh. In: Dubash NK, (ed) Handbook of climate change and India: development. Politics and Governance, Earthscan, Abingdon

Salick J, Ross N (2009) Traditional peoples and climate change. Glob Environ Chang 19:137–139

Sharma D (2017) Himachal farmers start growing pomegranate and cabbage, As Apple Lines Recede Due to Climate Change https://wwwdowntoearthorgin/news/climate-change/himachal-farmers-start-growing-pomegranate-and-cabbage-as-apple-lines-recede-due-to-climate-change-59244. Accessed 4 July 2017

Singh M, Wason GS, Bhan SC (2014) Trends of snowfall over catchments of Sutlej River in Himachal Pradesh. Presented at the National Conference on Glaciology, October 30–31, 2014

State Department of Horticulture Government of Himachal Pradesh (2016a) Overview. http://wwwhpagrisnetgovin/hpagris/Horticulture/Defaultaspx?SiteID=5&PageID=1034 &Language=En. Accessed 5 Aug 2017

State Department of Horticulture Government of Himachal Pradesh (2016b) Horticulture at a glance. http://wwwhpagrisnetgovin/hpagris/Horticulture/Defaultaspx?SiteID=5&PageID=1219. Accessed 10 Sept 2018

Swiss Agency for Development and Cooperation (2012) Why mountains matter in global sustainable development. file:///E:/Downloads/Why%20mountains%20matter%20in%20global%20sustainable%20 development%20(3).PDF. Accessed 4 Dec 2018

Usher PJ (2000) Traditional ecological knowledge in environmental assessment and management. Arctic 53:183–193

White S (2012) History and Climate change. Perspectives on History 50(7). https://www.historians.org/publications-and-directories/perspectives-on-history/october-2012/historians-and-climate-change

Xu J et al (2009) The melting Himalayas: cascading effects of climate change on water, biodiversity, and livelihoods. Conserv Biol 23:520–530

Check for updates

30 Climate Change Modeling for Ecosystem Services in Western Himalayan District: A Methodological Approach Towards Sustainability Science

Swarnima Singh

Abstract

The term Ecosystem Services (ES) concerned around profits offered by ecosystem functioning to the humanity for well-being. The changing global environmental condition, variable anthropogenic drivers, and ecological conditions impact the delivery of ecosystem services. These services include provisioning services, regulating, cultural, and supporting services. The purpose of the study pivots around changing climate impact quantification of ecosystem services in Kangra district. The meteorological measurements across western Himalaya are examined and analyzed for the past 43 years since 1970. Visible increase in temperature trends with a substantial variation during different seasons over the past quartile period have been noted. North-eastern and south-western part of the district depicts much significant variation in the mean minimum and mean maximum. The differential decadal and annual trend exhibit inconsistent signals of cooling in the high-altitude north-eastern block in the district as compared to other parts in the region that is affecting the quantity and quality of ecosystem services. Atmospheric Infrared Sounder (AIRS) satellite, Tropical Rainfall Monitoring Mission (TRMM), Providing Regional Climates for Impacts Studies (PRECIS) data together have been considered to evaluate ES status with changing climate.

Keywords

AIRS · Climate change · Composite capital vulnerability index · Downscaling · Ecosystem services · PRECIS · TRMM

30.1 Introduction

The extent and frequency of extreme climate related events are on continuous rise throughout India and the world. The continuous and persistent impact of climatic variability on the flora, fauna, and crops are much evident therefore, it could not be called as a myth. Consequently, it has awakened even the non-believers though the amplified vulnerability and their outcomes on food security (GCOS 2003; UNISDR 2007; Lal et al. 2011; Kumar et al 2006; DFID 2000). However, the increasing signals of climate variability in the cold and dry regions of Himachal Pradesh like the aftermath of June 2013 and 2014 floods and subsequent crop loss provided enormous indications regarding agricultural and hor-

S. Singh (✉)
Department of Geography, Deen Dayal Upadhyay Gorakhpur University, Gorakhpur, India

B. Thakur et al. (eds.), *Resource Management, Sustainable Development and Governance*, Sustainable Development Goals Series, https://doi.org/10.1007/978-3-030-85839-1_30

ticultural production losses in the state. These conditions are not just jeopardizing the crops but are also pushing incessant pressure on the marginal and small farming communities for out-migration. Indeed, it would lead to critical socio-economic implications to the future of the state. The current trends in GHGs emission and potential agricultural losses have induced alarm on livelihood sustainability (Fischer et al. 2002). Subsequently, the quantification of climate change impact on each Agro Ecological Zone (AEZ) has become the need of the hour, but since the scenario modeling capacity is still in elementary stage whenever a small region is concern, the potential impact of climate change might provide a general overview. Despite these limitations in the methodology it is still feasible to compute the plausible impacts of changing climate on the AEZs and blocks under different climate change scenarios with the help of integrated approach; combination of remote sensing, data archives, primary survey, public response, governmental and non-governmental information to study impact of climate change on ecosystem services at microlevel with the help of deductive and empirical methodologies (Singh and Singh 2011, 2014, 2016a, b).

The examined arrangement of three sets of potential indicators of the vulnerability of ecosystems services: (1) existing and observed changes based on past records to depict changing climate of twentieth century, (2) projected twenty-first century vegetation changes using the MC1 dynamic global vegetation model under two Intergovernmental Panel on Climate Change (IPCC) emissions scenarios, and (3) overlap of results from (1) and (2). The confidence levels for vegetation projections and observations related to changing climate, the estimation of probability density functions have been performed to classify areas into ecosystem services vulnerability classes based on capital assessment for treatment of uncertainty. The philosophy present is key to the past by James Hutton (Gould 1965) and is very much pertinent even in the case of climate change before outlining impact, adaptation, and mitigation strategies. The present climate variability needs a lucid understanding for adaptation and acclimatization even for the future investigation of climate change. The spatio-temporal variability (minimum and maximum) of precipitation and temperature on all time scales as intra seasonal, inter annual, and decadal etc., are needed to be examined for impact analysis through standard deviations and statistics of extremes (Bhutiyani et al. 2007; Bhutiyani et al. 2009; GCOS 2011; Singh and Singh 2011, 2014, 2016a, b). The continuous warming trend in the western part of the district and changing climate has been projected to analyze the increasing number of extreme temperature and rainfall events as already been explained previously, hence climatic variability is estimated to exhibit an upward trend in the entire region. The Climate Change Ecosystem Services Vulnerability Index (ESVI) has been developed for the impacted ecosystem capital services in the region to investigate how these changes will alter Kangra's future ecosystem services sustainability.

Box 30.1 Sustainable Development Goals

Ecosystem services which include provisioning (food, feed, fuel, and fiber), regulating (the role of an environment in maintaining larger ecological systems), cultural (the recreational, spiritual, and aesthetic services), and support (for example, as a sink for pollutants) services are all very crucial for human well-being. Both the UN-SDGs and the governance framework within India recognize that the key to sustainable development lies in achieving a balance between the exploitation of natural resources for socio-economic development and conserving ecosystem services. In this context the author studies the Kangra Himalayas of Himachal Pradesh in India using among others, climate change modeling (PRECIS) to find out change in climate over time and its impact on ecosystem services. Her study reveals an average decline in precipitation and increase in winter temperature which resonates with the general conclusions of IPCC or intergovernmental panel on climate change.

Building on this context, the author recommends that the challenges of vulnerable ecosystem services in Kangra can be managed by flexibility and adaptability as circumstances and conditions change. This requires that the local people self-regulate different uses and respond to incentives which demonstrate incomes or clear livelihood benefits, to support sustainable management of ecosystem services.

30.2 Materials and Methods

The sample units were scattered over nine different blocks (Indora, Fatehpur, Paragpur, Dehra Gopipur, Kangra, Nagrota Bagwan, Baijnath, Dharamsala, and Multhan) selected from each Agro Ecological Zones (AEZs) and in total 27 villages have been chosen (on an average 3 village from each block) based on total numbers of households (270), variable temperature and precipitation records agricultural production and accessibility, etc. Out of total 15 development block 9 blocks have been selected purposively based on different AEZs. There are altogether 8 AEZs fall in the district so from each AEZ the block has been selected like AEZ 1.1 (Indora, Fatehpur, Paragpur, Dehra), AEZ 1.2 (Dehra, Kangra, Nagrota), AEZ 2.2, 3.2 and 4.1 and 4.2 (Dharamsala), and AEZ 4.2 and 4.3 (Baijnath). It has been kept in mind that there should be at least one representative block from each AEZ. From each 9 block, numbers of Gram Panchayats (GPs) have been identified to select at least three villages having largest population and smallest population. Ten households were interviewed from each village (9 blocks, 3 village from each block and 10 households from each village means 27*10 = 270). Three blocks from the high hills wet sub-temperate (Dharamsala and Multhan) including 3 villages from each with 10 households, 3 blocks with 9 villages from valley region and Pong dam, 2 blocks with 6 village from Changar region, and one block with 3 village from Shivalik foot hill region. The valley region comprises three valley areas; Palam (include Baijnath), Kangra, and Nurpur. Changar region of the district comprises the areas falling between these valleys extending up to Bias basin. Grid sampling was used to select the households. Points, 1 km apart were drawn on the whole study transect. Point coordinates were saved on a handheld GPS receiver. The receiver was used to measure distance from the point when approaching a point.

30.2.1 Baseline Data Requirement

The mean monthly maximum and minimum temperature and precipitation together with annual minima and maxima from networks of 7 stations, for the period of 1970–2013 have been compiled because these are fundamental instrumental statistics for the analysis of climate dynamics in the micro-study area (Fig. 30.1). Primarily, the India Meteorological Department (IMD) has been considered with its monthly weather records that have been updated for the period of 1970–2013 with the help of downscaling Atmospheric Infrared Sounder (AIRS) satellite, TRMM, and PRECIS data to analyze the gap in IMD dataset. Consequently, several checks have been made for missing values in the previous data set to distinguish the regular pattern of regional precipitation and temperature variation in the district. The annual and seasonal temperature and precipitation progression for the winter months (December–January-February), pre-monsoon (March–April–May), monsoon (June–July–August–September), and post-monsoon (October–November) months have been calibrated. The ground station data has been acquired from the CSK University, Palampur in Kangra district and the satellite data have been acquired from AIRS, TRMM from National Aeronautics and Space Administration (NASA), and PRECIS from Climate Research Unit (CRU), United Kingdom.

30.2.2 Database and Methodology for Climate Change Modeling

The climate change modeling has been done based on 1970–2013 baselines to find out a defi-

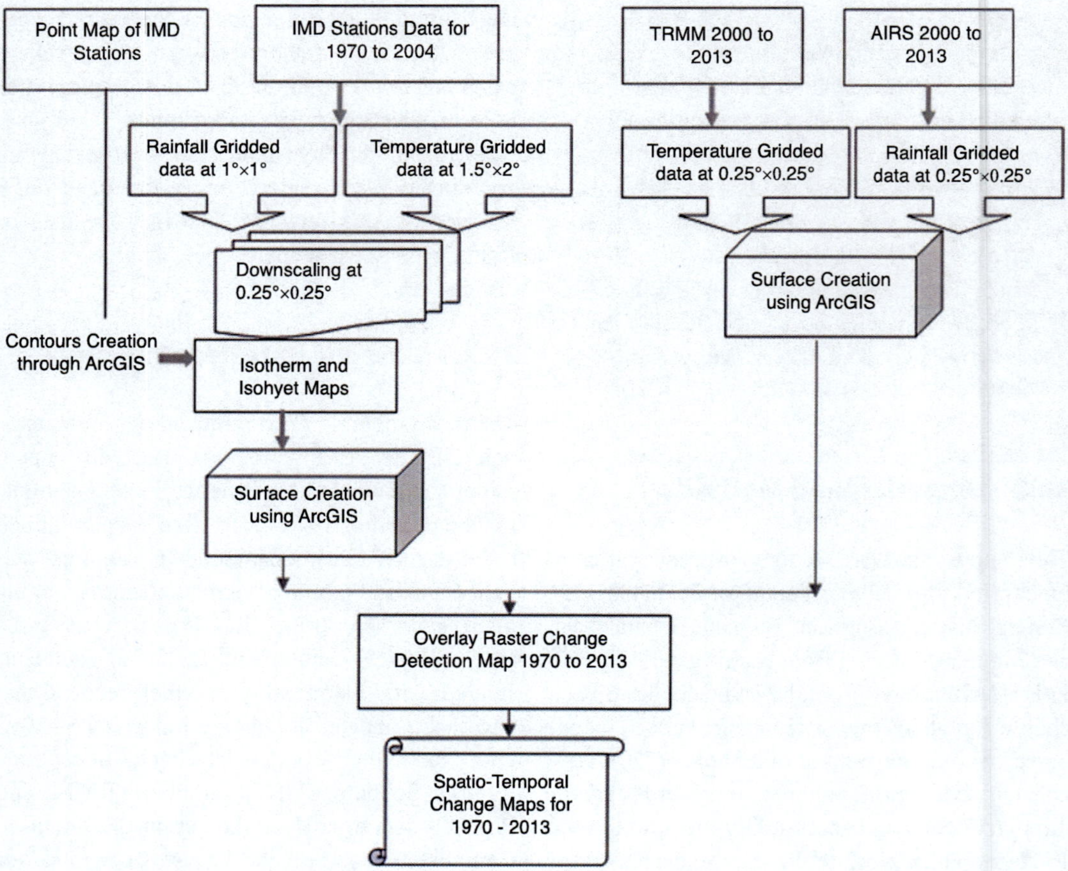

Fig. 30.1 Methodological Framework for Preparing Spatio-Temporal Maps

nite consciousness from infinite number of equally credible changes. Basically, it should be consistent with current understanding of real calibrations without assuming that the future will resemble the past and the present. PRECIS based on Hadley Regional Climate Model (HadRM) downscaling methodology (shepherd method) data have been used on IMD, AIRS, and TRMM combined gridded data to generate climate change with a resolution of 25 × 25 km approximately (1° = 102.3 km on tropics) (IMD 2013; Collins et al. 2001).

The regional monthly temperature trends are calculated by simple average of the fundamental grid point data of the respective block region. The temperature trend is computed through the slope of a simple linear regression fitted line against time to each of the series. The statistical significance of trend is assessed by means of the F-ratio subsequently, by annexing into account the autocorrelation. Therefore, both the ground and satellite data have been taken into consideration for the last one decade to minimize the gap/error in climate change simulations for the baseline study. The Mann–Kendall non-parametric test has been calculated for the 43 years (1970–2013) baseline temperature and rainfall data from:

$$T = \sum_{i=2}^{n}\sum_{j=2}^{i-1}(xi - xj)\, R = \sum_{i=2}^{n}\sum_{j=2}^{i-1}(xi - xj)$$

where n = data set length, while xi and xj = standard chronological data values. The independent and randomly distributed variables from Tmax Tmin and Tmean provide data set length $n \geq 3$ similarly for Rtotal and Rdays where $n \geq 2$. The non-parametric temperature statistic T and R is distributed normally with variance and neutral or zero mean. With the times-series length the standardized test statistic is calibrated from Zt score for temperature and Zr for rainfall value to test the null hypothesis as:

$$Zt = \left[T - \frac{1}{\sqrt{V}} \right] \quad Zr = \left[R - \frac{1}{\sqrt{V}} \right]$$

$$\text{if variance } V > 0, \text{or } V = 0 \text{ or } V \text{ is} < 0$$

The increasing temperature T_{max} and T_{min} trend indicates $a + Z$ value and a decreasing T_{max} and T_{min} temperature trend indicates -Z value. The significant T_{max} and T_{min} levels of 0.05, 0.01, and 0.001 have been converted into percent later. The temperature and rainfall are shaped by GHGs emissions. This is based on computing average, z-score, standard deviation, and moving average for temperature variability and identification of dry/wet years for rainfall. For downscaling, the regional adaptive responses for climate change have been computed as

$$Cs_c = f(C_L, Ps)$$

where Cs_c is the microscale climate, which is downscaled through the functional relationship,

C_L is the large-scale climate and Ps s is the micro-level physiographic details. These steps of downscaling from GCM to RCM have been adopted (Fig. 30.2) to fill the gap in IMD databases for the given missing years (1975 to 1981, 1987 to 1992, and 1998 to 2000).

30.2.3 Database and Methodology for Climate Change Scenario Modeling

The climate change scenario modeling has been done based on 1970–2013 baselines to find out a definite consciousness from infinite number of equally credible changes. The basic quality of the used simulated scenario is that it is consistent with current understanding of real calibrations without assuming that the future will resemble the past and the present. PRECIS based on Hadley Regional Climate Model (HadRM) downscaling methodology (shepherd method) data have been used on IMD, AIRS, and TRMM combined gridded data to generate climate change scenario for 2020, 2050, and 2080. The 2020, 2050, and 2080 scenarios for temperature and rainfall are shaped by GHGs emissions preceded from Intergovernmental Panel on Climate Change-Special Emission Scenario (IPCC-SRES) that are A_1 and B_2 (IPCC 2013; IMD 2010, 2014). The deviations in the simulated time are regarding the baseline of 1970 to 2013 and the maps have been generated for the past, present, and future scenarios accordingly. The baseline (A_1) and the emission (A_1B_2) climate change scenario have been developed for the present study to provide advanced alternative adaptation and mitigation strategies. The A_1 and B_2 scenarios have been referred for future conditions to precipitation, temperature, and CO_2 emission that have been used in assessing impact of climate change on ecosystem services security in the district. It is based on computing average, z-score, standard deviation, and moving average for temperature variability and identification of dry/wet years for rainfall.

30.2.4 Calculating the ESVI: With Impact Approach and Capital Vulnerability Indices

The re-gridding has been done after downscaling the data at same resolution, i.e., all data has been downscaled at 0.5°×0.5° latitude and longitude and then re-gridding to 0.1° lat. ×0.1° long has been done to generate blockwise data to study the climate variability at micro-level. AIRS gridded rainfall data for the study has been obtained for the distribution of rainfall, latent heating, and the variability of monsoon in terms of Outgoing Longwave Radiation (OLR), Length of Growing Period (LGP). To find out Ecosystem Services Vulnerability Index (ESVI) equation the following steps have been taken to calculate composite index of ESVI, lies between −1 and +1, where −1 denotes least vulnerable, 0 signifies vulnerable, and 1 designates highly vulnerable, the cal-

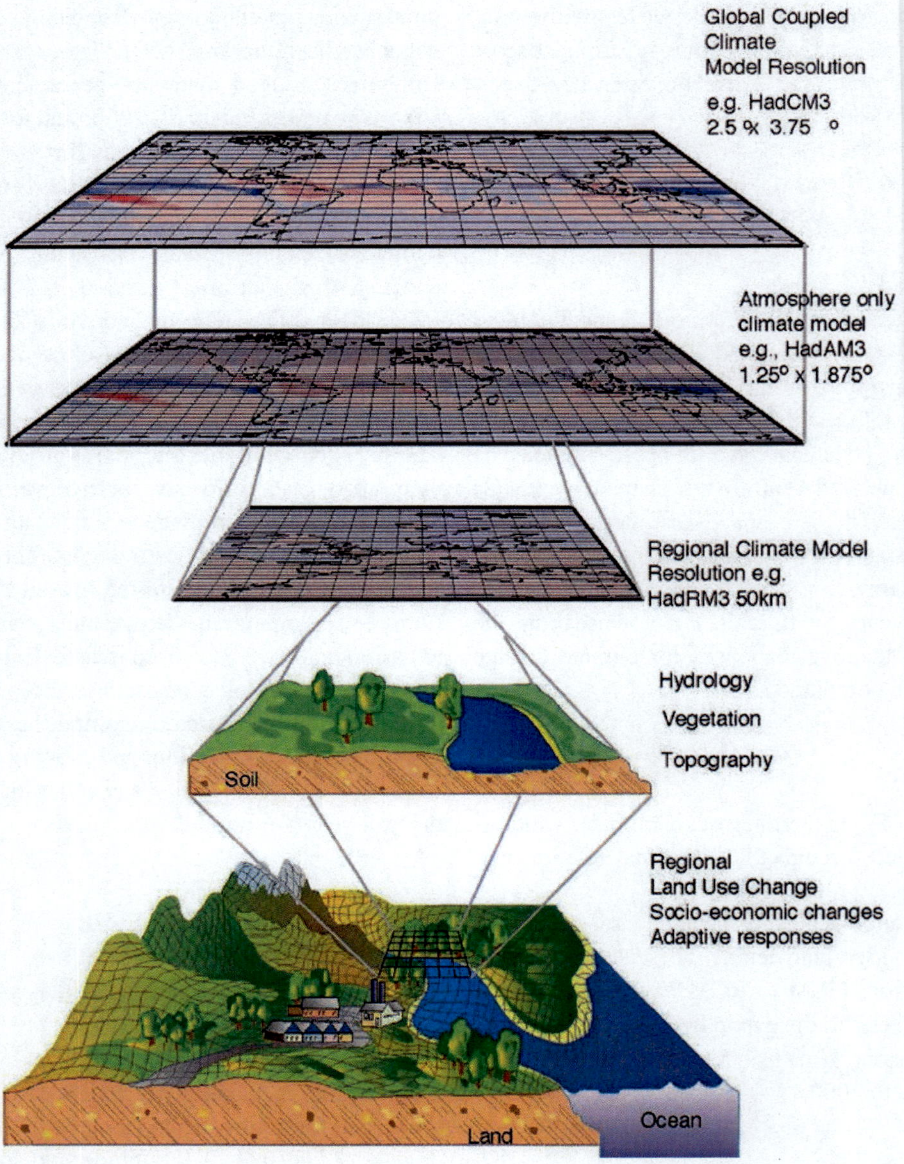

Fig. 30.2 Climate change modeling mechanism GCMs downscaling to regional adaptive responses of RCMs. Source: WMO

culation has been done through the following steps as.

$$VPr_t = \frac{TPr_{ob-}TPr_{min}}{TPr_{max-}TPr_{min}}$$

Step 1: (reiterate for all variables) where, VPr_t is productivity of a particular crop has increased with temperature variable, that has been computed based on 30 years data, $TPr_{ob}$is observed productivity in that temperature range, TPr_{min} and TPr_{max} are minimum and maximum productivity at that temperature range. In other words it is (observed - minimum)/(maximum - minimum) $VPr_t = \frac{30-12}{41-12}$ = 0.6206, then comes second step.

Step 2: (reiterate for all components) if for the temperature component.

$$T_I = \Sigma\left(\frac{V_{Ob+}V_{LU+}V_{OLR+}V_{Albedo+}\ldots\ldots\ldots V_{n+}}{N*}\right).$$

where T_I is temperature index component, the variable; V_{ob} is observed changes, V_{LU} is land use changes, V_{Albedo} is albedo changes with spatio-temporal temperature trend, where the temperature component is having 6 variables and the respective values of $V_{ob...}$ .... V_n are as:

$$= \Sigma\left(\frac{0.571 + 0.900 + 0.500 + 0.566 + 0.705 + 0.919}{6}\right)$$
$$= 0.693$$

Step 3: would follow Ecosystem Services Vulnerability Indices (ESVI) with summation of its all components, the variable values are multiplied by the assign weightages (where V_t = temperature, V_r = rainfall, V_{LU} = Land use) that are divided by their respective total weights as:

$$\text{ESVI} = \Sigma\left(\frac{W_1 Vt_{+} W_2 V_{LU+} W_3 V_r}{W_1 + W_2 + W_3}\right)$$

$$\text{Where ESVI} = \Sigma\left(\frac{4 \times 0.511_{+} 3 \times 0.532 + 4 \times 0.614}{4 + 3 + 4}\right)$$
$$= 0.554$$

Finally, the ESVI_s of all 27 villages have been clubbed together with their respective ESVI_s to show climate impacted ecosystem services for all the respective capitals: Natural capital index (NCI), Physical capital index (PCI), Social capital index (SCI), and financial capital index (FCI) across district.

For calculating the ES vulnerability and changing condition, 270 households from each sampled part of the district with semi-structured questionnaire were interviewed (individually and in groups). Village location has been taken as point coordinates and were saved on a handheld GPS receiver to map the interview track. The exploratory inductive research, methodological triangulation has been done for obtaining data from different sources to increase the validity, reliability of the findings and simplify data analysis ((Singh and Singh 2014; DFID 2000; Swaminathan 2010; Tubiello and Rosenzweig 2008; WDR 2010; WRI; 2008; Directorate of Agriculture Records (2008–2012); District Statistical Abstract-Kangra 2013; Department of Economics and Statistics 2013; Government of Himachal Pradesh 2013; GOI 2004). It has been done to cross check individual interview answers, where the questions were focused on past and current ES linked livelihood indicators for both supporting and provision services such as household composition, general household livelihood security as provision ES, net sown area availability, land use/change, crop production/combination, livestock, soil productivity, soil fertility, status of employment and earnings and other income, household assets, urban agriculture, fish production, environment, water and sanitation, socio-economic conditions and constraints to improved supporting ecosystem services. Above all the data was coded and seeded in Statistical Package for Social Scientists (SPSS) version 20.0 for calculation and analysis. Ecosystem services vulnerability concept can be operationalized mathematically as:

$$V = f\left(\frac{P_s - P_{\min}}{P_{\max} - P_{\min}}\right) + \frac{1}{n}\sum_{x=1}^{n} Di_x - \frac{1}{n}\sum_{x=1}^{n} IRi_x$$

Or can be summarized in the following way.

$$\text{Vulnerability}\, V = f\ (\varepsilon + S - AC)$$

In this equation, V is vulnerability of the block/village, $\mathcal{E}$ is the exposure to vulnerability (reflected in the amount a problem occurred like flood/drought, etc.), S is the sensitivity to perturbation (like sensitivity of harvest to perturbations), and AC is adaptive capacity of regions to cope with exposure, that can be determined using socio-economic proxy indicators. Thus, on this basis ESVI has been calculated in the range of −1 to +1 where −1 is the least vulnerable, 0 vulnerable, and + 1 is the most vulnerable without climate change factor.

$$\text{where Exposure}(\varepsilon) = \left(\frac{P_s - P_{\min}}{P_{\max} - P_{\min}}\right)$$

where $\mathcal{E}$ is the exposure in each season in the respective village, Ps is the latest year production or facility recorded, that is based on 3 year moving point average of increase or decrease in facility/production. P_{max} and P_{min} are the maximum and minimum value during the observation period.

$$\text{Sensitivity}(S) = \frac{1}{n}\sum_{x=1}^{n} Di_x$$

where average sensitivity is calculated by S, for the ith variable n is the number of drivers and Di_x driver of the sensitivity of each variable.

$$\text{Adaptive Capacity}(\mathcal{AC}) = \frac{1}{n}\sum_{x=1}^{n} IRi_x$$

where $\mathcal{AC}$ is the average adaptive capacity of ith variable, n is the number of variables, and IRi_x represents influence of resilience. This paper would provide further support to analyze provisioning and supporting ecosystem services without and with climate change impact where three interconnected methodologies have been derived and considered. First is to measure Ecosystem Services (ES) vulnerability index where the villages have been put in vulnerability order, second is to assess the climatic impact on provisioning and supporting ES for present and future and the third one is to analyze ES vulnerability, exposure, sensitivity, and adaptation. The researcher has visited each household 2 times during May 2012 to July 2014 to complete all sections of the questionnaire for ESVI. The sample selection was done carefully to assess the vulnerability of ES. The provisioning and supporting services of ecosystem have been analyzed based on 47 parameters (good fit with data collection). To avoid complexities of measuring too many indicators altogether the abbreviations have been used and the indicators have been grouped into five indices, i.e., Natural Capital Index (NCI), Physical Capital Index (PCI), Social Capital Index (SCI), Financial Capital Index (FCI), Human and Health Capital Index (HHCI), and Shock Faced and Coping Strategies Index (SCSI) (Table 30.1). This includes a set of factor / indicator variables that have been derived from the data acquired from baseline survey to calculate the indices that will principally differentiate household status.

Table 30.1 Selected Minor Variables and Major Components for Ecosystem Services Vulnerability Index in the District (based on what)

Minor Variables (in percent)	Abbreviation Minor Variables	Major Components
Households having agricultural land	HAL	Natural capital index (NCI)
Households having cultivated area under irrigation	HCAI	
Households that utilize a natural water source	HNW	
Households grow multiple crops in a season	HMCS	
Households having >50% net area sown of total holding	H50NSA	
Households having change in >10% net area sown of same crop 5 year back	H10BNSA	
Households forced to undergo crop change due to previous failure	HCH_F	
Household having surplus production at present	HSP_Present	
Household having surplus production 5 year back	HSP_Past	
Household home retained out of total production	HReTPro	
Households sell the product out of total production	HSoPro_Tpro	
Households dependent on family farm for food	DH_FF	
Households dependent on livestock for milk and other	DHL_M	
Average agricultural livelihood diversification index	AAgri_LD	
Average time spent in collecting firewood by an HHs	T_Fire	

(continued)

Table 30.1 (continued)

Minor Variables (in percent)	Abbreviation Minor Variables	Major Components
Fuel-wood used by household predominantly LPG more than 70–80%	FW_LPG	Physical capital index (PCI)
Households that have consistent water and electricity supply	WES_Regular	
Households having tap inside their house for drinking	Tap_in	
Households having shared tap for drinking water	Tap_Shared	
Households go for public hand pumps/bore well for drinking water	Pub_Pump	
Households who are dependent on well/ponds for drinking water	Pwell	
Households dependent on forests for NTFPs (honey, medicine etc.)	H_NTFPs	
Household getting help in case of crop diseases/dieback	Own_Equip	
Household facing plantation impact on agricultural land availability	Plan_LA	
Percent of households dependent solely on the family farm for food	D_FF	
Average number of months households struggle to find food	Av_Sfood	
Working population to the total population	%WP_TP	
Households dependent solely on agriculture as a source of income	D_AI	
Per person income (₹ per person per month)	PPIM	
Per person current value of land/house/animal shed/pond (₹)	PP_LaValue	
Per person current value of livestock asset (₹)	PP_LiValue	
Per person current value of machineries & equipment (₹)	PP_MeValue	Social capital index (SCI)
Per person current value of other asset (₹)	PP_OValue	
Active population ratio (15–59 years population/family size)	P_Ratio	
Proportion of 15–59 population in employment	Emply_N	
Household income earned by women (₹ /person)	HHI_F	
Per person current savings (₹)	PP_Saving	
Households that do not save crops	NoCr	Financial capital index (FCI)
Households that do not save seeds	NoSeed	
Average number of flood and drought events since 2000	Av_Dro/Flo	Shock faced and coping strategy index (SCSI)
Households that did not receive a warning about the pending natural disasters	Nwarn_Dis	
Households with an injury or death as a result of flood or drought since 2000	Injur_Dis	
Households forced to sell their assets in case of crop loss	Crlos_SA	
Households forced to sell their assets in case of loss of job	Joblos_SA	
Households sold their assets in case of income shortage	Inco_SA	
Households sold their assets in case of illness/death/theft	ill_SA	
Person forced to migrate due to crop failure/loss of job/other	Mig_crlos	

Source: Established from Primary Survey, 2012–2014

30.3 Result and Discussion

30.3.1 The Observed Spatio-Temporal Change in Temperature

The temporal change in temperature and rainfall has been plotted based on monthly decadal average for the region for four decades, namely 1970–1990 and 1990–2010. The summer (June) and winter (January) temperature in 1990–2010 has witnessed variable change. The maximum and minimum temperature increase ranges between 0 °C and 0.8 °C. While a noticed decline of almost 10 cm of rainfall for 3 months summer monsoon (June, July, and August) average from 49.3 cm (1970–1990) to 39.3 cm (1990–2010) has been perceived. Whereas the winter rainfall months; November, December, and January have seen decline of 4.5 cm from 7.6 cm during 1970–

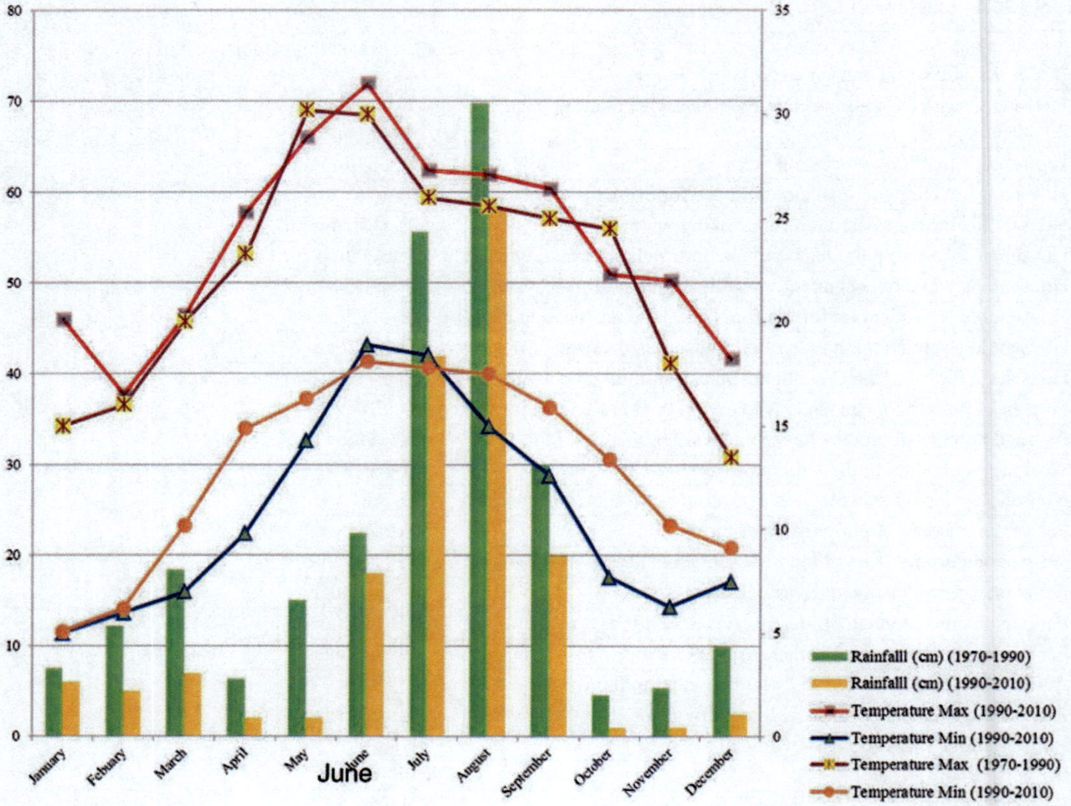

Fig. 30.3 Comparative status of average monthly decadal rainfall (cm), maximum and minimum temperature (°C) of Kangra district for 1970–1990 and 1990–2010

1990 to 3.1 cm during 1990–2010 (Fig. 30.3). Comparably, IPCC Vth Assessment Report observed global amplification in the average surface temperature of around 1.74 ± 0.18 °C (2.33 ± 0.32 °F) during the twentieth century steered by the increasing emission of GHGs ever since the last century (IPCC 2013). During the last 40 years, the temperature has shown a great variability with an average rise of about 0.5 °C for the month of June. The average temperature during 1970 to 1990s rose from 22.9 °C to 23.7 °C during 1990–2010 though the temperature graph shows a decline in temperature in the later part. The mean annual temperature of Kangra district shows significant warming trend of 0.8 °C per 10 years through the period 1970–2010 representing a considerable increase of rate of the warming in the last four decades (Fig. 30.4). Therefore, the aggregated data from seven meteorological stations; Nurpur, Dharamsala, Pong Dam, Palampur, Kangra, Malan, and Dehra have been computed on spatial scale in this Cwa type of climate, plotted through MATLAB R2009a.

The average annual temperature ranges between 10 °C and 26 °C in January month while in the summer month of June the temperature ranges between 14 °C and 36 °C. The district climate changes from sub-tropical in low hills and valleys to sub-humid to temperate in the mid- and the high hills, respectively. Southward and eastward latitudinal shifts in isotherms have been observed and it is markedly visible in central and southern section of the district over 1970 and 2010. Meanwhile, a marginal shift in the northern and extreme eastern section of the region is also witnessed. The January isotherm of 12 °C, 14 °C, and 20 °C has noticed strong shift similarly in the case of June isotherms. The major shift in June isotherms can be noticed in 20 °C, 22 °C, 30 °C, and 31 °C

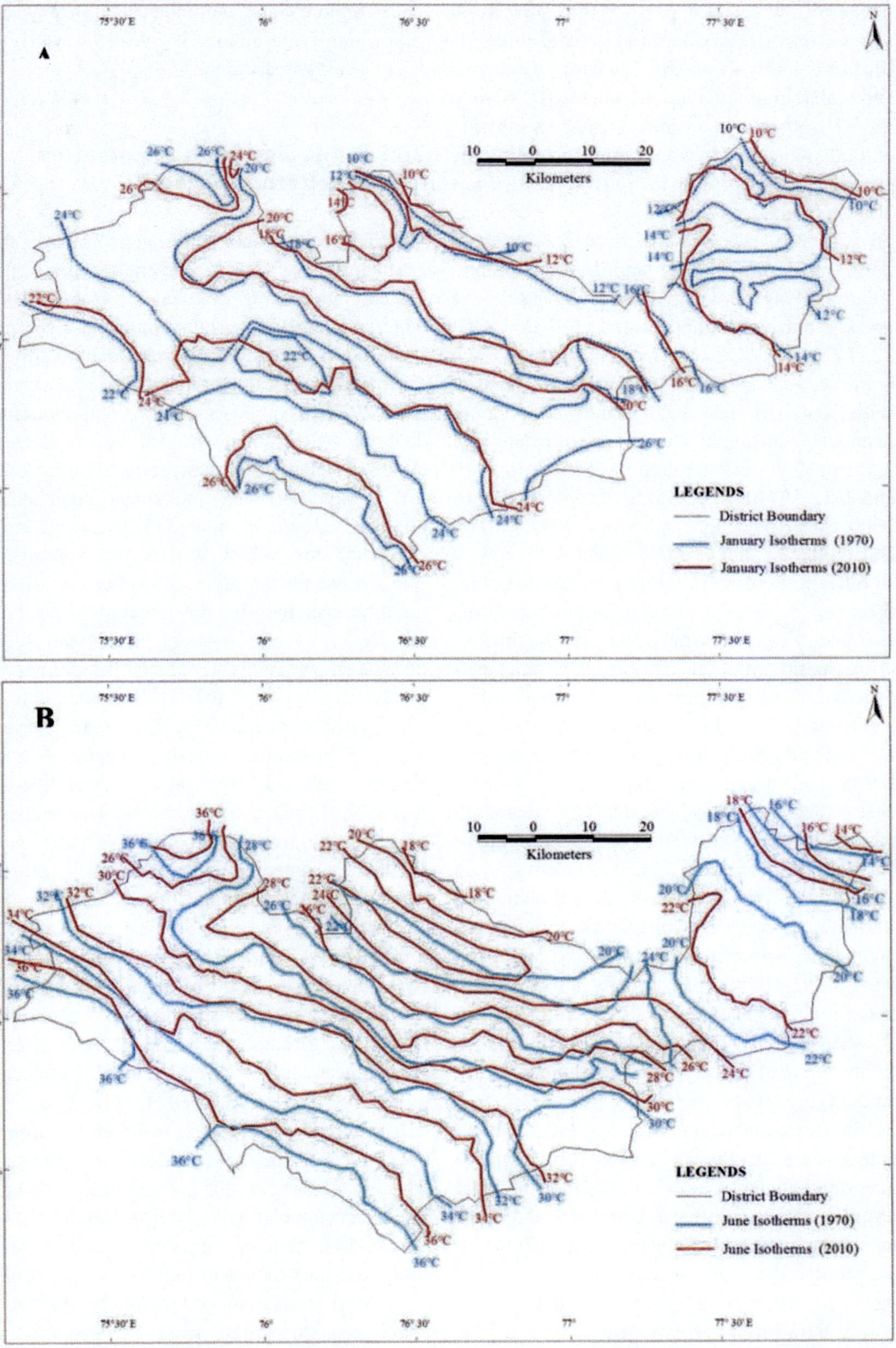

Fig. 30.4 (**a**) Spatio-temporal change in January isotherm and **4** (**b**) change in June isotherm for the year 1970 and 2010

isotherms, which are passing through central and western tracts where the general direction of shift is south-westward. The winter isotherms show maximum deviation where 10 °C and 12 °C isotherms are shifting much northward and 20 °C isotherm in intruding in central and eastern blocks of the district. Over the last 30 years Palampur and Baijnath have witnessed 16 °C and 18 °C isotherm shift towards more north ward during June and Dehra Gopipur, Kangra, Pong Dam have been witnessing the average north ward and central shift in 30 °C, 32 °C, and 36 °C isotherm which pass through western and southern parts in the district (Fig. 30.4a, b). The mean annual temperature displays substantial trend of warming of 0.06 °C/10 year all through the period till 1970 and 0.22 °C/10 year during recent decade of 2000 to 2010 (Singh and Singh 2011, 2014). During the last three to four decades, it shows a significant increase in the rate of the warming; therefore, it portrays a major turnaround in the asymmetry of the diurnal trends of temperature. From spatio-temporal analysis, it is quite evident that the temperature change ranges between 0 °C and 0.5 °C during summer months with western and central section in Nurpur, Fatehpur, Dehra and Kangra are experiencing highest change between 0.6 °C and 0.8 °C in the month of June over the years The least change in temperature of about 0–0.2 °C has been seen in eastern part of Bada Bhangal, Tarmehr, and Baijnath region. For winter season the maximum change noticed is of 1.0–1.2 °C and minimum of 0.2–0.4 °C, respectively, in western and eastern region. The trends of temperature rise in October–January months are just reverse of what one can observe during June to July months (Fig. 30.4a, b).

The increasing temperature T_{max} and T_{min} trend indicates a + Zt value and a decreasing temperature trend indicates −Zt value. On a regional level the variation in mean maximum and minimum temperatures over the period of 1970–2010 has observed a net escalation in temperature ranges from 0.86 °C ± 0.04 °C. The total population of the district has experienced 0.25–0.5 °C increase. Some 14.5% of the population acknowledged a more significant increase of 0.75 °C and more, signifying almost 30.1% of the total district's geographical area.

30.3.2 The Simulated Temperature Pattern and Trend

To circumvent minor anthropogenic and other problems in the global temperature, the signals are computed using an ensemble of the RCMs. The modeled RCM results constitute a comprehensive representation of current identification of the methods of climate variability and change. The PRECIS based simulation on combined TRMM, AIRS, and IMD grid data is showing slow and steady increase in temperature for 2020, 2050, and 2080. The interannual temperature component is also computed to observe the accurate point from where the departure in temperature has been taken place. It has been plotted by the difference between black and green traces in the figure for the residual. The trend (red), decadal (green), and inter annual (blue) lines are marked to display the time scale as applied to an averaged area of each block (Fig. 30.5). The simulated temperature computations for Nurpur, Kangra, and Bada Bhangal have been used to regress and project the scenarios. The temperature scenario processing consists of three steps. Firstly, the screening of the individual grid box values for the chronological data, secondly, the de-trending; to extract slow changes and lastly, filtering to isolate the high and low frequency constituents. The result of simulation modeling for the 2020s signifies swelling GHG concentrations and throughout warming. The overall annual mean surface air temperature is projected to rise by 0.86–1.2 °C in 2020s (Fig. 30.6). The winter months are getting warmer by around 0.25 °C near the end of 2020s except Bada Bhangal. This region displays contrasting results. The mean temperature variability during the winter months is more as compared to summer, monsoon, and post-monsoon months. As per result, the annual mean temperature of the district is anticipated to increase from 1.2 ± 0.5 °C to 2.3 ± 0.69 °C during 2020s and 2050s, respec-

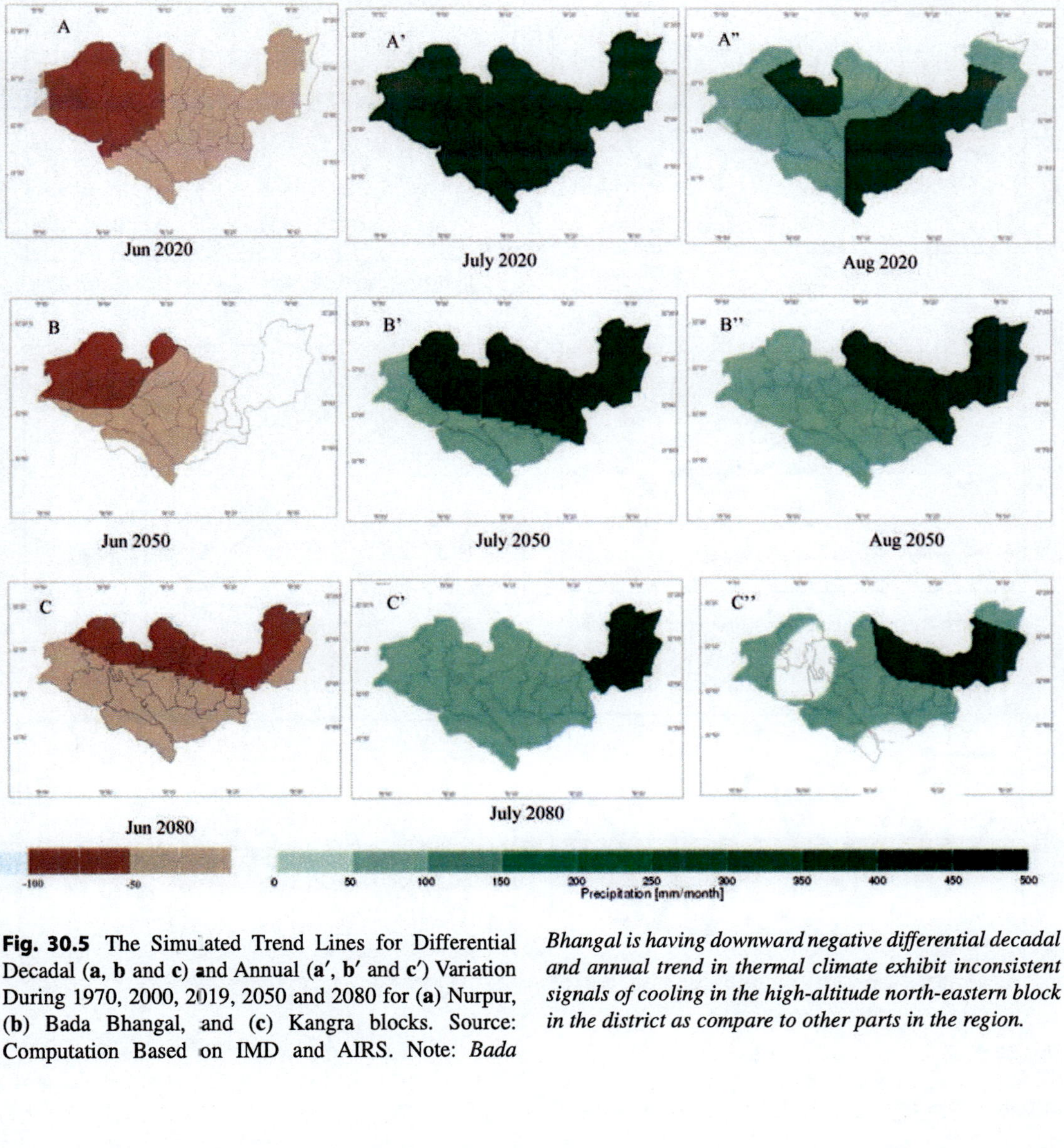

Fig. 30.5 The Simulated Trend Lines for Differential Decadal (**a**, **b** and **c**) and Annual (**a′**, **b′** and **c′**) Variation During 1970, 2000, 2019, 2050 and 2080 for (**a**) Nurpur, (**b**) Bada Bhangal, and (**c**) Kangra blocks. Source: Computation Based on IMD and AIRS. Note: *Bada Bhangal is having downward negative differential decadal and annual trend in thermal climate exhibit inconsistent signals of cooling in the high-altitude north-eastern block in the district as compare to other parts in the region.*

tively. The annual mean temperature in the baseline scenario of 2020s to 2050 explains that the temperature will increase from 0.25 to 0.75 °C while total precipitation will decrease from 22.4% to 14.3%, the seasonal winter temperature may rise from 0.79 to 2.89 °C. The winter precipitation will decrease from 13.6 in 2020s to 10.2% by 2080. Similarly, the monsoon temperature may rise from 1.01 to 4.37 °C during the A_1 to A_1B_2 with total monsoon precipitation decrease from 37.0% to 28.7%. The summer temperature will rise from 1.28 to 4.37 °C by the year 2080, and the precipitation will be decreased from 17.5% 10.7% until 2050, then further it will increase by 16.26% by the year 2080 due to increasing emission. The IPCC, GHG emission scenarios for A_1B_2 and the baseline has been compared for the trend analysis for temperature and precipitation parameters. The general trend in both the scenario is almost same and only difference is that amount of change. The high intensity of warming of temperature is observed in central and Palampur, whereas the precipitation exhibits the decreasing pattern (Table 30.2).

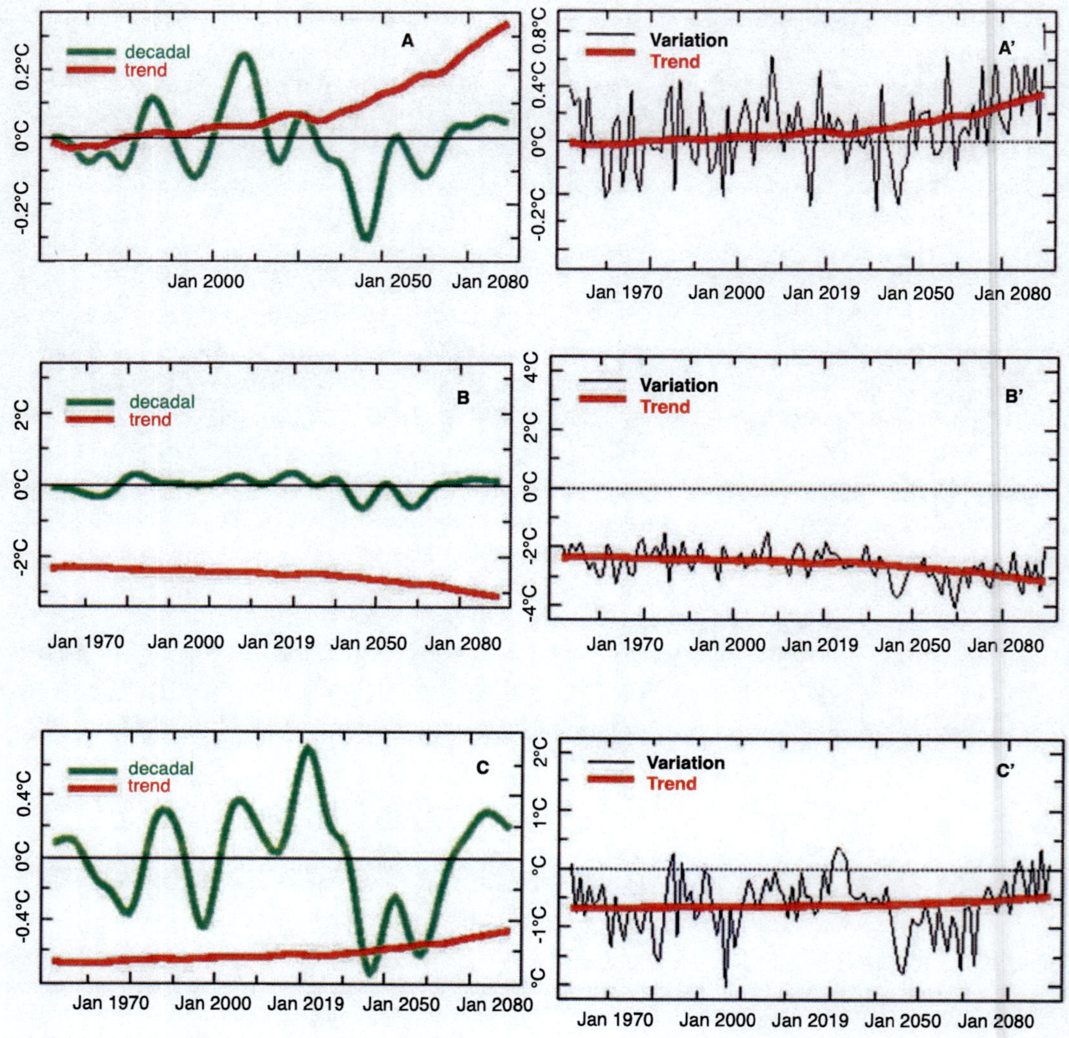

Fig. 30.6 Simulated Climate Change in **a**, **b**, and **c** (Observed); **a′**, **b′**, and **c′** (Baseline); **a″**, **b″**, and **c″** (for Emission) Scenario Rainfall Pattern for the Months of June, July, and August, respectively, for the Year 2020, 2050, and 2080. Source: Downscaled and Simulated by the Researcher

Table 30.2 Seasonal temperature (T°C) and precipitation (R in percent) change in Kangra district for the baseline (A_1) and emission scenario (A_1B_2)

	A_1						A_1B_2					
	2020		2050		2080		2020		2050		2080	
Seasons	T	R	T	R	T	R	T	R	T	R	T	R
Annual	1.23	22.4	2.81	21.50	4.76	14.30	1.52	25.20	2.43	21.10	3.49	11.6
Winter	0.79	13.60	1.24	18.70	2.89	10.20	0.96	13.50	1.73	24.40	2.48	44.7
Pre monsoon/ summer	1.28	17.50	2.61	10.66	4.37	16.26	1.38	15.45	1.97	10.88	3.11	14.54
Post monsoon/ autumn	1.37	12.40	2.71	7.79	4.91	5.67	1.52	16.25	2.7	6.13	3.71	11.71
Monsoon	1.01	37.20	2.61	36.80	4.37	28.70	1.38	26.13	1.97	21.9	3.11	17.6

Source: Primary Calculations for Representative Concentration Pathways (RCP)

30.3.3 Impact of Temperature and CO_2: Growth and Productivity of Crops

The climatic conditions in the district with wide range of altitudinal and temperature difference in the region provide a range of potentialities for growing vegetable and horticultural crop as onion, chilly, brinjal, bhindi, plums, apricot, little-apple, dry fruits, and other citrus fruits. The important wet season crops (green gram and *rajmah*) and two dry season crops (wheat and maize) have been showing early flowering and maturity due to temperature increase. The marked reduction in Leaf Area Index (LAI) that in turn expresses gradual reduction in the total biomass and yield of the crop increase in Length of Growing Period to more than 9 days in the study area.

It has been observed that with increase in CO_2 level (380 μmol/ mol) from 1995 to high CO_2 level in 2013 (440 μmol/mol) level the biomass of crops has initially increased then decline sharply with decreasing yield (Fig. 30.7). The same happens when the crops are exposed to temperature increase that has been observed in the district (+0.6 to 1.2 °C) (Fig. 30.8). After exposure to high temperature and CO_2 for more than a month, these crops showed maximum detrimental effect on their reproductive growth. The crop production fuctuates over time showing similar trend as rainfall, whenever the rainfall is less the production decreases and vice versa.

The production of cereal crops and horticultural crops has shown decreasing trend for 30 years. Therefore, the statistics exhibit that all types of farming is getting impacted and it is just not limited to cereal or horticultural or other food grain for marginal farmers, but it does show same trend across with low yield and production.

30.3.4 Composite Capital Index for Ecosystem Services Vulnerability

The average Ecosystem Services Vulnerability Indices (ESVI) with socio-economic parameters score is 0.387 that is making Kangra moderately vulnerable to climate change scenario, but the composite indices impacted ecosystem services shows 0.587, which is quiet high therefore, the possible impact of climate change and environmental degradation has been assessed based on composite scores and other identified indexing factors have been discussed below distinctly. Natural capital (NC) (provided directly from nature; water, land, vegetation, etc.), physical (PC) (forest, NTFPs, and other infrastructural produce), social (SC) and financial vulnerability index (FCVI) have been measured to find out extent of ecosystem services getting impacted by climatic variability. For the energy sources the entire high-altitude villages in Multhan sub-tehsil, Baijnath, and Dharamsala block rely largely on the forest and forests products

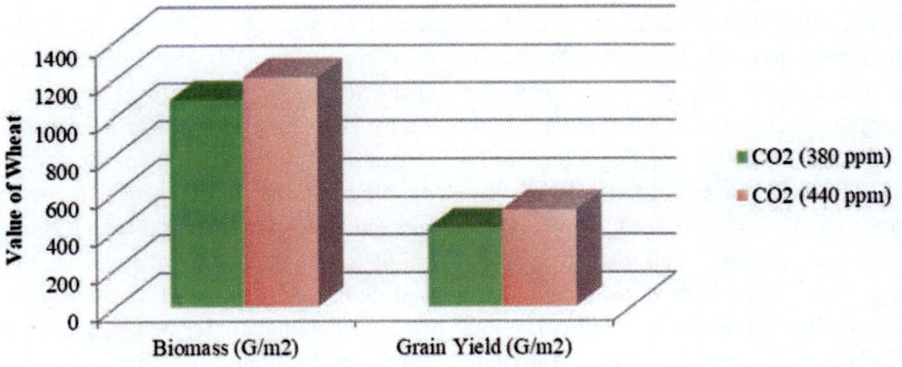

Fig. 30.7 Simulated Effect of Elevated CO_2 on Growth and Yield of Wheat Crop

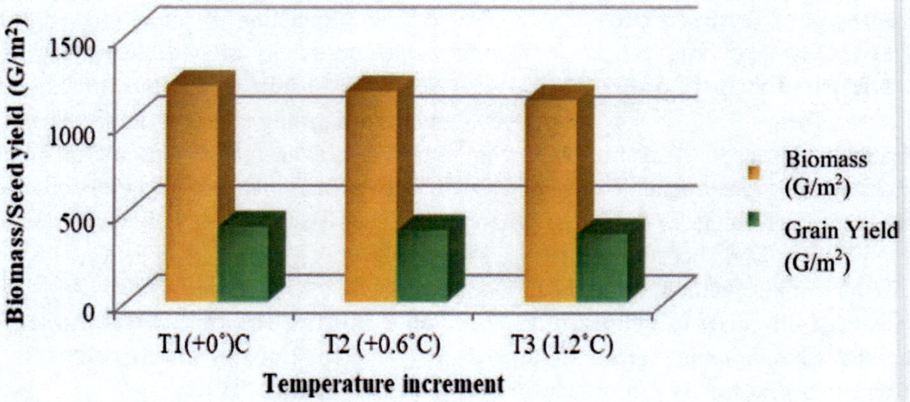

Figure 30.8 Simulated Effect of Elevated Temperature on Growth and Yield of all Cereal Crop

(NTFPs). It has been made clear by the respondents that these high-altitude mountainous forest trees take more time to grow as compared to low lying areas although the rate of depletion is very rapid. A respondent said that nowadays it takes almost on an average 7. 30 h to fetch firewood, which used to be done by 2–3 h, and if any alternative is not developed soon the entire region must face consequences of acute energy crisis. This might threat the ecosystem sustainability. Similarly, land resources have been found as one of the most vulnerable natural resource. Its vulnerability is quiet high at district level, i.e., 0.748, together with very low Inverse of Per hectare Yield Index (IYI) which is 0.593 that exhibit less land production capacity, while the respondents in Dehra block have said land productivity is declining day by day because it is becoming impossible for a marginal or small farmer to leave their land fallow for more than a month. It has also been strongly stated by the respondents that the fertile soil of these lands is being removed due to recurrent disasters like, drought, flood, upper surface erosion, and landslides. Another major component is forest resource, on which entire livelihood of 63.76% villages are fully dependent. The forest based natural vulnerability index varies from 0.412 to 0.513 in Kangra, Dehra, Paragpur, Nagrota, and Fatehpur block while more than 0.621 in high-altitude blocks and villages like Dharamsala, Baijnath, and Multhan (Fig. 30.9). While overall, 81.46 per cent HHs have observed depleting firewood species as well as numbers. The *kuhls* (indigenous canal) are major sources of irrigation in the entire Himalayan region and Kangra is no exception to that. The emerging influence of diversified large economy Kangra is undergoing remarkable socio-economic transformations due to that *kuhl* regimes, it is facing declining interests among present day farming community. Decreasing participation, increasing conflict, declining legitimacy of customary rules and authority structures are making *Kuhl* less reliable and manageable (Baker 2005). These plain fertile villages (Sohara, Aweri, Jhalot, Re Khas Gabli Dhar, etc.) are characterized by abrupt soil salinity, infertility, poor productivity, and small to marginal farm sizes. Mundla, Ulharli, Jhalot, Garh, Gangath, Indora, Raja Khas, Abdulla Pur, Puling, Tarmehr, Bada Bhangal, Samkar are having very high vulnerability index (more than 0.69) (Table 30.3). The physical capital vulnerability index (PCVI) is based on the infrastructural gap or absence in an area. HHs who are dependent on well/ponds for drinking water, HHs dependent on forests for NTFPs (honey, medicine, etc.). It takes minimum 2 h for HHs in Puling and Tarmehr to reach to the nearest bus stop while it takes 4 days for HHs in Bada Bhangal to do the same. The average time spent in collecting firewood for a HH in the same is 7.30 h, The PCVI varies between 0.28 (Zamana Bad, Sohara Gangath and Mundla) to more than 0.71 (Bada Bhagal, Tarmehr, Puling and Uparli Barol) in Multhan sub-tehsil and Dharamsala block.

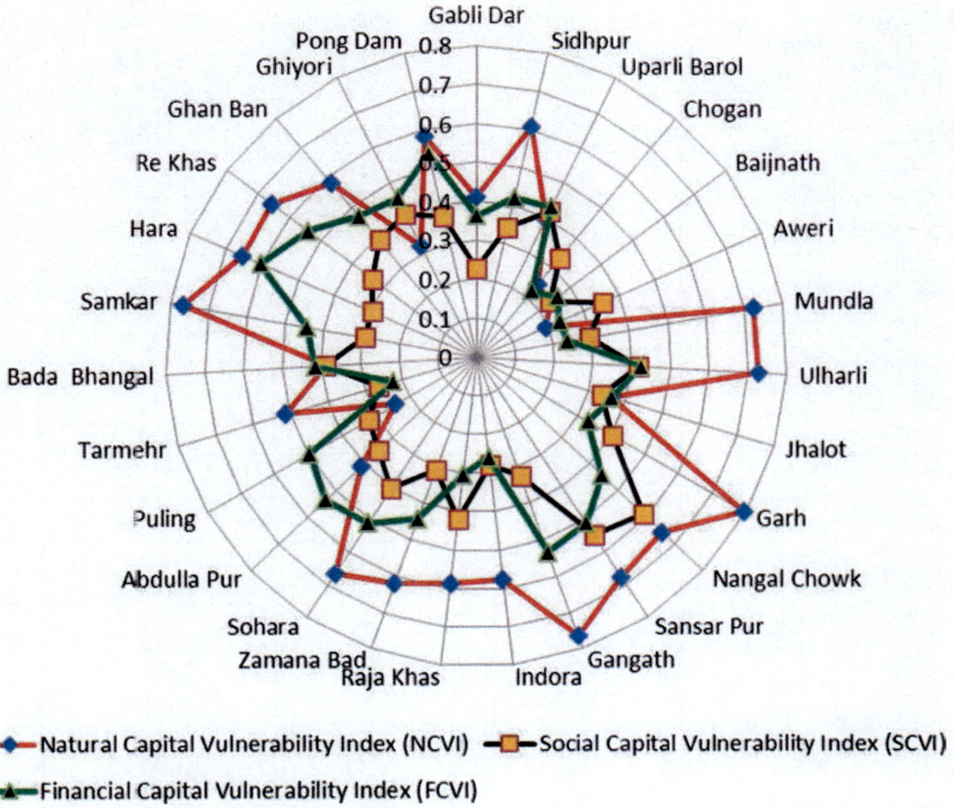

Fig. 30.9 Vulnerability Radar Diagram of NCVI, SCVI and FCVI for selected Villages of Kangra. Source: Primary Survey, 2013

Water vulnerability index is quiet high in Kangra, Mundla, Jhalot, Abdullapur, Chogan (varies between 0.45 and 0.51) as compared to inaccessible areas where it is less than 0.25 because of glacier and rainfed supply system. The financial capital is fundamentally associated with the sustainability, sufficient financial security can help to support and overcome during risks and shocks in the family. On an average this district is financially less vulnerable except few agricultural villages in central and western part. The FCVI is computed based on parameters, analyzed in monetary values (₹ /person HH or Rs. /person). It is focused on per person income and current savings (₹ per person/month), current value of land/house/animal-shed/pond (₹ per person), current value of livestock asset, value of machineries and equipment (₹/person), active population ratio (15–59 years population/family) and household income earned by women (₹). FCVI is comparatively high in lower plains as compared to high hilly terrain villages where economy is more focused on livestock, transhumance, or horticulture. Almost 38% HHs in Dehra, Fatehpur, Panchrukhi, Paragpur and 78.4% in Baijnath, Puling, Aweri, Tarmehr and Mundla are having family member working outside the respective villages at relatively developed place in these high lands from which they buy ration to eat for those 3 months of frost. Based on overall capital vulnerability a detail summery has been outlined to explain into five vulnerable categories from very low to very high. Very high ESVI is ranked between 1 and 4, high is between 5 and 9 and low varies between 15 and 19 (Table 30.3).

Table 30.3 ESVI results summary for all 47 sub-components, 13 components, and 5 ecosystem capitals in 27 sampled villages of Kangra district

Block Name	Villages	NCVI	PCVI	SCVI	FCVI	HCVI	Final Capitals Average	Overall Rank
Dharamsala	Gabli Dar	0.41	0.57	0.23	0.36	0.56	0.40	21
	Sidhpur	0.61	0.47	0.34	0.42	0.53	0.48	17
	Uparli Barol	0.40	0.41	0.42	0.43	0.55	0.44	18
Baijnath	Chogan	0.24	0.59	0.33	0.22	0.54	0.39	25
	Baijnath	0.24	0.66	0.24	0.26	0.55	0.39	24
	Aweri	0.19	0.50	0.35	0.23	0.52	0.38	26
Nagrota Bagwan	Mundla	0.73	0.72	0.29	0.24	0.42	0.48	16
	Ulharli	0.73	0.61	0.42	0.42	0.54	0.54	8
	Jhalot	0.36	0.42	0.34	0.36	0.53	0.40	23
Paragpur	Garh	0.79	0.50	0.40	0.33	0.38	0.48	15
	Nangal chowk	0.66	0.43	0.59	0.44	0.56	0.54	10
	Sansar Pur	0.68	0.54	0.55	0.51	0.51	0.56	4
Indora	Gangath	0.76	0.53	0.33	0.53	0.72	0.58	2
	Indora	0.58	0.82	0.28	0.26	0.71	0.53	12
	Raja Khas	0.59	0.76	0.42	0.30	0.64	0.54	9
Kangra	Zamana bad	0.62	0.88	0.31	0.44	0.43	0.54	11
	Sohara	0.67	0.69	0.41	0.51	0.51	0.40	22
	Abdulla Pur	0.41	0.77	0.35	0.53	0.54	0.55	7
Multhan	Puling	0.24	0.51	0.32	0.50	0.55	0.60	1
	Tarmehr	0.51	0.35	0.26	0.22	0.33	0.55	6
	Bada Bhangal	0.39	0.42	0.39	0.42	0.41	0.56	3
Fatehpur	Samkar	0.76	0.72	0.29	0.44	0.52	0.53	13
	Hara	0.65	0.85	0.29	0.60	0.60	0.56	5
	Re Khas	0.66	0.83	0.34	0.54	0.41	0.42	20
Dehra Gopipur	Ghan ban	0.58	0.84	0.39	0.47	0.46	0.52	14
	Ghiyori	0.32	0.78	0.41	0.46	0.67	0.42	19
	Pong dam	0.58	0.65	0.37	0.54	0.64	0.34	27

Source: Primary Survey, 2012–2014

30.4 Conclusion

Kangra occupies one of the largest percentages of Himachal Pradesh in terms of population as well as agricultural output. The important perennial river Beas provides water for drinking, irrigation, and hydropower to nearly 80% of its total population directly or indirectly. The climate of the district has shown changes. There is an average rise of 0.5 °C in temperature over the region. The rise in winter temperature is almost the double of rise in summer temperature. Spatially, the western part of the region has experienced greater change in temperature than the eastern part during summer months, whereas this trend is reversed during winter months. The rise in temperature experienced over the years in winter season is greater in eastern parts than in western. The rainfall pattern shows dual characteristics. The annual average rainfall has increased in western part but at the same time it has decreased in eastern part of the region. The precipitation trend over past 40 years has been very variable but an average decrease has been noticed in the annual precipitation. The underlying mechanism behind these changes is blamed on the rising GHGs concentration through it fragile forests and global circulation of rising GHGs concentration. The local factors affecting the rise in temperature, increased cloud cover, and decreasing rainfall over the region can be studied in more integrated and elaborative manner. The effects of changing climate are likely to be exacerbated, which may adversely impact the ecosystem through increased temperature, altered precipitation patterns, episodes of drought, and anthropogenic land use/land cover change and influences. It would not only impact the very sustenance of the indigenous communities in uplands but also alter the life of downstream dwellers across the region, country, and beyond. Therefore, there is an urgent need for giving special attention to sustain this fragile ecosystem. The results presented above exhibit that the geo-drivers have been playing pivotal role in sustainability science research with spatiotemporal perspective due to its interdisciplinarity, transdisciplinariy and multidisciplinarity approach.

References

Baker MJ (2005) The Kuhls of Kangra: community-managed irrigation in the Western Himalaya state. University of Washington Press, London

Bhutiyani MR, Kale VS, Pawar NJ (2007) Long-term trends in maximum, minimum and mean annual air temperatures across the northwestern Himalaya during the twentieth century. Clim Chang 85:9185–9196

Bhutiyani MR, Kale VS, Pawar NJ (2009) Climate in the northwestern Himalaya: 1866–2006. Int J Climatol 30:535–548

Collins M, Tett SFB, Cooper C (2001) The internal climate variability of Hadcm3, a version of the Hadley Centre coupled model without flux adjustments. Clim Dyn 17:61–81

DFID (2000) Sustainable livelihoods guidance sheets. Department for International Development, London

Directorate of Agriculture Records (2008–2012) Annual Administrative Report. Government of Himachal Pradesh, Shimla

Economic and Statistics Department Government of Himachal Pradesh (2013) District Statistical Abstract-Kangra,Shimla

Fischer G, Shah M, Velthuizen HV (2002) Climate change and agricultural vulnerability. IIASA, Laxenburg

GCOS (2003) The second report on the adequacy of the global observing Systems for Climate in support of the UNFCCC. GCOS-82, WMO/TD no. 1143, p. 74

GCOS (2011) Systematic observation requirements for satellite based data products for climate. GCOS no.154, pp. 1–127

Gould SJ (1965) Is uniformitarianism necessary. Am J Sci 263:223–228

Government of Himachal Pradesh (2013) Statistical outline of Himachal Pradesh. Department of Economics and Statistics, Himachal Pradesh, pp 86–89

Government of India (2004) India's initial National Communication to the UNFCCC, chapter 3: vulnerability and adaptation. Ministry of Environment and Forests, GOI, New Delhi, pp 59–71

India Meteorological Department (2010) Climatological tables for (1931–2010). Indian Meteorological Department, Pune

India Meteorological Department (2013) Climatological tables for (1970–2013). Indian Meteorological Department, Pune

India Meteorological Department (2014) Climatological tables for (1931–2013). Indian Meteorological Department, Pune

IPCC (2013) Summary for policy makers. In: Stocker TF, Qin D, Plattner GK, Tignor M, Allen SK, Boschung J, Nauels A, Xia Y, Bex V, Midgley PM (eds) Climate change 2013: the physical science basis. Contribution of working group I to the fifth assessment report of the Intergovernmental Panel on Climate Change. Cambridge University Press, Cambridge, pp 1–30

Kumar RK et al (2006) High-resolution climate change scenarios for India for the 21st century. Curr Sci 90(3):334–344

Lal R et al (2011) Climate change and food security in South Asia. Springer, New York

PRECIS (2010) Providing regional climates for impacts studies. Met Office. http://precis.metoffice.com/. Accessed 15 July 2010

Primary Census Handbook (2011) Government of India, New Delhi

Singh RB, Singh S (2011) Anthro-Biome: Integrated Global Climate Change Modeling, in Proceedings of 3rd International Indian Geography Congress on Sustainable Natural Resource Management Under Changing Climatic Scenarios, Kozhikode, Kerala, India 6–8 May, 1: 95–101

Singh RB, Singh S (2014) In: Singh RB, Hietala R (eds) Human induced biome and livelihood security, in livelihood security of northwestern Himalaya: case studies from changing socio-economic environments in Himachal Pradesh. Springer, New York, pp 53–66

Singh P, Singh S (2016a) Anthropogenic climate change and critical assessments of contemporary climate policy discourse. In: Raman VAV (ed) Geoanthropogenic environment: an appraisal. A.K. Publications, Delhi, pp 13–31

Singh RB, Singh S (2016b) Assessing climate change signals in Western Himalayan District using PRESIS data model. In: Singh RB, Mal S (eds) Climate change, glacier response and vegetation dynamics in Himalaya. Springer, Basel, pp 103–115

Swaminathan MS (2010) From green to Evergreen revolution, Indian agriculture: performance and challenges. Academic Foundation, New Delhi

Tubiello FN, Rosenzweig C (2008) Developing climate change impact metrics for agriculture. Integr Assess J 8(1):165–184

United Nations International Strategy for Disaster Risk Reduction (UNISDR) (2007) Gender perspective: working together for disaster risk reduction. Good practices and lessons learned. UNISDR, Geneva

WDR (2010) Development and climate change. World Development Report, World Bank, Washington, D.C.

Web Sources

AIRS (n.d.). airs.jpl.nasa.gov/

IRS (n.d.). bhuvan.nrsc.gov.in/data/

Landsat (n.d.). glcf.umd.edu/data/landsat/

MODIS (n.d.). http://modis.gsfc.nasa.gov/

TRMM (n.d.). http://disc.sci.gsfc.nasa.gov/

WRI (2008) Climate Analysis Indicators Toolkit (CAIT) http://cait.wri.org/

31 Urban Sustainability and Good Governance in Delhi Metropolitan Region

Shweta Rani

Abstract

Cities are the engines of economic growth. With the process of globalization, the entire world has shrunk to a common platform, whereby the dual process of economic growth and rapid urbanization largely determines the growth of cities and measures their ecological sustainability, with "Delhi Metropolitan Region" being no exception to this. A prime mover and nerve center of ideas and actions, Delhi mega city, however, stand at the crossroads, today. Delhi, a symbol of rich and ancient tradition, a seat of mesmerizing Indian culture, a growing hub of contemporary educational system, the seat of national governance and a center of business, is assuming increasing eminence among the great cities of the world. With an unprecedented pace of growth, Delhi needs to be able to integrate its elegant past as well as modern developments into an organic whole keeping pace with the future generations. The paper attempts to unravel the intricate relationships between urban infrastructure and development of urban environment ensuring its sustainability. Such an approach, if handled with care coupled with good governance practices, is likely to lead a movement towards making Delhi Metropolitan Region, a world-class sustainable city in the decades to come. In addition, some of the best practices in good governance pertaining to this mega city have been highlighted to depict and validate the cities move on the crossroads of sustainable development.

Keywords

Ecological Sustainability · Educational hub · Governance · Mega City · Planned urban development

31.1 The Background

There is a close, intricate, and most revered relation between the urban development process and environment. Urbanization will be the defining trend over the next several decades, especially in East Asia, South Asia, and sub-Saharan Africa, where the bulk of extreme poverty is concentrated. Cities, in these and other regions, will play a central role in the ability of nations to achieve sustainable development. Urbanization and economic growth are strongly associated, and, hence, it has been established that "cities are the engines of economic growth and agents of change" (Mohan and Dasgupta 2005). With the process of globalization, the entire world has shrunk to a common platform, whereby the dual process of economic growth and

S. Rani (✉)
Department of Geography, Dyal Singh College, University of Delhi, New Delhi, India

B. Thakur et al. (eds.), *Resource Management, Sustainable Development and Governance*, Sustainable Development Goals Series, https://doi.org/10.1007/978-3-030-85839-1_31

rapid urbanization largely determines the pace of growth of cities. These two processes are strongly correlated and dynamic in nature. Their combined effects have both positive and negative implications. On the one hand, it results in overall growth and development of cities with better infrastructural facilities but has its effect on the natural environment as well. Under the lure of developmental activities, natural resources of the regions are interfered with and exploited to the extent that it reaches the point of concern for the ecologists and environmentalists. Hence, the theme of the paper tries to unravel the intricate relationships between urban infrastructure, development process and development of the urban environment and ecology ensuring its sustainability with "Delhi Metropolitan Region" being no exception to this.

31.2 Cities as Analytical Unit of Study

Scholars, like Pincetl and Chester, are of opinion that cities are socio-ecological systems (SES) either of interacting biophysical and socio-economic components, or social and technical components. Within the SES, dynamic interactions between human endeavor and ecosystem function have many consequences for sustainability and resilience. Here, cities become the analytical unit for research (Gallopín 2006), as they affect level of environmental pressure on urban populations exerting on ecosystems and their natural resource base (Alberti and Susskind 1996). In fact, SES concept proves to be useful in understanding urban impacts and environmental feedback. A definition of urban areas as a socio-ecological system (Folke et al. 2005), representing five (SETEG) domains (Romero-Lankao and Gnatz 2016) which are listed as Socio-demographics (S), Economy(E), Technology (T), Environment (E), and Governance (G), is discussed later in the study.

Delhi Metropolitan Region (DMR), the focus of political and socio-economic life of India, a symbol of ancient values and aspirations, and capital of the largest democracy, is assuming increasing eminence among the great cities of the world. In fact, this mega city represents a symbol of rich and ancient tradition, a seat of mesmerizing Indian culture, a growing hub of contemporary educational system, etc. Growing at an unprecedented pace, the city needs to be able to integrate its elegant past as well as modern development into an organic whole, which demands a purposeful transformation of socio-economic, natural, and built environment. A prime mover and nerve center of ideas and actions, seat of national governance and a center of business, culture, education, and sports, Delhi, stands at the crossroads today. The choice is between either taking a road to indiscriminate uncontrolled development leading to chaos or a movement towards making Delhi a world-class city if handled with vision and care.

Box 31.1 Sustainable Development Goals

Megacities like Delhi have become the drivers of growth and so-called indicators of progress. However, unprecedented urban growth has implications on its environment and human development. With a population exceeding 30 million, the rapid urbanization of Delhi urban agglomeration comes with challenges of high carbon emissions exerting pressure on fresh water supplies, sewage, the living environment, and public health. This metropolis also experiences congestion, lack of basic services, a shortage of adequate housing and declining infrastructure. The UN 2030 Agenda recognizes making cities and human settlements inclusive, safe, resilient, and sustainable. In this context, the author observes that Delhi needs to transform into an engine for sustainable development, where its urban space offers opportunity for people to prosper in an evolving urban ecosystem, and not be overwhelmed by its growth, unplanned urban sprawl, as the city spills beyond their formal boundaries contributing to climate change. To this end, while SDG—11 is a stand-alone goal and formally acknowledges sustainable development of cities, several other SDGs recognize goals such as ending poverty, food security, health, education, water and sanitation, sustainable energy, resilient infrastructure, inclusive economic growth and productive employment, gender equality, and climate change action, all of which make cities like Delhi livable. These goals can be achieved by stakeholders coming together to strengthen urban governance and compliment collaboration and partnership between constituents.

31.2.1 Historical Evolution of Delhi: "Dilli"

According to India folklore, Delhi (*Dilli*) was the site of the magnificent and opulent Indraprastha, the capital of Pandavas in the Indian epic Mahabharata, founded around 3500 BC. eighth to sixth century onwards Delhi gained importance under Tomar dynasty. Thereafter, Delhi was annexed and conquered times to times by different dynasties during Mughal times, Akbar shifted the Capital to Agra, resulting in decline in fortunes of Delhi. In the mid-seventeenth century, Shah Jahan shifted the Capital back to Delhi and built the planned city of Shahjahanabad (today's Old Delhi). A treaty signed in 1752 made Marathas the protector of Mughal throne at Delhi, resulting in Hindu Maratha Empire rising to prominence.

In 1803, during the Second Anglo-Maratha War, the forces of British East India Company defeated the Maratha forces in the Battle of Delhi, ending the Maratha rule over Delhi. Shortly after Indian Rebellion of 1857, Calcutta was declared the capital of British India. But, in 1911, at Delhi Durbar, held at Coronation Park, King George V announced the shifting of capital back to Delhi. New Delhi, a monumental new quarter of the city designed by the British architect Edwin Lutyens to house the government buildings was inaugurated in 1931. Thus, Delhi remained a political, commercial, economic, and social hub for the whole of the North India.

After the Post-Independence period, New Delhi was officially declared as the seat of the Government of India in 1947. But the high influx of people from Pakistan and Bangladesh and other migrants from different parts of the country to Delhi led to increase in population from 7 lacs to 17 lacs by 1950; mixing of the properly planned commercial centers with haphazard and uncontrolled growth, increased pressure on land and resources, unbalance in the demand and supply in commerce in city, rise in unemployment rates, virtual collapse of civic services, etc. Delhi Improvement Trust and Municipal Body, the two local bodies at that time, failed to cope up with the changing scenario.

To plan Delhi and check its rapid and haphazard growth, the Central Government appointed a Committee under the Chairmanship of G.D. Birla in 1950, which recommended a Single Planning and Controlling Authority for all the urban areas of Delhi. Consequently, Delhi Development (Provisional) Authority—DDPA was constituted by promulgating the Delhi (Control of Building Operations) Ordinance, 1955 (replaced by Delhi Development Act 1957) with the primary objective of ensuring development of Delhi in accordance with a plan. Then, on 30th December 1957, Delhi Development Authority acquired its present name and its role as the ninth builder of the grand city of Delhi.

31.2.2 Delhi at the Crossroads of Urban Development

Cities have become unlikely but crucial zones for the survival of humanity. They are currently spaces for the most consequential attempts at human adaptation and sustainability. They provide a possible focus for the flourishing of future life on this planet. However, for this to take place in more than an ad hoc way, we need substantial rethinking, a new paradigm for urban development (James et al. 2015). Cities are responsible for the bulk of production and consumption worldwide.

Delhi is one of the most ancient cities in the world, characterized by its historical monuments, forts, and buildings. The city is a home to many significant government agencies, including the Parliament of India, making Delhi the commercial and political hub of India. Being the capital of the largest democracy, Delhi receives constant and rapid influx of migrants. Also, there are major changes in the economic structure due to liberalization of the economy, entry of multinational companies, increased per capita income, and the purchasing power of the people. As a suction force, they tend to pull in the pool of migrants which further gives an air to the process of urbanization. High influx of people from Pakistan and Bangladesh and different parts of the country led to increase in population of Delhi from 7 lakhs to 17 lakhs by 1950.

31.2.3 Delhi: The Growing Mega City

A spur in population growth occurred during the 1940s because of the migration of displaced Sikhs, Hindu Punjabis, and Sindhis. The capital city of Delhi is regarded as one of the largest forced resettlements in human history, and this movement continued into the following decades too. The intensification of Delhi's population has continued to be notably high in the last few decades with a total projected population as high as19.5 million in 2018*. The number of people projected to be living in Delhi by 2026 is around 30 million. Such a rapid urbanization has in conjunction, intensified the major Problems and challenges of environmental degradation, placed pressure on infrastructure, housing availability, and the spread of slums. It also led to dramatic and tremendous change in the land use/land cover of the metropolis of Delhi. The fertile grounds and water bodies and agricultural lands have been converted into built-up environment registering the phenomenon of urban sprawl.

The greater sprawl of metropolitan Delhi consumes an area of 1484 square kilometers, an expanse flanked by the rocky hills of the Aravalli Range and the Yamuna River. Neighbored by the territories of Uttar Pradesh and Haryana, Delhi is a largely dry zone, with significantly hot summers, transitioning into a monsoon season with the most of the city's annual rainfall recorded before winter begins. With climate change, seasonal change seems to be becoming more variable (James et al. 2015).

In 2016, the city registered a population of 16.8 million, making it the second most swarmed city after Mumbai and most packed urban agglomeration in the country. The projected population for this mega city as per estimated data is 19.5 million. The below line graph shows the trend in Delhi's population growth depicting an increasing and momentous trend for a period of 110 years (1901–2011) (Fig. 31.1).

The population density is 12,591 persons per sq. km. It is one of the snappiest growing cities in the world. Also, the mega city has registered one of the highest per capita income in India, i.e., 3.3 lakh (Economic Survey 2017–2018). Estimates show that urban concentrations of sulfur dioxide and particulates increase as income rises and then decline (World Bank 1992). Such relationships have been described by so-called Kuznet curves (inverted U-shaped curves) (Grossma and Krueger 1994). However, other environmental indicators worsen as income rises per capita generation of solid waste, nitrogen oxides, and carbon dioxide emissions. This suggests that as income increases, urban environmental problems tend to shift from the local to the global (Alberti and Layton 1996).

With this brief background, the paper aims at a concerted effort to outline, explore, and investigate the process of urbanism and urban growth prospects, on the one hand, and an assessment of quality and quantity of urban environment, on the other, thus transforming "Delhi"–as a green world-class mega city. It also attempts at critical understanding necessary to plan long-term development process in perfect blend with a purposeful transformation of the socio-economic, natural, and built environment in parity with environmental laws.

Based on preceding discussions, it is established that there is tremendous expansion of population in Delhi. Such a quantitative and momentous increase in the population results not only in affecting the quantity and quality of basic services available, but also puts a threat to the environmental and ecological surroundings of the mega city. Many a times, it has exceeded the carrying capacity of land as traditionally used. Improved natural resource utilization without

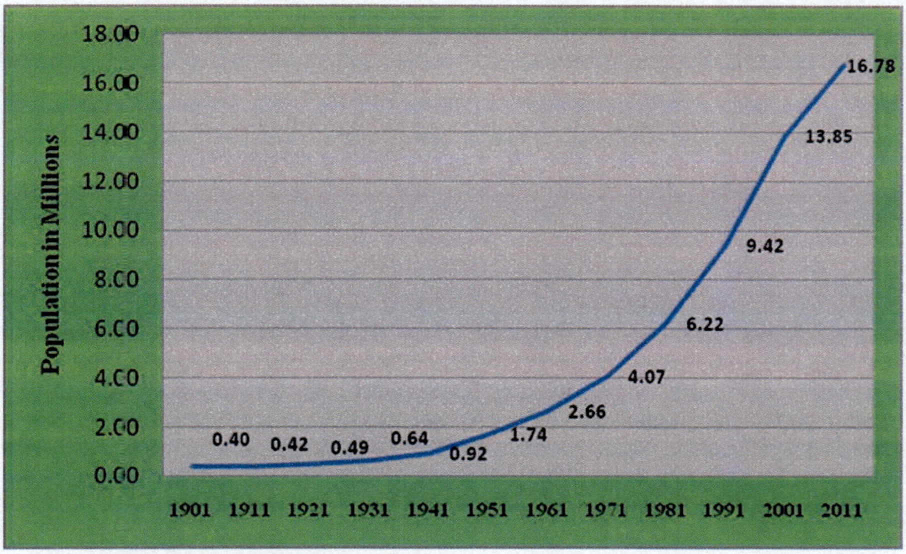

Fig. 31.1 Trends in population growth in Delhi (1901–2011). Source: Census of India (2011)

excessive environmental damage is necessary to provide an acceptable quality of life for current and future generations. And, thus, there is a need to adopt the approach of sustainable development to achieve sustainability.

31.3 Definitions and Conceptual Framework

Generally, developmental experts agree that a sustainable city should meet the needs of the present without sacrificing the ability of future generations to meet their own needs. The term "*Sustainability*" is used in ecology to indicate a condition that can be maintained indefinitely without progressive diminution of quality (Holdren et al. 1995). Applied to economic development, sustainability implies maintaining the capacity of natural ecosystems to support the human population over the long term. Arrow et al. (1995) hold that what decides whether they are compatible is the content or composition of growth, i.e., Sustainability. In the words of the Brundtland Report: "Sustainability is not a fixed state of harmony, but rather a process of change, in which exploitation of natural resources, direction of investments, orientation of technological development, and institutional change are made consistent with future as well as with present needs" (WCED 1987). Sustainable development: "economic development that satisfies current and future needs for resources and employment while minimizing the impact on biological diversity" (Lubchenco et al. 2015). Sustainable development is the development that meets the needs of the present without compromising the ability of the future generations to meet their own needs (WCED 1987). A condition of balance, resilience, and interconnectedness that allows human society to satisfy its needs, while neither exceeding the capacity of its supporting ecosystems to continue to regenerate the services necessary to meet those needs nor by our actions diminishing biological diversity (John Morelli, Environmental Sustainability: A Definition for Environmental Professionals). Ecological sustainability is: "the maintenance of life support systems and the achievement of a 'natural' extinction rate." It is an amalgamation of 4R's which is regarded as the key to "ecological sustainability," i.e., reduce, reuse, recycle, and recover. In fact, the definitions of sustainability are explored in terms of their orientation to the ontology of nature, substitutability of resources, economic growth, population growth, role of technology, and social equity (Vos 2007).

31.4 Need for Sustainability in Delhi Metropolitan Region

Delhi has a strikingly diverse multicultural population of about 16.8 million people according to Census of India (2011) and projected to rise to 19.5 million in 2018 with high density, high-rise buildings, restaurants, parks, and its metropolitan footprint radiates outwards into a region of ever-stretching car-dependent suburbs, mixed-use peri-urban zones, and a hinterland, where most of the trees have been cut down. It is a global city with well-educated populations who have a growing and sophisticated public consciousness about climate change, recycling, and water consumption and transportation issues, thus lying on the brink of sustainable urban development.

However, the public sensitivity to ecological sustainability issues in the city like the use of resources (land and energy consumption), civic amenities (water, sewage, and transportation, etc.), and carbon emissions continue to grow, in this mega city on a per capita basis that needs to be discussed in detail and remedial measures to be provided. In an endeavor to achieve sustainability, cities increasingly depend on natural resources and in the process are affected by various facets of environmental problems with Delhi metropolitan region being no exception to this.

31.5 Dimensions and Vagaries of Environmental Problems

Mega cities are primarily a phenomenon of the developing world experiencing rapid urban growth and tremendous pressure on available land existing other resources and Delhi presents a classic example to this. In fact, the environmental and ecological implications of the population pressure are reflected in terms of interfering with and deteriorating the available natural resources. In connection to this, the mega cities are faced with multitude problems, such as depleting natural resources, environmental degradation, loss of cultivated lands, disappearance of forests, soil erosion, energy crisis, dearth of civic amenities like water poverty in terms of availability of quality and quantity of clean drinking water, unhygienic sanitation and sewerages, crisis of efficient means of public transportation, etc. The main problems are highlighted as: (1) unprecedented growth in population leading increased demand for available resources, (2) increased pressure on land and resources, (3) unbalance in the demand and supply in the city, rise in unemployment rates, and virtual collapse of civic services, etc., (4) constant interference with nature without sound knowledge, (5) unsustainable use of resources like water, sanitation, and transport (the three basic civic amenities to life). The key urban concern is the growing gap between demand and supply of these basic amenities.

31.5.1 Water Related Problems

Due to rising population and increased demand DMR is facing profound challenges in securing and managing adequate and reliable water resources. Increasing urbanization and population pressure and major changes in hydrologic and climatic conditions impact water supplies and water quality. It is to be noted that warmer climate, deficit rainfall, and heavy urbanization have led to the perpetuating water shortages in the capital (52% of water gets wasted due to leaks in pipelines). There can be various management techniques for water problems like by conserving and judicious utilization of water based on graded pricing system, by recycling of wastewater and sewage for non-domestic use, and by encouraging water harvesting techniques.

31.5.2 Sanitation, Sewerage, and Waste Disposal

Delhi metropolitan region is characterized with many low-income communities and development of slums (e.g., the areas like Seelampur, Bawana J. J. Colony, Jahangirpuri slums) due to rapid influx of migrants which are not mainstreamed into the productive city life. They face serious public health and environmental problems due to lack of availability of safe sanitation and proper sewage treatment plants. Municipal and industrial wastewaters containing human pathogens, refuse and toxic wastes are directly dumped into Yamuna River. Almost 22 "*nalas*" drain into Yamuna River turning into a big "*mahanala*" *(big drain)*.

The city lacks proper waste disposal facilities with poor sewage treatment plants. Half a kilogram of waste is created per capita in Delhi, with 70% of this being collected and disposed of through formal means. It, therefore, implies that 30% of waste is disposed of through the streets or in illegal dumping places. It has led to piles of garbage and other litter across the city being increasingly common. It creates not only environmental and health issues, but dramatically affects and adds to the emissions of greenhouses gases, especially methane gas which is released from decaying of dumped and piled wastes over years. Presently, there are three major sanitary sites for the city of Delhi: Ghazipur, Bhalswa, and Okhla. The use of these sites as landfill locations is rapidly moving towards operational completion, which means there is an increased demand for the government to initiate new and safe alternatives.

31.5.3 Methods to Improve Sanitation Services

The following methods can be adopted to improve sanitation services:

- Sustainable integration of water supply and wastewater management by participatory planning, i.e., health institutions, municipal governments, and external aid agencies.
- Adoption of the conventional sewerage and treatment system would be preferred.
- Strict laws and heavy penalty to industries polluting rivers.
- Behavioral change in sanitation practices through awareness campaign.
- Concept of Green Toilets: Wastes are disposed in natural ways (Bio-toilets) and (zero-discharge toilets). In the bio-toilet system, human waste is converted into liquid and gases. The gases get mixed with air and liquid is discharged on the track. In the zero-discharge system, solid and liquid waste are separated using special type of solid-liquid separator, the liquid is recycled after proper treatment and waste is converted into manure.

31.5.4 Transportation Problems

Transportation in mega cities in the developing nations is in a state of crisis with Delhi metropolitan region being no exception to this. Heavy urbanization has massively led to problems of transport-related services like public transportation, noise, congestion, and provisions for non-motorized transport. Traffic-generated pollutants, such as particulates, nitrogen oxide, and carbon dioxide are an increasing problem as cities sprawl. Photochemical smog is a serious threat in most cities and the recent occurrence of smog in Delhi particularly during winter month (November) is a classic example to this. Increasing and choking levels of air pollution act as a threat to environment in the form of vehicular emission (NO2, CO2, CO). Also, it gives rise to increasing health hazard due to air-borne respiratory diseases like asthma, bronchitis, lung cancer, etc. (60 million cases per year reported).

31.5.5 Transportation Problems Faced

- "Premature Congestion" due to high levels of traffic and declining mobility.
- Long commute times in traffic which otherwise could have been productively utilized.
- Rising and choking levels of air pollution–a threat to environment, due to large amount of vehicular emission (NO_2, CO_2, CO).
- Increasing health hazard due to air-borne respiratory diseases like asthma, bronchitis, lung cancer, etc. (60 million cases per year reported).

31.5.6 Methods to Improve Efficiency of Transportations

- Planning for reduced motorized, comfortable, and "greener" transport like Delhi Metro and Delhi Transport Corporation (DTC) eco-friendly buses in Delhi. The Delhi Transport Corporation runs the world's largest number of eco-friendly transports. These vehicles continue running on CNG and are less harmful by manifolds compared in relation to those that continue running on diesel or petrol.
- Efforts to reduce congestion and travel time with fast means Rapid Transit Metro in Delhi (NCR) without affecting public spaces, for example in Gurugram.
- Affordable transportation that can enhance economic mobility by connecting the poorer segments like Delhi Metro.
- Discouraging private vehicles on roads in terms of parking, gasoline, etc. on odd-even rationale (last digit of license plate and quite successful in USA); for example, odd number on Monday, Wednesday, and Friday) and even number on Tuesday, Thursday, and Saturday for better traffic management and ensuring safety.

Hence, with an unprecedented pace of growth, and the vagaries of environmental problems discussed above, the city needs to be able to integrate its elegant past as well as the modern developments into a "green" organic whole keeping pace with the future generations so that it can attain the tag of a "Sustainable City." Such dynamism in Delhi represents a major sustainable development opportunity. By getting urban development right, cities can create jobs and offer better livelihoods; increase economic growth; improve social inclusion; promote the decoupling of living standards and economic growth from environmental resource use; protect local and regional ecosystems; reduce both urban and rural poverty; and drastically reduce pollution. Sound urban development will accelerate progress towards achieving SDGs.

An emerging group of scholars has started to position the problem of local government sustainability action within a system of multilevel governance (Bulkeley and Betsill 2005; Corburn 2009; Bulkeley 2010; Homsy and Warner 2013). Municipalities decide upon the appropriate action for each local situation. In this co-production approach, knowledge and policy innovation flow up from local governments, down from higher authorities, and horizontally across networks of municipalities (Homsy and Warner 2013).

31.6 How we Achieve Environmental Sustainability

Research shows that smaller and rural communities adopt general sustainability policies at a lower rate (Conroy and Iqbal 2009; Lubell et al. 2009; Homsy and Warner 2012) are three times less likely to protect their drinking water (Phoenix 2002) and tend to resist land use planning more than larger places (Wolensky and Groves 1977). These municipalities have lower rates of government service provision (Johnson et al. 1995; Warner 2006) and are less likely to experiment with service reforms (Hefetz et al. 2012). Smaller places also demonstrate lower quality plans for environmental protection (Tang 2009).

In an endeavor to make this mega city ecologically sustainable, efforts are made to check the rapid and haphazard growth of population, and optimally utilize the available resources through "Go Green" concept and "Green Innovation."

Studies have identified a multitude of factors that contribute to local sustainability for policy adoption. These include concerns with preserving home values (Fischel 2001), city governments interest in achieving certain co-benefits (e.g., cost savings) (Bulkeley and Betsill 2003; Kousky and Schneider 2003; Svara et al. 2011), competition for economic development (Jochem and Madlener 2003), fiscal capacity (Lubell et al. 2009), and citizen's advocacy (Portney and Berry 2010). But the ironical situation is such that despite the enthusiasm among researchers and practitioners for bottom-up policymaking, most municipalities do not take action to promote environmental sustainability (Svara et al. 2011; Saha and Paterson 2008; Conroy and Iqbal 2009).

Resources are needed for policy action and these resources include funding and technical skills (Thompson 1965). To add public services, such as environmental protection, local governments must have the capacity to seize new opportunities (Watson 1997) and, as problems become increasingly complex, the need for capacity increases (Honadle 2001).

31.7 Diffusion of Green Innovation (Ecological Sustainability) over Space

As a geographer, we are more concerned with the process of diffusion, i.e., spatial diffusion. There are five basic and interrelated stages, which are passed by some very quickly and while others take time to pass the stages (Fig. 31.2).

Every innovation or to be precise "new idea" has a particular point of origin, i.e., a source. Once the innovation is evolved, it needs to spread, diffuse, or pass on from one person to the other and from one space to another for it to be understood and, thus, adopted. This process of spread of innovation may be in the field of environment to make it ecologically sustainable thus emerges the diffusion of green innovation.

An innovative and educated person has thorough knowledge about the environment in which he/she lives and, thus, takes less time to adopt any new innovations which are ecologically sustainable. The person can very well analyze the pros and cons and can rationally interpret the results favorable to environment and which are certainly sustainable.

31.8 Good Governance and Green Innovations: Delhi Going Green

Urban sustainability is the idea that a city can be organized without excessive reliance on the surrounding countryside and be able to power itself with renewable sources of energy. The aim of this is to create the smallest possible ecological footprint and to produce the lowest quantity of pollution possible, to efficiently use land, compost-used materials, recycle it or convert waste-to-energy, and to make the city's overall contribution to climate change minimal.

Technology plays the biggest role to achieve the motto of "Minimum Government and Maximum Governance." Governance has become a "hot" topic as evidence mounts on the critical role it plays in determining societal well-being. Governance is "the process of decision-making and the process by which decisions are implemented (or not implemented)." The governance of urban service delivery in mega cities of the world plays a pivotal role. Institutions and governance can go a long way in for effective delivery of the basic services of water management and improved and hygienic system.

Good governance is an indeterminate term used in international development literature to describe how public institutions conduct public affairs and manage public resources. The major characteristics of Good Governance are shown in Fig. 31.3. The Secretary General of the United Nations, Kofi Annan, reflects a growing consensus when he states that "good governance is perhaps the single most important factor in eradicating poverty and promoting development." It is commonly agreed that effective water and sanitation rely on strong institutions and good governance, especially so that all people, even the poorest, receive water effectively and can use it to serve their needs (Bakker and Kooy 2007, Hardoy et al. 2005).

Formation of National Green Tribunal (NGT) presents an example of good governance in Delhi. The Green Court (December 11, 2015) was formed for environmental protection and conservation of forests and other natural resources including enforcement of any legal right relating to environment. National Green Tribunal (NGT) plays a pivotal role in safeguarding the environment and provides the inhabitants an opportunity to live in a clean, green, and free Delhi. In addition to this, there are some best practices in "green innovations" adopted in Delhi Metropolitan Region that needs worth mentioning. Some of them are discussed below.

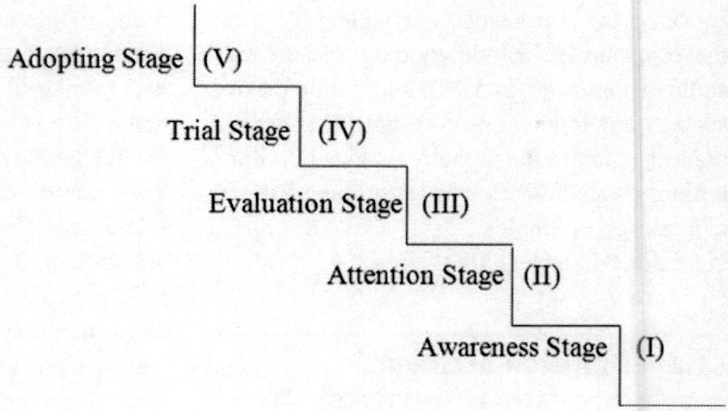

Fig. 31.2 Stages in Adoption and Diffusion

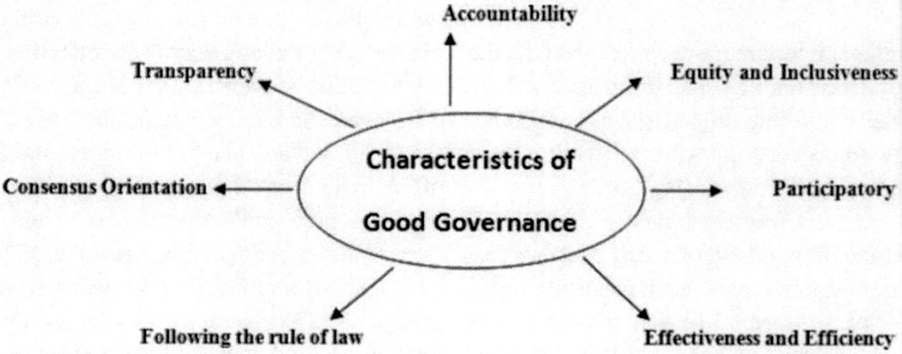

Fig. 31.3 Characteristics of good governance

31.9 Best Practices in Green Innovations: A Reflection of Sustainability in Delhi

Technology has a large part to play in enabling positive outcomes for cities. The solutions being developed today will be critical in ensuring that the cities of tomorrow represent a positive and sustainable development.

1. Paperless Computerized Electricity Bill Collection and Online Remittance (TPDDL).
2. Use of "Green Cycles" at metro stations for students to ply to Colleges and University.
3. Low floor CNG buses in Delhi to enable the differently abled citizens"keep moving."
4. Green Buildings: It aims at using natural building materials and adopts greener methods, such as rainwater harvesting for its water needs, solar energy for the electrification, wastewater treatment plants, etc. Hence, making the building self-sufficient ensuring sustainability with lower carbon footprint at every stage (Example: Civic Centre, Asaf Ali Road: With an area of 1.16 lakh m^2).
5. Delhi Metro bags the title of "World's First Green Metro" for its eco-friendly initiatives, i.e., solar power installations, initiatives for green depots, stations, and buildings.
6. Development of Vertical Gardens (Green Metro Pillars): Eight pillars of the elevated Blue Line (Dwarka Sector 21-Vaishali/ Noida City Centre) between Mandi House and Pragati Maidan Stations are sporting vertical gardens, an initiative by NDMC (Fig. 31.4). The main aim is to increase green space in Delhi and anti-pollution plants to reduce smog and to produce oxygen.
7. Green Toilets: An initiative taken by Indian Railways. In the bio-toilet system, human waste is converted into liquid and gases. The gases get mixed with air and liquid is dis-

Fig. 31.4 Panoramic view of green vertical gardens (Pragati Maidan)

charged on the track. In the zero-discharge system, solid and liquid waste is separated using special type of solid-liquid separator, the liquid is recycled after proper treatment and waste is converted into manure.

8. E-Rickshaw (Battery Driven Rickshaw): It is being widely accepted as an alternative to petrol/diesel/CNG auto-rickshaws and is anti-pollution in nature. They are widely used in Delhi and other parts of India. In Delhi, as per government official's figures in April 2017, the number was over 2, 50,000.
9. The wider use of Green Tanks: It proves to be most economical and environmentally sustainable storage solution. Modular in design, it is relocatable, re-useable and can be built to store capacities of 50 megaliters of water.
10. Using mobile phones to check traffic rule violations: The Third Eye Project, initiated by Gurgaon Traffic Police, utilizes innovative mobile phone application to capture images and track traffic rule violators in the city. (Example In Delhi, Ring Road on the way to ITO).
11. Sensitizing people for segregation of wastes using different colored Bins at public places.

31.10 Conclusion

In a nutshell, the mega city of Delhi thus represents a symbol of rich and ancient tradition, a seat of mesmerizing Indian culture, a growing hub of contemporary educational system, etc. A prime mover and nerve center of ideas and actions, the seat of national governance and a center of business, culture, education and sports, Delhi, however, stands at the crossroads today.

Innovation and technology along with social welfare reforms are regarded as an agent to fulfill human needs and aspirations and they present creative ideas facilitating the vision of "Good Governance" in Delhi metropolitan region and in India in general. Such an approach if handled with care is likely to lead a movement towards making Delhi Metropolitan Region, a world-class sustainable city in the decades to come.

References

Alberti M, Layton D (1996) Sustainability and the urban factor. Paper presented at the fourth biennial meeting of the International Society for Ecological Economies, Boston, 4–7 August 1996

Alberti M, Susskind L (1996) Managing urban sustainability: an introduction to the special issue. Environ Impact Assess 16:213–221

Arrow K, Bolin B, Costanza R, Dasgupta P, Folke C, Holling CS, Jansson BO, Levin SG, Mler K, Perrings C, Pimental D (1995) Economic growth, carrying capacity, and the environment. Science 268:520–521

Bakker, K. and Krooy, M. (2007) 'Governance Failure: Rethinking the Institutional Dimensions of Urban Water Supply to Poor Households', World Development, 36(10). https://doi.org/10.1016/j.worlddev.2007.09.015

Bulkeley H (2010) Cities and the governing of climate change. Annu Rev Environ Resour 35(1):229–253

Bulkeley H, Betsill M (2003) Cities and climate change: urban sustainability and global environmental governance. Routledge, London

Bulkeley H, Betsill M (2005) Rethinking sustainable cities: multilevel governance and the 'urban' politics of climate change. Environ Polit 14(1):42–63

Conroy MM, Iqbal A (2009) Adoption of sustainability initiatives in Indiana, Kentucky, and Ohio. Local Environ 14(2):109–125

Corburn J (2009) Cities, climate change and urban Heat Island mitigation: localizing global environmental science. Urban Stud 46(2):413–427

Fischel WA (2001) The Homevoter hypothesis. Harvard University Press, Cambridge, MA

Folke C, Hahn T, Olsson P, Norberg J (2005) Adaptive governance of social-ecological systems. Annu Rev Environ Resour 30:441–473

Gallopín GC (2006) Linkages between vulnerability, resilience, and adaptive capacity. Glob Environ Chang 16:293–303

Grossma G, Krueger AB (1994) Economic growth and the environment. NBER, Cambridge, MA

Hardoy A, Hardoy J, Pandiella G, and Urquiza, G. (2005) 'Governance for Water and Sanitation services in lowincome settlements: experiences with partnership-based management in Moreno, Buenos Aires' Environment and Urbanisation, 17(1):183–200

Hefetz A, Warner ME, Vigoda-Gadot E (2012) Privatization and Intermunicipal contracting: the US local government experience 1992–2007. Environ Plan C 30(4):675–692

Holdren JP, Daily GC, Ehrlich PR (1995) The meaning of sustainability: biogeophysical aspects. In: Munasinghe M, Walter S (eds) Defining and measuring sustainability, the biogeophysical foundations. The World Bank, Washington, DC

Homsy GC, Warner ME (2012) Off the beaten path: sustainability activities in small towns and rural municipalities, in municipal yearbook 2012, 53–61. ICMA Press, Washington, DC

Homsy GC, Warner ME (2013) Climate change and the co-production of knowledge and policy in rural USA communities. Sociol Rural 53(3):291–310

Honadle BW (2001) Theoretical and practical issues of local government capacity in an era of devolution. J Reg Anal Policy 31(1):77–90

James P et al (2015) Urban sustainability in theory and practice. Routledge, London. ISBN: 978-1-138-02572-1

Jochem E and Madlener R (2003) The forgotten benefits of climate change mitigation: innovation, technological leapfrogging, employment, and sustainable development. In: Workshop on the Benefits of Climate Policy: Improving Information for Policy Makers

Johnson KM, Pelissero JP, Holian DB, Maly MT (1995) Local government fiscal burden in nonmetropolitan America. Rural Sociol 60(3):381–398

Kousky, C. and Schneider, S. H. (2003) 'Global climate policy: will cities lead the way?' Climate Policy, 45(4): 1–14. https://doi.org/10.1016/j.clipol.2003.08.002

Lubchenco J et al (2015) Sustainability rooted in science. Nat Geosci 8(10):741–745

Lubell M, Feiock RC, Handy S (2009) City adoption of environmentally sustainable policies in California's Central Valley. J Am Plan Assoc 75(3):293–308

Mohan R, Dasgupta S (2005) The 21st century: Asia becomes urban, special articles. Econ Polit Wkly 40(3):213–223

Phoenix LE (2002) Introduction: small municipalities and water supply. Water Resources Impact, March 2002

Portney KE, Berry JM (2010) Urban Advocacy Groups, City Governance, and the Pursuit of Sustainability, in American Cities, presented at the 2010 Meetings of the American Political Science Association, September 5, Washington, DC

Romero-Lankao P, Gnatz DM (2016) Conceptualizing urban water security in an urbanizing world. Curr Opin Environ Sustain 21:1–7

Saha D, Paterson RG (2008) Local government efforts to promote the 'three Es' of sustainable development: survey in medium to large cities in the United States. J Plan Educ Res 28(1):21–37

Svara JH, Read A, Moulder E (2011) Breaking new ground: promoting environmental and energy programs in local government, conserving energy and the environment. IBM Center for the Business of Government, Washington, DC

Tang Z (2009) How are California local jurisdictions incorporating a strategic environmental assessment in local comprehensive land use plans? Local Environ 14(4):313–328

Thompson VA (1965) Bureaucracy and innovation. Adm Sci Q 10(1):1–20

Vos RO (2007) Defining sustainability: a conceptual orientation. J Chem Technol Biotechnol 82:334–339

Warner ME (2006) Market-based governance and the challenge for rural governments: U.S. Trends Soc Policy Adm Int J Policy Res 40(6):612–631

Watson DJ (1997) Climate for innovation. In: Watson DJ (ed) Innovative governments: creative approaches to local problems. Praeger Press, Westport, CT

Wolensky RP, Groves DL (1977) Planning in the smaller city: major problems and a possible solution. Socio Econ Plan Sci 11(1):37–41

World Bank (1992) World development report 1992. The World Bank, Washington, DC

World Commission on Environment and Development (1987) Our common future. Oxford University Press, Oxford

Investigating the Variability in Sanitation and Sustainability Issues: Evidence from Resettlement Colonies of Delhi

32

Pooja Yadav and Subhash Anand

Abstract

Sanitation is the hygienic means of promoting health by avoiding the human contact with the hazardous waste. The intervention of separating hazardous waste from humans is termed as sanitation. The sanitation interventions could be in form of hardware including building up of infrastructure, e.g., hygienic latrines or software in form of hygiene practice such as hand washing with soap. The sanitation is the safe disposal of human excreta defining "Safe Disposal" as the hygienic containment or treatment of the excreta to avoid adversely affecting human health. The present chapter focused on the status of sanitation issues related to the basic amenities (water, sanitation, drainage) in the four resettlement colonies of Delhi. The present study is based on primary and secondary data. To understand the situation of sustainable sanitation in the resettlement colonies of Delhi multiple variables are observed and studied. The study concludes that a radical shift in the behavior of the local people is required to achieve sustainable sanitation in urban areas. The urban sanitation approach also requires engagement of all the engaged stakeholders for the sustainable sanitation management of the resettlement colonies in Delhi.

Keywords

Resettlement colonies · Solid waste · Sustainable sanitation · Wastewater disposal

32.1 Introduction

Sanitation is not a new concept in India. Its importance dates to ancient times which were attached to cleanliness, sanitation, and drainage. The Indus valley civilization is an important example. The urban population of India is increasing rapidly with the pace of time and exerting huge pressure on present resources, i.e., urban services. It has been observed that present urban infrastructure is unable to keep pace with the increasing population. Challenge lying ahead in front of Indian cities is to make cities livable making every citizen accessible to basic services of acceptable quality. Lack in basic services further affects their daily life by affecting health status, economic condition, and social status. The sole authority for providing the basic services lies in the hand of urban local governments, but they are often strapped of funds and unable to discharge their duties satisfactorily. Sanitation can be defined as endowment of facilities and services aimed at safely disposing human feces and

P. Yadav · S. Anand (✉)
Department of Geography, Delhi School of Economics, University of Delhi, Delhi, India

B. Thakur et al. (eds.), *Resource Management, Sustainable Development and Governance*, Sustainable Development Goals Series, https://doi.org/10.1007/978-3-030-85839-1_32

urine. Improving sanitation or provision of sustainable sanitation has a positive impact over health of communities, thus inadequate sanitation is chief cause for diseases across world-wide.

The foremost challenge for countries seeking to work out on the problem of checking out the accessibility of sanitation defines "sanitation." The second challenge is to decide what aspects are crucial while working on sanitation. In other words, focus-based aspect of the problem is going to be dealt with according to the priority. The reported problem is not an easy task as many of the professionals confuse within the two steps. With respect to defining sanitation, most of the experts would reach agreement that "Sanitation" overall is an "Umbrella Term" but, as per the World Health Organization (WHO) sanitation covers inter alia:

1. Safe assemblage, storing, handling, and dumping of human and animal waste,
2. Management of waste, re-use, and recycling of solid wastes,
3. Drainage of water from discharged various sources, treatment, and disposal of sewage effluents, and
4. Discharge and management of industrial waste products.

Sanitation is the hygienic means of promoting health by avoiding the human contact with the hazardous waste. The intervention of separating hazardous waste from humans is termed as sanitation. The sanitation interventions could be in form of hardware including building up of infrastructure, e.g., hygienic latrines or software in form of hygiene practice, such as hand washing with soap. The sanitation is the safe disposal of human excreta defining "safe disposal" as the hygienic containment or treatment of the excreta to avoid adversely affecting human health (Mara et al. 2010). As per report, 2.6 billion people, mainly in the developing countries lack access to improved sanitation. It also showed that about 1.2 billion people lack even an unimproved sanitation facility and practice open defecation (WHO and UNICEF 2010). Present chapter focused on the present status of sanitation issues related to the basic amenities (water, sanitation, drainage) in the four resettlement colonies of Delhi.

Box 32.1 Sustainable Development Goals

Sanitation in Delhi's resettlement colonies is a challenge as it is transformed into the fastest urbanizing center in the nation. Even though sanitation has made considerable progress, yet as the authors in this chapter observe, the city administration still grapples with the enormous challenge of providing improved and adequate sanitation facilities to its emerging population in the resettlement colonies. While UN Agenda 2030 provides a framework and road map for capacity building with water and sanitation, the same is also complimented by the efforts of the Government at national, regional, and local scales that influence not only SDG-6, but also other related goals, targets, and indicators. According to the authors in the chapter, this not only places water and sanitation at the core, but also establishes synergies between SDG-6 and other goals such as poverty eradication (SDG 1), ending hunger (SDG-2), ensuring healthy lives, and promoting well-being (SDG- 3), education for all (SDG-4), gender equality (SDG-5), and inclusive cities (SDG-11). Achievement of the above goals will need adequate capacity building to facilitate collaboration and synergy between stakeholders to ensure empowerment of people and their participation. Finally, the authors recommend behavioral change as a key component towards achieving safe sanitation. They argue that governance process needs to be supplemented and synergized with awareness generation, for achieving intended outcomes.

32.2 Review of Literature

Inequality in the access to a satisfying water supply from the public network takes the typical geographic shape where peripheral areas get lower volumes per residents, but the main determinant of the inequality is the fragmented administrative status of the housing stock in Delhi. As is the case in many developing cities (UN-Habitat 2003) the occurrence of slums and other forms of informal housing is a continuing concern in Delhi. This trend is largely attributed to the failure of the public agency in charge of land development, defining and implementing urban policies and housing promotion in the NCT, the Delhi Development Authority (DDA). The DDA's ambitions in controlling urban growth are denounced as being disconnected from reality, and the political debate is particularly fierce between those for asking strict compliance with the plan and others demanding its amendment (Verma 2002). The sanitation is the safe disposal of human excreta defining "safe disposal" as the hygienic containment or treatment of the excreta to avoid adversely affecting human health (Mara et al. 2010). Sanitation is basic part of life for health and well-being, overcrowded cities facing challenges in providing sustainable sanitation, i.e., safe, functional sanitation systems, and affordable. Unsustainable sanitation system and lost chances to handle the urban challenges are result of factors such as restricted political will, insufficient technical, financial and institutional capabilities (Andersson et al. 2016). The larger cities like Delhi have reached exorbitantly colossal population size, further moving towards the detained level of urban services like poor housing or housing shortage, lack in proper water supply and quality of life, sanitation and other basic amenities, thus, at the end proliferation of slums (Ray 2017).

32.3 Issues and Challenges in Sustainable Sanitation

Sanitation is basic part of life for health and well-being, overcrowded cities facing challenges in providing sustainable sanitation, i.e., safe, functional sanitation systems, and affordable. Factors such as limited political will, inadequate technical, financial, and institutional capacities, and failure to integrate safe sanitation systems into broader urban development have led to a persistence of unsustainable systems and missed opportunities to tackle overlapping and interacting urban challenges (Andersson et al. 2016).

The poorly controlled waste also refers to invitation to an unpleasant environment. Discharging the untreated wastewater and human excreta into the environment affects human health by different modes, i.e., polluting of drinking water, allowing entering in the food chain through food stuffs, by bathing and other recreational activities (Andrianou et al. 2019). These activities are further supporting for breeding grounds to flies and insects which spread disease. Therefore, to avoid or minimize the socio-economic and environmental impacts, there is need to work upon various indicators of sanitation. The conceptual framework presented provides a way to accumulate information with the help of knowledge acquired and shared by the respondents during the field survey. Response of the respondents and observation during the field surveys have been used for "illustrating relations, identifying patterns, presenting an overview, and details" regarding various complex issues that are arising in sanitation (Fig. 32.1). The key points related to institutions that are accomplished from the interviews taken are presenting results; coordination among the multiple departments like solid waste, infrastructure development, environmental health, water and sanitation is foremost requirement for sustainable planning.

There encountered a matter of concern about accountability in decision-making and communication among the various departments. Another related issue came across as the insufficient flow of information among the service providers and users, i.e., the municipal departments or the Non-Government Organizations and the private sectors involved in providing service to residents.

There exist an unresolved and politicized query relating to the sustainability of the service, i.e., whether there are progress or upgradation of plans for the informal settlements, and how the

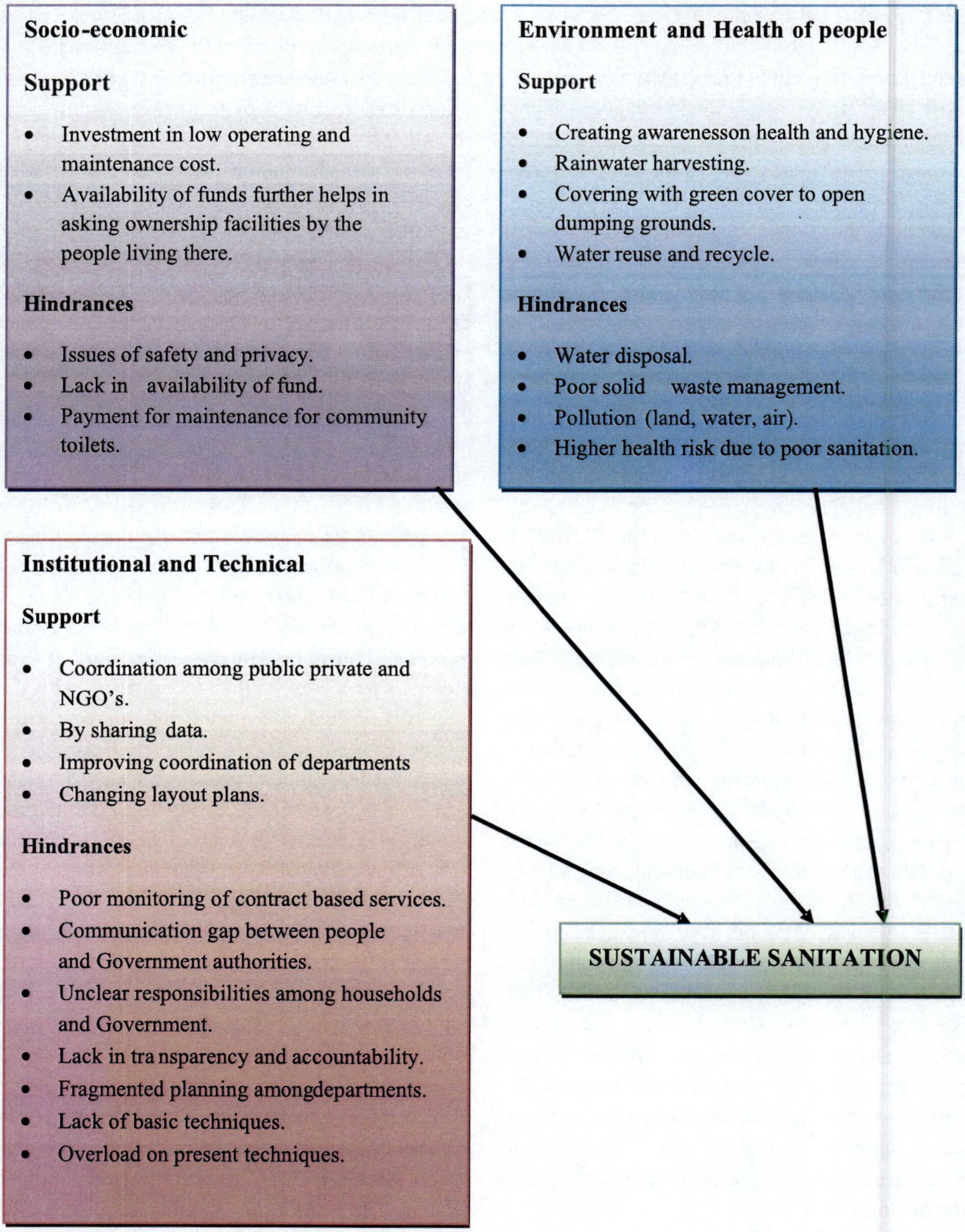

Fig. 32.1 Conceptual framework of sustainable sanitation. *Source:* Prepared by Authors

planning and design of sanitation system process are influenced by these plans. For example, more operating and maintenance costs due to high cost of chemical and other toilets cleaning stuffs demarcated as a barrier to sustainability. One among the major technical challenges (Fig. 32.1) recognized for increasing amenities in informal settlements is the lack of water accessibility and the not well-connected sewer facilities. These areas characterized of both areas, i.e., rural and urban land use but they lack in sanitation-based services. The safety for women and children is the main social issue identified for the prerequisite of sustainable sanitation. The discouraging factor for not using sanitation facilities at night and early morning is greater walking distance to community-based facilities. Besides the social and economic concern, the major one is regarding the health. To make sanitation sustainable, there is need to create awareness regarding health and hygiene.

Therefore, there are methods which are sustainable and can assist in walking on the footstep's equity based and sustainable nature of services. The same approach can be followed for taking decision of ongoing projects and at various levels of work undergoing for the informal areas. These approaches and tools would be more beneficial to take into consideration during the planning and design approach-based stage. At these stages, decision making associated with making priority of sanitation-based issues like technology based, layout, and the number of constructed toilets is to be influenced. Overall, the sustainable sanitation, which is a holistic concept, taking into consideration the institutional factors, economic, social, environmental, and then walking in the shoes of sustainability-based approach. To achieve the targets of Sustainable Development Goals (SDGs) there is need to involve every section of society, be it the women, children, old aged, and disabled person.

32.4 Data Source and Methods

The present study is based on both primary and secondary data. To understand the situation of sustainable sanitation in the resettlement colonies of Delhi, multiple variables are collected and studied. Both the qualitative and quantitative research methods are used based on nature of information required. The secondary data was collected from government offices but some of the Government offices were consulted such as Delhi Urban Shelter Improvement Board, Delhi Development Authority (DDA), Ministry of Urban Development, Census of India, Directorate of Health Services, and National Institute of Urban Affairs (NIUA). The background information pertaining to extent, area, demographic characteristics, urbanization, and socio-economic profile was compiled from district census handbook, and census of India. Also, the research work of School of Planning and Architecture also taken into consideration.

For the primary data, a questionnaire was prepared regarding the various aspects of sustainable sanitation, i.e., quality of water, sources of water, disposal of waste generated, etc. The total number of respondents selected for survey were based on purposive random sampling and study area is chosen based on stratified sampling. For the primary survey, a total sample of 400 respondents were selected and out of which 100 respondents from each resettlement colony. The data presented with tables; percentage analyzed quantitatively give clear picture of the topic quantitatively. Various descriptive statistics methods like measure of central tendencies are used vastly. Mapping and graphical work is made with the help of MS-excel and GIS techniques.

32.5 The Study Area

Delhi is situated on the banks of the Yamuna River. The National Capital Territory (N.C.T.) of Delhi spreads over 1483 km^2 between the latitudes of 28°25′ North to 28°53′ North and the longitudes of 76°50′ East to 77°22′ East. The region occupies a significant position in the Indian subcontinent. It has extreme climate due to its continental situation which is very cold in winter and terribly hot in summer (Directorate of Economics and Statistics 2012). The N.C.T. of Delhi at the time of Census of India (2011) comprises 9 districts with 3 tehsils in each district, but now Delhi is divided into 11 districts as two more are added, i.e., Shahdara and South-East Delhi. Delhi has experienced phenomenal population growth immediately after declaration as the capital of India during the year 1911. In 1901, Delhi had total population of 0.4 million and 16.78 million population in 2001, Delhi is the second largest city in India according to 2011 census. Due to the migration of people from across the country, Delhi has grown to be a cosmopolitan metropolis.

The resettlement colonies of Delhi were developed during the period from 1962 to 1977, which can be divided into two categories, i.e., (1) Those established before 1975 called *jhuggi-jhopri* clusters and (2) those that came up during the national emergency period of 18 months called resettlement colonies. Presently, there are 47 resettlement colonies in Delhi. The present study is based on the selected four resettlement colonies of Delhi, i.e., Seelampur, Hastsal, Madanpur Khadar, and Jahangirpuri (Fig. 32.2).

Jahangirpuri Resettlement colony lies in the Northwest District of Delhi. It is situated along the G.T. Road, very near Adarsh Nagar. The number of plots allotted to the people was 18,938, size of the plots was 25 m^2, and the area was 988 acres as per the Delhi Urban Shelter Improvement Board (2015). Seelampur mainly falls in the East district of Delhi, and it is further divided into different blocks. Seelampur is basically a resettlement colony formed through the collusion of state violence and elite-middle class interests. The number of plots allotted to the people living here was 5976 with 25 m^2. The area covered was 59.2 acres. Madanpur Khadar is a relocation colony in southeast Delhi built by the government to relocate people from unauthorized slums from Nehru Place and Alaknanda. The area covered was 9.615 hectares and the size of the plots allotted was 12.60 m^2 (1071 plots) and 18 m^2 (1304 plots) (Delhi Urban Shelter Improvement Board 2015). The Hastsal resettlement colony falls under the West district of Delhi. The number of plots allotted to the Hastsal colony were 3969, the covered area included 17.43 hectares at the time of relocation.

32.6 Result Analysis and Discussion

Every person needs 20–50 L of potable water a day for their basic needs: drinking, cooking, and cleaning, but more than one in six does not have access to such amount of potable water. Africa has the lowest total water supply coverage of any region, with only 62% of the population having access to improved water supply (UNESCO 2006). China and India combined currently accounts more than a third of world's population. Progressive development in water and sanitation sector in China and India are of great significance for bringing change in the global scenario, as these two countries comprising a vast proportion of the regional as well as world's population share. To some level, these two have immense influence on the global trend, and world's ability to achieve the Sustainable Development Goals (SDG) target is highly dependent on the performance of these two countries. Subsequently, it is affecting the health benefits accrued from improvement in the water and sanitation conditions.

The big cities like Delhi have reached a level of over population, moving towards provision of detained level of urban amenities, poor accommodation facilities, lack of water supply, and poor quality of life. The capital city is characterized by lack in housing space, multiplication of slums, inadequate water supply, poor sanitation, medical and institutional arrangements. The resettlement colonies are the products of population explosion, the people from the selected slum areas being rehabilitated to the other areas referred to as "Resettlement Colonies."

32.6.1 Water Accessibility

Access to improved water sources is also among the counted indicators taken by the Joint Monitoring Program to determine the achievements or target taken for Millennium Development Goals (MDGs) for water and sanitation. The share of population using improved sources of drinking water are those who get water from many of the following types of water supply: piped water (into dwelling yard or plot tube well or borehole), public tap or standpipe, protected well, protected spring and rainwater collection while unimproved sources are unprotected dug well, vendor-provided water (cart with small tank or drum, tanker truck), unprotected spring,

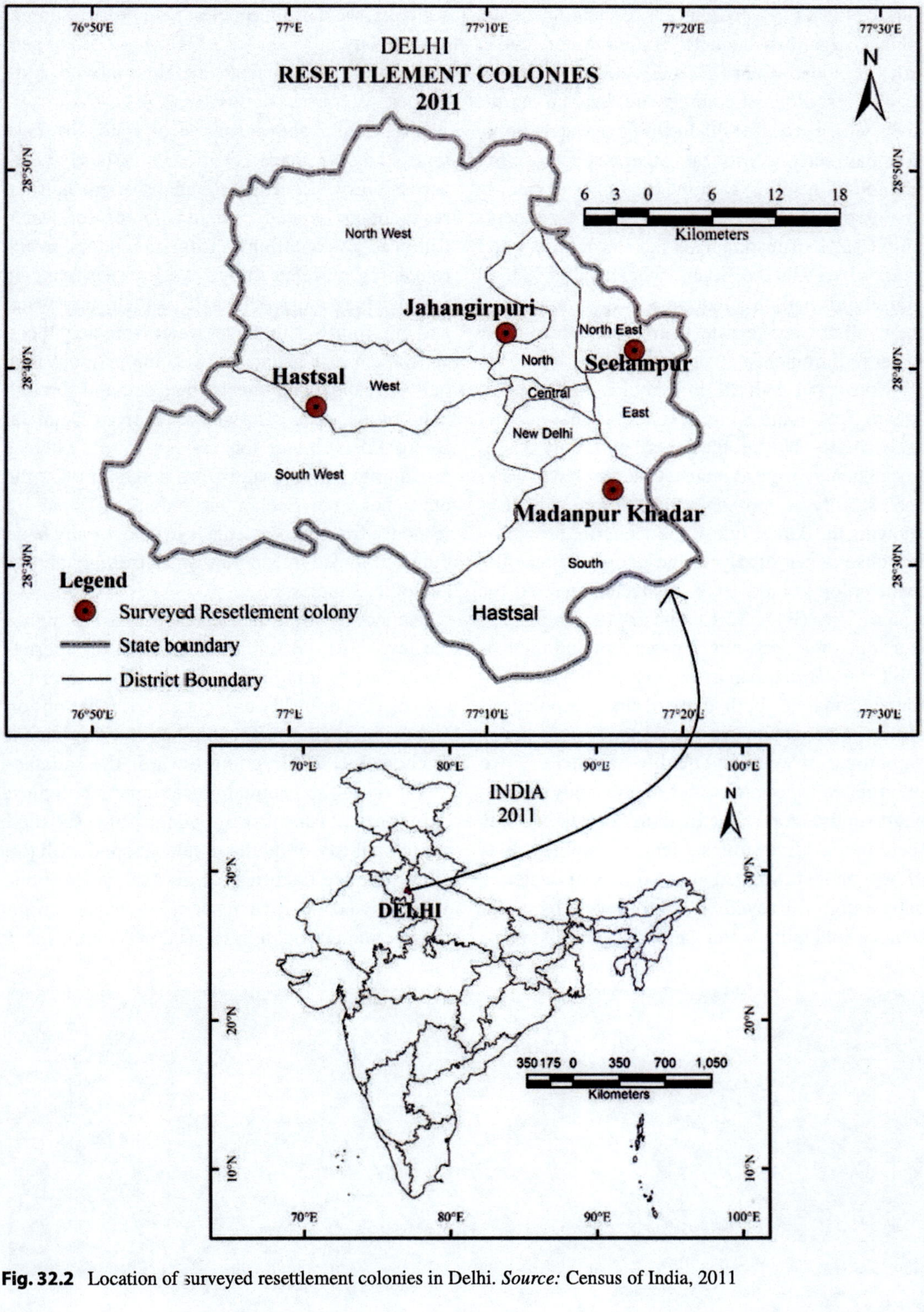

Fig. 32.2 Location of surveyed resettlement colonies in Delhi. *Source:* Census of India, 2011

surface water (river, dam, lake, pond, stream, canal, irrigation channel), bottled water, tanker truck provided water (UNICEF 2006).

The resettlement colonies are also among the areas which are affected due to poor water supply that indirectly affects the other aspects of life including health, well-being, and degrading quality of life. As per the response of respondents only 61.2% of respondents have access to water supply by Delhi Jal Board, whereas 38.8% total households had no direct piped water supply and they relied on alternate sources for their daily water requirements.

More than half of households surveyed in Jahangirpuri and Hastsal found to have water accessibility, but, on the other hand, only 3% of respondents in Madanpur Khaddar have water accessibility to proper water source which is showing the darker side of the scenario. Providing in house water supply to the people of resettlement colony could be a great challenge to the water utility (Plate 32.1). The urban water sector is a zone of serious mismanagement and the reason being large urban areas representing concentrated demands, both due to large populations and large per capita use and waste.

Sources of water and quality of water are two interrelated aspects of water; in the study area no better option for water in terms of quality was seen. In the surveyed resettlement colonies, 44% of respondents argued that source of water is hand pump surveyed, it is followed by other sources including water tanker as one of major sources, water from pipeline, etc., other sources include buying water, i.e., 20.05%, 17.5% of people getting water from neighbors who have proper water accessibility.

Figure 32.3 shows sources of water for those who do not have Delhi Jal Board water accessibility. In Hastsal, more than half of respondents depend on hand pumps for water, followed by one-fourth from neighbors, while remaining 17.7% buy water, whereas in Jahangirpuri more than half of them buy water and one-fourth of them get water from neighbors. In Madanpur Khadar and Seelampur more than half of respondents meets their demand through hand pump, remaining get water from Delhi Jal Board (DJB) water tankers for water, whereas remaining 1% get water from neighbors. On the other hand, in Seelampur maximum share of response from respondents is in category by hand pump followed by sharing water from neighbors and other sources.

Hence, it would be expected that the higher the level of a desirable attribute in an alternative, other factors held constant, the greater the satisfaction or utility associated with that option and the more likely, it would be for a respondent to choose it. While going towards the satisfaction level of the people in resettlement colonies, only one-fourth of respondents are satisfied while majority of them are not satisfied with the present water facilities. In Hastsal, no respondent showed positive response towards strong satisfaction although in Hastsal maximum share

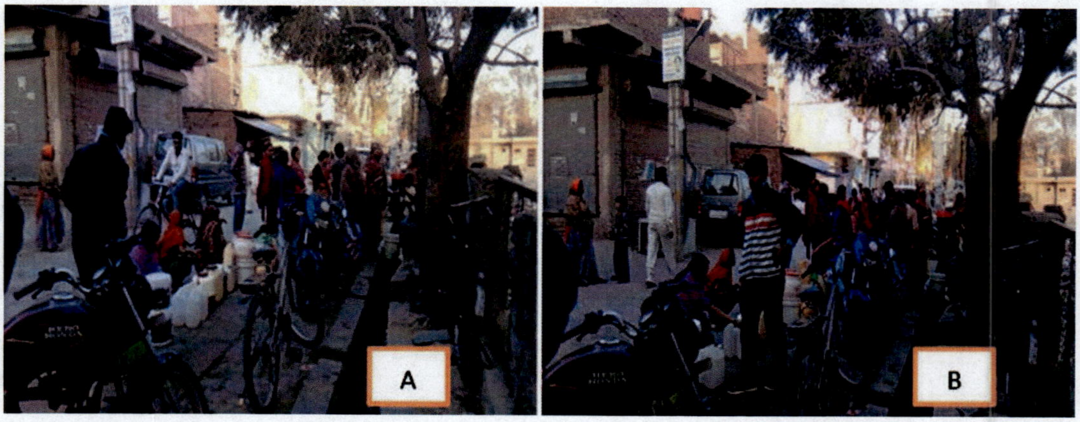

Plate 32.1 Water supply and accessibility in Hastsal (A&B)

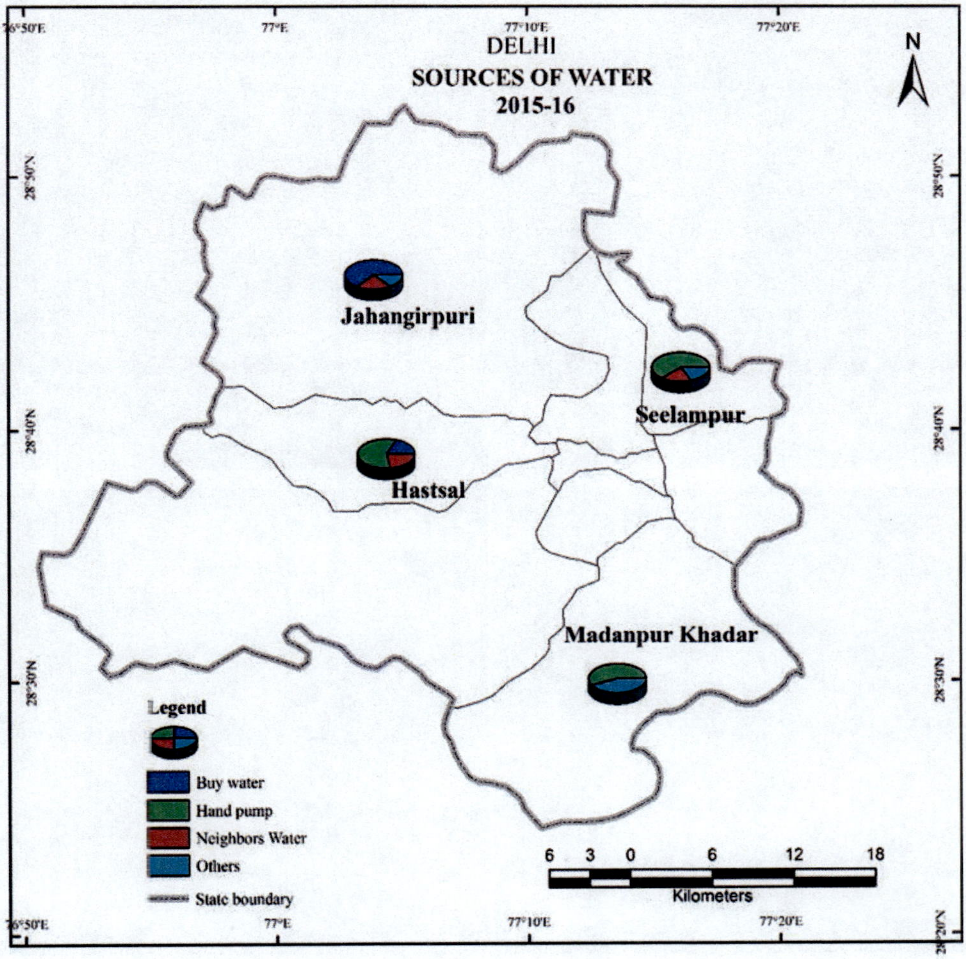

Fig. 32.3 Different sources of water in four resettlement colonies of Delhi. *Source:* Primary Survey, 2015–16

of population have access to Delhi Jal Board tap water. Somehow their satisfaction level is good in comparison to other colonies. In Jahangirpuri only 23.0% of respondents are satisfied, whereas in Seelampur the proportion for satisfaction is only 21%.

32.6.2 Wastewater Disposal

Wastewater refers to water whose properties have been modified by domestic, industrial, agricultural, or other uses. It also includes water (sewage) discharged with it during dry weather as well as the run-off and collected water (rainwater) from built-up or paved areas due to rainfall. Sewage also includes the liquids discharged and collected from waste treatment and storage plants. But in the surveyed resettlement colonies, water was drained out after using from the domestic purpose not industrial or agricultural wastewater.

During survey, the information was collected on the system of disposal of wastewater and other liquid wastes from the house consisting of kitchen wastewater, bath, and wash water. However, it did not include the wastes from latrines. Water disposal is one of major problems that all the urban areas are facing mainly the informal settlements as shown above (Plate 32.2). In Hastsal, majority of respondents highlighted that the water from their household discharges into com-

Plate 32.2 Water disposal in four resettlement colonies (**a–d**), (**a**) (Hastsal), (**b**) (Madanpur Khadar), (**c**) (Jahangirpuri), (**d**) (Seelampur)

mon public drainage, i.e., 68.5% though the condition of the drains observed to be choked and over-flowed, one-fourth of them responded that water is standing near their house. Among the four colonies, Jahangirpuri is having better condition in terms of wastewater management as 66.7% of people stated that water goes to public drainage system. But in CD Park area of Jahangirpuri condition observed was not satisfactory, with no constructed public drainage system and water standing in front of the house. In Madanpur Khadar, 11.4% of household's wastewater discharges into public drainage, rest of 86.4% of the household suffering from poor drainage system as water standing outside (Plate 32.2). In Seelampur, the houses are built in such a way that they are having narrow drains outside their house which are chocked by vegetables as the wastewater includes water after washing utensils as well.

32.6.3 Accessibility of Toilet Facilities

It has been expressed that more people in India have more access to cell phone than to a toilet and improved sanitation. The United Nation experts published a 9-point prescription for achieving the world's Millennium Development Goal for sanitation by 2015 (Hamilton 2014). The same statement is supported through results of the survey conducted in the four resettlement colonies of Delhi: people prefer more luxury items than necessity goods.

In the surveyed resettlement colonies, 91.75% of the people responded that they have mobile and only 8.25% people do not have mobile phones, more than 90% household have Television and Fridge, but, in terms of toilet accessibility, half of the households do not have. The availability of bathroom facility in a house clearly defines the hygiene and health condition of the people living in the house. It has also been brought under light through primary survey that 60.4% households have bathroom access and 39.6% do not have. On the other hand, the availability of safe sanitation system is one of the important factors to gauge socio-economic and cultural development of any country. Availability of toilet is symbol of human development; the regions which are lagging in terms of toilet access are also lagging human development. Thus, the percentage of households without latrine was arrived at as the percentage of households using open area for defecation or using the *Sulabh Shauchalaya*.

In Madanpur Khadar, irony is that 79.5% households do not have toilets and 93% having mobile phone access. In Seelampur, 77.3% households do not have toilets but with more than 90% of them have mobile phones. Hastsal is on better side than the other colonies surveyed, with 82.5% of surveyed houses having toilet accessibility (Fig. 32.4). In Jahangirpuri, 50% of households are having toilet and half using *Sulabh Shauchalaya* or defecating openly.

The reasons for not having access to improved sanitation facilities can be many including major factors, such as cultural, economic, technological, topography, and plot size. The questionnaire survey identified that economic is of the major factor following small plot size, and technological.

Economic factor: Poverty is a strong barrier to improved sanitation facilities. Poverty phenomenon hits different areas with varying degrees of acuteness. In the resettlement colonies of Delhi, majority of living population have family income less than rupees 10,000 per month which is insufficient for the entire family. Plot size is another factor identified which claims to be not enough to facilitate improvement of sanitation facilities. The size of the plot allotted to each family is very small, whereas the family size has also increased over time.

In Jahangirpuri, the reason for not having access to improved sanitation found to be economic followed by lack of space. With half of them stating, 53% for economic reasons including higher cost for construction and maintenance, followed by other reasons including lack of awareness, lack of space are the major reasons that came across and for cultural reasons like no toilet near kitchen. Moreover, 2% responded for lack of sewerage system.

Lack of sewerage system is found to be dominating reason in Seelampur and Madanpur Khadar, whereas in Jahangirpuri, 2% of respondents noted lack of sewerage system and 59% in Madanpur Khadar. As stated earlier, 49.50% of household have access to toilet facility and 51.5% do not. Among those who do not have toilets in their households 71.25% of them have access to public toilets, whereas 28.75% still prefer to go for open defecation (Table 32.1).

The share of open defecation is higher in Madanpur Khadar and Seelampur. In Madanpur Khadar, those who do not have toilets in their house and openly defecates more than half and in Seelampur, it is 38%, lesser open defecation is found in Jahangirpuri, i.e., only 4% of the households openly defecates. The public toilet usage is highest among the people of Jahangirpuri, i.e., 96%, followed by Hastsal with 86% of respondents.

In Jahangirpuri during survey it was found that females are most affected in using public toilets, as they are not allowed to use the public toilet "Sulabh toilets" after 11:00 am in the morning. Problems have been found in some or the other study area regarding toilet usage. Areas where people are using toilets are not allowed after a specific time. The condition of toilets is either unhygienic or poorly maintained. Still, people visit for it. The reasons for using present facility can be categorized under, convenient, not expensive, environment friendly, etc. Figure 32.5 indicates the reasons of using

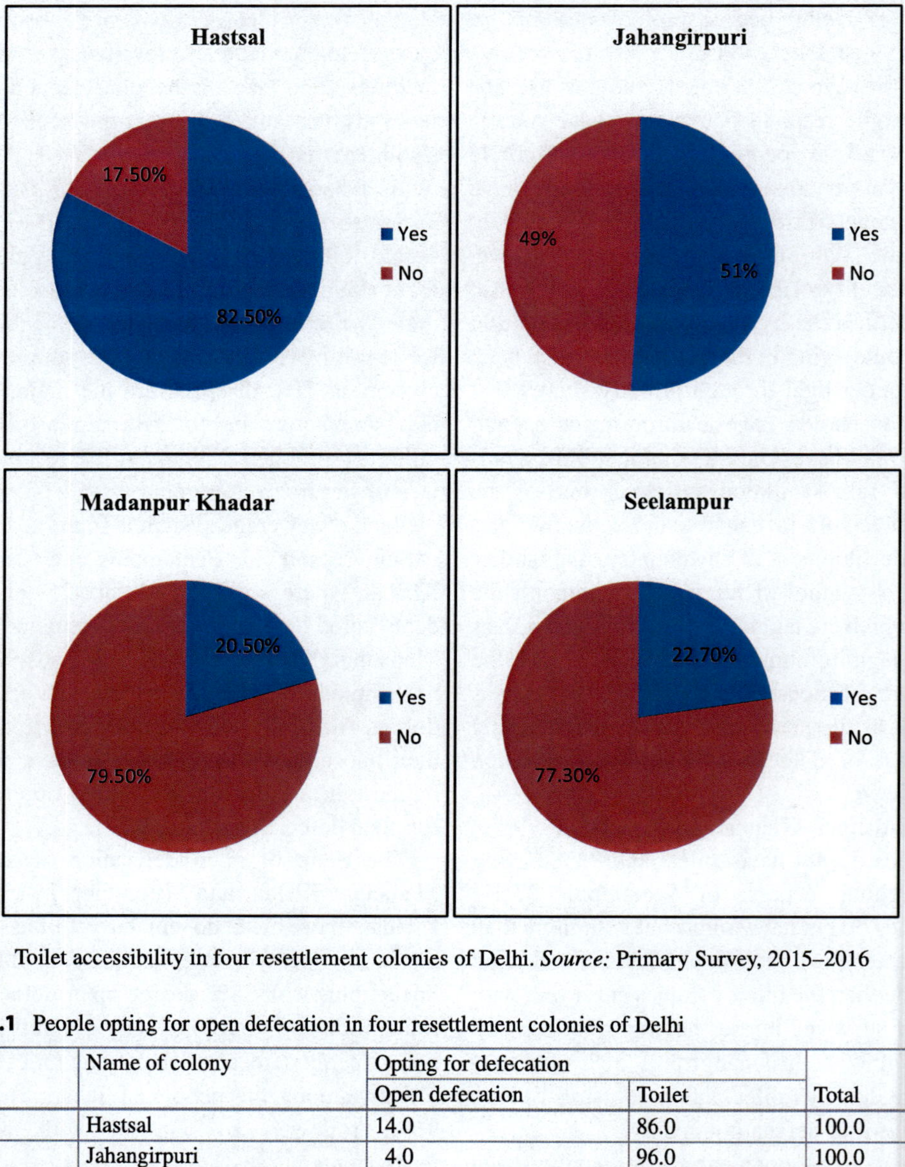

Fig. 32.4 Toilet accessibility in four resettlement colonies of Delhi. *Source:* Primary Survey, 2015–2016

Table 32.1 People opting for open defecation in four resettlement colonies of Delhi

S. no.	Name of colony	Opting for defecation		Total
		Open defecation	Toilet	
1.	Hastsal	14.0	86.0	100.0
2.	Jahangirpuri	4.0	96.0	100.0
3.	MadanpurKhadar	59.0	41.0	100.0
4.	Seelampur	38.0	62.0	100.0
5.	Total	28.75	71.25	100.0

Source: Primary Survey, 2015–2016

the present facility whether it is open defecation or the public toilets. One-third of the people responded for convenience, while one-fourth of them for environment friendliness of present facility, 20.15% for lack of space those who do not have toilets and are openly defecating, whereas 13.62% for not expensive, and 5.05% for other reasons.

In Hastsal, majority of them affirmed that present facility they are using is per their own convenience, 17.2% emphasized that it is eco-friendly, only 10.3% responded for lack of space if they are not having toilets in their house, so they are using public toilets. More than half of the people in Jahangirpuri identified reason for using available facility as it is environment

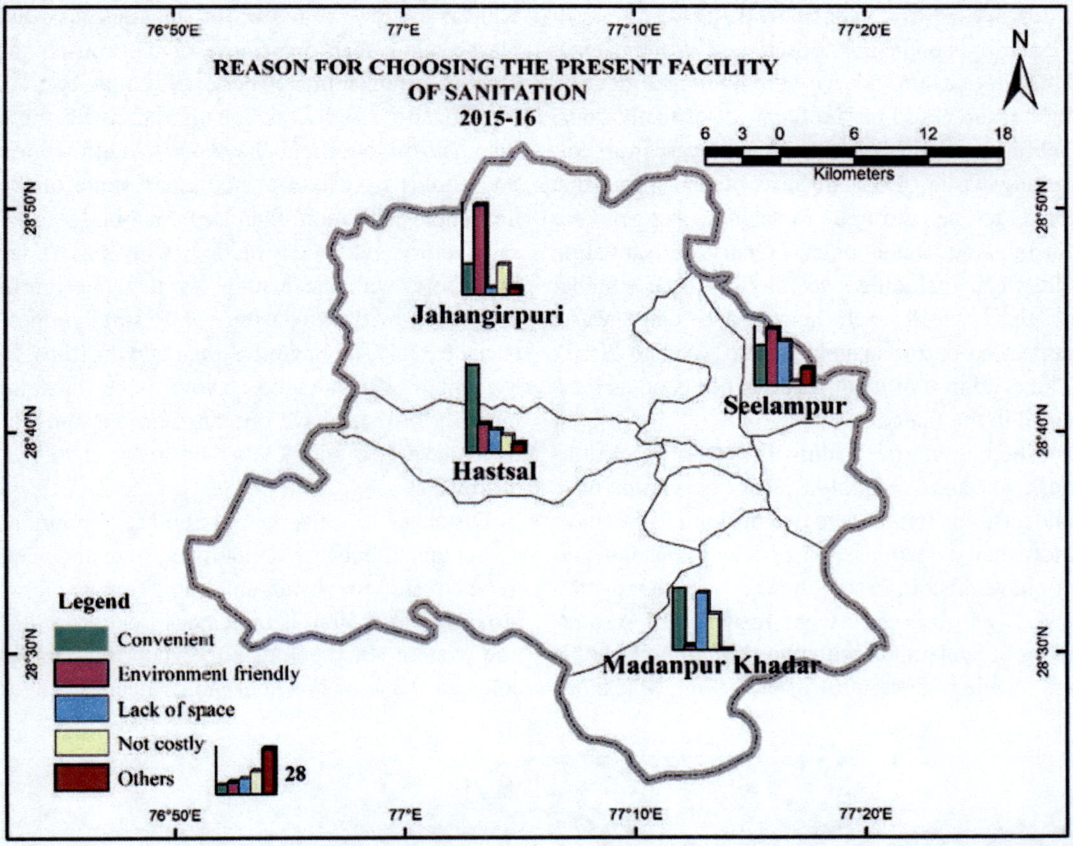

Fig. 32.5 Reason for choosing the available facility of sanitation. *Source:* Primary Survey, 2015–2016

Table 32.2 Willingness to participate for initiative of sustainable sanitation

S. no.	Name of colony	Willingness to participate (per cent)				
		Strongly Agree	Agree	Do not Agree	Strongly Disagree	Total (In %)
1.	Hastsal	22.8	66.7	10.5	–	100.0
2.	Jahangirpuri	33.3	66.7	–	–	100.0
3.	MadanpurKhadar	75.0	25.0	–	–	100.0
4.	Seelampur	70.0	26.7	1.7	1.7	100.0
5.	Total	49.8	46.4	3.3	0.5	100.0

Source: Primary Survey, 2015–2016

friendly while in Seelampur the weightage is of 32.1% and Madanpur Khadar is only 2.6% of for environment friendly. The residents of the four colonies are not happy with the present available sanitation facilities so the level of willingness to participate is high as shown in Table 32.2.

In Madanpur Khadar, the residents are highly motivated to participate, i.e., two-third of respondents strongly agree to participate in developing the sustainable sanitation system, followed by 70% of people in Seelampur, while in Hastsal less than one-fourth of respondents strongly agreed. But on an average maximum share of people in Hastsal and Jahangirpuri have responded that they agree to participate, i.e., 66% as an average but in Seelampur 1.7% of the respondents strongly disagree regarding their responsibility make sustainable sanitation policy or program successful.

32.6.3.1 Solid Waste Management

Generally, people are not satisfied with the available services provided by the municipalities and the reason could be the financial or institutional debility over the municipalities. Apart from collecting waste, there are also other aspects that need to be covered to achieve targets for Sustainable Development Goals on sanitation and water including sweeping of house, number of dustbins in each house, and daily waste collection and management, etc. (Anand 2010). The condition of solid waste is not good as indicated in the Plate 32.3.

The number of dustbins is symbol of cleanliness. On an average 49.8% of households have one dustbin, 16.9% have two dustbins, 3.8% have more than 2 dustbins, and 29.5% of household do not have dustbin in their house. Less satisfaction level was observed as per response of respondents in Seelampur, with two-third of households not having dustbins, whereas in Madanpur Khadar owing to 6.8%. In Jahangirpuri and Hastsal, less than one-fourth of the households do not have dustbin, i.e., better than other two. As mentioned in Fig. 32.6, Jahangirpuri with more than half of households have one dustbin, 19.6% households have two dustbins and none of the household with more than two dustbins. Hastsal satisfactory place on ladder, with 65.5% of households with one dustbin, less than one-fourth households with two dustbins and a small proportion, i.e., 3.4% having more than two dustbins. In Madanpur Khadar, those who have dustbin, majority of them have one dustbin, one-third of them have two and 9.1% have more than two dustbins.

Distance to dustbin is another important aspect that affects the sanitation system and need to be covered for sustainable sanitation as closer proximity to dustbin is indicator of better sanitation system. In the four surveyed resettlement colonies 9.3% of the households need to cover

Plate 32.3 Solid waste management status in four resettlement colonies of Delhi (**a**–**d**), (**a**) (Hastsal), (**b**) (Madanpur Khadar), and (**c**, **d**) (Seelampur)

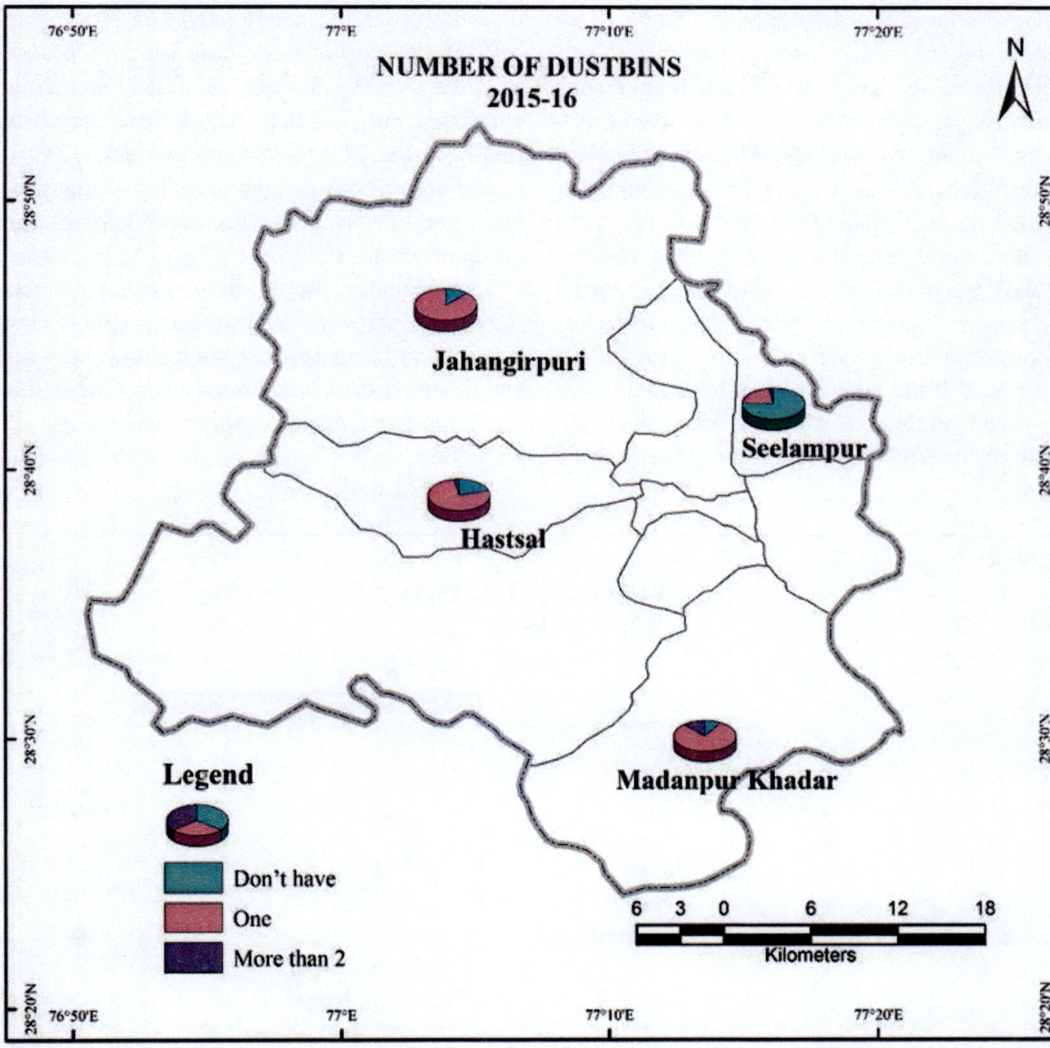

Fig. 32.6 Number of dustbins per household in four resettlement colonies of Delhi. *Source:* Primary Survey, 2015–2016

more than 500 m of distance to throw domestic waste, whereas nearly one-third of them walk 50 m. In Seelampur, observed during field survey, people throw waste outside their house and later being collected by concerned authorities. While the people who are having dustbins, they cover around 50 m to throw waste. In Jahangirpuri, half of the households cover around 200 m to throw their household waste.

One of the important factors in ensuring cleanliness in the micro-environment of the households is the garbage collection system. Garbage collection arrangement means the arrangement which usually exists to carry the refuse and waste of households to a final dumping place away from the residential areas (Biswas et al. 2010). One of the major problems most of the urban centers facing is of waste collection and management. Delhi is among the fastest growing cities of country, facing the problem of waste generation and poor waste management. The indiscriminate dumping of waste results in huge dumps of waste in front of residential areas, directly or indirectly affecting the health of the people and decreasing the aesthetic value of the environment (Latif et al. 2012, Aydin 2017).

On the ground of evidence, it can be noted that, still there are areas where waste is not collected on daily basis or collected once in a week. As per the response of the people of colonies, on an average 4% of the respondent indicated that no waste collection or sometimes as they throw of their own, and one-fourth of the household responded for once or twice a week. Figure 32.7 serves as a backdrop to the analysis on Madanpur Khadar with lesser frequency of daily waste collection. But nearly two-thirds of the households exhibit that waste has been collected weekly. Area wise, in Jahangirpuri 78% of the households responded that daily waste is collected, one-fourth of them indicated for waste collection done once or twice a week, whereas 2% stated that no one collects waste, they themselves throw the waste (mainly the CD park area in Jahangirpuri, where no proper waste collection and management practices are observed and where a large dump of waste is seen.

In Seelampur, people throw waste on streets rather than throwing in dustbins. People do not throw every bit in dustbins but most of the cases people throw their waste once a day. The drains are found to be choked with the solid waste on inspection.

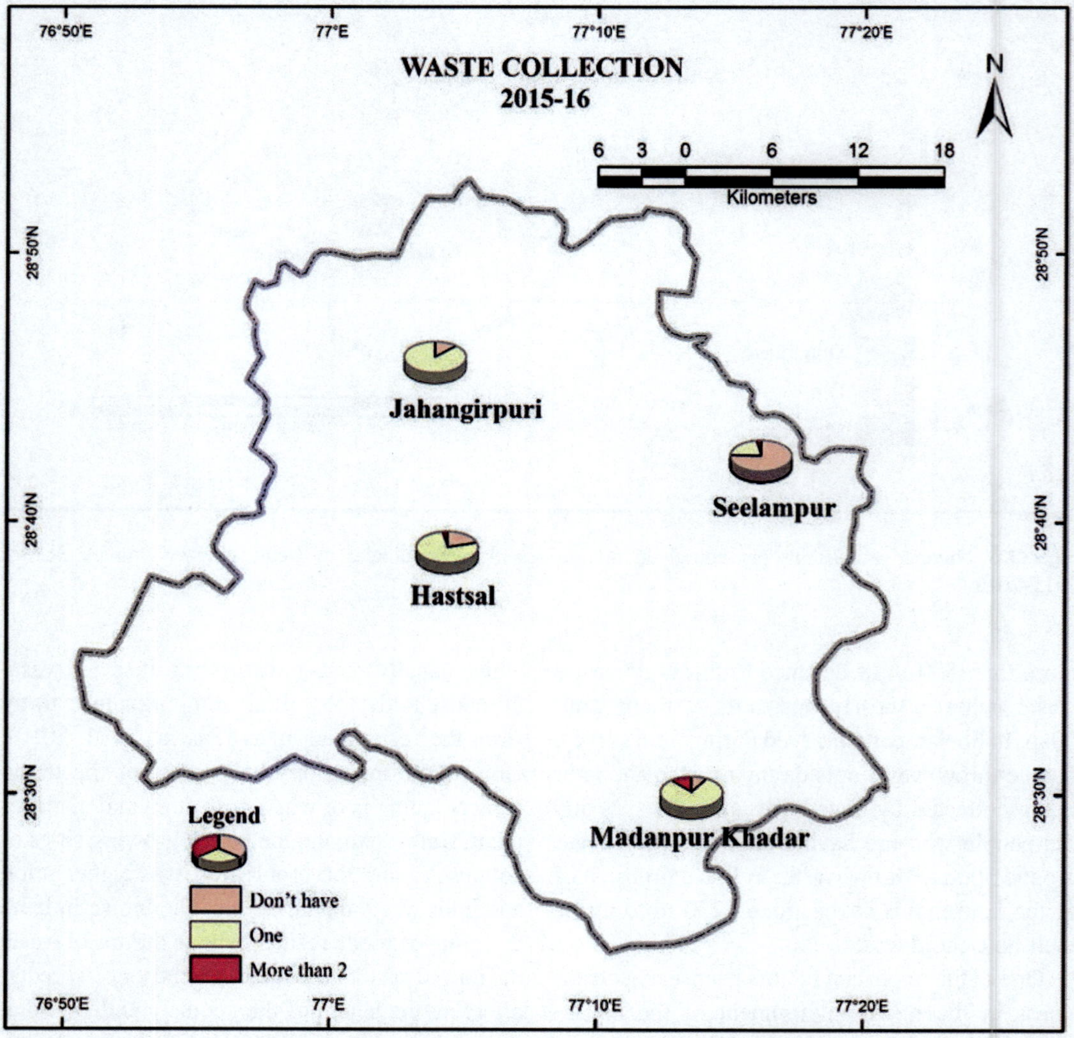

Fig. 32.7 Waste collection in four resettlement colonies of Delhi *Source:* Primary Survey, 2015–2016

32.7 Conclusions

During the field visit in the resettlement colonies, it was found that in case of Seelampur and Madanpur Khadar where the respondents were facing more severe problems regarding water supply, sewerage, toilet accessibility, and waste collection. It is equally important to note that majority of the respondents in Hastsal had no fixed time for water supply while this was not in case of Jahangirpuri. The irregular water supply has always been a problem for the residents. And as per results discussed above in the study conducted, show that the people were not satisfied with the basic services especially water supply, drainage system, community latrines, and solid waste management.

Now what is required is a radical shift in the behavior of the local people and practices to achieve sustainable sanitation in urban areas. The other requirement is urban sanitation approach which impacts more on funding, designing and management of sanitation, etc., and will also require engagement of all the engaged stakeholders for the sustainable sanitation management of the resettlement colonies in Delhi. Equally important is to support the principles of equity and sustainability and community-based sanitation approach found to be operative in encouraging change.

There is need of integrated models like role played by Sulabh International Social Service Organization known for achieving success in the field of cost-effective sanitation and behavior change of people from open defecation to use of Sulabh Shauchalaya, Ruchika Social Service Organization played significant role in sanitation management and rainwater harvesting is praising and worthy by involving the school children. Their work based on approaches on Child-to-child interaction, Child to family, Child to community. More number of governments funded sanitation and water-based programs are Swachh Bharat Abhiyan, Jal Jeevan Mission and the Atal Mission for Rejuvenation and Urban Transformation, and the Jal Jeevan Mission.

References

Anand S (2010) Solid Waste Management. Mittal Publication, New Delhi

Andersson K, Dickin S, Rosemarin A (2016) Towards 'sustainable' sanitation: challenges and opportunities in urban areas. Sustainability 8(12):2–14

Andrianou XD et al (2019) Mapping drinking water and quality of life aspects in urban settings: an application of the urban exposome. BioRxiv 401927

Aydin, N. (2017) Review of municipal solid waste management in Turkey with a particular focus on recycling of plastics. Energy Procedia, 113:111–115

Biswas AK, Kumar S, Babu SS, Bhattacharyya JK, Chakrabarti T (2010) Studies on environmental quality in and around municipal solid waste dumpsite. Resources, Conservation and Recycling, 55(2):129–134

Census of India (2011) Administrative atlas. National Capital Territory of Delhi. Government of India, New Delhi

Delhi Urban Shelter Improvement Board (2015) Retrieved JJ cluster details, Resettlement Colonies. Weblink: http://delhishelterboard.in/main/?page_id=3644

Directorate of Economics and Statistics (2012) Delhi statistical handbook. Government of National Capital Territory of Delhi, New Delhi

Hamilton (2014) Greater Access to Cell Phones Than Toilets in India. United Nations University. Weblink: https://unu.edu/media-relations/releases/greater-access-to-cell-phones-than-toilets-in-india.html

Latif SA et al (2012) Environmental problems and quality of life: situational factor as a predictor of recycling behavior. Procedia Soc Behav Sci 35:682–688

Mara D et al (2010) Sanitation and Health. PLoS Medicine, 7(11). Weblink: https://www.ncbi.nlm.nih.gov/pmc/articles/PMC2981586/?tool=pmcentrez. Accessed on 23 Mar 2017

Ray B (2017) Quality of life in selected slums of Kolkata: a step forward in the era of Pseudo-urbanization. Local Environ 22(3):365–387

UNESCO (2006) Water A Shared Responsibility. The United Nations World Water Development Report 2. Berghahn Books, New York, USA

UN-HABITAT (2003) The challenge of slums. Global report on human settlements. Earthscan Publication, London

UNICEF (2006) Meeting the MDG Drinking Water and Sanitation Target: The Urban and Rural Challenge of The Decade, Switzerland

Verma GD (2002) Slumming India. Penguin Books, New Delhi

WHO and UNICEF (2010) Progress on Sanitation and Drinking Water: 2010 Update World Health Organization, Geneva?

33 Water Governance in Thiruvananthapuram City, Kerala: Existing Practices and Alternative Paradigms

Srikumar Chattopadhyay and K. N. Harilal

Abstract

Urban water governance is a challenging issue, globally. Questions are often raised about the viability of the existing practices and a new paradigm of integrated urban water management is proposed. Overcoming spatial differentiation in service delivery, providing quality service, and devising measures for source sustainability are emerging challenges of urban water management in Thiruvananthapuram City. Despite strong commitment to decentralization, water management is fragmented and centralized with little role for the city authority. Applicability of integrated urban water management concept has been examined in case of Thiruvananthapuram City. It is suggested that participatory polycentric governance may be developed considering the city and the hinterlands and necessary space may be created to debate and evolve alternative urban water management practices suitable for the city.

Keywords

Decentralization · Integrated Urban Water Management · Polycentric Governance · Urban Water Governance

33.1 Introduction

Urban water governance is emerging as a critical development challenge particularly in the developing countries which are experiencing high rate of urban growth (UNESCO 2015). Despite several initiatives, most of the cities in India perform poorly in the matter of availability and reliability of drinking water supply and maintenance of water quality (Nastar 2014). The situation is quite alarming when other aspects of urban water services like providing sewer system, drainage services, quick dispersal of storm water, and combating negative impact of cities on water resources within and outside urban areas or environmental wholesomeness are considered (Gopokumar 2010; Nadhamuni 2012). The existing technology centred water governance model is not adequate to address all these concerns. The concept of Integrated Urban Water Management (IUWM) is considered as an alternative paradigm (Barga 2001; Mitchell 2006). It accords equal priority to technical as well as non-technical solutions and stresses on sustainability, good governance, and empowerment.

S. Chattopadhyay (✉)
National Centre for Earth Science Studies, Thiruvananthapuram, India

K. N. Harilal
Centre for Development Studies and Member, State Planning Board, Government of Kerala, Thiruvananthapuram, India
e-mail: harilal@cds.ac.in

B. Thakur et al. (eds.), *Resource Management, Sustainable Development and Governance*, Sustainable Development Goals Series, https://doi.org/10.1007/978-3-030-85839-1_33

Thiruvananthapuram, capital city of Kerala, located along the southwest coast of India spreads over an area of 250 km^2 and houses 7.4 lakh people distributed in 100 Wards (Census of India 2011). It enjoys relatively better services in the matter of drinking water supply and continuity of supply among all State capital cities in India. However, in all other sectors of water services and overall environmental management, performance of the city is far from satisfactory. The present paper proposes to analyse water service sectors and emerging water governance challenges in Thiruvananthapuram City and examines applicability of IUWM framework to overcome these problems. This study differs from earlier attempts to deliberate on urban water management, which concentrated mostly on service delivery of drinking water and sanitation and the focus was often on technology and financial allocation for provisioning of the services. In this paper we argue for a holistic approach and try to discuss all aspects of water regime in the city and surroundings including governance challenge to integrate the services and their effective delivery. This study is expected to facilitate further analysis of political and economic processes and contribute to policy formulation.

Box 33.1 Sustainable Development Goals
Water governance in Thiruvananthapuram City remain a challenge given the combined impact of biophysical, socio-economic, and policy issues. These challenges are further compounded through the impact of sea-level rise, river flooding, increased frequency, and magnitude of extreme rainfall, water scarcity, water pollution, ageing infrastructure for water, wastewater, and solid waste in an expanding metropolis. Recognizing these challenges is important to articulate the UN SDG framework (goals # 6 & 11) and the recommendations of policies and programs of the Government of India through its ***Smart City Mission,*** among others. The authors of this chapter study water governance challenges and present solutions in the form of integrated water management practices and water governance strategies. Their research also confirms the need for long-term strategy and a multi-level approach as a way forward. Integrated sustainable water governance can be achieved in Thiruvananthapuram City with the involvement of government, business, and civil society early in the process of collaboration within the city's rapid implementation of sustainable development goals.

33.2 Integrated Urban Water Management (IUWM) and Challenges of Urban Water Governance

33.2.1 Integrated Urban Water Management (IUWM)

Urban water management witnesses shift in approach and content. The emerging new paradigm is comprehensive, multidimensional and advocates perceptional change in management practices of urban water (Table 33.1). The International Water Resource Association introduced the concept of Integrated Urban Water Management (IUWM) a couple of decades back (Barga 2001). However, it gained serious consideration since 2000 and now there are several initiatives across the world to adopt this new approach (Mitchell 2006; GWP 2013; Whitler and Warner 2014). Country level experiences indicate that IUWM has brought perceptible positive change in urban water management (Maheepala 2010; Closas et al. 2012) although there are criticisms about its long-term viability, macro-scale applicability (Biswas 2008), and difficulty in transforming entrenched institutional practices particularly in large cities (GWP 2013). Nevertheless, IUWM provides a framework that is "flexible, participatory, and iterative process which integrates the elements of urban water cycle with both the city's urban development and river basin management to maximize economic,

Table 33.1 Old and new paradigm of urban water systems

Old paradigm	Emerging paradigm
• Human waste is a nuisance. It should be disposed of after treatment. • Storm water is a nuisance. Drain it out of the city as rapidly as possible. • Demand is a matter of quantity. Amount of water required or produced by different end users is the only parameter relevant to infrastructure choices. Treat all supply side water to potable quality and collect all wastewater for treatment. • One use (throughput). Water flows one way path from supply, to a single use, to treatment and disposal to the environment. • Grey infrastructure. Infrastructure is made of concrete, metal, and plastic. • Bigger/centralized system is better for collection and treatment plants. • Limits complexities and employs standard solutions. Small number of technologies by urban water professionals defines water infrastructure. • Integration by accident. Water supply, wastewater, and storm water may be managed by same agency as a matter of historical happenstance. Physically, however, three systems are separated • Collaboration = public relation. Approach other agencies and public when approval of pre-chosen solution is required.	• Human waste is a resource. It should be captured and processed effectively, used to nourish land and crops. • Storm water is a resource. It should be harvested for water supply, retain for supporting aquifer recharge, waterways, and vegetation. • Demand is multifaceted. Infrastructure choice should match the varying characteristics of water required or produced for different end users in terms of quantity, quality and level of reliability, etc. • Reuse and reclamation. Water can be used multiple times, by cascading from higher to lower quality needs and reclamation treatment for return to the supply side of infrastructure. • Green infrastructure. It is made of concrete, metal, plastic, and also of soils and vegetation. • Small/decentralized is possible, often desirable for collection system and treatment plants. • Allow diverse solutions. Decision makers are multidisciplinary. Allow new management strategies and technologies. • Physical and institutional integration is by design. Linkages must be made between water supply, wastewater, and storm water, which requires highly coordinated management. • Collaboration = engagement. Enlist other agencies and public in search for effective solutions.

Source: Mitchell (2006) (adopted from Pinkham 1999, p. 591)

social and environmental benefits in an equitable manner" (World Bank 2012).

The thrust is to consider all aspects of urban water as components of an integrated physical system, position this physical system within an organizational framework of governance and a broader natural landscape like river basin. On the one hand, IUWM recognizes hydrological interconnectedness of all water bodies within and outside the urban area, and, on the other hand, it stresses on designing location-specific solutions by encouraging tapping of locally available water sources and innovation. Initiatives may begin with an overarching national policy on integrated water resources management to set the natural landscape scale/ watershed/ river basin level contextualization of the urban area followed by effective legislations at appropriate levels to guide concerned authorities, engage local communities, and empower them to solve the problems of water management through collaboration of all stakeholders.

Government of India has shown interest in IUWM as the enabling framework for efficient management of vast challenges and needs of urban India (Brikke 2015). The Mihir Shah committee constituted by the Government of India to restructure the Central Water Commission (CWC) and Central Groundwater Board (CGWB) recommended unification of these two organizations to form National Water Commission (NWC) which will be responsible for water policy, data generation and storage, and water governance in the country. The new paradigm proposed by this committee envisaged fundamental changes in the existing system. It suggests suiting interventions to the contour of nature, partnership, multidisciplinarity, demand management and sustainability as central focus, emphasis on equity in access to water, transparency, and national water framework law (Shah 2016). Arghyam, (a public charitable foundation based in Karnataka) introduced IUWM in Mulbagal town in mid-2008 and the results are quite encouraging (Nadhamuni 2012). Taking note of this experience, Government of

Karnataka proposed Integrated Urban Water Management programme for funding by the Asian Development Bank (ADB 2014). Given the nature of Thiruvananthapuram City—population size, geo-environmental settings, and water regime—and the overall commitment of the State of Kerala to decentralized governance the IUWM frame provides good opportunity to address urban water management problems.

33.2.2 Challenges of Urban Water Governance

The World Water Forum held at The Hague in 2000 accorded high priority to water governance for action (Cooley et al. 2013) and failures in water sectors are often attributed to the failure in governance (Biswas 2004; Bakker 2010; Pahl-Wostl et al. 2012). Traditionally, urban water governance refers to technical decision-making process following a fragmented departmental demand-supply cycle. Inadequacy of this approach is well documented. The emerging approach of urban water governance considers technical as well as non-technical issues and enlarges its ambit to address water security and source sustainability, reduction of impact on environment controlling human dimension of water environment change, improvement of performance of water services across socio-economic groups, and allocation and reallocation of water among resource rich and resource poor areas. Case studies across the World indicated that there are three essential roles of urban water governance: first, to manage the environmental dynamics, including climate change to always provide water for cities; second to ensure justice and fairness in the distribution and access to water in cities; and third to ensure quality in terms of human health and environmental pollution (Olsson and Head 2015). As there are competing demands and multiplicity in management authorities, the challenge of urban water governance is also to resolve conflicts among techno-scientific, market, policy administration, ecological and socio-political actors. The global pattern manifests that water for urban and industrial uses to go up substantially and, as a result, reallocation of water between urban and rural areas is potent to create social tensions and even conflicts. There is a need to look beyond the city boundary and consider the broader territory, establish relationship with the surrounding rural areas, evolve reciprocity with the hinterlands, and operate in the frame of co-management with other administrative units surrounding the city.

The risk and uncertainties associated with various changes that the urban centres trigger in hydrological cycle are difficult to comprehend; however, the emerging challenges for water governance due to these changes form part of the urban water governance. The drivers of water use and abuse are location-specific, so the insights on water management should arise from local level experiences. A careful analysis of the water governance system, its actors, interests, values, and processes in each locality is necessary to bring out required change in the present governance practices. Going beyond the ‘instrumental and idealistic’ notion of governance and an attempt to depoliticization (Castro 2007) as nowadays being advocated in some parlance, it is necessary to strike a balance among different aspects of water management activities, ensure convergence between research and practice in water management science, and help the society to evolve a proper governance system through democratic means of debate and stakeholder participation in policy making.

33.2.3 Water Service Sectors and Emerging Management Issues in Thiruvananthapuram City

Thiruvananthapuram City receives rain almost in every month. Both the monsoons active in the city account for 76% of average annual rainfall of around 1700 mm. The river Karamana, its tributary Killi Ar, and couple of small streams drain the city. Water service sectors in Thiruvananthapuram City cover drinking water supply, sewage system, and storm water drainage. There is no separate provision for wastewater drainage. So far as drinking water supply is concerned, Thiruvananthapuram City is well placed among all the capital cities in India. Around 80% of households in the city enjoy piped water connection and the continuity of water supply averages at 18 h a day (Jacob 2012a; b) in contrast to all India figure of <50% of urban dwellers having access to tap water by 2010 (Nastar 2014), and average continuity of water supply limited to 4–5 h a day. However, there are problems of intra-city variations and core periphery differences in access to drinking water as pointed out from time to time (Chattopadhyay and Harilal 2017). Even there are some areas facing acute shortage of drinking water. In the matter of sewage system, wastewater disposal, and storm water drainage, performance of the city is far from satisfactory. We discuss here status of all three service sectors and the emerging management issues in each case.

33.2.3.1 Drinking Water Supply

Piped water supply project started in Thiruvananthapuram City as Wellington Water Works in the year 1933 covering an area of about 30 km^2 and with capacity to serve a population of 1.35 lakh, (projected population for 1961). There were arrangements for water quality checks at various points. The scheme was considered as one of the best in the Country in those days. At present, drinking water supply in Thiruvananthapuram City is augmented through Aruvikkara reservoir (storage capacity of 2 Mm^3) located in the outskirt of the city drawing water from the Karamana river and a new installation at Peppara (storage capacity of 70 Mm^3) located further upstream on the Karamana river at the foothills. Kerala Water Authority manages drinking water supply in the city. Around 81% households receive piped water supply at their premises under KWA. Among the rest 19%, wells cater to the need of drinking water supply for 10% households followed by 4% depending on public tap water, 3% households receive water from tankers, and rest 2% manage from various sources (Fig. 33.1). However, the supply is not adequate as is evident from frequent shortage reported by the people particularly during summer months. At present, the Thiruvananthapuram Water Supply Scheme has the capacity of 273 MLD to cater to the need of 9.57 lakh people. The gap between production and supply is estimated at 137 MLD, whereas the water loss due to leakage has been estimated to be between 35% and 43%. Cutting across physiographic barriers the problem areas are distributed in various parts of the city (Fig. 33.2). Most of these locations along the peripheral areas and suburbs face acute water shortage due to fall in well water level and non-availability or infrequent supply of piped water administered by KWA (The Hindu 2015). Only 58% of slum dwellers have access to tap water (PRIA 2014). In many places poor people depending entirely on piped water supply or well water are compelled to purchase water from private vendors. Vizhinjam in the coastal tract off Kovalam coast is one such locality where people must purchase drinking water (Field investigation).

In order to augment the water supply scheme and meet up all demands till 2036, KWA has taken up a project with loan assistance from JBIC (Japan Bank for International Cooperation) and proposed to follow a three-tier system like (1) augment the supply by 76 mld, (2) bring down the unaccounted water losses to the range of 15%, and (3) cut-off some boundary areas from the city water supply system (WSS) by providing small WSS exclusively for those areas.

33.2.3.2 Emerging Management Issues

Our experience suggests that even with augmentation water services are not reaching to all sections and all parts of the society, there are significant losses of water due to leakage. The basic settlement structure in the city is linear and dispersed. With horizontal expansion of urbanization more and more rural areas along the periphery come under city administration. This type of spatial growth impacts drinking water facilities of the original city, which used to enjoy better services compared to their rural counterparts but now struggling to keep up with population growth and sprawl. Similar trend has been reported from many parts of the world (WHO-

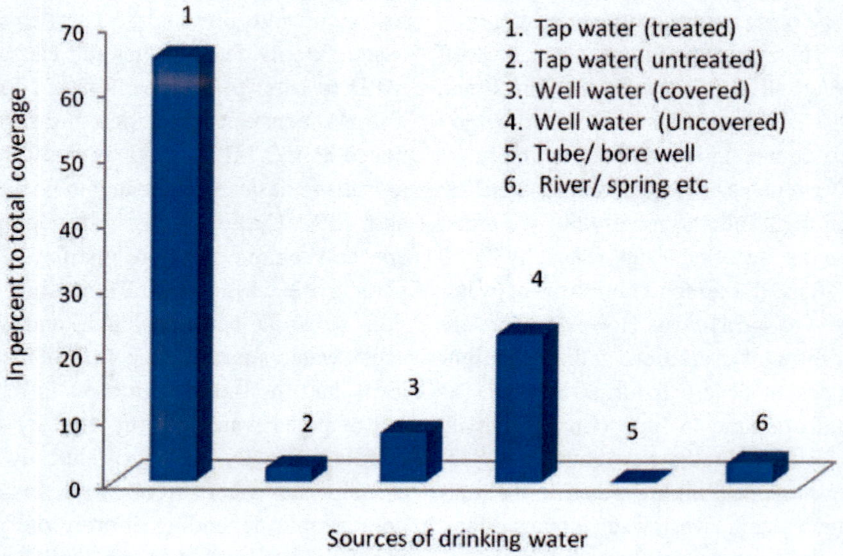

Fig. 33.1 Sources of Drinking water in Trivandrum Urban area

UNICEF JMP 2010). As a result, poor people in the peri-urban areas mostly rely on informal practices that lie beyond formal support strategies and mechanisms, whether centralized supply policies or market-based approaches (Bahri 2012). Thiruvananthapuram faces similar situations with growing chasm between core periphery service deliveries.

At present, the city water supply depends on surface water sources, which varies with rainfall and is subject to climate change vagaries. It is important to recognize that the conventional surface water-based water supply system as existing today in the city has limitations and rather than disregarding all other sources like bore well, well, spring, etc., it is necessary to assess their viability and regulate them for quality. State regulations to this effect may be expanded and, where appropriate, may provide operational management and financial assistance, so that all water supply activities within the urban environments may form part of a coordinated effort of urban water management under public utility development.

There are reforms in water governance sector in Kerala from time to time like formation of KWA, and subsequent start of Jalanidhi, which is now in phase-II of its operation. KWA is mainly looking after the urban sector. However, these reforms have not yielded desired results for various reasons from official apathy to take advantage of 74 constitutional amendments and decentralize the operation with necessary capacity building to fragmented reductionist approach to tackle the water issue. Performance of the schemes under KWA has been questioned (Chakrapani 2014). Water authority has been taken beyond public scrutiny and social auditing although KWA would not survive but for public funding. What reform can do is globally best demonstrated by Phnom Penh experience from Cambodia. It succeeded to install one of the three most globally referred successful drinking water supply programme in urban sector under public sector/institution (Das et al. 2010). The other two are Porto Alegro of Brazil and Kampala in Uganda. In all three cases, it has been demonstrated that it is public utilities themselves that have been able to bring much needed change through certain reforms. The culture of change, implemented by Phnom Penh Water Service Association, focused first and foremost on employees, particularly on education, motivation and, when necessary, sanctions. It can bring transparency, accountability, operational efficiency through internally driven incentives and succeeded to engage the consumers and civil society. These lessons may be useful for future

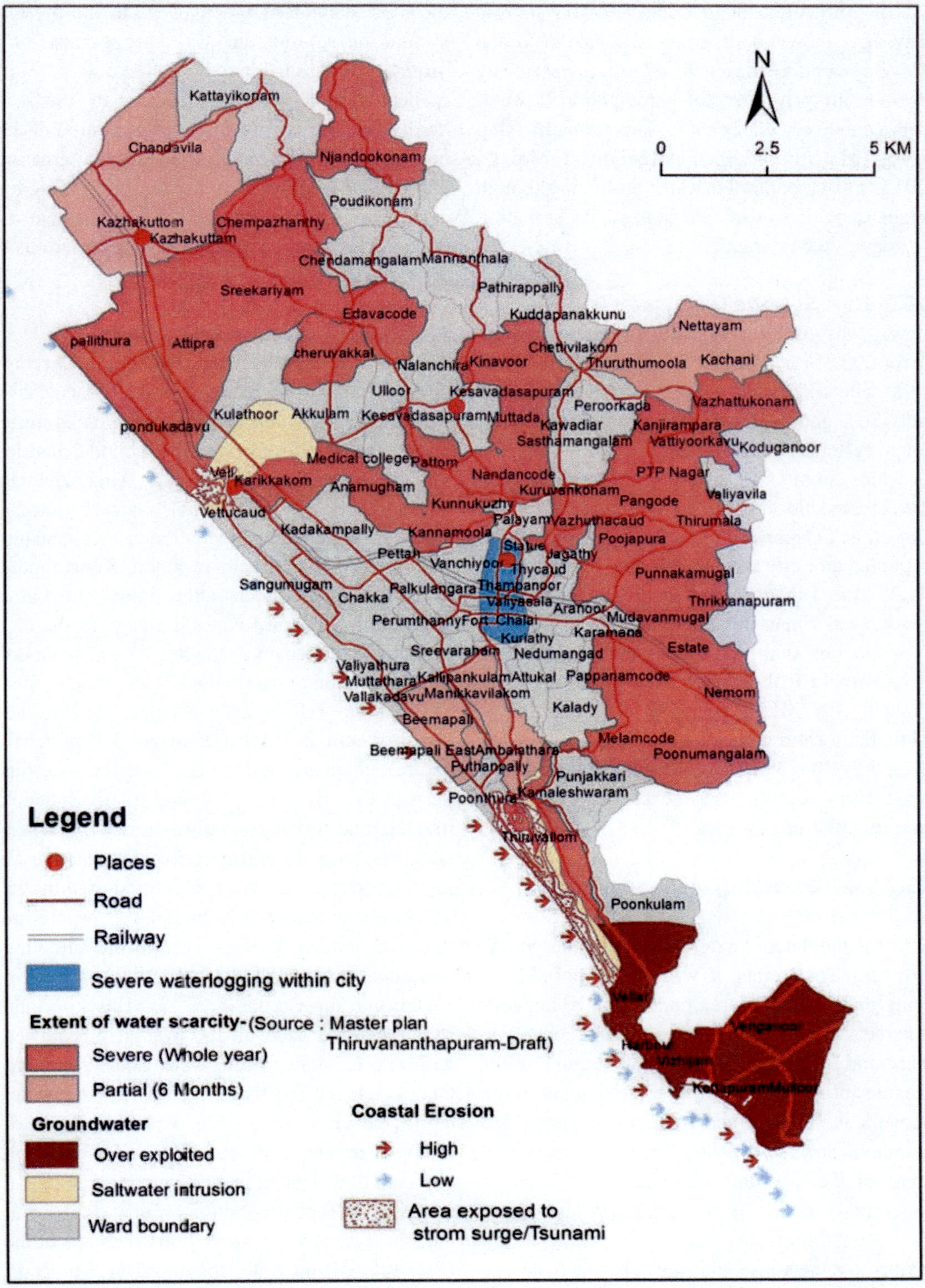

Fig. 33.2 Spatial distribution of water related problems in Thiruvananthapuram City
Souces: CESS, 2009, CGWB Thiruvananthapuram master plan (Draft)

restructuring of water governance for Trivandrum city. It is more important as Ministry of Urban Development, Government of India is seriously considering private sector participation in water service delivery under public–private partnership (PPP) (Ministry of Urban Development (MoUD) 2012) to overcome the water problem although global experiences of PPP indicate its limitation to address the problems.

33.2.3.3 Sewage Disposal

Sewage disposal is a big challenge although State record 98.3% households in urban sector to possess toilets. Trivandrum City Corporation has only 30% geographical area covered under sewerage system. The sewerage network is available in 5 blocks out of 18 sewerage blocks in the city. The covered blocks under the Thiruvananthapuram Sewerage Scheme are part of the original corporation area or city core spreading over an area of 74.93 km^2 (Dileep Kumar 2017). Sewerage networking continues in parts of the city under ADB program. On completion, this scheme is expected to cater to 45% of city area. A Sewage Treatment Plant of 107 MLD capacities is functioning at Muttathara from November 2013 onwards; however, only 30–40 MLD of sewage is reaching the plant every day through the sewerage network catering 30% of city area.

33.2.3.4 Emerging Management Issues

Due to undulated topography and other infrastructural constraints, it will not be possible to bring the entire city area under sewage network scheme. At present, 70% area depends on septic tanks and leach pits, which will continue in foreseeable future. Moreover, the existing sewerage network is very old and frequent overflows are common. Sewage overflow from the unsewered areas of the city and adjoining areas into the water body creates water pollution and unhygienic conditions. As the leach pits used by majority of the households are not sealed, there is leakage affecting groundwater. The present practice is to collect the septage using vacuum suction into tankers which are then emptied into open spaces and even into water bodies, one of the most dangerous practices. With the present system of sewage disposal both surface and groundwater are liable to be polluted. Expansion of network for centralized sewage treatment is hardly feasible as evident after installation of the new plant. It is necessary to look for alternative approach. Use of water recycled from sewage for agriculture and environmental management as done in many countries may be seriously considered.

33.2.3.5 Storm Water Drainage

The river Karamana and its tributary Killi Ar flow through the city. There are natural drainage channels, manmade drains and lakes serving as storm water drainage. These are Ulloor thodu, Pattom thodu, Pazhavangadi thodu, Amayizhanjan thodu, Thekkenekara canal, Kariyil thodu, Tettiyar thodu, Pangappara thodu, Kaimanam-Azhamkal thodu, Amathara thodu, Koori thodu, and Vattakkayal Thodu. Other important drains contributing to storm water drainage to the city canals are Choozhampala thodu, Anathana thodu, Edanada thodu, Arayalloor Ela thodu, and Thiruvallom Pallathukadavu thodu. Parvati putnar also facilitates storm drainage. Several small natural channels and canals draining various parts of the city carry water and finally debouch into the mainstream/river to be carried to the sea. Based on location and severity of problems, 59 high priority areas were identified within the Thiruvananthapuram Municipal Corporation area and another 53 sites were in the adjoining panchayat areas (KUDA 1992).

Trivandrum Corporation has 180 km long streamlets and each of 100 Wards is bestowed with one or more natural water courses/streamlets, which are mostly converted into rainwater drains, concrete/stone lined. Water accumulates in these drains from individual plots through overland flow. Due to undulating topography and relatively elevated position water drains out quickly in most of the areas, however, parts of the CBD suffer from water logging 5 to 6 times in each year and depth of water reaches up to 1.2 m in some cases. Genesis of the problem can be deciphered from a detailed investigation in one of the micro watersheds, the Amayizhanjan thodu following the events of waterlogging on 21st April and 28th June 2015 with rainfall of 161 mm and 120 mm, respectively (Shaji and Ravindran 2017). It is found that more than 78% of 3.94 km^2 watershed area is under built-up category or paved, which does not allow percolation. Instantaneous flow has increased, and capacity of drainage net has reduced due to encroachment. Frequent waterlogging in some parts of the city particularly in the CBD area during monsoon months has raised serious question about storm water management in the city.

33.2.3.6 Emerging Management Issues

Storm water drainage in the city is in the form of conveyance, e.g., to drain out the water at a faster rate. This trend prevails for traditional storm water management almost in all urban centres is now found inadequate. Factors contributing to waterlogging are expansion of settlements, growth of surfaced area, reduction of flood cushioning areas due to occupancy of floodplains, encroachment along drainage channels and obstructions, dumping of solid wastes within the water bodies, inadequate and or absence of drains, low level of plinth of houses vis a vis road/drain level, lack of clear-cut outlet to primary/secondary drains, inadequate cross drainage, overtopping of banks, inadequate inlets, poor maintenance level, and tides/ea surge in the coastal areas. Majority of the causes are of anthropogenic origin. In the event of climate change there will be an increase in high intensity rainfall with intervening prolonged dry periods. Probability of water logging will increase. At present there is no clear-cut responsibility among the Government departments in the matter of storm water management. It is also not particularly effective at public engagement, at the same time people's interest in storm water management may be difficult to sustain. For long-term solution and sustainability, there must be a balance between infrastructure development, its performance, and environmental goals. It is unlikely that cities can build distributed storm water management infrastructure to cope with the eventuality like the rainfall of 161 mm that occurred on 21st April 2015. The changing storm water management proposes hybrid infrastructure which combines structural measures that facilitate conveyance with distributed measures that promote infiltration (Porse 2013). Plot level intervention by individuals, business houses, and institutions is necessary to increase in-situ infiltration and reduce surface run off. Capacity of the existing ponds/wetlands may be enhanced to store water. These interventions may not be possible through legislation. It requires awareness among the people and voluntary action at the individual/family level. Storm water management warrants people-centric decentralized approach and appropriate system design of distributed infrastructures.

33.2.4 Fresh Water Sources of Thiruvananthapuram City: Problems with Availability and Sustainability Challenge

Sustainability of fresh water sources is one of the issues that emergent urban water governance is expected to consider. Thiruvananthapuram City uses both surface and groundwater. While the piped water supply in the city fully depends on Karamana River, many households along the periphery of the city continue to depend on dug wells. Even within the city, there are several households, still maintaining dug wells as additional source. There are instances of people within the city sinking tube wells in their premises. Thiruvananthapuram City is facing problem in both surface and groundwater concerning quantity and quality. We plan to discuss some of these issues here and highlight governance challenges in the matter of source sustainability.

33.2.4.1 Reservoir Siltation and Sand Mining in Karamana River

Peppara and Aruvikkara are two reservoirs impounded on this river to cater to the drinking water need of Thiruvananthapuram City. The first stage of construction of Aruvikkara reservoir started in 1928 and completed by 1933 and the second stage was completed in the year 1972. The water spread area at the time of impounding was 48 hectares and the maximum storage capacity was 2 million cubic metres (Mm^3). Survey conducted by Kerala Engineering Research Institute (KERI) in May 2009 brought out that water holding capacity of the reservoir had come down from 2 Mm^3 at the time of final commissioning in 1972 to 1.137 Mm^3 in May 2009. There is a reduction of 19 ha in water spread area and the storage loss of 0.863 Mm^3 for 37 years. The Peppara reservoir with a water spread area of 5.82 km^2 and storage capacity of 70 mm^3 was commissioned in 1982 to augment water supply capacity of Thiruvananthapuram City at the level of 24/7. However, this reservoir is facing problem of siltation and reduced lean season flow due to various interventions in its catchments. In dry years, water storage in Peppara reservoir is also not sufficient to meet the demand of the city. Water is being diverted from Neyyar river to supplement city water supply.

Sand mining from riverbed is another problem that impacts on source sustainability. Assessment of sediment deposit conducted under the aegis of Revenue Department; Government of Kerala indicated that in a 32 km stretch of Karamana River there are sediment deposits of 0.18 Mm^3 within a depth of 2 m below summer water level (River Management Cell 2015). However, all the sediments above summer water level have been removed. The riverbed level helps maintaining the piezometric level of water and controls water levels in the surrounding wells. Removal of sands from riverbed causes not only lowering of riverbed level but also reduces water holding capacity. This is an important issue in the matter of source sustainability.

33.2.4.2 Water Quality of Karamana River

Analysis of selected parameters indicated that water quality of Karamana is in general not good, and it deteriorates as river enters Thiruvananthapuram Corporation Area. Comprehensive monitoring of river water quality by Centre for Water Resource Development and Management (CWRDM) under the aegis of KSCSTE highlighted that the maximum deteriorated condition is noted in the downstream side of the Trivandrum Corporation towards the coast manifesting effluent load from the urban areas. Nutrient pollution is attributable to domestic use and agricultural runoff along the upstream stretch. Water of Karamana River is not potable at any stretch at any time of the year (HariKumar 2016). Even the monsoon rainfall is not sufficient to flush out the pollution load. Water cannot be used without treatment. During summer months, when river discharge comes down substantially, but other waste inputs continue to be at the same level, the colour of river water records high deterioration. Domestic sewage discharge, increased surface run off from paved urban areas, urban effluents, and waste dump in combination contribute to the deteriorating condition of the Karamana River and other surface water bodies draining the city. Impact of urbanization and human action are well evident. Deterioration of river water quality is a serious issue in the matter of source sustainability, and it requires public action to ameliorate the situation.

33.2.4.3 Groundwater

Groundwater is also used in many households. Now, there are reports of falling water table and quality deterioration. Water is tapped mostly by dug wells for domestic or irrigation purposes. Thiruvananthapuram City area experiences high fluctuation of water table (Rani et al. 2011). According to the CGWB estimates, the net groundwater availability of the Thiruvananthapuram Block is 23.74 MCM and the current stage of development is 81% (CGWB 2013). Doubts have been raised about this estimate particularly on account of local geology and aquifer characteristics. Ground reality hardly matches the CGWB estimated resource availability (Soman 2016). Due to hard rock terrain groundwater occurs in patches and there are wide spatial variations. Some parts around the city show overexploitation with more than 100% development. In these areas, groundwater extraction is more than recharge resulting in fall in groundwater level. Tapping of groundwater has increased significantly with increase in population and water use. Number of private tube wells and bore wells has increased tremendously.

Majority of the households in the panchayat area and even in the urban area have their own dug wells or bore wells The groundwater, alternative source for drinking water is under stress. While certain pockets are overexploited, the quality of water is an issue warranting serious consideration. Besides, quality of groundwater is an emerging problem. The coastal aquifers reported high salinity, which is further aggravated due to overexploitation. At present, for most of the water quality parameters, the groundwater in major part of Thiruvananthapuram City area is still within the limit. However, bacteriologically, contamination is an issue for all the seasons (HariKumar 2016).

This brief discussion brings out that the city is facing problems due to fall in quality of surface water, which affects quantity of availability also, reduction in storage capacity, fluctuation of resource availability due to vagaries in climate, and overexploitation and pollution of groundwater. Therefore, the challenges of water governance so far as source sustainability in Thiruvananthapuram City concerned will be to address both surface water and groundwater sectors. Revenue department, Water Resources Department, particularly, Irrigation department, and Groundwater Department are involved to administer these two sectors of surface and groundwater. Governance challenge here is more of coordination among the departments and to enhance in-situ water harvesting within the city limit.

33.3 Present Water Governance System in Thiruvananthapuram City

The discussion in Sect. 33.3 has brought out the emerging challenges of water governance in all three water service sectors and Sect. 33.4 deliberated on problems of fresh water that impact sustainability of water resources in Thiruvananthapuram City. In this section we intend to deliberate present water governance system in Thiruvananthapuram City, and decentralized planning and urban water management.

33.3.1 Water Governance in Thiruvananthapuram City

Water governance in Thiruvananthapuram City falls under the purview of more than one Government departments and the City Corporation has hardly any role (Table 33.2). In fact, urban self-governments in Kerala play no role or a very marginal role in providing basic water services of drinking water, sewage water, and storm water drainage to the city dwellers although there is evidence in other States in India where local urban governments manage all these basic services (Jacob 2012a; b). Thiruvananthapuram Corporation with 100 wards is administered through six zones for operational purposes. There are eight standing committees, namely Finance, Development, Welfare, Health, Public Works, Town Planning, Tax appeal, and Education & sports. However, there is no standing committee looking after water. Only during emergency supply of water through tankers in some of the stressed areas is managed by the city corporation.

Kerala Water Authority, an autonomous body established by Government of Kerala in 1984 for the development and regulation of water supply, wastewater collection and disposal manage drinking water supply and sewage treatment in the city. The two dams on Karamana River, principal source for harnessing water for the city are under the administrative control of Water Resources Department. The upper catchment area of the Karamana River comes under the administration of Forest Department. Any activity within the Forest Department is subject to the control of Ministry of Environment, Forests and Climate Change, Government of India. Even desiltation of the reservoir falling within the forest boundary needs clearance from Government of India. Downstream part of the Karamana river from the Peppara reservoir till its confluence is under the jurisdiction of Revenue Department, Government of Kerala and most lands are privately owned. The drainage channels within the city are partly controlled by City Corporation although some of the construction activities are undertaken by Minor Irrigation Department.

Table 33.2 Actors in relation to Urban Water Governance of Trivandrum City

Administrative level	Department/Organization	Responsibilities/Actions
Government of India	Ministry for Water Resources	National Water policy, 1987, 2002, and 2012
	Ministry of Urban Development	National Urban Policy
Government of Kerala	Department of Water Resources	Looking after the dams and reservoirs
	Department of Local Self-Government (Urban)	Policy formulation for local self-government activities
	Thiruvananthapuram Development Authority (TRIDA)	Supervision of development work in TRIDA area
	Town and Country Planning Department	Preparation of perspective plans for urban development
	KSUDP	Working out plans for urban development
	KWA	Drinking water, sewage, and wastewater disposal of urban centres
	Thiruvananthapuram Corporation	–
	Thiruvananthapuram District Administration	Emergency management of storm water

There is no specific agency to manage storm water drainage separately. To initiate the programme of drainage rejuvenation following 2015 waterlogging of the city Government of Kerala pressed seven organizations into the service and the provision of disaster management law was invoked. The Central Groundwater Board, Government of India and State Groundwater Board, Government of Kerala are entrusted with the job of exploring groundwater. The coastal zone (tidal affected areas) is under the purview of Coastal Regulations, promulgated by Government of India from time to time. What emerges from this brief discussion is that urban water governance is fragmented, and urban self-government virtually plays no role.

To communicate clearly about the important aspects of urban water governance and their interrelationships, a three-layer model of governance consisting of Content, Institutional and Relational aspects has been proposed (Hofstra 2014; OECD 2015). It serves as a checklist and can be used to assess the prevailing condition. We use this checklist in the context of Thiruvananthapuram (Table 33.3), and it is found that the city lacks in almost all front and urban water governance is yet to be conceived properly.

33.4 Decentralization and Urban Water Management

The 73rd and 74th Constitutional Amendment set the motion for decentralized government in the country. While 73rd amendment was for strengthening Panchayati Raj Institutions, the 74th amendment aimed to strengthen municipal level governance. The 11th and 12th schedules of 73rd and 74th constitutional amendments included water supply as one of the transferable responsibilities to LSGs. The Ministry of Urban Development (MoUD) initiated a series of institutional and financial reforms under this Act to increase the efficiency of urban local bodies for service delivery. This Act also provided a basis for state governments to delegate responsibility to urban bodies for providing a host of services including water supply. The Kerala Municipality Act 1994 envisaged various reform activities covering transfer of certain institutions and giving responsibility to prepare plans and implement those plans. Providing urban basic services, including water supply, sanitation, storm water drainage, and urban roads (excluding those provided/maintained by the State Public Works Department) is one of those. Urban Local Body was expected to manage the "water regime" and provide water services. To facilitate this devolu-

Table 33.3 Three-layer model of governance and Position of Thiruvananthapuram

Layers	Questions	Thiruvananthapuram City
Content	*Do we have sufficient and relevant information?	*Partly
	*Do we have the necessary knowledge and skills?	*Yes
	* Is there a clear policy and planning for the water management?	*Partly
Institutional	*Are the roles and responsibilities clear?	*Partly
	*Do we have the necessary tools?	*No
	*Is functioning of the financing system ensured?	*No
Relational	*Is the water policy well connected with other policy fields?	*No
	*Are all stake holders involved in decision-making in water management?	*No
	*Is there transparency in water management?	*No
	*Is there enough trust to work together?	*No

tion, the Kerala Water Authority (KWA) Act was amended permitting any local governments to take over from KWA an existing water supply scheme or to establish a new stand-alone water supply project (Government of Kerala 2009a; b). However, so far, no LSG has taken over any of the existing schemes from KWA, which continues to hold the responsibility for piped water supply and sewage treatment. There are enormous challenges of implementing decentralization in water service delivery (Nastar 2014).

Despite all initiatives since 1992 Act and introduction of People Plan Campaign in 1996 (Thomas Isaac and Franke Richard 2000) facilitating substantial decentralization and providing greater functional autonomy to municipal and local self-governments, which has resulted in "increased legitimacy of democratic local government and widened the political space for local politicians and civil society" (Heller and Harilal 2007), the centralized administration continues its hegemony in management of urban services. Continuation of central control through regulation and funding by the federal and state governments, inadequate capacity to perform decentralized functions effectively by local self-government and non-inclusion of the stake holders in decision-making are some of the issues that hindered effective decentralized governance and required institutional changes (Nastar 2014). The City Corporations and Municipalities operate within a policy framework designed at the state level and executed locally by the District Collector. There is a tendency of recentralization of all urban services through various authorities and government departments, besides bureaucratic and technocratic apathy to devolve authority to the LSG institutions. As a result, the authority of urban self-government has reduced to merely issue city registry certificate, management of solid waste and to perform some mundane duties. Thiruvananthapuram Corporation is neither technically equipped nor hold the authority to manage the water regime. Redesigning of urban governance is considered a key element in effecting decentralization and improving basic services (Jacob 2012a; b). The elected urban local self-governments are mere implementing agency with little scope for creativity. In this context one of the observations "The Municipal governance in Kerala is seen by some state level elite as nothing more than a costly exercise in keeping the local government democratic" (Gopokumar 2010) merits serious consideration.

33.5 Coping with Future Governance Challenges

To cope with future water governance challenges, we propose some structural/ institutional changes, which require legislative as well as technical interventions and people's participation at various levels. Our discussion specifically covers polycentric governance, bringing water back to urban life and waste tax/ effluent tax.

33.6 Polycentric Governance

Intra-city variations, especially core periphery differences in access to water services, growth of urban sprawl without commensurate increase of urban service facilities, encroachment, and diversion of natural water bodies for other purposes, effluent discharge from urban settlement and other wastes, and allocation and reallocation of water between the city and the surrounding rural areas are emergent issues of serious concern. Available water resource base is shrinking, whereas demand for water is increasing. The City authorities need to pay simultaneous attention to improve the existing water availability and access to sewage, and at the same time respond to environmental problems like storm water management, quality deterioration of surface and groundwater, and sustainability of freshwater resource catering the city today. The existing urban water governance has limitations to address these issues and a new governance system is necessary to cope up with future challenges. Interdepartmental coordination has emerged as a serious impediment to integrate all three water service sectors.

In the macro-scale country level analysis, it was suggested that performance improvement requires polycentric structures, which create possibilities to respond at different spatial scale as well as dealing with heterogeneity in impacts and capacities among different places and sub-basins (Pahl-Wostl et al. 2012). This is perhaps significant for Thiruvananthapuram City; however, the challenge will be how to evolve polycentric water governance structure both at the level of Karamana River basin and at the city level as the activities, priorities, actors, and aspirations are different at these two levels. To initiate the process, it is imperative to evaluate the nature of distribution of power among the State Government, District administration, Corporation, and the surrounding panchayats in management of hydrological regime. Proper assessment of the interplay among different governance modes —line departments and bureaucratic hierarchies is also necessary. Differing power structures between the urban and rural areas within the basin and among various socio-economic groups within the urban and rural areas are important contending factors.

The most promising level of intervention is perhaps at the bottom, building local governance capacity at the level of local self-government and civil society through formal and informal process, and the lessons from Jalanidhi experiment should be internalized in such an attempt. Participatory processes have great potential to integrate governance within and across scales, however, the power positions of actors at various scales influence the success of participation and, therefore, the scale perspective is an important issue in multi-level aspects of water management and governance in view of democratic legitimacy, efficiency, and equity (Moss and Newig 2010). Often the impediment of a bottom-up approach is competing relations between formal and informal institutions, short-sightedness at the local level and lack of technical support necessary to sustain the structures developed through informal process. There are several barriers including lack of involvement of all stakeholders, particularly from economically weaker groups, in the process of decision-making.

To address these issues, it is suggested to have multiple level interactions between governments and all other stakeholders/actors, all of whom must be drawn into continuous dialogue and negotiations, which may turn into conflicts and uncertainties. However, all these as part of democratic water governance may progressively lead to concentrative process to reach agreement to move on to implementation and evolve an iterative multi-level governance processes that continually progress through social learning and create a broader conceptual space for wide ranging debate over urban water governance (Gupta and Pahl-Wostl 2013; Bakker 2010). To facilitate this process and to evolve polycentric governance for water management in Thiruvananthapuram City a three-tier activity level has been envisaged as elucidated here:

Firstly, landscape level or macro-level perspectives at river basin scale: The forces operating at this level are mostly ex-situ and beyond direct control of the city authority, however, they need to be factored for urban water governance. This level consists of larger hydrological regime, climate change related aspects, demographic pressure, rural–urban water resource sharing, sustainability of resources, environmental dimensions, and political economic issues linked to the national/ regional development domain. Governance at this level refers to the integration between river basin and the urban area.

Secondly, city region or regime level: This level concerns about the existing situation within the city. The institutional arrangements, prevailing technology, local users, customs, markets, local politics, rules, and regulations as effective at the city level are part of this stratum. The governance at this level is essentially interdepartmental coordination, integration of service sector-drinking water, wastewater, sewage, and storm water drainage as part of an integrated water system, and integration of urban planning and water bodies through zonation of urban landscape.

Thirdly, local level or niche level: This level refers to the resident levels, which are building blocks. They act in micro-scale and facilitate innovation. People are directly involved in action. The units are small, and decisions are in the form of commonly agreed principles. Governance at this level is to ascertain direct involvement of local people, accord them ownership, honour local value system, internalize views of water user's group, which are mostly heterogeneous, nurture innovation and facilitate up-scaling of good practices.

Institutional repositioning is necessary both at the Karamana river basin level and at the city level to initiate the process to reorient governance system as proposed here. Water regime in Thiruvananthapuram City cannot be managed by the corporation alone in the present condition even if it is given all the required support-administrative, financial, and technical, as provenance of the surface water courses and recharging area for groundwater extend well beyond the jurisdiction of the Corporation. In view of this reality, it is important to plan for the transition from KWA and or Corporation centric management to decentralized management and create necessary socio-political space for a wider debate involving all the stack holders and society at large.

33.7 Bringing Water Back to Urban Life

An important way to evolve more integrated as well as democratic/participatory system of governance is to bring back water resources and water related infrastructure to the public sphere. One major reason for the crisis in the water sector is lack of social visibility of water resources and infrastructure in the urban areas. They are now dumped in the back yard of urban life, far removed from public gaze or intervention. Rivers and water bodies have turned into dumping grounds for urban waste of all sorts. They have also become sites of illegal activities, such as brewing of illicit liquor, sand mining, etc. As abodes of illegal gangs and waste dump riverbanks are not approachable, and usually do not attract people or even the media. In many locales, even the law enforcing authorities hesitate to visit these places. Obviously, this was not the case in the past. Rivers and other water bodies occupied a pride of place in social life and public activities. As such they were remarkably visible. Water continues to be critical for daily life even today; but its sources have generally become too remote and obscure from everyday social life.

To accomplish this task of bringing water bodies back to the centre stage of public life, it is necessary that people belonging to the Corporation wards should engage in healthier competition to take care of water bodies in the respective wards. Students and institutions should be encouraged to participate in the integrated water management programme. There could be "Know Your River/Water Resources" campaigns to cultivate public interest in the rivers/ streams/ water bodies. Research Institutions and the media may be encouraged to undertake periodic survey of the water bodies. The water bodies should become public spaces like public parks that the locals proudly present to the visitors as examples of ideal river management and waste mitigation programme, and cleanliness.

33.8 Tax on Waste/Effluent Tax

It is also possible to have a system of incentives to promote best practices and disincentives to check activities that adversely affect water resources. Producing waste is not a crime; but not owning up is. An important idea that we intend to put forward in this connection is a local tax on waste generation. All big economic units should be encouraged to submit waste tax return every year to the local government. The returns should give a clear estimate on the generation of waste by the unit. It should also say how the waste is treated or disposed. Such a system of waste accounting will not only help mobilize resources; but in addition, it would also act as a major disincentive for those who hide and take free ride on its disposal. Indonesia introduced a programme of voluntary disclosure of waste generation and environmental performance of the industrial units as part of river cleaning programme, which has been partly effective (Afsah et al. 1996).

Effluent tax, one of the economic instruments for environmental protection has been widely discussed in terms of impact and frequency of application (EEA 1996). The idea germinated in mid-1960s when there was proposal to study a tax-like system in which all polluters would be subject to effluent charges in proportion to their contribution to pollution (Boyd 2003). From the environmental point of view, pricing pollution is useful to assess full social cost of polluting activities, besides, it can achieve something that is not possible through presently followed command-control system and technology centred practices. Case studies on effluent tax in Germany indicated that a policy mix consisting of regulatory and economic instruments can be very powerful in implementing and enforcing policies to address direct effluent emissions in water bodies and in this context, setting up of right incentive structure, which is often hampered by interest groups assuming significance (Moller-Gulland et al. 2011). So far, eco-taxes have not been introduced in India, but there are deliberations to introduce environment taxes from time to time and the state of Karnataka introduced forest development tax and Sikkim collects ecology and environment tax (Verma 2016). The Kerala State Environment Policy-2009 proposed introduction of polluters pay principle to create a separate corpus fund for meeting expenses exclusively to protect environment and tackle environmental hazards (Government of Kerala 2009a; b). The Working Group on Environment constituted by State Planning Board, Government of Kerala also suggested introduction of polluters' pay principle and collect eco-taxes. These proposals are yet to be implemented.

Implementation of effluent tax is fraught with various challenges. The first and foremost is about measurement of effluents. While measurement is somewhat possible for point sources, in case of non-point sources, it appears not so realistic. Thiruvananthapuram City is primarily an administrative town with concentration of tertiary sector activities dominated by residential land use. Around 2 lakh households in the city produce wastewater and therefore the sources are largely non-point. In such a situation, the suggestion to impose tax on input resulting in pollution may be useful. Such practice is already in place in some of the European countries like Germany, where tax for wastewater (wastewater user fee) is collected at the same rate levied for water. Non-point source pollution and its pricing are a major issue throughout the World and have drawn considerable attention among the concerned academicians and policy makers. Besides, there are challenges of governance in introduction of effluent tax—institutional mechanism, execution, operation, and utilization.

In India, there is paucity of serious studies on this issue. However, available studies on Forest Development Tax introduced in Karnataka and Ecology and Environment Tax introduced in Sikkim highlighted certain drawbacks related to project selection, transparency, technical expertise, and monitoring mechanism while executing environmental safeguard projects for which these taxes were primarily introduced (Verma 2016). All these issues will crop up in case of Thiruvananthapuram. Nevertheless, the idea is significant and warrants serious considerations. With little effort effluent tax can be introduced in the urban areas at the household level. KWA manages water supply and sewage disposal in Trivandrum. As both the sectors are under single department it may not be difficult to administer this tax. However, the major impediment may be people's willingness to pay such tax. Massive awareness campaign at the ward level may be required to take people into confidence and apprise them about the necessity to introduce wastewater tax to serve the city better and proper provisioning for sewage management. A proper study can be initiated in Kerala to introduce this concept. The effluent tax generated at the local level as proposed in our study can be used for waste treatment and to meet administrative costs for waste management as is being practiced in many of the developed countries.

33.9 Conclusion

Problem of water management in urban areas is a growing concern across the world and it is being stressed that the problems emanate not so much from resource scarcity but due to governance failure. Integrated Urban Water Management approach is being advocated as an alternative paradigm. This study covering the water management issues in Thiruvananthapuram City argues that there is need to reconsider the existing water management practices and posit urban water issue in the larger context of water regime management covering surface as well as groundwater. The city is facing serious challenges. Due to growth of city, there is already core periphery differentiation in urban services. Despite various augmentation plans there are limitations and unsuitability of networked supply of drinking water to reach all parts of the city. Technological approaches in water service delivery may not yield desired results. The poorer sections are often deprived which is partly geographical due to location and partly infrastructural. Non-networked water supply alternatives, on which many people, both poor and non-poor depend, may be properly regulated, quality checked, and developed to supplement the piped water supply. In many of the scarcity areas local groundwater sources could be developed as alternative. Source sustainability is an emerging challenge as quality and quantity of surface and groundwater are deteriorating alarmingly. The water management of Thiruvananthapuram City cannot be considered in isolation. The Karamana basin and the underground water regime cutting across administrative boundary may be considered. This will certainly call for interacting with the surrounding panchayats within the basin. Even within the city the fragmented departmental approach in handling different component of water is one of the impediments and the responsibilities are sometimes not well defined, particularly in storm water management. Instead of draining out storm water there may be provision to recharge groundwater and store as part of rainwater harvesting. This will also contribute to source sustainability. Use of various water harvesting techniques could be a viable innovative option and for which there is a need for suitable and facilitating institutional set up.

The real challenge in water governance is integration, rural urban or upstream downstream and inter-agency partnership and people's participation. Urban water is part of a total water regime, and it cannot be governed in isolation and fragmented manner. Although the 74th Constitutional Amendment and Kerala Government Municipal Act created provisions for decentralization of water service delivery it has not been effective. There are bureaucratic-technocratic apathy and tendency to recentralize. Thiruvananthapuram Corporation has virtually no role in water management be it drinking water, sewage, or storm water. The Corporation is also not technically, financially, and administratively equipped to perform any role. To face the emerging situation, it is necessary to initiate reform of the existing public institutions, capacity building and restructuring the institutions involving all stakeholders across the society for managing all aspects of water through polycentric/ multi-level governance. This is important as there is a tendency already underway to involve private bodies to solve water shortage problem to overcome governance failure. The pressure will continue to grow with involvement of international funding agencies.

Experience of PPP is not without question. Water is a heritage resource, and all citizens are bestowed with the right to water, therefore, management of water should be through public institutions and public utility services. There is a need for coordinated efforts of local people, civil society, and political action which can create an arena to promote alternative visions of development in the water sector that are different from the existing practices. An important element of the new system of water governance would be to bring the water bodies to the centre stage of social and public life, so that their visibility is enhanced. It is also argued to have an appropriate system of incentives and disincentives to manage externalities related to water use. Focusing on bottom-up approaches, it is necessary to explore the possibility of building continuous pressure and create socio-political space for such experiments. Adopting IUWM framework may open the opportunities to innovate new alternatives however the key elements are decentralization, cooperation, and participation. For which Kerala has successful niche experiments that may be upscaled and improved through social learning and create a broader conceptual space to address the challenges of urban water governance. The discourse on Thiruvananthapuram city will be helpful in this context and shall provide a broad contour for the policy makers to deliberate on alternative urban water governance both in theory and practice drawing from empirical studies. The lessons are also useful for the cities of similar size in other parts of the country to examine existing urban water governance and scope for reorientation. In a broader context, this paper contributes to societal quest for designing appropriate urban water governance for sustainable urban water management.

Acknowledgement This paper has been completed while the first author was serving as Fellow in the HWK Institute for Advanced Studies, Delmenhorst, Germany. The first author (SC) is grateful to the HWK management for awarding the fellowship. This paper is part of the project "River restoration in Kerala: developing a co-evolutionary framework and river restoration action plan for Thiruvananthapuram City" funded under Research Unit for Local Self Government, Centre for Development Studies, Thiruvananthapuram. Authors are grateful to the Director, Centre for Development Studies, Trivandrum for providing facilities to take up this project. We also thank Prof. U S Mishra, Fellow, CDS for constructive suggestions in the matter of water governance. We also thank an anonymous referee for constructive suggestion to improve the earlier version of the paper. Mrs. Vandan M helped in preparation of maps and Mrs. Resmi P S assisted in data collection and analysis.

References

ADB [Asian Development Bank] (2014) Report and recommendation of the president to the Board of Directors, project no. 43253. Proposed Multi-Tranche Financing Facility India: Karnataka Integrated Urban Water Management Program, p. 18

Afsah S, Laplante B, Makarim N (1996) Program-Based Pollution Control Management: The Indonesian PROKASIH Program. Policy Research Working Paper No 1602, The World Bank, Policy Research Department, Environment, and Infrastructure Division

Bakker KJ (2010) Privatizing water: governance failure and the World's urban water crisis. Cornell University Press, Ithaca

Barga BPF (2001) Integrated urban water resources management: a challenge into 21st century. Int J Water Resour Manag 17(4):581–599

Biswas AK (2004) From Mar Del Plata to Kyoto: an analysis of global water policy dialogue. Glob Environ Chang A 14(Supplement 1):81–88

Biswas AK (2008) Integrated water resources management: is it working? Water Resour Dev 24(1):5–22

Boyd J (2003) Water pollution taxes- a good idea doomed to failure. Discussion paper 03-20. Resources for the Future. Washington, DC, p. 33

Brikke F (2015) Urban Water Management in India. Water and Wastewater International. Weblink: www.waterworld.com

Bahri A (2012) Integrated urban water management. Global Water Partnership Technical Committee (TEC) Background Paper No 16. Global Water Partnership, Stockholm 89p

Castro JE (2007) Water governance in the twentieth-first century. Ambiente Soc 10(2). https://doi.org/10.1590/S1414-753X2007000200007

Census of India, (2011) Kerala-District census handbook, Series-33, part XII B, Thiruvananthapuram-Village and town wise primary census abstract, Directorate of Census Operations, Kerala

CGWB (Central Ground Water Board) (2013) Ground water information booklet of Thiruvananthapuram District, Kerala state, technical report series D. CGWB, Kerala region, Thiruvananthapuram

Chakrapani R (2014) Domestic water and sanitation in Kerala: a situation analysis. Water aid, Chalakudy puzha Samrakshana Samity and forum for policy dialogue on water conflicts in India, Pune. (With inputs and guidance from K. Madhavan Namboodiri)

Chattopadhyay S, Harilal KN (2017) Emerging challenges of urban water governance-case of Thiruvananthapuram City, India. RULSG occasional papers no 1/2017, RULSG, Centre for Development Studies, Thiruvananthapuram

Closas A, Schuring M, Rodriguez D (2012) Integrated urban water management: lessons and recommendations from regional experiences in Latin America. Central Asia, and Africa. Water partnership program case profile no 1/November, World Bank, Washington, DC

Cooley H, Ajami N, Ha M, Srinivasan V, Morrison J, Donnelly K, Christian-Smith J, (2013) Global water governance in the 21st century. Pacific Institute, California. http://www.pacinst.org/publication/global-water-governance-in-the-21st-century/

Das B, Chan ES, Visoth C, Pangare G, Simpson R (eds) (2010) Sharing the reform process: learning from the Phnom Penh water supply authority (PPWSA). Mekong Water Dialogue Publication No 4, Gland

Dileep Kumar M (2017) Achieving Total sanitation in Thiruvananthapuram City. Unpublished background note, RULSG, Centre for Development Studies, Thiruvananthapuram

EEA [European Environment Agency] (1996) Environmental Taxes-Implementation and Environmental Effectiveness. Environmental Issues Series No 1. Copenhagen

Gopokumar G (2010) Transforming water supply regimes in India: do public-private partnerships have a role to play? Water Altern 3(3):492–511

Government of Kerala (2009a) Kerala state environment policy. Thiruvananthapuram

Government of Kerala (2009b) Report of the Committee for Evaluation of decentralized planning and development. Thiruvananthapuram

Gupta J, Pahl-Wostl C (2013) Editorial-global water governance. Ecol Soc 18(4):54

GWP [Global Water Partnership] (2013) Policy brief-integrated urban water management (IUWM): towards diversification and sustainability. Global Water Partnership, Global Secretariat, Stockholm

Harikumar PS (2016) Water quality status of Thiruvananthapuram District, Kerala. Unpublished background note, RULSG, Centre for Development Studies, Thiruvananthapuram

Heller P, Harilal KN (2007) Building local democracy: evaluating the impact of decentralization in Kerala, India. World Dev 35(4):626–648

Hofstra M (2014) Water resource management and governance-the Dutch experience. Water Governance Centre, Dutch Water Authorities, Ministry of Infrastructure, and the Environment

Jacob B (2012a) Rethinking development in an age of scarcity and uncertainty-new values, voices and alliances for increased resilience. Paper presented in the University of York, 19-22 September 2011

Jacob S (2012b) Infrastructure for urban water supply and disposal of sewage in Kerala: present and future. Indian Plumbing Today. May Issue: 14–22

Kerala Urban Development Project (KUDA), (1992) Kerala Urban Development Project consultancy servcies for Syrface Water Drainage, Final Report for Trivandrum, Kirloskar Consultants Ltd. Pune, Unpublished report

Maheepala S (2010) Towards the adoption of integrated urban water management for planning. Proceedings of 2010 International congress on environmental modeling and software modeling for Environment's sake, firth biennial meeting, International Environmental Modeling and Software modeling Society, Ottawa, Canada

Ministry of Urban Development (MoUD) (2012) Urban Reforms. URL: http://www.urbanindia.nic.in/urbanscene/urbanreforms/urbanreform.htm

Mitchell GV (2006) Applying integrated water management concepts: a review of Australian experience. Environ Manag 37(5):589–605

Moss T, Newig J (2010) Multi-level water governance and problems of scale: setting the stage for broader debate. Environ Manag 46(1):1–6

Moller-Gulland J, McGlade K, Lago M (2011) Effluent Tax in Germany. Ecological Institute, Berlin, 47p

Nadhamuni S (2012) An approach to integrated urban water management (IUWM)— the Mulbagal experience. Arghyam, Bengaluru

Nastar A (2014) What drives the urban water regime? An analysis of water governance arrangements in Hyderabad, India. Ecol Soc 19(2):57

Olsson L, Head BW (2015) Urban water governance in times of multiple stressors: an editorial. Ecol Soc 20(1):27

Organisation for Economic Co-operation and Development [OECD] (2015): Water governance for future cities, Concept Note for OECD Projects. Weblink: www.oecd.org/gov/water

Pahl-Wostl C, Lebel L, Knieper C, Nikitina E (2012) From applying panaceas to mastering complexity: toward adaptive water governance in river basins. Environ Sci Policy 23:24–31

Porse EC (2013) Storm water governance and future cities. Water 5:29–52

PRIA (Society for Participatory Research in India) (2014) Thiruvananthapuram study report, 2014: government led exclusion of urban poor: a greater contributor through a lesser recipient. PRIA, New Delhi

Rani VR, Shyam A, Nandakumaran P (2011) Thiruvananthapuram City, Kerala in groundwater scenarios in major cities in India. Central Groundwater Board, Ministry of Water Resources, Government of India, New Delhi, pp 223–229

River Management Cell (2015) Riverbank mapping and sand auditing of Karamana River, Thiruvananthapuram. Institute of Land and Disaster Management (ILDM), Government of Kerala, Thiruvananthapuram

Shah M (2016) Water governance: the way forward. Econ Polit Wkly L1(52):56–62

Soman K (2016) Status report on groundwater potential and use in Thiruvananthapuram District with special emphasis on Thiruvananthapuram City. RULSG, Centre for Development Studies, Thiruvananthapuram

Shaji J, Ravindran (2017) Waterlogging in parts of Thiruvananthapuram City and an Overview of 'Operation Anantha'. Background Note III, Research Unit, Local Self Government, Centre for Development Studies, Thiruvananthapuram 26p

The Hindu (2015) Daily Newspaper dt. 5th April 2015, Thiruvananthapuram edition

Thomas Isaac TM, Franke Richard W (2000) Local democracy and development: People's campaign for decentralized planning in Kerala. Leftword, New Delhi

UNESCO (2015) The United Nations world water development report, 2015: water for a sustainable world. United Nations World Water Assessment Programme, Paris

Verma R (2016) Ecotaxes: a comparative study of India and China. Working paper 353, Institute of Social and Economic Change, Bangalore

Whitler JY, Warner J (2014) Integrated Urban Water Management for Planners. Planning Advisory Services (PAS) Memo – September/October 2014, Water Research Foundation, American Planning Association, Chicago, pp. 1–13

World Bank (2012) Integrated urban water management case study: Buenos Aires. Washington, DC, World Bank

WHO (World Health Organisation), UNICEF (United Nations Children's Fund)-JMP (Joint Monitoring Programme) (2010). Progress on Sanitation and Drinking Water: 2010 update. United Nations Children's Fund, New York, World Health Organisation, Geneva

Bhitarkanika Wildlife Sanctuary, Odisha: People's Attitude Towards Its *Conservation*

34

S. C. Rai and Arpita Panda

Abstract

Mangrove forests are one of the most productive and bio-diverse wetlands covering an area of about 190,000–240,000 km^2. The Bhitarkanika Mangroves were designated a Ramsar Wetland of International Importance in 2002. Bhitarkanika in Odisha is India's second-largest mangrove ecosystem in terms of area and hold 63 of the 110 species of mangroves identified worldwide. About 9 lakh people in 900 revenue villages and hamlets live in and around the Bhitarkanika Biosphere Reserve and are directly or indirectly dependent on this diversified ecosystem for their livelihoods. However, encroachment, timber smuggling, and prawn farming are proving to major threats to its survival. The continued exploitation of mangroves has led to habitat loss, changes in species composition, loss of biodiversity, and a shift in dominance and survival ability. The present study suggests that ecosystem services provided by mangroves are of importance to the coastal communities as they increase their economy and employment opportunities.

Keywords

Livelihoods · Ecosystem services · Mangrove ecosystems · Odisha

34.1 Introduction

Mangroves are salt-tolerant plants of tropical and subtropical intertidal regions of the world. The specific regions where these plants occur are termed as "mangrove ecosystems." These are highly productive but extremely sensitive and fragile ecosystems (Bhomia et al. 2016). Mangrove resources are available in approximately 117 countries, covering an area of 190,000 to 240,000 km^2. Indonesia, Nigeria, and Australia have the largest mangrove areas. The total mangrove area in India was 4628 km^2 mangrove forests (covers about 3% of the total world mangroves vegetation), of which 60% are found along the eastern coast, 27% western coast, and 13% on the Andaman and Nicobar Islands (Bhomia et al. 2016; FSI 2013). Mangrove ecosystem has the highest level of productivity among natural ecosystems and performs several ecosystem services. The continued exploitation of mangroves has led to habitat loss, changes in species composition, loss of biodiversity, and a shift in dominance and survival ability. Mangrove forests are one of the most productive and bio-

S. C. Rai (✉) · A. Panda
Department of Geography, Delhi School of Economics, University of Delhi, Delhi, India

B. Thakur et al. (eds.), *Resource Management, Sustainable Development and Governance*, Sustainable Development Goals Series, https://doi.org/10.1007/978-3-030-85839-1_34

Box 34.1 Sustainable Development Goals

Mangrove forests are some of the most productive and biologically diverse ecosystems. The Bhitarkanika Mangroves were designated a Ramsar Wetland of international importance in 2002 and with the implementation of Sustainable Development Goal 14 in 2015, its significance and crucial role in maintaining biodiversity and preservation of ecosystem functions have been better realized. The challenge with Bhitarkanika Mangroves located in the northeastern coast of Odisha in India is its vulnerability because of the impact of anthropogenic pressures on the ecosystem on the one hand, and the benefits arising from eco-restoration programs on the other. The multiple functions of mangroves are that they provide protection from the impacts of climate change by acting as a barrier to coastal storms, adapting to rising sea levels, & soil conservation. They also support sustainable fisheries, provide medicinal products, fuelwood, fodder, habitat for wide range of flora & fauna. It is a great source of carbon sequestration and play an important role in climate mitigation also. Bhitarkanika Mangroves also provide food security and livelihoods (livelihoods, ecosystem services, climate regulation, fisheries, tourism, water filtration, and wood) for local communities. The marginalized population around this mangrove take shelter in the natural defenses against the impacts of natural disasters. The challenges with Bhitarkanika Mangroves can be converted into opportunities as has been displayed by local women taking to innovative practices such as mangrove nurseries and floating gardens. The various stakeholders of mangrove ecosystems can also contribute by mapping and understanding the landcover change to better articulate SDGs. This will help local, regional, and national policymakers understand and navigate changes with mangroves.

diverse wetlands. The mangrove ecosystems are widely recognized as providers of a wide variety of goods and services to people.

Bhitarkanika in Odisha is India's second-largest mangrove ecosystems in terms of area, and hold 63 of the 110 species of mangroves identified worldwide. However, encroachment, timber smuggling, and prawn farming are proving to major threats to its survival. Bhitarkanika Mangrove Conservation Area (BCA) comprises the Bhitarkanika National Park and Gahirmatha Marine Sanctuary that covers about 2107 km^2. The mangrove forests in the BCA consist of 145 km^2 of intact forest notified as a national park, and 385 km^2 of degraded forests comprising the Bhitarkanika wildlife sanctuary (Hussain and Badola 2010). About 1.5 lakh people depend on fuel, fodder, and other non-timber forests produce from this ecosystem. Bhitarkanika mangroves are home to 55 of India's known mangrove species (BCA 2016) besides harboring one of India's largest populations of saltwater crocodiles in the Gahirmatha marine sanctuary.

The Bhitarkanika Mangroves were designated a Ramsar Wetland of International Importance in 2002 (BCA 2016) to conserve the complex and fragile mangrove ecosystem and the endangered flora and fauna associated with it. However, overexploitation by the surrounding populace and other anthropogenic activities has contributed to the severe degradation of this mangrove ecosystem. Therefore, the main purpose of this paper is to examine the factors responsible for causing threats to mangroves and to suggest measures for sustainable management of mangroves.

34.2 Study Area: Bhitarkanika Wildlife Sanctuary

Bhitarkanika wildlife sanctuary is situated between 20.50°N and 86.42°E. Bounded on the north by river Baitarani and Dhamra estuary, which separates it from Bhadrak district on its south, Jagatsinghpur district and Cuttack district on its west is known as Kendrapara district. In 1975, an area of 672 km^2 was declared the Bhitarkanika wildlife sanctuary (Fig. 34.1). The

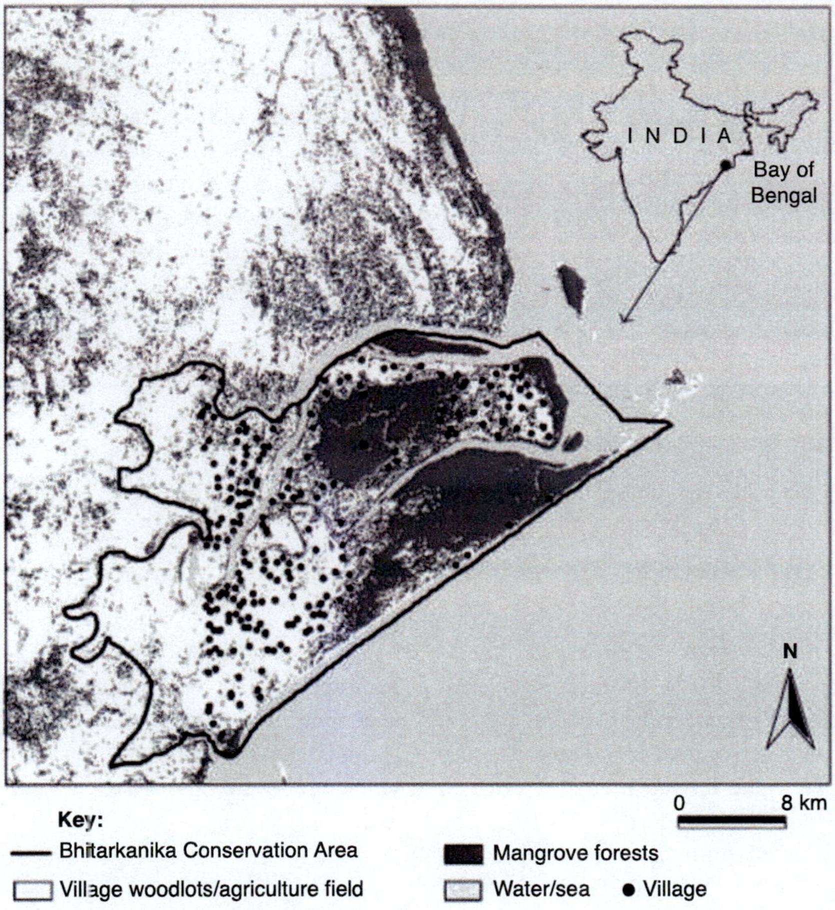

Fig. 34.1 Bhitarkanika Wildlife sanctuary, Odisha

Table 34.1 Characteristics of agro-ecosystem of Bhitarkanika conservation area

Parameters	Total
Average village size (ha)	113.3
% Cultivated area irrigated	8.7
Mean family size (nO)	7.6
No. of months of employment	6.3
% People engaged in agriculture	60.3
% People engaged in labor	25.5
Income of family (US$/annum)	488.9
% People literate	48

Source: Panda (2016)

core area (145 km^2) of the sanctuary was declared Bhitarkanika National Park in September 1998. The Gahirmatha Marine Wildlife Sanctuary, which bounds the Bhitarkanika wildlife sanctuary to the east, was created in September 1997. The sanctuary is surrounded by 336 villages having 200,000 population.

34.3 Ecosystem Services

About 9 lakh people in 900 revenue villages and hamlets live in and around the Bhitarkanika Biosphere Reserve and are directly or indirectly dependent on this diversified ecosystem for their livelihoods (Panda 2016) (Table 34.1). The Bhitarkanika mangrove wetland encompasses a host of ecosystems—an estuarine brackish water ecosystem, a riverine, and a forest ecosystem (Table 34.2). The value of forestry and fisheries products harvested by the locals estimated in an independent study is INR 18,000/– per house-

Table 34.2 Overall vegetation structure of Bhitarkanika mangrove forests

Tree density (trees/ha)	1376.93
Sapling density (sapling/ha)	83.33
Seedling density (seedling/ha)	45.79
Total tree basal area (m²/ha)	220.26
Total canopy cover	33.25%

Source: Hussain and Badola (2010)

Table 34.3 Use of different products by local people in Bhitarkanika mangrove forests

Resources	Uses	Mean quantity (kg/hh/annum)	Monetary value (Rs./hh/annum)
Fuelwood	Food cooking	2205	5500.08
	Livestock food preparation	312	800.00
	Others	21	102.08
Fish	For own consumption	98	4483.01
	For marketing purposes	115	6800.00
Timber	As thatching materials	343	1019.61
	As roof supports	27	294.12

Source: Panda (2016)

hold per year (Panda 2016). Of the 25 species of fish reported from Bhitarkanika (Chadha and Kar 1999), the total catch of in-shore fishery was estimated as 3.77 kg/ha having a market price of US$2.25 (Hussain and Badola 2010). The resources extracted from the mangrove forests contributed to more than 14.5% of the total income of the household (Panda 2016). This can form a far more substantial chunk of the income earned by the poor and marginalized residing near the forests. Since almost all households use fuelwood for cooking, the total fuelwood requirement in the 36 surrounding villages amounts to 2205 ± 104.2 kg/household/annum. The average fuelwood requirement per household is 14 kg per day, with the mean consumption standing at 312 kg per annum, with around 20, 000 people engaged in collecting fuelwood from the forests daily (Panda 2016). (Table 34.3).

Mangroves can help clean waters by trapping metal ions and other pollutants in their root systems and storing them in their stems. The concentration of lead across plant species varies from 0.03 to 4.21 g. The maximum concentration of copper was recorded in the stem of *Avicennia alba* (10.03 g), while the leaves of *Ceriops decandra* (8.05 g) accumulate the highest amount of copper (Chadha and Kar 1999).

Bhitarkanika has been a major tourist destination for years and is subject to heavy biotic pressure, with the area witnessing a maximum inflow during the winter months of October to March. In 2010–2011, about 18000 foreign and 18500 domestic tourists visited this area. In recent times, several aquaculturists from distant areas have started buying land along the Gahirmatha area, for instance, Gupti area has been acquired for conversion into prawn farms. These farms are a serious threat to mangroves, which are even otherwise, subject to several other depredations.

34.4 Sustainable Management

To protect these mangroves, sustainable management strategies are necessary, viz. to reassess the value of the mangrove ecosystem, develop a mangrove management plan, rehabilitate degraded mangrove ecosystem, improve community information, enhance community participation, develop alternate income generation activities, and develop an effective public awareness program, etc.

It would also help to fall back on the historical legacy of mangrove conservation followed in the past. In 1951, under the Kanka Raj Jungle Mahal Niyamawali rule, these forests had been designated as class I, prohibiting entry herein. Subsequently, in 1961, this area was declared a Protected Forest under section 29 of the Indian Forest Act 1927. In 1975, the Government of India declared 672 km² area of Bhitarkanika as a Wildlife Sanctuary, which was later upgraded to a National Park in 1998. In 1997, Gahirmatha Marine Sanctuary was set up on 1435 km^2 area with a core area of 725 km^2.

People's attitude towards conservation also plays a major role in management. Education plays a significant role in influencing the attitude of the people. Almost all literate respondents are in favor of eco-development and about 47.05% literate respondents are willing to cooperate with the forest department in this regard. Only 18.30% of the respondents feel that there is a violation of their rights after declaration as a protected area (Primary survey by second Author, 2015). About 90% of the respondents of the area are aware that Bhitarkanika mangrove forests have protected status. A high percentage (84%) of people felt responsible for the conservation of flora and fauna, while 93% were in favor of an integrated conservation and development program. About 43% of the respondents were willing to cooperate with the forest department in mangrove restoration.

During the last century, mangroves from these areas were destroyed or degraded by people, making these areas vulnerable to damage caused by cyclones. The vulnerability of many coastal human communities to cyclones is heightened by the removal of mangroves for development, agricultural expansion, and human habitation, etc. Mangrove forests are natural buffers against storm surges protecting tropical shores from erosion by tides and currents. It is suggested that a mangrove strip at least 100 m wide should be left as a buffer zone on more exposed shores. The awareness and appreciation of the local people of the functions performed by the mangrove forest are a positive sign for the conservation of the area.

Community participation at all stages of development is imperative for the success of any venture. Hence, public information networks need to be strengthened. The Government must strictly adhere to the coastal land use plan for the protection of mangroves to conserve biodiversity. Engaging the community in ecotourism related activities could provide a welcome source of income and yield. It could also help prevent unabated construction activities by realtors, and in-shore fisheries using mechanized vessels in the area.

In short, it is important to deal with the problems and issues spatially rather than addressing these using a sectoral perspective. There is a need for more research both on the productivity of mangrove dependent systems and their contribution to the local economy. A community-based mangrove management plan alongside an awareness program for forest guards and locals is the need of the hour. It is suggested that a Bhitarkanika Conservation Area Management Authority, which also involves outreach programs the forest guards, be set up to conserve this fragile ecosystem for posterity.

34.5 Conclusion

Bhitarkanika Mangroves are a mangrove wetland in Odisha. This is widely considered to be India's one of the largest mangrove in terms of area. Out of the 110 species of mangroves identified worldwide, 63 species are found in this mangrove. Goods and services provided by Bhitarkanika Wild Life Sanctuary are high as compared to other land use in the area, significant for low-income groups. Mangroves forests contributed more than 14% of total family income. The Bhitarkanika Mangroves are facing steady degradation due to population pressure, urbanization, and unplanned developmental activities. This mangrove is deteriorating very fast due to encroachment by the surrounding community. They extract fuelwood, timber, and other products from these mangrove forests. This mangrove faces a major threat from prawn farming and timber smuggling. It is suggested that to deal with the problems and issues on a spatial scale rather than addressing sectorally. The conservation initiative of creating protected areas has halted the loss of mangroves but simultaneously affected the livelihoods of local communities. To stem the decline of this important ecosystem, a spatial approach is essential wherein a mangrove management plan is formulated taking communities living within the area into confidence.

Acknowledgments The authors are thankful to the Head, Department of Geography, Delhi School of Economics, University of Delhi for providing facilities.

References

BCA [Bhitarkanika Conservation Area] (2016) UNESCO World Heritage Centre. Whc.unesco.org. Accessed on 19 Aug 2016

Bhomia RK, Mackenzie RA, Murdiyarso D, Sasmito SD, Purbopuspito J (2016) Impacts of land use on Indian mangrove Forest carbon stocks: implications for conservation and managements. Ecol Appl 26(5):1396–1408

Chadha S, Kar CS (1999) Bhitarkanika, myth and reality. Natraj Publishers, Dehradun

FSI [Forest Survey of India] (2013) State of Forest report. Forest Survey of India, Dehradun

Hussain SA, Badola R (2010) Valuing mangrove benefits: contribution of mangrove forests to local livelihood in Bhitarkanika conservation area, East Coast of India. Wetl Ecol Manag 18(3):321–331

Panda A (2016) Sustainable Management of Mangrove Ecosystem of Bhitarkanika, Odisha. Unpublished M. Phil Dissertation, Department of Geography, University of Delhi, Delhi

35 Socio-Economic Development in Highway Corridor Zones of Bihar: Case Study of Kuchai Kot-Muzaffarpur Section

Krishna Kumar and Anjan Sen

Abstract

Highways are an important tool to bring socio-economic and infrastructural development in the society. It is regarded as engine of growth and development as it brings development with itself. It is very much required in expanding the production base, expanding trade, and linking together resources and markets into an integrated economy. It also helps in connecting rural areas with towns and market centers and in bringing together underdeveloped and developed regions closer to one another. Road transport, therefore, forms an essential input for production processes, and adequate provision of transport infrastructure and service helps in increasing productivity and lowering production costs. In this study a sub-section, i.e., Kuchaikot–Muzaffarpur sub-section (159 km) of East-West Highway Corridor in Bihar has been taken as study area, then zones along highway have been delineated to examine the spatial dimensions of socio-economic development. Use of GIS (Georeferencing and digitization), Observation, Questionnaire and Purposive Stratified Random Sampling, normalization (UNDP method for obtaining scale free unit), composite index, and t-test are used in the study for analyzing the data for result. In the present study it has been found that highway corridor is playing a great role in the development of socio-economic status of the region. It can be concluded that highway corridor is providing better connectivity and accessibility followed by the frequency to travel to the urban area to avail better health and educational facilities. Land price is also increasing which is followed by coming up of new settlements along the highway. It has been also found that villages of zone one have better opportunities in the context of connectivity and accessibility. It has been found that the highway is creating positive environment for bringing development in the region.

Keywords

Accessibility · Connectivity Development · Highway Corridor · Socio-Economic Development

K. Kumar (✉)
Discipline of Geography, School of Sciences, Indira Gandhi National Open University (IGNOU), New Delhi, India

A. Sen
Department of Geography, Delhi School of Economics, University of Delhi, Delhi, India

B. Thakur et al. (eds.), *Resource Management, Sustainable Development and Governance*, Sustainable Development Goals Series, https://doi.org/10.1007/978-3-030-85839-1_35

35.1 Introduction

Every day we read, talk, and debate about different concerns of socio-economic development like unemployment, poverty, development of roads and bridges, and facilities like educational institutions, hospitals, availability of safe drinking water, and latrine facility in newspapers, magazines, on television, etc. Even during elections these socio-economic indicators have been discussed by our political leaders, political parties, electorates, and media. Whenever there are debates and discussions on the issue of development and, particularly on socio-economic development, the issue of empowerment of disadvantaged groups and backward regions in our society is generally referred. What does this mean? Why and how should we understand the relationships between socio-economic development and empowerment of disadvantaged groups and backward regions?

Today, all community-based studies focus on socio-economic stratification, as this is the key to understand affordability of health services, amenities, and purchasability (Dudala SR 2013). In the present-day situation, social scientists and researchers need the economic revision of income variable in socio-economic status. Almost all community-based studies give importance to the socio-economic stratification, which is considered as a key parameter for proper understanding of the affordability of community of health services, amenities, and their purchasing capacity. When it is taken as a totality of education, occupation, and income, it reflects the value system expected for that level of education and occupation (Kumar BPR et al. 2013). Socio-economic status influences on the incidence and prevalence of various health-related conditions. It also influences social security in terms of accessibility, affordability, acceptability, and actual utilization of various health facilities (Agarwal AK 2008).

Thus, to understand the concept of socio-economic development, we must first understand the concept of development. Generally, development is a process of societal transformation from a traditional society to a modern society, also known as modernization. Development is a process of improving the capability of a nation's institutions and value system. Development means economic growth with social change. Core values underlying the concept of development are human dignity, equality and social justice are interrelated and interdependent. Another principle implied in social development is the emphasis on cooperation against competition and collectivity as against rugged individualism. Social development means something is more than the economic, political, and environmental development. Thus, development is based on principles of human dignity, equality, and social justice and can be defined as integrated, balanced, and unified (social and economic) development in the society.

Socio-economic development, therefore, is a process of social and economic development in a society. It is measured with indicators, such as gross domestic product (GDP), life expectancy, literacy, levels of employment, housing conditions, etc. For the better understanding of socio-economic development, we may understand the meaning of social and economic development, separately. Social development is a process which brings positive transformation in every section and community of the society in a manner which improves the capacity of the society to fulfill its needs and desires. It involves a qualitative change in the way the society transforms itself and carries out its activities, such as through more progressive approach and behavior by the population, the adoption of more efficient processes, or more advanced technology. Therefore, the purpose of this paper is to evaluate the socio-economic developmental disparities in the highway corridor zone by taking the example of Kuchai-Kot-Muzzaffarpur section in Bihar with the help of selected indicators.

Box 35.1 Sustainable Development Goals

Movement of people is fundamental to economic and social prosperity everywhere. The development of transport corridor is crucial not only to improve efficiency in the transport and logistics processes but also to generate economic development which further impacts sustainable human well-being in the form of improved income and employment, reduced poverty, and promotion of equity and inclusion in society. It is in this context that the authors study the Kuchaikot- Muzaffarpur sub-section (159 km) of East-West Highway Corridor of Bihar in India. The UN Sustainable Development Goals (SDGs) identify growth in sustainable transport infrastructure as key solution to achieving the other dimensions of SDGs. SDG-9 is about establishing infrastructure and connectivity for trade in goods and services and between people which is very critical. But these also are paramount for achieving SDG-8 (decent work and economic growth), SDG-10 (reduce inequality within and among regions), and SDG-11 (sustainable cities and communities). Thus, in a sense, growth in transport infrastructure is critical for achievement of all other sustainable development goals. AS has been laid out at various platforms of the UN, stakeholders can come together through the lens of "shared value" to identify opportunities in addressing infrastructural challenges.

35.2 Study Area

The study area is a sub-section, i.e., Kuchaikot-Muzaffarpur (159 km) of East-West Highway Corridor in Bihar (Fig. 35.1). The total length of East-West Highway Corridor is 3300 km which runs from Porbandar (Gujarat) to Silchar (Assam). The total stretch of the corridor in Bihar is 517 km, which has 3 sub-sections: (1) The Kuchaikot–Muzaffarpur section of NH-28 (Barauni to Lucknow) and some stretch of NH-57, whose length is 159 km; (2) The entire stretch of NH-57 (Muzaffarpur–Darbhanga–Purnia), whose length is 310 km; and (3) The Purnia–Dalkhola and Islampur–Galgalia sections of NH-31 (Barhi in Jharkhand to Jalukbari in Guwahati, Assam), whose length is 48 km.

The Corridor has been constructed under phase I and Phase II of the National Highway Development Project (NHDP). The NHDP was started in 2000 and is being managed by the National Highway Authority of India (NHAI), which works under the Ministry of Road, Transport and Highways (Government of India).

35.3 Selection of Indicators

The selection of indicators is of utmost significance. According to Nayak and Narayankar (2009) a good indicator should have the following properties:

1. It should be closely related to the situation and always be objective oriented.
2. It should be unique and clearly understandable.
3. It should be sensitive, such that small changes in the areal unit could be reflected in the value of the indicator.
4. As to feasibility, value of the indicator should be easily workable and information should not pose any problem, and
5. It should be stable, such that comparable values can be obtained over time and space.

Thus, the indicators selected for this study pertain to have various socio-economic aspects. In total, 27 indicators have been selected from Household Amenities of Census of India, 2011 and Village Directory of Census of India, 2011. Out of total 24 indicators, we have clubbed similar indicators to make an index to represent the level of socio-economic development. The grouping is mentioned in Table 35.2.

35.4 Methodology

35.4.1 Database

The study has used secondary database. For construction of socio-economic development index

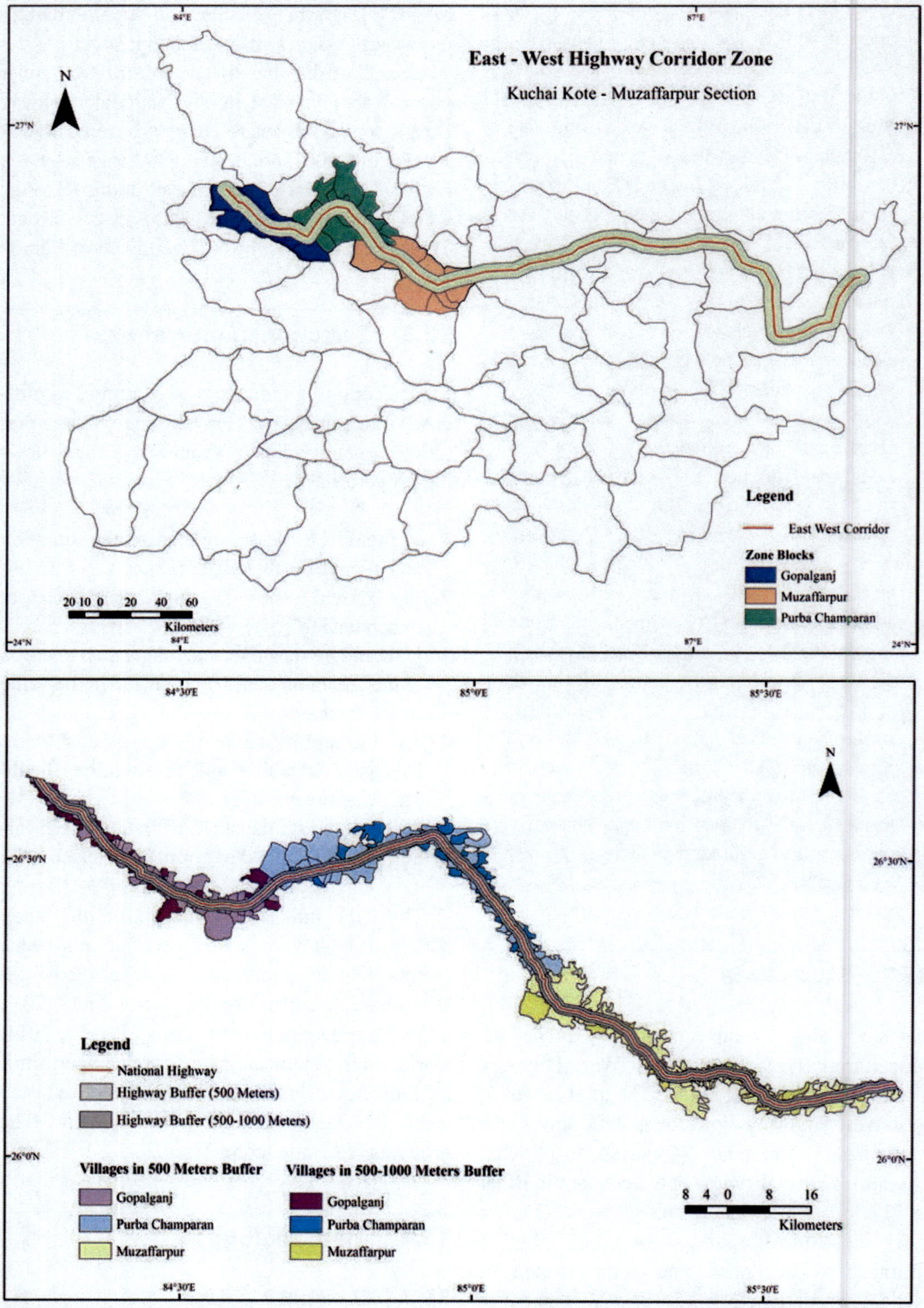

Fig. 35.1 Study Area

Table 35.1 Villages covered in the zone district wise

S.No.	District	No. of Villages Zonewise	No. of Villages District wise
1.	Gopalganj	Zone 1: 46	82
		Zone 2: 21	
		Zone 3: 15	
2.	Purba Champaran	Zone 1: 32	66
		Zone 2: 23	
		Zone 3: 11	
3.	Muzaffarpur	Zone 1: 56	125
		Zone 2: 44	
		Zone 3: 25	
	Total		173

Source: Calculated by authors
Note: Same blocks can come under buffer zone 1, buffer zone 2, and buffer zone 3 depending on their respective administrative boundary size

Table 35.2 Indicators used for construction of socio-economic index

S. no.	Groups	Indicators
1	Good housing conditions (such census houses which do not require any repair and are in fairly good condition).	Concrete roof
		Burnt brick wall
		Numbers of covered drainage
		Numbers of exclusive rooms
		Bathroom facility
2	Potable water	Tap water facility
		Drinking water
		Hand-pump connection
3	Electricity facilities	Electricity connection
4	Health facilities	Hospitals, PHC, DHC,
5	Sanitation facilities	Latrine
6	Banking facility	Co-operative Bank, Nationalized Bank, Private Bank
7	LPG facilities	LPG connections

Source: Compiled by authors

data related to good housing conditions, two rooms or more, dilapidated house, electricity connection, latrine facility, liquid petroleum gas (LPG) connection, banking facility, and two-wheeler have been collected from household amenities, Census of India, 2011. The data on agricultural laborers, cultivators, household workers, other workers, and literacy rate have been collected from Village Directory, Census of India, 2011 (Fig. 35.2).

35.4.2 Data Analysis

For calculating level of socio-economic development, the first step was to determine the indicators. The indicators have been grouped in six categories, i.e., good housing conditions, potable water, electricity, health, sanitation, banking, education, and LPG facilities. The collected data has been arranged in the form of a rectangular matrix with "" rows representing villages and "j" columns representing indicators, where Xij be the value of the indicator j corresponding to village i. In the second step, normalization of indicators use for functional relationship has been done. It is obvious that the indicators have different units and scales. The methodology used in UNDP's Human Development Index (HDI) is followed to normalize them. That is, to obtain figures which are free from the units and to standardize their values; first, they are normalized so that they all lie between 0 and 1. Before doing this, it is important to identify the functional relationship between the indicators and development. Two types of functional relationship are possible: development increases with increase in the value of the indicator. Assume that higher the value of the indicator, more is the development. In this case the normalization is done using the formula:

$$x_{ij} = \frac{X_{ij} - \min_i\{X_{ij}\}}{\max_i\{X_{ij}\} - \min_i\{X_{ij}\}}$$

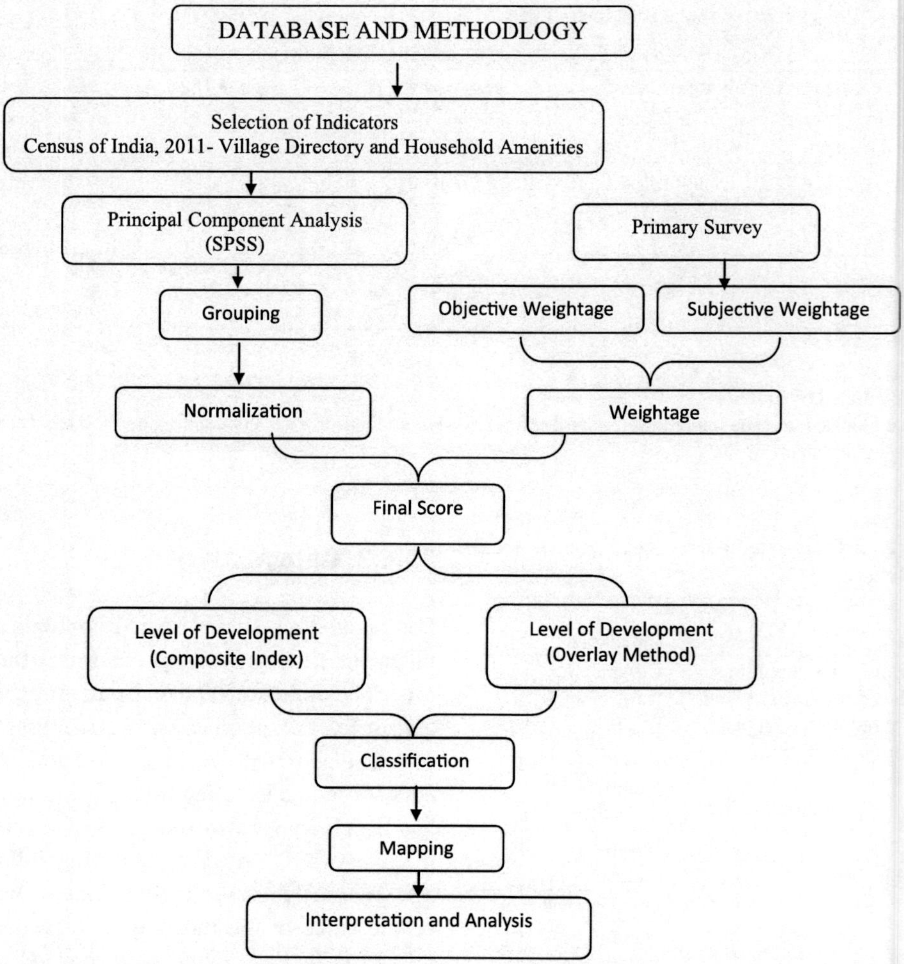

Fig. 35.2 Data Source and Methodology

All these scores lie between 0 and 1. The value 1 corresponds to that village with maximum value and 0 corresponds to the village with minimum value. On the other hand, the second functional relation corresponds to increase in development with decrease in value of indicator. For this case, the normalized score is computed using the formula:

$$y_{ij} = \frac{\max\limits_{i}\left\{X_{ij}\right\} - X_{ij}}{\max\limits_{i}\left\{X_{ij}\right\} - \min\limits_{i}\left\{X_{ij}\right\}}$$

It can be easily checked that Xij + Yij = 1 so that Yij can be calculated as Yij = 1−Xij. After normalization, the Average Index (AI) for each source of vulnerability was worked out and then the overall vulnerability index was computed by employing the following formula:

$$VI = \left[\sum_{i=1}^{n}\left(AI_i\right)^{\alpha}\right]^{1/\alpha} \quad n$$

where n is the number of sources of vulnerability and $\alpha = n$. The vulnerability indices can be worked out for each period, and they can be compared to assess the changes in vulnerabilities over the period.

It is assumed that there are M villages, K indicators of development and Xij, i = 1,2,...M; j = 1,2,...K are the normalized scores. The level or stage of development of ith zone, yi, was assumed to be a linear sum of Xij as.

$$\bar{y}_i = \sum_{j=1}^{K} w_j x_{ij}$$

where ws $\left(0 < w < 1 \, and \sum_{j=1}^{K} w_j = 1\right)$ are the weights.

The weights are assumed to vary inversely as the variance over the regions in the respective indicators of vulnerability. That is, the weight wj was determined by

$$w_j = \frac{c}{\sqrt{\underset{i}{\text{var}}\left(x_{ij}\right)}}$$

where *c* is a normalizing constant such that

$$c = \left[\sum_{j=1}^{j=K} 1 \sqrt{\underset{i}{\text{var}}\left(x_{ij}\right)}\right]^{-1}$$

The choice of the weights in this manner would ensure that large variation in any one of the indicators would not unduly dominate the contribution of the rest of the indicators and will distort inter-regional comparisons. The development index so computed lies between 0 and 1, with 1 indicating maximum development and 0 indicating no development at all. For classificatory purposes, a simple ranking of the regions based on the indices, viz. yi is enough. However, for a meaningful characterization of different stages of development, suitable fractal classification from an assumed probability distribution was used. A probability distribution which is suitable for this purpose is the Beta distribution, which is generally skewed and takes values in the interval (0,1). This distribution has the probability density given by

$$f(z) = \frac{z^{a-1}(1-z)^{b-1}}{\beta(a,b)}, 0 < z \langle 1 \, and \, a,b \rangle 0$$

where β (a, b) is the beta function defined by

$$\beta(a,b) = \int_0^1 x^{a-1}(1-x)^{b-1} \, dx$$

The two parameters a and b of the distribution were estimated using SPSS software packages. The Beta distribution is skewed. Let (0, z1), (z1, z2), (z2, z3), (z3, z4), and (z4, 1) be the linear intervals such that each interval had the same probability weight of 20%.

The database has been represented and analyzed by using maps. Techniques like regression and correlation have also been used wherever necessary. For such analysis SPSS software has been used.

35.5 Development Index

The selection of indicators in construction of socio-economic index is of utmost significance. The indicators need to represent all aspects of socio-economic development. The indicators selected for the study pertain to various socio-economic aspects. A total of 16 indicators have been used to show the spatial pattern of socio-economic development as discussed below:

35.5.1 Good Housing Index

According to the Census of India, 2011, such census houses which do not require any repair and are in good condition are considered as a house with good condition. Good housing condition shows both the social and the economic conditions of the resident, thus it is an important indicator for evaluating the socio-economic conditions of any individual or region. The status of good housing condition is not satisfactory which is clearly visible from Fig. 35.3. It has been found that the condition is just satisfactory as most of houses fall above the value of 0.31.

Among the three districts, the status of good housing condition is much better in Gopalganj district, which is followed by Purba Champaran and Muzaffarpur district. In Gopalganj out of 145 villages, 63 villages (43%) fall above the value of 0.46 in comparison to 51 villages (26%) and 35 villages (29%) of Muzaffarpur and Purba Champaran, respectively.

The reasons for Gopalganj district having better housing condition are that the per capita income of the villagers is much higher as large numbers of youth have migrated from here to

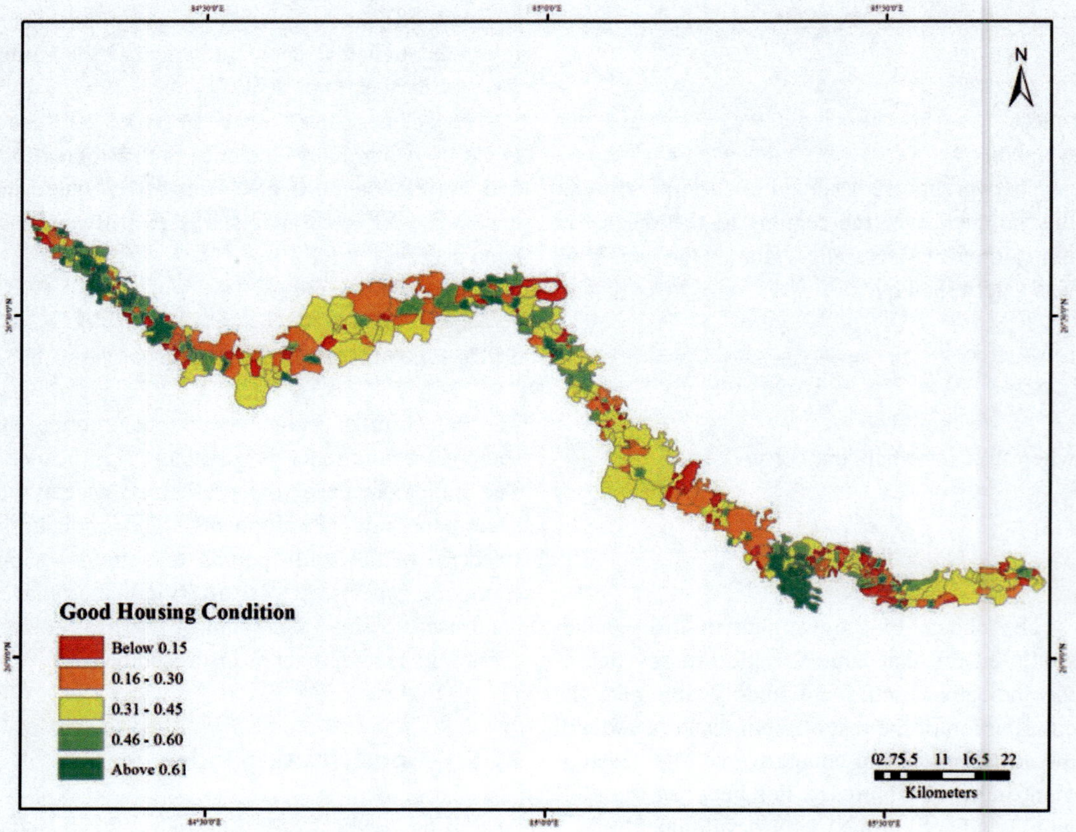

Fig. 35.3 Good Housing Conditions

the Gulf countries and metropolitan cities in India to earn. The Purba Champaran district is affected by naxalism, while the Muzaffarpur district gets affected by annual seasonal floods. These are other reasons for low numbers of good houses.

35.5.2 Latrine Facility

Availability of latrine facility represents the socio-economic condition of a household. According to the United Nation estimates, approximately 600 million people (i.e., 55%) of Indians openly defecate even after 60 years of independence (The Times of India, Nov.19, 2013). According to Mr. Jack Sim, founder, and President of the World Toilet Organization (WTO), India has a lot of work to do to improve sanitation.

The overall condition of availability of latrine facilities is very poor in the study area. In Purba Champaran, the percentage of availability of latrine is below poor. In most of village's households the availability of latrine is below 30%. The percentage of availability of latrine in Gopalganj district is good in comparison to Purba Champaran. The easternmost part of the district has worst percentage of availability of latrine. Even there are three villages which do not have any latrine in any one of its household. In Muzaffarpur district, the availability of latrine is not good. Out of 189 villages, only 53 villages have more than 41% of latrine in the house, while rest of villages has below 20% (Fig. 35.4). This shows the poor condition of sanitation. The condition of latrine facility is also very poor. Thirty percent of households of the villages have no latrine facilities which show the poverty condition and the lack of awareness for sanitation.

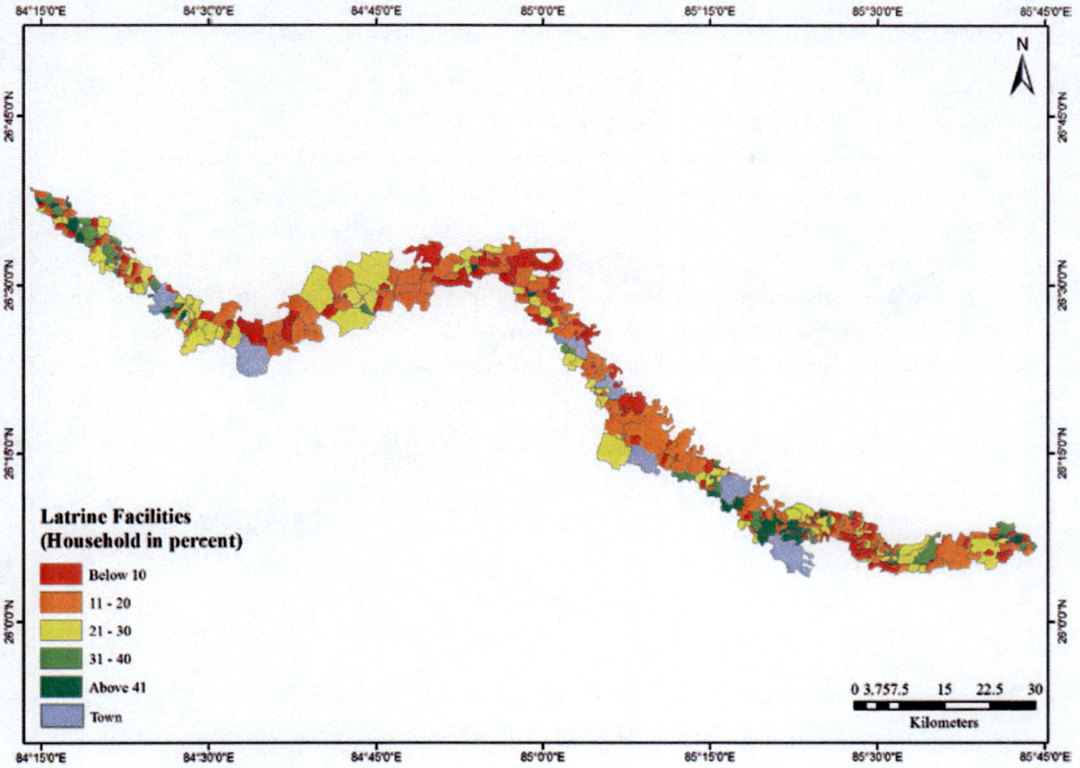

Fig. 35.4 Latrine Facility

35.5.3 Electricity Connection

Power sector is an essential infrastructure required for the smooth functioning of the development process and the economy of any country or region. An efficient, reliable, affordable power helps in the rapid agriculture, industrial, and overall economic development of any region. Besides, it is also essential for growth and poverty reduction.

Since the beginning of planned economic development, rural electrification has been a high priority in India. Both the central and the state governments have been attempting to improve the accessibility, availability, and quality of electricity, especially in rural areas. The central government has started the Rajiv Gandhi Grameen Vidyutikaran Yojana (RGGVY) in April 2005, with a vision to electrifying all villages, providing access to all rural households and free connections to all below-poverty-line (BPL) households by 2010 (South Asia Energy Unit Sustainable Development Department, The World Bank 2010).

In 2015, the Prime Minister of India, launched a new scheme—Deen Dayal Upadhayaya Gram Joyti Yojna (DDUGJY) for electrifying all the villages of India phase wise. Till February 2016, 5855 villages have been electrified under this scheme. The condition of electricity connection is very poor in the whole corridor zone. It is worst in the district of Purba Champaran. The electricity connections are uniform in Gopalganj district, while in Muzaffarpur district the electricity connection is below 10% in the eastern part. The condition of electricity connection is very poor in Purba Champaran district. In majority of the villages only 10% of the households have the electricity connection, while only 15 villages, above 41% of households have electricity connection. Its connection in Gopalganj district is uneven and it varies from below 10% to above 41% (Fig. 35.5). It is difficult to explain this pattern. The condition of electricity connection in

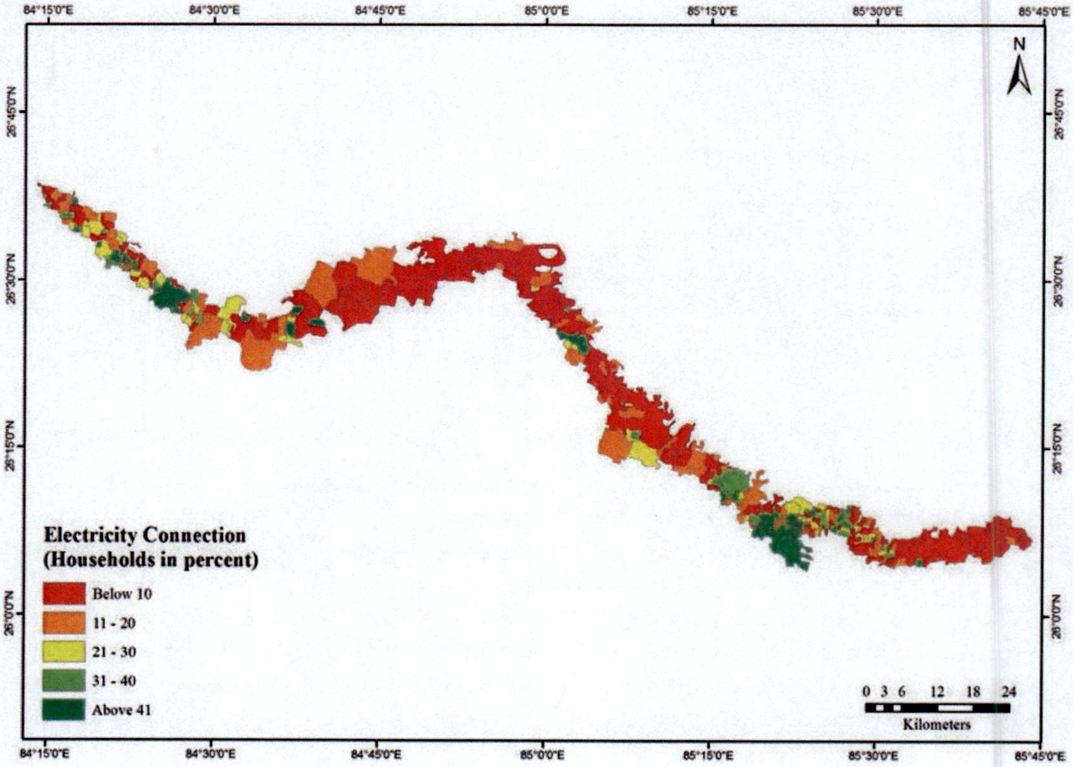

Fig. 35.5 Electricity Connection

Muzaffarpur district is also uneven but its condition is worst in the eastern part followed by the western. Its connection condition is better in the central part of the district of the corridor.

35.5.4 LPG Connection

The connection of LPG in household indicates both the social and economic conditions of the society or an individual. It is an accepted fact that access to energy is a critical input for the process of sustainable development. Access to clean and economic sources of energy is also considered as an essential need for different challenges world is facing in the form of rapid urbanization, globalization, rising inequality, and global environmental changes.

According to the National Sample Survey 68th round, in rural India, the dominant cooking fuels still consist of firewood and dung, with around two-thirds of the households still depending on them. However, the transition to cleaner cooking fuels has largely materialized in urban India, with 68.4% of the households using LPG. More than 70% of all households in India still use solid fuels mostly biomass and, thus, causes lots of health issues in rural parts of India, whereas only 16% uses gaseous fuels. This shows that India still must go long ways in providing clean and affordable cooking energy to its people (Government of India 2015).

It has been highlighted by several studies conducted in the past that the choice of household fuel is a complex process and depends upon a variety of factors. It has been found that income is a key determinant of adopting new energy sources and total energy demand. It determines the affordability of different types of fuel for households (Ouedraogo B 2006).

The condition of LPG is very poor in the study area. In almost whole region LPG connection was less than 10%. There are 13 villages where households had above 41% of LPG con-

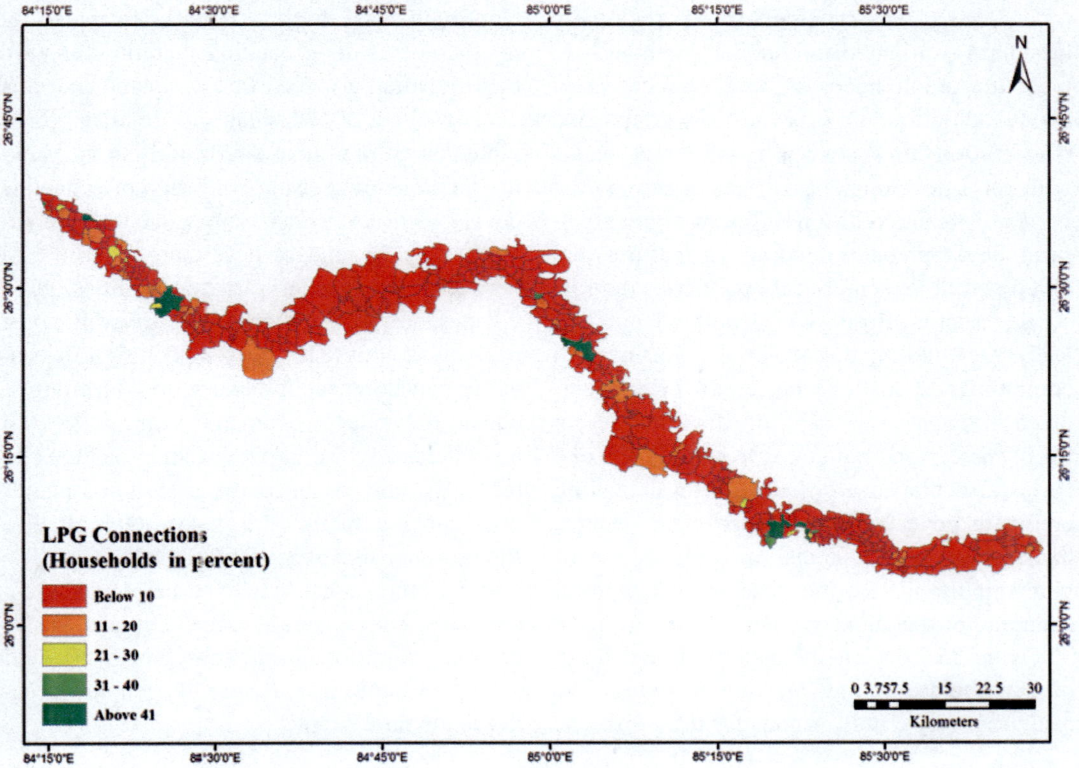

Fig. 35.6 LPG Connection

nections, while in 35 villages connections were between 11% and 20%. Six villages had 21 to 30% of LPG connection. The condition of LPG in Purba Champaran is very poor. Except 9 villages, all villages household have LPG connections below 10%. Only 2 villages have households with above 41% of LPG connections. In Gopalganj district also, the status of LPG connection is very poor. Most of the households of the villages have LPG connection with below 10%. Only 5 villages have above 41% of LPG connections (Fig. 35.6).

35.5.5 Banking Facility Index

The role of banking sector in the economic development of a country cannot be ignored as it helps in saving and provides credit at the time of needs. When we talk stated about rural India, then, the role of banks becomes more important as Mahatma Gandhi - "India is a country of villages and real India lies in village" and village economy is the backbone of Indian economy. Without the development of the rural economy, the objective of economic planning cannot be achieved in India. Thus, banks and other financial institutions are considered to play a critical role in the development of rural economy in India.

Many reviews on the development of rural economy are available, but there are very few studies, on the role and impact of banking sector in the economic development of rural India. Burgess and Pande (2005) indicated that the branch expansion in unbanked rural area in India has a positive impact on the reduction of poverty and rural inequality. Ketkar and Ketkar (1992) emphasized that post-nationalization of banks evidenced an increase in saving bank amount, investment, productivity of capital and GDP; however, the desired goals were not met with satisfaction. Yadav (2005) depicted on the role of Regional Rural Banks (RRB) on rural development. He also pointed that RRB has now become the main plank of organized rural credit structure.

The banks are state-sponsored, regionally based, rural-oriented and functioning mainly to meet the credit needs of weaker sections of rural population. Das SK (2007) established the relationship between commercial banks with rural development. He has examined the impact of bank credit on different segments of rural development including agriculture and allied activities, small-scale industries, and its service sector. Bose (2005) pointed out that RRBs are enabling the weaker sections of rural community to institutional credit on a subsidized interest. The bulk of the loans from RRBs have been granted to the priority sectors. It has also played a critical role in extending credit for poverty alleviation schemes. Hence, banks and other financial institutions are of vital importance for the development of rural economy of the country.

Figure 35.7 depicts the pattern of percentage of households availing banking facilities. The results are very erratic. Among the three districts, Gopalganj has the maximum percentage of households availing banking facilities followed by Muzaffarpur and Purba Champaran. In Gopalganj out of 140 villages, 116 villages have more than 80% of households availing the banking facilities. It shows a good sign of economic development. In Purba Champaran out of 118 villages, only 23 villages have more than 60% of households are availing banking facilities. In 61 villages, only 60% of households are availing the banking facilities. It shows a poor sign of economic development. Similarly, in Muzaffarpur, out of 202 villages, only in 7 villages, 80% of households are availing the banking facilities. In rest of the villages, the percentage of households with banking facilities is below 60%. It also shows a poor sign of economic development. The state government needs to focus on providing co-operative banks, rural banks, Gramin banks, branches of nationalized banks, etc. in the rural parts of Bihar to give a sense of economic security to the rural people.

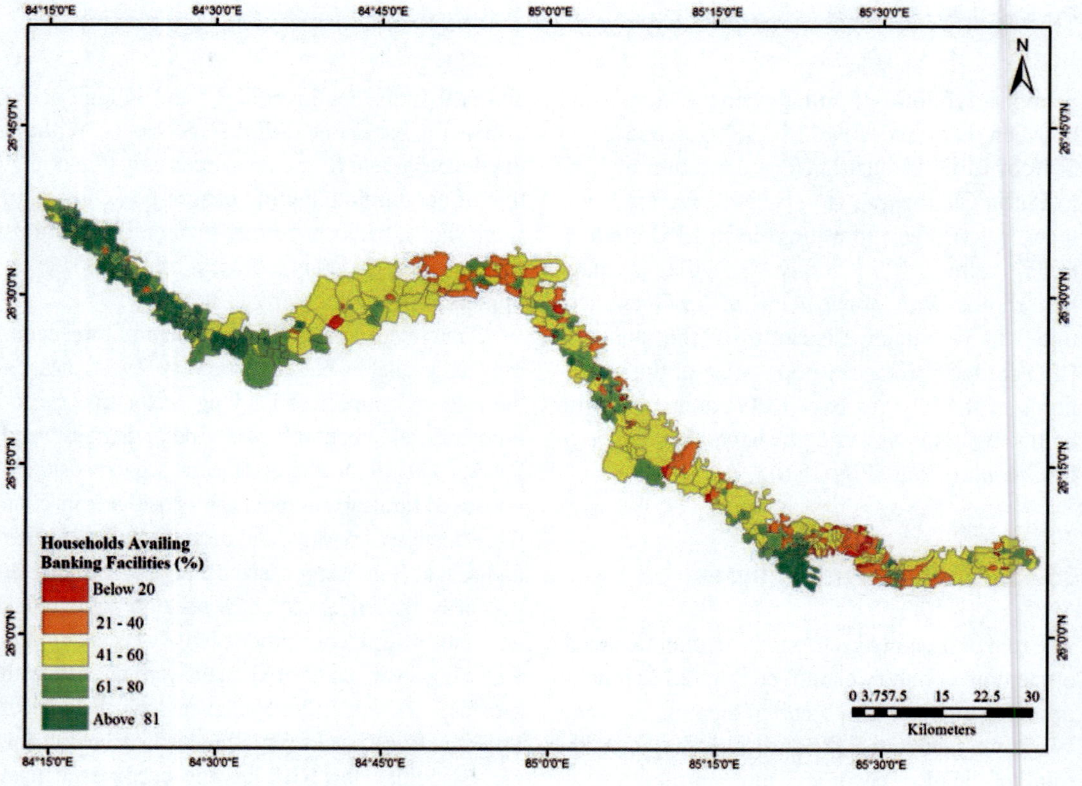

Fig. 35.7 Households Availing Banking Facilities

35.5.6 Drinking Water Facility Index

Water is one of the most important resources for any country, and most of this water is obtained from rivers and underground water to meet the needs of irrigation, cattle rearing, and various other sanitation purposes by our states. There is a provision in our constitution (Article 47) which ensures that government needs to provide clean and drinking water to all its inhabitants for improving public health standards to the state. Abide by the constitution the government has started various programs since independence to provide safe drinking water. According to the estimate by the Planning Commission a total sum of Rs. 1105 million was spent on providing safe drinking water in India till the tenth Plan. One would argue that the expenditure is huge, but it is also a fact that despite such a huge investment, lack of safe and secure drinking water continues to be a major hurdle and a national economic burden.

India, which has 16% of the world's total population, has roughly 4% of the world's water resources and 2.45% of the world's land area. Even in the distribution of the available freshwater resources in the country, there are great variations in space and time, i.e., between different parts of the country and in different periods in a year.

In India, there are many villages which are facing either scarcity of water supply or without any source of water. In many rural areas, women still must walk a long distance of about 2–5 km to fetch potable water or any source of water. The hardship of their life can just be imagined that after fetching water from 2 to 5 km in a heavy pot, they do not take rest but start doing other household works like cooking, washing, cleaning, caring for children, and looking after livestock. Here, to show, the status of the availability to drinking water facility, we have taken different source of drinking water like hand-pump, tap water, and tube well to prepare an index. The status of drinking water facility is given below.

It is clear from Fig. 35.8, that the availability of drinking water facility is satisfactory in all the three districts. In Gopalganj district,

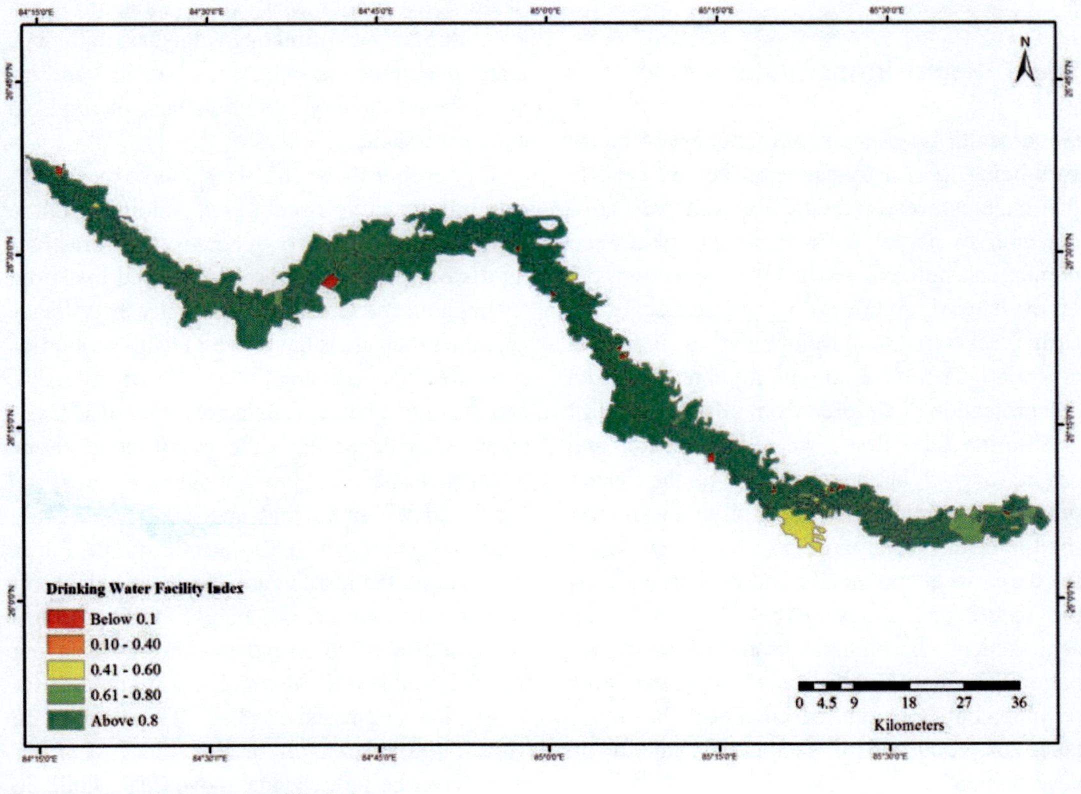

Fig. 35.8 Drinking Water Facility Index

there is only one village "Bhatwa Rup" in Kuchaikot district where the availability of drinking water facility is critical, while out of 140 villages, in 137 villages the availability of drinking water is good as they show the index value more than 0.8.

Similarly, in Purba Champaran district the availability of drinking water facility is also good. Out of 118 villages, the status of availability of drinking water facility is good in 111 villages whose index value is above 0.8. Only in 6 villages its availability is not satisfactory as the index value is below 0.1.

In Muzaffarpur district also, its availability is good as out of 220 villages, 173 villages have an index value above 0.8. Only in 6 villages (Haraulia Mansundi, Panapur, Chak Mundi, Bhatgawan, Chak Ghazi, and Raghopur Manjhouli) its availability is very poor as their index value is below 0.1 which is critical. Even in the municipal area of Muzaffarpur district the availability of drinking water facility is just satisfactory. Local authority needs to investigate the matter seriously and urgently.

35.5.7 Health Infrastructure Index

Good health is an integral component of human well-being. It is a fundamental human capacity that enables every individual to achieve her/his potential to participate actively in social, economic, and political processes. A growing body of evidence highlights the significance of the early years in the development of individual potential. Therefore, optimum care, nutrition, and protection of children from infection, at birth and during their first 3 years of life, not only ensure survival, but importantly form the foundations for lifelong development. Overall, improving the health of its large population, especially for the most economically and socially vulnerable sections of the society, is central to the achievement of human development of any nation. The entire approach of socio-economic development is human based. Thus, the importance of good health sector is vital for its development.

Health care infrastructure is an important indicator of development. A better healthy nation is an indicator of its development and vice versa. As we know that poor health condition and health infrastructure mean underdevelopment as it requires lots of amount from the public fund. Making health care affordable and accessible to all its citizens is one of the important agenda of our country today. It is an immense challenge as nearly 73% of the county's population lives in rural areas, and approximately 26.1% is below the poverty line.

In Bihar, the population density is 880 km^2, and it is the third highest populated State of India. Approximately, 40% of its population lies below the poverty line. The major health and demographic indicators of the state like infant mortality rate (IMR), maternal mortality ratio (MMR), total fertility rate (TFR), etc. are much higher than the all-India level and reflect a poor health status in the State.

State government needs not only to provide a well-structured public health care system but as well as the staffs that are required to provide the health care services. Many rural residents are not even able to obtain treatment for basic ailments either due to the non-presence of health care services in the vicinity or due to lack of funds to access the same.

Figure 35.9 shows that the status of the healthcare infrastructure is not good. Among the three districts, the availability of health care infrastructure is better in Muzaffarpur followed by Purba Champaran and Gopalganj, respectively. Villages near the urban areas have better health care infrastructure. In Gopalganj, the availability of health care facility is below satisfactory. Out of 140 villages, 86 villages have the index value below 0.40 which shows its poor condition. Only 10 villages and the municipal area are showing the index value above 0.80. Similarly, in Purba Champaran, the health care facility is below satisfactory. Out of 118 villages, 57 villages have index value above 0.66 and 36 villages are having the index value of 0.40 which is a far better situation in comparison than Gopalganj. In Muzaffarpur district, out of 202 villages, 127 villages have the index value above 0.80, while 36

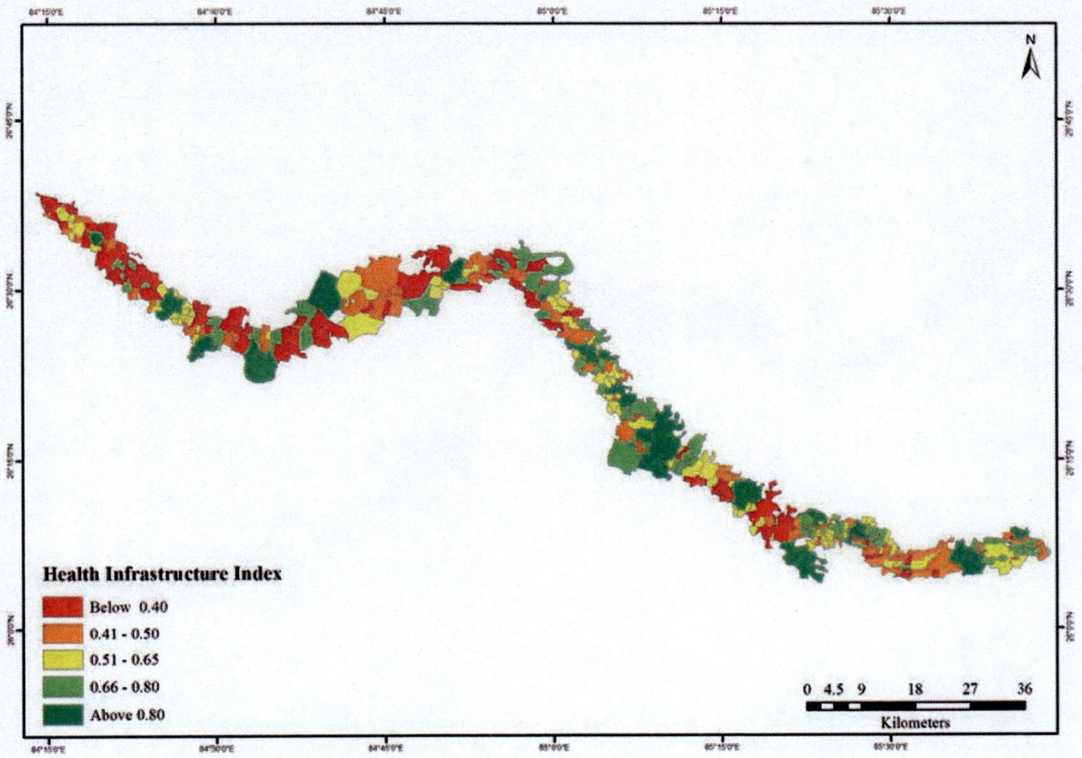

Fig. 35.9 Health Infrastructure Index

villages have index value below 0.40. They show much better condition of the availability of health care facility. No doubt that the state government is doing good work in this sector, but they need to build more primary health center, community health center, government hospitals, and make ensure to provide sufficient staffs so that villager or people will be served in much better way. After all, sound health leads to sound future of a nation.

35.6 Level of Development

There are different methods to measure level of development of any indicators. Here, we have used both composite index method and overlay method to determine the level of development of socio-economic infrastructure in the highway corridor.

Figure 35.10 illustrates that level of development tends to decrease as one move from the urban areas and vice versa. In all the three districts the level of development is high near the towns and tends to decrease as one moves away from towns. It is also found that the level of development in zone 1 is higher compared to zone 2 and 3. Our findings show that highway corridor has positive impact on the socio-economic development. The study area has analyzed that villages of zone I, which are at 500 meters each side of the highway, show high level of development in terms of socio-economic infrastructure. It is demonstrated that as we move away from zone I to zone II to zone III, level of development tends to decrease. This result also affirms that "the socio-economic status is higher among settlements located along the highway corridor, compared to those located away from the highway corridor." Conclusively, settlement nodes, particularly towns and cities provide the benefits of highway accessible to the inhabitants along the highway corridor.

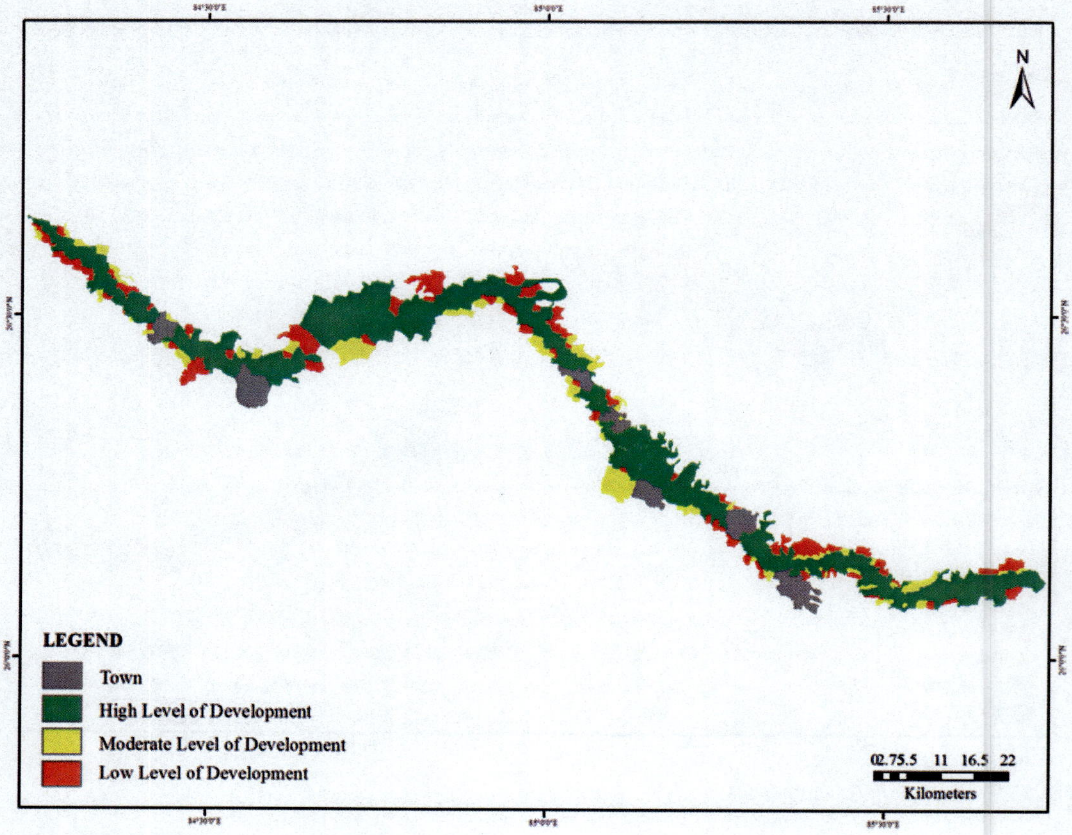

Fig. 35.10 Level of Development

35.7 Conclusion

The study shows the socio-economic development in highway corridors zones of Bihar with a case study of Kuchai Kot-Muzaffarpur section. The selected indicators used to measure the development provided to be very significant in analyzing the regional patterns of disparities in development. The study concludes that the highways development contributes to the development of activities in the surrounding areas which promotes the development in the area and in nearby areas also. Due to this, there is direct impact on the improvement in housing conditions, sanitation facilities, communication network, and other services like banks or water availability. All these are necessary to bring socio-economic transformation of a region. Therefore, it is recommended that highway corridor zones should also been seen as a major driving force behind the development of a region. The rapid development in the region and India in general can contribute to the rapid economic development of the country.

References

Agarwal AK (2008) Social Classification: The need to update the present scenario. Indian J Community Med; 33:50–1

Bose S (2005) Regional Rural Banks: The Past and the Present Debate, http://www.macroscan.com/fet/jul05/pdf/RRB_Debate.pdf

Burgess R, Pande R (2005) Do Rural Banks Matter? Evidence from the Indian Social Banking Experiment. Am Econ Rev 95(3):780–795

Dudala SR (2013) Updated Kuppuswamy's socioeconomic scale for 2012. J Dr NTR Univ Health Sci 2:201–2

Government of India (2015) Energy Sources of Indian Households for Cooking and Lighting, 2011–12. National Sample Survey Organisation Report No. 567, Government of India, New Delhi

Ketkar KW, Ketkar SL (1992) Bank nationalization, financial savings, and economic development: a case study of India. J Dev Areas 27(1):69–84

Kumar BPR et.al (2013), Kuppuswamy's Socio-economic status scale-a version of economic parameter for 2012, International Journal of Research and Development of Health, 1(1):2–4

Mekonnen A, Kolhin G (2008) Determinants of household fuel choice in major cities in Ethiopia. Working papers in economics 399, Department of Economics, University of Gothenburg

Nayak LT, Narayankar DS (2009) Identification of regional disparities in levels of development in Bellary District – Karnataka. Indian J Reg Sci 41(1):35

Indian households defecate in open: World Bank says on World Toilet Day , https://timesofindia.indiatimes.com, https://timesofindia.indiatimes.com/india/53-indian-households-defecate-in-open-world-bank-says-on-world-toiletday/articleshow/26032829.cms (2013).

Ouagadougou BF (2006) Household energy preferences for cooking in urban Ouagadougou, Burkina Faso, Internet, https://www.researchgate.net/publication/222513612_Household_energy_preferences_for_cooking_in_urban_.

Role of commercial banks in rural development - A case study of Keonjhar district of Orissa, https://shodhganga.inflibnet.ac.in/handle/10603/129654, http://hdl.handle.net/10603/129654 (2007).

Yadav BS (2005) Role of regional rural banks in rural development. Shree Publishers & Distributors, New Delhi

Part VII

Future Directions

36 Sustainable Resource Governance: Lessons for the Future

Rajiv R. Thakur, Baleshwar Thakur, Srikumar Chattopadhyay, and Rajesh K. Abhay

This volume was conceived, planned, and coordinated over the last 3 years by the editors. It sheds light on the connections and interdependence between resource management, sustainable development, and governance in the context of India and many other countries. In the process, chapter contributors have analyzed the spatial, temporal, ecological, environmental, economic, political, and social relationships between resource management, sustainable development, and governance. In most cases, chapter authors examine natural resource management issues by integrating the same through the lens of Sustainable Development Goals (SDGs) making use of geospatial technologies as well as innovative approaches. In the perspective of chapter authors, resource management and sustainable development play a critical and decisive role in intelligent garnering of natural resources. Given this overview, it is argued that the existing development practices have not brought meaningful results and there is an urgent call for action within the broader framework of the UN-SDGs both in India and elsewhere to ensure that policies and programs synchronize with realities and dangers of climate change and associated vulnerabilities. Therefore, governance at different scales and places need to transition by developing innovative and adaptive practices and techniques to protect their regions and better cope with new and worsened risks even as population expands and so does carbon growth threaten the natural environment. The authors come together in their recommendation that there is a need to act now, together, and differently adopting an integrated sustainable natural resource management approach as the dominant inertia. Hence, stakeholders need to understand the ever-evolving interactions between natural and social processes which will determine how natural resources will be governed, produced, managed, and allocated to society.

The following three research questions were addressed: (1) how can we integrate resource management, sustainable development, and governance, (2) how can we boost human resilience and move towards sustainability, and (3) how can we achieve livelihood security without stopping the development process? Taken together, the 36 chapters were organized into seven sections.

R. R. Thakur (✉)
Department of Geosciences, Missouri State University, West Plains, MO, USA
e-mail: rajivthakur@missouristate.edu

B. Thakur
Department of Geography, Delhi School of Economics, University of Delhi, Delhi, India

S. Chattopadhyay
National Centre for Earth Science Studies, Thiruvananthapuram, India

R. K. Abhay
Department of Geography, Dyal Singh College, University of Delhi, New Delhi, India
e-mail: rkabhay@dsc.du.ac.in

B. Thakur et al. (eds.), *Resource Management, Sustainable Development and Governance*, Sustainable Development Goals Series, https://doi.org/10.1007/978-3-030-85839-1_36

They include (1) introduction consisting of 2 chapters, (2) understanding conceptual foundations comprising 5 chapters, (3) unpacking problems with 3 chapters, (4) how resource management, sustainable development, and governance work: case studies made up of 6 chapters, (5) exploring human dimensions consisting of 7 chapters, (6) response to national, regional, and global change comprising 12 chapters, and (7) future directions comprising the concluding chapter.

36.1 Sustainable Resource Governance: Lessons

While the introductory part of the volume outlines the ambitious goals of the central theme—resource management, sustainable development, and governance and outlines the contributions of Bruce Mitchell in whose honor this volume is a festschrift, the second part of the book quickly dives to deal with conceptual issues around shifting paradigm change in natural resource management, the need to appreciate collaborative "turn" in water governance, institutionalization of disaster risk governance in some Asian countries, China's attempt at reshaping natural resource management, and the need to understand regional sustainable development and natural resource decision-making in the context of India. The authors in chapter three while examining the paradigm shifts in natural resource management take a longitudinal, descriptive, and analytical perspective to deconstruct the literature that has emerged between 1935 and 2005. Important paradigms identified in early period include ecological, integrated, economic, and technological-I and in the modern period decision-making and technological-II further divided into four sub-paradigms are dominant. The research has concluded the dominance of non-technological paradigms in the early period and ruling technological sub-paradigms in the modern period. From the discussion of paradigm shifts in this section, readers will find the investigation of collaborative "turn" in water governance as not only meaningful but also insightful from a normative perspective, for how collaborations in water governance need to emerge, develop given benefits and cost, as well as factors which seem to affect performance and success. With respect to collaborations the author draws our attention towards six key lessons for policy and practice, namely (1) collaboration is not a panacea for all resource-related problems or management situations, (2) collaborations do not happen overnight, (3) collaboration requires credible, trusted leadership to work well, (4) skepticism is often associated with collaboration and collaborative initiatives, (5) collaboration should not be entered lightly, and finally, (6) collaboration can generate a range of benefits related to knowledge sharing, joint learning, trust, and conflict resolution. The notion of collaboration is embedded in the success of disaster risk governance too, which is significant in the current climate of frequent natural disasters. In Asia and the Pacific, in recent years disaster risk governance has gained focus as it becomes institutionalized through focal point agencies, establishment of national platforms, promotion of legislation, and development of synergy for cooperation between national and local governments. The authors (Pal and Routray) conclude on the need for devolution of power, responsibilities, and resources to local governments and support decentralized multilevel planning. This can be achieved by strengthening the coordination role of the national government through an enhanced legal framework as well as by building a flexible cooperation system among concerned organizations and with local governments. It is recognized that disaster governance is in the evolutionary process and is a work in progress. All of this raises question about natural resource management and decision-making in the context of China and India, the world's two largest populated country. Today, more than ever before China needs to rethink natural resource management as it evolves into a high-consumption economy. Thus, policy making in China is now aware and recognizes the need to incorporate a co-benefits approach as it deals with a variety of regional ecological dilemma. On the other hand, in the context of India it is recognized that natural resources are important for economic development as it provides agricultural output, input for production process, sources of foreign exchange, and recreational amenity. The author (Sudhir K. Thakur) of chapter seven raises by implication questions and concerns about economic growth without end. It is argued that economic and environmental systems are interdependent and are characterized by stability and resilience properties. In addition, natural capital is regenerative but also depletes and deteriorates when over-used. Thus, utilization of natural capital in the production process, and its depletion and degradation lead to depletion, and then the GDP must be adjusted. Put simply, the favored argument is the need to operationalize sustainable development in terms of regional sustainable development which further implies attaining social welfare for regional population both for the current and future generations subject to the attainment of sustainability at regional scale.

Part III of the book unpacks some problems associated with development imperatives. In chapter eight the authors (Mathur and Sikka) appear to present that development prescriptions advocate market solutions to all manner of social and ecological problems excluding the displaced and dispossessed. They recommend the need for policy and programs to pay attention both in India and elsewhere (Brazil, China, and the USA) to those excluded and impoverished by development. This section is more about distilling powerful new approaches to counteract challenges of human needs in urban ecosystem. In a compelling research the author (Georgina Drew) of chapter eight brings out the need to rethink the problem of water stress in Delhi through innovative approaches such as rainwater harvesting. She recognizes the challenges of water governance in world's second largest metropolis as water gets commodified and the power structure becomes ever more complex with politicians, municipal employees, engineers, contractors, and private water suppliers making competing demands. Each of these groups plays a role in stalling, derailing, or misguiding the efficient and equitable management of water resources based on the narrow institutional, and sometimes even personal, interests. This chapter's assertions use Bruce Mitchell's integrated approach to water management embracing the importance of understanding interconnections among variables and relationships in a relevant socio-ecological system. This chapter is a grim reminder of how the urban population lives, struggles, negotiates, and endures their situations. In the next chapter the author (Shabana Khan) continues to engage with water resource governance in the context of Delhi. It is argued that Water problems in Delhi are not merely the result of inadequate water provision, but the complex water governance where each stakeholder is aware of different reality and addresses them in their own way creating further complexity in this domain. It is essential that all the stakeholders realize the risk that city faces and respond accordingly within their domain of understanding and power. It implies that existing practices and capacity building at the city scale need a rethinking and research for a more holistic water risks management.

Part IV of the book demonstrates how resource management, sustainable development, and governance work through six case studies. In this section of the book, except for chapter eleven (by Kapil Gavsker), all chapters are concerned with water resource management. In chapter eleven the author argues that the ecosystems in the Eastern Ghats are fragile to degradation owing to fragmented and narrow distribution and heavy anthropogenic pressure. It is suggested that regional environmental governance is a key driver for the achievement of sustainable development. In this chapter the author deconstructs the inextricable links between environment and the social and economic dimensions of sustainable development which relies on decision-making processes, effective institutions, policies, laws, standards, and norms. If anything, water resource management anywhere and everywhere calls for employing new approaches. In Haryana, groundwater depletions have become a matter of concern and challenge. The author (Inder Jeet) of chapter twelve recognizes that in the state of Haryana, India, water has been pumped faster than recharge, as a result, water tables are falling. Unfortunately, the widespread practice of the state government to appease the farmers lobby with zero electricity tariff has led to unsustainable groundwater exploitation. The enormous amount of water being pumped from aquifers—mainly to irrigate crops—is having a significant effect on the state's hydrological cycle. The author laments that the need of the hour is promotion of artificial recharge methods which can replenish the aquifers, and adoption of micro-irrigation techniques, especially for the plantation crops. In respect of water, its development and management are key for preserving ecosystems including wetlands. In chapter thirteen the authors (Bhagabati and Deka) analyze wetlands in the Brahmaputra Valley of Assam and conclude that they contribute directly to the local environment and the associated rural economy. However, they are concerned that the rural development plans and programs followed in Assam do not pay due to attention to the status and problems of wetland ecosystems. As a result, most of the programs adopted for conservation and development of wetlands do not bear expected fruits because of the indifferent attitudes on the part of implementing agencies towards the need and aspiration of the people directly sharing the concerned micro-ecological settings. The authors recommend that further research needs to be carried with a view to understand traditional management practices, and ways to integrate ecological needs and economic benefits of the local people. The last three chapters in this part of the book remind us that the word crisis is perhaps over-used in the context of river basin management and managing transboundary waters. Thus, in the context of Kosi, Teesta, and the Rapti river basins, the authors recognize that managing water is a challenging task, particularly in shared river basins with a dense population. These chapters acknowledge the unique physical characteristics of these river basins and explore both threats and opportunities presenting the need for governance that analyze the multi-faceted and dynamic nature of the interplay between domestic and international water security. It is concluded across the board, that if concerns, left unchecked, can derail progress towards the UN-SDGs, exacerbating on-going and emerging international water disputes between Nepal and India (with both Rapti and Kosi rivers) on the one hand and between India and Bangladesh with reference to the Teesta River on the other.

Part V is about exploring and understanding human dimensions of resource management, sustainable development, and governance. While the authors of the chapters in this section have repeatedly stressed the need for sustainable resource management through case studies, the common theme which emerges include the significance of the relationship of ecosystems and land use dynamics, as well as their implications on sustainable resource management and livelihood options with special reference to vulnerable population such as tribal women. The reality is that there are numerous, interrelated, and growing ecological problems arising from the impact of the introduction of neoliberal economic policies which has led to unintended consequences, such as outmigration of working population from the hilly areas of Himalayan regions both in Nepal and India. Land uses connect us directly with land cover, biodiversity, agriculture, ecosystems, economic development, and climate change, among others. Land use change, whether in the hills of the Uttarakhand, the highlands of Orissa or the plains in the rest of India, is constantly bringing about drastic transformation of the environment by creating ecological footprints which are often detrimental and unsustainable for people's livelihoods. Often, the cost of social and economic transformation on the landscapes is irreversible, which has led to soil and land degradation and major depletion in vegetation. Evidence from studies displays that because of neoliberal policies, the change in land use has caused irreversible loss in the ecosystem, such as loss of productive land to housing, or urban development. Several authors suggest that changes in land use impact directly on the ecosystem and are intimately linked with the issue of sustainability while having significant consequences for food security and sustainable livelihoods. Collectively, the dominant message from this part of the book is that the major natural resource management issues that are being experienced—of which climate change is only one—

cannot necessarily be solved by technological fixes or market-based solutions. It needs a transformation in community, culture, and economy, in how people relate to each other and to the ecosystems that they belong.

Part VI of the volume acknowledges priorities which includes acting now, together, and differently by taking an integrated approach in responding to a wide variety of issues, such as challenges of sanitation in developing countries, compelling need to address solid waste management, prioritizing sustainable water governance, addressing urban sustainability challenges, need for natural resource conservation, and energizing development without compromising the climate. In addressing these issues and challenges there cannot be tailor made solutions. However, it is important to understand and appreciate how unsustainable development is contributing to changing the climate, thus making it imperative that we understand the links between climate change and development. Unmitigated climate change is incompatible with sustainable development. Addressing sanitation needs and reducing solid waste would result in reducing human vulnerability which requires adaptive management, and managing physical, financial, and social risks. Across the board, this means balancing competing objectives, integrating development into climate regime, closing the funding gap, accelerating innovation and technology diffusion, and finally harnessing individual's behavioral change.

36.2 How the Book Contributes to Sustainable Resource Governance

This research took up the challenge of integration by cutting across ecological, economic, social, and governance dimensions. Throughout the volume, it has been argued that a new field of research has emerged emphasizing integration among resource management, sustainable development, and governance. It also demonstrated that governance is a growing contemporary paradigm in the field of natural resource management. During the last two decades, the focus of enquiry has been changing from government to governance. Water management and analysis of governance system have become the strongest interest areas, which are reflected in the research. The book provides a fresh viewpoint to research in resource governance like capacity building, collaborative turn, and disaster risk which are expected to contribute to the resource management and sustainable development processes. Several chapters have furnished critical insight towards achieving resource sustainability in different regions. It also contributes to initiate discussion among policy makers to frame policies towards achieving sustainability at different levels. Finally, the book has largely reflected some dominant sustainable development goals and problems of various countries.

36.3 Future Research Directions

Based upon the research in this volume by contributors on resource management, sustainable development, and governance several issues, challenges, and opportunities emerge which suggest prospective research trajectories.

1. Resource management under environmental uncertainty.
2. Model building in resource management, sustainable development, and governance.
3. Panchayat resource mapping.
4. Resource management decision-making: role of perception, attitudes, and public participation.
5. Understanding perceptions of climate change to devise climate-smart development policy.
6. Traditional ecological knowledge (TEK) and resource management.
7. Community-based resource management in tribal, marginal, and culturally diverse areas.
8. Role of governance in resource management: capacity building and collaboration.
9. Resilience and resource management.
10. Political economy of resource depletion.
11. Impact of climate change on resource sustainability.

12. Understanding land use change dynamics to facilitate agriculture, forestry, ecosystem services, gender, and sustainable rural livelihood.
13. Institutional constraints on sustainable resource use.
14. Application of geospatial technologies in resource management.
15. Water scarcity, risk, and vulnerability issues in peripheral regions.
16. Water challenges in megacities and need for rainwater harvesting.
17. Transboundary water management and governance issues.
18. Solid waste management challenges in urban areas.
19. Challenges arising from the vast deficit in sanitation.
20. Understanding and facilitating local, regional, and global cooperation to achieve integrated sustainable resource management.

Index

A
Acceptability, 564
Accessibility, 564, 571
Adaptation, 435, 476
Adaptive environmental assessment management (AEAM), 113
Adaptive practices, 583
Adaptive strategies, 483
Adaptive water management, 435
"Ad hoc water tribunals", 432
Affected Citizens of Teesta (ACT), 257
Affordability, 564
Agrarian economy, 367
Agrarian practices, 286
Agricultural activities, 284
Agricultural development, 213
Agricultural diversity, 378
Agricultural expansion, 365
Agricultural farming, 343
Agricultural lands, 28, 308–310, 374
Agricultural Non-point Source Pollution (AGNPS) model, 20
Agricultural pump (AP) sets, 200
Agricultural sector, 193, 302
Agricultural sustainability, 367
- alternative cash crops, 374
- crop rotation, 373
- irrigation efficiency, 373
- multi-dimensional challenges, 372
- soil conservation measures, 373
- space and time dimensions, 374
- yield improvement, 372

Agricultural use of land, 365
Agriculture, 31, 32, 201, 325, 333
- diversification, 379
- production, 289
- resources, 107
- in Surveyed Villages, 378

Agriculture land, 378
Agro-based products, 451
Agro-climatic and socio-economic dynamics, 353
Agro-climatic conditions, 382
Agro-climatic zones, 477
Agro Ecological Zone (AEZ), 488, 489
Agroecosystem, 326
Agro-techniques, 390
Agro-Techniques of Medicinal Plants, 392
Air pollution, 447, 448
Alpine Convention, 189
Alternative cash crops, 374
Alternative pro-poor sanitation solutions (APPS), 409–412
Amazon forests, 180
Ambala, 197
Amenities, 564
Amenity resources, 107
American Society for Photogrammetry and Remote Sensing (ASPRS), 48, 52
Ammonia-based bag, 408
Anaerobic vermi-composting, 409
Annapurna Conservation Area (ACA), 294
Anti-hail net, 480, 481
Anti-India policies, 263
Anti-pollution regulatory frameworks, 447
Apple cultivation, 483
Apple farming communities, 476–483
Appropriate transboundary water management, 226
Aravalli hills, 194
Area-based development (ABD), 449
Asian Development Bank (ADB), 227, 433
Asia-Pacific region, 76
Assam
- British rule in, 206

Assimilates waste, 109
Assured irrigation facilitates, 370
Atal Mission for Rejuvenation and Urban Transformation (AMRUT), 416, 457
Atmosphere, 35
Audali, 353, 355, 356
Average Index (AI), 568
Ayu geriatric regional care centers, 393
Ayurveda personnel, 386
Ayurvedic healers, 380
Ayurvedic spas, 380

B. Thakur et al. (eds.), *Resource Management, Sustainable Development and Governance*, Sustainable Development Goals Series, https://doi.org/10.1007/978-3-030-85839-1

B
Baitarini, 367
Bali Sustainable Development Project (BSDP), 17
Balrampur, 274
Bangladesh, 250–256, 261–266, 408
Bangladesh Government, 259
Bangladesh Nationalist Party (BNP), 263
Bangladesh's biggest Teesta project, 261
Bangladesh's Outline Perspective Plan, 81
Banking facility index, 573, 574
Bansi, 274
Barren and unculturable land, 306
Basin/watershed-wise management of water, 33
Beekeeping, 342, 343
Behavioral geography, 49
Beijing's pollution, 96
Below-poverty-line (BPL), 571
Bengal Province, 251
Beta distribution, 569
Bhitarkanika in Odisha, 558
  mangrove forests, 558
Bhitarkanika Mangrove Conservation Area (BCA), 558, 559, 561
Bhitarkanika Mangroves, 558
  Bhitarkanika National Park, 559
  Bhitarkanika Wild Life Sanctuary, 561
  challenges, 558
  ecosystem services, 559, 560
  products use by local people, 560
  sustainable development goals, 558
  sustainable management strategies, 560–561
  vegetation structure, forests, 560
  wildlife sanctuary, 558
Bhitarkanika Wildlife sanctuary, Odisha, 559
Bhorer Alo, 259
Biligiri hills, 186
Billion cubic meters (BCM), 251, 252
Biodegradable sanitation, 408
Biodiversity, 476
  unsustainable exploitation of resources and threats, 187, 188
Bioengineering, 343
Biofertilizers, 343, 390
Biofuels, 182
Biogas, 408
Bioindustrial Watershed mechanism, 389
Biological/ecological environment, 467
Biopesticides, 343
Biophysical conditions, 391
Bioremediation, 421
Biosphere reserve, 41
Body Shop, 456
Boston (book), 51
Brahmaputra Valley (BV), 586
  agricultural development, 213
  biological diversity and productivity, 206
  conservation efforts, 211–213
  development, 213
  distribution of wetlands, 209, 210
  ecological context, 208
  ecosystems, 206
  floodplain, 207
  in LULC, 206
  physiography, 208, 209
  rainfall, 206
  rural landscape, 216
  shapes, sizes and characteristics, 205–206
  and tributaries, 208, 209
Brazil
  good practices, 138, 139
Brick factory encroaching wetland, 209, 211
Broom cultivation, 388
Buddhist-Hindu Teesta basin, 252
Build Back Better, 87
"Buri Teesta", 263

C
Cadastral map, 214
Campylobacter jejuniin, 69
Canada
  good practices, 138
Canal, 219
Capacity building, 85
Capacity reduction, 96, 97
Carbon sinks, 30
Cardiovascular disease, 380
Careful analysis, 101
Cartography, 45, 46
Catchment Restoration Fund (CRF), 68
The Central Ground Water Board (CGWB), 273
Central Rural Sanitation Programme (CRSP), 407
Centre for Water Resource Development and Management (CWRDM), 546
Cereal production, 290
Channel migration zones, 275, 276
Chemical fertilizers, 197
Children's Club (CC), 288
China, 584
  GDP, 90
  good practices, 139
Cholai, 341
Circular economy, 453, 457
Circular metabolism, 101
Citizens Forum of Sikkim (CFS), 257
Civil Défense Force, 83
Civil society, 139
Civil society organizations (CSOs), 82
Clean Water Act, 69
Clean water provision, 408
Climate change, 172, 332, 583
  and adaptation strategies, 476, 488
  AEZ, 488
  apple farming communities, 479, 480
  ecosystem services, 488
  farmers understanding and response, 480, 481
  geo-climatic environments, 476
  hail, 479
  local/traditional knowledge system, 476
  mitigation strategies, 488

physical dimensions, 476
rainfall, 478
and rising temperatures, 476
scientists' climate change knowledge, 481, 482
snowfall, 478
social-ecological processes, 476
social-political connections, 475
socio-economic dimensions, 476
study area and methodology, 477
temperature, 478
Climate change mitigation, 390
Climate change modeling, ecosystem services in Western Himalayan district
baseline data requirement, 489
climate change scenario modeling
database, 491
methodology, 491
CO_2
crop growth, 501
crop productivity, 501
composite capital vulnerability index, 501–503
database, 490, 491
ESVI, 491–495, 504
FCVI, 503
GCMs, 492
GHGs concentration, 504
Kangra, 504
materials, 489
methodological framework, spatio-temporal maps, 490
methodology, 489–491
NCVI, 503
precipitation, 500
rainfall, 496, 504
river, 504
SCVI, 503
seasonal temperature, 500
simulated temperature pattern, 498–500
temperature
growth and productivity, crops, 501, 502
spatio-temporal change, 495–498
winter season, 504
trend, 498, 499
winter temperature, 504
Climatic variability, 487, 488
Climatological disasters, 76
Coal, 36, 91–93
Co-benefits approach, 102
Cold-Temperate Zone, 336
Collaboration
benefits, 61–63
costs, 61–63
definitions, 61–63
implications
active capacity-building support, 73
benefits, 73
farm sector organizations, 73
organizations, 72
policy and practice, 72
stakeholders, 72
technical support, 73
process
complex and cross-scale institutional settings, 64
cooperation/coordination, 63
design elements, 63
governance, 63
groups, 64
initiatives and arrangements, 64
iterative cycle, 63
learning and negotiation, 64
policy outcomes, 64
relationships and interorganizational arrangements, 64
rules, 63
self-interest, 63
Collaborative catchment management, UK
antecedents and motives, 65
dialog and trust, 66, 67
interdependency and reciprocity, 67, 68
organic and inorganic nutrients, 65
outcomes, 68
outputs, 68
power and power relations, 66
Colonial economy, 446
Columbia River Treaty Agreement, 228
Commercial cultivation, 384
Commercial energy sources, 36
Commercial farming, 284
Common pool resources (CPR), 323, 351
Common property resources (CPRs), 349, 351
Communicable diseases, 448
Communication network, 368
Community-based forest management, 339–341
Community-based mangrove management plan, 561
Community-based micro-financing and insurance, 435
Community-based organizations (CBOs), 289
Community bio-degradable treatment plants, 425
Community capacity-building, 348
Community-driven projects, 411
Community education, 407
Community expectations, 277, 278
Community Forest User's Group (CFUGs), 289
Community forests, 323–325
Community health, 327
Community initiatives, 348
Community participation, 407
Community partnership programs (CPP), 435, 439–441
Community resource utilization, 325
Community response, 378
Compendium of Environmental Statistics and Environmental Statistics of India, 109
Compensation, 137
dimensions, 138
Complementary and Alternative Medicine (CAM), 341
Composite capital vulnerability index, 501–503
Composite index, 304
Composite water management index (CWMI), 195
Compost plants, 419
Computer-aided cartography, 47
Concerned Lepchas of Sikkim (CLOS), 257
Conflict-prone/vulnerable regions experience, 351

Connectivity development, 565
Conservation, 561
Construction and demolition (C&D) waste, 419
Consumer market, 286
Contaminated water, 448
Contested nations, 254
Cool-temperate zone, 335–336
Cooperative management, 250
Cost-benefit analysis (CBA), 116–118, 127
COVID-19 pandemic, 38, 55
Creative destruction, 285
Crop diversification, 389, 481
Crop rotation, 373
Crop sustainability and profitability, 373
Cropping intensity, 325, 371
Cropping pattern, 196
CSR initiatives, 354
Cultivated land, 325
Cultivation, 335
    sustainable livelihood diversification, 343, 344
Culturable wasteland, 302, 306
Cultural attractions (CA), 467, 468
Cultural services, 292–294
Current fallows, 306, 307
Current land use, 322, 324, 325
Cyclone Nargis disaster, 83
Cyclones, 34

D

Dalia Barrage, 259, 261
Damodar River Valley (DVC) project, 438
Damodar Valley Corporation (DVC), 43
Dams, 438
Data discrepancy, 99
Data inventory, 237, 238, 244
Data reliability in China, 99
Data source and methods
    qualitative and quantitative research methods, 523
    questionnaire, 523
Data statistics system, 99, 100
Dauk Nagar Main Canal, 257
Dead channel, 218, 219
Decentralised waste management, 457
Decision-making, 20, 49, 50, 352
Deen Dayal Upadhayaya Gram Joyti Yojna (DDUGJY), 571
Deficit of water, 237
Deforestation, 31, 326
Degradation, 209, 211–213, 216
Degraded groundwater quality
    on human health, 437
Delhi, 523
    adaptation and sustainability, 509
    Battle of Delhi, 509
    co-production approach, 514
    Delhi Development Authority, 509
    diverse multicultural population, 512
    eco-friendly, 530
    ecological sustainability, 512
    economic structure, 510
    environmental and ecological surroundings, 510
    environmental degradation, 512
    fortunes, 509
    humanity, 509
    Jahangirpuri, 529
    mega city, 510
    metropolitan, 510
    monuments, 509
    policy adoption, 514
    population, 509, 510
    poverty, 529
    quantitative and momentous, 510
    rainwater harvesting (*see* Rainwater harvesting)
    resettlement colonies, 524, 528–529
    sanitation, 513, 529, 531
    sewerage system, 529
    solid waste, 534
    sustainability, 511
    sustainable development, 516
    toilet facilities, 528
    transportation, 513
    waste disposal, 513, 527
    water resources, 512
    water supply, 526
Delhi Jal Board (DJB), 526
Delhi Metropolitan Region (DMR), 11, 508, 512
Delhi Transport Corporation (DTC), 514
Demand-side constraints, 78
Democratization of Water Management project, 439
Dense forest, 336
Department of Geography and Environmental Management (GEM), 14
Dependent territory, 251
Descriptive statistics, 46, 47, 523
Desertification, 29
Desirable land use dynamics, 313
Deterioration, 196
Developing countries, 399–412
Development
    opportunity, 135–141
Development agenda, 350
Development of Humane Action (DHAN), 435
Development-induced displacement, 136
Development opportunity, 136
Development strategy, 135
Dhamdaha Kosi, 238
Dhaus Kosi, 238
Digital interpretation techniques, 321
Dimensions of governance, 272
Disaster Countermeasure Basic Act, 81, 82
Disaster governance, 77, 78, 81, 86
Disaster management, 19
Disaster Management Act, 273
Disaster management authority, 270
Disaster management cycle, 274
Disaster management paradigm, 77
Disaster risk governance, 76
Disaster Risk Governance and Management, 81
Disaster risk management (DRM)

development plans and policies, 77, 80
governance, 76
governmental organizations, 82
institutions, 80
in Japan, 81
legislation, 82
multi-sectoral approach, 81
national development strategies, 85
organizations, 85
policies and countermeasures, 83
strategies, 82
tasks, 81
Disaster Risk Management and Reduction, 76
Disaster risk reduction (DRR)
approach, 77
disaster management, 77
framework, 80
governance process, 78
mainstream, 77
measures categories, 76
policies, 77
resource management, 76
sustainable development, 81
types, 78
Displacement, 137
District Irrigation Plan (DIP), 368
District-level distribution of pump sets, 200
Domestic consumption, 386
Domestic natural gas, 91
Domestic sewage discharge, 546
Dominant caste, 28
Drinking water facility index, 575, 576
Drinking Water User's Groups (DWUGs), 289
DRM-related budgets, 86
Drought, 34
Drought-prone areas, 34
Dry season, 261

E
Earth-orbiting satellite, 48
Earthquakes, 34, 79
Earth systems, 112
Eastern Ghats, 186, 189, 191
biodiversity, flora and fauna, 186, 187
ecological treasure and natural heritage, 185
E-charak portal, 386
Ecological audit, 441
Ecological community, 207
Ecological economic analysis, 120
Ecological implication, 300, 302
Ecological interlinkages, 40, 41
Ecological paradigm, 40–42
Ecological sector, 302
Ecological sustainability, 400, 407–409
Ecology and environmental impact assessment
current institutional capability, 436
groundwater management challenges, 436, 437
progressive assessment, 437, 438
Economic activities, 394
Economic analysis, 106
Economic development, 112, 113, 300, 307, 584–585
Economic dynamism, 446
Economic-ecological analysis, 121
Economic efficiency, 172
Economic growth, 106, 307
and natural resources, 106
and social development, 90
Economic paradigm, 44, 45
Economic sector, 300
Ecosystem, 4, 5, 8–10, 12, 38, 586
Ecosystem approach (EA), 212
Ecosystem services, 114, 284, 285, 289, 293, 294, 299, 326, 327, 557–560
climate change, 488
cultural, 488
ESVI, 488
Kangra, 489
provisioning, 488
regulating, 488
socio-economic development, 488
vulnerability, 488, 493, 494
Ecosystem Services Vulnerability Index (ESVI), 491–495, 503, 504
Ecosystem shift, 236
Ecosystems services, 488
Ecotones, 205
Ecotourism, 561
Ecotourism, Hong Kong
eco-site visits, 462
and geotourism, 462
inventory, 465
mass tourism, 462
methodology, 470, 471
natural areas, 462, 465
natural environmental potential, 465
natural resource, 463, 465
nature-based tourism, 462, 470
physical environment, 470
resource evaluation
adoptable definition, 471
cultural attractions (CA), 467, 468
environmental variables, 471
human impact (HI), 467, 469, 470
inventory studies, 468
natural areas, 470
natural attractions (NA), 467, 469
overall attractiveness (OA), 468
protected area, 470
tourism infrastructure (TI), 467, 469
simplicity, 470, 471
sustainable development, 463
sustainable tourism, 464
Tung Ping Chau (TPC)
adopted methodology, 466
natural areas, 468
nature-based recreation, 465
study area, 466, 467
supply component, 467
visitor survey, questionnaires, 471

Education, 561
Effluent tax, 552
EIA consultancy business, 436
Electricity biogas projects, 424
Electricity connection, 571, 572
Embankments, 276
Energy conservation/efficiency frameworks, 101, 102
Energy consumption
  coal, 92, 93
  natural gas, 91, 92
  oil industry, 92
  structure, 91
Energy crises, 35–37
Engendered good practices, 139–141
Engineering methods, 76
Enhanced legal framework, 85
Enviable ecology (EE), 302
Environment-adjusted domestic product (EDP), 107
Environmental Assessment (EA), 18
Environmental clearance (EC), 430
Environmental evaluation system (EES), 118
Environmental impact analysis (EIA), 119
Environmental impact assessment (EIA), 11, 18, 429, 436, 438, 441
Environmental impacts, 439
Environmental Justice, 18
Environmental Kuznets curve (EKC), 109
Environmental management, 17, 18
Environmental remote sensing, 52
Environmental sustainability, 112
Environmentally adjusted state domestic product (ESDP), 122
Environment degradation, 349
Environment impact analysis (EIA), 118, 119, 127
Environment Kuznets's curve (EKC), 126
Environment management, 278
Environment Protection Act, 212
Environment providing services, 106
E-Rickshaw, 517
Erodibility, 29
Erosivity, 29
EU Water Framework Directive, 65
Excess of water, 237
Expert observation, 285
Explosion of groundwater, 201
Ex-situ conservation, 392

F
Falling water table zone, 194
Fallow lands, 306, 307
Farm-based work, 357
Farm-based workforce, 354
Farmers financial capital measurement, 392
Farm management, 342
Farming, 357, 390
Farraka barrage, 261
Female workforce participation, 362
Feminine tasks, 356
Financial capital index (FCI), 493, 494
Financial services, 350
Financial vulnerability index (FCVI), 501, 503
First and second law of thermodynamics, 108
Fisheries management program, 50
Fishponds, 390
Five-Year Plans, 28, 182
Flexible cooperation system, 86
Flood cushioning, 213
Flood disaster, 237
Flood emergency plans, 274
Flood forecasting, 274, 275
Flood hazards, 238, 244
Flooding, 257
Flood Management Improvement Support Centre (FMISC), 227, 241
Floodplains, 206, 207, 209, 213, 214, 221, 269, 273–278
Floods and floodplain management in India, 273, 274
Flow resources, 106
Flying toilets, 408
Focal point agencies, 84–86
Food, 327
Food security, 348, 452, 487
Food sufficiency, 287
Forest based natural vulnerability index, 502
Forest consumption, 351
Forest covers, 371
Forest, land and water policy, 277
Forest Panchayats, 323
Forest products, 352
Forest resource
  community-based forest management, 339–341
  medicinal plant, 341, 342
Forests, 30, 304, 326
Free-flowing undammed rivers, 439
Frequent disruption, 260
Fresh water sources, 545
  groundwater, 546, 547
  Peppara reservoir, 546
  sand mining, 546
Fruit Production in Kangra, 379
Fruits diversification, 389
Fuel woods, 340, 341
Full cost recovery, 433

G
Ganga Action Plan (GAP), 117
Ganga Rejuvenation, 272
Ganges-Brahmaputra-Meghna (GBM) River basin, 43, 228, 255
Garbage collection arrangement, 533
Garlic, 390
Gasification, 423
Gazaldoba Barrage, 262
GB Pant Institute's intervention program, 356
Gender awareness, 140
Gender-based discrimination, 348
Gender-based inequality, 350
Gender-biased environment, 139
Gender differentiation, 350

Gender disparity, 351
Gender inequality, 140
Gender-related benefits, 140
Gender relation, 352
General Circulation models, 197
Generalization, 350
Geo-climatic environments, 476
GeoDa software, 109
Geo-garden, 394
Geographic information system (GIS), 53, 54, 394
Geographical and non-geographical factors, 355
Geographical environments, 285
Geographical fix, 480, 481, 483
Geographical Information System (GIS), 268, 343
Geographical location, 272
Geography, 20, 21, 37, 45, 51
Geoparks, 462
Geopolitics, 250–254, 262, 263, 265, 266
Geotourism, Hong Kong
  adoptable definition, 471
  biological diversity, 464
  concept, 464
  conservation, 464
  criteria, 465
  current development, 463
  definition, 462
  eco-site visits, 462
  environment, 464
  GIS, 471
  inventory, 465
  landscape evaluation research, 465
  methodology, 471
  natural area tourism, 462, 465
  natural resource assessment, 465
  natural resource management, 465
  nature-based tourism, 462, 464, 470
  physical environment, 470
  resource evaluation
    cultural attractions (CA), 467, 468
    human impact (HI), 467, 469, 470
    inventory studies, 468
    level of human impact (HI), 467
    natural areas, 470
    natural attractions (NA), 467, 469
    overall attractiveness (OA), 468
    protected area, 470
    tourism infrastructure (TI), 467, 469
  resource evaluation methodologies, 465
  SDGs, 463
  simplicity, 470, 471
  sustainability, 463, 464
  sustainable tourism, 464
  Tung Ping Chau (TPC)
    adopted methodology, 466
    checkpoints, 464
    natural areas, 468
    nature-based recreation, 465
    nature-based tourism, 463
    study area, 466, 467
    sustainable development, 463
  visitor survey, questionnaires, 471
German International Cooperation- Advisory Service for Environmental Management, Program of Government of India and Germany (GIZ ASER), 425
GHGs emissions, 488, 491
Global Climate Model (GCM), 235
Global disaster fatalities, 76
Global economy, 112
Global Environment Facility (GEF), 425
Global Gross Domestic Product (GDP), 80, 109, 113, 201
Global warming, 351
Global Water Partnership (GWP) toolbox, 431
Good governance, 77, 78, 85, 515, 516
Good housing index, 569, 570
Good practices
  Brazil, 138, 139
  Canada, 138
  China, 139
  engendered, 139–141
  global and local, 136
  Japan, 138
  land, 137
  stakeholders, 137
Governance, 77, 80, 112, 249, 250
  groundwater sustainability in Haryana, 201–203
  and river basin management in India, 272, 273
  in river basin planning in India, 268
  recommendations, 263–265
  sustainable water management (*see* Sustainable water management)
Governance lessons, 75
Governance mechanism, 78
Governance theory, 244
Government Ayurveda college, 386
Government-led/supply-driven sanitation, 407
Government schemes, 377
Government statistics, 286
Gozoldoba Barrage, 257, 259, 261
Gozoldoba eco park, 259
Grain yields, 194
Gram Panchayat Adhikari, 270
Gram Pradhan, 270
Gram Vikas Adhikari, 270
Grazing lands, 304, 305
Great East Japan Earthquake (GEJE), 81
Great Himalayas, 231
Great Pacific Garbage Patch, 423
Green innovations, 514
  footprint, 516
  green toilets, 516
  natural building materials, 516
Green Revolution, 201, 437
Green toilets, 516
Green Vertical Gardens, 517
Greenfield land, 95
Greenhouse gas (GHG) emissions, 93, 102, 450, 456

Gross cropped area, 372
Gross domestic product (GDP), 106, 286, 564
Groundwater, 159, 546
Groundwater availability, 195, 196
Groundwater depletion, 32, 437
Groundwater-energy nexus, 437
Groundwater Estimation Committee, 201
Groundwater extraction, 195
Groundwater guzzling industries and enterprises, 436
"Groundwater hotspot" zones, 437
Groundwater irrigation, 437
Groundwater legislation, 202
Groundwater management, 436, 437
Groundwater reserve, 437
Groundwater resources, 123
Groundwater sustainability in Haryana
  Aravalli hills, 194
  availability, 195, 196
  and development, 194
  ecosystem, 195
  falling water table zone, 194
  food deficit state, 194
  governance, 195, 201–203
  grain yields, 194
  indicators, 195
  inland basin, 194
  irrigation, 198–201
  overexploitation and quality deterioration, 193
  population growth, 193
  quality, 195–197
  quality deterioration, 193
  quantity, 195
  rising water table zone, 194
  SDGs, 194, 195
  second-generation problems, 194
  Shiwalik hills, 194
  socioeconomic, 195, 197, 198
  southwestern region, 194
  WYC, 194
Guangdong Province, 99
Gurugram, 197
Gyre, 423

H

Hadley Regional Climate Model (HadRM), 490, 491
Hail, 479
Hand holding, 342
Handicraft work, 358
Hanging toilets, 408
Haryana, groundwater sustainability, *see* Groundwater sustainability in Haryana
Hazard's risk map, 332
Health and herbs, 380
Health care system (CAM), 393
Health infrastructure index, 576, 577
Health-related conditions, 564
Health risk assessments, 406
Health services, 564
Herbal cultivation, 386
Herbal medicines, 380, 394
Herfindahl Hirschman Index, 389
Herfindahl Index (HI), 389
Heritage City Development and Augmentation Yojana (HRIDAY), 416
High-consumption economy, 584
High population pressure on resources, 296
High rent staple trap model, 107
High Traded Medicinal Plants, 387–388
Higher temperature, 237
Highway corridor, 564–578
High-yielding varieties (HYV), 290, 291
Himachal Pradesh, 477
Himachal Pradesh Forestry policy, 380
Himalayan bioresources, 384
Himalayan mountains, 326
Himalayan regions, 586
Himalayans medicinal plant species, 389
Himalayas, 476, 477
HIMBALA (institute), 392
Hindu Kush Himalayan (HKH) region, 227
Hiran Kosi, 238
History of deforestation, India, 30
Holas, 219
Holistic health, 389
Horticultural State of India, 477
Horticulture Technology Mission (HTM), 389
Household-level heterogeneity, 351
Household responsibility, 351
Human and health capital index (HHCI), 494
Human capital, 110, 113
Human civilization, 300
Human development dynamics, 349
Human Development Index (HDI), 567
Human dignity, 564
Human-environment systems, 335
Human geography, 47
Human impact (HI), 467, 469, 470
Human-induced water pollution, 213
Humanitarian response and coordination
  Hyogo Conference, 83
  primary importance, 83
Human welfare, 120
Hydel projects, 260
Hydrometeorological/climate-induced disasters, 76
Hydrometeorological disasters, 76
Hydrometeorological events, 76
Hydro-politics, 253
Hydropower, 36, 226, 228, 251, 256, 257, 264, 265
Hyogo Framework for Action (HFA), 77, 78, 80, 81
Hyper-urbanisation, 447

I

Incineration plants, 422
Inclusive wealth index, 113
Index of desirable land use dynamics, 303
Index of Sustainable Economic Welfare (ISEW), 113
Index of undesirable land use dynamics, 303
India, 250–257, 261–266, 408, 583, 584

capitalistic society, 446
cities, 446
colonial economy, 446
formal urban structure, 446
hyper-urbanisation, 447
internal migration, 448
net annual availability of groundwater resources, 124
Net Annual Ground Water Availability, 127, 128
Per Capita Wealth, 111
public sector, 446
SDGs, 108
species, 183
SWM (*see* Solid waste management (SWM), India)
Total Forest Area, 123, 124
Total Renewable Installed Power, 123, 125, 126
urban development, 446
urban growth scenarios, 447
urbanisation, 448
urban-rural interactive events, 446
urban structure, 447
India's hydro-hegemony, 228
Indian agro-economy, 436
Indian economy, 390
Indian Himalaya, 390
Indian micro-irrigation program, 434
Indian Ocean, 34
Indian Ocean Tsunami, 82, 85
Indian Plains, 231
Indian Social Organization, 356
Indian Traditional System of Medicine, 380
Indian Union, 250
Indian villages, 348
Indigenous Health Care, 389
Indo-German Watershed Development Programme (IGWDP), 436
Indo-Nepal water cooperation, 227
Industrial/agricultural wastewater, 527
Industrial development, 307
Industrial Development Corporation of Uttarakhand, 354
Industrialization, 480
Industrial revolution, 209
Infant mortality rate (IMR), 576
Inferential statistics, 46, 47
Informal sector role, waste management industry
air and water pollution, 454
biodegradable waste, 454
health risks, 455
landfill sites, 454
livelihood challenges, 455
MSW mismanagement, 454
rag pickers, 454, 455
recycling, 454
reduction, 454
reuse, 454
social isolation, 455
Swachh Bharat guidelines, 455
SWM framework, 453, 454
3R strategy, 454
unsanitary working conditions, 455
urban institutional structure, 454
waste collector responsibility, 455
waste picker community, 454
waste recycling, women, 456
women waste pickers, India
body shop, 456
SEWA, Ahmedabad, 456
WIEGO, 456
Informal systems, 412
Information and communications technologies (ICT), 448
Information, Education and Communication (IEC), 426
Infrastructural development, 394
Inland delta, 231
Inland Waterways Authority of India (IWAI), 227
Input-output (I-O) analysis, 121
Institute of Himalayan Bio-resource Technology (IHBT), 392
Institute of Social Development (ISD), 354, 356
Institutional repositioning, 551
Institutional restructuring, 268
Institutional systems, 411
Integrated paradigm
Canada, 43
DVC, 43
endemic poverty, 43
GBM basin, 43
geography, 42
hazards, 42
natural hazards, 42
natural resource management, 42
river basin forms, 42
river basin planning, 43
social and economic change, 43
Subarnarekha River basin, 43
sustainable development, 43
types of human adjustments, 42
water, 42
Integrated resource management (IRM), 17, 43
Integrated Rural Nursery Program (IRNP), 344
Integrated Scheme of Oilseeds, Pulses, Oil palm and Maize (ISOPOM), 434
Integrated Teesta River Water Management (ITRWM), 264
Integrated urban water management (IUWM), 537–539
Integrated water resource management (IWRM), 17, 161, 253
Intergovernmental Panel on Climate Change (IPCC) emissions scenarios, 488
Intergovernmental Panel on Climate Change-Special Emission Scenario (IPCC-SRES), 491
Interlinking of rivers (ILR), 117
International Centre for Integrated Mountain Development (ICIMOD), 227
International Day of Rural Women, 348
International Decade for Natural Disaster Risk Reduction, 79
International financial institutions (IFIs), 433
International markets, 100
International water law, 228
Interrelationships, 355
Inter-sector coordinating mechanisms, 82

Inter-State Water Disputes (ISWD), 33
Iron-belt, 35
Iron-ore, 35
Irrigation, 438
development, 194
in Haryana
annual groundwater status, 198, 199
AP sets, 200
cropping pattern, 200
crop production, 198
district-level distribution of pump sets, 200
physical productivity markers, 198
private-owned shallow tube wells, 201
stage of development for selected states, 198, 199
subsidies, 201
tube wells, 198
water-rich, 201
ISD organized skill development programs, 357
Isotherms, 496

J
Jamuna SHG, 357, 359
Japan International Cooperation Agency-Official Development Assistance (JICA ODA), 425
Jawaharlal Nehru National Urban Renewal Mission (JNNURM), 416, 418
Joint decision-making, 354
Joint Forest Management (JFM), 324
Joint Rivers Commission (JRC), 261–263
Joint Standing Technical Committee, 242
"Jhum cultivation", 350
Juxtaposition, 357

K
Kajri, 238
Kakardhari Bhinga, 274
Kalagarh Forest Division, 333
Kali Kosi, 238
Kamla, 238
Kamla Kosi, 238
Kangra district Himachal Pradesh
Ayurveda, 383
herbal cultivation, 386
medical geography, 382
medicinal plants, 382
medicinal plants, altitudinal distribution, 383, 384
Panchakarma, 388
policy issues, medicinal plant sector, 384–386
symposium, 382
udara chikitsa, 382
Withania Somnifera, 382
Kendujhar District, land degradation study
agricultural sustainability, 368, 370, 372–374
aim and objectives, 367
block-wise status, 369
contributing factors, 370–372
data source and methodology, 368
geographic profile, 367, 368
mining activities and shifting cultivation, 372
Kerala Municipality Act, 548
Kerala State Environment Policy-2009, 552
Kerala Water Authority (KWA), 541, 547
formation, 542
KWA Act, 549
operation, 542
performance, 542
water supply and sewage disposal, 553
Key Informants Interview (KII), 285
khal, 218, 219
khal-beel, 216
Kharif season, 339
Kosi Project, 240
Kosi River basin
in Bihar, 227
canal irrigation, 242
climate, 231, 232
climate change and water demand
characterization, 235
NCVST, 235
in Nepal, 235, 236
NOAA, 235
precipitation, 235–237
runoff, 237
temperature, 235, 236
water-related hazards, 237
data inventory, 237, 238
development, 238–241
drainage system, 232, 233
economic activities, 234, 235
Ganga-Brahmaputra-Meghna, 228
geographical factors, 230, 243
Himalaya, 229
HKH region, 227
hydropower development, 228
Indian territory, 230
Indo-Nepal water cooperation, 227
location, 229
natural resources, 243
physiography, 229, 230
population, 234, 235
precipitation, 231, 232
relief, 230, 231
sustainable development, 226
temporal water availability, 233, 234
Tibetan region, 230
water discharge and irrigation, 241, 242
Kosi Treaty, 241
"Krishi Utpadak Sangh", 358
Kuchai Kot-Muzaffarpur Section, 564–578
Kuhls, 502
"Kutumb", 350

L
Labor migration, 285, 286, 296
Lacto-fermentation, 409
Lake District National Park Authority (LDNPA), 65

Land access, 28
Land-based groundwater, 202
Land-centred activities, 367
Land degradation, 29
 human interference, 366
 land and water resources, 366
 prime environmental issue, 365
 remedial measures, 366
 SDGs, 366
Landfilling dumping, 421
Land filling, 426, 454
Landfill island, 423
Landfill mining, 421
Landfill tax, 423
Land management, 366
Land reform policies, 28
Land resource, 300, 367
 sustainable development and management, 335
Land submergence, 228
Land transformation, 299, 300, 302
Land under miscellaneous tree crops, 305
Land use budgeting, 300, 311
Land use change, 300
 advantages, 332
 in agricultural lands, 308–310
 beekeeping, 342, 343
 climate change, 332
 community-based forest management, 339–341
 and hazard's risk map, 332
 human survival, 331
 interview and questionnaire, 334, 335
 livelihood culture, 332
 livelihood securities, 342
 LULC (*see* Land use/land cover (LULC) changes)
 medicinal plant, 341, 342
 migration, 332
 multiple regression model, 312
 natural resources management, 332
 PRA exercises, 335
 property resources, 332
 qualitative and quantitative methods, 334
 rural areas, 332
 satellite imageries, 334
 SDGs, 333
 socioeconomic and environmental conditions, 331
 study area, 333, 334
 substantial change, 310, 311
 substantial increase in nonagricultural lands, 307, 308
 sustainable livelihood diversification, 343, 344
 tourism, 332
 unprecedented population growth, 312
Land use changes, Kumaon Himalayas
 agricultural and wastelands, 327
 agricultural land, 325
 biomass-based subsistence agriculture, 320
 community forests, 325
 community health, 327
 cultivated land, 325
 current land use pattern, 322, 324, 325
 data collection, 321, 322
 ecological impacts, 328
 economic globalization, 319
 ecosystem services, 320, 326
 environmental changes, 320
 food, 327
 forests, 325
 Himalayan watersheds, 320
 land use policy, 328
 livelihood security, 327
 methodology, 321–323
 mountain landscapes, 319
 natural ecosystems, 319
 natural resources, 320
 population growth, 319, 320, 327, 328
 resource use dynamics, 325
 rural water supplies, 328
 SDGs, 320, 321
 tourism, 320
 traditional subsistence agriculture, 320
 Upper Kosi Catchment, 321, 324, 325
  biomass supply, 327
  fodder, 327
  food, 327
  fuelwood deficit situations, 327
  irrigation potential, 327
  rural livelihood, 328
  water availability, 327
  water resources, 326
 wasteland, 325
 water, 328
Land use dynamics
 agricultural sector, 302
 categories, 300
 component, 299
 composite index, 304
 desirable, 313
 in different sectors, 302
 ecological implications, 302
 ecological sector, 302, 311
 geographical area, 300
 index of desirable, 303
 index of undesirable, 303
 land use changes, 307–310
 land use pattern, 304–307
 linearly additive land use changes, 301–302
 rapid transformation, 300
 SDGs, 301
 sectoral, 310–312, 314
 social and cultural development, 304
 state-wise, 314
 statistics in India, 301
 time periods, 301
 undesirable, 313
Land use/land cover (LU/LC) changes, 300, 301
 BV, 206
 livelihood sustainability, 335
 natural resource management and action planning, 332
 in Pasolgad watershed, 335, 336
 in Upper Pasolgad watershed, 336–339

Land use pattern
area under nonagricultural use, 306
barren and unculturable land, 306
culturable wasteland, 306
current fallows, 306, 307
fallow lands, 306, 307
forests, 304
grazing lands, 304, 305
land under miscellaneous tree crops, 305
net area sown, 307
permanent pastures, 304, 305
total reported land, 304, 305
trends, 304, 305
Land use planning, 343
Land use statistics in India, 301
Landslide, 34
Latrine facility, 570, 571
Laws of thermodynamics, 106
Leaf Area Index (LAI), 501
Leasehold Forest Users Groups (LFUGs), 289
Length of Growing Period (LGP), 491
Leopold matrix, 118
Level of human impact (HI), 467
Liberal economic policies, 271
Libri Kosi, 238
Life expectancy, 564
Light detection and ranging (LiDAR), 53
Limestone, 35
Liquid Gold, 36
Liquid petroleum gas (LPG) connection, 567
Livelihood culture, 332, 335
Livelihood options, 339–341
Livelihoods, 558, 559, 561
activities, 349
context, 360
decision-making process, 349
diversification, 351, 359
environmentally sustainable, 348
framework, 349
human beings, 349
opportunities, 349
people, 347
Rapti River basin, 276, 277
related issues, 352
research, 349
resource-dependent, 352
rural, 350
Livelihood securities, 327, 431, 432
resource and skill management, 342
Livelihood sustainability, 335
Livestock, 320
Livestock productivity, 327
Local economy, 351
Local governments, 85, 237
Local *haats* (markets), 357
Local knowledge, 379
Local knowledge/traditional knowledge system, 476
Local self-governance, 432
Locally Unwanted Land Uses (LULUs), 18
Low rent competitive diversification model, 107
Low yield cash crops, 388
Loweswater Care Project (LCP), 66
Loweswater Improvement Group (LIP), 65
LPG connection, 572, 573

M

Macro-scale country level analysis, 550
Madurai-based civil society organization, 435
Mahananda Main Canal, 257
Mainstreaming DRM, 81
Make in India initiative, 393
Man and Biosphere Programme (MAB), 41
Management, 172, 249, 250, 366
Mangrove conservation, 560
Mangrove ecosystem
Bhitarkanika Mangroves, 558
conservation of flora and fauna, 561
ecosystem services, 559–560
endangered flora and fauna, 558
productivity, 557
Mangrove resources, 557
Man-made pond, 218, 219
Mann–Kendall non-parametric test, 490
Maoist's insurgency, 283
Maoist's Revolutionary Forces (MLF), 284
Maosikhali jungle, 339
Map overlay analysis, 127
Map use analysis, 51, 52
Marginalization, 351
Marine environment, 466
Market-based approaches, 354
Market-driven model, 409
Marsh, 218, 219
Mass tourism, 462
Maternal mortality ratio (MMR), 576
Mcleodganj and Namlang Himal Ayurvedic resort, 383
Mcleodganj region, 388
Measures for the Environmental Information Disclosure, 100
Medical geography, 382
Medical tourism, 388, 389, 393
Medicinal plants, 394
demand and supply, 385
forest resource, 341, 342
Medicinal plants in Kangra district
agricultural development, 379
climate change, 393
collection, 381
commercial basis, 381
crop diversification measurement, 389
demand and supply, 386
domestic demand, 380
global export market, 379
health purposes, 380
Himachal Pradesh, 380
indigenous medicine, 392, 393
medical tourism, 393
occupation, 380
prices, 388

reduced cultivation 386, 388
richness, 380
suatainable cultivation, 392
utilization, 380, 381
Mega cities, 162, 510, 512
Mental health, 380
MGNREGA (program), 343
Mica, 35
Micro Units Development and Refinance Agency Limited (MUDRA Yojana), 425
Micro-irrigation (MI), 433–434, 586
Micro-level case study, 360
Micro-level experiences, 361
Microscale climate, 491
Middle or Lesser Himalayas, 231
Migration, 332
Millennium Development Goals (MDGs), 5, 80, 272, 399, 402, 404, 405, 412, 524
and sanitation challenges, 401–403
Millennium Ecosystem Assessment (MEA) report, 4
Mineral extractions, 106
Mineral resources, 107
Minerals, 35–37
Ministry of Drinking Water, 272
Ministry of Environmental Protection, 100
Ministry of Irrigation and Power, 272
Ministry of Natural Resources, 96–97
Ministry of Urban Development (MoUD), 548
Ministry of Water Resources, 272
Mitchell, Bruce
awards, 15
disaster management, 19
environmental change, 21
Environmental Justice, 18
environmental management, 17, 18
ethics, 20
field of resource and environmental management, 13
fishing, 20
geography, 20, 21
honors, 15
management, 21
natural resource management, 16, 17
professional contribution, 16
recognition, 15
resource analysis, 21
soil erosion, 20
sustainable development, 18, 19
University of Waterloo, 14
Waterloo, 14, 15
Mitigation, 435
Modern agriculture, 365
Modernization, 564
Modernized mixture model, 408
Monsoon rainfall, 34
Moral hazard, 79
Mother's Group (MGs), 287, 288
Mountain ecosystems, 326
MS-Excel and QGIS software, 368
Muktapur village, 213, 214, 216, 217
Multi-criteria analysis, 127
Multidisciplinary research, 28
Multi-hazards and disasters, 86
Multiparty democratic system, 285
Multiple criteria evaluation (MCE), 119, 120
Multiple linear regression model, 312, 314
Multiple stakeholders, 82
Multi-stakeholder collaboration, 400, 407, 408, 412
Municipal solid waste, 415, 417
cities, 420–421
generation, 417
Municipal Solid Waste Management (MSW), 417
Muslim Teesta basin, 252

N

Nagar Tangon Main Canal, 257
National Accounting Matrix including Environmental Accounts (NAMEA), 121
National and local disaster management institutions, 76
National Bank for Agriculture and Rural Development (NABARD), 354, 356
National Bureau of Statistics of China (NBS), 92
National Commission on Flood (NCF), 438, 439
National disaster management agencies, 76
National Disaster Prevention and Protection Agency, 82
National DRM systems, 82
National economic growth, 107
National Environmental Policy (NEP), 430
National Fire Agency, 82
National Food Security Mission (NFSM), 434
National governments, 85, 584
National Green Tribunal (NGT), 515
National Highway Authority of India (NHAI), 565
National Highway Development Project (NHDP), 565
National Mission on Micro-irrigation (2010) program, 434
National property, 261
National Resettlement and Rehabilitation Act (2007), 429
National Rural Livelihoods Mission (NRLM), 343
National Rurban Mission (NRuM), 457
National Trust (NT), 65
National Urban Livelihoods Mission, 457
National Water Development Agency (NWDA), 33, 273
National Water Policy (NWP), 242, 434
National Wetland Conservation and Management Rules, 212
Natural areas, 468, 470
Natural attractions (NA), 467, 469
Natural biodegradable fertiliser, 451
Natural capital (NC), 106, 107, 109, 110, 113, 125, 501, 585
Natural capital index (NCI), 493, 494
Natural disasters, 80
Natural environment, 462, 583
Natural gas, 91, 92
Natural hazards, 42, 80
Natural mega-disasters, 83
Natural pond, 218
Natural reservoirs, 213

Natural resource decision-making
in India
CBA, 116–118
ecological economics, 115
EIA, 118, 119
MCE, 119, 120
natural resource accounting, 120–122
predictive performance/realism of assumptions, 116
quantitative measurement, 127
research Methodology, 109
Natural resource economics
in India, 108
Natural resource management, 16, 17, 584
in China
co-benefits approach, 102
energy conservation/efficiency frameworks, 101, 102
policy implications, 102
policy integration, 102
provincial level, 98, 99
reliability in data statistics system, 99, 100
resource consumption, 90–95
state level, 95–98
strategic investments towards resource-efficient cites, 100, 101
early paradigms (1935–1975), in India
ecological, 38, 40, 42
economic, 44, 45
integration, 42–44
Technological paradigm-I (*see* Technological paradigm-I)
modern paradigms (1976–2005), in India
decision-making, 49, 50
technological paradigm-II (*see* Technological paradigm-II)
multidisciplinary research, 28
paradigm analysis, 38
paradigm development
global and indian experiences, 54
Natural resources, 45, 105, 181, 184, 190, 243, 250, 320, 348, 352, 463
accounting, 120–122
analysis and decision-making (*see* Natural resource decision-making)
and assets, 110
and development, 107
ecological stability and resilience, 106
EDP, 107
and economic growth, 106
flow resources, 106
in India, 300
mineral extractions, 106
national economic growth, 107
natural capital, 107
nonrenewable, 105
renewable, 105
rent curse theory, 107
and SD, 109, 110, 112
resource curse thesis, 106
stock resources, 105
Natural resources management, 332
Natural wetlands, 223
Nature-based resource inventory, 463
Nature-based tourism, 462, 470
Nature-society interaction, 27
Nepal
agricultural activities, 284
citizens, 285
collective consciousness, 285
creative destruction, 285
development financing, 286
government, 284, 294
hill regions, 284
"People's Movement I", 285
rural area, 284
Nepal Climate Vulnerability Study Team (NCVST), 235
Nepalese water resources, 228
Net present value (NPV), 116, 117, 120
Net price method, 120
Net sown area, 310
Net state domestic product (NSDP), 122
Network power, 62
Niche Extrapolator, 391
Nile River water resources, 253
Nilgiri District of Tamil Nadu, 354
Nisargruna biogas projects, 419
Nonagricultural land use, 301, 303, 306, 312
Nonagricultural lands, 307, 308
Nondominant caste, 28
Non-farm sectors, 356
Nongovernmental organizations (NGOs), 77, 237, 276, 277, 324, 409, 424
Non-household activities, 351
Non-irrigation revenue, 438
Nonlinear transformation, 285
Nonparametric statistics, 46
Non-standardized agro-technologies, 386
Nonstructural/flood emergency measures, 274
Non-technological paradigm, 55, 584
Non-threatening exercise, 359
Normalized Difference Vegetation Index (NDVI), 321
North Pacific Sub-Tropical Gyre, 423
North-West Himalaya, 380
Not in My BackYard (NIMBYs), 18
Novel Drug Delivery System (NDDS), 380
Nuclear energy, 36
Nuclear minerals, 37

O

Occupational structure, 361–363
Ocean landfilling, 423
Odisha
Bhitarkanika Mangroves, 558
Bhitarkanika Wildlife sanctuary, 559
Off season vegetable cultivation, 377
Oil industry, 92
Oilseeds, 290
On-site application, 409
Open door policy, 435

Organization for Economic Cooperation and Development (OECD), 15, 415
Orissa R&R policy, 140
Outgoing Longwave Radiation (OLR), 491
Overexploitation and quality deterioration, 193

P

Paddy/chick-pea cultivation, 437
Panchayat resource mapping, 587
Panchkarma, 388
Pan-City developments 449
Panipat, 197
Paradigms
  analysis, 55
  beliefs, 38
  definition, 37
  development of science, 37
  exemplar, 37, 38, 40, 45–55
  geography, 37
  Kuhnian model, 37
  natural resource management, 38
  phases, 37
  sampling frame, 39–40
  scientific enquiry, 37
  stratification, 39–40
Parametric statistics, 46
Participatory approach, 356
Participatory forest management, 342
Participatory rural appraisal (PRA), 214, 285
Partyless Panchayat governance system, 283
Pasolgad watershed, 335, 336
Payment for ecosystem services (PES), 113
Pen culture technology, 213
Perception, 49
Perennial lakes, 269
Peri-urban sanitation systems, 409
Permanent pastures/meadows, 304, 305
Petroleum, 36
Physical capital index (PCI), 493, 494
Physical capital vulnerability index (PCVI), 502
Physical input-output accounts, 120
Physical science methods, 76
Picroside biosynthesis, 392
Piped water supply project, 541
Policy implications, 102
Policy integration, 102
Policymakers, 341
Political ecology, 41, 42
Political economy, 44, 45
Political elements, 433
Pollutant emissions, 95
Pollutee-pay-principle, 113
Pollution, 32
Population dynamics, 285
Population growth, 287, 307
Post-audits, 435
Poverty, 36
Power dynamics, 355
PRA exercises, 335
Pradhan Mantri Awas Yojana, 457
Pradhan Mantri Krishi Sinchai Yojana (PMKSY), 273
PRECIS based simulation, 498
Present value (PV), 116
Primary data, 162
Principal Component Analysis (PCA), 321
Principles of ecological resilience, 108
Prioritization, 385
Private-owned shallow tube wells, 201
Private sector engagement, 411
Privatization, 351
Produced capital, 110, 113
Production-oriented livelihoods, 342
Profit-maximization
  principles, 206
Property resources, 332
Protected agriculture, 390
Provincial region, 98, 99
Provisioning services, 289–292
Public and private institutions, 275, 354
Public owned agencies, 436
Public Private Participation (PPP), 424
Public sector, 446
Public utility management
  in channel migration zones, 275, 276
Public Works Department (PWD), 451
Pulmonary diseases, 448
Punai, 214, 218
Punjabi settler community, 355
Pyrolysis techniques, 423

Q

Quality deterioration, 193, 202
Quality, groundwater, 196–197
Quality of life, 300
Quantitative measurement
  natural resource decision-making, 127
Quasi-sovereign kingdom, 251

R

Rabi season, 339
Rag pickers, 454, 455
Rainfall, 206, 478, 491, 495
Rainwater harvesting, 585
  "downscale" challenges, 149–152
  future resource self-sufficiency, 148, 149
  implementation, 144
  methodological and empirical context, 145
  municipality and government, 144
  political ecology and IWRM, 145, 146
  potential, 149
  resource inadequacy, 147, 148
  "upscale" challenges, 153, 154
Rajiv Gandhi Grameen Vidyutikaran Yojana (RGGVY), 571
Ramsar Convention, 207

Rapti River basin, 586
community expectations, 277, 278
concept of governance, 271, 272
data and methods, 270, 271
demographic and land use data, 270
embankments, 276
flood forecasting, 274, 275
floods and floodplain management in India, 269, 273, 274
Ganga plain, 269
integration of sectorial plans, 274, 275
livelihood issues, 276, 277
location, 269, 270
NGOs, 277
perennial lakes, 269
physical and socioeconomic characteristics, 273
prevention of sand mining, 276
primary and secondary data sources, 270
public utility management, 275, 276
SDGs, 268, 269
southwest monsoon, 269
synchronization of forest, land and water policy, 277
warning, 274, 275
waterlogging crops, 270
Rasayana trantra, 393
Raw material-based resources, 352
Rayadaga district of Odisha, 349
Reclamation, 212
Recovery operations, 78
Recycle nutrients, 109
Recycling, 457
Recycling human waste, 408, 409
Reduce, Reuse and Recycle (3Rs), 422, 423
Refuse Derived Fuel (RDF) plants, 419, 421
Regional environmental governance
Alpine Convention, 190
biodiversity and ecological degradation, 182, 183
development effect on natural resources, 183–185
licensing protocols, 189
sustainable development goals, 180
vulnerability and challenges, 189
Regional Rural Banks (RRB), 573
Regional sustainable development (RSD)
in India
conceptualized, 114
economic development, 113
ecosystem services, 114
forest area, 122–124
goals, 114
Net Annual Ground Water Availability, 127, 128
regional distribution of forest coverage, 114, 115
rural population, 114
SD, 113
societies grow, 114
Total Renewable Installed Power, 125, 126
research methodology, 109
Regulating services, 291, 292
Rehabilitation, 137
dimensions, 138
Rehabilitation and Periphery Development Advisory Committee (RPDAC), 140
Reliability
in data statistics system, 99, 100
Relief, 230, 231
Remittance, 284, 286, 293, 296
Remote sensing, 48, 49, 52, 53, 268, 394
Rent curse theory, 107
Reporting Land, 304
Research methodology
natural resource decision-making, 109
RSD, 109
Reserved forests, 322
Resettlement
best practice, 136
cost, 135
definition, 137
and development-induced displacement, 136
development opportunity, 136
development strategy, 135
dimensions, 138
gender-related benefits, 140
good practices (*see* Good practices)
SDGs, 136
spatial displacement, 135
Resettlement colonies, 526, 530
Resident Welfare Associations (RWAs), 424
Resilience, 4, 10, 101, 366
Resource allocation, 44
Resource consumption in China
direct resource consumption sectors, 93, 94
economic growth and social development, 90
energy consumption, 90–93
environment, 90
GDP, 90
humans, 90
pollutant emissions, 95
SDGs, 90
steel industry, 93, 94
Resource curse thesis, 106
Resource deficit, 327
Resource degradation, 351
Resource-efficient cites, 100, 101
Resource extraction, 284, 285, 293, 294
Resource management, 46, 463, 583, 584, 586, 587
content organization, 6, 7
customary practices, 5
economic development, 4
ecosystem, 4, 8
geopolitical issue, 9
geo-system, 3
governance, 5, 12
groundwater sustainability, 8
growth, 4
human dimensions, 9, 10
human society, 4
indicators, 4
informal norms, 5
institutional interventions, 5
instrumental and idealistic, 6
land, 6
local-level experiences, 5
management practices, 5

microlevel, 6, 9
multi-scale and nonlinear, 6
national, regional and global change, 10–12
policy formulation, 5
policymaking, 6
pressure and socioeconomic growth, 4
process of transformation, 3
regional environmental governance, 8
science, 4
sectors, 5
social learning, 6
sustainability issue, 4
sustainable development, 4, 5
systems, 4
transboundary river basins, 9
understanding conceptual foundations, 7
unpacking problems, 8
wetland environments, 8
Resource management decision-making, 49
Resource management problems in India
deforestation, 30, 31
degrading atmosphere, 34, 35
desertification, 29
energy crises, 35–37
groundwater depletion, 31, 32
hazards, 34, 35
human-environment relationships, 28
ILR, 32–34
land degradation, 29
land-man ratio, 28
minerals, 35–37
redistribution, 28, 29
shifting cultivation, 30, 31
soil erosion, 29
soil salinization, 30
transboundary water resources, 32–34
unequal land access, 28, 29
waterlogging, 30
water quality, 31, 32
water scarcity, 31, 32
Resource management sector, 437
Resource sustainability, 284
and skill management, 342
Resources and environmental policy, 102
Resource use dynamics, 325
RESPECT model, 50
Respiratory problems, 380
Restoration, 212
Revenue department, 547
Rice cultivation, 209, 211
Ringnal jungle, 339
Rising water table zone, 194
Risk and vulnerability assessment
chronic poverty, 433
livelihood security issue, 431, 432
poor accessibility, 433
water disputes, 432, 433
Risk governance
in Asia, 80–82
adaptation and urban environmental management, 78
communication and participation, 78
comprehensive means, 78
demand-side constraints, 78
dynamics, structures and functionality, 78
HFA, 78
institutional structure, 78
international scenario, 79–80
pro-poor governance, 79
risk-prone settlements, 79
UNISDR, 78
Risk Interpretation and Action (RIA) framework, 8, 160, 162
Risk management legislations, Asia
Civil Défense Force, 83
DRM policies, 82
flood management, 83
focal point agencies, 83
HFA, 82
Japan's DRM plan, 82
lines and policy responsibilities, 82
national disaster management agency, 83
River basin management
in India, 272, 273
river development, 272
Rivers
and wetland, 206–208
River Valley Projects, 438, 439
River water management, 252
Rural areas, 287, 332
Rural economy, 350, 352
Rural Economy and Land Use Programme (RELU), 66
Rural geographical domain, 286, 287
Rural industry, 122
Rural jobs, 201
Rural livelihood security, 377
Rural Non-Farm Employment (RNFE), 377
Rural poverty, 351
Rural small landholders, 354
Rural socio-economy, 384
Rural spaces, 361
Rural systems, 206
Rural village *Audali*, 348
Rural women, 351

S

Safe Drinking Water Act, 69
Safe drinking water and sanitation, 433
Safe drinking water supply systems, 201
Salinization, 202
Salt-affected soils, 202
Samuel Evans Stokes, 477
Sand mining, 276
Sanitary landfilling, 421, 455
Sanitation, 519–521
approach-based stage, 523
community-based facilities, 523
conceptual framework, 521
definition, 520
in Delhi's resettlement, 520
facility and practice, 520
interventions, 520

Sanitation (*cont.*)
  SDG, 520
  sustainable, 521
  inequality, 521
  wastewater and human excreta, 521
Sanitation challenges, 587
  accessibility, 406
  APPS, 409–412
  ecological sustainability, 408, 409
  facilities, 407
  government-led/supply-driven sanitation, 407
  health risk assessments, 406
  human rights, 403, 404
  and hygiene, 405, 406
  innovation and enterprise in sanitation provision, 409
  interventions, 407
  and MDGs, 399, 401–403
  multi-stakeholder collaboration, 407, 408
  partnerships, 407, 408
  policy options, 407
  provision, 400
  recycling human waste, 408, 409
  and SDGs, 400, 404–406
  and vulnerability, 400, 401
  and water, 401–406
  WHO, 406
Sanitation provision
  innovation and enterprise, 409
Sanitation wheel, 412
Sapt (seven) Kosi, 229
Sapt-Kosi, 232
Satellite imageries, 334
Savara and Lodha tribal community, 352
Savara tribe, 352
Scale of production, 358
Scientific-based policymaking, 99
Scientists climate change knowledge, 481, 482
Scrub forest, 336
Seasonal requirement, 254
Seasonal variation, 236
Second-generation problems, 194
Sectoral land use dynamics, 310–312, 314
Sectorial plans, 274, 275
Self-help groups (SHGs)
  alternative livelihood options, 357
  Audali village, 356, 358
  female members, 357
  hesitation, 358
  Jamuna, 356
  share reserves, 358
  socio-economic hierarchies, 356
  training sessions, 356
  tribal women, 361
  in Udham Singh Nagar district, 356
Self-Help Women's Association (SEWA), 456
Semi-structured interviews, 355, 477
Sendai Framework, 79, 80, 86
Sendai Framework on Disaster Risk Reduction, 81
Sense of security, 62
Service level benchmarking (SLBM), 434, 435
Service-skill based livelihoods, 342
Sewage disposal, 544
Sex ratio, 352
Shanxi Province, 99
Shifting cultivation, 31
Shimla district, 476–478, 481–483
Shiwalik hills, 194
Shiwalik/outer/sub-Himalayas, 231
Shock Faced and Coping Strategies Index (SCSI), 494
SIA in Indian water resource management
  community-led management ventures, 441
  economic development, 430
  EIA programs, 441
  NEP, 430
  performance level, 430
  SDGs, 430, 431
  social *vs.* political undercurrents, 430
  structural planning and management philosophy, 430
SIA in Indian water and allied policies
  assessment instruments tool, 431
  ecology and environmental impact assessment
    agricultural power consumption and subsidies, 437
    ESIA consultancy farms, India, 436
    institutional capability, 436
    groundwater management challenges, 436, 437
    progressive assessment, 437, 438
  economic efficiency
    community partnership programs, 439–441
    flood mitigation efficiency, River Valley Projects, 438
    irrigation performance, River Valley Projects, 438
  GWP toolbox, 431
  risk and vulnerability assessment
    accessibility, 433
    chronic poverty, 433
    livelihood security issue, 431, 432
    social and economic stability, population, 431
    water disputes, 432, 433
  social assessment
    farmers' perception, micro-irrigation tools, 434
    social impacts, current policies, 434
    social monitoring framework, 435, 436
    water audit-plans, 434, 435
SIDCUL Ltd., 354
Sikkim Association for Environment, 257
Sikkim State Government, 257–260
Silsako Beel, 211
Simple additive weighting method (SAWM), 466, 468
Simpson Diversity Index, 390
Single-sector development planning, 81
Sino-Indian border, 251
Sitarganj Tehsil, 353
Skill management, 342
Slash and burn, 31
Slope, 34
Small production units, 342
Small-scale development activities, 287
Small-scale engineering, 343
Small-scale entrepreneurship, 409

Small villages, 343
Smart Cities Mission, 449
Smart city
  challenge, 449
  circular economy framework, 450
  definition, 449
  efficiency, 449
  future urban sustainability, 449
  India, 449
  infrastructure elements, 449
  municipalities, 449
  recycling, 450
  Smart Cities Mission, 449
  stakeholders, 450
  urbanisation, 448, 449
Smart village, 348
Snowfall, 478
Snow-laden Himalayan splendor, 334
Social and cultural development, 304
Social and economic transformation, 586
Social auditing, 436
Social capital index (SCI), 493, 494
Social development, 564
Social distribution, 161
Social-ecological processes, 476
Social Forestry Project, 343
Social impact assessment (SIA), 11
  characterization, social aspects, 429
  EIA, 429
  Indian water and allied policies (*see* SIA in Indian water and allied policies)
  management efficiencies *vs*. livelihood security, 429
  neutral information, 430
  sound science, 430
  water resource management (*see* SIA in Indian water resource management)
Social impacts, 429
Social inclusion, 112
Social interactions, 112
Social learning, 173, 435
Social monitoring framework, 435, 436
Social-political connections, 475
Social transformation
  agrarian practices, 286
  CBOs, 289
  CC, 288
  CFUGs, 289
  cultural services, 292–294
  distribution of survey locations, 285
  DWUGs, 289
  ecosystem services, 289 (*see also* Ecosystem services)
  factors, 284
  field data, 285
  food sufficiency, 287
  HYV, 291
  labor migration, 285, 286
  LFUGs, 289
  location of study area, 285, 286
  Maoist's insurgency, 283
  MGs, 287, 288
  MLF, 284
  multiparty democratic system, 285
  Nepal (*see* Nepal)
  nonlinear transformation, 285
  Partyless Panchayat governance system, 283
  population change, 287, 289
  population growth, 287
  provisioning services, 289–292
  regulating services, 291, 292
  remittances, 284
  on resource extraction, 285
  rural areas, 287
  rural geographical domain, 286, 287
  SDGs, 284
  supporting services, 292, 293
  sustainability, resource supply, 293–296
  traditional practices, 284
  war against feudalistic society, 285
Social welfare, 110
Societal transformation, 564
Socio-ecological systems (SES), 508, 585
  concept, 508
Socio-economic development, 83, 405
  community-based studies, 564
  concept, 564
  data analysis, 567–569
  database, 565, 567, 568
  data source and methodology, 567, 568
  development index (*see* Socio-economic index)
  disadvantaged groups and backward regions, 564
  GDP, 564
  in Haryana, 197, 198
  highway corridor (*see* Highway corridor)
  human dignity, 564
  indicators, 564, 565, 567
  level of development, 577, 578
  life expectancy, 564
  nation's institutions and value system, 564
  present-day situation, 564
  SDGs, 564, 565
  social development, 564
  study area, 565, 566
Socio-economic development index, 565
Socio-economic dimensions, 476
Socio-economic dynamics, 361
Socioeconomic impacts, 260, 261
Socio-economic index
  banking facility index, 573, 574
  drinking water facility index, 575, 576
  electricity connection, 571, 572
  good housing index, 569, 570
  health infrastructure index, 576, 577
  indicators, 565, 567, 569
  latrine facility, 570, 571
  LPG connection, 572, 573
Socioeconomic landscape, 253
Socioeconomic life, 259
Socio-economic monitoring framework (SMF), 435
Socio-economic parameters, 355

Socio-economic processes, 484
Socio-economic profile, 357, 359
Socio-economic status, 564
Socio-economic structure, 356
Socio-economic sustainability, 354
Sociopolitical stakeholders, 255
Softwood, 340
Soil and Water Assessment Tool (SWAT), 228
Soil conservation, 373
Soil erosion, 20, 29, 30, 45, 343
Soil regeneration, 109
Soil salinization, 30
Soils, 197, 480
Solid waste, 415, 416
Solid waste disposal commitments, 422
Solid waste generation, 419
Solid waste landfilled, 423
Solid waste management (SWM), India
 advanced technologies, 426
 AMRUT, 416
 community participation, 457
 decentralised waste management, 457
 endemic air pollution, 447
 generated municipal solid waste collection, 418
 GHG emissions, 456
 HRIDAY, 416
 Indian economy, 447
 informal sector (*see* Informal sector role, waste management industry)
 inter- and intra-country best practice
  Biotech, Kerala, 424
  dumping waste, oceans, 423
  eco-towns, 422
  incineration plants, 422
  incineration technology, Japan, 422
  Japan, 422, 423
  Ministry of Chemicals and Fertilizers, 424
  municipalities, 424
  Netherlands, 423
  NGOs, 424
  Okhla compost plant, Delhi, 424
  SBM, 424
  Singapore, 423
  SWM Rules 2016, 424
  3Rs, 422
  USA, 423
 JNNURM, 416
 MSW Rules, 2000, 417
 municipal areas, 417
 municipal solid waste generation, 417
 open and unsanitary landfills, 447
 processing facilities, 426
 recycling, 457
 responsible consumption and production, 447
 SBA, 457
 SBM, 416, 426
 schemes, 457
 segregated storage of waste, 426
 smart city, 416 (*see also* Smart city)
 sustainable waste management, 425, 426
 technological interventions, 457
 2030 agenda, sustainable development, 456
 3R system, 446
 treatment, collected waste, 418
 treatment efficiency, 417–419
 un-encroached landfill sites, 447
 urbanisation, 447 (*see also* Urbanisation and SWM in India)
 waste collection, 417–419
 waste generation, 417–419
  cities/towns (2004–05 to 2010–11), 419
  cities/towns (2010–2011), 418
 waste handling (*see* Waste handling, India)
Sonepat, 197
Source water protection
 antecedents and motives, 69, 70
 dialog and trust, 70, 71
 interdependency and reciprocity, 71
 outputs and outcomes, 71, 72
 power and power relations, 70
 regulatory and policy framework, 69
 regulatory environment, 69
 treatment and distribution systems, 69
 water supply, 69
South Asia, 253
Southwestern region, 194
Spatial analysis, 51
Spatial displacement, 135
Spatial equity, 114
Spatio-temporal analysis, 498
SPSS software packages, 569
Stability, 125
Stage-of-life identities, 357
Stakeholders, 137, 261, 264, 407, 583
Staple food, 378
State/civil administration, 77
State-sponsored and supported industrial development drive, 354
Statistical analysis, 107, 261
Statistical Package for Social Scientists (SPSS), 493
Steel industry, 93, 94
Stereotypical restrictive norms, 360
Stock resources, 105
Stockholm Resilience Centre, 4
Storm water drainage, 545
Strategic investments towards resource-efficient cites, 100, 101
Stratified Random Sampling (SRS), 334
Stress-capability framework, 19
Strong sustainability, 110, 126
Structural flood mitigation, 274
Subject-oriented/project-oriented organization, 273
Subsistence agriculture, 320, 326
Substantial change
 LU/LC cover categories, 310, 311
Substantial reduction of disaster, 79
Subtropical Forest Zone, 335
Super cyclone, 34
Supply-driven strategies, 407
Supply-Side Structural Reforms (SSSR), 96

Supporting services, 292, 293
Surface water resources, 36
Surface water sector, 432
Sustainability, 250, 349
community-based forest management, 339–341
of economy, 108
and livelihood potential, 341, 342
livelihood securities, 342
Susceptibility-Resilience Matrix (SRM), 29
Sustainable development (SD), 44, 95, 99, 101, 126, 226, 320, 583, 584, 586
accumulation, 112
adaptive cycles of growth, 112
complex system, 112
concept, 110
earth systems, 112
ecosystems, 112
GDP, 113
global economy, 112
governance, 112
inclusive wealth index, 113
intergenerational, 110
ISEW, 113
Kosi River basin, 226
measurement of economic progress, 113
and natural resources, 108–110, 112
PES, 113
pollutee-pay-principle, 113
renewal, 112
restructuring, 112
RSD, 113–115
social interactions, 112
transboundary water management, 243
Sustainable Development Goals (SDGs), 5, 10, 33, 77, 80, 160, 229, 272, 320, 321, 348, 378, 379, 400, 404, 430, 431, 447, 524, 532, 583
China, 90
climate change, 508
goals, 509
groundwater in Haryana, 194, 195
in India, 108
land use change, 333
land use dynamics, 301
Rapti River basin, 268, 269
resettlement, 136
and sanitation challenges, 404–406
social transformation, 284
socio-economic development, 564, 565
Teesta River, 250
transboundary water management, 226, 227
UN 2030 Agenda, 508
SDG-3 of Agenda 2030, 378
SDG 5, 349
SDG-13, 81
wetland, 207
Sustainable future, 100
Sustainable livelihood diversification, 343, 344
Sustainable management, 558, 560
Sustainable natural resource, 278
Sustainable resource governance, 6, 584–587
Sustainable sanitation, 531
Sustainable tourism, 463, 464
Sustainable waste management, 425, 426
Sustainable water management
at basin scale, 278
evaluations of policies, 268
governance, 278
imbalances, 268
institutional framework and governance, 268
Integrated River Basin Planning and Management, 268
interdependencies, 268
Rapti River (*see* Rapti River basin)
Swachh Bharat Abhiyan (SBM), 402, 416, 424, 426, 430, 457
SWM Rules 2016, 417, 419, 424–426
SWOT analysis, 253
System of Integrated Environmental and Economic Accounting (SEEA), 120, 121
System of national accounts (SNA), 121

T

Tamil Nadu Water Supply and Drainage Board (TWAD), 441
Tata Energy Research Institute (TERI), 122
Tea cultivation, 391
Technological fix, 480, 483
Technological paradigm-I
cartography, 45, 46
computer-aided cartography, 47
descriptive statistics, 46, 47
inferential statistics, 46, 47
natural resource management, 45
remote sensing, 48, 49
Technological paradigm-II
categories, 51
geographic information system, 53, 54
map use analysis, 51, 52
mathematical and statistical analytical techniques, 51
remote sensing, 52, 53
spatial analysis, 51
Technology Mission on Cotton (TMC), 434
Teesta Barrage (TBPB), 259, 261
Teesta-Jaldhaka Main Canal, 257
Teesta-Mahananda Link Canal, 257
Teesta River
BCM, 251, 252
definition, 251
discharge, 261
equitable and reasonable utilization, 255
hydropower, 251
India and Bangladesh, 250, 251, 255
linguistic region, 252
literature, 252, 253
location, 251, 252
Maor dams, 260
methodology, 254, 255
objectives, 254
research, 253, 254

Teesta River (*cont.*)
SDGs, 250
Teesta River Treaty, 261–263
water resources (*see* Water resources)
water-sharing, 255
in West Bengal, 251
Teesta River Basin Organization (TRBO), 264
Teesta River water resources
discharge characteristic, 255, 256
governance and management, 255
Bangladesh Government, 259
Sikkim State Government, 257–260
socioeconomic impacts, 260, 261
state subject, 256
transboundary river, 256
West Bengal State Government, 257, 259
nature of conflict, 255, 256
space and time, 255
Teesta River Water Treaty, 261–263, 266
Teesta valley, 254
Temperature, 478, 491, 495, 498, 499
Temperature index component, 493
Temples of modern India, 261
Temporal water availability, 233, 234
Tera petra sanitation, 409
Terrace farming, 373
Terraces, 373
Terrain, 343
Tharu tribe, 356
Thermal power plant, 197
Third Pole Ice, 236
Third World Conference, 79
Thiruvananthapuram Corporation, 549
3-bin storage, 425
3R, 446, 454
Tibetan Plateau, 231
Tibetans, 381
Tilawe Kosi, 239
Total fertility rate (TFR), 576
Total reported land, 304, 305
Total wealth, 110
Tourism, 332, 462, 464
Tourism development, 213
Tourism infrastructure (TI), 467, 469
Tradable water rights, 433
Traditional ecological knowledge (TEK), 587
Traditional knowledge system in Kangra, 394
Traditional medicine and building capability, 386
Traditional wisdom communication, 394
Transboundary river, 229, 256
Transboundary water governance, 255, 261
Transboundary water management
description, 225
governance, 226
India, 243
international cooperation, 229
Kosi River basin (*see* Kosi River basin)
Nepal, 243
per capita availability, 226
per head availability, 226
SDGs, 226, 227
study, 227
sustainable development, 226, 243
water availability, 229
Transboundary water resources, 32–34
Trans-Himalayan Highlands, 231
Transport network, 367
Transportation, 307, 450
Tribal women challenges, 359
Trimohanighat, 274
Tsunami, 35, 79
Tung Ping Chau (TPC), 11
Twelfth Five-Year Plan, 202
2010 International Recovery Forum, 77
2011 Interim Deal, 254
Types of tourism, 463

U
UN International Strategy for Disaster Reduction (UNISDR), 78
UN sustainable development goals, 348
Undesirable land use dynamics, 313
UNDP-World Bank Water, 409
Unenviable ecology (UE), 302
Unilateral physical control, 253
United Nations Development Program (UNDP), 349
United Nations Development Programme-Small Grants Programme (UNDP-SGP), 425
United Nations Economic and Social Commission for Asia and the Pacific (UNESCAP), 227
United Nations Economic and Social Commission for Western Asia (UN-ESCWA), 237
United Nations General Assembly, 79
United Nations International Strategy for Disaster Reduction (UNISDR), 76
Universal Soil Loss Equation (USLE), 20
University of Waterloo, 14, 15
Upper Pasolgad watershed
agriculture, 333
Bironkhal block and Thailisain tehsil, 333
cultivation, 333
dense forest and scrub forest, 336
developmental activities, 336
forest areas, 337, 339
FSI, 336
LULC, 336, 337
Maosikhali jungle, 339
Ringnal jungle, 339
study area, 333
topography, 334
Van Panchayat, 338
village community, 338
Urban atmospheric resources, 35
Urban development, 201
Urban ecosystem, 585
Urban India, 448
Urban Infrastructure Development of Small and Medium Towns (UIDSMT), 416

Urban Local Bodies (ULBs), 457
Urban population, 448
Urban rainwater harvesting
  "downscale" challenges, 149–152
  "upscale" challenges for, 153, 154
Urban sustainability, 515
Urban water and sanitation sector, 435
Urban water governance, 537, 548
  decision-making process, 540
  IUWM, 538
  policy formulation, 538
  risk and uncertainties, 540
  sewage network scheme, 544
  social tensions, 540
  spatial growth impacts, 541
  water recycle, 544
  water sectors, 540
Urban water systems, 539
Urbanisation, 448, 449
Urbanisation and SWM in India
  agriculture-based organic forms, 450
  circular economy, 453
  collection efficiency, 453
  colonial period
    agro-based products, 451
    Black Towns' wastes, 452
    charitable interventions, 451
    concepts, 451
    epidemic, 451
    large-scale fossil fuel, 451
    postcolonial period, 452
    PWD, 451
    societal framework, 451
    stakeholders, 451
    waste characteristic, 451
    waste segregation, 451
    western industrial revolution, 451
    White Towns' wastes, 452
  consumer behavioural change, 450
  food, 450
  fuels use, 450
  innovative business models, 450
  lack of awareness and education, 452
  lack of source segregation, 453
  land requirement, disposal, 453
  maedieval era, 450
  medicines, 450
  municipal solid waste disposal, 450
  paint, 450
  poverty, 453
  pre-colonial era, 450, 451
  public health, 452
  recycling, 452
  reuse, 452
  traditional houses, 450
  transportation, 450
  waste characteristics, 450, 452
  wastewater management, 452, 453
Urbanization, 307, 480, 507
User cost allowance, 120
Uttarakhand, 348, 353, 354
Uttarakhand State Rural Livelihoods Mission, 354

V

Value-added medicinal plants-based materials, 393–394
Van Panchayat, 338, 340
Vegetable production, 290
Vegetation, 488
Vermicompost plants, 419
Village commons and forests, 350
Village community, 338
Village Councils, 350
Village spatial economy, 355
Vocational Skill Development Centre (VSDC), 342
Vulnerabilities, 348, 431
  and sanitation challenges, 400, 401

W

War against feudalistic society, 285
Warm-Temperate Zone, 335
Warning, 274, 275
Waste assimilative capacity, 106
Waste collection, 417, 418
Waste generation, 417, 418, 426
Waste handling, India
  bioremediation/landfill mining, 421
  C&D waste, 419
  chain of events, 419
  community bins, 419
  compost, 419
  Deonar landfill site, 422
  Kerala, 421
  land acquisition, landfilling, 422
  landfilling dumping, 421
  landfills of Delhi, 422
  RDF Plants, 419, 421
  recycling, 419
  sanitary landfills, 421
  segregation, 419
  small-scale biogas plants, 421
  solid waste disposal commitments, 422
  Tamil Nadu, 421
  Uttarakhand, 422
  vermicompost plants, 419
  waste processing, 419, 421
  WtE plants, 419, 421
Waste management, 425, 426
Waste recycling, 456
Waste segregation, 423, 450, 451
Waste segregators, 455
Wasteland, 325
Waste-to-energy (WtE), 419, 421, 423, 426, 455
Wastewater management, 452, 453
Water audit-plans, 434, 435
Water availability, 227, 229
Water bodies, 217
Water cooperation, 228
Water disputes, 432, 433

Water governance, 228, 268, 584
  climate change, 60
  collaboration, 60
  conventional approaches, 160
  cross-scale interagency, 60
  decision-making, 59
  economic and social welfare, 161
  environmental management, 60
  in India, 242, 243
  institutional and organizational arrangements, 59–60
  Karamana River, 547
  multiparty collaboration, 60
  Nepal, 242, 243
  Risk Interpretation and Action framework
    complexity, 169
    influence, 171
    learning, 168, 169
    past experiences, 168
    scale of, 169
    social context, 170
    trust, 167
    uncertainty, 167
  and stakeholders, 163–166
  sustainable development goals, 160
  in Thiruvananthapuram City, 547
  types of interaction, 60
Water management (WM), 33, 161, 439, 585, 587
  See also Transboundary water management
Water policies and programs, 430
Water pollution, 480
Water quality, 32, 33, 431
Water resource management, 585
Water Resource Regulation, 242
Water resources, 228, 326, 438
  governance and management, 251, 263
  Teesta River (*see* Teesta river water resources)
Water risks, 160
Water scarcity, 31–33, 326
Water shortages, 431
Water storage, 196
Water supply, 31, 32
Water supply systems (WSS), 441
Water vulnerability index, 503
Waterbodies, 325
Waterborne diseases, 270, 327
Waterlogged area, 219
Waterlogging, 30, 257, 270
Waterpower, 272
Water-related hazards, 237
Water-retaining capacities, 274
Water–Sanitation–Hygiene (WASH), 405
Watershed, 325
Watershed Development Fund (WDF), 436
Weak sustainability, 110
Wealth shares, 106, 108
Welfare generation, 120
West Bengal's largest Teesta project, 261
The Western Yamuna Canal (WYC), 194
Wetland
  brick factory encroaching wetland, 209, 211
  in BV (*see* Brahmaputra Valley (BV))
  canal, 219
  change in number and area, 216, 218
  checklist of common birds, 213, 216
  checklist of common fish, 213, 215
  checklist of common plants, 213, 214
  conservation efforts, 211–213
  dead channel, 218, 219
  degradation, 209, 211–213, 216
  development, 213
  distribution, 209, 210
  diversity, 206
  ecosystems, 206
  ecotones, 205
  in floodplain (*see* Floodplain)
  human pressure, 209, 211
  khal, 219
  khal-beel, 216
  management, 212
  man-made pond, 218, 219
  marshes, 218, 219
  Muktapur village, 214, 216, 217
  natural pond, 218
  natural wetlands, 223
  ponds and fish consumption pattern, 223
  protecting, restoring and reviving, 206
  resources, 221–223
  rice cultivation, 209, 211
  and rivers, 206–208
  SDGs, 207
  size and distribution, 219–221
  traditional fishing, 209, 211
  transitional location, 205
  traditional use, 206
  types and sizes, 205, 209, 221
  utility pattern, 217
  in Village, 216, 218
  waterlogged area, 219
Women in Informal Employment Globalising and Organizing (WIEGO), 456
Workforce participation, 361
World Commission on Environment and Development (WCED), 4, 226
World Conferences, 79
World Toilet Organization (WTO), 570

Y

Yamunanagar, 197
Year wise Crop Price, 392
Yoga-themed retreats, 393